The aquatic and semi-aquatic bugs (Heteroptera: Nepomorpha & Gerromorpha) of Malesia

René Cobben, 1925–1987
Nils Otto Møller Andersen, 1940–2004

We dedicate this book to our late teachers, colleagues and friends Prof. Dr. René Cobben and Prof. Dr. Nils Møller Andersen, who both passed away prematurely leaving their enormously important work unfinished. Their contributions to general morphology, classification, biology, zoogeography and phylogeny of semi-aquatic bugs have inspired and guided us, and will inspire coming generations of heteropterists. As the leading scientists in the field of water bugs they are also respected and admired for their serious attitudes and modest personalities.
Andersen encouraged the first author to take on the challenge to write this handbook. His support and advice have enabled us to finally present it.

FAUNA MALESIANA HANDBOOKS

Series editor A. J. de Winter
Project editor G.O. Keijl

Fauna Malesiana
National Museum of Natural History Naturalis
P.O. Box 9517
2300 RA Leiden
The Netherlands

Fauna Malesiana received financial support from the Royal Netherlands Academy of Arts and Sciences, and the National Museum of Natural History Naturalis, Leiden.

FAUNA MALESIANA HANDBOOK 5

The aquatic and semi-aquatic bugs (Heteroptera: Nepomorpha & Gerromorpha) of Malesia

Ping-ping Chen, Nico Nieser & Herbert Zettel

BRILL
LEIDEN · BOSTON
2005

Library of Congress Cataloging-in-Publication Data
The Library of Congress Cataloging-in-Publication Data is also available.

This book is printed on acid-free paper.

ISSN 1388-3895
ISBN 90 04 14768 3

Cover design by John Stelck (Biskaje), Leiden
Typesetting Fritschy opmaak & redactie, Leiden

Printed in The Netherlands.

Contents

Foreword

This fifth volume of the Fauna Malesiana Handbooks treats the families and genera of aquatic and semi-aquatic bugs of Malesia. For the first time since the classic work of Lundblad (1933), the only existing general overview of this group, it brings together a large amount of information on identification and biology of this group for the entire region up to generic level. Lundblad listed 169 species in 55 genera for Malesia, numbers that increased to the almost 1000 species in more than 100 genera in the present volume. According to the authors, only half of the species may as yet have been described. To put these numbers in perspective: in the vast Palaearctic region, an area much better studied than Malesia, the Nepomorpha and Gerromorpha fauna consists 'only' of 484 species in 85 genera (Catalogue of the Heteroptera of the Palaearctic Region, 1995). So, although this volume is a beautiful and important milestone, a lot of work remains to be done. This clearly underlines the need for this book!

Since the exact distribution of genera (and species) is still poorly known, the authors decided to include information on all known genera in the Indo-Australian Region, not only those reported from Malesia *sensu stricto*, thereby increasing the scope and usefulness of the book considerably.

Waterbugs inhabit a wide range of terrestrial, freshwater and marine habitats, as well as the transitional zones in between. They may be beneficial to mankind, for instance in mosquito pest control, as bio-indicators, or even as condiment in food. This book is thus of interest to a wide audience. With this state of the art overview, the authors aim to stimulate further research, which this work will undoubtedly do.

Berend Aukema,
Heteropterist

Acknowledgements

The authors are deeply indebted to I. Lansbury (Oxford, UK) for his generous help of reading the first draft of the manuscript, correcting the English, giving many suggestions for improvement, and permitting us to use some of his illustrations. We are also deeply indebted to J.T. Polhemus (Englewood, Colorado, USA) who critically read the final draft of the manuscript, pointing out various inacurracies and omissions and thus considerably improved the text. Any remaining errors are the responsability of the authors.

The great help for loan of specimens from the following colleagues are sincerely acknowledged: N.M. Andersen (Copenhagen), J. Duffels (Amsterdam, The Netherlands); I.A. Kobke (München, Germany); D. Kovac (Frankfurt, Germany); D.A. Polhemus (Honolulu and Washington DC, USA); J.T. Polhemus (Englewood, USA); J. van Tol (Leiden, The Netherlands); T. Vásárhelyi (Budapest, Hungary); M. Webb (London, UK), and C.M. Yang (Singapore).

We are grateful to the Library of National Museum of Natural History (Leiden, The Netherlands) and the Library of The Netherlands Entomological Society, especially to G. Tweehuysen and R. van Kampen (Amsterdam, The Netherlands), for helping with literature requests which always consisted of long lists of journal volumes; and to W. Hogenes (Amsterdam, The Netherlands), D.A. Polhemus (Washington DC, USA) & C.M. Yang (Singapore) for helping with some publications on the fauna of continental Malaysia.

We appreciate the discussion with J. T. Polhemus and D.A. Polhemus on some taxonomic problems. Also, they made a part of one of their manuscripts on Naucoridae available to us. We thank L. Cheng (La Jolla, USA) for her generous offering of important data on the genus *Halobates*, and also for reading the text on Halobatinae and giving useful comments.

We thank The Royal Society of South Australia for permission to use illustrations no. 121, 127 and 157, which were originally published in the Transactions of the Royal Society of Australia.

We acknowledge deeply the help from Dr. G. Thirumalai (Chennai, India) and Dr. A. Bal (Calcutta, India) for confirming the situation of the type of *Ventidius distanti* Paiva, deposited in the Zoological Survey of India.

We are deeply indebted to Ms. V. D. Thompson (The Netherlands) for her generous help with the linguistic control of part of the manuscript.

We thank the editors of this handbook series A. de Winter and G. Keijl for their great involvement and pleasant cooperation. We would like to express our great gratitude to J. van Tol for discussion on various matters of this work and the strong logistic support.

Finally, special thanks are due to N.M. Andersen for his encouragement and support to this work. This book benefited largely from his important book of 1982 (Andersen, 1982a) and his further work on semi-aquatic bugs.

Ping-ping Chen
c/o Department of Entomology
National Museum of Natural History Naturalis
P.O. Box 9517
2300 RA Leiden
The Netherlands

Nico Nieser
Hertog Eduardstraat 16
4001 RG Tiel
The Netherlands

Herbert Zettel
Natural History Museum in Wien
2. Abteilung, Postfach 417
Burgring 7
A-1014 Vienna
Austria

chapter **1**

Introduction

This handbook provides an introduction to the genera of adult Gerromorpha (semi-aquatic bugs) and Nepomorpha (true aquatic bugs) known from Malesia and adjacent areas. It deals with the classification and nomenclature, distribution, morphology, and adaptations to aquatic life, and it presents identification keys of families, subfamilies and genera of both Gerromorpha and Nepomorpha. As the exact distribution of many genera and species in Indo-Australia is not yet known, and some of them are not recorded from Malesia but may be found there in the future, all Indo-Australian genera are included in the keys.

Malesia

Malesia (fig. 1) is the area between the isthmus of Kra in southern Thailand and the islands east of New Guinea: the Bismarck and Solomon Islands. The data in this book refer to taxa occurring in the Malesian area. However, some genera or species from adjacent areas are included since they may be found in Malesia in the future.

'New Guinea' refers to the entire island (including surrounding islands), 'Papua (Irian Jaya)' to the western half and 'Papua New Guinea' to the eastern half.

FIGURE 1
Map of the area.
The black area is Malesia, with Wallace's Line and Weber's Line indicated. The part of Wallace's Line north of Sulawesi, running between the Sulu Islands and Palawan and the remainder of the Philippines, is also referred to as Dickerson's Line.

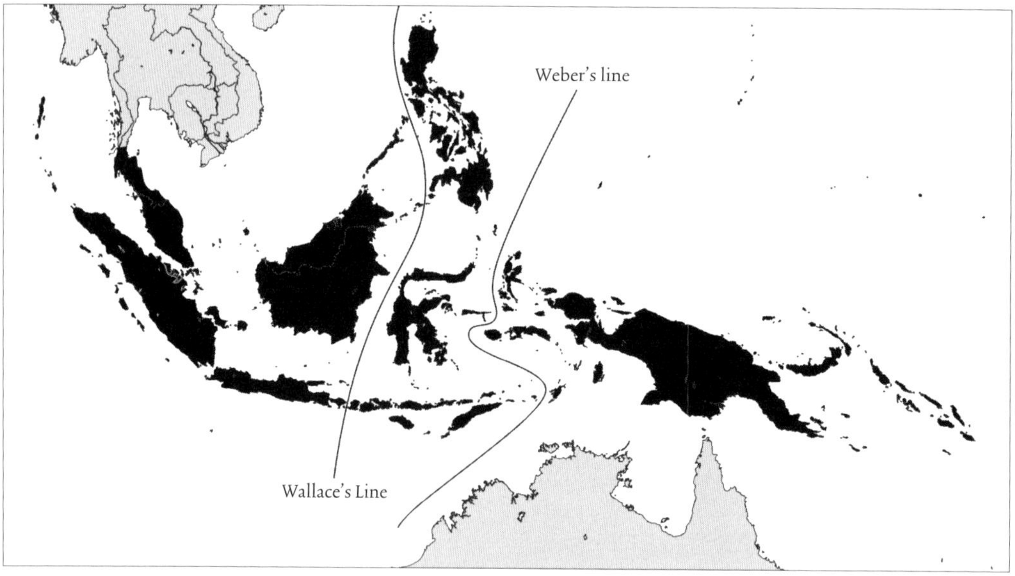

About this book

Water bugs are divided in two infraorders (Popov, 1971a): Nepomorpha (true water bugs), mostly living in the water, and Gerromorpha (semi-aquatic bugs, waterstriders, sea skaters and their relatives), mostly living on the water surface. Leptopodomorpha, specifically Saldidae, which is also considered to be semi-aquatic by many authors as they always occur at the edge of water bodies, is not included.
Over 90% of studies on biology of water bugs deals with species from temperate regions. The distinct seasons in these areas have a clear impact on their behaviour and biology: during winter most or all activities stop. Therefore, a number of excellent studies (e.g. Ekblom, 1926, 1930; Young, 1965; Andersen, 1973; Vepsäläinen, 1974) are unsuitable to describe water bug biology in Malesia, as they deal with adaptations to totally different climatological conditions. Habitat preference of each genus is mentioned, to encourage collecting in the field. We hope that the relatively limited information on biology and habitat preference offers a base for further investigations on populations and species in Malesia.
The main part of this book consists of an overview of families and genera of water bugs. In the family accounts morphological features are discussed, sometimes also on specific level. Subgenera or species groups within genera are discussed in cases where they are easily recognizable or have distinct distribution patterns. Distributional data based on undescribed species are occasionally mentioned under the genus but not in the checklist. Biological notes, if available, are added under each genus. At the end of each subfamily or family references on general information and identification of Malesian species are given. Names of authors of species and higher taxa occurring in Malesia are only given in the checklist.

This book was written jointly by the first two authors, except chapter 4, which has been divided among the three of us as follows: Chen – Gerromorpha except Hebridae, Perittopinae and Rhagoveliinae, and all the habitus illustrations; Nieser – Nepomorpha except Helotrephidae; Nieser and Chen – Microveliinae (Gerromorpha); Zettel – Helotrephidae (Nepomorpha), Hebridae (Gerromorpha); Zettel and Chen – Ptilomerinae (Gerridae); Zettel and Nieser – Rhagoveliinae, Perittopinae and Veliinae (all in Veliidae).

History of research in South-east Asia

There are several general introductory works on Heteroptera, a few of which are briefly mentioned here. An excellent, although somewhat antiquated, book on biology is Weber (1933; in German). The book by Miller (1971) is more superficial but still useful as a general introduction. Schuh & Slater (1995) focused on classification by presenting keys to all subfamilies of Heteroptera, but some brief notes on natural history were also offered. Finally there are three works dealing with biology of aquatic bugs, all three centred on temperate fauna: Wesenberg-Lund (1943) contains a wealth of observations on the Danish fauna, but also includes references to worldwide literature; Hungerford (1920) presents a taxonomic survey of North American species, but the book also contains many observations on biology of the North American, especially the Kansas fauna; J. Polhemus (1984, in Merritt & Cummins) gives keys to genera of North American aquatic insects, with tables on general ecological characters of the genera.

Esaki (1923-1930) provides good information on water bugs from the early twentieth century. He describes species from Malesia and adjacent areas, and gives additions to the higher classification, which are still partly followed today. Lundblad is one of the founders of the modern study of aquatic Heteroptera: he wrote the first and so far only general overview of water bug species for the whole Malesian area (Lundblad 1933a), with checklists of species from India to New Guinea and Japan (169 species in 55 genera for Malesia, including 42 newly described species and five new genera). He redescribes a large number of species, and is the first who presents excellent illustrations of both habitus and detailed structures. Although the book is written in German, it is still a basis for study of water bugs on species level in the Malesian area.

Between 1933 and 1960 little was published on the Malesian water bug fauna. Only Hungerford & Matsuda (1957-1965) published a series of papers, mainly on Gerridae. The significant systematic work of Matsuda (1960) provides the base for modern systematics of Gerridae on generic level.

Cobben (1946-1987) published extensively, but mostly on Leptopodomorpha of the Old World. In addition, he wrote two major and beautifully illustrated publications on the evolution of Heteropteran families (Cobben, 1968, 1978). Both publications mainly deal with structure and development of eggs, structure of mouthparts and feeding strategies, but several other structures and larval development are also discussed. From 1962 on, Lansbury studied various families of water bugs in Malesia; he published many significant papers, and continues to do so until today. Cheng & Fernando (1969) and Fernando & Cheng (1974) published their survey of Gerridae of continental Malaysia. Cheng

(1965-1966, 1976) later described a number of new species from this area and published several papers on ecology and biogeography of *Halobates*.
The most important work for modern phylogenetic and systematics on semi-aquatic bugs was written by Andersen (1982a). His first revision of *Limnogonus* and *Neogerris* (Andersen, 1975) provides a character analysis and reclassification of the Gerrinae. It was followed by a number of studies as preparations for his master piece, in which he presents a splendid survey of morphology, phylogeny, biogeography, adaptations and classification on generic level of the entire infraorder Gerromorpha (Andersen, 1982a). This work has been acknowledged worldwide as the modern classification of Gerromorpha, and has stimulated a considerable number of revisions of Oriental gerromorphan genera. Moreover, many genera were described by him, as outcome of his treatment of phylogeny, biology and other aspects in general. In recent years Andersen started to use modern techniques in phylogenetic studies (e.g. Andersen et al., 2000, Damgaard et al., 2000a, b). His untimely passing away in 2004 has unfortunately checked his activities in this field.
J. Polhemus started to study paleotropical fauna in 1976 (a paper on Philippine aquatic bugs; J. Polhemus & Reisen, 1976). He was later joined by D. Polhemus. At present, both are among the most prolific writers on Malesian water bugs. Many publications, including a number of revisions covering systematics, biogeography and phylogeny, have contibuted greatly to the knowledge of Malesian water bugs (e.g. D. Polhemus 1990, 1994b, 2001; D. Polhemus & J. Polhemus, 2000a, b, 2001; J. Polhemus, 1990a, 1997; J. Polhemus & Lansbury, 1997; J. Polhemus & D. Polhemus, 1993, 1994b, 1995b, 1996, 2000a). They now focus on the New Guinean fauna, especially on Naucoridae, the veliid subfamily Microveliinae and the gerrid subfamily Trepobatinae.

The authors of this book started the study of Malesian water bugs in the early 1990's by dealing with the faunas of Sulawesi, Continental Malaysia, North Borneo, the Philippines and adjacent areas. A considerable number of publications, including several revisions, have since been published. Besides, many other authors have contributed to the knowledge of the Malesian water bugs as well; they are referred to under the various families. Up till now, there has been no comprehensive up-to-date reference dealing with identification and biology of the water bug fauna of Malesia up to generic level, especially on true aquatic bugs (Nepomorpha). This handbook intends to fill this gap. It is our hope that the handbook will provide a sound basis and stimulation for further research on aquatic and semi-aquatic bugs in Malesia.

chapter **2**

Introducing Nepomorpha and Gerromorpha

Bugs are characterised by (usually) a sturdy, dorso-ventrally flattened body, a wide thorax, sucking mouthparts, and fore wings with a leathery basal and a membranous distal part. So-called 'water bugs', with the exception of Hebridae, differ from other bugs by completing their life cycle entirely on or under the water surface. Water bugs consist of two major groups: Nepomorpha, which are usually found under water (althoug many species can fly), and Gerromorpha, which live on the water surface. Because of the adaptations to living under water, Nepomorpha possess a highly specialised breathing system. Most water bug species are predators and feed on (living or dead) insects or other invertebrates, but a number of Nepomorpha are partially or completely vegetarian.

Terms printed in **bold** can be found in the glossary.

Morphology

Discussion of morphology in this paragraph will be restricted to those aspects used in identification or correlated with ecological and biological adaptations. For a more detailed treatment of morphology the reader is referred to other publications.

In this book length refers to total body length, unless otherwise stated. It is measured – with the animal oriented with its dorsum upward with the main body axis horizontal – from (anterior) tip of head to (posterior) tip of abdomen, except in macropterous Nepomorpha, where it is measured from (anterior) tip of head to (posterior) tip of hemielytra. Sizes given for families, genera, et cetera, are relative sizes: a large gelastocorid has the size of a small belostomatid. In Nepomorpha a length less than 12 mm is small, 12-40 is medium and over 40 mm is large.

Eggs

An elaborate study on embryology and morphology of the eggs of Heteroptera, and copiously illustrated, is given by Cobben (1968). Eggs are generally suboval with the dorsal side usually more convex. Eggs of Nepidae possess respiratory horns (figs. 30, 31; see the introduction on Nepidae).

Most eggs of Gerromorpha and Nepomorpha are simply glued to the substrate with the flattened side. Some eggs which are glued upright to the substrate have a basal adhesive disc (e.g. in Corixidae and Hydrometridae; figs. 122, 291). Many species insert their eggs in plant tissue. Both ways of oviposition may occur within the same genus, suggesting an evolutionary plastic, mainly ecological feature. As to terminology, one must be aware of the so-called revolution: during development the embryo makes a half turn, so that the prolarva lies with its dorsal side against the ventral side of the egg shell.

Nymph

Heteroptera are hemimetabolous insects: the development after the egg stage is a gradual process and the juveniles, called nymphs, resemble the adult.

First instar larvae of Gerrinae have a strongly underdeveloped abdomen compared with adults. In other groups the differences, especially in later instars, are mainly restricted to differences in wing development and reproductive organs. In addition, there are various small differences, e.g. most larvae have single-segmented tarsi (tarsal formula 1-1-1), but a few taxa have two-segmented middle and hind tarsi (tarsal formula 1-2-2), e.g. *Metrocoris*. This is useful in separating larvae from apterous adults (especially in some tropical Gerridae with reduced abdomen), which always have at least two-segmented tarsi on some legs. For example, adult Microveliinae have a tarsal formula of 1-2-2. Adult Trepobatinae have fused segments on the hind tarsi.

Apart from this tarsal character and the development of the genitalia, apterous or micropterous adults can be told apart from larvae by the structure of wing pads (fig. 164): wing pads are absent in larvae of truly apterous Gerromorpha. Larval wing pads are broadly connected with the mesothorax and metathorax and never possess veins. Their degree of development (usually in combination with measurements like head width) can be used in determining the instar stage. It is usually difficult, or even impossible, to tell from the development of the larval wings pads whether the adult will have fully developed or reduced wings.

The majority of Heteroptera has five larval instars. A number of species in various families are reported to have an aberrant number of instars, nearly always four. In Nepomorpha this has been confirmed only for the American species *Nepa apiculata* Uhler. All other Nepomorpha investigated have five instars. In Gerromorpha there seems to be more variation: four instars are verified in one species each of the genera *Mesovelia*, *Rhagovelia* and *Hermatobates*. Two species of *Microvelia* have four instars, and another species of *Microvelia* may have four or five instars. Other species of *Mesovelia* and *Microvelia* studied have five instars. The number of instars in other species of *Rhagovelia* or *Hermatobates* are unknown. So far no Malesian species is reported to have four instars (Štys & Davidová-Vilimová, 1989).

Adult

The division of the insect body (head, thorax and abdomen) is easily recognizable in Heteroptera (fig. 2). In dorsal view in most Nepomorpha and Gerromorpha with well developed wings the abdomen and posterior part of the thorax are covered by the wings. In dorsal view one thus sees the head, the anterior part of the thorax (pronotum) and the wings (figs. 44, 285). In apterous specimens one recognizes the head, thorax and abdomen (fig. 2). In most Nepomorpha and in some Gerromorpha there is a triangular sclerite, the **mesoscutellum**, behind the pronotum between the basal parts of the hemielytra, usually simply named **scutellum** (figs. 3, 44).

2

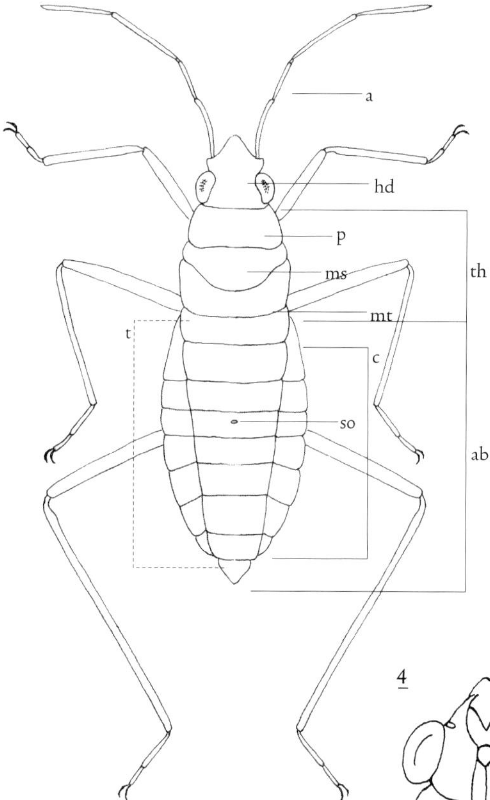

3

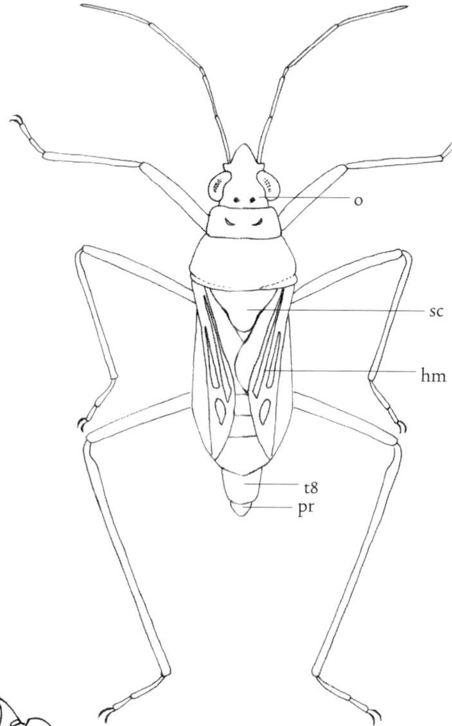

4

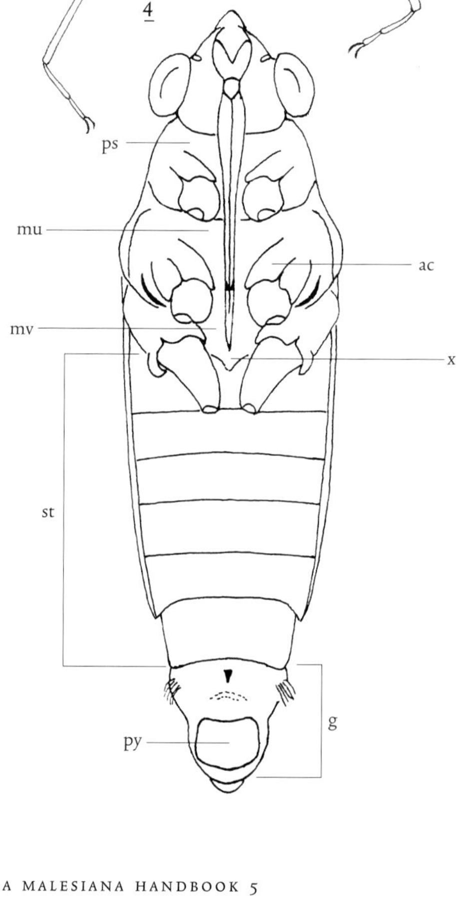

FIGURES 2 – 4
Mesovelia vittigera
2 apterous female, dorsal view – 3 macropterous dealate male, dorsal view – 4 macropterous male, ventral view (appendages removed); a = antenna, ab = abdomen, ac = acetabulum, c = connexivum, g = genital segments, hd = head, hm = hemi-elytron, ms = mesonotum, mt = metanotum, mu = mesosternum, mv = metasternum, o = ocellum, p = pronotum, pr = proctiger, ps = prosternum, py = pygphore, sc = scutellum, so = scent gland orifice, st = abdominal sternites, t = abdominal tergites 1 – 8, t8 = eighth abdominal tergite, th = thorax, x = metasternal xiphus.

Head

The head segments are so intimately fused that without study of embryonal development the segmentation is not recognizable. Parsons (1964) gives a good outline of how the transformation from a generalized insect head with chewing mouthparts into a sucking head is thought to have come about. In an earlier study (Parsons, 1959), she studies the skeleton and musculature of the head of *Gelastocoris* and finds it to be similar to the head skeleton of other nepomorphan families (Naucoridae, Notonectidae). Rieger (1976) presents an extensive study of the head skeleton, based on *Ochterus*, and compares it to the other Nepomorpha and many other Heteropterous families. The gerromorphan head structure is discussed by Andersen (1982a).

The **compound eyes**, usually simply termed eyes, are generally large, especially in some nepomorphan families (Corixidae, Micronectidae, Naucoridae, Notonectidae), in which they occupy most of the surface in lateral view (fig. 5). In other nepomorphan families and in Gerromorpha the eyes are somewhat smaller but, with a few exceptions, well developed (figs. 6-8). The exceptions are *Nepa anophthalma* Decu et al. (Nepidae), a recently discovered troglodyte from Romania, and *Cavaticovelia aaa* Gagné & Howarth (Mesoveliidae), a Hawaiian troglodyte with vestigial eyes. In two small apterous species of *Cryptovelia* (Mesoveliidae; fig. 250), from the Amazon basin and Sarawak respectively, each compound eye is composed of three or four facets.

Ocelli are absent in most families with exception of the Hebridae, Gelastocoridae and Ochteridae. Mesoveliinae have ocelli in macropterous specimens and lack them in apterous specimens (figs. 2, 3).

Behind eyes the head narrows to form the neck which inserts into the pronotum, the thickened hind rim of the head around the occipital foramen (the hole where nerves and the alimentary canal enter the thorax) is called the postocciput. The dorsal part of head is called **vertex** or **cranium** (fig. 10, v).

In some groups with relatively small eyes the vertex may extend in front of and behind the eyes. The area between the eyes is often referred to as **interoculus**. In most species of water bugs interoculus and vertex roughly refer to the same area. The frontal part of the head is formed by the **clypeus**. It is more or less distinctly divided into two sections: a smaller anterior part, the **anteclypeus** (often simply called clypeus; figs. 288, 289) and the larger **postclypeus** (or **frons**; fig. 10, f). The ventral part of the head is the gula, which is usually well developed in Heteroptera (fig. 8, g) and absent in Homoptera (fig. 9). However, in e.g. Notonectidae and various genera of Veliidae the gula may be reduced (figs. 5, 7). The remaining lateral part of the head is formed by the genae, in front of the compound eyes. Other front parts have a few plates which are generally poorly recognizable. Of these the maxillary plates, which lie at the base of the rostrum, are sometimes useful in identification. In some gerromorphan groups the **ventral lobes**, or **bucculae**, are sutured off ventrally from the

FIGURES 5 – 9
Head and prothorax in lateral view.
5 *Anisops kuroiwae* – 6 *Aphelocheirus robustus* – 7 *Rhagovelia skoura* – 8 *Limnogonus hungerfordi* – 9 *Bothrogenia farinosa* (Fabricius) (Homoptera); a = antenna, a1 = first antennal segment, e = eye, g = gula, l = labrum, p = pronotum, r = rostrum, rp = rostral prong, t = trichobothria, t4 = additional pair of trichobothria, v = ventral lobe.

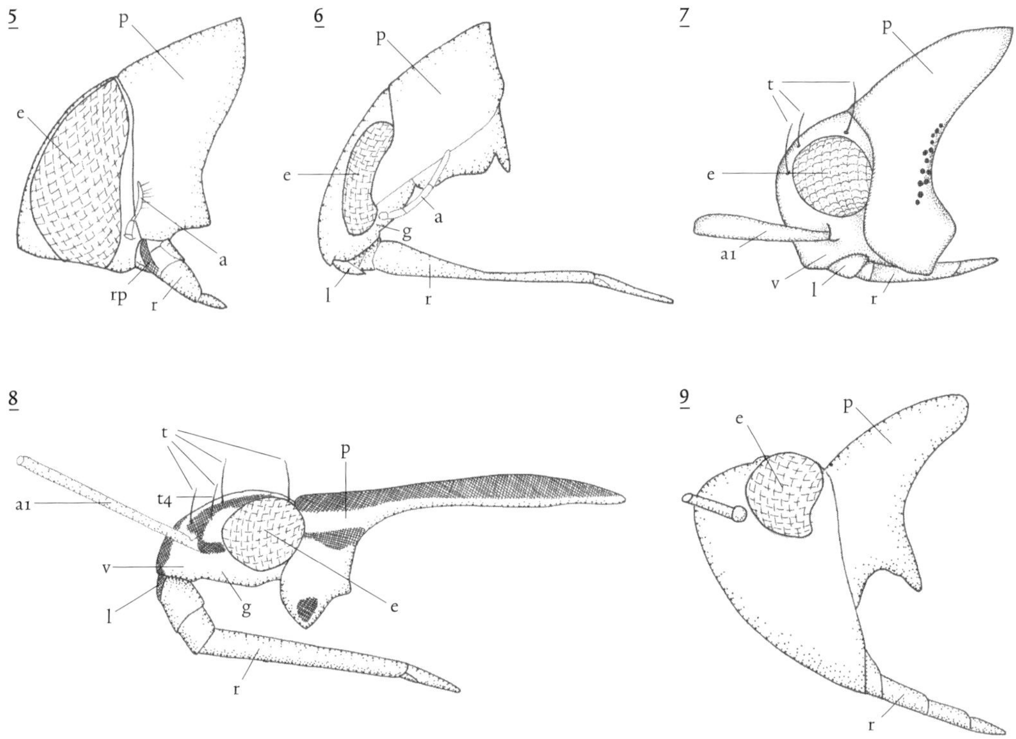

FIGURE 10
Angilia bispinosa Andersen, head in dorsal view; e = eye, f = frons, g = medio-dorsal impression on vertex, p = ocular pit, t = tricho-bothria, v = vertex.

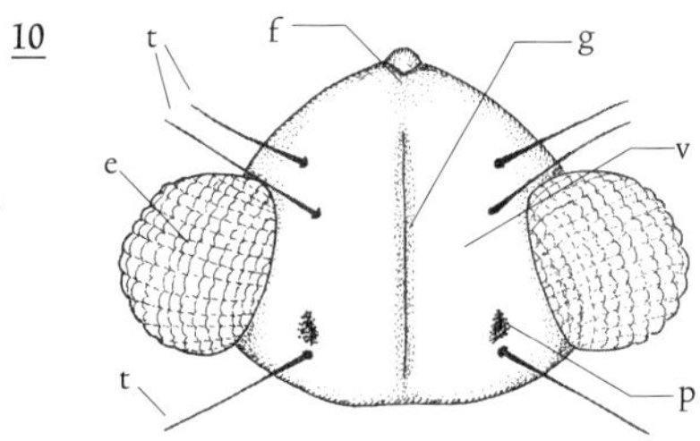

maxillary plates. These are especially important in identification of Hebridae (figs. 272, 276).

Apart from the eyes the head has two sets of appendages: the **antennae** and the mouth parts. In Nepomorpha the antennae are inserted beneath the eyes (figs. 5, 6, a), are usually shorter than the head and in most families concealed under the eyes, hence the old name Cryptocerata ('with concealed threads').

This reduction of antennae is supposed to attribute to the streamlining of the animals. The tactile function and most of the olfactory function are lost. However, in most species they function as an organ of equilibrium (see 'Johnston's organ'). Antennae of Gerromorpha are well developed, filiform and four-segmented. The genus *Hebrus* has secondarily five-segmented atennae. The relative length of antennal segments provides taxonomic characters in some cases, usually at the specific level. In recent literature there is a tendency to call the segments of appendages 'meres', e.g. antennomeres and tarsomeres for antennal segments and tarsal segments respectively.

The mouthparts are fused to form the **rostrum** or beak, which in Hemiptera shows a characteristic modification. An excellent and profusely illustrated study dealing with mouthpart structures and feeding strategies of Heteroptera was given by Cobben (1978).

At the base of the rostrum in front of the anteclypeus lies the **labrum**, a usually more or less triangular sclerite (figs. 6, 8 l). In Nepoidea, however, the labrum is fused with the anteclypeus and difficult to recognize. The outer sheath of the rostrum proper consists almost exclusively of the labium (the most posterior mouthpart), which is elongated, secondarily segmented and curled up anteriorly (dorsally). These modifications result in an apically closed tube with a wide posterior and a narrow anterior channel. There are four rostral segments with a small intercalary segment between segments 3 and 4, which, however, in most taxa of Gerromorpha and Nepomorpha are more or less fused with segment 3. Species with a long moveable rostrum have an elongated third segment (Ochteridae, Aphelocheiridae; fig. 6). The mandibulae and maxillae lie in and move through the narrower anterior channel. Figure 11 (based on Ekblom, 1926) presents a diagrammatical cross section of this narrower channel about halfway down the rostrum. The mandibulae and maxillae are modified into slender needles which are usually called **stylets**. The mandibular stylets lie anteriorly or laterally, and the maxillary stylets posteriorly or medially. The mandibular stylets have a cutting function and are often adorned apically with minute barbs or teeth. The maxillary stylets form two canals, an anterior one through which fluid food is pumped into the alimentary canal, and a much narrower posterior one, through which saliva is pumped into the prey (fig. 11, mxa, mxp). Proximally the labial tube is more open and covered by the labrum. The maxillary palps are absent and the stylets represent only the apical part of the maxillae. The basal part forms the so-called maxillary plates which are part of the head capsule. Inside the head the stylets diverge and are connected to the head wall by levers.

The anterior part of the muscular pharynx (cibarium, pharynx- or food-pump) pumps fluid through the maxillae. It is formed from epipharynx and hypopharynx. The deviations in the unsegmented rostrum of Corixidae are external only (Benwitz, 1956). The the central part of the pump is modified into chitinous structures in Corixidae, but present also in the remaining Nepomorpha; it

11

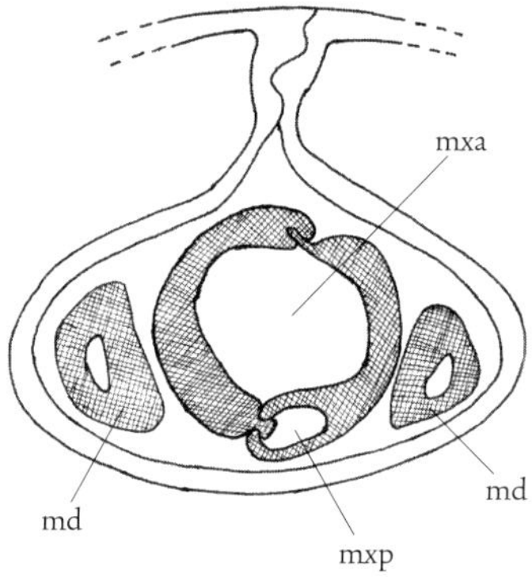

FIGURE 11
Diagrammatical cross section through the narrow channel in the rostrum of a heteropteron (after Ekblom, 1926); md = mandibulary stylet, mxa = anterior canal of maxillary stylets, mxp = posterior canal of maxillary stylets.

sieves and grinds the food. Similar structures occur only sporadically in other Heteroptera. These chitinous structures, almost absent in *Ranatra*, are considered to be an evolutionary reduction by Marks (1958). Parsons (1969b) on the other hand considers the less developed chitinous structures in Belostomatidae and Nepidae as original.

Trichobothria are large sensory bristles, almost always distinctly longer and tapering more gradually than the surrounding hairs. They occur in both larvae and adults of many groups of Heteroptera and other insects. The trichobothria of special interest are the (normally) three pairs of cephalic trichobothria: a posterior pair at the back of the head, usually close to the hind corners of the eyes, and two anterior pairs on the front of the head, usually with one pair just behind the other (fig. 10). A fourth pair, present in most adult Gerridae, is inserted laterally between the two anterior pairs (fig. 8, t4) and usually of the same structure as in the larva. In larvae the seta or 'trich' (from the Greek tricha = hair) arises from a dome-like elevation of the cuticula, called bothrium. In gerromorphan adults the bothrium is located at the bottom of a depression. The function of the trichobothria is apparently sensory, but to which stimulus they respond has not been established. A survey of trichobothria in various Heteroptera is described by Schuh (1975).
Trichobothria should not be confused with **macrotrichia**, which are simply longer setae. Notably in Gerromorpha there is a double layer of hairs, which in Gerroidea cover the whole body. In other families they occur on head, thorax and basal part of the abdomen. The dense layer consists of very short (length 2-10μ) cuticular outgrows, the **microtrichia**, and a layer of much more widely set and longer (length 20-100 μ) pointed hairs (macrotrichia), flexibly inserted in sockets. The latter are divided in two or three length classes on a single specimen. The function of the microtrichia layer, which shows similarities with the plastron of some Nepomorpha (see respiration), is supposed to avoid the animal becoming wet when accidentally submerged (Andersen, 1977a). Length and structure of both macrotrichia and microtrichia differs between genera and species.

Thorax

The thorax consists of three segments: **prothorax**, **mesothorax** and **metathorax**. These are most easily recognized in several apterous Gerromorpha (fig. 2). Their exoskeleton consists of sclerites with the following nomenclature: the dorsal sclerite (tergum or tergite) is named **notum**, the lateral sclerite **pleurite** or **pleuron** and the ventral sclerite (sternite) is named **sternum**. They are distinguished in the same way as the segments: pronotum, mesonotum, and metanotum/propleuron, mesopleuron and metapleuron/prosternum, mesosternum and metasternum.

In macropterous specimens the main part of the visible thorax in dorsal view is the pronotum (fig. 3). In most Nepomorpha (excluding Corixidae) there is a visible triangular sclerite behind the pronotum, the **mesoscutellum**, usually called **scutellum** (fig. 44). The mesoscutellum is a part of the mesonotum. In Hebridae the apparent scutellum is part of the metanotum, whereas the mesoscutellum is a short transverse plate at its base (fig. 285). The remainder of thorax and abdomen is usually covered by the hemielytra. Most Gerromorpha, with the exception of Hebridae and Mesoveliidae, lack the scutellum.

There are ten pairs of (tracheal) spiracles, three of which are situated on the thorax after development of the embryo. In the embryo the spiracles develop one segment posterior of where they appear in the larva and adult. Thus, the metathoracic spiracle in adults and larvae is the one belonging to the often greatly reduced first abdominal segment. Contrary to terrestrial Heteroptera and Gerromorpha, in Nepomorpha there are no structures to close the spiracles, as these closing mechanisms are meant for water retention. In addition, due to the specialized respiration in various Nepomorpha, some spiracles may be more or less distinctly reduced. See also the paragraph on adaptations to aquatic life (p. 24).

Each thoracic segment has a pair of legs, while the mesothorax and metathorax have a pair of wings each. The wings may be reduced or absent.

The legs consist of **coxa**, **trochanter**, **femur**, **tibia** and **tarsus** (figs. 12-15). In Gerromorpha, the tarsi have more than one segment in at least one pair of legs. In larvae all tarsi are usually one-segmented. The fore legs are mainly used in prey capturing and/or handling. Otherwise, function and specialization of the legs is different between the infraorders.

In Gerromorpha the positioning of claws on the second tarsal segment is an adaptation to semi-aquatic life. Species living on the water surface have subapical tarsal claws (fig. 13), such as in most Gerridae and Veliidae. Taxa mainly living on land or on water plants often have apical claws, e.g. Hebridae, Hydrometridae and Mesoveliinae. Males of many Gerromorpha have grasping fore legs for holding the female during copulation. This is often restricted to a small comb of pegs (fig. 13) but it can also result in modifications of the femur and tibia, notably in Gerridae (figs. 453, 532).

In many Nepomorpha the fore legs have a strongly developed femur (fig. 12), usually with an anterior groove, against which the tibia and tarsus can be pressed to capture and hold prey. Although fore legs of e.g. *Enithares* are also raptorial, **'raptorial fore leg'** is reserved for the type with the strongly incrassate femur. In some taxa the non-raptorial fore legs show structures used in sound production. These are sometimes restricted to males (in Notonectidae: Anisopinae), but may also occur in both sexes (various Corixidae). During mating the males of most groups use both fore and middle legs for clasping the females. The exception is formed by Corixidae, and possibly Micronectidae, where only the fore legs are used. Excepting semi-terrestrial Nepomorpha, such as Ochteridae which have legs for walking, legs and body shape of Nepomorpha are adapted to life under water.

Wing

In Heteroptera the fore wings or **hemielytra** (alternatively spelled as hemelytra) are leathery basally and membranous distally. In most Heteroptera the hemielytra when fully developed can be divided in three parts: **clavus**, **corium** and **membrane** (fig. 16). In Gerromorpha there is hardly a structural difference between the basal and apical part of the wings. Exceptions are notably Hebridae, Mesoveliidae and *Perittopus* (fig. 295, Veliidae), in which the veins are strongly chitinized and restricted to the basal half of the hemielytron, whereas the apical part is membranous. Autotomy (tearing off the apical part of the wings after dispersion flight) is observed in various genera of several families of the Gerromorpha (Mesoveliidae, Veliidae, Gerridae) and the nepomorphan family Aphelocheiridae. Specimens with autotomous wings are called dealate (fig. 3). The hemielytra of taxa showing autotomy have fixed break lines: they are situated at the base of the wings (e.g. Ptilomerinae); distally of the closed cells of the hemielytron (e.g. *Mesovelia*, Trepobatinae), or behind the corium (e.g. Aphelocheiridae). Within the genus *Rhagovelia*, break lines at the base of the wing or distally of the closed cells may occur according to species group.
In most Nepomorpha the membrane is recognizable by the membranal suture, although the difference in texture of membrane and basal wing part may be small. Within the aquatic families, Notonectidae have the most typical membranes, whereas in Helotrephidae and Pleidae they are absent or modified. In Nepomorpha the venation of the hemielytra is not or hardly recognizable except for a vein running through the clavus, which is distinct in several families. In addition, the membranes of Nepoidea and Ochteroidea show a number of veins and cells (figs. 44, 65). In macropterous Gerromorpha there is usually a number of well developed longitudinal and cross veins in the hemielytra. Cells are formed by cross veins connecting longitudinal veins, or by fusion with the longitudinal veins (fig. 18).
In both Gerromorpha and Nepomorpha the hind wings are membranous and show a variable number of well developed longitudinal veins. Andersen (1982a)

FIGURES 12 – 15
Legs of Nepomorpha and Gerromorpha.
12 *Naucoris scutellaris*, fore leg, dorsal view – 13 *Microvelia douglasi* fore leg of male and tibia and tarsus of female, lateral view – *Enithares intricata* 14 middle leg – 15 hind leg, ventral view; cl = claws, co = coxa, f = femur, ta = tarsus, ta1 = first tarsal segment, ta2 = second tarsal segment, tc = tibial grasping comb, ti = tibia, tr = trochanter, va = ventral arolium.

12

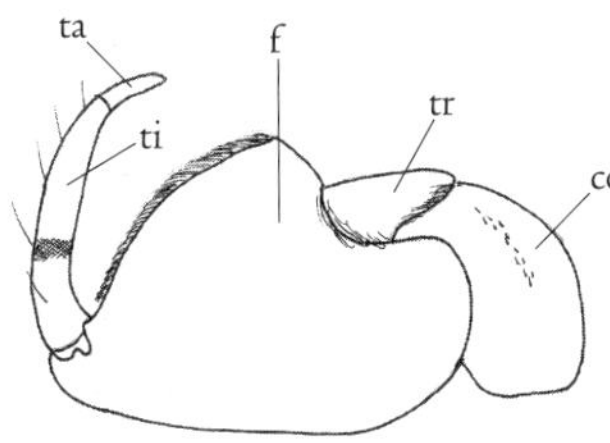

13

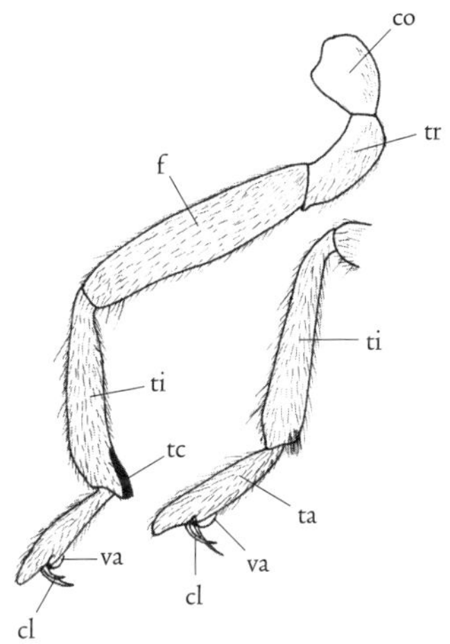

14

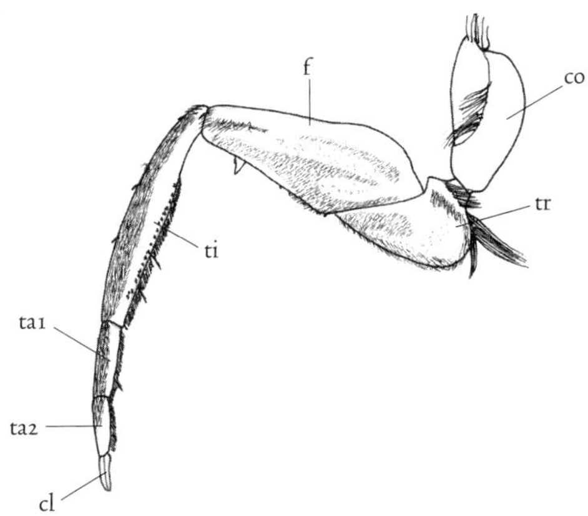

15

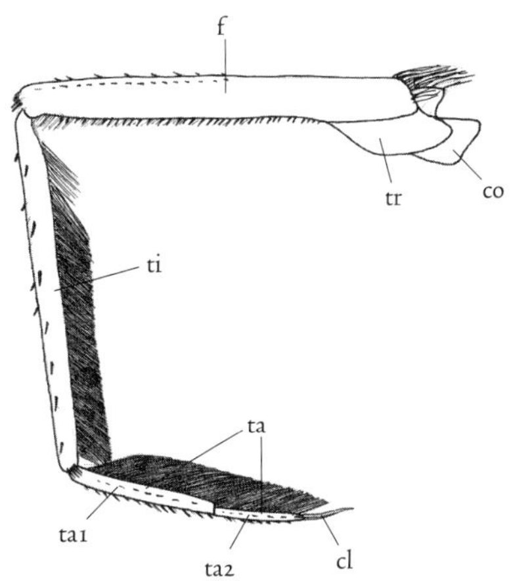

discusses the wing venation of gerromorphan families. For Nepomorpha there is no survey, although the hemielytra of several families are depicted in Mahner (1993). In other cases basic morphological papers mentioned under some family accounts have to be consulted. Longitudinal veins of the insect wing are considered to have derived from a single ground plan containing much more veins than found in Heteroptera. The names applied to the veins are similar in all insect orders and are considered to represent homologous veins. Although differences of venation may be diagnostic for subfamilies or sometimes genera, the names of the veins are hardly used in keys within the Gerromorpha or Nepomorpha. Therefore, we provide only one example of each of the infra-orders: in most insects the vein forming the anterior wing margin is called the costa (figs. 16-19). This vein seems to have vanished in Heteroptera, but the term **costal margin** for the anterior margin is still used.

Wing polymorphism

Wing polymorphism, the various differences in wing length, occurs in many species. The **macropterous** morph has fully developed wings (fig. 368) and is the only one which may be able to fly. The capacity to fly depends on the development of the indirect flight muscles which, even in morphologically macropterous morphs, may not be functional. This is why a division is often made between flying and non-flying specimens of a particular species. In **apterous** specimens wings are completely absent (fig. 367). In **meiopterous** specimens wings are reduced (figs. 283, 287). They are further divided into **brachypterous** (wings roughly half as long as in the macropterous morphs) and **micropterous** (vestiges of wings) specimens. In such individuals, notably various Gerrinae, the wings can be hidden under the posterior lobe of the pronotum, the specimen looking apterous at first sight. Some authors use the prefix 'sub' to indicate intermediate states. Hence, a submacropterous specimen has somewhat reduced wings, but not as strongly as in a brachypterous specimen. If such detail is needed it is better to indicate how far the hemielytra reach respective to the tergites, e.g. hemielytra 4 means they reach to the caudal margin of tergite 4, and $4^1/_2$ means they reach halfway tergite 5. The use of terminology is sometimes rather loose and may differ somewhat among families. In Nepomorpha usually only macropterous and brachypterous morphs are encountered. Exceptions are found in Aphelocheiridae and some Naucoridae, which have plastron respiration. In *Aphelocheirus* the species with strongly reduced hemielytra qualify as micropterous, although they are called brachypterous by many authors. In Nepomorpha the term submacropterous is sometimes used for brachypterous morphs (distinctly reduced hind wings, only slightly reduced hemielytra). We prefer to restrict submacropterous to morphs with slight reduction of both hemielytra and hind wings. These morphs are rare and have not yet been reported in any Malesian species.

Wing polymorphism and thorax development are often correlated with life cycle. In temperate regions there may be two generations annually. One generation with fully developed wings overwinters, while the next generation, of which a considerable fraction of specimens has reduced wings, is active in summer. In the tropics generations are not annually regulated, and wing polymorphism is more related to habitat choice. Species living in temporary or otherwise unstable habitats are partly or entirely macropterous; species living in stable habitats have as a rule predominantly reduced wings. Strictly marine species (e.g. *Halobates*, some genera of Haloveliinae) live in very stable habitats

FIGURES 16 – 19
Wings of Nepomorpha and Gerromorpha.
Enithares sp. **16** hemielytron – **17** hind wing – *Limnogonus* sp. **18** hemielytron – **19** hind wing; capital letters indicate longitudinal veins: al = anal lobe, 1An = first anal vein, 2An = second anal vein, cl = clavus, co = corium, Cu = cubitus, m = membrane, M = media, n = nodal furrow, Pcu = postcubitus, R = radius, r+m = cross vein between radius and media, Sc = subcosta; stippled = folding lines.

16

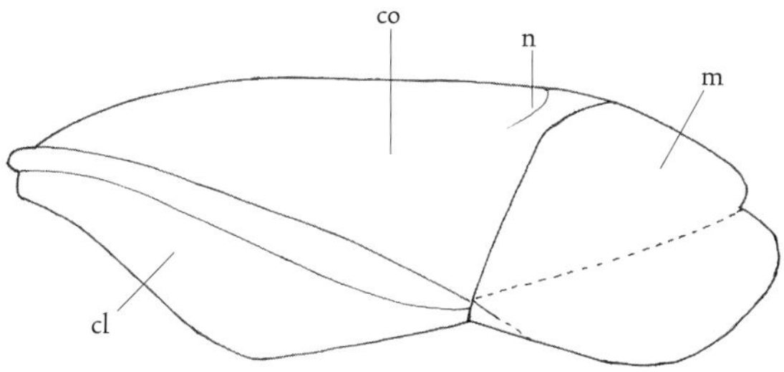

17

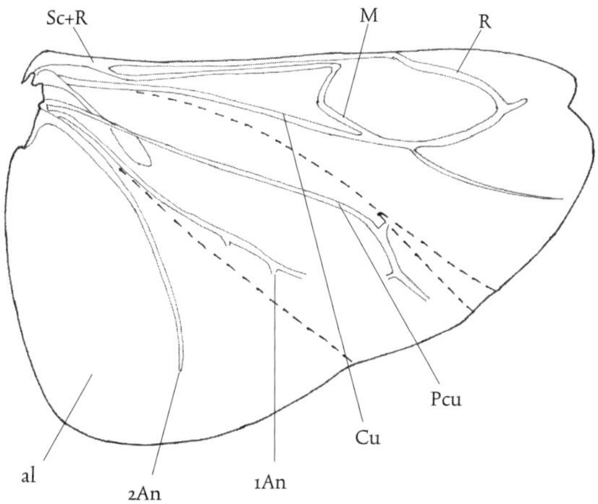

18

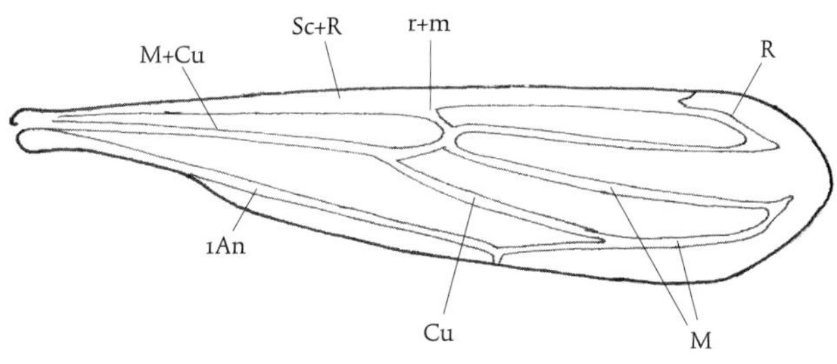

19

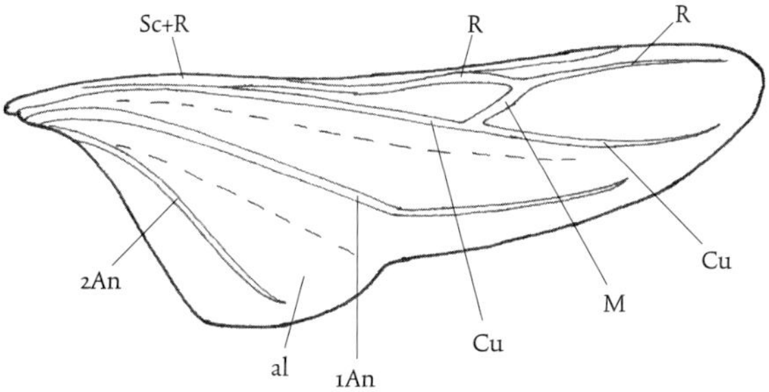

and are only known in apterous morphs. There may be a correlation between wing polymorphism and seasons (e.g. dry and wet seasons), but this has thus far not been investigated.

In Nepomorpha there are various structures keeping the hemielytra in place when closed. These are called **co-aptatations**. The main function seems to be to prevent air lifting the wings while under water. The co-aptatations usually consists of folds and rims along the costal margin of the hemielytra, margin of the abdomen, claval margin, margin of the scutellum and along a depression in the costal margin of the hemielytra; the knob on the mesopleuron fits in this depression. This structure is rather similar in Nepomorpha, except for Corixoidea (Rieger, 1973), in which the knob lies on the costal margin and the depression on the pleura.

Abdomen

The dorsal sclerites of the abdomen are called (abdominal) **tergites**, the ventral sclerites of the abdomen are called **sternites**. Roman numerals are used for reference to the the thorax segments, and arabic numerals for the ones on the abdomen. TII thus refers to the mesonotum, T2 to the tergite of second abdominal segment. We use full names (e.g. pronotum, metasternum) for the thoracic sclerites.

The abdominal **pleurites** (lateral sclerites) are considered to have fused with the sternites, which carry the abdominal tracheal spiracles. In Gerromorpha the spiracles are directed straight forward (visible in lateral view; fig. 22, sp). In Nepomorpha some or all spiracles may be hidden in some families (Nepidae), or show special adaptations (Aphelocheiridae, fig. 73). In unmodified abdomina they are visible on the venter (figs. 51, 79). The distinct lateral regions of the abdomen in dorsal view (fig. 2) are named **connexiva** (one connexivum at each side of the abdomen). In Gerromorpha connexiva are formed by dorsal sclerites, called laterotergites. The central dorsal sclerite represents only a part of the tergite and should formally be called mediotergite. In taxonomic literature it is however usually called tergite. In Nepomorpha the situation is somewhat more complex. In most families the laterotergites are folded ventrad so that they are also visible in ventral view, but they are indistinct or absent in Aphelocheiridae, Corixidae and Micronectidae. In Helotrephidae the narrow ventral part of laterotergites is usually hidden beneath a rim of the hemielytra. In families which have distinct ventral parts of the laterotergites there is often an additional secondary division in the abdominal sternites, resulting in a division in median sternites flanked by a pair of parasternites (fig. 50, 79). Especially in Gerromorpha the connexiva furnish quite often useful identification characters.

The first abdominal segment is often fused with the metathorax and difficult to recognize. In other instances abdominal tergite 1 is well defined (often larger

than subsequent tergites), but the sternopleurite is fused with the metathorax, except in the Pleidae (Papáček, 1987).
In Gerromorpha abdominal segment 7 is often larger, but apart from that there is usually little external differentiation between segments 2-7. Again, in Nepomorpha the situation is more complex; it will be discussed under the respective families.
The genitalia proper are situated in segment 9. Segment 8 is often modified while segment 10 (supposed to be a fusion of segments 10-11) is strongly reduced and forms an, often rather small, dorsal sclerite (**proctiger**, **anal lid** or **anal cone**), covering the aedeagus and ending in the anus. In some male Nepomorpha (Aphelocheiridae, Corixidae, Ochteridae) the posterior part of the abdomen is asymmetrical. The chitinous parts of the genitalia are often important for specific identification.

The genitalia, especially in males but sometimes also in females, often provide good characters for specific identification. The genital morphology should both promote fertilization results and function as a barrier against interspecific mating. It is interesting in this respect to note that in some groups using sound during courtship (*Anisops*, and some species groups in Corixidae) the secondary characters on the sound producing apparatus (notably the fore leg in males, fig. 154) show more distinct specific characters than the parameres. An extensive study of genitalia and mating behaviour of Nepomorpha is given by Larsén (1938), of Gerromorpha by Andersen (1975, 1982a). Scudder (1959) offers a good treatment of female genitalia of Heteroptera in general, and Dupuis (1955), apart from a morphological description, presents a useful bibliography and inventory, including definitions of terminology (in English, German and French).
In male Nepomorpha the genitalia are retracted in the abdomen. Segment 8 is modified in such a way that the true genital segment 9, the **pygophore** or **genital capsule**, can protrude during copulation. Generally, the genital capsule has a strongly developed (pleuro)sternite which is elongated caudally. The genital capsule encloses a hollow space, the genital chamber (atrium genitale), in which the copulatory organ (penis, phallus or **aedeagus**) rests (fig. 81). Tergite 9 however is very small and hardly recognizable. The membrane between segments 9 and 10 is large and forms the membranous lining of the genital chamber. The genital chamber opens either caudally or dorsally. In the latter case the lateral and caudal margins of sternite 9 are curved dorsad. Attached to the margin of sternite 9 at the sides of the opening of the genital chamber is a pair of **parameres** (genital claspers) which are of different size depending on (sub)family or genus (figs. 34, 35, 81, 150, 151). These are of various, often species specific, form and can operate independently from the aedeagus during copulation.

In male Gerromorpha segment 8 is cylindrical and narrower than the preceding segments, and although partly retracted into segment 7, it is usually visible at the apex of the abdomen. Segment 9 is more strongly developed ventrally than dorsally. Its proximal cylindrical part lies inside segment 8, whereas the open distal part is exposed. The proctiger covers the dorsal opening of segment 9 as a lid. The parameres in some taxa, notably the Gerrinae and the tribe Metrobatini of the Trepobatinae, are strongly reduced and of little use in identification. The aedeagus of Gerromorpha is primitive. It consists of a **basal plate** (or basal articulatory apparatus) which is attached to the pygophore, to which the parameres are attached. Apically of the basal plate are the sclerotized **phallotheca** and the **endosoma**, consisting of soft tissues. The basal part of the endosoma is called the **conjunctivum** and the apical part the **vesica**. Within the vesica lie sclerites which, especially in genera with reduced or uniform parameres, are important in identification. After removing the proctiger from a pygophore the vesica becomes visible, lying folded back in the phallotheca (figs. 20, 384). When the basal plate is detached and the vesica folded out, the entire aedeagus is visible (fig. 21). To study the vesical sclerites they have to be cleared with KOH or enzymes. Alternatively the phallotheca with the vesica inside can be cleared until the vesical sclerites are visible. In some cases the pigmentation and/or form of the tip of the phallotheca is of taxonomic importance (J. Polhemus & D. Polhemus, 1993). The vesical sclerites consist of four pairs of median sclerites: the anterior (= apical), dorsal, basal and ventral sclerites (figs. 372, 429). However, some of the sclerites may be fused or reduced, sometimes leaving only one pair of median sclerites (figs. 373, 374). Moreover, the median sclerites may be fused medially, suggesting unpaired median sclerites (fig. 385). In addition, the pair of lateral sclerites (figs. 429, 435) may also be reduced or absent. To

20

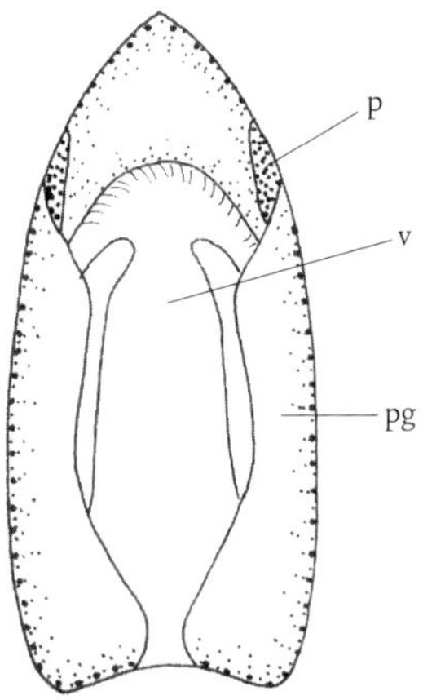

21

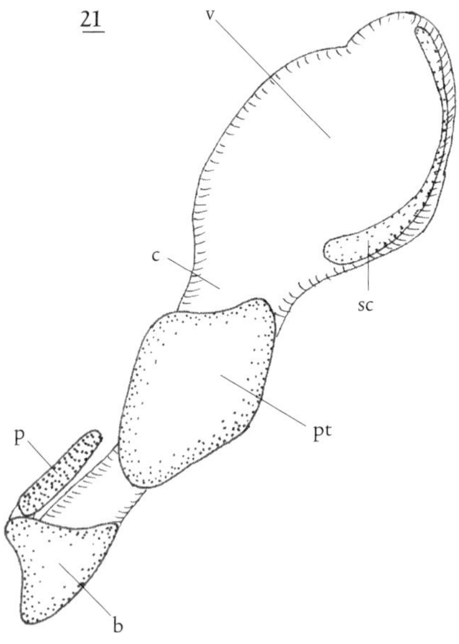

FIGURES 20 – 21
Male genitalia semi-diagrammatically, based on *Cylindrostethus persephone*. 20 pygophore and vesica, dorsal view, after removal of proctiger; 21 extended aedeagus; b = basal plate, c = conjunctivum, p = paramere, pg = pygophore, pt = phallotheca, sc = vesical or endosomal sclerite, v = vesica.

complicate matters there may be additional vesical sclerites, which are called **accessory sclerites** (fig. 429), and/or sclerotized structures on the wall of the vesica, which are called plates. The definition of vesical plates and vesical sclerites is given by Andersen (1975, 1982a). Matsuda (1960) suggested that these vesical sclerites were originally separate and have been fused in more specialized groups.

The female genital apparatus is more uniform (Scudder, 1959), although in Nepomorpha it is retracted within the abdomen, whereas in Gerromorpha it is often well visible from the outside. The ovipositor is formed by segments 8 and 9. Abdominal sternite 8 is divided into two plates: the first gonocoxites or **gonocoxae**. They are usually medially connected by a membrane. In most Gerromorpha these cover the rest of the genital apparatus (fig. 22, 23). Segment 9 also forms a pair of (second) gonocoxites (absent in Gerridae except the Rhagadotarsinae), which are, in resting position, nearly always retracted into segment 8. The ovipositor proper is formed by two pairs of **gonapophyses**. These are the apical parts of sternites 8 and 9 and connected to the gonocoxites in an articulating way, except in most Gerridae, where the second gonocoxites are lacking. Species depositing eggs in plants have more strongly developed ovipositors compared to species which glue the eggs onto the substrate (figs. 24, 239, 247). Tergite 9 in some Veliidae is converted into an anal lid which closes the apex of abdomen more or less completely. In other taxa of Gerromorpha tergite 9 is reduced. In some Nepomorpha the genital apparatus is covered ventrally by a **subgenital operculum**, often called genital operculum, either in both sexes (sternite 7 in Nepidae and Belostomatidae; figs. 37, 51; sternite 8 in Pleidae) or in females only (sternite 7 in Aphelocheiridae and Naucoridae; fig. 80).

22

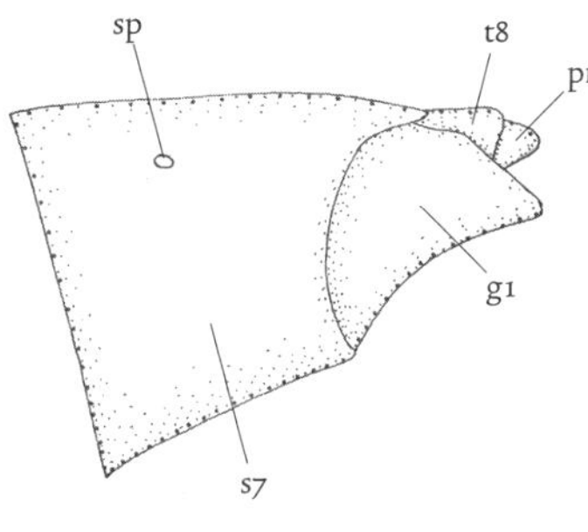

23

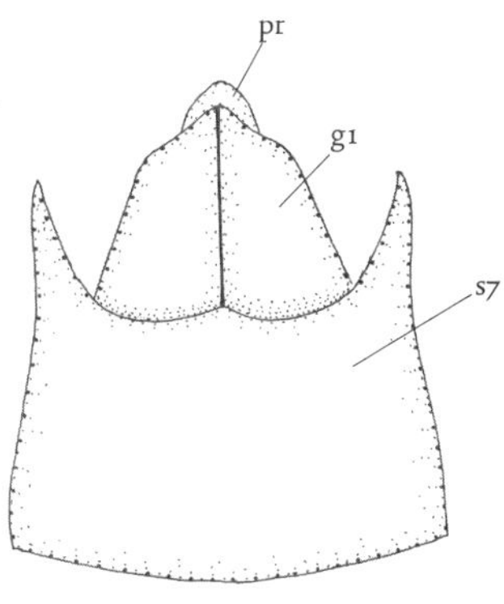

FIGURES 22 – 23
Apex of abdomen of female *Cylindrostethus persephone*.
22 lateral view – 23 ventral view; g1 = first gonocoxa, pr = proctiger, s7 = sternite 7, sp = spiracle, t8 = tergite 8.

24

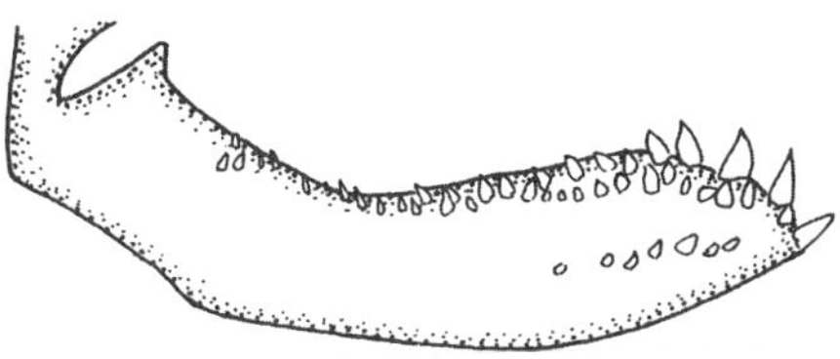

FIGURE 24
Ovipositor of *Anisops rhomboides*.

Adaptations to aquatic life

In this section a few adaptations of Nepomorpha are discussed. Adaptations of gerromorphan species to life on the water surface, like hydrofuge pilosity, long to very long middle legs adapted for rowing, subapical claws, receptors for detecting surface disturbances, and other tarsal modifications, are extensively dealt with by Andersen (1982a).

Respiration

Insects breathe through a system of tubes. Near the organs gas exchange is realised through extremely small tubes. The larger tubes are called tracheas, the very narrow ones tracheoles. The larger tracheal trunks usually consist of a pair of main longitudinal tracheas running along the sides of the abdomen; another major longitudinal pair runs dorsally through the abdomen and thorax. The dorsal tracheas are connected with the lateral ones by segmental tracheas. The lateral trunks are usually connected with each other by transverse tracheas and with the exterior by (tracheal) spiracles, or stigmata. Most insects also have thoracic spiracles. Gas is exchanged by diffusion, aided by some osmosis in the tracheoles near the organs. It has clearly developed in terrestrial ancestors.
One of the precarious aspects of the life of most aquatic insects is to obtain sufficient oxygen. The two principal reasons are: 1) the respiratory system evolved to be used under atmospheric conditions, and 2) under normal conditions the amount of oxygen available in water is about 10,000 times less than in air, and can under unfavourable conditions drop to zero. Gerromorpha use atmospheric air, as do most adult Nepomorpha; in this way they avoid the problem to get oxygen from the water. The drawback is that these Nepomorpha are obliged to surface regularly to replenish their gas stores. The following types of respiration are found in Nepomorpha (Thorpe & Crisp 1949; Kovac 1987): tracheal gill or cutaneous respiration, reservoir breathers, air bubble respiration, and plastron respiration. They are each discussed below.

Tracheal gill or cutaneous respiration

In this type of respiration oxygen diffuses into the closed tracheal system through the thin cuticula. The gas volume in the tracheal system is constant, which causes a continuous diffusion of oxygen in the direction of the lower partial pressure in the system. This type of respiration is found in larvae of *Aphelocheirus* and *Micronecta*, and apparently also in first instar larvae of Corixidae. Therefore, even when there are no gill-like structures (as seen in for instance various Ephemeroptera and Zygoptera larvae), this type of respiration is called 'respiration by tracheal gills'. Wesenberg-Lund (1943: 134) mentioned that in various aquatic bugs rectal respiration plays a role, referring specifically to larvae of *Ranatra*. To our knowledge, detailed study of this aspect is not known.

Reservoir breathers

Aquatic bugs catergorised as reservoir breathers almost exclusively use atmospheric air to cover their oxygen need. When diving they use oxygen from air stores on their body, which has to be replenished regularly. The air is stored under the hemielytra or in ventral grooves covered by hair and is not, or only very limited, in contact with the surrounding water. This type of respiration is found in Nepidae, Notonectidae and Belostomatidae. The genus *Anisops* (Notonectidae) is a special case: it has specialised haemoglobin cells in the abdomen in which the oxygen reserve is stored.

Air bubble respiration

Here atmospheric air is also used, and the animals have to surface regularly to renew the air store. The air is not concealed but at least partially transported in a bubble which is in contact with the water. When oxygen is used, the partial oxygen pressure in the bubble lowers and oxygen from the water diffuses into the bubble. At the same time carbon dioxide is moving outward. However, the partial pressure of nitrogen in the bubble tends to be relatively high and although nitrogen diffuses more slowly than oxygen, the nitrogen gradually disappears from the bubble. When the nitrogen level becomes too low, the bubble collapses, so the water bugs have to get a new supply of nitrogen instead of oxygen from time to time. It is calculated that under favourable circumstances an insect can obtain eight times as much oxygen from a bubble in one dive compared to the amount of oxygen originally available. The system is known as a **physical gill**. This type of breathing is found in Corixidae, Helotrephidae, Micronectidae, Pleidae and many genera of Naucoridae. *Notonecta*, and probably also *Enithares* (both Notonectidae), seem to have an intermediate position: the ventral bubble and the air under the hemielytra is supposed to function more or less as a reservoir. The air film on the outer side of the hemielytra functions a physical gill (Ruiter et al., 1951). Corixidae have a smaller air film on the basal part of the hemielytra, but in these species the bubble on the venter functions as physical gill. According to Ruiter et al. (1952) the higher concentration of carbon dioxide inside the air bubble stimulates movements of the legs. These movements are especially numerous in Corixidae and cause a current of fresh water to flow over the bubble, thus accelerating the oxygen intake.

Plastron respiration

A plastron consists of a layer of very fine, closely set hairs with curved tips which covers a large part of the body (most developed in *Aphelocheirus*, with 43,000/mm^2). A permanent air film is maintained between the hairs where gas exchange takes place. *Aphelocheirus* can stay submerged permanently. This breathing type is sometimes referred to as 'non-collapsable physical gill', to distinguish it from the 'normal' air bubble, which is then referred to as 'collapsable physical gill'. In species using this type of breathing, the stigmata are sometimes surrounded by specialised areas with a rosette of small grooves, which are supposed to facilitate air exchange into the interior of the tracheal system (fig. 73). Experiments with the European species *Aphelocheirus aestivalis* (L.) have shown that the air film in the plastron stays present even in completely gas-free water, and that a pressure of four to five atmospheres is needed to collapse the film (Thorpe & Crisp, 1947). Apart from *Aphelocheirus* some bottom-dwelling Helotrephidae and Naucoridae apparently show plastron respiration as well. This has been demonstrated for some American genera of Naucoridae. In Malesia especially some genera inhabiting the bottom of fast flowing streams possibly use plastron respiration.

Parsons (1970) shows that function and structure of (collapsable) air bubbles and their associated spiracles varies between families. Apart from the supra-alar air film in Corixidae and Notonectidae, there are usually also air stores on the abdominal and part of the thoracic venter, under the hemielytra, and between the first two thoracic segments in Naucoridae, Notonectidae and Corixidae. The Corixidae have an additional air store between head and prothorax. These air stores are interconnected, sometimes indirectly. The important spiracles in respiration are the mesothoracic and metathoracic and first abdominal pairs. They are intersegmental, lie in air-filled spiracle chambers and are distinctly larger than the remaining spiracles.

The spiracle chambers are directly connected with parts of the air stores, but are only indirectly connected to each other. Apparently the first abdominal pair of spiracles is used for inhaling and the thoracic spiracles for exhaling, except in Corixidae, where the metathoracic spiracles are used for both in- and exhaling. The intake of fresh air in the air stores also varies between (sub)families. Naucorinae and Laccocorinae take in air at the apex of abdomen, from where it is directly stored under the hemielytra. Air exchange with water takes place in the ventral bubble. Air from the metathoracic spiracles is exhaled into the ventral bubble on the thorax. The ventral air store of Notonectinae is divided into two channels by a median keel. Fresh air also enters at the tip of the abdomen. Gas exchange takes mainly place in the supra-alar air film. Species of Corixidae take in atmospheric air through the space between head and pronotum and the mesothoracic spiracle chamber, and further air exchange predominantly both in the ventral and supra-alar air stores.

Sensory organs

Scolopal (chordotonal) organs

In the air chambers of the thorax and first abdominal segment of Nepomorpha lie the so called **tympanal** organs (Wotzel, 1933) or **scolopal** (or scolopophorous) organs. They are found in most insects. These organs are characterized by a group of specialized sensors, each containing a relatively conspicuous sensory rod (the scolops or scolopale). The nerve ending is drawn out to a slender prolongation enclosed by an envelope cell and a cap cell. The scolops is formed by the envelope cell and is distally nearly always thickened to form an end knob. The entire rod lies in a liquid medium in which it can move freely. The cap cell attaches the organ to the inner body wall beneath the cuticle. At the other end the organ is attached to the cuticle by connective tissue which forms a sheath around it.

Scolopal organs are thought to react to pressure or change in pressure (including sound waves). They were initially found in tympanal (hearing) organs and were thought to be sound receptors only. However, they are stimulated by tension and are used for various purposes such as to detect pressure changes, mechanical vibrations, and sound. Apart from Johnston's organ (see below), which also contains cells with a scolops, scolopal organs in Nepomorpha especially occur in the air stores of mesothorax and metathorax and the first abdominal segment. In some families at least the mesothoracic scolopal organ reacts to sound and thus seems to be a true tympanal organ. The function of other scolopal organs is still unclear, but it is supposed to either register the orientation of the insect in the water, or to be pressure receptors checking the depth of a dive.

Parsons (1969a) drew attention to the fact that in micropterous *Aphelocheirus* the scolopal organs of mesothorax and metathorax and the first abdominal segment are situated in the subalar air bubble. An extensive study of these organs in Nepomorpha was published by Larsén (1957). He suggested that they were most likely sensory organs related to equilibrium. The hearing function, which is widely attributed to these mesothoracic and metathoracic sensory organs, has not been proven. This is why he preferred the neutral indication scolopal organs over tympanal organs. However, Arntz (1972, 1975) proved that the mesothoracic scolopal organs react to airborne sound in the genera *Nepa* and *Notonecta*. In addition, Popham (1961), Prager (1973, 1976) and Prager & Streng (1981) have shown that in adult Corixidae the mesothoracic organ reacts to sound. Hence, in recent literature the term tympanal organ is used again. The microscopic structure of the tympanal organ of *Callicorixa praeusta* (Fieber, 1848) was studied by Michel (1977).

Johnston's organ

One of the more important sensory organs in the antennae of Pterygota (winged insects) is known as Johnston's organ. It is an organ located in the second antennal segment. The antennae and their sensory organs in Nepomorpha were studied by Hüfner (1939). He found that in Nepomorpha Johnston's organ is also located in the second antennal segment, thus refuting the opinion of some authors that the first antennal segment in this infraorder evolved from fusion of two segments. The fundamental function of Johnston's organ is the perception of movement at the joint between the second and third antennal segments. As a consequence, an insect can respond to various stimuli such as sound, gravity, and/or speed of air current. In true aquatic nepomorphans it is part of the equilibrium organs. The antennae lie against an air bubble at the back of the head. When the insect changes position the bubble moves because of its buoyancy, displacing the antennae and thus causing Johnston's organ to respond (Rabe, 1953). In the families studied in this respect, Johnston's organ is best developed in Corixidae and Notonectidae, and least developed in Nepidae, whereas in Aphelocheiridae and Naucoridae the development is intermediate. There are various additional means to detect pressure change. Species which carry an air store, such as in Corixidae and Notonectidae, have sensory hairs among the hairs lining the air store. When pressure changes the airstore is compressed or expanded, which moves the sensory hairs. Bottom dwelling species usually have a specialized set of organs, called static organs, which react on pressure changes and check the orientation in the water. As these are located differently in the various families they are mentioned in the general parts of the families Aphelocheiridae, Naucoridae and Nepidae.

Secretion grooming

Most Heteroptera possess metathoracic secretory glands, usually called scent glands. They are integumentary invaginations oriented ventrally in the metathorax. There may be a single gland with a single or a paired opening (**ostiole**), or paired glands opening into closely or widely spaced ostioles. The structure of the glands differs from simple to complex. The simple ones consist of a single gland which produces and stores scent fluid. The complex ones consist of one or more glands secreting into a reservoir which may have an accessory gland attached to it. The accessory gland probably delivers enzymes which are important in production of the final scent gland chemicals. Older hypotheses postulate that the secretion, because of bad smell or taste, functions as a defense against predators. This is true for several terrestrial Heteroptera but not for true water bugs. Moreover, in terrestrial bugs the function of the secretion varies as well. A survey of structure and function of these glands in Heteroptera is given by Staddon (1979).

The Gerromorpha, with the exception of the Hydrometridae, possess metathoracic scent glands with a single opening on or near the posterior margin of the metasternum. The area around the opening is sometimes slightly differentiated (e.g. somewhat elevated or with a hair fringe) and called **omphalium**. In Veliidae and some Gerridae there are channels from scent glands to the sides of the body. In Veliidae hair tufts at the end of these channels are considered to aid in evaporation of the secretion (fig. 25). The function of the secretion in Gerromorpha is unclear. The hypothesis of a waterproofing agent has been refuted in experiments with the European gerrid *Aquarius najas* (de Geer) by Staddon (1972).

In Nepomorpha there is more variation in metathoracic glands. They are lacking in Anisopinae, Aphelocheiridae, Belostomatinae and Nepidae. The simplest metathoracic glands are found in *Lethocerus*. These consist of two simple separate

25

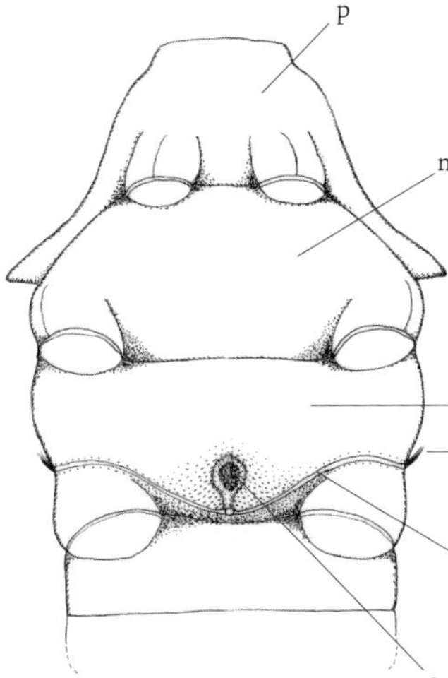

FIGURE 25
Thorax and second abdominal sternite in ventral view of *Angilia bispinosa* Andersen, legs removed; c = evaporation channel, ms = mesosternum, mt = metasternum, o = metathoracic scent gland orifice, p = prosternum, t = tuft of hairs at end of evaporation channel.

tubes which open at the base of the metaxiphus. In Corixidae simple tubes open in a common median reservoir which has paired openings ventrally between the hind coxae. In Notonectidae the structure is the same as in Corixidae, but the tubes open on the lateral side of the thorax. Finally, in Naucoridae and Pleidae the metathoracic scent gland consists of branched tubes with a single median reservoir with paired openings on the sides of the thorax. In Corixidae the secretion fluid is transported through ventral channels to the sides of the thorax. In Helotrephidae metathoracic glands are present but have not been studied in detail.

Many adult nepomorphan species actively spread the secretion from the metathoracic glands over the body. The insects crawl out of the water (Naucoridae, Notonectidae, Pleidae) or lie on the water surface (Corixidae) to do so. The secretion, often containing considerable amounts of hydrogen peroxide, has an antimicrobial effect. Its function is primarily to counter colonization of bacteria on hairs containing air for respiration. Bacteria would diminish the hydrophobic properties and thus efficiency of the hairs. Secretion grooming has indeed been found only in species with a collapsible physical gill (Kovac, 1987; Kovac & Maschwitz 1989, 1990, 1991). Nepidae species preen the eyes and the respiratory siphon only. Preening behaviour has not been observed in genera *Aphelocheirus* (Aphelocheiridae) and *Anisops* (Notonectidae). *Micronecta* has glands but lacks the preening behaviour.

Kovac (1987) provided a survey of preening behaviour and analyzed factors instigating secretion preening in *Plea minutissima* Leach (Kovac, 1993). He found that either rising temperature or increasing light intensity, or both simultaneously, leads to preening activity. This is correlated to spring in temperate climates. It is meant to prevent bacterial pollution and is not instigated by pollution itself. European species were studied under European climatological conditions, but due to the general physiological purpose it may occur in tropical species too. Which factor eventually stimulates this behaviour under tropical conditions remains to be investigated.

Another scent gland system is situated dorsally on the abdomen. It has paired openings in larval Nepomorpha. In Gerromorpha it is present in Hebridae, Mesoveliidae, and in the primitive genera of Hydrometridae in both larvae and adults. In all these genera it is visible as a single opening situated medially on tergite 4 (fig. 2, so).

Locomotion

Nepidae are poor swimmers. They usually walk, and hence the hind legs are hardly modified. The legs are moved alternately: one middle and one hind leg on opposite sides at a time. A distinct adaptation to the submerged way of life is the stronger development of a fringe of swimming hairs on the hind legs in the better swimmers of Nepidae. In the other families the hind legs are the propulsive organs: the two hind legs are pushed backwards simultaneously. They become more flattened distally and beset with thicker fringes of swimming hairs towards the tips. In addition, adaptations in the joints allow for horizontal movement. Aphelocheiridae and Naucoridae have slender hind legs with a moderately developed fringe. In *Lethocerus* the hind legs are widened and flattened with dense fringes (fig. 59). In the best swimmers, such as Notonectidae and Corixoidea, the hind legs are widened, flattened and beset with thick fringes, and proportionally very long, adding to the thrust (fig. 15). Belostomatidae, Helotrephidae, Naucoridae and Pleidae are reasonable to very good swimmers which can crawl quite well on land, although very slowly in the case of Pleidae. In Corixoidea and Notonectidae the hind legs are incapable of vertical movement and cannot be used for walking. Representatives of these families, when put on a dry surface, will clumsily crawl or hop.

In water bugs without specialized hind legs the middle legs usually assist in swimming and are clothed with swimming hairs as well, with the exception of Aphelocheiridae. In Notonectidae (fig. 14) the middle legs assist in catching and handling prey. In Corixoidea they are strongly specialized, mainly for anchoring the animals when resting at the bottom, and during grooming.

Gerromorpha display three types of locomotion on the water surface (Andersen, 1976). The first one, exhibited by *Microvelia* and its relatives, is walking as in terrestrial insects. Their middle legs are usually slightly shorter than the hind legs. Fore and hind leg of one side and the middle leg on the other side move at one step. This pattern of locomotion is shared by at least the more generalized representatives of Hebridae, Mesoveliidae and Hydrometridae. Species of these families walk about as easily on water as on land and many live on water plants or at the edge of water bodies.

The second method is rowing, exhibited by Veliidae (genera *Rhagovelia* and *Velia*), most Gerridae, and apparently also in Haloveliinae (Veliidae). Their

middle legs are – often considerably – longer than the hind legs. Fore and hind legs support the insect, while both middle legs are pushed forward simultaneously without contact with the water and then backward while resting on the water. The support of fore and hind legs is much less, to virtually zero, during the thrust. Quick turns are executed by giving a push with the middle leg opposite to the turning direction, assisted by rudder movements of the hind legs. The long middle legs are less suited to walking on land, and especially many Gerridae hop around erratically when placed on dry ground.
Finally, many Gerridae jump to avoid predators. By lowering the body supported by fore and hind legs, and with a vigorous thrust of the middle legs, the insect lifts itself free of the water. If the thrust has a considerable forward vector they slide a considerable distance over the water surface after coming down again due to the low friction. The fore legs are used intermittently for support, the hind legs are used mainly in steering.
Several Veliidae use a method of locomotion in which the legs are not providing the thrust. It is called expansion skating and will be discussed in the introduction to Veliidae.

Feeding

An important aspect of efficiency in feeding in predacious Heteroptera is the amount of time they spend with a single prey. In a feeding *Diplonychus rusticus* most food is consumed during the first two minutes; after ten minutes only twice this weight is extracted. When prey densities are high a prey is discarded sooner than when prey densities are low, thus enabling the insect to ingest more food when prey density is high (Dudgeon, 1990). The functional response in which predators consume more prey at higher densities at an decelerating rate towards an asymptote is common amongst invertebrate predators. It has even been called 'the invertebrate response' by Holling (1966).

Economic importance of aquatic bugs

Water bugs can be beneficial to humans in many ways. They are, with a few possible exceptions, predacious and may thus occasionally be harmful in fish ponds, not only for eating fish fry but also for competing for food with the fish. On the other hand, several taxa (*e.g.*, many species belonging to Aphelocheiridae, Naucoridae, Helotrephidae, Notonectidae) contribute to the control of insects which are harmful to mankind, notably mosquito larvae (especially Culicidae, but for instance in the case of Aphelocheiridae Simuliidae; Hoffmann, 1927; Williamson, 1949; Laird, 1956; Andersen, 1975; Takagashi et al.,

1979; Chesson, 1984; Sites, 2000), which are an important part of the diet of a number of water bugs. Some species of Limnotrephini (Helotrephidae: *Limnotrephes, Idiotrephes* and *Tiphotrephes*) are known to be able to live in small water bodies under extreme conditions. *Idiotrephes asiaticus* can survive even in artificial water bodies like ceramic bonsai bowls (Papáček, 1993). Such small stagnant waters are breeding places for culicids, vectors for diseases as dengue fever. As dipteran larvae are accepted for prey by Helotrephidae, some helotrephids may be considered for control of mosquitos.

Water bugs may also play a role in the control of leaf hoppers (Homoptera) in rice fields: there are records of Veliidae (*Microvelia*; Laird, 1956; Spence & Andersen, 2000) attacking pests. See further under Veliidae.

The glands in *Lethocerus* contain a pleasantly aromatic substance (trans-hex-2-enyl-acetate). This is why *Lethocerus* is valued in South-east Asian cooking: they are made into a paste which is used as a condiment. In Cantonese shops in San Francisco's Chinatown (Usinger, 1956) and in various Thai markets (Pemberton, 1988) conserved *L. indicus* and extracts of *Lethocerus* spec. were available. In Mexico, where some species of Corixidae are extremely abundant in lakes, eggs are used for human consumption and eggs and larvae are sold as fish food.

The tolerance to water pollution and habitat disturbance is quite different among species, so some species may be expected in the future to be used as indicator organisms. For instance, benthic genera with high demands on their habitats like *Fischerotrephes, Trephotomas* and *Distotrephes* may be used as indicators for highly oxygenated water. In addition, some *Anisops* species have been used for evaluation of toxic effects of various pesticides used against locusts in Senegal (Lahr, 2000). Also *Halobates* could possibly be used as an indicator of heavy metal pollution (e.g. cadmium) in the open sea (see under *Halobates* for references). At present however, our knowledge of habitat requirements and tolerance to adverse conditions of Malesian species is insufficient.

Most water bugs are not harmful for mankind. As under natural circumstances they occur in or on almost every type of water, and the number of species in a given area is usually rather low, they form an ideal component in biodiversity studies.

Collecting and preservation

Collecting

For qualitative collecting of aquatic bugs a net is required. For a quick scan a kitchen sieve with a diameter of 15 cm will do, although various taxa move too fast to be collected more than occasionally by this method. A larger net will catch most species. The net must be able to be dragged forcefully through aquatic vegetation and thin branches, and hold sand and pebbles while collecting bottom fauna. The frame must be solid (e.g. steel bar, diameter of about 5 mm), the front straight and not too narrow, the net not too shallow (e.g. 0.4-0.5 m wide, 0.5 m deep). It can be triangular or square. The handle should be about 1 m long. Good material for the net is nylon mosquito screening, to be covered on the rim by canvas or leathery material. Mesh size can be 1 mm^2, although species like *Micronecta* will pass through this. However, at least a few will usually remain in the net. With a mesh size of 0.25 mm^2 even the smallest species are captured. The resistance in such a net is much higher, resulting in strong swimmers to easily escape and the net gets clogged more easily. Some species occur in very small waters and at the shallow edges of larger water bodies. They require a small net with fine mesh. For this purpose a tea-strainer with a handle or a small net as used in aquarium shops can be used.

Most species of Nepomorpha can be collected by dragging the net quickly through the water. As they tend to dive when disturbed a net should be dragged over the bottom in more or less open water. When vegetation is present, the net should be forcefully dragged through it. Several species stay at the edge of the water, including overhanging banks, or cling to the (under)side of logs. Finally, *Aphelocheirus* and some helotrephid and naucorid genera occur on or in the bottom substrate of streams. They can be collected by kicking up the substrate in front of the net so that the current will wash them into the net.

Many species of Gerromorpha and Nepomorpha are attracted to light during darkness. As they are often not able to cling to a cloth there must be a surface beneath it upon which the insects are conspicuous. Alternatively, a light trap where the light source is placed in a smooth funnel leading to a container with killing vapour or fluid can be used. The result of light catches is usually better as the fraction of ultraviolet light in the light source of the trap increases.

The best way to sort out a catch is to put the contents in a white tray with some water. After some time the bugs will crawl or swim around. A large tray for developing photo-posters is very useful, but any tray slightly larger than the net will do. If the tray is small the amount of material scooped up in a drag must be restricted. Alternatively, the contents of a net can be spread out on a flat part of the bank. However, small hardly moving specimens such as *Paraplea* may be largely overlooked. Attention must also be paid to various Nepidae, which tend to stay immobile for some time after disturbance.

When not handled carefully, *Aphelocheirus* may occasionally sting. The effect is somewhat painful and part of the hand may be numb for a short time, but after-effects are unknown. Most Naucoridae can sting painfully.

Killing

Aquatic bugs are killed with any of the vapours used for killing insects. Among these, ethyl acetate is one of the better as it keeps the specimens soft when kept in a closed tube. The drawbacks are that it will extract fat from the specimens, is volatile, and it dissolves some plastics. Killed specimens can also be kept in chlorocresol (4-chloro-3-methylphenol). They will stay in good condition for at least three months when the boxes are not cooled (with temperatures rising up to 30 °C during daytime) or a year when the boxes are kept cool (Tindale, 1961). Chlorocresol is sold by the chemical company Merck under the name 4-Chlormethylphenol.

Alternatively, specimens can be put in 70% (ethyl) alcohol, in which they can be kept for years. Methylated spirit will also do. If there is a large amount of specimens the alcohol should be changed after a few days. Some species of Anisopinae may loose their colour when put in alcohol. Isopropyl alcohol is less suitable as the specimens become hard.

Preservation

Larger specimens are mounted with stainless insect pins when they are still soft. If done later, the appendages will be damaged or the specimen could break. After drying they are kept in well-closed boxes. Very large specimens such as *Lethocerus* may take two months to dry completely. The boxes are stored in a dry place and contain an insect killing agent to avoid damage by storage pests. Alternatively, specimens can be kept in 70% alcohol in tubes with a tightly closing stopper. Especially when series of the same species are to be kept, alcohol storage is an option. If specimens are dry before being prepared they can be softened by putting them in acetic acid vapour put on filter paper. After one day most specimens are soft again.

Smaller specimens are glued venter downward on small pieces of cardboard. In this way the ventral side is not visible, but the specimen is protected by the cardboard and the appendages can be arranged nicely. If a series of specimens is avaible a male and a female could be glued on their back. The use of water-soluble glue is strongly advised. Specimens can also be glued with their side (usually the right side) to the tip of triangular or pointed pieces of cardboard. The advantage is that dorsal, ventral and a lateral side of the specimen are visible, the disadvantage is that it is less well protected.

Labelling

It is very important to properly label collected insects. The label must contain at least the locality, date and name of the collector. Additional information, such as habitat type, is useful. Notes on habitat should be taken in the field. In the field labels written with pencil on good quality paper will do. Dry labels accompanying mounted specimens can be written with Indian ink on thick paper or thin carton. Computer-produced labels are sufficent as well. Labels for specimens stored in alcohol should be written with Indian ink on trace paper.

Preservation and study of genitalia

For reliable identification it is often necessary to study the genitalia, notably of the male. For preparation of genitalia of dried specimens, the abdomen is

relaxed in acetic acid (vinegar) vapour or in water with some alcohol or detergent. After that the pygophore can be pulled out with an insect needle with bent tip.
Especially in small specimens the easiest way is to detach the entire abdomen and extract the genitalia in a petri dish. The abdomen can be glued on the carton with the specimen, or on a separate carton which is pinned with the specimen. Muscle tissue is removed from the genitalia by hydrolysis in a 5-10% solution of potassium hydroxide (KOH) or sodium hydroxide. The cleaned item is thoroughly rinsed with dilute acetic acid and water, and preferably dried, before it is put in conservation fluid. If acetic acid is not available rinsing well in water will do. Large genitalia can be glued to a carton, but small ones are better kept in glycerine in microvials (small polythene tubes) which are also pinned with the specimen. Very small structures have to be studied in microscopic slides. Temporary mounts can be made in glycerine. In cooler climates slides made in glycerine jelly can be kept for several years. Permanent slide mounts are better and allow observation at high magnification. The drawback is that reorientation of the object is impossible.
To expose the apex of the paramere of fresh specimens (see Gelastocoridae) it can be pulled out before pinning. To do this, the apex of the abdomen of a fresh or relaxed specimen is gently pulled backward and to the left with a dissecting needle. The paramere should now come clear of the abdomen. Then allow the segments to pull back to their normal position, but direct the paramere to the ventral side of the abdomen. It will stay in this position and the apex of the paramere can be studied easily afterwards.
The scanning electron microscope is perfectly suited for observing details of the exoskeleton; it is frequently used in the study of insects.

chapter 3

Taxonomy, phylogeny and distribution

At the time of Fabricius most insects with sucking mouthparts, including 'look-alikes' (e.g. the present-day Mallophaga) were united in an order Rhyngota or Rhynchota (Fabricius, 1803). During the 19th century various groups were removed and by the early 20th century the Rhynchota included the Heteroptera and Homoptera, and sometimes still the Siphunculata, at that time usually named Pedunculata (real or sucking lice). Halfway the 19th century however the name Hemiptera, with subdivisions Heteroptera and Homoptera, was already in use (Amyot & Serville, 1843), and the name Rhynchota gradually disappeared.
About 4000 species of Nepomorpha and Gerromorpha have been described worldwide, nearly equally divided between these two infraorders (appendix 1), but the actual number of species may be twice as high. Aquatic Heteroptera are distributed throughout Malesia and almost a quarter of the world's known species occurs there. Many widespread genera occur from South-east Asia to New Guinea and Australia, but do not reach the Bismarck and Solomon Islands (fig. 565).

Taxonomy above family level

Nowadays most authors define the order **Hemiptera**, containing the suborders **Heteroptera** and **Homoptera** (e.g. Schuh & Slater, 1995). Homoptera exclusively contains herbivorous taxa such as aphids, cicadas, scale insects, leafhoppers and white flies. Heteroptera contains the true bugs, many of which are also herbivorous, but a considerable number, including all or nearly all aquatic and semi-aquatic species, are carnivorous.
Heteroptera are presently divided into seven infraorders, summarized first by Štys & Kerzhner (1975) and followed by Schuh & Slater (1995). Two infraorders are considered in this book: Nepomorpha or true aquatic bugs, containing the bugs living under water such as waterscorpions, water boatmen and backswimmers together with a few shore-inhabiting groups, and Gerromorpha or waterstriders, containing e.g. sea skaters, pond skaters, marsh treaders and watercrickets.
Older subdivisions of Heteroptera are: 1) Cryptocerata (= Nepomorpha) and Gymnocerata (= the remaining infraorders), or 2) Hydrocorisae (= Nepomorpha), Amphibicorisae or Amphibiocorisae (= Gerromorpha), and Geocorisae (remaining infraorders). Both are considered incorrect at present, as Gymnocerata and Geocorisae are polyphyletic at the level where Cryptocerata or Hydrocorisae and Amphibicorisae are considered monophyletic. However, Mahner (1993) revived Cryptocerata as he considers them of equal rank with Gymnocerata, but this has not been followed by subsequent authors (e.g. Schuh & Slater, 1995).
Some recent publications deal with taxonomy on the level between infraorders and genera. A splendid book on Gerromorpha was written by Andersen (1982a). Štys & Jansson (1988) present a checklist of names of supraspecific taxa in

Nepomorpha (valid taxa with synonyms). The catalogue of Palearctic Heteroptera (Aukema & Rieger, 1995) also gives extensive treatment of supraspecific classification in accordance with the above mentioned papers. Finally, Mahner (1993) gave a complete system of Nepomorpha based on cladistic analysis, but did not deal with synonyms.

In this book, we follow the taxonomic order of Andersen (1982a) for Gerromorpha and Mahner (1993) for Nepomorpha (figs. 26-27). However, in our arrangement of Nepomorpha in the text of chapter 4 the Corixoidea have been placed between Naucoroidea and Notonectidae instead of placing it between Nepoidea and the remaining Nepomorpha which he gave the new name Tripartita (fig. 27) as in Mahner (1993).

Phylogeny

The first evolutionary tree which agrees well with present views was given by China (1955). He combined physiological (respiratory) and morphological characters (mainly associated with respiration and swimming) of recent taxa. Better adaptation to aquatic life was considered the progression within this evolutionary line. This idea led him to believe that Ochteridae is close to the ancestral stock of Nepomorpha and Aphelocheiridae at the top. Nowadays, Nepoidea are considered to have branched off earlier (fig. 27). Moreover, in 1955 it was not yet known that several naucorid subfamilies (Cheirochelinae, Cryphocricinae and Limnocorinae) also use plastron respiration. In view of the plesiomorphic head and fore leg characters it is plausible that the Aphelocheiridae and Potamocoridae branched off from the (proto)-naucorid stock quite early, and that plastron respiration developed independently in Aphelocheiridae and Naucoridae. This view is supported by the analysis of Rieger (1976), who considers Aphelocheiridae as a sister group of Notonectidae and Pleoidea (fig. 28).

Combining fossil records with comparative morphology, Popov (1971b) came to quite similar conclusions, but added time scales. From an protoheteropteran stock the line leading to the aquatic and semi-aquatic bugs should have branched off at the beginning of the Triassic (about 250 million years ago), soon followed by a split in Protochteridae and Protosaldidae. The Protochteridae developed into Nepomorpha, and the Protosaldidae split into Leptopodomorpha and Gerromorpha. In a recent phylogenetic tree by Andersen (1998), the Gerromorpha branch off from Nepomorpha, Leptopodomorpha and the two main infraorders of terrestrial bugs Cimicomorpha and Pentatomorpha. Subsequently the Nepomorpha split from the other three. It is estimated that this happened in the early Triassic, which coincides with Popov's model.

FIGURES 26 – 28
Cladograms indicating the relationship between families.
26 Gerromorpha according to Andersen (1982a) – *27* Nepomorpha according to Mahner (1993) – *28* Nepomorpha according to Rieger (1976). N.N. = unknown taxon.

26

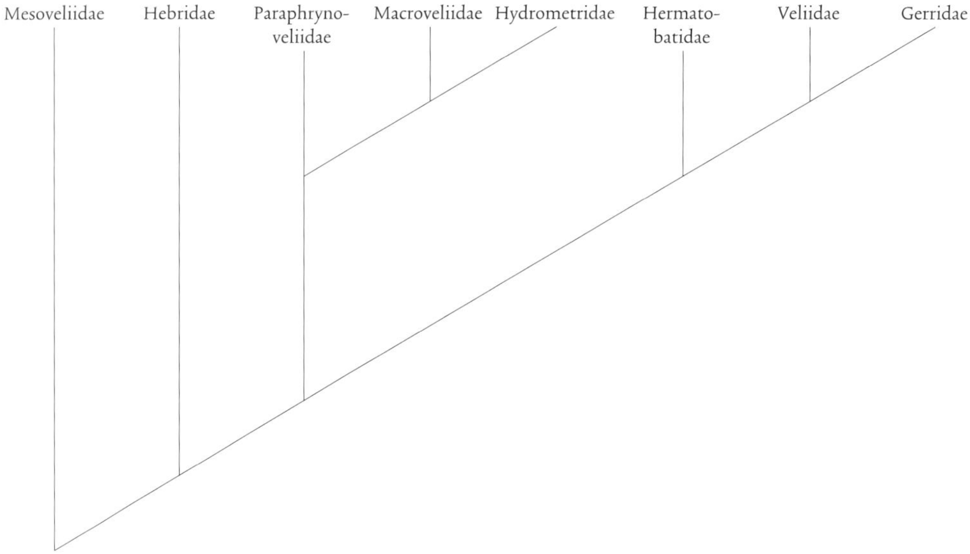

27

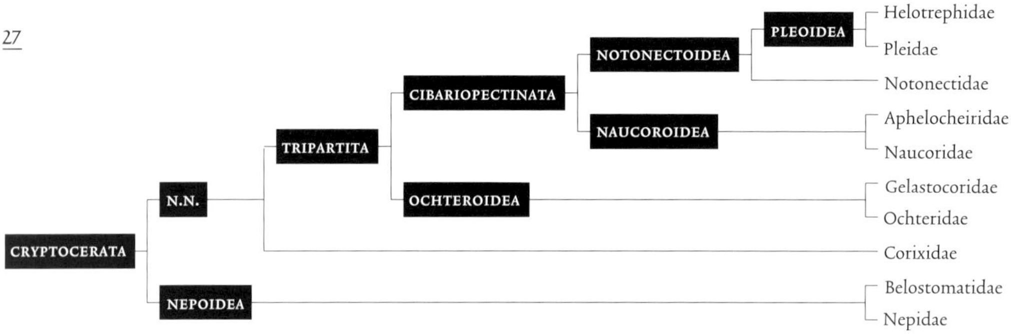

28

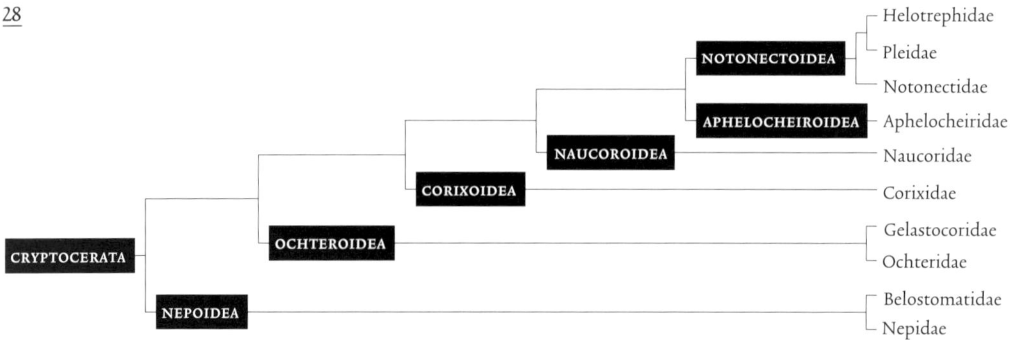

There are supposed to be three main branches from the lineage leading to the following current groups: 1) Ochteroidea, 2) Nepoidea (Nepidae and Belostomatidae) and Corixoidea (Corixidae, Diaprepocoridae and Micronectidae) early on, and 3) Naucoroidea, Notonectoidea and Pleoidea somewhat later (around the Middle Triassic period). At the beginning of the Jurassic period there seems to have been an explosion of differentiation, except for Pleidae, which are supposed to have arisen in the early Tertiary. All recent families and several subfamilies are supposed to have arisen at that time. Nepoidea seemed to have had palmy days during the second part of the Jurassic. This can partly be attributed to their size combined with strongly chitinized structures, which favours fossilization over smaller, lighter built forms.

Other comparative studies, of egg structure by Cobben (1968), and food pump by Parsons (1969b), supported the scheme proposed by China (1955). A recent cladistic analysis by Mahner (1993, fig. 27) sets Nepoidea more apart and isolates Aphelocheiridae from Naucoroidea as a separate family.

As stated above, the Gerromorpha are supposed to have branched off from other Heteroptera in the early Triassic. A fossil from the Lower Jurassic, *Engynabis tenuis* Bode, was placed in the Gerromorpha by Popov & Wootton (1977). Andersen (1998) pointed out that this fossil has several characters which do not occur in any recent Gerromorpha, and thus strongly doubted that this fossil belonged to this infraorder. Fossils of Mesoveliidae, Hydrometridae and Veliidae have recently been identified from the Lower and Middle Cretaceous, up to 120 million years ago (Jell & Duncan, 1986; Andersen, 1998; Nel & Popov, 2000; Andersen & Grimaldi, 2001). Especially the Lower Cretaceous veliid fossil (Jell & Duncan, 1986) is important. As Gerridae and Veliidae are still closely related it seems reasonable to assume that they were the last to split. This would mean that the families of Gerromorpha had already evolved into separate taxa. This supports the hypothesis that most gerromorphan families probably originated in the early Mesozoic, and is consistent with a branching-off of Gerromorpha in the early Triassic (the beginning of the Mesozoic). Especially Hydrometridae are comparatively well represented by early Cenozoic fossils (Andersen & Grimaldi, 2001). By the Middle Eocene (50 million years ago) the genera presently occurring in the northern temperate region were already fully differentiated (Andersen et al., 1994). Gerromorphan fossils from Eocene baltic amber (50-40 million years ago) belong to extinct genera which are, however, assignable to recent subfamilies (Andersen, 2000b). A gerrid from upper Oligocene (26 million years ago) in Dominican amber belongs to an extinct subfamily (Andersen & Poinar, 1992). This suggests differences in patterns of extinction between temperate and tropical faunas. Other fossils from Dominican amber belong to recent genera (Andersen & Poinar, 1998; Andersen, 2000b; 2001) or genera closely related to extant ones and clearly assignable to recent subfamilies.

China (1955) also constructed a diagram of probable evolutionary lines for Gerromorpha, using mainly characters correlated with habitat adaptation, such as modifications of the middle and hind legs including reduction of tarsal segments, and the subapical position of the tarsal claws. In addition, he emphasized modifications of the metatoracic scent glands. His diagram agrees well with the recent diagram by Andersen (1982a), which is based on cladistic analysis of an extensive list of characters (fig. 26). However, there are differences between the two interpretations. The main difference is that Andersen (1982a) considers the Mesoveliidae to be a sister group of all other gerromorphan taxa, and the Hebridae to be a sister group to other gerromorphan taxa except for Mesoveliidae, whereas China (1955) considers Mesoveliidae and Hebridae together to be a sister group to the remaining gerromorphan taxa. This is still disputable, mainly because some of the synapomorphies of the Gerromorpha exclusive of Mesoveliidae are supposed to have been secondarily lost again in some families, and some are only visible in macropterous specimens which are unknown in some groups. China (1955) considered the well developed ovipositor ofMesoveliidae as an apomorphy, whereas Andersen (1982a) considered it to be a plesiomorphy. Other differences refer to small exotic families which are not shown in the diagrams.

Until very recently cladistic phylogenetic studies were based on comparison of morphological characters only. Recently a few studies using DNA-sequences have been published. Most of these studies deal with the phylogeny of a single genus. The analysis based on DNA of *Aquarius* fully supports the former phylogeny based on analysis of morphological characters (Andersen, 1990; Damgaard et al., 2000b). In a study of *Halobates* there were some discrepancies between phylogenies based on DNA and morphological character analysis respectively (Damgaard et al., 2000a). Muraji & Tachikawa (2000) published a phylogenetic analysis based on DNA sequences for several families of Gerromorpha. Their preliminary results agree with the phylogeny of Andersen (1982a; which was based on morphological characters), notably the position of Hermatobatidae as a sister group to Gerridae and Veliidae. According to one method, however, the subfamily Microveliinae (Veliidae) sorted out in the Gerridae. This stresses the close relationship of the families Gerridae and Veliidae, and led to the observation that analysis of sequences of DNA from different sources is advisable (Muraji & Tachikawa, 2000).

An attempt for a complete survey of fossil Gerromorpha and Nepomorpha is given by Nel & Paicheler (1992, 1993a, b). It contains descriptions of some new species and an extensive bibliography.

Distribution

The distribution of genera in Malesia, except for the genera distributed throughout the entire Malesian region, is given in figures 565-601. Apart from the summarized information on distribution of waterbugs presented in appendix 1, we also give the distribution of taxa worldwide, and the distribution in Indo-Australia and Malesia, under each taxon. When there is a risk of textual overlap, information is given only under the lowest taxon.

The last ice age, which ended about 20,000 years ago, has become an important factor for determining the present distribution of many genera. Because much water was fixed in the polar ice caps and glaciers, the sea level was so much lower that continental Malaysia, Sumatra, Java, Borneo and Palawan formed a contiguous land mass called Sundaland (fig. 563). Likewise, New Guinea was connected with Australia. Especially the existence of Sundaland made it easier for taxa living on the Asian mainland to disperse to what are now the islands Sumatra, Java and Borneo. The sea between Bali-Borneo-Palawan and the smaller Sunda Islands (= Nusa Tenggara)-Sulawesi-Philippines was much deeper and did not dry up. Therefore there is a sudden drop in the number of Asian taxa from Borneo and Java eastward. Figure 564 shows the extent to which various genera with an apparent origin in the Asian mainland have dispersed eastward into Malesia. This pattern of a distinct diminishing number of Asian taxa east of Java and Borneo is observed in many animal groups and is known as Wallace's Line. Originally Wallace's Line ran east of the Philippines. In figure 1 we present the modification of Wallace's Line by Dickerson et al. (1928), wich runs west of the Philippines excepting Palawan. Another important dividing line is Weber's Line (fig. 1), which gives roughly the division between the Oriental and Australian plus Oceanic fauna's (Weber, 1902; see also Oosterbroek, 1998).

Natural expansion of taxa originating on mainland South-east Asia follows a fixed pattern: Sumatra and Java are first colonized, then Borneo and the Lesser Sunda Islands, followed by the southern Philippines, Sulawesi and the remainder of the Philippines, and finally the Moluccas and New Guinea. The first to present a survey of genera of water bugs and their distribution in Malesia and adjacent areas were J. Polhemus & D. Polhemus (1990).

The difference in flora and fauna between Sundaland and Sulawesi and the remaining Philippines is partly attributed to geological origin. Sundaland and apparently South-west Sulawesi were connected with Eurasia already 50 million year ago, whereas East Sulawesi was lying on the Australian plate west of Australia and quite close to it. The Philippines were lying on the North New Guinea plate, quite far north of New Guinea (Hall, 1998). This may account for the fact that there are faunistic connections between the Philippines and New Guinea (in water bugs notably the Sagocorini and a species-group of *Rhagovelia*), whereas the faunistic connections between Sulawesi and New Guinea has thus far been restricted to one genus (Gerridae: *Andersenella*). The Sangihe arc of islands running from North Sulawesi to Mindanao is apparently a more recent phenomenon (D. Polhemus, 1995). However, as far as water bugs are concerned, most of the colonization of Sulawesi and the Philippines apparently happened by active dispersal of the insects and not by rafting of pieces of continents. The main argument against rafting is that most taxa seem to be of continental Asian origin and dispersed eastward, against the direction of the rafting (J. Polhemus & D. Polhemus, 1990).

chapter 4

Families and genera of Malesian bugs

Aquatic and semi-aquatic bugs belong to the suborder Heteroptera of the order Hemiptera and are related to the suborder Homoptera. The main character separating Hemiptera from other insects are the sucking/piercing mouthparts: the labium forms a sheath around the mandibulae and maxillae. In Heteroptera the rostrum originates anteriorly on the head, and they have a gula ('throat'), whereas in Homoptera the rostrum originates ventrally and more backward on the head, hence eliminating the gula (figs. 8, 9).

Key for recognition of Gerromorpha and Nepomorpha within Hemiptera

1 Rostrum originating anteriorly on the head; ventral part of head usually forming a distinct gula (fig. 8, Heteroptera) ▸ 2
– Rostrum originating ventrally and backward on the head; gula absent (fig. 9) ▸ **Homoptera**

2 Head with three pairs of conspicuous trichobothria, inserted in deep pits and placed near the inner margin of compound eyes (figs. 7, 10); bothrium concealed in the pit. Many species with a fourth pair of trichobothria, which is usually not inserted in a pit (fig. 8). Hemielytra when fully developed not divided into clavus, corium and membrane ▸ **Gerromorpha**
– Head without trichobothria inserted in deep pits. In species with fully developed hemielytra these are usually divided into clavus, corium and membrane ▸ 3

3 Antennae inserted ventrally on the head, usually shorter than head and concealed in a groove beneath the eyes. Eyes well-developed (figs. 5, 6; except in one European species of *Nepa*). In Aphelocheiridae, Naucoridae and Ochteridae apical part of antennae sometimes visible from above ▸ **Nepomorpha**
– Antennae inserted anteriorly, usually longer than head, clearly visible from above over the entire length or nearly so. One specialized family, the Termitaphididae, has antennae shorter than head; not visible when seen from above, eyes are lacking. Short legs (not visible from above); small (2-3 mm), looks like scale-insects and lives in termite nests ▸ **remaining Heteroptera**

Key to Indo-Australian families of Nepomorpha

1 Ocelli present. Usually living at the edge of water ▸ 2
– Ocelli absent. Usually living under water ▸ 4

2 Rostrum short and broadly triangular, not segmented but with transverse grooves. Posterior margin of head cover the anterior part of pronotum. [Australia and New Zealand] ▸ **Diaprepocoridae**
– Rostrum slender and of variable length, distinctly segmented and more or less parallel-sided. Posterior margin of head not covering anterior part of pronotum ▸ 3

3 Fore femur not thickened. Rostrum very long and slender, reaching hind coxae (figs. 6, 64) ▸ **Ochteridae, p. 77**
– Fore femur strongly thickened. Rostrum short and stout, not or hardly reaching beyond the posterior margin of the prosternum (figs. 85, 68) ▸ **Gelastocoridae, p. 82**

4 Rostrum short and broadly triangular, not segmented but in nearly all species with transverse grooves. Posterior margin of head covers anterior part of pronotum ▸ **(Corixoidea) 5**
– Rostrum distinctly segmented and more or less parallel-sided. Posterior margin of head not covering anterior part of pronotum ▸ 6

5 Scutellum exposed. Length less than 4 mm (fig. 145) ▸ **Micronectidae, p. 141**
– Scutellum entirely or almost entirely covered by pronotum. Length 4 mm or more (figs. 133, 135) ▸ **Corixidae, p. 130**

6 Respiratory siphon present, usually long and non-retractable; the halves of the siphon not entirely fused; the siphon may split, especially in dry specimens (figs. 44, 47) ▸ **Nepidae, p. 51**
– Respiratory appendages absent, or paired and retractable ▸ 7

7 Eyes pedunculate, in dorsal view strongly mesocaudally emarginated. Legs without swimming hairs (fig. 68) ▸ **Gelastocoridae, p. 82**
Six species of *Nerthra* without ocelli, of which only *N. macrothorax*, with a body length of 8-11 mm, occurs in Malesia.
– Eyes sessile, in dorsal view not mesocaudally emarginate. Hind legs with fringe of setae and usually more or less flattened (except in Pleidae, which are ≤ 3 mm) ▸ 8

8 Body dorsoventrally flattened, oval to nearly circular. Swimming with venter downward. Fore legs inserted on the anterior margin of prosternum ▸ 9

– Body boat-shaped. Swimming with venter upward. Fore legs inserted on the posterior margin of prosternum ▸ 11

9 Rostrum long and narrow, reaching hind coxae. Head elongate, as long as or slightly longer than wide. Apical part of antennae often visible from above. Fore femur hardly thickened (figs. 72, 74) ▸ **Aphelocheiridae, p. 86**

– Rostrum short and stout, not reaching halfway of mesosternum. Head distinctly transverse. Antennae fully concealed under the head. Fore femur distinctly thickened (fig. 12) ▸ 10

10 Anteclypeus in dorsal view laterally delimited by a pair of sutures (fig. 52). Membrane of hemielytron with distinct veins. Abdomen apically with a pair of retractable respiratory appendages (figs. 58, 59) ▸ **Belostomatidae, p. 67**

– Anteclypeus in dorsal view not recognizable. Membrane of hemielytron without veins. Apex of abdomen without respiratory appendages (figs. 106, 110, 117, 119) ▸ **Naucoridae, p. 95**

11 Head and pronotum more or less fused; in most species head and pronotum recognizable by a shallow W-shaped groove. Antennae with one or two segments. Small: 3.5 mm or distinctly less (figs. 228-234) ▸ **Helotrephidae, p. 166**

– Head and pronotum separate. Antennae with three or four segments. Size variable ▸ 12

12 Small, length about 2.5 mm or less. Antennae with three segments. Membrane of hemielytra apparently absent (fig. 242) ▸ **Pleidae, p. 190**

– Size variable but nearly always over 3 mm. Antennae with four segments. Membrane of hemielytra present, large in macropterous morph, very small in some brachypterous morphs (figs. 16, 159, 161, 165) ▸ **Notonectidae, p. 148**

Key to Indo-Australian families of Gerromorpha[1]

1 Macropterous (wings fully developed) ▸ 2
– Apterous (wingless) or meiopterous (wings reduced) ▸ 6

2 Mesoscutellum exposed, forming a subtriangular (fig. 3) to transverse (fig. 255) plate behind the pronotum ▸ 3
– Mesoscutellum not exposed, covered by the posteriorly extended pronotal lobe ▸ 4

3 Head ventrally with a pair of prominent vertical plates (bucculae) covering the base of rostrum (fig. 272). Tarsi two-segmented, first segment very short (fig. 277) ▸ **Hebridae, p. 204**
– Head not as above. Tarsi three-segmented (fig. 254) ▸ **Mesoveliidae, p. 194**

4 Head behind eyes distinctly elongate (fig. 293), length of postocular part longer than diameter of one eye. Anterior lobes of head covering lateral part of first and sometimes also second labial segments. Claws inserted apically on last tarsal segment ▸ **Hydrometridae, p. 220**
– Head not elongated behind the eyes, which are situated at or close to the hind margin of head. Anterior lobes of head not as above. Claws nearly always inserted distinctly before the apex on last tarsal segment ▸ 5

5 Head with distinct longitudinal median impressed line on dorsal surface (fig. 10). Fore tibia of male nearly always with comb of short spines along inner margin (fig. 13), hind femora usually stouter than middle femora ▸ **Veliidae, p. 230**
– Head without a median impressed line on dorsal surface. Fore tibia of male without comb, hind femora usually more slender than middle femora ▸ **Gerridae, p. 292**

6 Abdominal scent gland orifice absent ▸ 7
– Abdominal scent gland orifice present on tergum 4 (fig. 2) ▸ 8

7 Head more than three times longer than wide; eyes far removed from posterior margin of head (fig. 293) ▸ **Hydrometridae, p. 220**
– Head three times longer than wide at most; eyes situated close to or at posterior margin of head ▸ 5

---

1 Adapted from Andersen (1982a).

8 Pronotum very short, exposing both mesonotum and metanotum; base of rostrum in lateral view exposed. Tarsi three-segmented ▸ 9

– Pronotum longer, covering at least most of mesonotum; head ventrally with a pair of prominent vertical plates (bucculae) covering the base of rostrum in lateral view (fig. 272). Tarsi two-segmented, with first segment very short (fig. 277) ▸ Hebridae, p. 204

9 Head longer than wide; mesonotum and metanotum separate, together shorter than abdomen (fig. 251). All claws apical ▸ Mesoveliidae, p. 194

– Head shorter than wide; mesonotum and metanotum fused, together much longer than abdomen (fig. 294). Fore claws subapical, middle and hind claws apical. (Exclusively marine species) ▸ Hermatobatidae, p. 227

Family Nepidae – waterscorpions

Diagnosis

Medium-sized to large (15-60 mm), elongate oval to tubular. Dorsally more or less uniformly medium-brown; legs yellowish to light brown, fore femur often annulated with darker brown. Antenna shorter than head, hidden in groove beneath eyes, usually three-segmented with segment 2, and sometimes segment 3 as well, often with finger-like projection (fig. 29). Ocelli absent. Rostrum four-segmented, short, the first segment represented by short ventral segment. Metathoracic scent glands absent. Fore legs raptorial, femur widened and with anterior groove to receive tibia and tarsus. Tarsal formula 1-1-1. Fore tarsus with single blunt claw, middle and hind tarsi with two equal claws. Membrane of hemielytra well developed with numerous veins in most species. Apex of abdomen with slender, non-retractable, respiratory siphon of varying length (from rather short to longer than body).

29

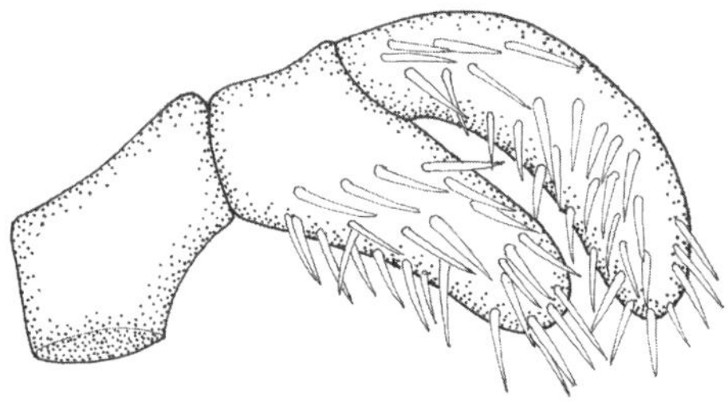

FIGURE 29
Antenna of *Ranatra* sp.

Characteristics

Eggs of Nepidae, studied by Hinton (1961), have characteristic respiratory horns on the anterior pole. In Nepinae there is variation between the number of horns between species, within species and between eggs of a single female. Eggs are deposited in a wet terrestrial substrate, including decaying plant debris, with the horns sticking out. In Ranatrinae there are always two horns (fig. 30), in Nepinae the number varies from 4-26 (fig. 31). The horns are tube-like, narrow slightly towards the apex, and contain a fine mesh which, if the egg becomes submerged, acts as a plastron. This function is restricted to the apical part of the horn, roughly the apical third in Nepinae, and varying from nearly the entire length of the horns to the apical third in Ranatrinae. The part with plastron function has small holes in the outer wall for gas exchange. The inner meshwork runs through the horns and is connected with a similar meshwork in the egg shell. The egg shell has many pores to let air in, but it lets water vapour out when the substrate dries. Eggs of Nepidae are thus susceptible to drought. The plastron can withstand excess pressures of over a half atmosphere (*Nepa*) or over two atmospheres (*Ranatra*) in clean water. These values are dependent of the surface tension of the water, which tends to drop in polluted

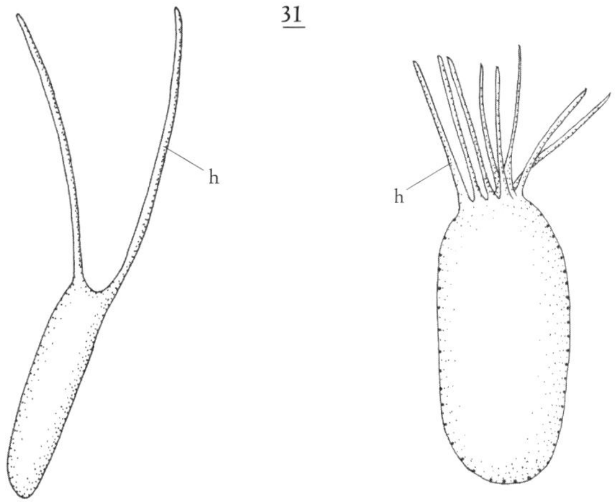

FIGURES 30 – 31
Eggs of Nepidae.
30 *Ranatra* sp. – 31 *Laccotrephes* sp.; h = respiratory horn.

water. When the surface tension is lower, the plastron becomes wetted and finally collapses.

The first thorough paper on morphology of Nepidae (Hamilton, 1931) is based on *Nepa cinerea*. Nepidae species are reservoir breathers. When undisturbed they keep contact with the atmosphere by using their siphons. Only some *Ranatra* species are reported to habitually perch in deeper water. There is apparently only one functional pair of tracheal stigmata, the eighth abdominal pair, dorsally opening into the proximal plates of the anal armature, which are partly surrounded by the basal plates of the respiratory siphon. However, at least in *Ranatra* the first abdominal pair may have an exhalant function. The respiratory siphon consists of strongly elongated laterotergites of abdominal segment 8, which in their apical parts form the two halves of the tube.

The abdominal apex and the genitalia are symmetrical in both sexes. In Nepinae the external difference between the sexes is small; females are somewhat larger on average and their genital operculum has a more acute apex. In *Cercotmetus* and *Ranatra* the female operculum has a ventral keel and sharply pointed apex, whereas in the male the genital operculum is rather flat and its apex obtuse (figs. 32, 33). The operculum is used by females to make slits in the substrate (usually decaying plants) to deposit eggs. The male parameres (fig. 35) and the shape of the genital capsule (fig. 34) give clues to specific identity in some cases, although in several genera this is not sufficient. Morphology of the genital capsule is quite uniform and only one *Ranatra* species is depicted here. Lansbury (1974b) evaluated the internal sclerites of the aedeagus for *Ranatra*, and J. Polhemus & Keffer (1999) for *Laccotrephes*, as an aid to specific identification. The results are, however, not promising. Especially *Laccotrephes* species are very difficult. In Asian *Ranatra* species there are a number of quite well defined species-groups (Lansbury, 1972a). Previously used characters appear to be unreliable, although J. Polhemus & Keffer (1999) reintroduced refined antennal characters in their treatment of the *Laccotrephes grossus*-group.

32

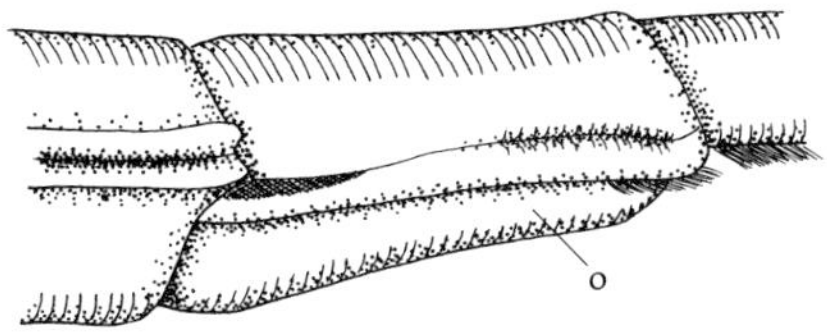

33

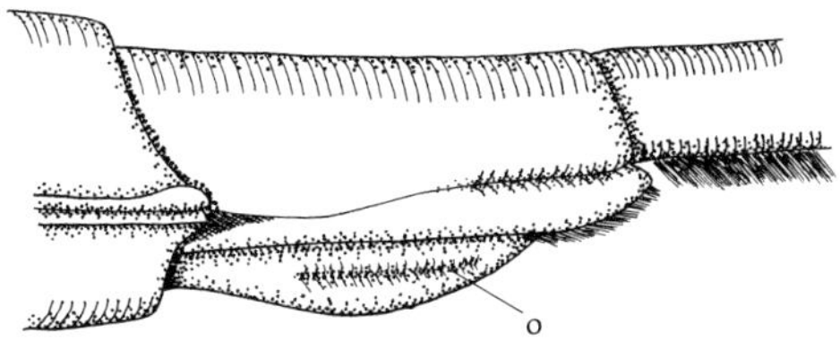

FIGURES 32 – 33
Apex of abdomen of *Ranatra longipes longipes* in lateral view.
32 male – 33 female;
o = genital operculum.

34

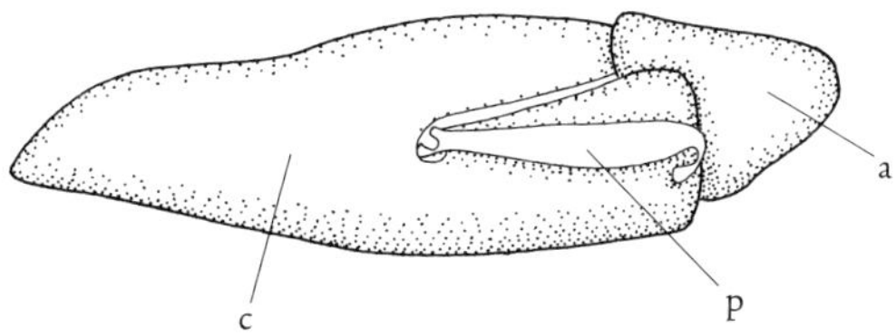

35

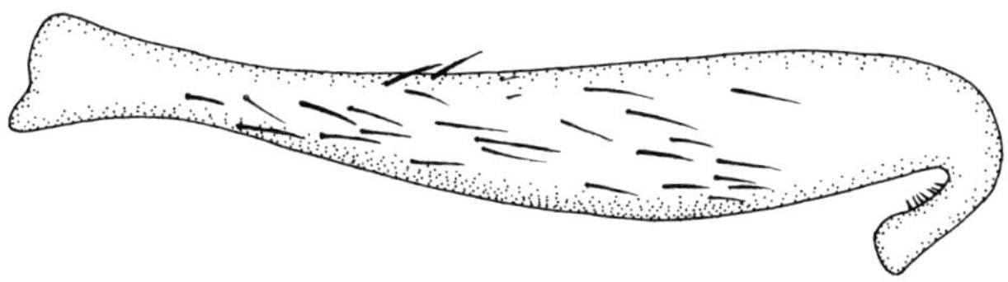

FIGURE 34 – 35
34 Genital capsule of *Ranatra longipes longipes*
35 Paramere of *Ranatra longipes longipes* (enlarged);
a = anal cone, c = capsule, p = paramere.

Adult Nepidae have rather large subcircular organs on the ventral laterotergites of segments 4-6 (figs. 36, 37, s). These look like large openings of tracheal stigmata, but they are closed, consist of a strong chitinous ring with a central field and function as **static organs**. The real stigma lies inside the ring at the median edge. It is very small, apparently non-functional and hidden by a chitinous fold. In larvae these subcircular organs are present on segments 3-6 and visible as small grooves, with the stigmata lying further towards the apex on the sternites. In larvae of Nepidae there are medially directed outgrows ventrally on the ventral paratergites, forming a respiration channel. These outgrowths are more strongly developed in Nepinae than in Ranatrinae, and are especially large in *Laccotrephes*. They obscure both stigmata and sensory grooves. In the adult organ the central field is a membrane with sensory bristles. Dorsally of it is an air bubble. Through the tracheae all six air bubbles are connected by transverse channels in the main longitudinal tracheal tubes. If an individual is not horizontally oriented the membranes of a higher situated organ are slightly convex due to the lower pressure relative to the lower situated ones, which are slightly concave. Sensory hairs detect these slight changes; in this way an individual can check its orientation. Electrophysiological research has confirmed that the system works as a differential manometer system (Bonke, 1974). Nepidae appear to have this static organ to compensate for their poor swimming ability. In other aquatic Nepomorpha, with the possible exception of Belostomatidae, such abdominal static organs are lacking. Aphelocheiridae and bottom dwelling Naucoridae possess static organs on the thoracic venter. Details on structure and function of the static orientation are described by Baunacke (1912).

Fossil record

The preferred habitat of Nepinae – shallow muddy waters – is prone to fossilization, but there are surprisingly few fossils of the entire Nepidae family. Popov (1971) described *Laccotrephes incertus* from the Upper Jurassic, about 150 million years ago, from Germany. He concluded that the split between Nepidae and Belostomatidae took place during the Lower Jurassic, about 190 million years ago. Nel & Paicheler (1992) questioned the assignment to a recent genus because of the poor condition of the specimen. Other fossils are much more recent: some species of *Nepa* are from the Upper Eocene, about 40 million years ago, and some *Ranatra* from North America are about 7000 years old, among them a species still occurring today (Nel & Paicheler, 1992).

Distribution worldwide

The family is represented in all zoogeographical regions, but the number of species per area is highest in the tropics. Judging from some recent papers the tropical Asian fauna probably still harbours a good deal of undescribed species (Nieser, 1996, 1997; Nieser & Chen, 1996; J. Polhemus & Keffer, 1999).
Most authors recognize two subfamilies with fourteen recent genera (appendix). Nepinae contains ten genera. The dominant genus is *Laccotrephes*, widely distributed in the Old World tropics with about 80 described species (although several names are probably synonyms). Africa, with 36 species, has the greatest species density (Poisson, 1965). Ranatrinae contains four genera, all of which occur in the Indo-Australian region. The dominant genus, the cosmopolitan *Ranatra* with about 150 species (see the key below), is the only genus occurring outside the Indo-Australian region. Greatest species density of *Ranatra* lies in South America with nearly 50 species (De Carlo, 1964).

Distribution in Indo-Australia

Of Nepinae four genera occur in the entire Indo-Australian region and three in Malesia with fourteen species. Of Ranatrinae also four genera occur in Indo-Australia and two in Malesia with seventeen species.

Biological notes

The number, length and density of hairs along middle and hind tibia increases in the following order: *Laccotrephes, Ranatra, Cercotmetus*. When disturbed at the water surface these bugs dive, for which they have to discharge air from their air storage. Especially *Nepa* has only a few hairs along the middle tibia and a thin fringe along the hind tibia. In glass jars under laboratory conditions *Nepa cinerea* cannot swim up at all and will perish on the bottom. *Ranatra linearis* will swim up from depths less than 75 cm. However, Larsen (1949), who analyzed flying and swimming in *Ranatra*, suggests that *R. linearis* should be able to swim up from greater depths and concludes that *Ranatra* is a moderately good swimmer. *Laccotrephes* may behave like *Nepa*. *Cercotmetus*, a better swimmer than *Ranatra*, may have less trouble, but experimental results with *Cercotmetus* and *Laccotrephes* are not available. Other aquatic Nepomorpha are stronger swimmers and dive against the upward force of the air store. If they want to surface they do so passively. Nepidae crawl to reach the surface.

Eggs of Nepidae are too large to obtain sufficient oxygen when simply deposited under water and left alone. Also, they lack the wax layer and complicated chorionic egg shell structure present in terrestrial bugs to prevent water loss, so if laid outside the water they will desiccate. Nepidae have solved this problem by depositing eggs in a waterlogged environment with the respiratory horns sticking out above the surface.

Nepidae species are ambush predators: they wait for prey hidden in the mud or between vegetation until they can grab it with the fore legs. The fore tibiae have an apical sense organ which apparently senses vibrations caused by a potential prey; visual clues however seem to be more important.

Laccotrephes has been used with moderate succes in mosquito control in China (Hoffmann, 1927). One species of *Cercotmetus* is reported to feed almost exclusively on mosquito larvae (Williamson, 1949).

Key to subfamilies and genera of Nepidae in Indo-Australia[1]

1 Body broad and flat; parasternites visible; abdominal sterna divided into six more or less distinct longitudinal zones (figs. 36, 38) ▸ **Nepinae 2**

– Body usually cylindrical; parasternites concealed by the ventral laterotergites (figs. 37, 39) ▸ **Ranatrinae 5**

2 Small, length excluding respiratory siphon not exceeding 14 mm. Siphon short, less than half of length of genital operculum [one species, *Montonepa erutus* (Montandon), from Assam, see Lansbury (1973b)] ▸ ***Montonepa* Lansbury**

– Large, length over 16 mm. Respiratory siphon longer than genital operculum ▸ 3

1 Adapted from Lansbury (1972b, 1974c).

FIGURES 36 – 39
Diagrams to illustrate the difference between Nepinae and Ranatrinae.
36, 37 abdomen in ventral view – 38, 39 cross section of abdomen – 36, 38 Nepinae – 37, 39 Ranatrinae; m = mediosternite, o = genital operculum (sternite 7), p = parasternite, s = static organ, v = ventral laterotergite.

36

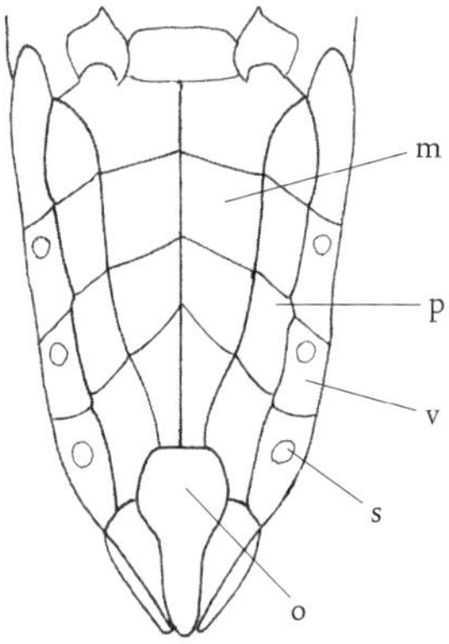

37

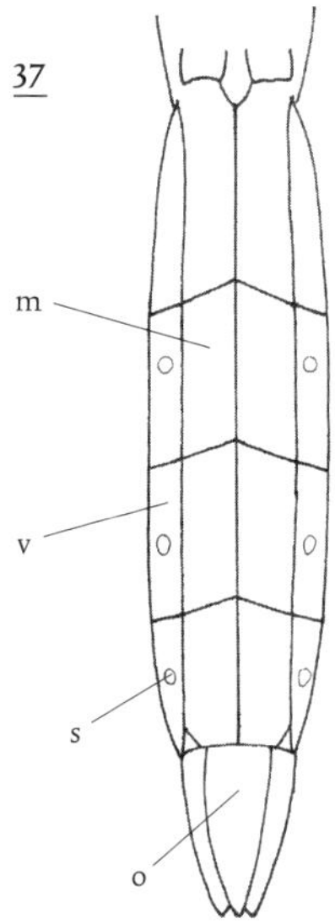

38

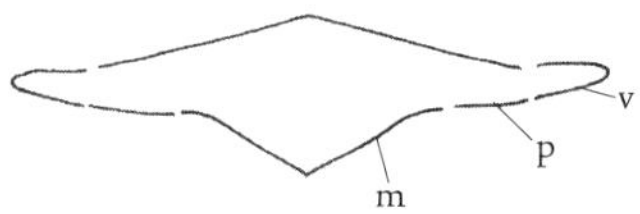

39

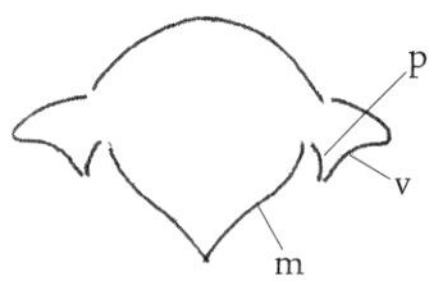

3 Sulci on prosternum caudally of fore coxa distinct (fig. 40, s). Siphon less than half of length of inner margin of hemielytra (measured from apex of scutellum to membrane; fig. 45) [South America, South-east China, North Borneo] ‣ ***Telmatotrephes* p. 61**

– Sulci on prosternum caudally of fore coxa vestigial or absent. Siphon more than 2/3 of length of inner margin of hemielytra ‣ 4

4 Membrane of hemielytra with distinct venation and differentiated from corium. Siphon longer (usually much longer) than inner margin of hemielytra (fig. 44) [Paleotropical] ‣ ***Laccotrephes* p. 59**

– Membrane of hemielytra with indistinct venation and not differentiated from corium. Siphon shorter than inner margin of hemielytra (fig. 43) [one species, *B. mayri*, from North Luzon; see Lansbury (1973b)] ‣ ***Borborophyes* p. 58**

5 Body subcylindrical. Distance between middle and hind coxa narrower than width of one coxa ‣ 6

– Body flattened. Distance between middle and hind coxa broader than width of one coxa ‣ 7

40

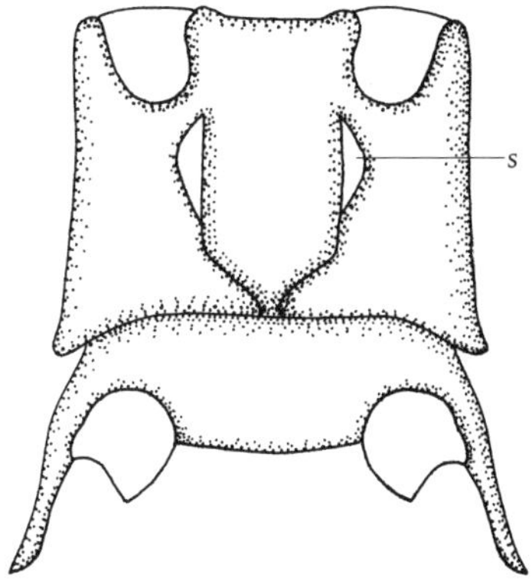

FIGURE 40
Prosternum and mesosternum of *Telmatotrephes breddini*; s = sulcus on prosternum.

6 In lateral view outer margin of eyes reaching downward to ventral margin of head (fig. 41). Fore femur shorter than pronotum [tropical Asia]
▸ *Cercotmetus* p. 63

– In lateral view outer margin of eyes not reaching downward to ventral margin of head (fig. 42). Fore femur longer than pronotum [cosmopolitan]
▸ *Ranatra* p. 64

41

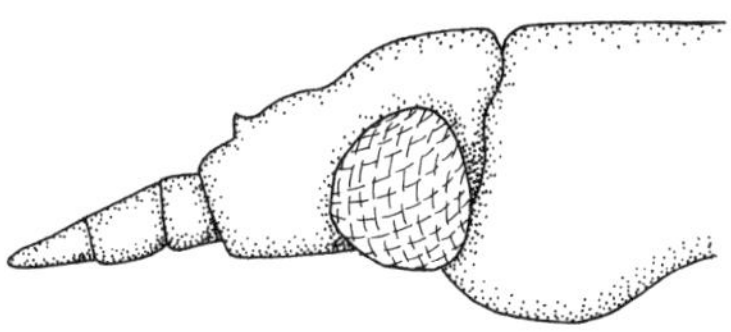

42

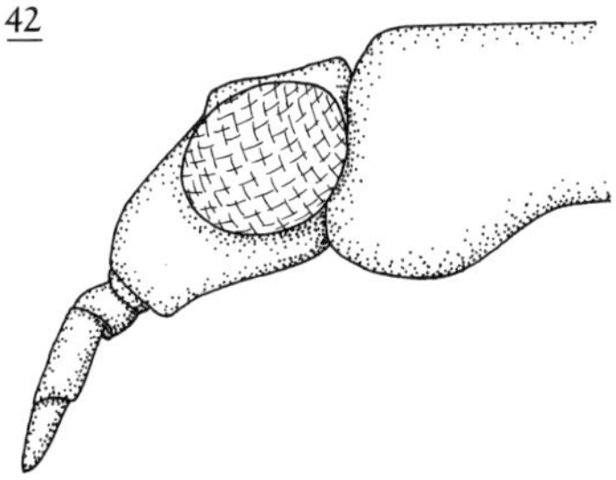

FIGURES 41 – 42
Head and anterior part of pronotum of Ranatrinae. 41 *Cercotmetus* – 42 *Ranatra*.

7 Humeral width of pronotum larger than median pronotal length. Groove in anterior femur extending about 4/5 along inner margin [known by three species from North Australia (Lansbury 1974c, 1978b)] ▸ *Goondnomdanepa* Lansbury

– Humeral width of pronotum less than median pronotal length. Groove of fore femur short, extending about halfway along inner margin [known by one species, *A. angusta* (Hale), from Australia (Menke & Stange, 1964)]
▸ *Austronepa* Menke & Stange

SUBFAMILY NEPINAE – waterscorpions

Diagnosis

Small to large (10-45 mm), flattened, parasternites not folded inward. Head narrower than anterior width of pronotum; anterior angles of pronotum embracing posterior part of head up to halfway eyes. Fore coxa short and stout, less than twice as long as wide (except in Neotropical genus *Curicta* Stål). Venter divided in six longitudinal zones (fig. 36). Eggs with a variable number of horns, but always more than two.

Genus *Borborophyes*

Figure 43

43

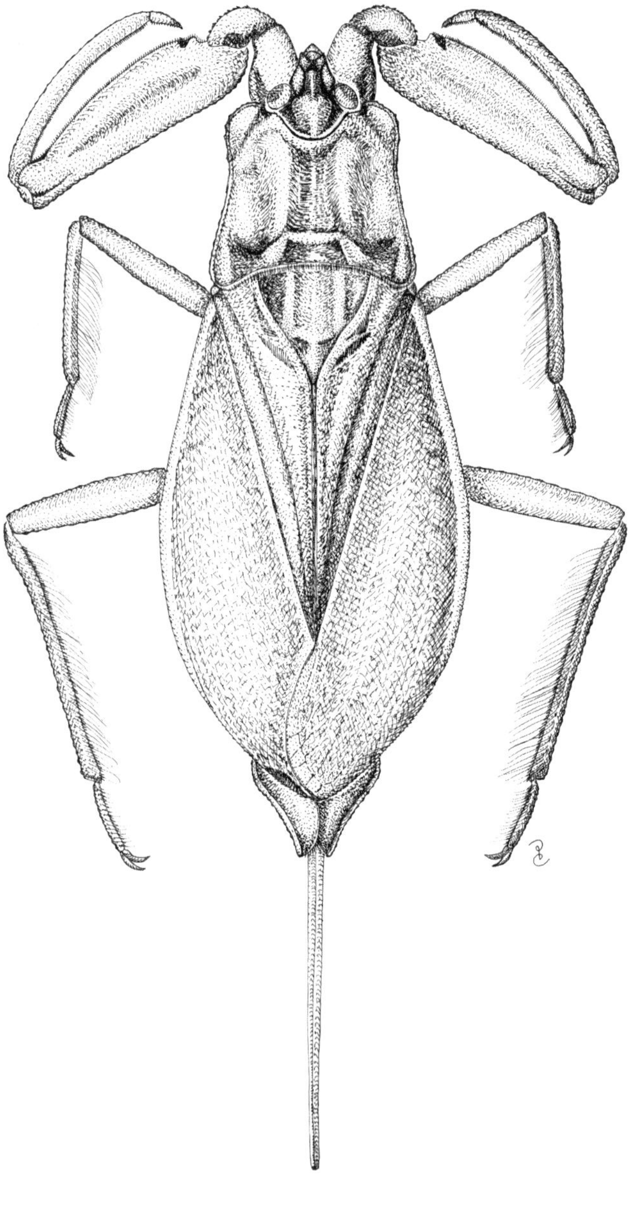

FIGURE 43
Borborophyes mayri,
female, length 23.5 mm.

Diagnosis

Medium-sized (21-25 mm) with respiratory siphon about as long as 1/3 of body length. Antenna three-segmented, the second without finger-like projection. Lateral margin of pronotum slightly concave but not constricted at the transverse groove. Prosternum without sulci posteriorly of fore coxa. Hemielytra smooth, membrane coriaceous with obscured venation.

Distribution in Malesia (fig. 565)

There is one species, *B. mayri* (fig. 43) in this genus. It is known from a few specimens from northern Luzon only.

Biological notes

A number of specimens were taken from the margins of a clear rocky stream and from leaves and trash in a concrete-lined spring-fed ditch in very shallow water (J. Polhemus, pers. comm.). Out of ten specimens seen by Lansbury (1973b) two were taken at light.

Genus *Laccotrephes*

Figure 44

Diagnosis

Small to large (11-45 mm) with respiratory siphon at least 2/3 of body length, but usually as long or slightly longer. Dorsum of abdomen (under the wings) in several species pale reddish to bright crimson. Antennae three-segmented, second segment with finger-like projection. Lateral margins of pronotum slightly concave but not constricted at transverse groove. Prosternum without sulci posteriorly of fore coxae. Hemielytra smooth, distal part membranous with distinct venation.

Distribution in Malesia (fig. 565)

Distributed throughout Malesia except for the Bismarck and Solomon Islands. There seem to be no areas of high species density in Malesia, although nearly half (five out of twelve species) occur in Sumatra.

Biological notes

The data on *Laccotrephes* presented below are based on Hoffmann (1927, 1933). Biology is comparable to that of *Nepa*. *Laccotrephes* species live at the water's edge, usually in very shallow muddy places; pools, ponds, ditches and small streams, including small irrigation canals in paddy fields, are preferred. Immatures occur in shallower water, away from adults. *Laccotrephes grossus* survives in polluted water. It often covers its back with mud and waits in concealed position, head downwards. *Laccotrephes tristis* is common in rockpools in Australia, up to 5 m deep with very clear water (Lansbury pers. comm.). The adults position

44

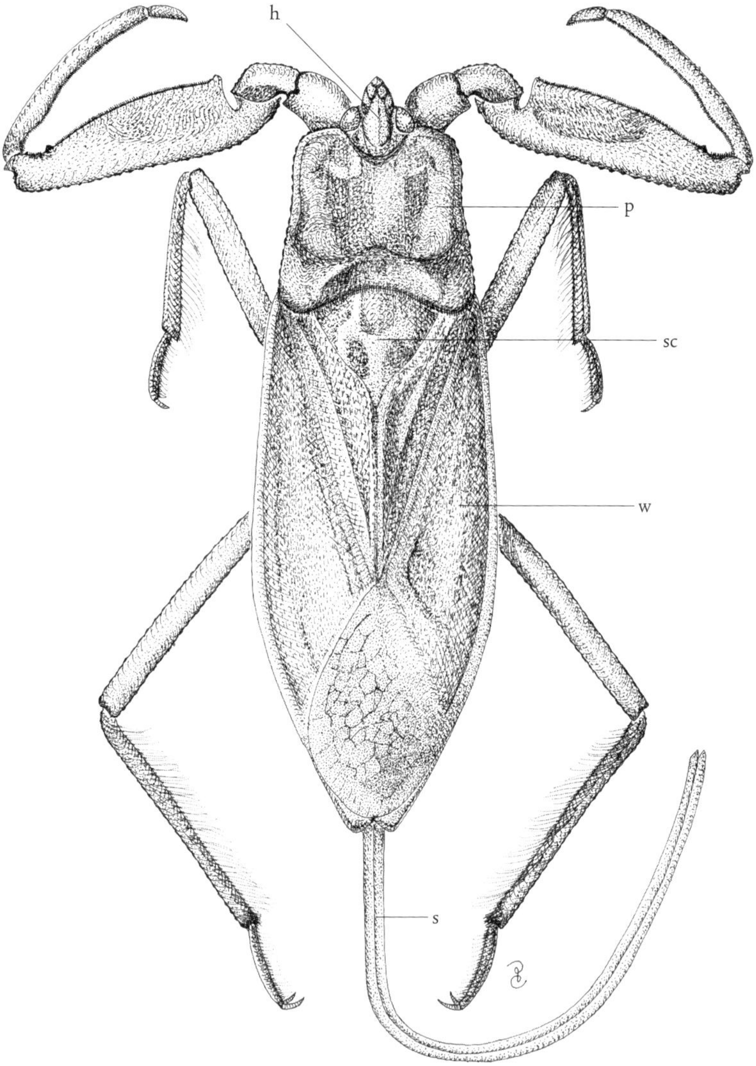

FIGURE 44
Laccotrephes celebensis, male, length 35 mm; h = head, p = pronotum, sc = scutellum, s = respiratory siphon, w = wing.

themselves with heads down, siphons penetrating the surface and fore legs stretched outward. Some species walk on land during daytime on cloudy days after rain, and perhaps also at night.

During copulation the male grasps the female with the fore legs over the prothorax and with his middle legs just behind hers. Other legs are used to secure footing in the vegetation (observed in an aquarium). When a more stable support is available, the male possibly grasps the female thorax, as in *Nepa*, with both fore legs. When secured the male curves the tip of its abdomen beneath the abdominal tip of the female and the genital capsule is pushed upward between the bases of the respiratory siphon. There is one record of possible

courtship: a male was above a female rapidly moving his anterior legs, somewhat like shadow boxing, part of the time only making the motions, part of the time touching the female on the back or legs.
Eggs (fig. 31) are laid in mud outside the water, preferably on a steep slope, and are covered with mud until only the filaments are exposed. Oviposition seems to take place at night only. The larvae enter the water after hatching and start feeding after about 10 hours.
Laccotrephes species usually wait until prey comes within reach, after which the fore legs are suddenly closed. When hungry they will stalk prey. The diet consists of various kinds of small invertebrates and tadpoles; mosquito larvae appear to be favourite. Especially nymphs can be used for mosquito control in small garden ponds. Hoffmann (1927) reports success in mosquito control on Lingnan University campus, Guandong, China, using *Laccotrephes* and *Diplonychus*. Later he succesfully used larvae of a smaller unidentified species of *Laccotrephes* to control mosquitoes in small garden ponds and pots with marsh plants, notably those in which lotus plants are grown (Hoffmann, 1933).

Genus *Telmatotrephes*

Figure 45

Diagnosis

Medium sized (23-32 mm), siphon short, up to about 1/4 of body length. Antennae three-segmented, second segment without a finger-like projection. Lateral margins of pronotum slightly concave but not constricted at transverse groove. Prosternum with distinct sulci posteriorly of fore coxa (fig. 40). Hemielytra smooth, membrane coriaceous with obscured venation. Hind wings reduced, coriaceous.

Distribution in Malesia (fig. 565)

There is one species, *T. breddini* (fig. 45), in North Borneo (Sarawak, River Baram, near the border with Brunei) and another in Thailand (Sites & J. Polhemus, 2000).

Biological notes

In North-central Thailand a species of *Telmatotrephes* was found in 'a stream in dense forest at the base of a steep incline. The water was clear and approximately 3 m wide at the greatest width, substratum was gravel and sand, and flow was sluggish with occasional riffles. The specimens were collected by sweeping a D-net through rootmats and leafpacks along stream margins' (Sites & Polhemus, 2000).

REFERENCES – Nepinae. Study of static organs (Baunacke, 1912); biology of *Laccotrephes* (Hoffmann, 1927, 1933); general morphology based on *Nepa*

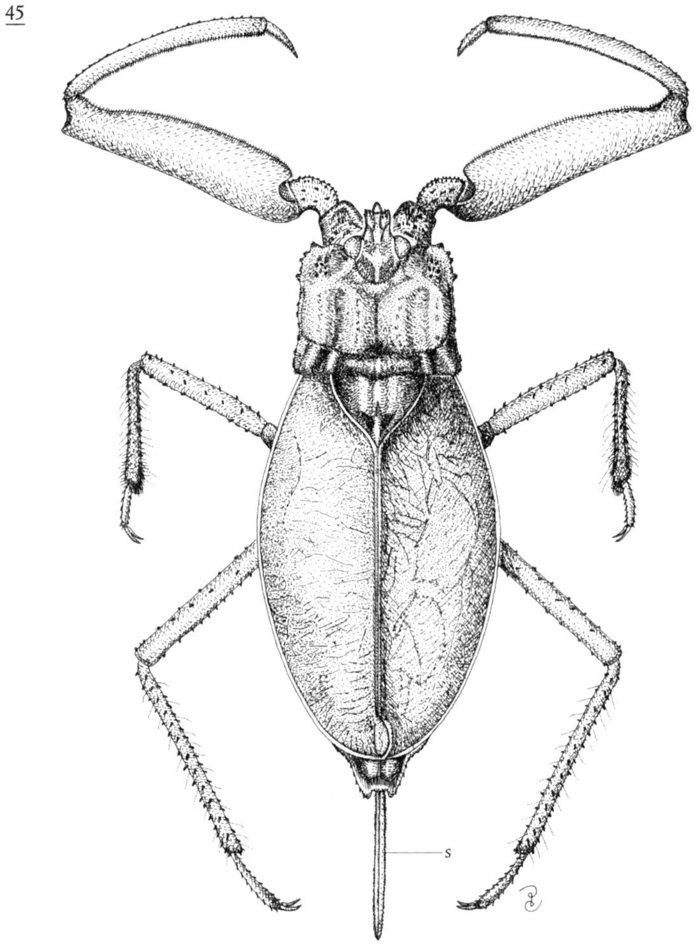

FIGURE 45
Telmatotrephes breddini, male, length 29 mm; s = respiratory siphon.

(Hamilton, 1931); eggs (Hinton, 1961); fossil Nepomorpha (Popov, 1971); revision of *Telmatotrephes* (Lansbury, 1972b); *Borborophyes*: redescription and generic position within Nepinae (Lansbury, 1973b); function of static organs (Bonke, 1974); fossil Nepomorpha (Nel & Paicheler, 1992); structure of genital capsule in *Laccotrephes* and key to species of *Laccotrephes grossus*-group (J. Polhemus & Keffer, 1999).

SUBFAMILY RANATRINAE – water stick insects

Diagnosis

Medium to large (20-60 mm); body cylindrical to subcylindrical. Head usually wider than anterior width of pronotum, if narrower anterior angles of pronotum not embracing posterior half of head. Fore coxa elongate, about twice as long as wide in *Goondnomdanepa* (in other genera much longer). Parasternites folded inward, covered by ventral laterotergites; venter of abdomen thus divided into four longitudinal zones (figs. 37, 39). Eggs with two horns (fig. 30). (*Austronepa* and *Goondnomdanepa* with one and three species respectively are so far only known from Australia.)

Genus *Cercotmetus*

Figure 46

46

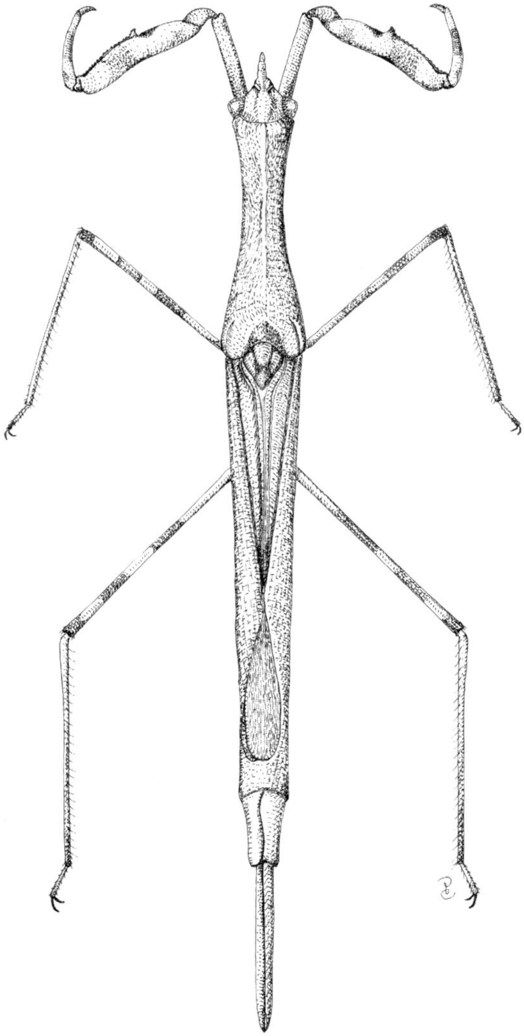

FIGURE 46
Cercotmetus brevipes, male, length 34 mm.

Diagnosis

Large (32-60 mm), subcylindrical, with comparatively short respiratory siphon about $^1/_4$ of body length. Antennae three-segmented, second segment with or without finger-like projection. Ventral eye margin reaching below venter of head, in lateral view obscuring ventral head margin (fig. 41). Fore femur shorter than prothorax, with ventral tooth halfway but without sharp constriction at the same level. Middle and hind tibia with well-developed hair fringes.

Distribution in Malesia (fig. 565)

Cercotmetus occurs throughout Malesia with the exception of Bismarck and Solomon Islands. Sumatra is relatively rich, with three out of five Malesian species.

Biological notes

Biology virtually unknown. Due to the strongly developed fringes of hairs on middle and hind tibiae *Cercotmetus* species can swim much better than *Ranatra*. Although they are often found in streams they rather float among plant debris in virtually stagnant bays (pers. obs.).

Egg filaments (respiratory horns) may be either shorter or longer than in *Ranatra* (Lansbury, 1973a).

Cercotmetus asiaticus is reported to feed almost exclusively on mosquito larvae in peninsular Malaysia, and an adult may keep an area of about five square metres virtually clear of larvae (Williamson, 1949). Laird (1956) suggested to introduce it in permanent habitats on Pacific Islands.

Genus *Ranatra*

Figure 47

Diagnosis

Medium-sized to large (20-53 mm), body subcylindrical; respiratory siphon $^1/_3$ of body length to longer than body. Antenna three-segmented, in most species second segment with finger-like projection (fig. 29). In lateral view ventral eye margin not obscuring ventral head margin (fig. 42). Fore femur longer than prothorax with one or two teeth at or distally of the middle of ventral margin, with sharp constriction at level of tooth or teeth. Middle and hind tibia sparsely beset with swimming hairs. Also differs from *Cercotmetus* in poorer development of hair fringes on middle and hind tibia.

Distribution in Malesia (fig. 565)

Distributed throughout Malesia, except for the Bismarck and Solomon Islands, with twelve species and three subspecies. The species are rather evenly distributed, with two or three on most larger islands. Three species occur in the Philippines.

47

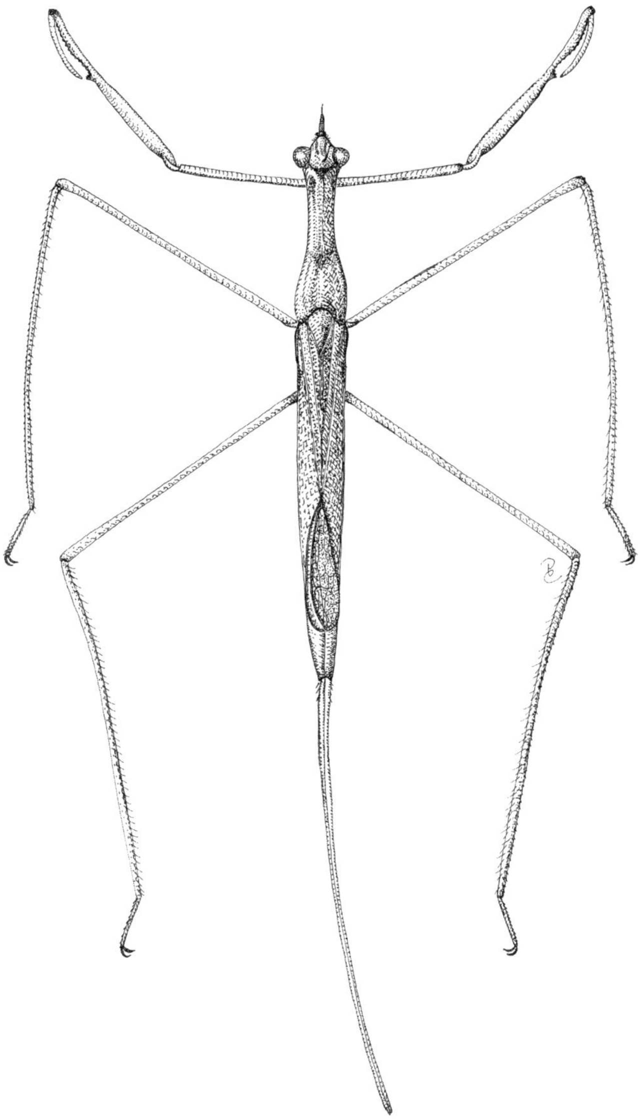

FIGURE 47
Ranatra longipes longipes, male, length 22.7 mm.

Biological notes

Contrary to Nepinae, *Ranatra* species are reported to hide between vegetation while waiting for prey, with head and fore legs pointing downward. This is apparently true for most species, but some live in streams in dense forest which are virtually devoid of vegetation and must thus hide somewhere else. A few species may burrow in mud like Nepinae and may catch prey through the surface film. This behaviour has been observed in hungry specimens living between plants in the laboratory. Due to the long fore coxa and femur the reach of the raptorial fore leg is considerably increased, especially compared to Nepinae.

Cloarec (1972) and Cloarec & Joly (1988) studied behaviour of European *R. linearis* and confirmed the general results of Radinovsky (1964) for the American *R. fusca*. They found that first instar larvae tend to stay in the upper few centimetres of the habitat, preferably perching between *Lemna* and *Elodea*. During development the preferred sites lie gradually deeper and plant preference shifts to plants like *Calamagrostis* and *Phragmites*. Younger larvae prefer green plants, older larvae and adults brown plants to perch on. These preferences probably aid in keeping the different instars separated and thus to some extent avoid cannibalism.

Adults walk slowly backwards to the surface, moving the respiratory siphon up and down until they make contact with the air. There are no detailed observations on the behaviour of larvae in this respect. Personal observations on *R. linearis* and some Neotropical species suggest that at least the first to third instars tend to stay more closely to the water surface. Gas exchange usually takes 30-60 seconds. They then walk down again until they reach roughly the same place. In temperate conditions *Ranatra* stays submerged for on average 40 minutes.

Copulation position and intrusion of male genitalia into the female is similar to that in Nepinae, only the angle between the longitudinal body axes of the pair is larger. The eggs of *Ranatra* are reported to be deposited in floating plant debris. This is true for the Palearctic *R. linearis* and has been verified in at least four more species, but *R. chinensis* Mayr is reported to deposit eggs in mud (Hoffmann, 1927). Egg deposition is unknown in most other species. In aquaria some species deposit their eggs also in floating polystyrene foam.

The orientation while waiting for prey is head down and fore legs stretched forward, with the body axis under normal circumstances is held in an angle of 30° or less. First instar larvae catch small prey such as water fleas. Although the maximum prey size increases with the size of the bug, even adults will still catch and suck water fleas when available.

REFERENCES – Ranatrinae. Study of static organs in Nepidae (Baunacke, 1912); movement in *Ranatra* (Larsen, 1949); structure and function of eggs (Hinton, 1961); biology of *Ranatra* (Radinovsky, 1964; Cloarec, 1972; Cloarec & Joly, 1988); revision of Oriental species of *Ranatra* (Lansbury, 1972a); revision of *Cercotmetus* (Lansbury, 1973a); function of static organs (Bonke, 1974); structure of genital capsule in *Ranatra* (Lansbury, 1974b); concise review of biology and ecology of Nepidae (Sites & Polhemus, 1994); key to Malesian species of *Cercotmetus* and Sulawesian species of *Ranatra* (Nieser & Chen, 1991); economic importance of Nepidae and extensive bibliography (Keffer, 2000).

Family Belostomatidae – giant water bugs

Diagnosis

Medium to very large (9-110 mm), ovoid to elongate-ovoid. More or less uniformly medium brown; legs yellowish to light brown, often annulated with darker brown. Antenna usually four-segmented, at least segments 2-3 with dorsal projections, shorter than head and lying in groove beneath eyes. Rostrum three-segmented, short to moderate in length. Fore femur raptorial (except in the African genus *Limnogeton* Mayr). Tarsal formula 2-1-1 or 3-1-1. Membrane of hemielytra usually well developed with numerous veins. Apex of abdomen with pair of retractable respiratory straps (fig. 58). With exception of *Limnogeton*, which has swimming hairs on slender hindtibia only, Belostomatidae species have middle and hind legs broadened, flattened and fringed with swimming hairs.

Characteristics

The antennae have projections on two or three segments, similar to Nepidae (figs. 54-56). In Nepidae these are exposed at rest, lying flat under the head, while in Belostomatidae they lie in a groove under the head, pointing dorsally, and are invisible unless the antenna is dissected. Hence, the antennal projections are called lateral projections in Nepidae, and dorsal projections in Belostomatidae. Instead of a single respiratory siphon, Belostomatidae have two retractable respiratory straps (fig. 58, r) which originate from tergum 8 and are present only in adults. These flaps are densely beset with long bristles, especially medially. Together they form the channel through which air passes the subalar air store. Apparently the first abdominal pair is the functional respiratory tracheal stigma. All stigmata are covered by a network of fine chitinous rods, and all except the first pair are closed by a membrane, which may show some slits. The transverse tracheal connections are similar to those in Nepidae. Especially the eighth abdominal spiracle may have a exhalant function. The subalar air store is partially exposed and there is an additional fully exposed ventral abdominal air store (Parsons 1972a, b, 1973). The abdominal stigmata are superficially similar to the static organs of Nepidae, with the chitinous outer ring lying internally. Möller (1920) concluded that adult *Lethocerus* static organs are associated with the abdominal stigmata 3-6. These have less sensory bristles which are less efficiently arranged compared to adult Nepidae. Conclusive proof that these structures indeed function as a static organ is still lacking. Abdomen and genitalia are symmetrical in both sexes. The abdominal sternum shows ventral laterotergites as in Nepinae. The pilose ventral laterotergites contain the abdominal stigmata. The pilosity is variable in both thickness and extension and may obscure the suture between laterotergites and sternites. In Lethocerinae (and the American Horvathiniinae Lauck & Menke) the apical sternites show an extra fold dividing them in median- and parasternites (fig. 50).

In males the apex of the genital operculum is more or less pointed (fig. 49), in females it is apically truncate and has a pair of small tufts of bristles sublaterally in the apical part (fig. 48). The parameres are of little use in identification, except in the genus *Appasus* Amyot & Serville. In *Lethocerus* and American Belostomatinae the aedeagus provides clues for identification, in *Diplonychus* this has not yet been verified.

There are no recent publications summarizing morphology of belostomatids only.

48

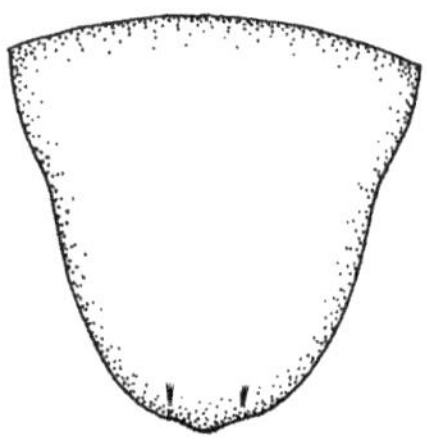

49

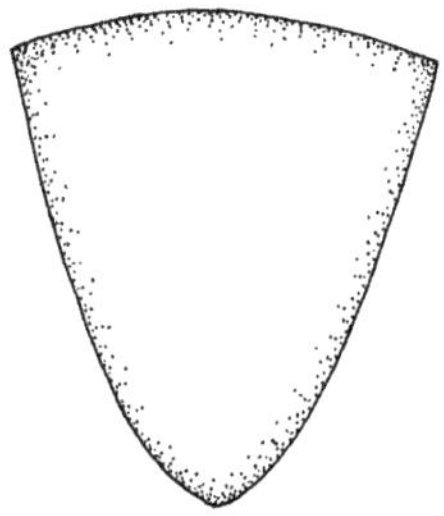

FIGURES 48 – 49
Genital operculum of *Diplonychus rusticus*; 48 female – 49 male.

Fossil record

At present, eight extinct Mesozoic genera of Belostomatidae are known, ranging from the Upper Jurassic to the Upper Cretaceous. At least one of these belongs to Lethocerinae and one to Belostomatinae, and there are two genera belonging to the extinct subfamily Stygeonepinae (Popov, 1971b; Nel & Paicheler, 1992). This diversification supports the hypothesis of Popov (1971b) that Belostomatidae became a separate taxon at least in the early Jurassic (190 million years ago). As Nepidae is the sister group of Belostomatidae their age is estimated to be the same, although definite Mesozoic fossils belonging to Nepidae are still lacking.

Distribution worldwide

Three extant subfamilies with nine genera and about 140 recent species are recognized (table 1). Belostomatidae is rather poorly represented in Asia (twelve species) and Africa (just over 20 species), whereas the Americas have over 100 species.

Distribution in Indo-Australia

There are three genera in Indo-Australia: the mainly African genus *Appasus* has one species in Thailand, *Diplonychus* and *Lethocerus* have six and five species respectively in Indo-Australia.

Biological notes

This account is mainly based on Smith (1997). Belostomatidae live in stagnant water or in sluggish parts of streams, in which oxygen is in lower supply than in fast flowing streams. The largest species are specialized in feeding on vertebrate prey, especially fish and amphibians. *Lethocerus* will even attack snakes and wading birds (Smith, 1997). Adults and larvae wait in dense vegetation or at the waters' edge for prey, keeping the respiratory straps in contact with the air.

Many species, like many other Heteroptera, come at light during the night, frequently in large numbers. As such large numbers of large insects are striking, they have been named 'electric light bugs'. A monthly or lunar periodicity has been observed by several naturalists. Duviard (1974) for instance, using light traps, found correlated factors for some African *Appasus*: over 75% of her catches were taken during the two weeks around full moon and 75% of the bugs were caught in the first three hours of darkness. After 03.00 hrs flight activity stopped entirely. Rain does not seem to influence nocturnal flight activity, but relative humidity and temperature seem to be limiting factors.

After oviposition, the males attend the eggs until hatching. As in Nepidae, the eggs are big, and Belostomatidae therefore face the same problems of lack of oxygen or dessication. Belostomatidae have solved this problem by the male keeping the eggs wet and in contact with air. An additional advantage is that the eggs are defended against predators. Laboratory experiments have shown that eggs which are not attended will perish or show a hatching percentage of only 20-30%, compared to over 90% in guarded eggs. In addition, larvae from unattended eggs have difficulties in hatching.

Many species eat aquatic snails; the African genus *Limnogeton* Mayr is even an obligate snail eater. No representatives of other Nepomorpha, with the possible exception of *Aphelocheirus*, eat snails. There are indications that Belostomatidae may alternate between waiting and actively foraging (e.g. *Diplonychus*, see below). As behaviour differs between subfamilies it will be discussed there.

Key to Indo-Australian subfamilies and genera of Belostomatidae

1 Rather large (50-70 mm). Caudal abdominal sternites with additional fold (fig. 50) (Lethocerinae) [pantropical] ▸ *Lethocerus* p. 74

– Rather small (usually 15-20 mm, occasionally smaller). Caudal abdominal sternites without additional fold (fig. 51). (Belostomatinae) ▸ 2

2 Lateral margin of eye in line with anterolateral margin of head (fig. 52). First antennal segment equal to or longer than lateral prolongations of segments 2 and 3 (fig. 56). [Tropical Asia and Australia] ▸ *Diplonychus* p. 71

– Lateral margin of eye diverging more strongly than anterolateral margin of head (fig. 53). First antennal segment shorter than the lateral prolongations of segments 2 and 3 (fig. 55). [Africa, Continental Asia, Japan] ▸ *Appasus*

FIGURES 50 – 51
Diagram of the abdominal venter of Belostomatidae.
50 *Lethocerus* – **51** *Diplonychus*; stippled area represents the long lateral pilosity, m = mediosternite, o = genital operculum, p = parasternite, s = sternite, sp = spiracle and probable static organ, v = ventral laterotergite.

50

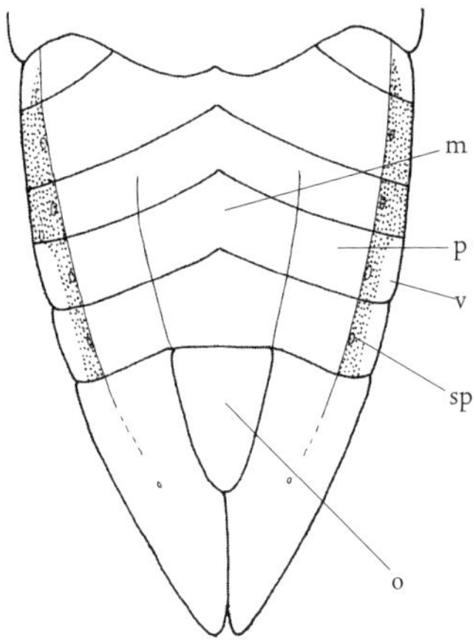

51

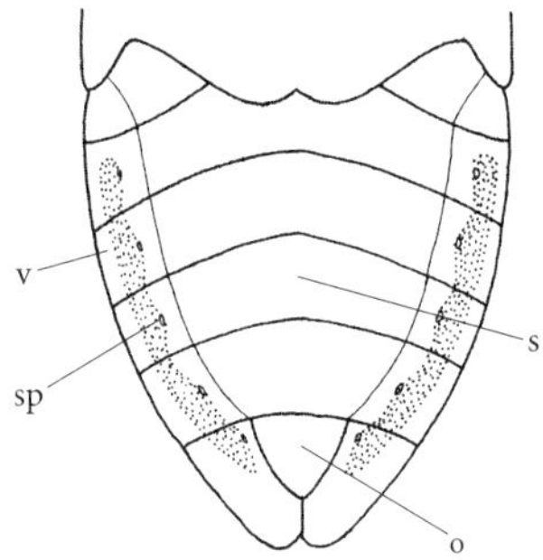

FIGURES 52 – 53
Head of Belostomatinae in dorsal view.
52 *Diplonychus* – **53** *Appasus*.

52

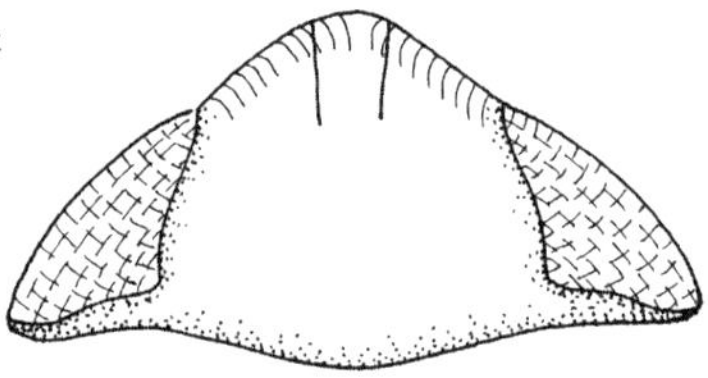

53

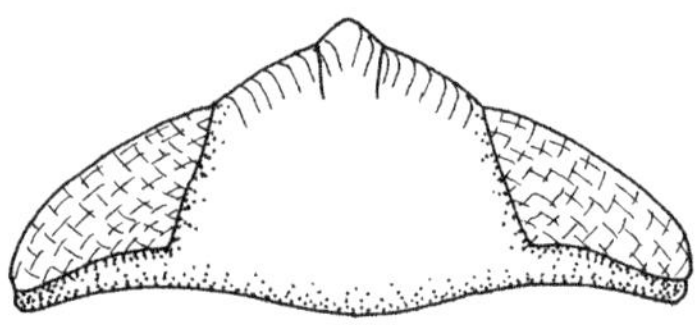

FIGURES 54 – 56
Antennae of Belostomatinae.
54 *Lethocerus* – **55** *Appasus* – **56** *Diplonychus*.

54

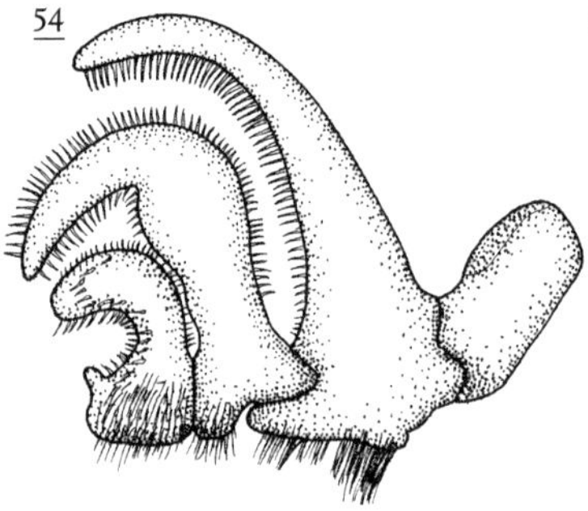

55

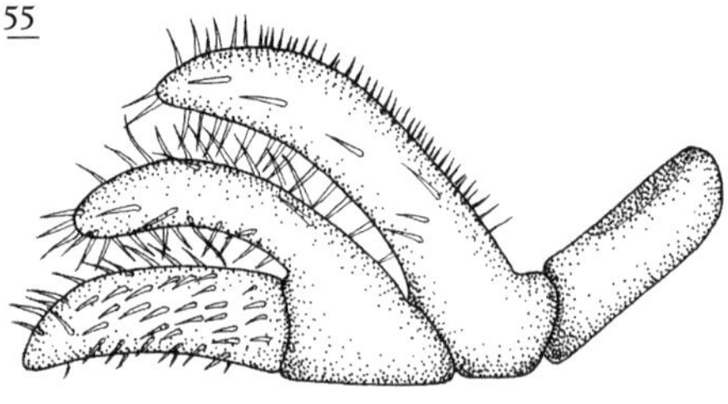

56

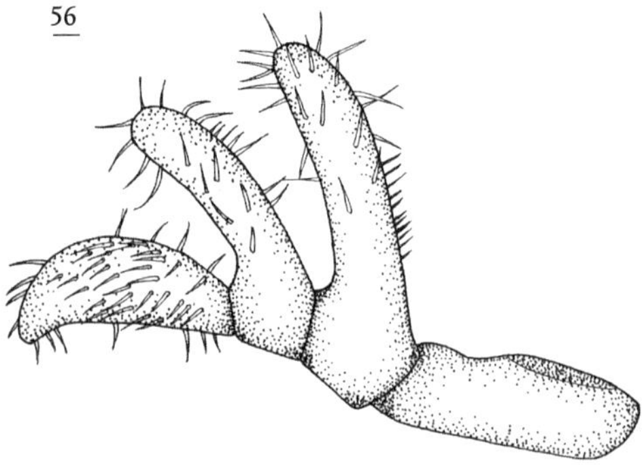

REFERENCES – Belostomatidae. Classification of the family to genus level (Lauck & Menke, 1961); economic importance of Belostomatidae, notably *D. rusticus* (Venkatesan & Rao, 1980; Venkatesan, 2000); biology, especially parental care (Smith, 1997); fossil Nepomorpha (Popov, 1971b; Nel & Paicheler, 1992).

SUBFAMILY BELOSTOMATINAE

Diagnosis

Small to large (9-70 mm). Antennal segments 2 and 3 bearing dorsal finger-like projections (except in some species of American genus *Abedus* Stål). Tibia and tarsus of middle and hind legs similar, flattened but not broadly dilated. Abdominal sternites without fold, divided in median and parasternites. Abdominal spiracles situated in centre of ventral laterotergites (fig. 51).

Genus *Diplonychus*

Figures 57-58

Diagnosis

Small to medium-sized (13-25 mm). First antennal segment equal to or longer than segment 4 and lateral prolongations of segments 2 and 3. Lateral eye margin in line with lateral head margin, not protruding laterally. Middle and hind tibiae flattened. Fore tarsi with two claws. Pubescence on ventral part of laterotergites 4 not reaching lateral margin. In *D. rusticus* claws of fore tarsus shorter than width of tarsus (fig. 57), in *D. molestum* longer.

57

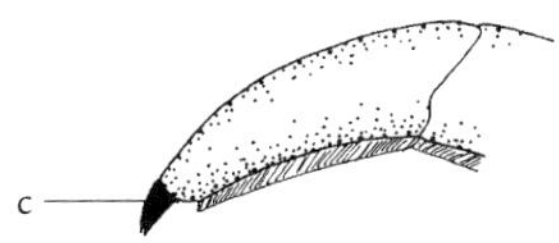

FIGURE 57
Foretarsus of *Diplonychus rusticus*; c = claws.

Distribution in Malesia (fig. 565)

Diplonychus is distributed throughout Malesia, excepting the Bismarck and Solomon Islands, with two species.

Biological notes

Most biological data mentioned here are based on observations by various authors of *D. rusticus* and not verified for other species. *Diplonychus* species live in stagnant or predominantly stagnant, usually shallow, vegetated water, including concrete tanks and ditches. They hide between the vegetation near the surface and seem to prefer floating plants such as *Eichhornia* and *Pistia*. However, in South Australia they hide in decaying vegetation at the bottom (I. Lansbury, pers. comm.).

58

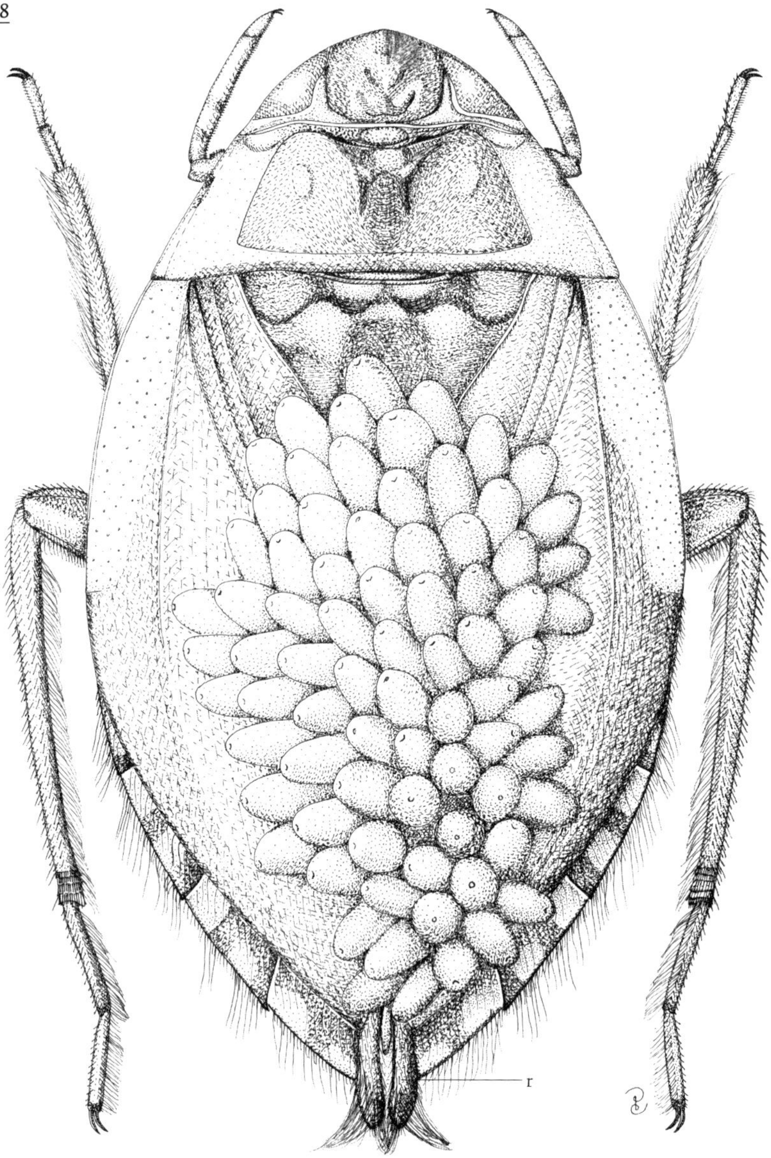

FIGURE 58
Diplonychus rusticus,
egg-bearing male,
length 16.5 mm;
r = retractable respiratory
straps.

Mating is initiated by the male, which makes up and down movements near the surface, called display-pumping. It is nearly always initiated by the presence of another displaying bug. This causes surface and subsurface wave movements which attracts females. When a female reaches the male there is usually some sparring with the fore legs, after which the female attempts to deposit eggs on the back of the male. The male always resists and attempts to start copulation. Only after – usually several – copulations the male allows the female to deposit eggs; she starts at the apex of the hemielytra and proceeds forward. Finally the

eggs uniformly cover the entire surface of the hemielytra (fig. 58). However, after every few eggs the male interrupts oviposition and copulates again with the female. The oviposition may go on for over 24 hours. Eggs of *D. rusticus* are when deposited oval, 1.65 mm long and 1.1 mm wide, and green. During development they increase sixfold and become greenish brown with the anterior (upper) end becoming darker. The brooding behaviour consist mainly of 'surface brooding': the male positions itself close to the surface so that the eggs are exposed to the air-water interface for protracted periods. This keeps the apices of the eggs in contact with the air, thus providing sufficient oxygen. This makes them vulnerable to predation and males dive immediately when movement from above is observed. Submerged males may display brood-pumping, which is similar to display-pumping but slower. Submerged males may also rhythmically brush and circulate water over the clutch. This brushing or 'grooming' of the eggs keeps the batches clean and helps to prevent infections by mould or bacteria, apart from aerating them. Brooding males do not actively hunt, but they will feed when prey comes within reach (Venkatesan, 1983; Smith, 1997).
In the laboratory *Diplonychus* alternate between waiting for prey and active foraging. There seems to be no fixed trigger for these alternations and they occur from first larval instars to adults. When sucking one prey they will strike at others coming close enough, but they do not eat them. This means that, especially in high prey density, they will kill more prey than is actually consumed. Up to three preys can be held at the same time (Cloarec, 1989-1992). Early instars will only catch small invertebrate prey, adults take anything they can master, including fish and amphibians. When fed mosquito larvae, adults prefer the older instars. In specimens which are regularly fed there are different feeding times between the sexes in the laboratory. Females show the highest predation rate between 08.00-17.00 hrs, after which there is a steep drop in feeding activity. Males show low feeding activity between 08.00-14.00 hrs but a peak around 20.00 hrs, when female activity is low. It is supposed that this pattern restricts intraspecific food competition Venkatesan & Rao, 1980).
Hoffmann (1927) used a species of *Diplonychus* successfully in mosquito control on the campus of Lingnan University in China. Venkatesan (2000), noting the absence of mosquito larvae from habitats with *D. rusticus*, suggests its potential for control of mosquitoes.

REFERENCES – Belostomatinae. Feeding rhythm in *D. rusticus* (Venkatesan & Rao, 1980); brooding behaviour of *D. rusticus* (Venkatesan, 1983); biology of *Diplonychus* (Cloarec, 1989-1992); redefinition of *Diplonychus* (J. Polhemus, 1995); biology, especially parental care (Smith, 1997); fossil Nepomorpha (Popov, 1971b; Nel & Paicheler, 1992).

SUBFAMILY LETHOCERINAE

Diagnosis

Large to very large (40-110 mm), elongate. Antennal segments 2-4 bearing dorsal finger-like appendages. hind tibia and tarsus very flat and broadly dilated, much broader than middle tibia and tarsus. Abdominal sternites 4-6, subdivided in median and parasternites by weak suture-like fold, originating at basal angle of genital operculum. Abdominal spiracles on mesal margin of ventral laterotergites.

Taxonomic note

Although in the past several different genera were assigned to this subfamily, it is currently considered to contain only one genus, *Lethocerus*, with two subgenera, of which *Benacus* Stål is American and *Lethocerus* pantropical. Apart from the generic name *Belostoma*, which was incorrectly applied to *Lethocerus* for a long time, the other commonly used synonym is *Kirkaldyia* Montandon. Although several nominal taxa were described as *Kirkaldyia*, these are now considered to be synonyms of a single species: *Lethocerus deyrolli* Vuillefroy, which does just not extend into Malesia. *Kirkaldyia* is also used in Lundblad (1933a) for *L. deyrolli*.

Genus *Lethocerus*

Figure 59

Diagnosis

As for the subfamily.

Distribution in Malesia (fig. 565)

Lethocerus occurs throughout Malesia with the exception of the Bismarck and Solomon Islands. Of the two species occurring in Malesia *L. indicus* is widespread in the area, and *L. insulanus* reaches New Guinea from Australia.

Biological notes

American species thrive in ponds and lakes and are occasionally found in quiet parts of streams. Several Asian species are common in paddy fields. Most specimens of this genus in collections were captured at light.

Breeding in *Lethocerus* is very different from Belostomatinae (Ichikawa, 1988-1995; Smith, 1997). Eggs are deposited on emergent vegetation or other objects above the water surface. Mating is induced by the male taking position at the base of emerging vegetation or other objects, with the tip of its abdomen held above water. In response to movement in the vicinity or, sometimes, spontaneously the male starts display-pumping, causing waves both on the surface and in the water column, which attracts receptive females. After a female arrives they will copulate once or several times. Copulations are interrupted by one or

59

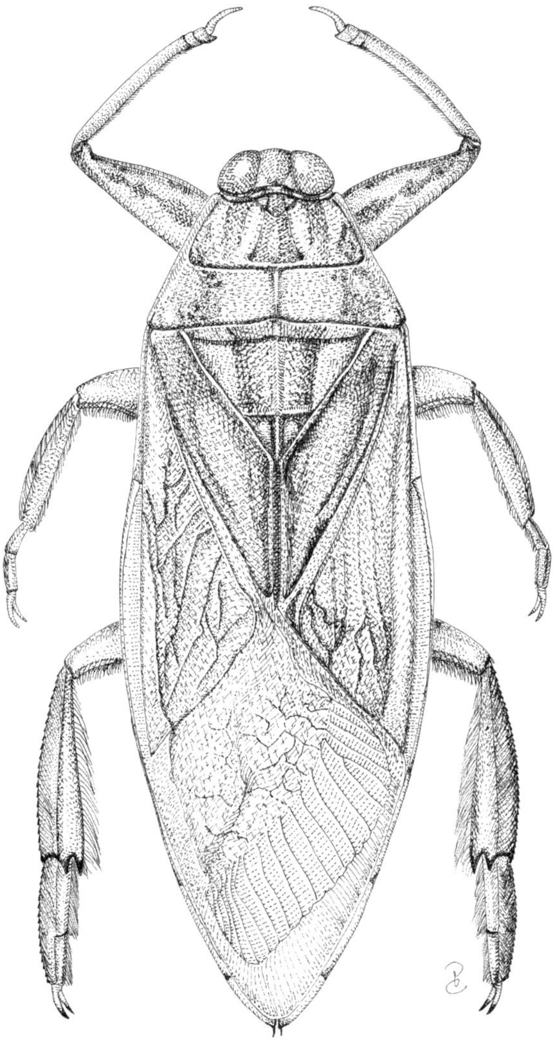

FIGURE 59
Lethocerus indicus,
male, length 72 mm.

both individuals climbing out of the water onto vegetation and then returning. Finally the female will climb out of the water to a place where she will deposit the eggs, with the male following her. As in Belostomatinae oviposition is frequently interrupted (after each 3-10 deposited eggs) by copulation. After the female has deposited all eggs she allows the male to assume the watering position (see below). She then jumps in the water and swims away, leaving the male in charge of breeding. The abandoning female never returns to the clutch. If males are removed during oviposition the female will stop laying eggs after five minutes and abandon the incomplete clutch if the male has not returned within one hour. The male stays in the water at the base of the clutch. At irregular intervals he climbs towards the clutch, inserts his rostrum between the eggs in the upper part of the clutch and lets water from his body and imbibed water flow into the clutch. Unattended eggs will perish both when submerged and when kept above the water. Apart from watering, males also defend the eggs against predators. If a clutch is tapped the male will emerge to check and chase away a predator. The most important predators of lethocerine eggs seem to be other females. As they are, on average, larger than the males it is usually difficult for the latter to defend a clutch against a female. After a clutch has been devoured by a female the male usually mates with this female and starts brooding the new clutch. One of the reasons may be that, as males are much longer occupied by an act of reproduction than females, there is scarcity of free males. It may thus be profitable for a female to take over an already brooding male. *Lethocerus* larvae are supposed to stay near the area of hatching for some time. Moreover, older larvae feed on younger ones. This means that receptive females would have to look for a new mate and new oviposition site if they do not destroy a previously laid clutch.

Lethocerus are the only Nepoidea with metathoracic scent glands being 10-25 times larger in males than in females. The glands in this genus contain a pleasant-smelling aromatic substance (trans-hex-2-enyl-acetate; see chapter 1: Economic importance). However, it is not useful in killing bacteria as do the contents of the metathoracic scent glands in other Nepomorpha. Smith (1997) hypothesized that it is used to mark a trail to the egg clutch. This would explain why the female gland is much smaller and measures only a tenth of the male's gland. It would also explain how males are able to go straight to the clutch they are brooding, even at night.

In spite of their considerable swimming power Lethocerinae are ambush predators. They prefer vertebrate prey, although they will take any aquatic animal. Contrary to Belostomatinae they do not seem to eat snails, although there is a single report of *L. indicus* in Vietnam attacking snails (Smith, 1997). In some waters *Lethocerus* is the main predator of tadpoles. Only moving prey is attacked; a dead animal will be ignored if gradually introduced, but when offered abruptly it will be seized immediately. There are several isolated reports of *Lethocerus* taking much bigger and heavier prey; once caught the bug will not let go, but even more important is the saliva, which, apart from powerful proteolytic and haemolytic enzymes, contains neurotoxins which affect the heart.

REFERENCES – Lethocerinae. Morphological notes on *Lethocerus* (Möller, 1920); key to Old World species of *Lethocerus* (Menke, 1960); classification of the family to genus level (Lauck & Menke, 1961); fossil Nepomorpha (Popov, 1971b); brooding behaviour of *Lethocerus* (Ichikawa, 1988-1995); fossil Nepomorpha (Nel & Paicheler, 1992); edible insects in Thailand (Chen et al., 1998).

Family Ochteridae – velvety shore bugs

Diagnosis

Small (4.5-9 mm), broadly oval, moderately dorso-ventrally flattened with soft velvety hemielytra. Predominantly blackish with yellowish to light brown markings; in addition, notably scutellum and hemielytra variably marked with bluish-grey pruinose spots. Antennae four-segmented, shorter than head but in dorsal view often not fully concealed beneath it. Ocelli present, rostrum four-segmented, very long. Fore legs not raptorial. Tarsal formula 2-2-3. Legs adapted for walking, without swimming hairs. Membrane of hemielytra well developed with either seven or over 20 cells.

Characteristics

Ochteridae show some superficial similarities to Saldidae (shore bugs), which live in comparable habitats. There are, however, many fundamental differences in morphology suggesting that similarity is due mostly to convergence. Ochteridae are of a more solid build than Saldidae, while the latter have much longer antennae. Male abdominal segments are asymmetrical distal of segment 6; in the female these are symmetrical (figs. 60, 61). The genital capsule is strongly modified, twisted to the right, with a large convex sternite and a reduced narrow tergite (fig. 62). The parameres are distinctly asymmetrical, the left one strongly reduced, nearly vestigial, the right one large. In addition to colour and size, the right paramere, together with the general form of the genital capsule, usually gives the best characters for identification.

Taxonomic note

Until recently the taxonomy of this family was little studied, partly because representatives live on the shore and are therefore usually not collected by hydrobiologists and somewhat neglected by students of terrestrial bugs. Also, these animals are apt to escape netting.

Distribution worldwide

Most species are tropical with only relatively few reaching subtropical or temperate regions. There are three genera with some 60 described species, nearly all belonging to the cosmopolitan genus *Ochterus*. There are two small genera, of which *Megochterus* Jaczewski with two species occurs in Australia.

Biological notes

See under *Ochterus*. Economic importance is negligible.

Key to the Indo-Australian genera of Ochteridae

1 Large (length 8.3-9.8 mm). Frontal plate strongly produced above base of rostrum. Antennae very short. Membrane with c. 20 cells (fig. 65). [Australia] ▸ *Megochterus*

– Smaller (less than 7 mm). Frontal plate slightly produced above base of rostrum at most. Antennae, especially third and fourth segments, elongate. Membrane with only seven cells (fig. 64). [cosmopolitan] ▸ *Ochterus*

Genus *Ochterus*

Figures 60-64

Diagnosis

Small (4.5-6.5 mm). Head declivent anteriorly, frontal plate not or hardly produced above base of rostrum. Antennae, especially third and fourth segments, elongate. Membrane with seven cells in two rows: four cells in basal row, three cells in apical row. Parameres in Old World species with an apical knob and two appendages (fig. 63).

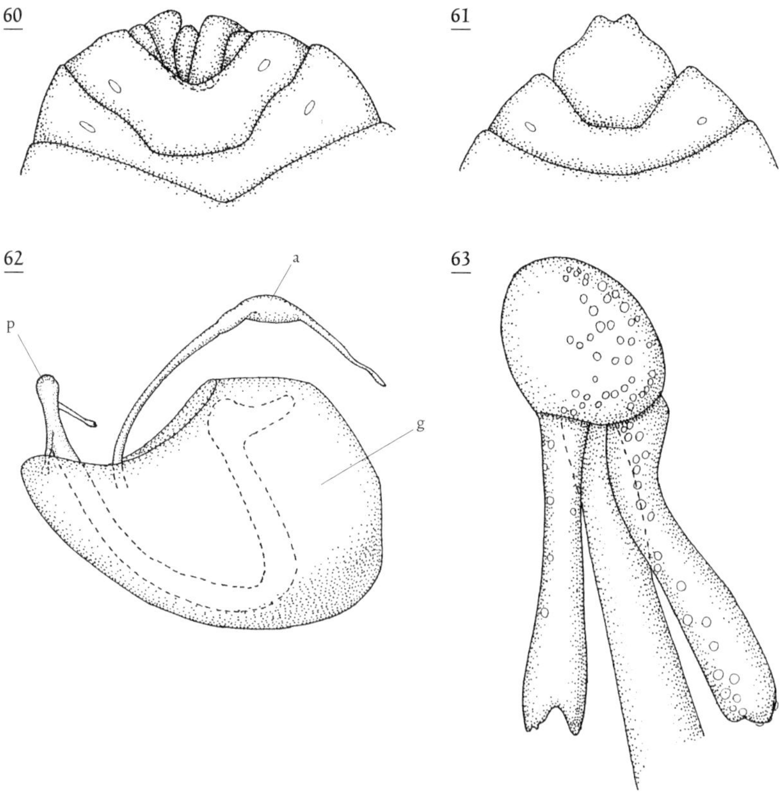

FIGURES 60 – 63 *Ochterus marginatus.* **60 – 61** apex of abdomen in ventral view: **60** male – **61** female – **62** genital capsule – **63** apex of right paramere; a = aedeagus, g = genital capsule, p = right paramere.

64

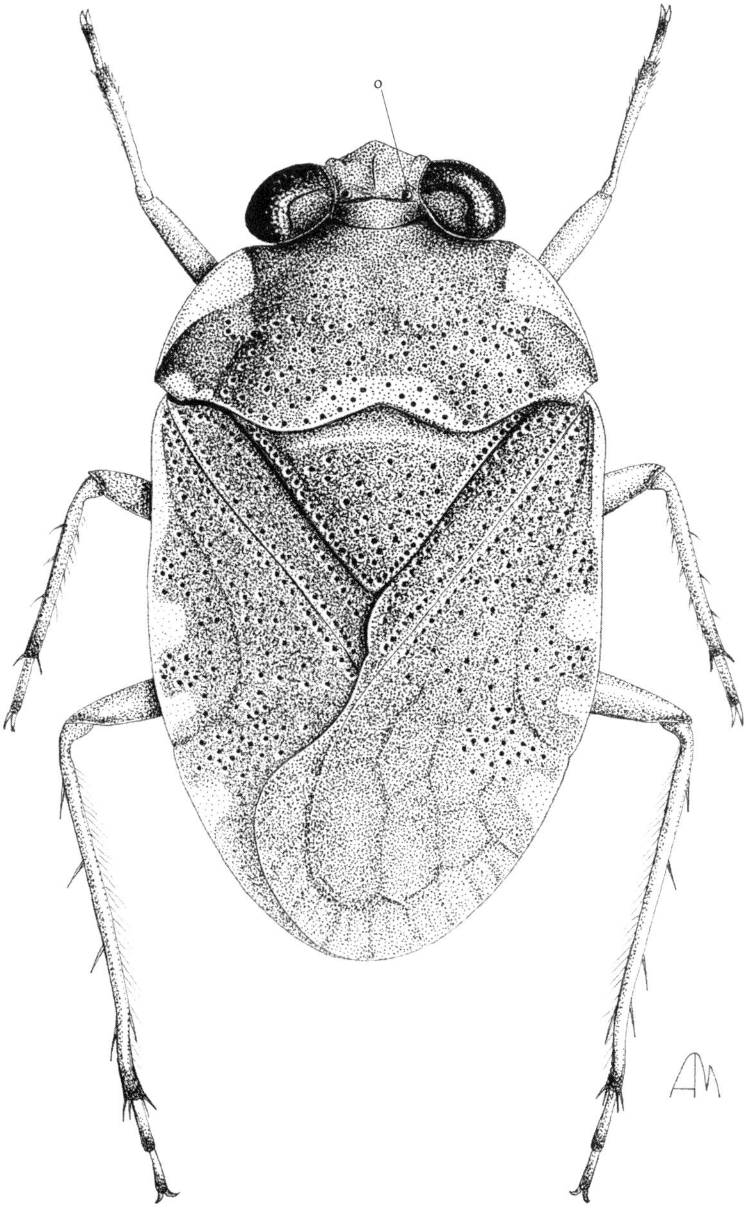

FIGURE 64
Ochterus marginatus, male, length 4.5 mm (from Nieser, 1982); o = ocellus.

Distribution in Malesia

Distributed throughout Malesia with 28 described species. One of these, *O. marginatus* (fig. 64) occurs from the Mediterranean area in Europe to Japan, Philippines, northern Borneo, Java and Sulawesi. However, a study by Chen and Nieser indicates that the current interpretation of this 'species' apparently comprises several closely related species, each occupying a part of this large area. In Malesia and adjacent areas there are apparently four areas with

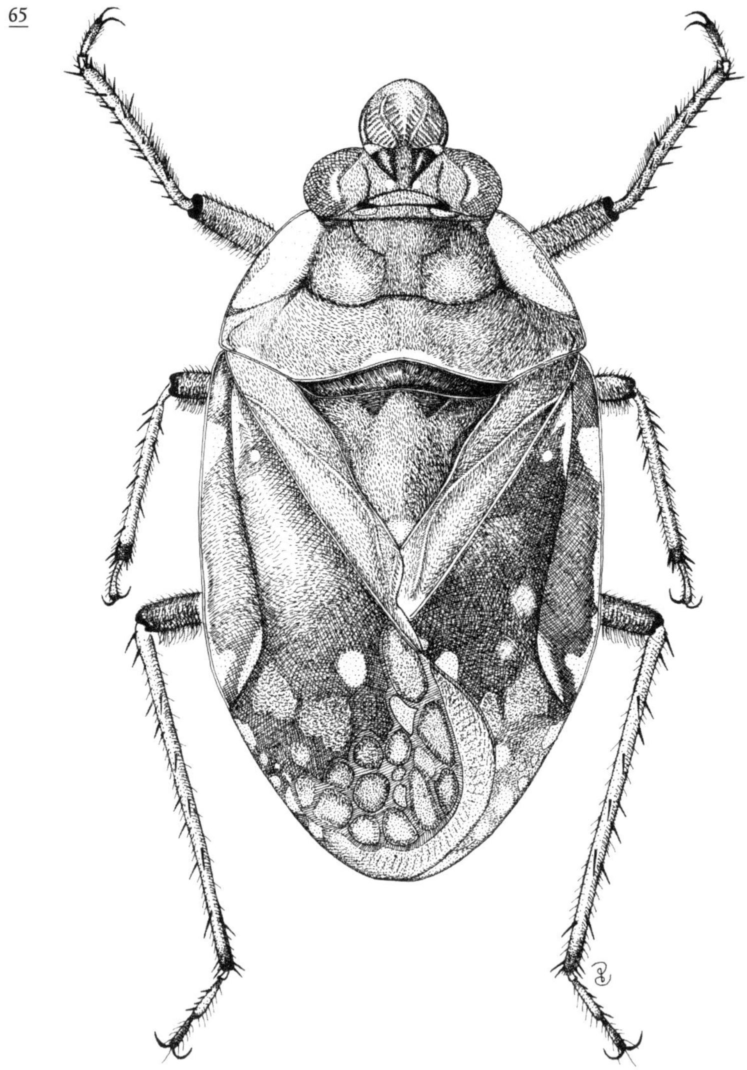

FIGURE 65
Megochterus nasutus,
female, length 9.5 mm.

comparatively high species density: New Guinea with ten species, Australia with nine species (of which only one shared), Sulawesi with six species and Luzon with five species (one shared). This compares favourably with Borneo and Sumatra with two species each, of which one shared. In continental South-east Asia only one species is recorded. The low species number for these areas is presumably due to insufficient study, as several additional species have already been recognized in continental South-east Asia (pers. obs.).

Biological notes

Most observations on biology, including those presented here, are based on European populations of *O. marginatus*. *Ochterus* species usually live at the edge of running water, including the base of waterfalls. They are mostly found at sandy or stony places with little vegetation and prefer sunny places, which is especially true for *O. marginatus* (pers. obs.). Recent reports on habitat however show that many species live in shaded places (Gapud & San Valentin, 1977; Nieser & Chen, 1999). In the Philippines it is quite common to find 2-4 species sympatrically, indicating that they have different habitat requirements. Adults fly away immediately when movement is observed. To collect *Ochterus* one should look around quietly and slap a net over them; they will fly up and into the net. One can also try to carefully catch them in a glass tube.

Eggs are suboval, one side being somewhat more convex than the other. They are deposited on the surface of small objects, not inside plant tissue.

Contrary to other Nepomorpha larval biology differs somewhat from that of adults. Apparently, larvae need sand to develop. Larvae, at least of *O. marginatus*, are nearly always covered with sand and loam, which is actively replaced when removed: sand or loam is scooped up by a small comb on the front of the head just over the labrum (the so-called pala frontalis, found in larvae only), and then put into place by fore and hind legs. The sand is apparently fixed with a glue-like compound. Prior to the imaginal moult the larva constructs a moulting chamber with a sandy or loamy roof with some holes in it. Moult takes two days (Boulard & Coffin, 1991).

Apart from the fact that they are carnivorous, adult biology is poorly known.

REFERENCES – Ochteridae. First revision of Oriental species (Jaczewski, 1934); revision of Ochteridae in the Oriental and Australian region (Kormilev, 1971); description of subspecies of *Ochterus marginatus* from the Philippines (Rieger, 1977); Philippine fauna (Gapud & San Valentin, 1977; Gapud, 1981, 1995, 2003); revision of *Ochterus* in Australia and New Guinea (Baehr, 1990b); larval development (Boulard & Coffin, 1991); key to Sulawesian species (Nieser & Chen, 1999).

Family Gelastocoridae – toad bugs

Diagnosis

Medium sized (4-15 mm), broadly built; dorsally often rough and warty. Colour variable. Antennae four-segmented, not visible in dorsal view. Eyes large with mesal emargination of dorsal surface appearing reniform (fig. 68). Adults with ocelli (excepting six species). Rostrum four-segmented, short and stout. Tarsal formula 1-2-3. Fore legs raptorial, femur strongly enlarged with anterior groove for reception of tibia and tarsus. Fore tarsus sometimes fused with tibia. Middle and hind legs adapted for walking, without swimming hairs. Membrane of hemielytra in some species reduced, when fully developed usually with numerous veins.

Characteristics

Male abdominal segments distal of segment 5 or 6 are asymmetrical (fig. 66). In females of Malesian species the apex of abdomen is ventrally symmetrical or slightly asymmetrical (fig. 67). The genital capsule is strongly modified with a well developed right paramere and vestigial left paramere. In *Nerthra* only the apex of the right paramere is used in identification, as it is difficult to extract the whole capsule without destroying the apex of the abdomen. Apart from the apex of the paramere (fig. 67) important characters for species identification are

66

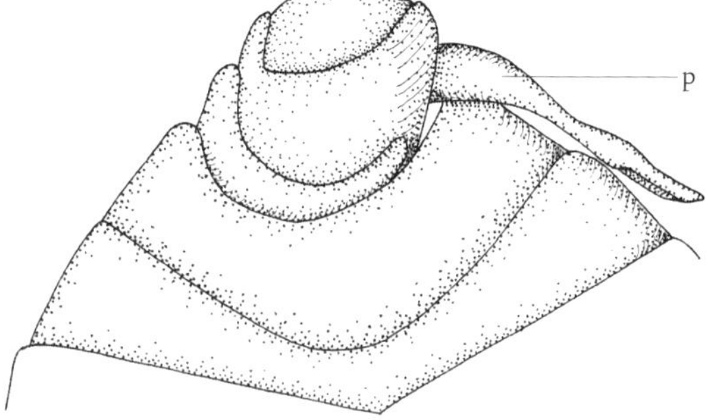

67

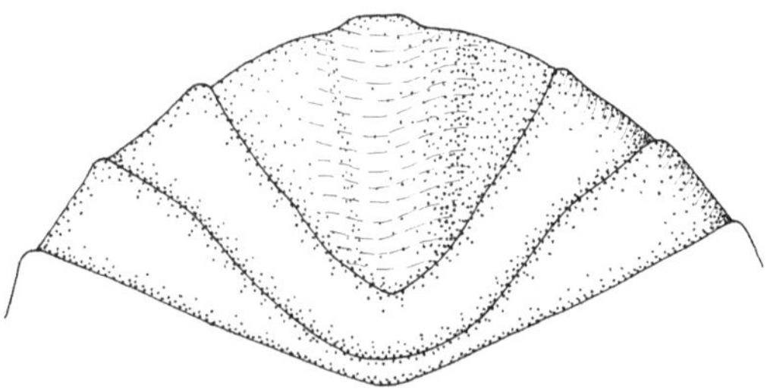

FIGURES 66 – 67
Nerthra mixta, apex of abdomen in ventral view. **66** male – **67** female; p = exposed right paramere.

size, development of the membrane, and shape of various body parts (e.g. lateral margin of the pronotum, shape of embolium), and various ratios. Females of some species are not or only hardly separable.

Distribution worldwide

There are 99 species of Gelastocoridae, of which 87 in the genus *Nerthra*. The three recognized genera are divided over two subfamilies. The two genera in Gelastocorinae are restricted to the Americas. Nerthrinae, with the single genus *Nerthra*, is distributed through the tropics and subtropics, excepting the Mediterranean area.

SUBFAMILY NERTHRINAE

Genus *Nerthra*

Figure 68

68

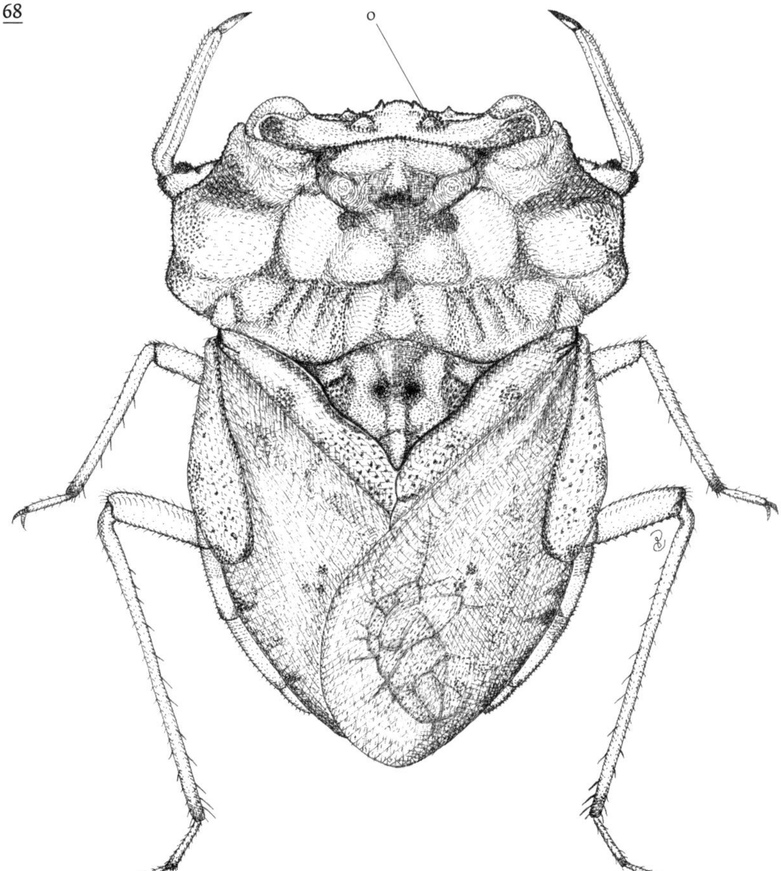

FIGURE 68
Nerthra laticollis, macropterous female, length 10.3 mm; o = ocellus.

Diagnosis

Small to medium-sized (4-15 mm), dorsally flattened. More or less uniformly medium to dark brown. Front of head usually provided with tooth-like tubercles; rostrum four-segmented but first segment concealed, apparently originating on the ventral surface of the head, apical part pointing ventrally or anteriorly. Fore tarsus fused with tibia, one well-developed claw. Ninth abdominal segment not invaginated in body cavity; ninth sternite visible externally. Left paramere absent, right paramere strongly developed and folded longitudinally. *Nerthra macrothorax*, widespread in the eastern part of Malesia, can be separated from all other Malesian species by absence of ocelli and reduced membrane of hemielytra, which are entirely coriaceous and fused together and with at least one longitudinal carina reaching to the middle of the hemielytron. It is regularly found near the seashore.

Distribution in Malesia

Sixteen of the 28 species recorded in Malesia occur in New Guinea, of which twelve endemic to New Guinea. The other twelve species are more or less evenly distributed over Malesia: the Moluccas have four species, Sulawesi three and most other Malesian areas two species. Twenty-two species are known from Australia. In continental Asia including India there are only six species.

Biological notes

Biology of *Nerthra* is mainly based on more or less isolated observations, mostly of American species. Many species seem to be nocturnal, hiding during daytime under stones, plant debris or in humid sand or mud. Although most species are still supposed to live at water edges, several species have been collected in banana plantations, cow dung or pitfalls away from water. On the other hand, at least three species have been found under water, crawling on plants or on the (under)sides of floating logs. On one occasion specimens were found under rocks in fast flowing water (La Rivers, 1953).
When threatened *Nerthra* species either jump away or press themselves against the substrate, depending on their coloration to escape predators. Once spotted they are easy to collect. Macropterous specimens are common, but in several species there are no records of flying individuals. Some species have fused hemielytra and are thus flightless. *Nerthra macrothorax*, with fused hemielytra, is widespread in the eastern part of Malesia and the Pacific. This species has been found digging into rotten *Pandanus* logs. Todd (1959) supposes that it has spread by drifting on such logs.
The life cycle is little known and no species have been reared from egg to adult in captivity. Eggs are broadly oval, slightly flattened at one side, and the shell has a rough hexagonal sculpture. The American *N. martini* Todd has been found to deposit eggs in small muddy holes under stones near water. The female apparently guards the eggs until they hatch (Usinger, 1956).

Recently a stridulatory mechanism has been discovered in males of all species inspected for this character (J. Polhemus & Lindskog, 1994). It consists of a rastrate area on the genital capsule near the base of the right paramere, against which the sclerotized posterior margin of the anal cone is rubbed. Stridulation in *Nerthra* however has apparently never been heard by humans.
As they take various small insects in captivity they are probably carnivores.

REFERENCES – Gelastocoridae. Biological notes (La Rivers, 1953); revision of family, biological notes (Todd, 1955); revision of species from Melanesia and adjacent islands (Todd, 1959); biological notes (Usinger, 1959); stridulatory mechanism (J. Polhemus & Lindskog, 1994).

Family Aphelocheiridae – benthic water bugs

Diagnosis

Small to medium-sized (3.5-11.5 mm), oval, strongly flattened. Colour in dorsal view varying from light brown or light greyish to dark brown or dark grey; most species more or less uniformly coloured but some with yellowish spots on a brown background. Head produced anteriorly and posteriorly, embraced by anterolateral angles of pronotum. Antennae elongate, slender, filiform, four-segmented, apices in dorsal view often visible. Rostrum very long, reaching well onto the metasternum, segment 2 very short, segment 3 very long (fig. 6). Metathoracic scent glands lacking, venter of thorax and abdomen with plastron. Fore femur only slightly thickened. Tarsal formula 3-3-3. Basal tarsal segment small, difficult to recognize. All tarsi with pair of equally well developed claws. Abdominal spiracles 2-7 surrounded by rosettes (fig. 73). Micropterous morph common. Membrane in macropterous morph well developed, without veins.

Characteristics

First abdominal segment strongly reduced (second segment is the first segment completely visible). Tergite 5 in males large, covering most of the caudal segments (fig. 76), slightly asymmetrical in many species, in some distinctly so. In ventral view sternites 6-8 slightly asymmetrical. Genital capsule prominent, can be extracted in fresh or relaxed specimens simply by pulling it caudally with a pair of tweezers. The asymmetrical parameres lie dorsally on the capsule and are species-specific (figs. 69, 70). Abdomen in female symmetrical, without enlarged segments. Sternite of segment 8 forms operculum which in many cases is species-specific. In addition, inner propleural projection (fig. 71) and absence or presence of small peg-like spines on the medial hind margin of some abdominal sternites may give clues to specific identity. All tarsi three-segmented, but the first tarsal segment always small and often partly covered

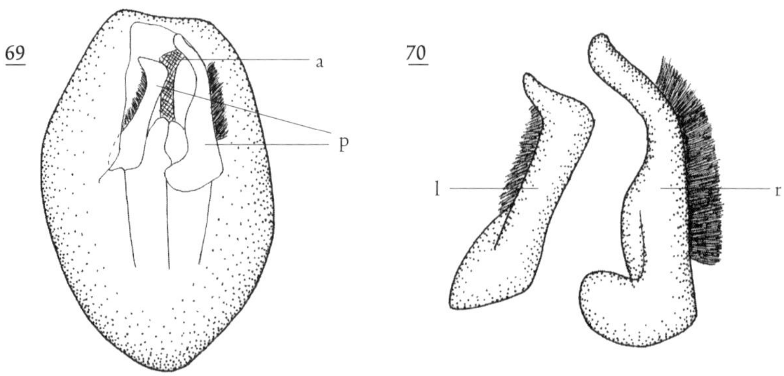

FIGURES 69 – 70
Genital capsule and parameres of *Aphelocheirus robustus*.
69 genital capsule in dorsal view – **70** parameres; a = aedeagus, l = left paramere, p = parameres, r = right paramere.

71

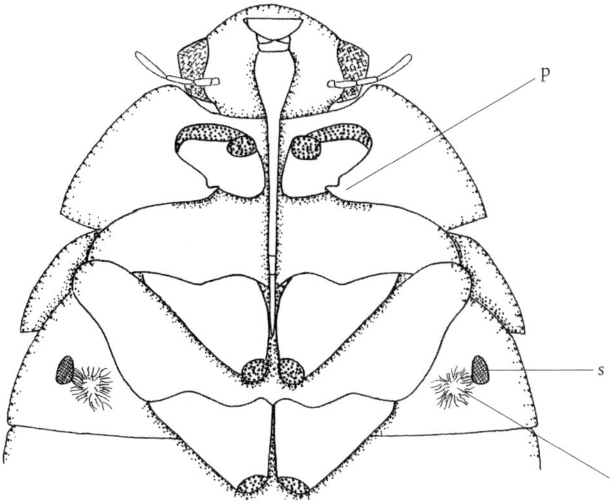

FIGURE 71
Head, thorax and abdominal sternite 2 of *Aphelocheirus robustus* in ventral view; p = inner projection of propleuron, r = rosette, s = static organ.

by the apex of the tibia. The first segment is often visible only after clearing the leg and only under high magnification. The fore leg is often described as non-raptorial, probably because it lacks the chitinous anterior groove on the femur. However, as seen in figure 72, there is a groove formed by the hairs on the femur. *Aphelocheirus* mainly catches prey by probing with the long rostrum in crevices and piercing prey with it.

72

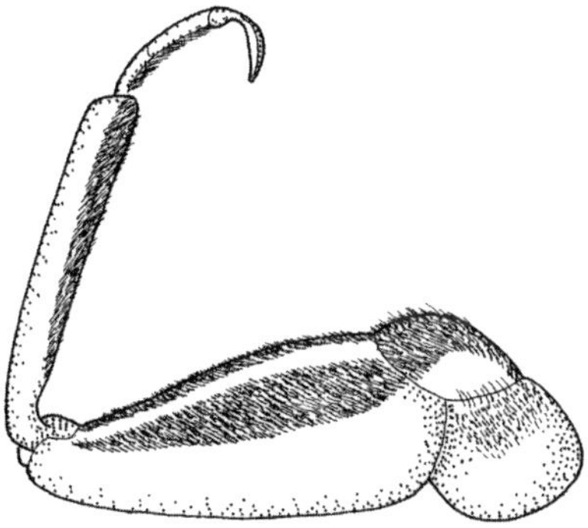

FIGURE 72
Right fore leg of *Aphelocheirus zamboanga* in ventral view.

Taxonomic note

Aphelocheirus, although considered a distinct family already by Fieber (1851b), has until recently been considered a subfamily of Naucoridae. Cladistic analyses by Rieger (1976) and Mahner (1993) support the view to consider it a distinct family. However, D. Polhemus & J. Polhemus (1988), in their revision of Asian *Aphelocheirus* including a survey of taxonomy of Naucoridae, still follow Hoberlandt & Štys (1979) in treating it as a subfamily of Naucoridae.

73

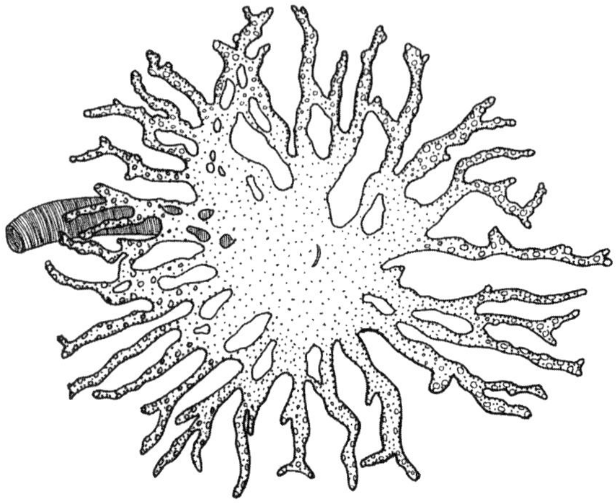

FIGURE 73
Aphelocheirus aestivalis, rosette at entrance of tracheal system (after Szabó-Patay, 1924).

Distribution worldwide and in Indo-Australia

There is only one recent genus divided into two subgenera, of which *Micraphelocheirus* is restricted to the tropical part of continental Asia and Borneo. *Aphelocheirus* s.str. is widely distributed in the Old World. The number of described species of *Aphelocheirus* s.l. is about 65 of which 23 occur in Malesia.

Biological notes

Biological studies have been performed only on *Aphelocheirus aestivalis* (Fabricius), a widespread species occurring in Europe eastward to Georgia. Tropical Asian *Aphelocheirus* are mostly found in well aerated streams, although they may occur in slow flowing streams and occasionally in stagnant channels or pools. *Aphelocheirus aestivalis* has occasionally been found in lakes; two African species have also been reported from lakes. In streams they tend to reach higher densities on exposed stretches compared to shaded parts. Exposed streams in country cleared for cultivation may occasionally contain *Aphelocheirus* in incredible numbers (provided there is sufficient current and not excessive organic pollution, which causes explosive growth of filamentous algae). They may be useful in controlling Simuliidae (Diptera, black flies), the larvae of which share the same habitat.

During daytime *Aphelocheirus* species hide under pebbles or in the sand of the stream bed; at night they are active and crawl over the substrate. This behaviour is common among stream-inhabiting insects and is supposed to be a defense against fish predation (Peckarsky, 1984). Because of their nocturnal and benthic habits they usually escape the attention of the casual collector. Until recently *Aphelocheirus* were thought to be rare. However, especially in tropical Asia, they are quite common in streams, especially mountain streams. Specific ecological demands include coarse versus fine substrate, and differences in current velocity and water depth. Two to three species may occur in the same stream. There are indications that various larval instars and imagines of a single species prefer different habitats (D. Polhemus & J. Polhemus, 1988).

During copulation the male sits on the dorsum of the female on the left side with its head above the left hemielytra (in micropterous specimens). The fore legs hold the lateral angles of the female's pronotum, the middle legs the margin of the abdomen at about the suture between tergites 4-5, the hind legs point caudally and the longitudinal axis of the male is virtually parallel with the longitudinal axis of the female. The abdominal apex of the male bulges out in the course of which a slit opens between the abdominal segments 5-6. The side of the female abdomen fits into this slit; the apical segments bulge out further downward until they are under the apex of the female abdomen whereupon the aedeagus makes contact with the female genital opening (Larsén, 1938).

Eggs are virtually cylindrical with broadly rounded ends and have a distinct hexagonal sculpture. They are glued to larger substrate particles, especially deeper in the substrate. They are deposited singly, in clusters or in rows. In *A. aestivalis* the development takes 9-10 weeks or more. Larval development takes two years. However, these observations are from Swedish populations (Larsén, 1927, 1932), and in tropical species the development is probably faster.

Respiration has been studied extensively by Thorpe & Crisp (1947a, b). The larvae have cutaneous respiration, the adults plastron respiration. The plastron is very stable, its gas layer persists even when specimens are put in gas-free water. It takes a pressure of 4-5 atmospheres to collapse the plastron gas film. The breathing system is refined compared to Naucoridae (also plastron breathers): the stigmata on the abdominal venter are surrounded by so-called rosettes (Szabó-Patay, 1924; Thorpe & Crisp, 1947a). These are areas of irregularly branching canals of chitin radiating out from and in open connection with the trachea (fig. 71, 73). The original stigma is closed or nearly so. The canals lie under the plastron and are connected with its gas film by small holes. Close to the trachea they run through the chitin, but are not connected with the surface. The function of these rosettes is apparently to maintain surface extension. There are also many small holes (diameter 7-8 μ) on the abdominal dorsum, venter and head (Messner et al., 1981). These lead to narrow channels through the cuticula and end in air sacs under it. The air sacs are lined with a membrane. Tracheoles connect the air sacs with the remainder of the respirational system. The plastron enables them to complete their entire life cycle under water. Although several species of Naucoridae living on the bottom of streams apparently do not need to surface in order to replenish their gas store, their plastron seems to be less suited to do this. D. Polhemus & J. Polhemus (1988) report that *A. kinabalu* survived being kept overnight in a completely filled jar, while the naucorid *Coptocatus oblongulus* died within two hours under similar conditions.

As a consequence of plastron respiration *Aphelocheirus* are among the few Nepomorpha having a micropterous morph. Apterous specimens do not occur. The air under the wings is in contact with the so-called scolopal (= chordotonal) organs, which are pressure receptors (Larsén, 1957; Parsons, 1969a). The pterothorax of micropterous *Aphelocheirus* is more heavily chitinized (especially the membranous parts) and more strongly fused than in macropterous specimens. This provides more solidity and is supposed, together with micropterism, to be an adaptation to a benthic way of life.

Adult *Aphelocheirus* have a pair of static organs on abdominal sternite 2 laterally of the rosettes (fig. 71). Each of the static organs consists of an oval sensory plate with its median edge cutting the lateral branches of the rosette. The sensory plates lack plastron hairs but are covered with a large number (± 60,000 per mm^2) of much larger hydrofuge hairs. Among these, thinly walled hairs are dispersed at intervals, each supplied with a sense cell in the proximal part from which a nerve end runs down through the cuticula. Below the cuticula they combine to a nerve. It is assumed that when pressure increases, the air film on the sense organ will be compressed. Since the hairs are strongly hydrofuge they themselves will be pressed down, thus moving the sensory hairs at the same time. When pressure is decreased the air film will expand and the hydrofuge hairs will resume their original position. The static organs thus react on increase or decrease of pressure. The main function seems to be orientation, especially during darkness. When one or both of the organs are destroyed and the animal is put on the surface in an aquarium in the dark, it swims either in spirals or erratically, whereas individuals with both organs intact swim straight to the bottom. When there is sufficient light they can orientate themselves well. When larvae are placed on the surface in the dark they behave as adults do with the static organs extirpated (Thorpe & Crisp, 1947c).

Food consists mainly of insect larvae such as Diptera, Ephemeroptera and Trichoptera. They might therefore be useful in controlling larvae of Simuliidae, which live in similar habitats. On the other hand, *Aphelocheirus* will also suppress populations of other predators, notably Trichoptera larvae, that feed on Simuliidae larvae too (Papáček, 2000). Adult Simuliidae (black flies) are vectors for several diseases, notably onchocerciasis, which can lead to river blindness. There are persistent reports of *Aphelocheirus* preying on fresh water mussels such as *Pisidium*. This report goes back on Larsén (1924). However, later he attributed this to unnatural behaviour due to lack of food under experimental conditions (Larsén, 1927). Unfortunately the sucking of mussels has been cited widely while the revocation is usually overlooked. Still there remains an observation by Brown (cited in Thorpe, 1966) that *A. aestivalis* eats *Viviparus* (L.) (Gastropoda). However, even if some *Aphelocheirus* would commonly attack molluscs, they would be of little use in biological control of the intermediate hosts of schistosomiasis, as this illness is associated with stagnant waters.

Key to subgenera of *Aphelocheirus*

1 Wing pads of micropterous morph covering at least lateral half of metanotum (fig. 74). Length more than 6.5 mm [Old World] ▸ *Aphelocheirus*

– Wing pads of micropterous morph covering up to $^1/_4$ of metanotum (fig. 76). Length less than 8 mm, usually less than 5 mm [India and South-east Asia] ▸ *Micraphelocheirus*

Genus *Aphelocheirus*

Figures 74-76

Diagnosis

See family diagnosis.

74

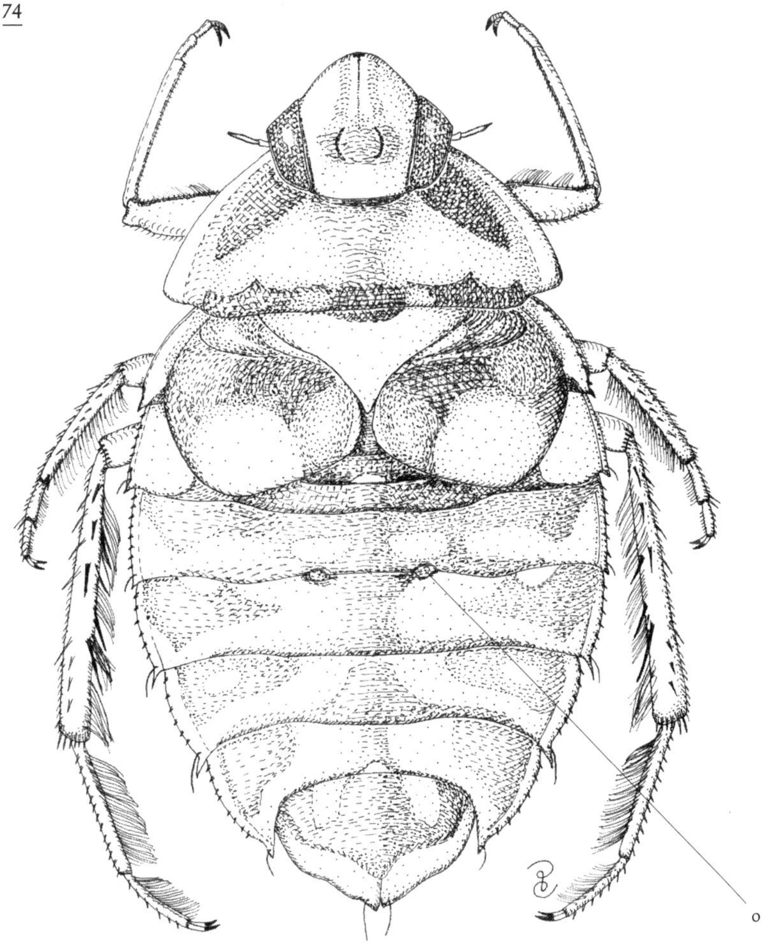

FIGURE 74
Aphelocheirus (*Aphelocheirus*) *robustus*,
micropterous female,
length 9.6 mm;
o = abdominal scent gland orifice.

75

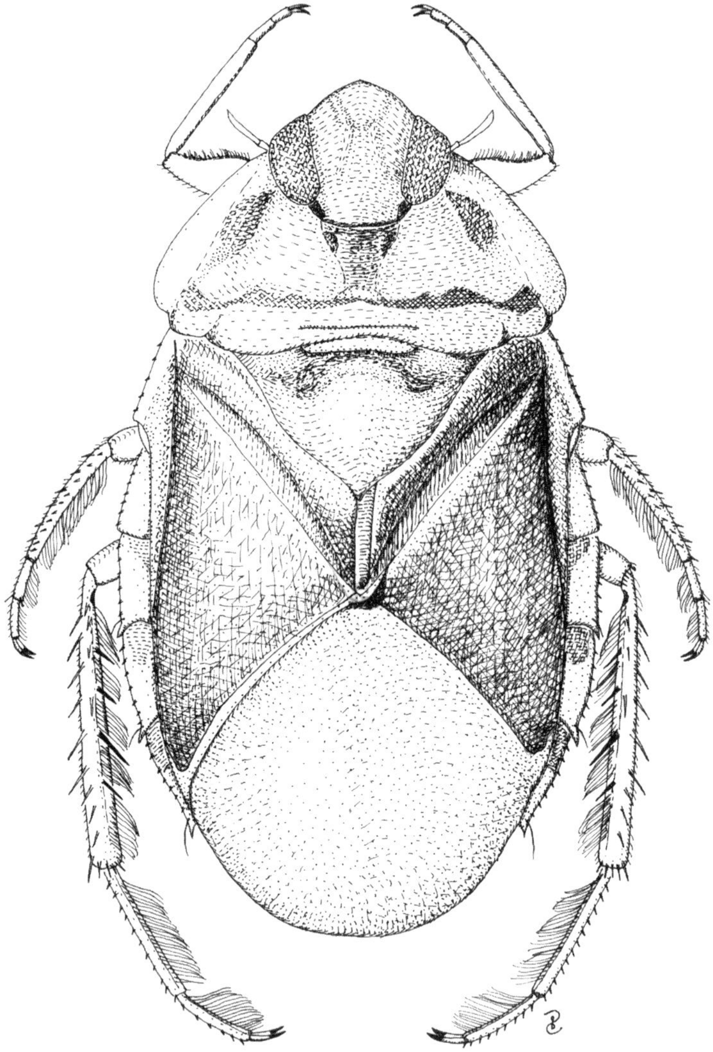

FIGURE 75
Aphelocheirus (*Aphelocheirus*) *robustus*,
macropterous male,
length 9.6 mm.

– *Aphelocheirus* **sensu stricto**
Figures 74, 75

Diagnosis

Small to medium-sized (6.5-11.5 mm). Inner propleural projections with apical notch (fig. 71). Wing pads of micropterous morph well developed, covering at least lateral half of metanotum (fig. 74). Openings of abdominal scent glands located closer to midline of abdomen than to lateral margin.

Distribution in Malesia (fig. 565)
Distributed throughout Malesia except for the Moluccas, Bismarck and Solomon Islands. Of 23 species of *Aphelocheirus* s.l. found thus far in Malesia 22 belong to this subgenus.

– *Aphelocheirus* **subgenus** *Micraphelocheirus*
Figure 76

76

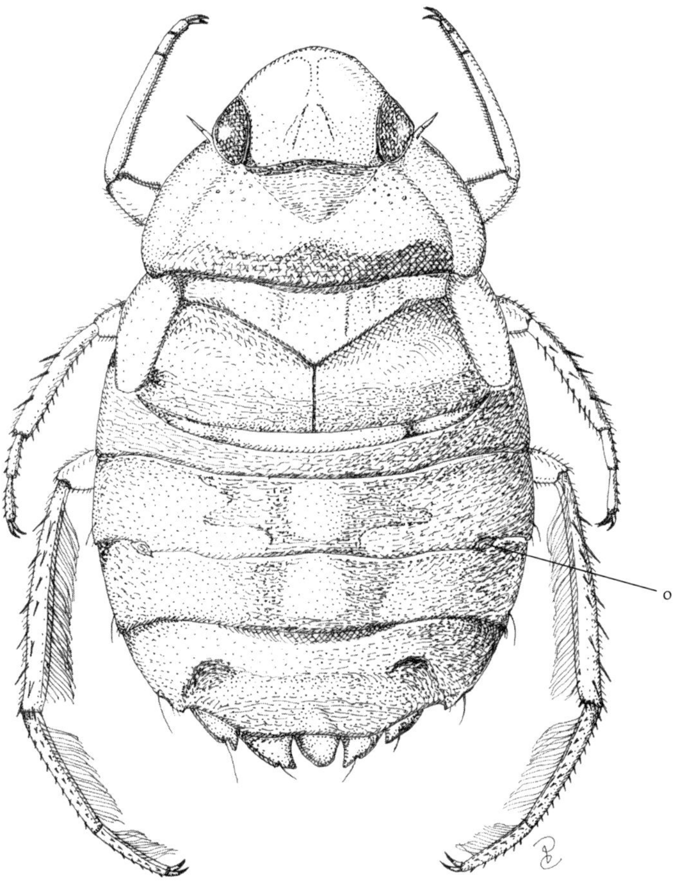

FIGURE 76
Aphelocheirus (*Micraphelocheirus*) *brevirostris*, micropterous male, length 4.2 mm; o = abdominal scent gland orifice.

Diagnosis
Small (3.5-8 mm). Inner propleural projections without apical notch. Wing pads of micropterous morph short, covering lateral 1/4 of metanotum at most (fig. 76). Openings of abdominal scent glands located closer to lateral margin of abdomen than to mid-line.

Distribution in Malesia

So far only one of the five described species of this subgenus has been found in Malesia: *A. (M.) kelabitensis* from Sarawak (fig. 566).

REFERENCES – Aphelocheiridae. Biology of *A. aestivalis* (Larsén, 1924, 1927, 1932); structure of respiratory system (Szabó-Patay, 1924); study of respiration (Thorpe & Crisp, 1947a–c); structure and function of scolopophorous organs (Larsén, 1957); biological notes on *Aphelocheirus aestivalis* (Thorpe, 1966); thorax morphology (Parsons, 1969a); systematics and morphology of *Aphelocheirus* (Hoberlandt & Štys, 1979); respiration (Messner et al., 1981); revision of South-east Asian species, biological notes (D. Polhemus & J. Polhemus, 1988); key to Sulawesian species (Nieser & Chen, 1991, 1996); key to Bornean species (Zettel, 1999a); key to Phillipine species (Zettel, 1999b; 2003b); economic importance (Papáček, 2000).

Family Naucoridae – creeping water bugs

Diagnosis

Small to medium sized (5-21 mm), ovoid to slightly elongate, flattened. Dorsally various shades of brown, often marked with yellowish, notably at lateral margins of pronotum and costal area of hemielytra. Antennae four-segmented, shorter than head, usually hidden below eyes (projecting laterally of eyes in Tanycricini). Eyes often overlapping anterolateral angles of pronotum. Rostrum three-segmented, short and stout. First femur widened (usually strongly) with anterior groove to receive tibia and tarsus. First tarsus usually fused with tibia, one- or two-segmented. Middle and hind tarsus with two distinct segments (basal segment very small or absent) and two equally developed claws. Hind leg usually modified for swimming. Metathoracic scent glands present. Membrane of hemielytra usually well developed without veins.

Characteristics

The family is morphologically diverse and there are few structural features which all genera have in common. The short rostrum and antennae are the main characters, separating them from their closest relatives Aphelocheiridae and the New World Potamocoridae.

The abdominal apex varies from symmetrical in both sexes – or nearly so – to distinctly asymmetrical, depending on subfamily or tribe (figs. 77-80). Sternites divided by folds into medial sternites and parasternites and the narrow connexiva or ventral laterotergites (figs. 79-80). However in many genera, especially in Cheirochelinae, this division is indistinct or absent (figs. 77-78). In ventral view the abdominal apex differs between sexes. In males the median parts of sternites 6-8 are short and the median part of sternite 8 is at least partly visible (figs. 77, 79). In females the median part of sternite 7 is enlarged, forming a (sub)genital operculum covering the median part of sternite 8, which forms the anterior gonocoxites (figs. 78, 80). The abdominal apex of the male is much more movable compared to that of the female, which relates to copulation (see below). In soft specimens the genital capsule can usually be pushed out by applying pressure on the abdominal sternites. Parameres are variably developed, and their use in identification varies between subfamilies. They lie dorsally in the genital capsule at base of the aedeagus (fig. 81). In Naucorinae and Cheirochelinae parameres are reasonably to well developed. In Laccocorinae the parameres are vestigial and not used in identification. Characters most frequently used in identification are size, pattern, caudolateral angles of connexiva, and various ratios. In addition, the shape of apical abdominal segments may give clues for identification.

FIGURES 77 – 80
Apex of abdomen in ventral view.
77 *Tanycricos acumentum* male – 78 *Tanycricos acumentum* female – 79 *Naucoris obscuripennis* male – 80 *Naucoris scutellaris* female; m = mediosternite, p = parasternite, s = spiracles, v = ventral laterotergite or connexivum, 7 = genital operculum (abdominal sternite 7), 8 = abdominal sternite 8.

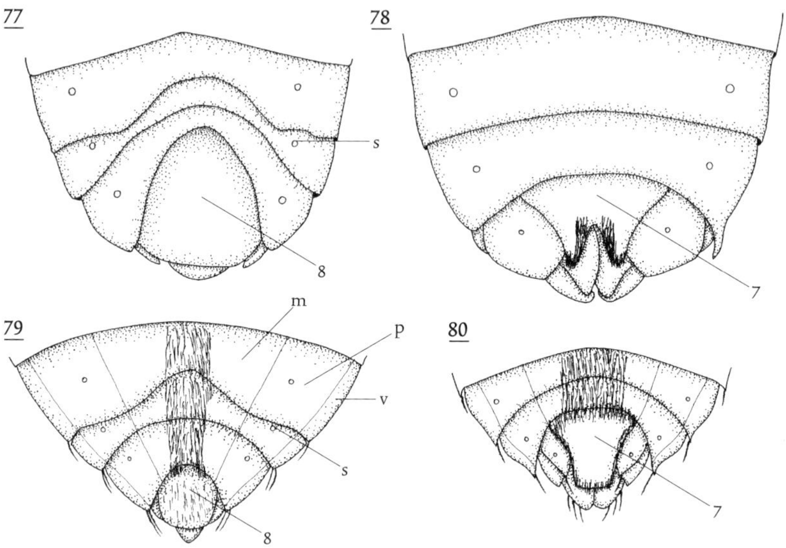

FIGURE 81
Genital capsule of *Naucoris scutellaris* in dorsal view; a = aedeagus, l = left paramere, r = right paramere.

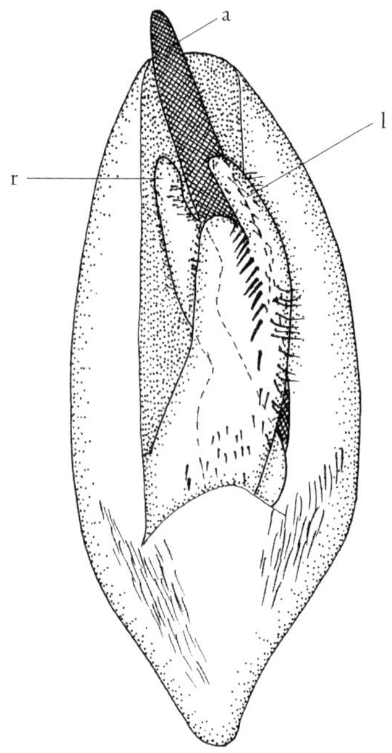

Fossil record

Similar to Belostomatidae, Naucoridae had a strongly diversified fauna already during the Mesozoicum. Three subfamilies are known, two have become extinct. The extant Naucorinae are represented by four different tribes. The Liadonaucorini Popov go back to the Lower Jurassic (about 180 million years ago). The recent tribe Naucorini is represented by three extinct genera from the Upper

Jurassic and Lower Cretaceous (about 135 million years ago) (Popov, 1971b; Nel & Paicheler, 1992). The strong diversity since the Lower Jurassic suggests that the radiation of the nepomorphan superfamilies took place in the Trias. The Cenozoic fossils of Naucoridae are not identifiable to genus, and it is often uncertain whether they belong to Naucoridae or Aphelocheiridae (Nel & Paicheler, 1992).

Distribution worldwide

Five or six extant subfamilies are recognized with 34 genera and about 350 described extant species (appendix 1), mostly in warmer regions. Subfamily Limnocorinae Stål contains only the genus *Limnocoris*, with 71 species (Nieser & Lopez-Ruf, 2001). Two subfamilies have a worldwide distribution and one is restricted to Indo-Australia. More details on these subfamilies are given below.

Biological notes

The habitat of Naucoridae varies between subfamilies, but most species live in streams. Although they are good swimmers, Naucorinae usually wait between vegetation until prey is close. The bottom dwelling species seem to crawl around with extended fore legs searching for prey. La Rivers (1951) reported both swimming and crawling forms of the American *Ambrysus*, depending on habitat. He stated that they have rather poor sight and only react to movement at less than two times a body-length distance.

The best known species regarding biology and morphology is *Ilyocoris cimicoides* (L.), a temperate Palearctic species. *Ilyocoris* is morphologically only slightly different from the more widespread *Naucoris*. Morphology is discussed by Rawat (1939). In dry specimens or specimens preserved in fluid the hemielytral membranes appear to be without veins. However, Rawat (1939) states that in glycerine mounts of freshly moulted adults a reticulate venation is visible in the membranes of the hemielytra. There are ten pairs of spiracles: three thoracical, placed posteriorly on the thoracic segments, and seven abdominal, placed ventrally on the parasternites (fig. 77). Although they are partly covered, a central opening remains, so they are probably all functional. Naucorinae are reported to surface from time to time in order to replenish the gas stores on the abdominal venter and under the hemielytra. The air stores are connected by an air chamber around the first abdominal spiracle. It is supposed that air passes through the dorsal air store and is inhaled by the first abdominal and the metathoracic spiracles. Air may also be inhaled occasionally by contacting the ventral abdominal air store with the atmosphere and inhaling air by the ventral abdominal spiracles. Air is exhaled through the rather ventrally situated mesothoracic spiracle into the ventral air store, where it is pushed caudally by stroking movements of the hind legs. So far none of the species with plastron respiration occurring in some other subfamilies have been studied in this respect. It is supposed that larvae of Naucoridae breathe through the skin (tracheal gill breathers). This is verified for some Naucorinae.

Copulation position is similar to that in *Aphelocheirus*. The male takes position on top of the female and clasps its fore legs around her so that the articulation between femur and tibia is at the anterior edge of the hemielytra. Then the male moves to the left side of the female and clasps with the claws of the right middle legs around of the right hemielytra of the female at the level of abdominal segment 5 or 6. With the left middle leg the male also supports himself. The hind legs point more or less caudally and do not participate in holding the female. The right part of abdominal segments 5-6 of the male are folded downwards and segments 6-8 are pressed against the ventral abdominal tip of the female so that the genital capsule is positioned against the females' abdominal apex. The genital capsule is then pressed out and into the female whereupon the aedeagus is extended and fertilization can take place. These data are based on *Ilyocoris* (Larsén, 1938) and *Naucoris*. Observations of copulation of Oriental genera in other subfamilies than Naucorinae are not available.
Eggs of Naucoridae are usually oval, elongate oval or more or less cylindrical with rounded ends. They are deposited on the surface of substrates, in Naucorinae mostly on the underside of leaves, in other subfamilies on pebbles or rocks (Sites & Nichols, 1999). The exception is *I. cimicoides*, which has an elongate curved egg with a flattened anterior pole. This species prefers to deposit its eggs inside submerged plant tissues with the anterior pole in contact with the water. The eggs of the Malesian *Naucoris* species are unknown.
Food consists of various kinds of invertebrates.

Key to Indo-Australian subfamilies of Naucoridae[1]

1 Rostrum arising from a deep excavation under the head, its insertion placed distinctly backward from the anterior margin of the head (fig. 84, 88, 89) ▸ **Cheirochelinae p. 100**

– Rostrum arising freely, not from an excavation (figs. 82, 85) ▸ 2

2 Fore tarsi not fused with tibia, with two claws, which may be small and inconspicuous (figs. 86, 87). In males fore tarsi two-segmented (fig. 86). Anterior margin of head strongly turned downward and usually backward, the rostrum consequently originating usually from well behind the anterior margin of the head (fig. 83) ▸ **Laccocorinae p. 120**

– Fore tarsi fused with tibia, although suture still visible. Tarsi one-segmented, with or without a minute scarcely distinguishable claw (fig. 12). Anterior margin of head only slightly turned down and backward. Rostrum originating at or near the anterior margin of the head (fig. 85) ▸ **Naucorinae p. 127**

---

1 Adapted from Usinger (1941).

82

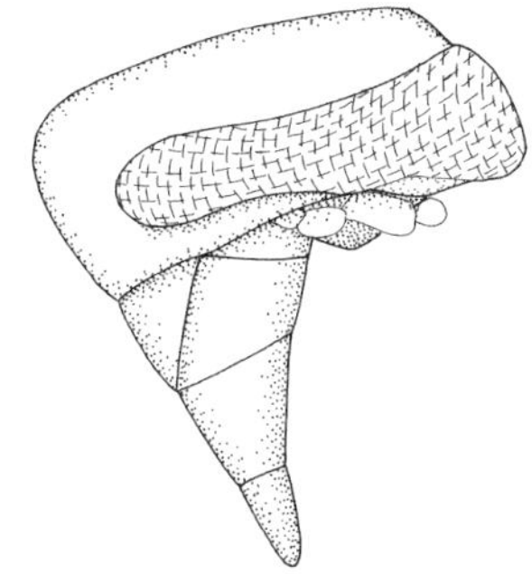

83

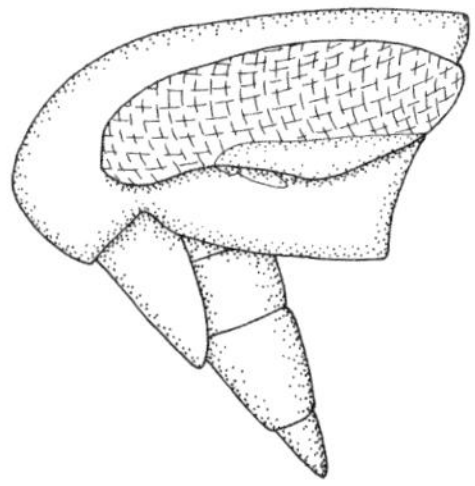

84

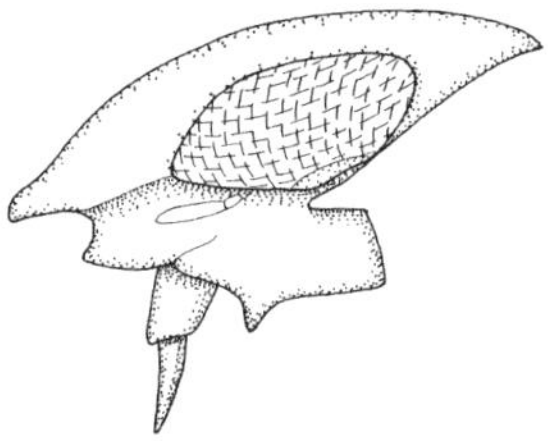

85

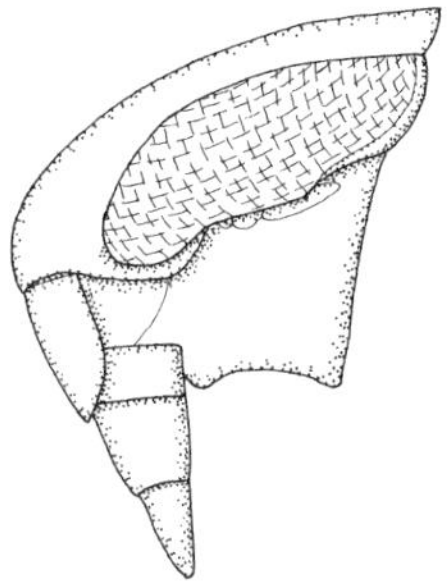

FIGURES 82 – 85
Head in lateral view.
82 *Ctenipocoris spinipes* Montandon – **83** *Heleocoris nebulosus* – **84** *Tanycricos acumentum* – **85** *Naucoris scutellaris*.

86

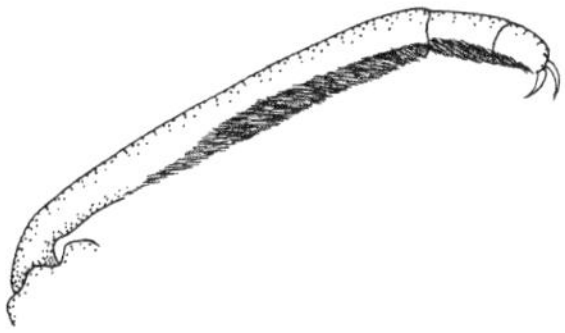

87

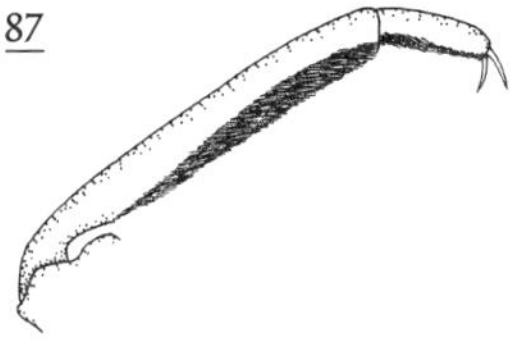

FIGURES 86 – 87
Laccocoris hoogstraali, fore tibia and tarsus.
86 male – **87** female.

88

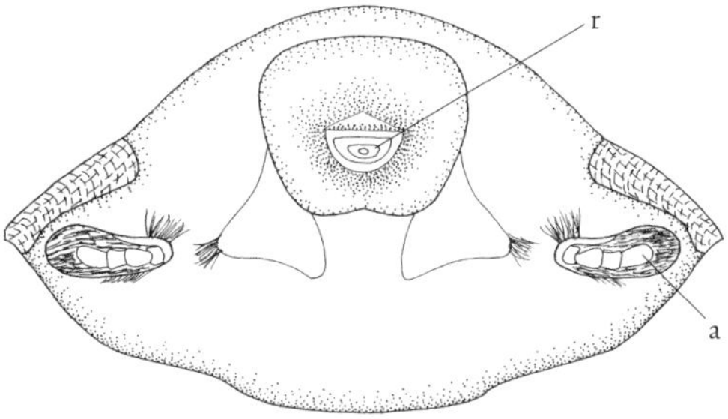

FIGURE 88
Gestroiella limnocoroides, head in ventral view; a = antenna, r = rostrum.

89

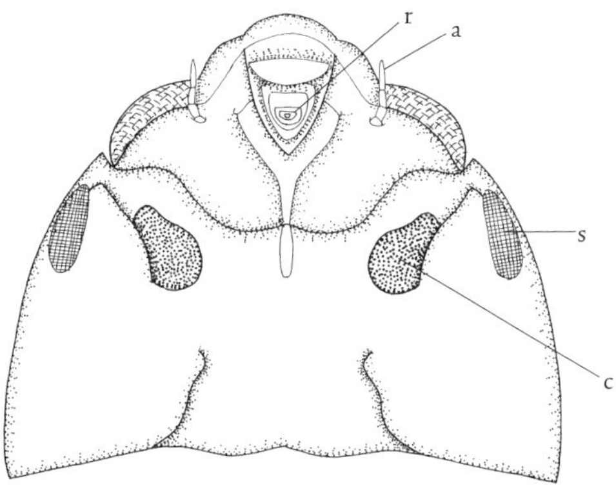

FIGURE 89
Tanycricos acumentum, head and prosternum in ventral view; a = antenna, c = insertion of coxa, r = rostrum, s = static organ.

SUBFAMILY CHEIROCHELINAE

Diagnosis

Small to large (6-21 mm), elongate to broadly oval, dorsoventrally flattened, usually brachypterous Naucoridae. Characterized by groove on ventral side of head in which the antenna fits (fig. 88), by reduced labrum and hair pads on middle and hindtibiae. Chitinous strip laterally of eyes, if present, not fitting over anterior angles of pronotum. Rostrum originating from ventral groove, distinctly posterior of anterior head margin (fig. 88). Gula well developed. Propleura not or only slightly medially prolonged, consequently prosternum fully exposed. First tarsus one-segmented, not fused to tibia, with one claw which may be very small and difficult to distinguish. Ventral surface of middle and hind femur smooth. Mesosternum and metasternum without or with poorly developed carinae and without apical fossa. Parameres well developed.

Taxonomic note

The position of Philippine genera presently placed in Cheirochelinae is under discussion (Zettel et al., 1999). They have been transferred to Naucorinae by Zettel (2001f). This is considered somewhat premature without a thorough cladistical analysis of at least all genera of Cheirochelinae and Naucorinae. The arrangement by La Rivers (1971b) is followed here awaiting further studies.

Distribution in Indo-Australia

Cheirochelinae are restricted to South-east Asia including New Guinea. There are 57 species in Malesia. The western distributional limit lies in Assam, India. Two clades are recognized (D. Polhemus, 1986). The tribe Cheirochelini probably evolved in South-east Asia from a laccocorine ancestor and subsequently expanded to North Borneo. At present there are thirteen described species in three genera: *Cheirochela* Hope with three, *Gestroiella* with four species restricted to continental South-east Asia, of which only one occurs in Malesia, and *Coptocatus* with six. Sagocorini and Tanycricini, of which the latter with three genera and seventeen species, are endemic to New Guinea. Sagocorini with seven genera and 33 species, are distributed from New Guinea along the eastern Philippines to Luzon.

Biological notes

Data on biology and habitat are scarce. Several species of Cheirochelinae inhabit sites with a strong current where it is difficult to surface. Hence they are supposed to have plastron respiration, which is supported by the presence of abdominal pilosity.

Key to tribes and genera of brachypterous Cheirochelinae

1 Antennae fitting in a groove under the head (fig. 88) ▸ **Cheirochelini 2**
– Antennae lying free against the underside of the head (fig. 89) ▸ 4

2 Claval commissure less than 1.5 times the median length of the scutellum. Hemielytra leaving apex of abdomen uncovered from tergite 4 onward (fig. 95) [Borneo] ▸ ***Coptocatus* p. 103**
– Claval commissure about two times the median length of the scutellum. Hemielytra reaching apex of abdomen ▸ 3

3 Lateral margins of pronotum convex (fig. 96) [South-east Asia] ▸ ***Gestroiella* p. 105**
– Lateral margins of pronotum concave [Assam, Myanmar and Laos] ▸ ***Cheirochela* Montandon**

4 Static sense organs usually large and conspicuous (fig. 89), present on ventral margin of prothorax. Body elongate (fig. 97, 107) [New Guinea] ▸ **Tanycricini and *Aptinocoris* 5**
– Prosternal static sense organs absent, body oval (fig. 101) [New Guinea and Philippines] ▸ **Sagocorini 8**

5 Static sense organ located just anterior of the middle of the lateral margin of the prothorax, situated away from the anterolateral angle of prosternum (several times its own length) ▸ ***Idiocarus* p. 115**
– Static sense organ located at or near the anterolateral angle of prosternum ▸ 6

6 Static sense organ rounded, placed at tip of anterolateral angle of prosternum ▸ 7
– Static sense organ elongate, at a distance of 1/2 to 3/4 of its length removed from the anterolateral angle of the prosternum (fig. 89) ▸ ***Tanycricos* p. 119**

7 Small, up to 10 mm. Static sense organ small ▸ ***Aptinocoris* p. 106**
– Large, over 12 mm. Static sense organs large ▸ ***Nesocricos* p. 117**

8 Middle tibia distally with five or six decreasingly sized transverse dense rows of short spines (fig. 90) [New Guinea] ▸ ***Cavocoris* p. 108**
– Middle tibia with maximally two transverse rows of distal spines (fig. 91) ▸ 9

9 Overhanging part of head anteriorly 1/2 to 3/4 as long as basal width of labrum [New Guinea] ▸ ***Warisia* p. 114**
– Overhanging part of head anteriorly at most 1/3 as long as basal width of labrum ▸ 10

90

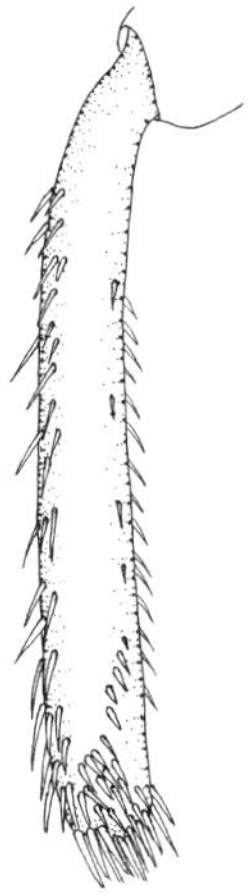

91

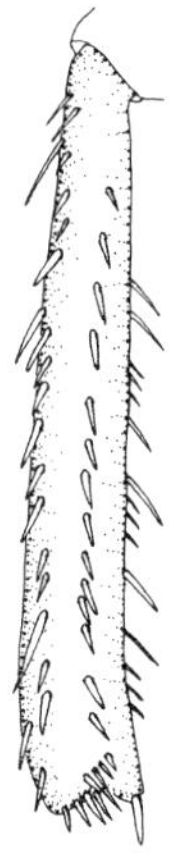

FIGURES 90 – 91
Middle tibia in ventral view.
90 *Cavocoris bisulcus* –
91 *Sagocoris biroi*.

10 Abdominal venter showing only four segments [New Guinea] ▸ *Sagocoris (Quadricoris)* **p. 111**
– Abdominal venter showing the usual five segments ▸ 11

11 Head anteriorly produced (in dorsal view), at least 1.2 times longer than pronotum measured along median line (fig. 98). In frontal view anterior margin of head evenly curved and covering labrum [Philippines] ▸ *Asthenocoris* **p. 107**
– Head not anteriorly produced (in dorsal view), subequal in length to pronotum along median line. Labrum visible in frontal view (fig. 92) ▸ 12

12 Length up to 7.4 mm [Philippines] ▸ *Stalocoris* **p. 113**
– Length over 7.5 mm; in Philippines length over 9 mm ▸ 13

13 Anterior margin of head in frontal view (anteclypeus) only slightly indented above labrum (fig. 92); labrum distinctly shorter than its basal width [Philippines] ▸ *Philippinocoris* **p. 110**
– Anterior margin of head in frontal view above labrum distinctly indented (fig. 93); labrum about as long as its basal width [New Guinea] ▸ *Sagocoris* **p. 111**

This key is mainly based on the work of La Rivers (1971a) for the New Guinean genera. D. Polhemus & J. Polhemus (2000b) revised and redefined *Aptinocoris* and *Sagocoris*, transferring some of La Rivers' (1971a) *Sagocoris* species to *Aptinocoris* and reducing *Quadricoris* to a synonym of *Sagocoris*. They present a simplified key to New Guinean Sagocorini, which is (slightly adapted) reproduced below.

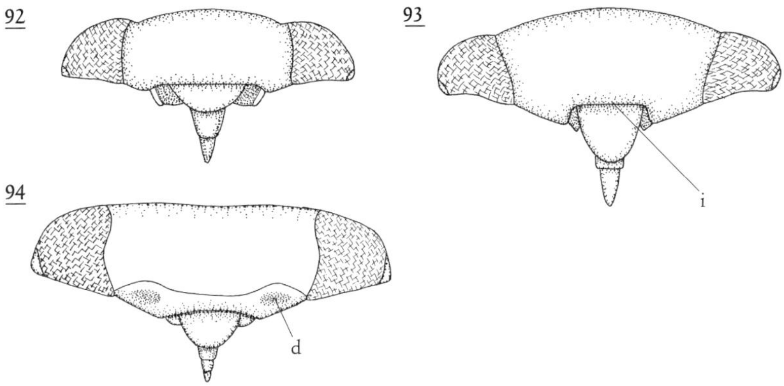

FIGURES 92 – 94
Head in frontal view.
92 *Philippinocoris usingeri* – 93 *Sagocoris biroi* – 94 *Warisia cavanceps*; d = depressed pit in anteclypeus, i = indentation of anteclypeus.

Alternative key to New Guinean Sagocorini

1 Anteclypeus with distinct notch above base of labrum (fig. 93) ▸ 2
– Anteclypeus lacking a distinct notch above base of labrum (fig. 92, 94) ▸ 3

2 Hemielytra in brachypterous morph posteriorly truncate (fig. 97). Anterolateral angle of pro-epimeron with a depressed patch of fine gold setae ▸ *Aptinocoris* p. 106
– Hemielytra in brachypterous morph posteriorly angulate (figs. 101, 102). Anterolateral angle of pro-epimeron without such patch ▸ *Sagocoris* p. 111

3 Anteclypeus with pair of shallow depressed pits to either side of the midline, flanking base of labrum (fig. 94). Hydrofuge pile of abdominal venter dark. Female abdominal paratergite 3 lacking a deep pit. Male left paramere compared to right paramere not markedly reduced ▸ *Warisia* p. 114
– Anteclypeus without pair of shallow depressed pits to either side of the midline. Hydrofuge pile of abdominal venter shining gold. Female abdominal paratergite 3 often with deep pit. Male left paramere compared to right paramere frequently reduced ▸ *Cavocoris* p. 108

– TRIBE CHEIROCHELINI

Genus *Coptocatus*

Figure 95

Diagnosis

Large (13-21 mm), elongate, oval. Antennae lie in groove of ventral part of head. Eyes distinctly raised above surface of interoculus, anteriorly parallel to slightly convergent. Anterior head margin projecting in front of curvature of lateral eye margin, in frontal view nearly straight. Labrum strongly reduced. Anterior margin of pronotum with broad indentation; lateral margins anteriorly straight

95

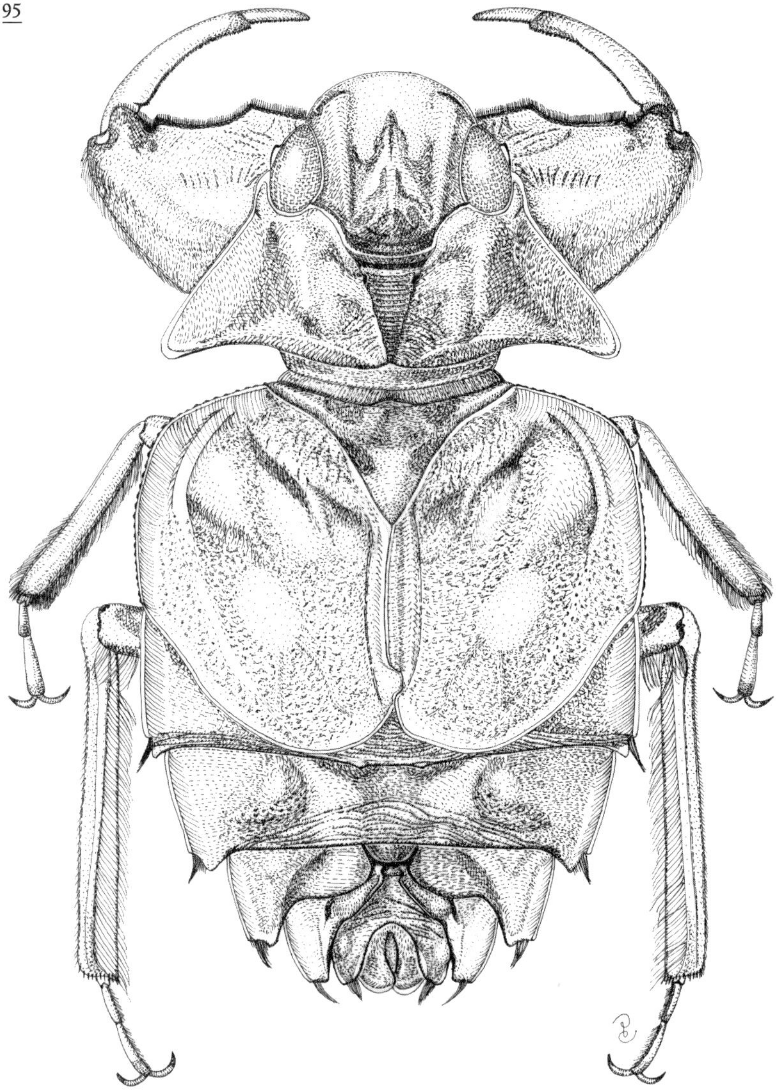

FIGURE 95
Coptocatus stereos, holotype, brachypterous male, length 20.4 mm.

to slightly concave, strongly convergent. Hair-lined pits at anterior tips of prosternum may be static sense organs. Embolium anteriorly only slightly narrowed, lateral margins of hemielytra anteriorly subparallel. Connexival angles of segments 3 and 4 with pair of blunt projections.

Distribution in Malesia (fig. 566)

Six species are known from North Borneo. One was collected close to the Indonesian border. There is little doubt that one or more species of this genus occur in Kalimantan.

Biological notes

All six species are known from streams with moderate to strong current and a gravel or cobble bottom, sometimes overlaid with rocks.

Genus *Gestroiella*

Figure 96

96

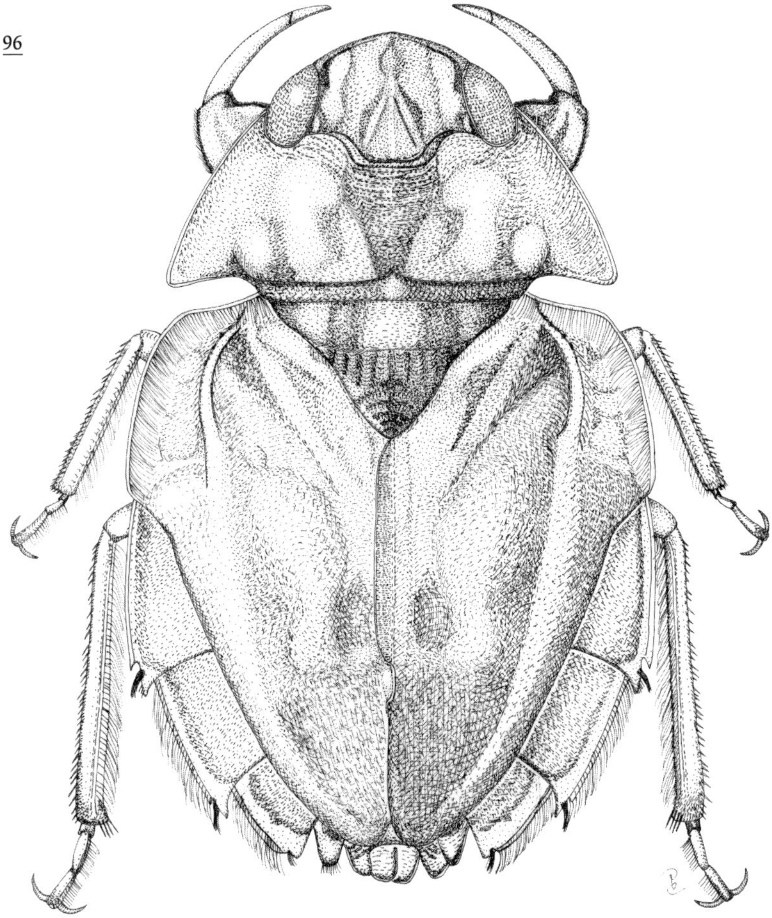

FIGURE 96
Gestroiella sp., brachypterous female, length 13.4 mm.

Diagnosis

Large (12.5-20 mm), broadly ovate to subcircular. Anterior head margin projecting in front of curvature of lateral eye margins; in frontal view nearly straight. Antennae lie in groove of ventral part of head. Eyes distinctly raised above interoculus surface; inner margins parallel. Labrum strongly reduced. Anterior margin of pronotum with broad indentation; lateral margin straight to very slightly convex, strongly converging. Anterior tips of prosternum without hair-lined pits. Embolium anteriorly only slightly narrowed. Lateral margins of anterior hemielytra subparallel. Connexival angles of segments 3 and 4 doubly spinose.

Distribution in Malesia (fig. 566)

Gestroiella limnocoroides has been recorded from the south of Thailand (Songkhla province; Sites et al., 1997). One specimen, apparently of another species, was collected near the Isthmus of Kra, Thailand (collection N. Nieser).

Biological notes

The habitat consists of gravel and rocky streams of various size, but specimens have also been collected in rock pools at the foot of a waterfall disconnected from the stream (Sites et al., 1997).

– TRIBE SAGOCORINI

Genus *Aptinocoris*

Figure 97

97

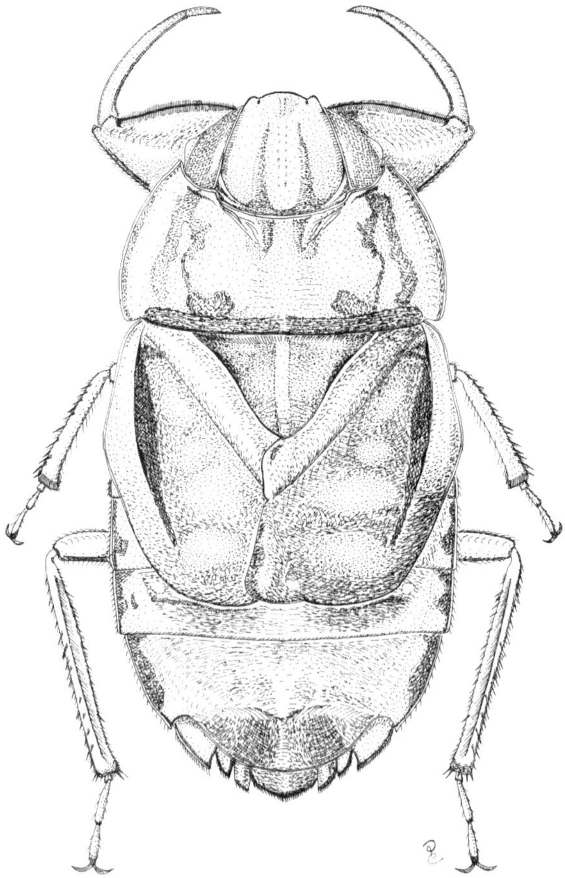

FIGURE 97
Aptinocoris sp., brachypterous male, length 6.6 mm.

Diagnosis

Small to medium-sized (5.4–10 mm), ovate to elongate ovate. Antennae lie free on underside of head. Eyes and interoculus separated by a broad groove. In posterior view dorsal surface of head and eyes not flat; eyes convergent anteriorly, lateral eye margins and anterior head margin forming a smooth curve. In frontal view anterior head margin with distinct notch above the labrum. Anterior pronotum margin shallowly concave, prosternum with small static sense organs. Hemielytra (in brachypterous morph) posteriorly truncate, leaving abdominal tergite 5 uncovered (fig. 97). Connexival angles right-angulate, not spinose.

Distribution in Malesia (fig. 565)

There are presently eight described species, all restricted to New Guinea.

Biological notes

Aptinocoris species inhabit the bottom of moderate to swift streams of various size. They apparently have plastron respiration (D. Polhemus & J. Polhemus, 2000b).

Genus *Asthenocoris*

Figure 98

Diagnosis

Small (5.9–8.4 mm), elongate, oval. Antennae lie free on ventral head surface. Eyes slightly raised above interoculus level, parallel to slightly convergent anteriorly. Anterior head margin projecting in front of curvature of lateral eye margins, in frontal view with distinct notch above the labrum. Anterior margin of pronotum strongly concave; prosternal static sense organs absent. Embolium anteriorly narrowed; lateral margins of hemielytra convergent anteriorly. Connexival angles on posterior segments in some species sharply angulate to broadly spinose.

Distribution in Malesia (fig. 566)

Four species are known, all restricted to the Philippines. Within this area they are widespread but do not occur in the Western Visayas (Negros, Panay), where *Stalocoris* is dominant.

Biological notes

Asthenocoris species live mostly on the bottom, usually of clean forest streams but sometimes also of larger exposed rivers. They hide under large stones or between coarse gravel in the current and are only rarely found at the edges. Of the Philippine genera of Naucoridae, *Asthenocoris* are the most adapted to life in fast flowing water. However, when found together with *Aphelocheirus* they prefer sections with less current (Zettel et al., 1999).

98

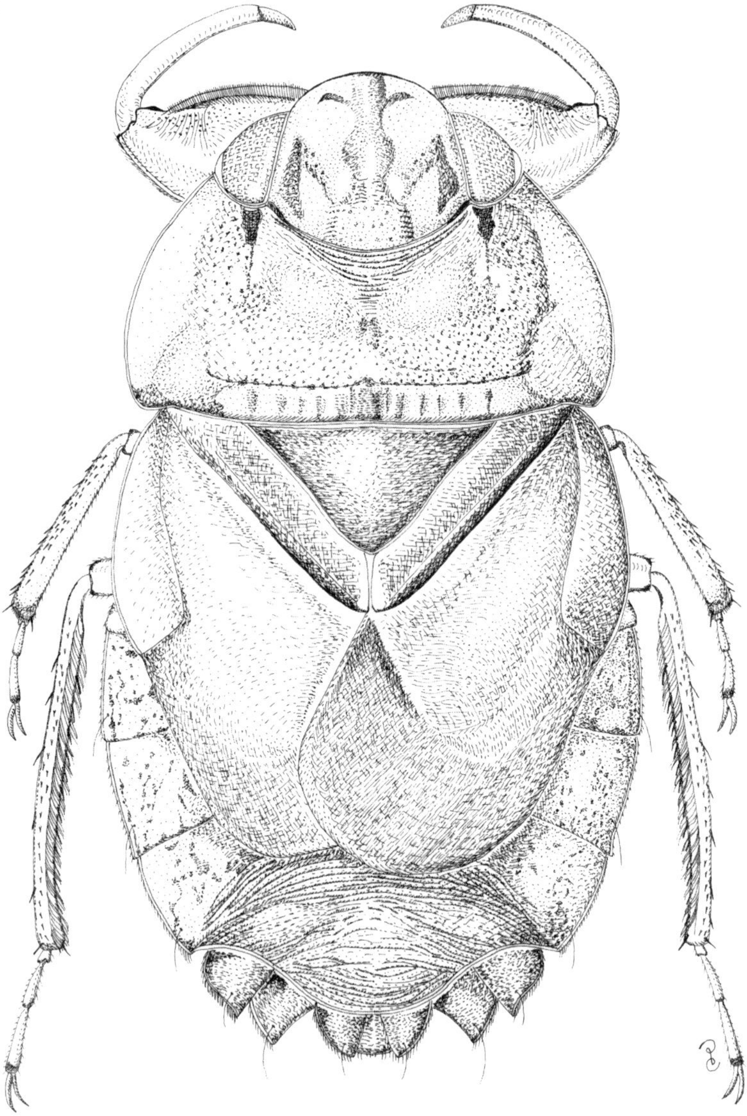

FIGURE 98
Asthenocoris luzonensis paradisianus, paratype, brachypterous male, length 7.0 mm.

Genus *Cavocoris*

Figure 99

Diagnosis

Medium-sized (9.1-12.5 mm), ovate to broadly ovate. Antennae lie free on ventral head surface. Eyes slightly and flatly raised above interoculus level, anteriorly divergent. Anterior head margin forms smooth curve with lateral eye margins, in frontal view only slightly concave, nearly straight above the labrum. Anterior margin of the pronotum slightly concave, prosternal static

99

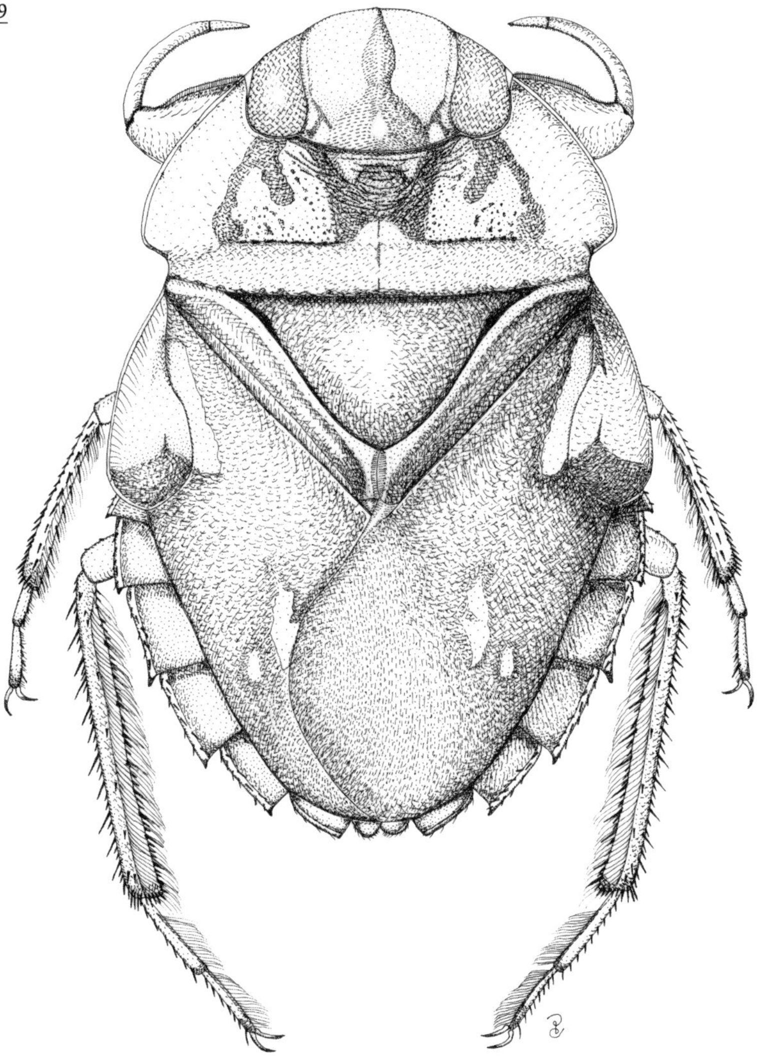

FIGURE 99
Cavocoris bisulcus,
macropterous female,
length 10.2 mm.

sense organs absent. Embolium anteriorly narrowed, lateral margins of the hemielytra convergent anteriorly. Connexival angles acute to spinose.

Distribution in Malesia (fig. 567)

The five known species are all endemic to New Guinea, from low altitudes (Lea) to high up in the mountains (Baiyer River).

Biological notes

Cavocoris species inhabit slow streams (D. Polhemus & J. Polhemus, 1989).

Genus *Philippinocoris*

Figure 100

100

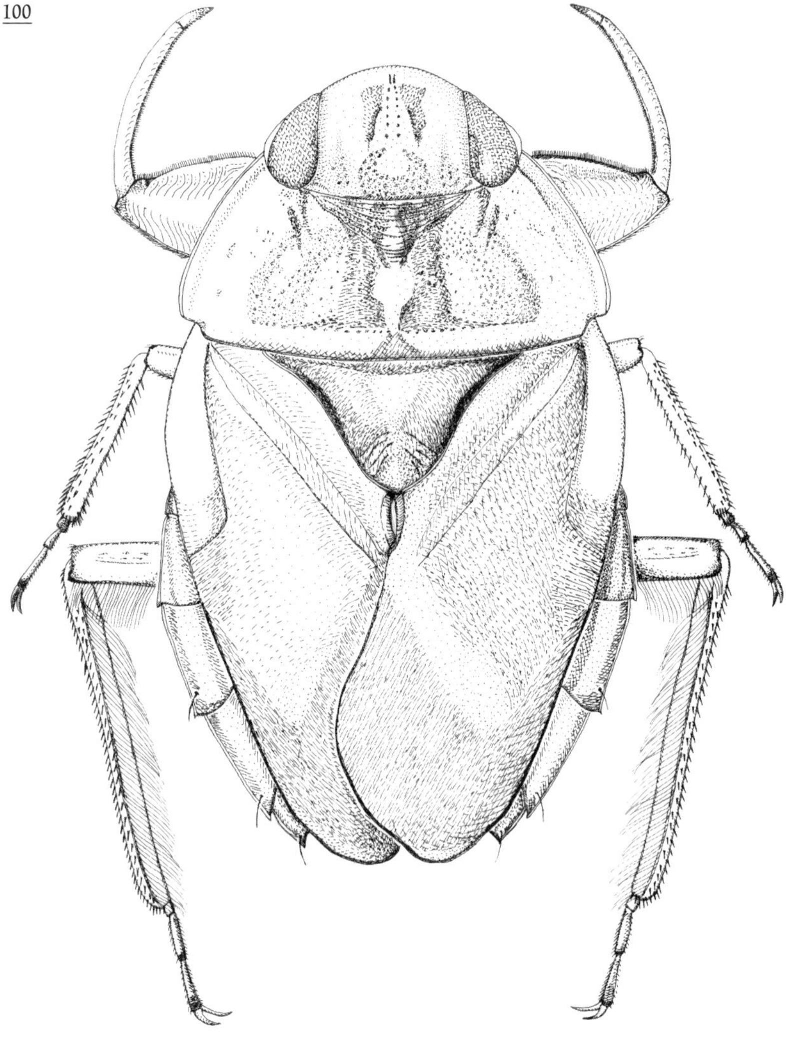

FIGURE 100
Philippinocoris usingeri, brachypterous female, length 9.8 mm.

Diagnosis

Medium-sized (9.5-12.5 mm), oval. Antennae lie free on ventral head surface. Eyes dorsally rather flat, only slightly elevated above level of interoculus, separated from it by shallow groove, parallel to convergent anteriorly. Anterior head margin following curve of lateral eye margin, in frontal view slightly indented above labrum. Anterior margin of pronotum slightly concave, prosternal static sense organs absent. Embolium anteriorly narrowed; lateral margin of hemielytra anteriorly convergent. Connexival angles rectangular to somewhat acute, non-spinose. Left paramere distally long and slender with one or two distal lobes. Female genital operculum triangular with deep and narrow apical emargination between two pointed tips.

Distribution in Malesia (fig. 567)

The two described species are endemic to the Cordillera Central of Luzon.

Biological notes

Philippinocoris species occur in unpolluted, cold, fast-flowing streams where they cling to submerged vertical rocks or roots along the edges of pools sheltered from the main current (D. Polhemus & J. Polhemus, 1987; Zettel et al., 1999).

Genus *Sagocoris*

Figures 101-104

101

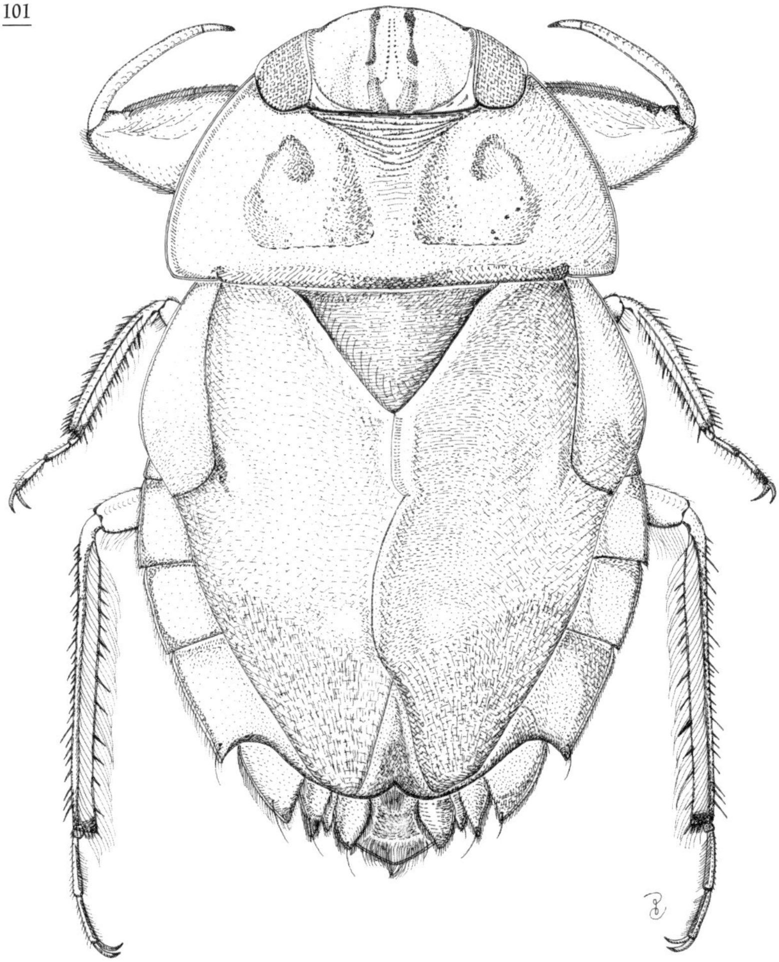

FIGURE 101
Sagocoris biroi, brachypterous male, length 10.3 mm.

102

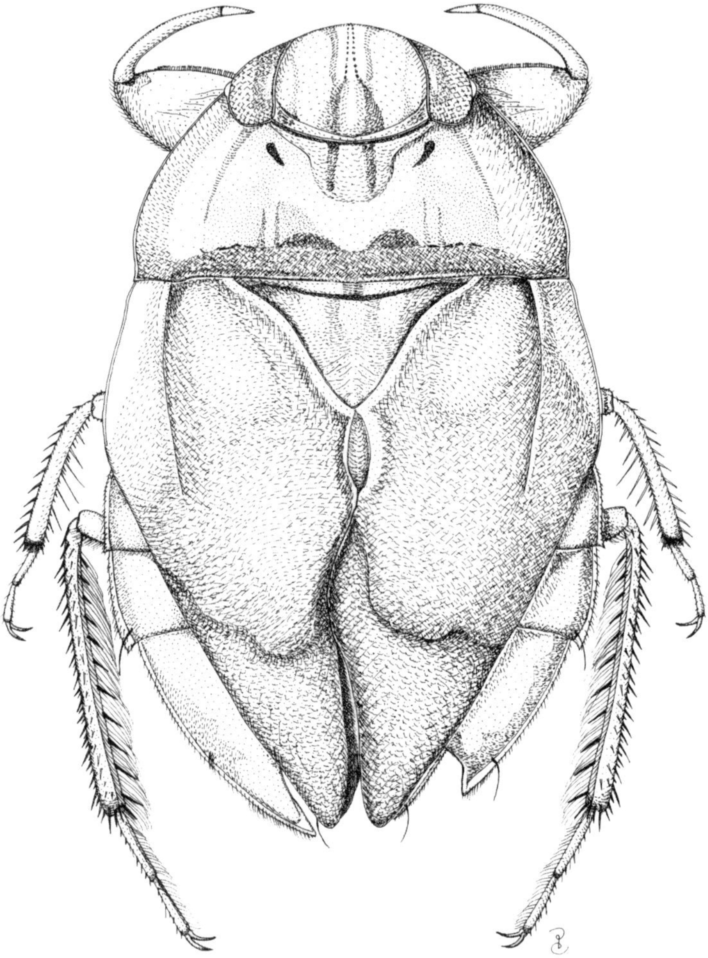

FIGURE 102
Sagocoris asymmetricus, brachypterous female, length 9.8 mm.

Diagnosis

Medium-sized (8-14 mm), oval. Antennae lie free on ventral head surface. Eyes dorsally flat, level with interoculus, separated from it by indistinct shallow groove, parallel to divergent anteriorly. Anterior head margin following curve of lateral eye margin, in frontal view distinctly indented above labrum. Anterior margin of pronotum slightly concave, prosternal static sense organs absent. Hemielytra of brachypterous morph caudally acutely rounded, reaching genital segments (fig. 101). Connexival angles rectangular to somewhat acute, non-spinose or with short broad spines; left connexival angle of segment 6 in female somewhat larger and more strongly produced than right angle.

Distribution in Malesia (fig. 566)

There are eight species (La Rivers, 1971a) , of which seven are endemic to New Guinea, occurring from sea level (Port Moresby, Wewak) to 1300 m altitude in various streams and rivers. *Sagocoris usingeri* was placed in *Philippinocoris* by D. Polhemus & J. Polhemus (1987). However, D. Polhemus & J. Polhemus (2000b) synonymized La Rivers' subgenus *Truncocoris* with *Aptinocoris*, thus removing four species. They describe three new species and synonymized *Quadricoris* (with the single species *Q. asymmetricus*; fig. 102) with *Sagocoris*. Hence, the present number of species is seven, and all are endemic to New Guinea. *Sagocoris asymmetricus* differs from other *Sagocoris* by having four instead of five abdominal segments (figs. 103, 104).

103

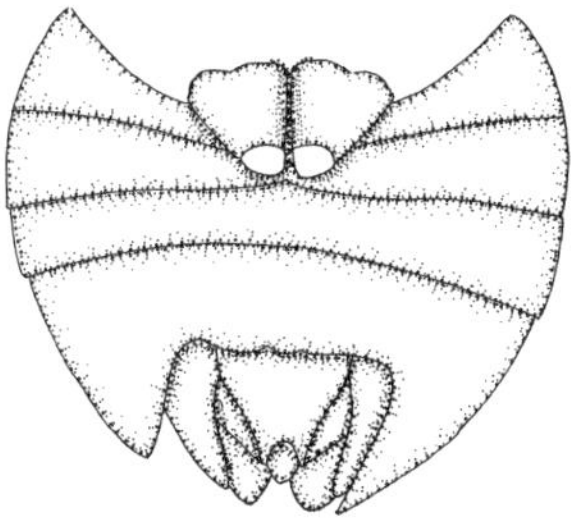

104

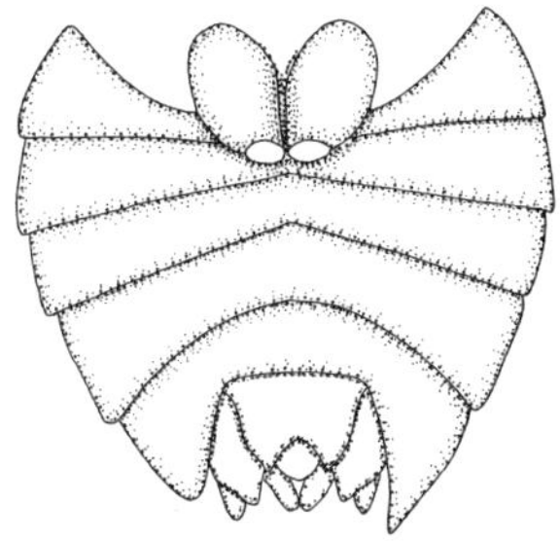

FIGURES 103 – 104
Abdomen in ventral view. 103 *Sagocoris asymmetricus* – 104 *Sagocoris biroi*.

Biological notes

Sagocoris species are found in shallow to moderately deep streams with medium to slow current. Some species prefer the edges of streams. They are occasionally encountered in the same stream as *Aptinocoris* species which, however, occur in the more swiftly flowing parts (D. Polhemus & J. Polhemus, 2000b).

Genus *Stalocoris*

Figure 105

Diagnosis

Small (4.9-7.4 mm), oval. Antennae lie free on ventral head surface. Eyes dorsally rather flat, only slightly elevated above interoculus level, separated from it by shallow groove, parallel to convergent anteriorly. Anterior head margin follows curve of lateral eye margin, in frontal view slightly indented above labrum. Anterior margin of pronotum slightly concave, prosternal static sense organs absent. Embolium anteriorly narrowed, lateral margins of hemielytra anteriorly convergent. Connexival angles on segments 5-6 rectangular to spinose. Left paramere distally stout with one distal lobe and deep subapical incision. Female genital operculum short, truncate, hind margin straight or broadly and weakly concave, in *S. breviceps* with shallow median emargination. *Stalocoris* is similar to *Philippinocoris*, but species of the latter are larger and differ in genital characters. Moreover, *Philippinocoris* has only been collected in Luzon, where *Stalocoris* has not been found yet.

105

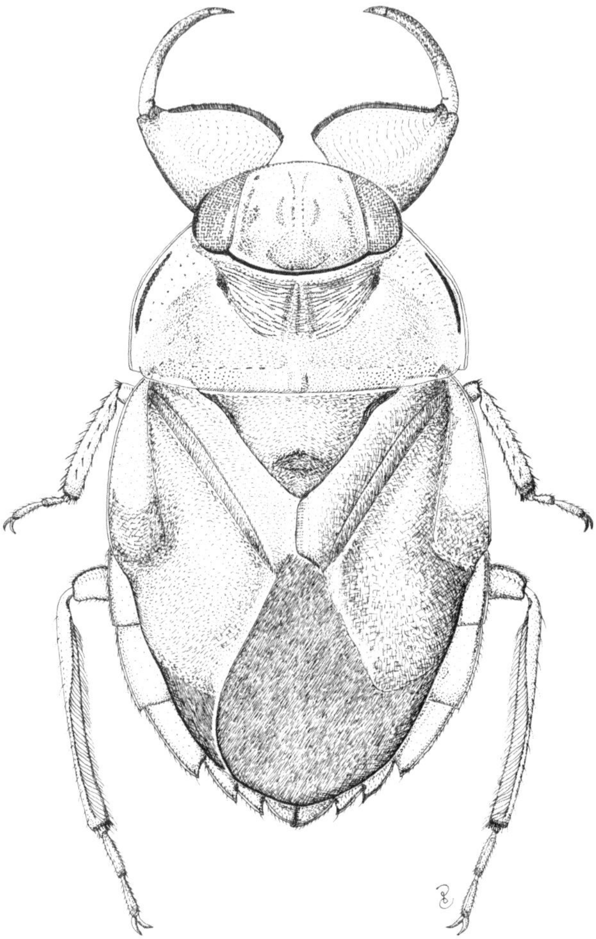

FIGURE 105
Stalocoris schoedli, paratype, macropterous male, length 6.4 mm.

Distribution in Malesia (fig. 565)

There are six species, endemic to the Philippines.

Biological notes

Stalocoris species live in streams under stones, sometimes together with *Aphelocheirus*, but they generally prefer stretches with less current. At least three of the six species show some resistance against pollution (Zettel et al., 1999).

Genus *Warisia*

Figure 106

Diagnosis

Medium-sized (10 mm), broadly ovate. Antennae lie free on ventral surface of head. Eyes flat but slightly above the level of the interoculus, divergent

106

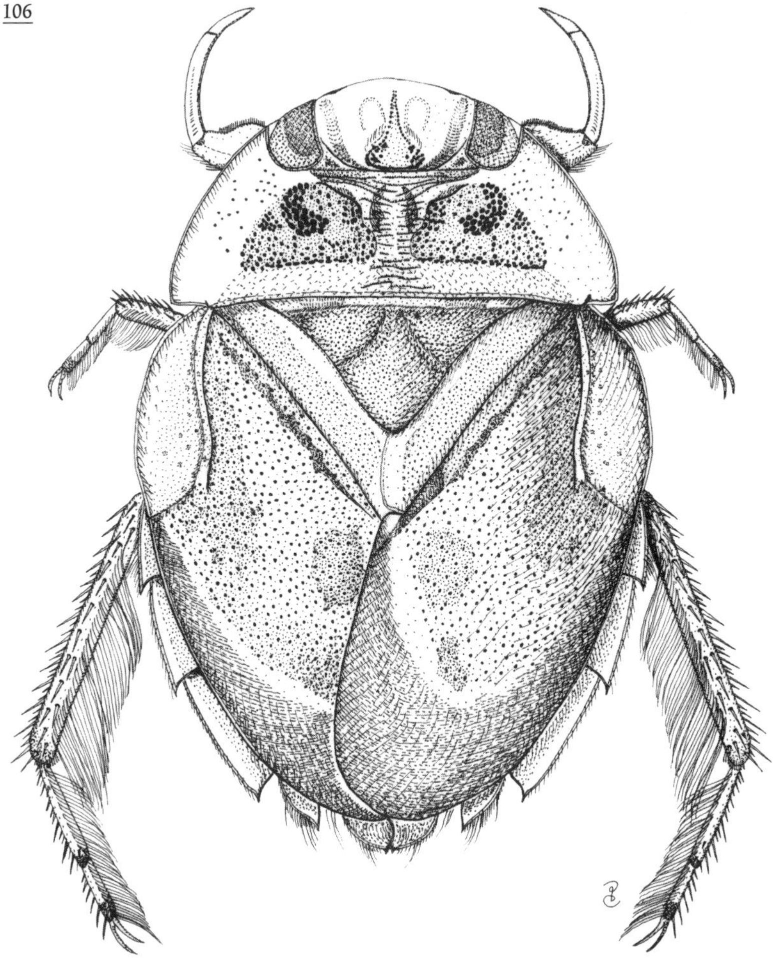

FIGURE 106
Warisia cavanceps,
macropterous female,
length 10.4 mm.

anteriorly. Anterior head margin follows curve of lateral eye margin, laterally with a pair of dimples. Anterior margin of pronotum slightly concave. Embolium anteriorly narrowed, lateral margins of hemielytra anteriorly convergent. Connexival angles rectangular to somewhat acute, non-spinose.

Distribution in Malesia (fig. 568)

A single species, *W. cavanceps*, endemic to New Guinea, has been collected at 350–750 m above sea level in swiftly flowing, rather narrow streams (La Rivers 1971a).

– TRIBE TANYCRICINI

Genus *Idiocarus*

Figures 107, 108

107

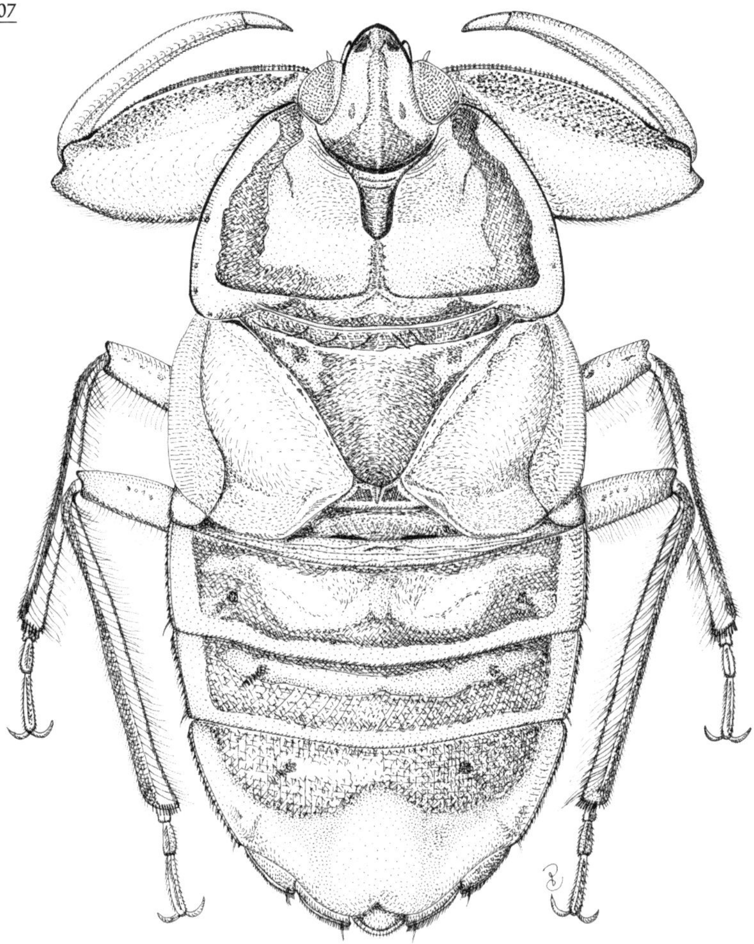

FIGURE 107
Idiocarus papuus, paratype, brachypterous male, length 10.6 mm.

Diagnosis

Small to medium-sized (7.7-12 mm), elongate. Antennae lie free on ventral head surface. Eyes dorsally convex, separated from interoculus by distinct groove, convergent anteriorly. Anterior head margin strongly projecting in front of eyes, in frontal view indented with small median projection above labrum. Anterior margin of the pronotum strongly concave, prosternal static sense organs present. Embolium anteriorly only slightly narrowed. Lateral margins of hemielytra only slightly convergent anteriorly. Connexival angles rectangular.

Distribution in Malesia (fig. 569)

Six species occur in New Guinea at elevations between 200-1700 m. Only one of these is known to extend into Papua (Irian Jaya).

108

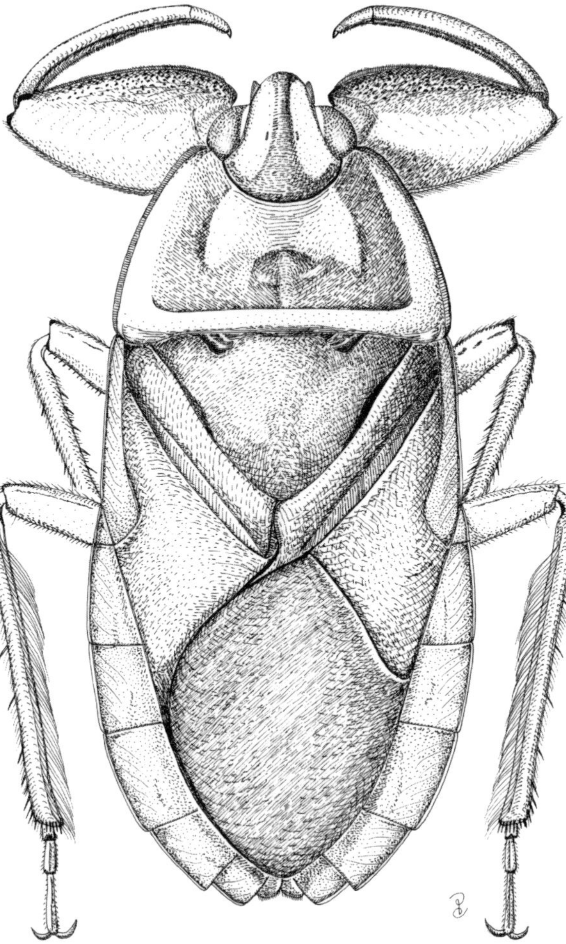

FIGURE 108
Idiocarus papuus, paratype, macropterous female, length 9.5 mm.

Biological notes

Idiocarus species are found under small to moderate-sized stones in cool, swift streams. At higher altitude this genus is replaced by *Tanycricos* (D. Polhemus & J. Polhemus, 1986a).

Genus *Nesocricos*

Figure 109

Diagnosis

Large (12.5-18.5 mm), oval. Antennae lie free on ventral head surface. Eyes dorsally convex, separated from interoculus by distinct groove, parallel to convergent anteriorly. Anterior head margin projecting in front of eyes, in frontal view slightly indented above the labrum. Anterior margin of pronotum slightly concave. Prosternal static sense organs present. Embolium anteriorly

109

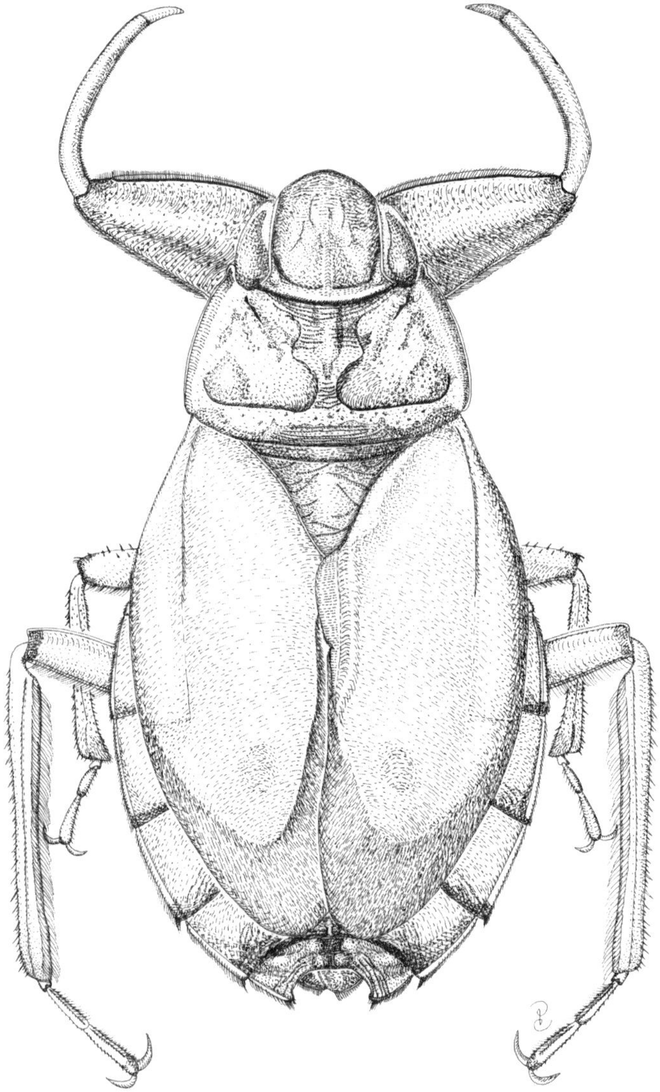

FIGURE 109
Nesocricos evops, paratype, brachypterous female, length 17.3 mm.

narrowed, lateral margins of hemielytra convergent anteriorly. Connexival angles rectangular.

Distribution in Malesia (fig. 570)

There are four species, endemic to the highlands of New Guinea. Only one of these is known to extend into Papua (Irian Jaya).

Biological notes

Nesocricos species occur at elevations above 800 m along margins of cold streams where they cling to overhanging vegetation, logs and undercut bank margins, frequently aggregating at favourable sheltered sites. In slower flowing water, e.g. side channels, they are also found under stones, usually close to the shore. They do not frequent the turbulent rushing midsections of streams favoured by the related genera *Idiocarus* and *Tanycricos*, with which they are often sympatric (D. Polhemus & J. Polhemus, 1985).

Genus *Tanycricos*

Figure 110

110

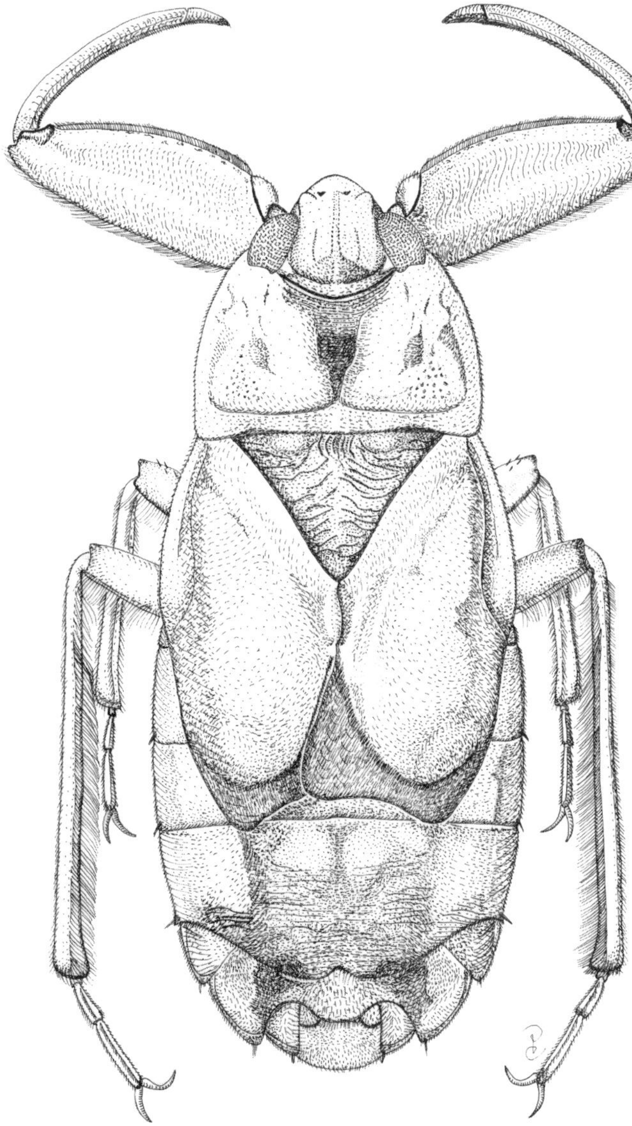

FIGURE 110
Tanycricos acumentum, paratype, brachypterous male, length 13.4 mm.

Diagnosis

Large (12-19 mm), elongate. Antennae lie free on ventral side of head. Eyes dorsally convex, about level with interoculus, separated from it by a distinct groove, parallel to convergent anteriorly. Anterior head margin slightly to strongly projecting in front of eyes, in frontal view not indented above labrum. Anterior margin of pronotum distinctly concave, prosternal static sense organs present. Embolium anteriorly narrowed. Lateral margins of the hemielytra anteriorly convergent. Connexival angles rectangular to somewhat acute, non-spinose or with short broad spines. Left connexival angle of segment 6 in females somewhat larger and more strongly produced than right angle.

Distribution in Malesia (fig. 571)

There are seven species, all endemic to New Guinea. They occur throughout the central highlands (D. Polhemus, 2000).

Biological notes

Tanycricos species are found at the bottom of cold mountain streams and rivers above 1500 m, sometimes with two species in one stream. In such cases the species show a preference for different current velocities. The preferred habitat lies under or behind large stones in swiftly flowing waters. *Tanycricos* often occur in high densities and represent the most common invertebrate predators, often preying on aquatic Lepidoptera larvae. Teneral (very young) specimens have been taken near the shore in relatively slack water among cobbles, suggesting that they may leave the midstream to moult (D. Polhemus & J. Polhemus, 1986b).

SUBFAMILY LACCOCORINAE

Diagnosis

Small to medium-sized (7.0-13.2 mm), oval, dorsoventrally flattened. Eyes laterally with a chitinous strip fitting over anterior angle of pronotum. Anterior head margin curved downward and usually somewhat backward ventrally. Rostrum originating somewhat posterior of anterior head margin, but not placed in a groove. Labrum well developed. Gula short. Anterior margin of pronotum nearly straight. Propleura not medially prolonged, consequently prosternum fully exposed. Fore tarsus not fused with tibia, with two claws which may be very small and difficult to distinguish; in males two-segmented. Middle and hind femur, in addition to two rows of bristles along posterior margin, ventrally with two distinct rows of bristles or spine-like setae, one near anterior margin, one somewhat remote from posterior margin; anterior row may be quite incomplete. Mesosternum and metasternum without or with poorly developed carinae, without apical fossa. Parameres very small.

Distribution worldwide

This subfamily has a circumtropical distribution. There are about 65 described species divided over seven genera. However, the status of several (sub)species is unclear. *Ctenipocoris* has seven species, of which one in South-east Asia. *Diaphorocoris* has three species in Sri Lanka and India, and extends into continental South-east Asia. *Heleocoris*, with about 25 species, has nineteen species in tropical Asia. *Laccocoris* has nearly 20 species in the Old World tropics, which are about equally divided between Africa and Asia.

Distribution in Indo-Australia

The species of the four laccocorine genera occurring in Asia are restricted to Indo-Australia except for one. For distribution of *Ctenipocoris* and *Diaphorocoris* see above and below. Of the eighteen species of *Heleocoris* in Indo-Australia ten are confined to India and two to continental South-east Asia. The remaining six species are Malesian. Of the eight species of *Laccocoris* occurring in Asia one is restricted to continental South-east Asia. One species of both *Heleocoris* and *Laccocoris* are reported from Australia (Cassis & Gross, 1995). These two species are not from Australia proper but from Christmas Island, south of Java (Izzard, 1936). However, *L. montandoni* Izzard is described from 'Indian Ocean, Tabameer, Propat, 1933'. This is not on Christmas Island but refers to Prapat at Lake Toba on Sumatra. Izzard (1936) himself already doubted that his specimen originated from Christmas Island. This species is considered to be a synonym of *L. nervicus*. The other species is *H. nebulosus*, which is also known from Sumatra and Java.

Biological notes

Biology and ecology of Laccocorinae is virtually unknown. The few literature data and personal observations suggest that, although they live in streams, Laccocorinae also stay at the edges or in quiet pools and avoid strong currents. Most of the more detailed habitat records mentioned below are from field notes by N. Nieser.

Key to Asian genera of Laccocorinae

1 Head anteriorly strongly folded downward but hardly backward. Vertical part wide (fig. 82). Chitinous strip laterally and posteriorly of eyes narrow [circumtropical] ▸ *Ctenipocoris* p. 122

– Head anteriorly folded downward and backward. Vertical part less wide (fig. 83). Chitinous strip laterally and posteriorly of eyes well developed ▸ 2

2 Eyes anteriorly distinctly divergent (fig. 116) [India, Sri Lanka and Thailand] ▸ *Diaphorocoris* p. 123
– Eyes parallel or anteriorly slightly convergent ▸ 3

3 Labrum broad, short and with bluntly rounded tip (fig. 114) [Paleotropical] ▸ *Heleocoris* p. 124
– Labrum long, usually with sharp or narrowly rounded tip (figs. 111-113) [Paleotropical] ▸ *Laccocoris* p. 126

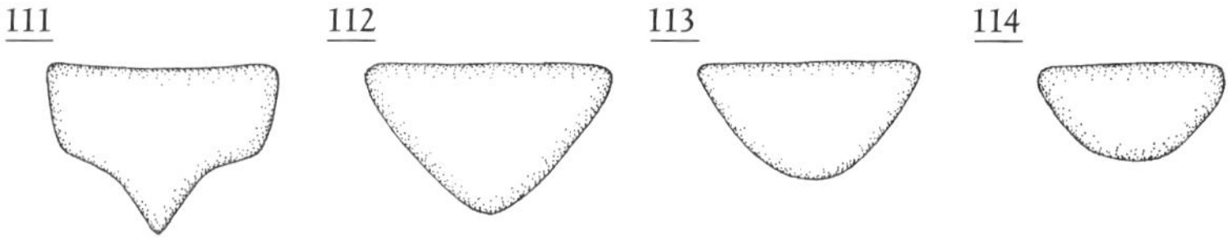

FIGURES 111 – 114
Labrum in Laccocorinae
111 *Laccocoris nervicus* – **112** *Laccocoris staudingeri* – **113** *Laccocoris hoogstraali* – **114** *Heleocoris nebulosus*.

Remark

The differences in the labrum are the only characters separating *Heleocoris* and *Laccocoris*. The difference is, however, not always very clear. Especially the labrum of *L. hoogstraali*, a common species in the southern Philippines, is very similar to that of *Heleocoris* (fig. 113).

Genus *Ctenipocoris*

Figure 115

Diagnosis

Rather small (7.5-9 mm). Eyes anteriorly convergent, head anteriorly high, frons not curved strongly backward (fig. 82). Anterior femur moderately thickened, middle and hind tibia strongly spinose.

Distribution in Malesia (fig. 568)

One species, *C. asiaticus*, occurs in South-east Asia including the extreme south of Thailand (Sites et al., 1997). D. Polhemus (1987) mentions that he has seen specimens from Malaysia, Singapore and Java.

Biological notes

Ctenipocoris asiaticus occurs in ponds or in vegetation along quiet margins of streams. A Brazilian species was collected (in Brazil) at a stagnant very shallow edge of a stream and seemed to burrow in soft mud.

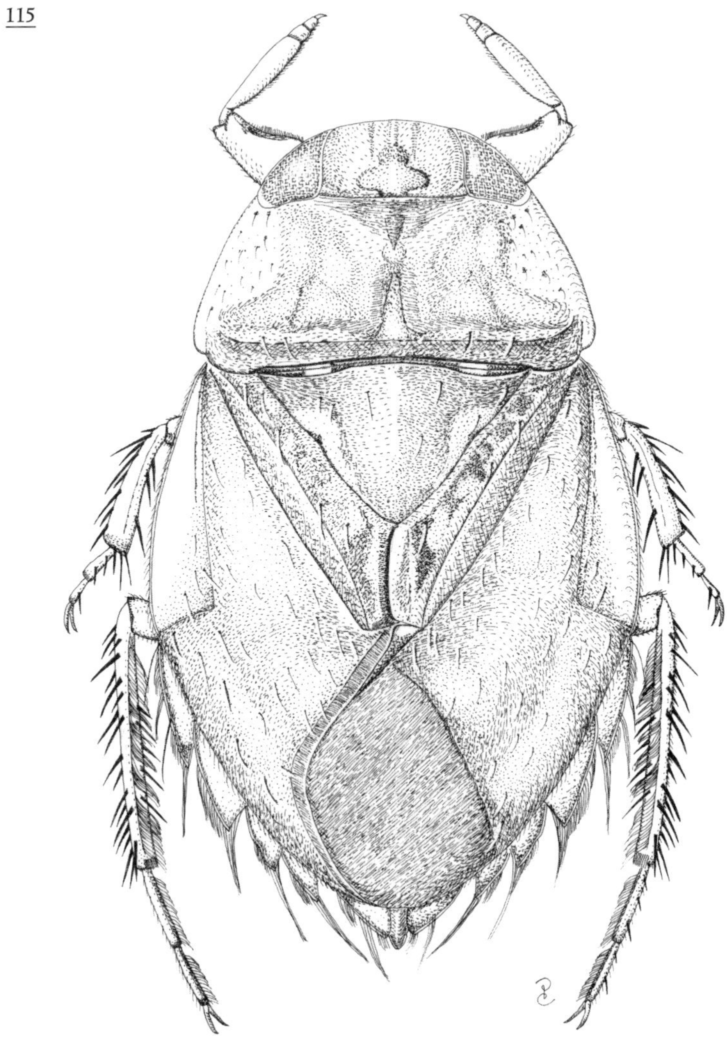

FIGURE 115
Ctenipocoris asiaticus, macropterous female, length 8.2 mm.

Genus *Diaphorocoris*

Figure 116

Diagnosis

Rather small (7-10 mm). Eyes anteriorly divergent, head anteriorly flat, frons moderately curved backward. Anterior femur moderately thickened, middle and hind tibia not strongly spinose.

Distribution in Malesia

Diaphorocoris does not occur in Malesia. Thus far it has only been recorded from India and Sri Lanka. One specimen of an unidentified species was collected in Chiang Mai, Thailand (collection N. Nieser).

116

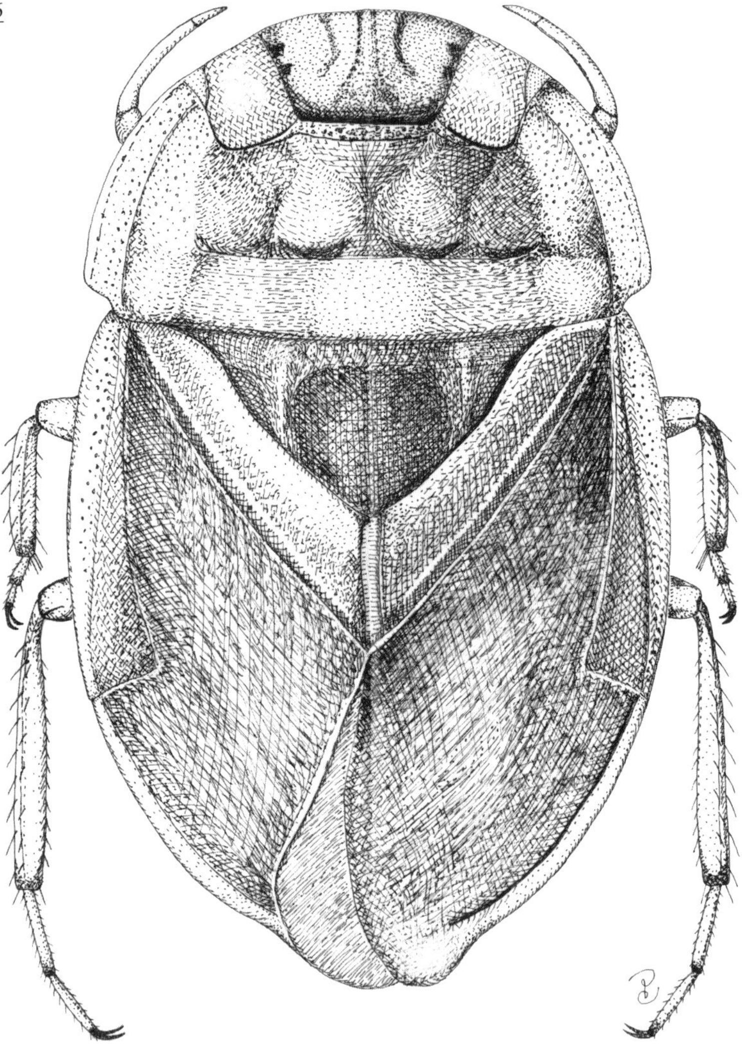

FIGURE 116
Diaphorocoris sp., brachypterous female, length 9.2 mm.

Biological notes

The Chiang Mai specimen was caught in a quiet pool with much silt of a small stream in disturbed forest.

Genus *Heleocoris*

Figure 117

Diagnosis

Small to medium-sized (6.7-11 mm). Eyes anteriorly parallel to convergent, head anteriorly flat, frons curved considerably backward. Labrum short (up to half as long as its basal width) and with a broadly rounded apex (fig. 114).

117

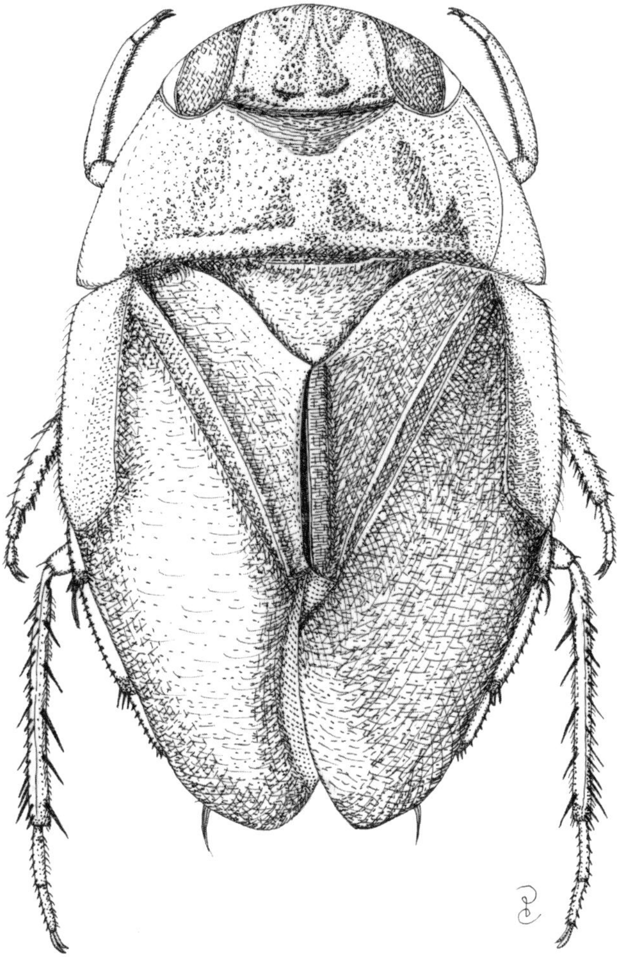

FIGURE 117
Heleocoris obscuratus, brachypterous female, length 9.8 mm.

Distribution in Malesia (fig. 565)

Six species occur in Malesia, notably in the southwestern part, which makes it the second-richest area after India (which has ten species).

Biological notes

Found in small streams. One *H. obscuratus* in Java was collected at the edge of a small stream under an overhanging rock.

Genus *Laccocoris*

Figure 118

118

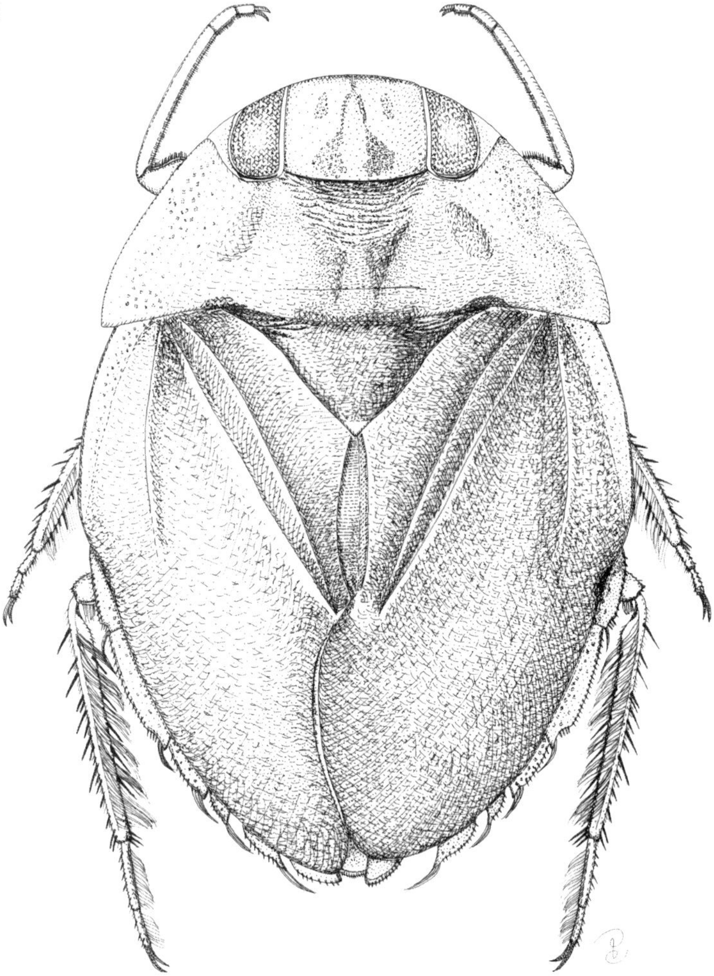

FIGURE 118
Laccocoris staudingeri, macropterous male, length 12.1 mm.

Diagnosis

Small to medium-sized (7.3-13 mm). Eyes anteriorly parallel to convergent, head anteriorly flat, frons curved considerably backward. Labrum long ($^2/_3$ as long as its basal width, or longer) with sharp or acutely rounded tip (figs. 111-113).

Distribution in Malesia (fig. 567)

Seven out of eight Asian species occur in Malesia, five of which in Borneo, one in Sumatra and one in the Philippines.

Biological notes

Laccocoris species inhabit small to large streams. *Laccocoris hoogstraali*, the only species known from the Philippines, occurs in small to medium-sized streams, often in agricultural areas, notably under overhanging banks and similar localities. One *L. horvathi* from Brunei was caught in rainforest at the edge of a very small and shallow stream under pebbles (N. Nieser).

SUBFAMILY NAUCORINAE

Diagnosis

Small to medium-sized (5.5-16 mm) broadly oval, not strongly flattened. Inner eye margin anteriorly convergent; eyes laterally without chitinous strap. Rostrum originating ventrally at anterior head margin. Labrum well developed; gula well developed. Anterior margin of pronotum nearly straight; propleura not prolonged medially and consequently prosternum fully exposed. Fore tarsi one-segmented with or without one minute claw; middle and hind femur ventrally smooth, without rows of bristles or spines. Mesosternum and metasternum usually without longitudinal carinae or at most with small thin plate-like carinae, without median groove. Parameres well developed.

Distribution worldwide

There are 60 extant species divided over seven genera. Species-rich areas are the Americas with 20 species in two endemic genera, and Africa with 28 species in three genera, of which two endemic. *Naucoris* has about 20 species in the warmer regions of the Old World, of which ten in Africa and eleven in Indo-Australia, five of which in Malesia.

Genus *Naucoris*

Figure 119

Remark

Zettel (2001f) reinstated the genus *Thurselinus* Distant for *Naucoris scutellaris*. From his text it is clear that he did not include *N. rhizomatus* in his considerations. *Thurselinus* is here considered synonymous with *Naucoris*.

Diagnosis

Small to medium-sized (5.5-15 mm) with flat or slightly convex dorsum. Eyes anteriorly convergent. Anterior margin of pronotum not excavated behind head, lateral margin of pronotum without a sublateral groove. Middle and hind tibia not widened, hind tarsus shorter than hind tibia. Parameres (in male) well developed. *Naucoris obscuripennis* and *N. pumilus* differ from the widespread *N. scutellaris* by having the posterior angles of the connexiva of abdominal segments 4-6 rectangular to weakly acute (fig. 79), whereas in *N. scutellaris* these are

119

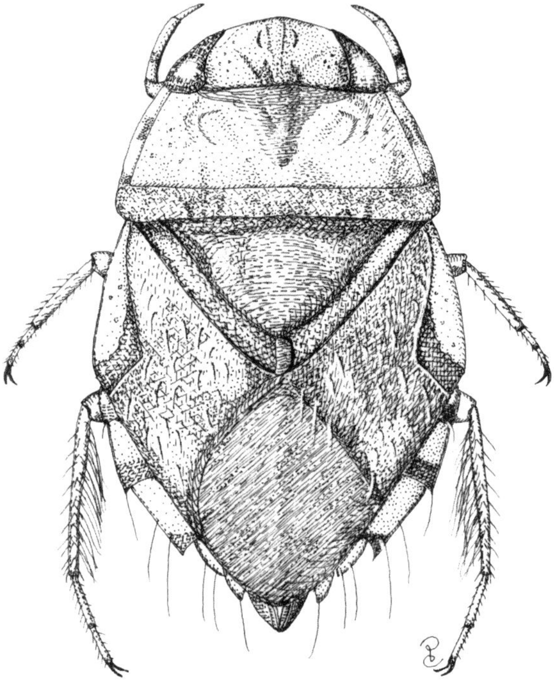

FIGURE 119
Naucoris scutellaris, macropterous female, length 5.7 mm.

spinose (figs. 80, 119). A key is provided by Zettel et al. (1999). The differences in parameres given by Usinger (1937) are incorrect: the legend and figures 1 and 2 have been interchanged (see Zettel et al., 1999a). *Naucoris rhizomatus* is similar to *N. scutellaris* but somewhat smaller on average; the female has a much longer genital operculum.

Distribution in Malesia (fig. 566)

Of the five described Malesian species only *N. sumatranus* has never been collected after the first specimen. Judging from the figure and the few data in the description (Fieber, 1851b) it may be *N. scutellaris* (the type nor fresh material of the latter from Sumatra is presently availiable). *Naucoris obscuripennis* and *N. pumilus* are known only from the Philippines. *Naucoris rhizomatus* occurs in Australia and continental Malaysia.

Biological notes

Most *Naucoris* species occur in stagnant water or in sluggish bays of streams with rich vegetation. *Naucoris rhizomatus* has, as far as known, an aberrant habitat preference: it clings to plant roots at the edge of streams (J. Polhemus, 1984a). Although they are good swimmers *Naucoris* species hide between vegetation rather than to chase prey.

REFERENCES – Naucoridae. Synonymy and (re)description of Indonesian taxa (Lundblad, 1933a); genital morphology of *Ilyocoris* (Larsén, 1938); morphology of *Ilyocoris* (Rawat, 1939); deposition of eggs (Lebrun, 1960); respiration (Parsons, 1970); Revision of Cheirochelinae of New Guinea (La Rivers, 1971a); catalogue of Naucoridae (La Rivers, 1971b); revision of *Nesocricos* (D. Polhemus & J. Polhemus, 1985); review of *Coptocatus* (D. Polhemus, 1986); revision of *Idiocarus* (D. Polhemus & J. Polhemus, 1986a); revision of *Tanycricos* (D. Polhemus & J. Polhemus, 1986b); erection of *Philippinocoris* (D. Polhemus & J. Polhemus, 1987); revision of *Cavocoris* (D. Polhemus & J. Polhemus, 1989); key to Indonesian genera (Nieser & Chen, 1991); records of Naucoridae in southern Thailand (Sites et al., 1997); key to Philippine genera and species of Naucoridae (Zettel et al., 1999; Zettel, 2003b); distribution of *Tanycricos* (splendid figure of *T. jaetipi*; D. Polhemus, 2000); revision of *Aptinocoris* and *Sagocoris* (D. Polhemus & J. Polhemus, 2000b).

Family Corixidae – water boatmen

Diagnosis

Small to medium-sized (3-15 mm), dorsoventrally flattened, usually elongate and parallel-sided. Mostly dorsally medium brown with yellowish markings. Antennae four-segmented. Eyes large, ocelli absent. Posterior margin of head covering anterior margin of pronotum. Rostrum broadly triangular, not segmented but usually with transverse grooves. Scutellum hidden by pronotum; hemielytra of uniform texture, membrane without veins. Tarsal formula 1-1-2. Fore tarsus (pala) usually broadened into spoon-like form with long bristles; palar claw in both sexes bristle- or spine-like. Middle tarsus with pair of striking long claws. Male abdomen usually with strigil, distally strongly asymmetrical.

Characteristics

Corixidae could be mistaken for Notonectidae because of their long hind legs with strongly developed fringes of swimming hairs and relatively large eyes. They are distinguished by the dorsoventrally flattened body and long middle legs with large claws. Corixoidea have several autapomorphies and were considered a separate infraorder by Börner (1934) under the name Sandaliorrhyncha. However, later analysis (Cobben, 1968; Mahner, 1993) include them in Nepomorpha, in spite of their obvious specializations.

The rostrum is unsegmented and broadly and immovably attached to the head. Instead of segments there are transverse grooves (fig. 123), except in the Cymatiainae (fig. 124). The one-segmented fore tarsus (pala) is enlarged and ventrally widened except in Cymatiainae (fig. 125). The ventral palm is bordered by palmar bristles. Apart from Cymatiainae males have pegs on the inner surface of the pala which are usually arranged in a row (fig. 126, 127). Stridulation is performed by rubbing fields of fine pegs on the fore femora along the sides of the head (maxillary plates). In several species both sexes stridulate. In some species males produce sound during mating by rubbing the middle femora against the edges of the female's hemielytra. It also functions in species recognition. Stridulation with the fore femora is usually a male's signal which in some species is answered by a female. In some species the male signals function in territorial definition (Jansson, 1986, 1989). Sounds are perceived by the tympanal organ, situated in the mesothorax. Observations on stridulation in Corixidae are only available for Northern Hemisphere temperate species (Jansson, 1973, 1986).

Adults have a strongly hydrofugous body. They take in air by bending head and pronotum forward and downward, thus creating a slit between pronotum and mesonotum (Popham, 1960). The air passes through channels along the side of the mesothorax to the subalar air store and the venter of both abdomen and thorax. Some air is kept between head and prothorax, between prothorax and mesothorax, and on top of the hemielytra (Parsons, 1967).

Larvae of Corixidae and Micronectidae are recognized by the lack of hemielytra. Third to fifth instars are recognized by their distinct wing pads. The first and second instars have closed spiracles and are tracheal gill breathers, but from the third instar on the spiracles open and atmospheric air is used.

The strigil, present in most male Corixidae, has been named wrongly as it has nothing to do with stridulation but keeps the subhemielytral air store open during copulation (Popham et al., 1984a). The male mounts the female from behind, presses his head against her pronotum and grips her around the mesothorax with the palae. The male then extrudes his aedeagus on the side where the strigil is and curves the genital segments sideways, downward and forward, so that he can insert the aedeagus in the vagina. Copulation can last long and a pair may need to surface to replenish gas stores. As the male is firmly attached to the female the pair rolls over once at the surface with the strigil pointing downward. The male, which is the only one in contact with the air, lifts its hemielytra to allow air to pass between the wings and the dorsal abdominal surface. The strigil prevents the water pressure to reduce the space between wings and abdomen at the submerged side. Since the bugs are strongly hydrophobous, air passes downwards to the strigil and then through the indentation between abdominal segments 5-6 to the ventral side. As male and female are close together, air passes from the ventral side of the male to the dorsal side of the female and from there to the whole surface of her body. When the gas stores have been replenished the pair goes back to the bottom.

Especially in *Sigara* only males can be reliably identified. Diagnostic characters lie in the pala, abdominal dorsum, and the right paramere (or both; figs. 127, 120, 121). Males are easily recognized by their asymmetrical abdominal apex,

120

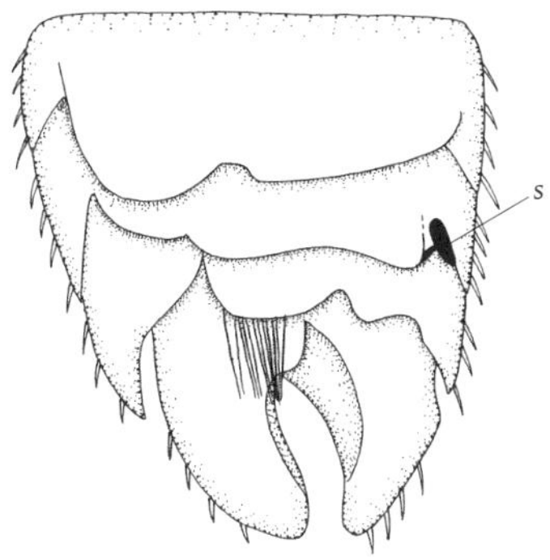

FIGURE 120
Sigara paivai male, apex of abdomen in dorsal view, semidiagrammatical; s = strigil.

121

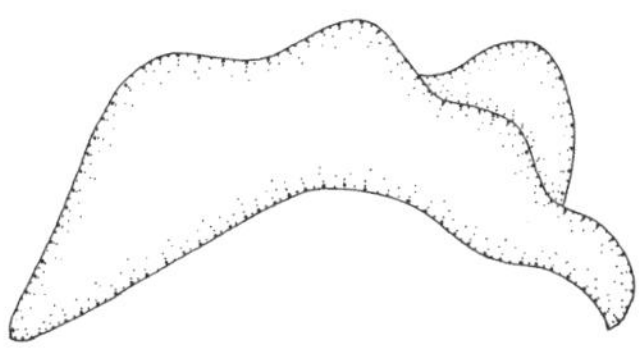

FIGURE 121
Right paramere of *Sigara paivai*.

and by palar pegs in Corixinae. In addition the frons of the male is flattened or indented. This is apparently used as a suction pad during copulation. In some species abdominal asymmetry is dextral, in others sinistral. In a few species however abdominal asymmetry may be on either side of the body.
Eggs of Corixidae have a disc with which they are glued to the substrate (usually plants), and a characterisitic tip. Between the disc and the main body of the egg there is a – usually short – stalk (fig. 122).

122

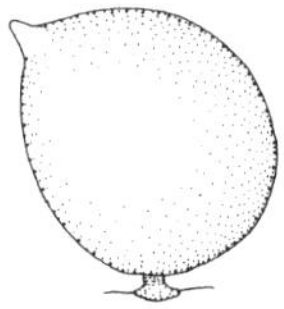

FIGURE 122
Egg of a species of *Sigara* subgenus *Tropocorixa.*

Fossil record

The fossil record of Corixoidea is comparatively rich, but many records are doubtful (Nel & Paicheler, 1993a). Popov (1971b) erected the family Shurabdellidae, represented by fossils from the Lower and Middle Jurassic, about 190-165 million years ago, as a sister group to recent Corixoidea. This family has, reluctantly, been accepted by Nel & Paicheler (1993a). All other mesozoic fossils placed in Corixoidea are either too fragmentary or need redescription. The only mesozoic fossils which can be identified with accuracy are assigned to the Diaprepocoridae Lundblad, which are usually considered a subfamily of Corixidae (see under Micronectidae). Three Mesozoic genera have been found in China and Siberia, ranging in time throughout the entire Jurassic era. An unnamed Miocene fossil belonging to the Diaprepocoridae was found in France (Nel & Paicheler, 1993a). The only extant genus in Diaprepocoridae, *Diaprepocoris* Kirkaldy, is restricted to Australia and New Zealand (Knowles, 1974; Wise, 1965). Cenozoic fossils of Corixidae are also abundant; most of them belong to three species recently collected at two sites in France and one site in Turkey. Most Cenozoic fossils can be identified with reasonable certainty and belong to Corixinae. However, the morphological structures used to classify modern genera of Corixidae are not recognizable in the fossils and therefore comparison with recent genera is impossible (Nel & Paicheler, 1993a).

Distribution worldwide

Corixidae has the largest number of species of the nepomorphan families. The four subfamilies contain 32 extant genera with about 450 species. Corixinae are predominantly distributed in temperate and subtropical areas; the number of species in the tropics is low. Cymatiainae has six species in two genera and are mostly Palearctic, but one species occurs in Australia and New Guinea.

Distribution in Indo-Australia

There are two subfamilies with six genera and 35 species in Indo-Australia. Thirteen of these species are, within Indo-Australia, restricted to India. Australia has thirteen endemics. Ten species divided over three genera are known from Malesia .

Biological notes

Life cycles have only been studied in temperate regions. In the wet tropics there are probably several generations in one year; in regions with distinct dry periods the life cycle is probably synchronized with these. In Europe, notably England and Poland, the relation between Corixidae and habitat choice has been studied extensively (Macan, 1954a, b; Midak, 1965; Wróblewski, 1939, 1952). When a recently dug pond develops into a marsh there is a succession of corixid species (Macan 1954a, b).

Most species stay close to the bottom except when replenishing air stores, but some hunt for prey in open water. Recent results show that most species feeding on the bottom eat mainly animal food, either as predators of small living prey or as scavangers. Animal food is necessary for reproduction and improves longevity, as shown by rearing experiments (Jansson, 1986). However, Bakonyi (1978) found that food preference varies in different genera and that even species that chiefly consume animal food may eat algae at certain times of the year. Popham et al. (1984b) analysed gut contents of 21 British species and found that there are not only differences between genera but also between species of the same genus, in one species even between sexes. Some genera, e.g. *Cymatia*, are strictly predatory, and correlation between pala shape and food preference has been described (Popham et al., 1984b). No specific data are known from tropical Asian Corixidae, but *Cnethocymatia* and *Agraptacorixa* are probably entirely or predominantly carnivorous (Popham et al., 1984b). British *Sigara* of comparable size to Asian *Tropocorixa* tend to be mixed feeders (Popham et al., 1984b). With some exceptions, Corixidae are not economically important (Popham et al., 1984b).

REFERENCES – Corixidae. Morphology of mouthparts and genitalia (Ekblom, 1930); structure and function of the genital apparatus (Larsén, 1938); general morphology (Griffith, 1945); skeleton, muscles and organs of the head of *Corixa* (Benwitz, 1956).

Key to Indo-Australian subfamilies and genera of Corixidae

1 Rostrum without transverse grooves (fig. 124). Pala elongate, nearly cylindrical, without palm and in males without palar pegs (fig. 125) (Cymatiainae) ‣ *Cnethocymatia* p. 136

– Rostrum with transverse grooves (fig. 123). Male pala more or less spoon-shaped with ventral palm. In males palar pegs present (fig. 127) (Corixinae) ‣ 2

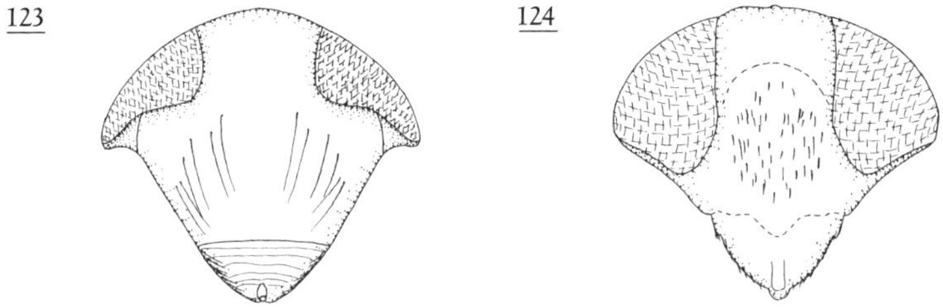

FIGURES 123 – 124
Head in frontal view.
123 *Sigara truncatipala*, female – **124** *Cnethocymatia nigra*, male (adapted from Lansbury, 1983).

2 Length to 5 mm. Male with strigil on left side and apex of fore tibia produced over base of pala (fig. 126). Female with apices of clavi not surpassing a line drawn through margins of hemielytra at nodal furrows (fig. 128) ‣ *Trichocorixa*
Only one species: *T. verticalis verticalis* (Fieber, 1851c), originating from the Atlantic coast of North America and the Antilles, introduced in New Caledonia (Jansson, 1982).

– Length 4.5 mm or more. Males with strigil on right side and apex of fore tibia not produced (fig. 127). Female with apices of clavi surpassing a line drawn through margins of hemielytra at the nodal furrows (figs. 133, 135) ‣ 3

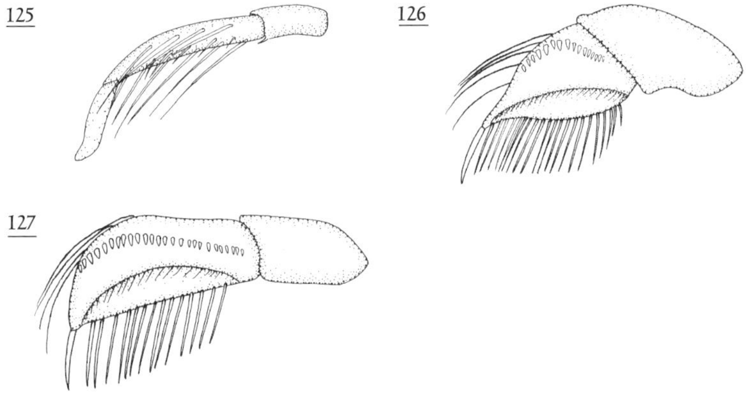

FIGURES 125 – 127
Right fore tibia and pala of male.
125 *Cnethocymatia nigra* (adapted from Lansbury, 1983) – **126** *Trichocorixa verticalis verticalis* – **127** *Sigara paivai*.

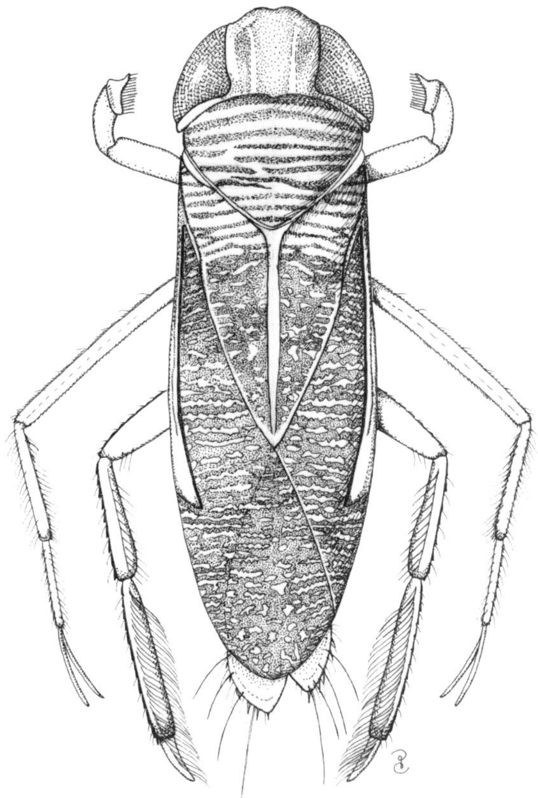

128

FIGURE 128
Trichocorixa verticalis verticalis, male, length 4.3 mm.

3 Pronotum and hemielytra concolorous or with small dark dots (at the base of hairs). Length 6.5-10 mm; head width 2.4-3.4 mm, body shape rather short (fig. 133) ‣ *Agraptacorixa* p. 137

– Pronotum and hemielytra brown with yellowish transverse bands which are interrupted on hemielytra. Length 4.5-8.0 (Malesian species not over 7.0 mm), head width 1.5-2.2 mm, body shape more slender (fig. 135) ‣ *Sigara* p. 139

SUBFAMILY CYMATIAINAE

Diagnosis

Small to medium-sized (3-8 mm). Rostrum without transverse sulcations, embolar groove present, nodal furrow absent. Pala elongate, nearly cylindrical, without palm in either sex. Claw of hindtarsus inserted apically. Male without strigil. Media vein curves apically towards the costal margin of the hemielytra (fig. 131), suggesting a nodal furrow.

Distribution in Malesia

Only one genus with one species occurs in Malesia.

129

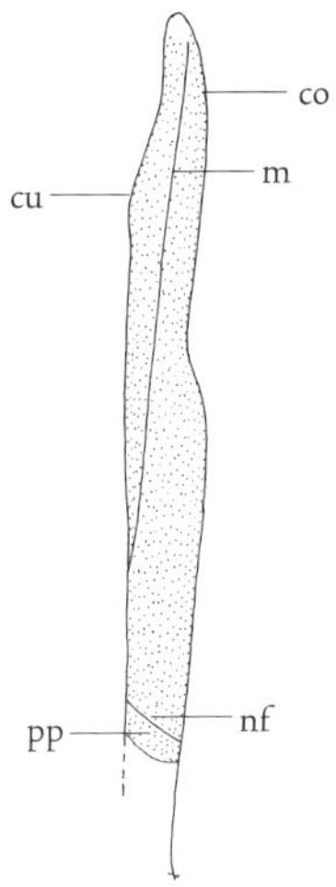

130

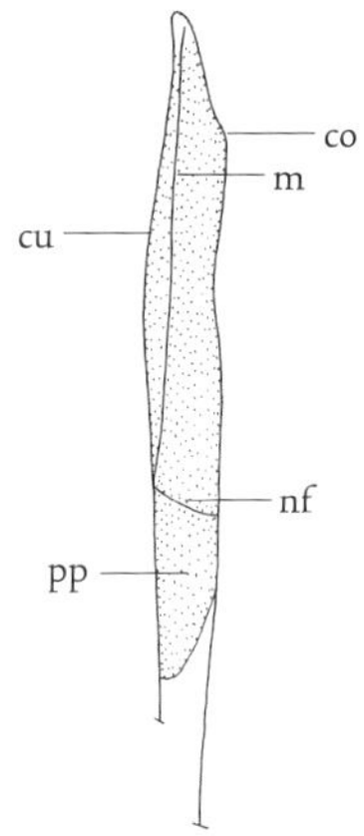

131

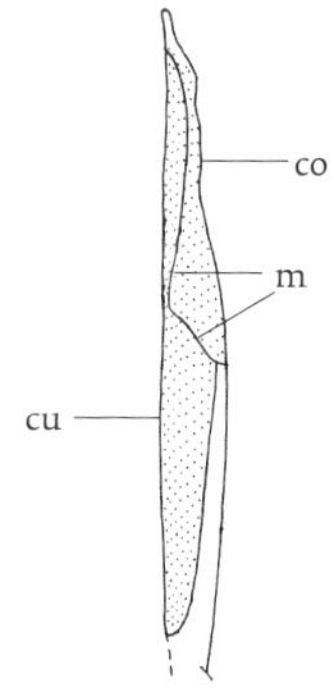

FIGURES 129 – 131
Embolar area of right hemielytron in lateral view. **129** *Agraptacorixa hyalinipennis* – **130** *Sigara truncatipala* – **131** *Cnethocymatia nigra* (131 adapted from Lansbury, 1983); co = costa, cu = cubitus, m = media, nf = nodal furrow, pp = pruinose area posterior of nodal furrow; dotted = pruinose area.

Genus *Cnethocymatia*

Figure 132

132

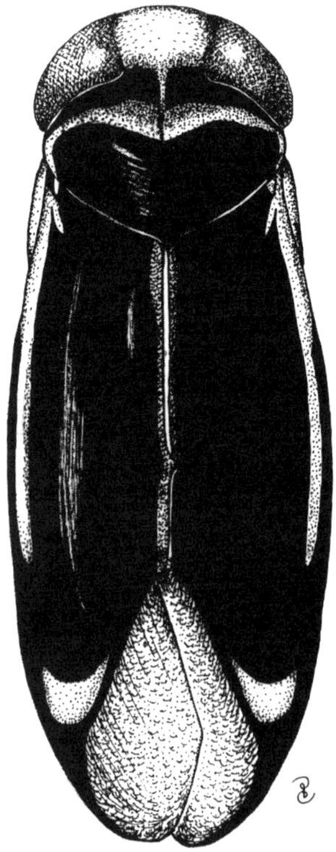

FIGURE 132
Cnethocymatia nigra (Hungerford), female, length 4.6 mm (adapted from Hungerford, 1947).

Diagnosis

Rather small (4.6-5.0 mm). Pronotum and hemielytra dark mahogany-brown, nearly black, except for an orange-yellow semi-lunar shaped spot at the distal apex of the corium. Pronotum without median longitudinal carina. Claval pruinose area half a long as pruinose area of embolar groove, posteriorly of the point where the media vein curves to costal margin (fig. 131). Male with finger-like projection anterior of seventh abdominal tergite. Fore tarsi elongate, cylindrical. Lansbury (1983) considered *Cnethocymatia* not sufficiently different from *Cymatia* s. str. and gave it subgenus status.

Distribution in Malesia (fig. 572)

Cnethocymatia nigra occurs in North Australia and New Guinea.

Biological notes

Most of the few specimens known from collections are caught with light traps; habitat data on other specimens are lacking. Only Lansbury (1983) studied some specimens in the field, but as these were from a pond with and a pond without macrophytes, the data give few clues about habitat preference. Representatives of *Cymatia* Flor, the other genus in the subfamily occurring in the Northern Hemisphere, are reported to actively catch mosquito larvae and other small invertebrates. Prey is grasped with the fore tarsi (Lansbury, 1983).

REFERENCES – Cymatiainae. Description of *Cnethocymatia*, redescription of *C. nigra* (Jansson, 1982); comments on *Cnethocymatia* and redescription of *C. nigra* (Lansbury, 1983).

SUBFAMILY CORIXINAE

Diagnosis

Small to large (4-15 mm). Rostrum with transverse sulcations; embolar groove and nodal furrow present. Pala in both sexes with palm, in females spoon-shaped, in males variable in shape; not elongate, nearly cylindrical. Claw of hind tarsus inserted subapically. Male usually with strigil.

Distribution worldwide and in Malesia

There are 27 extant genera with about 425 species. Two genera and nine species occur in Malesia.

Genus *Agraptacorixa*

Figures 133-134

Diagnosis

Medium-sized (6.5-10 mm). Pronotum and hemielytra concolorous or with dark spots at base of short hair-like setae, which form a uniform pubescence over both. Media of hemielytra fuses with cubitus before nodal furrow, at a distance equal to or greater than the length of the nodal furrow (figs. 129, 134).

133

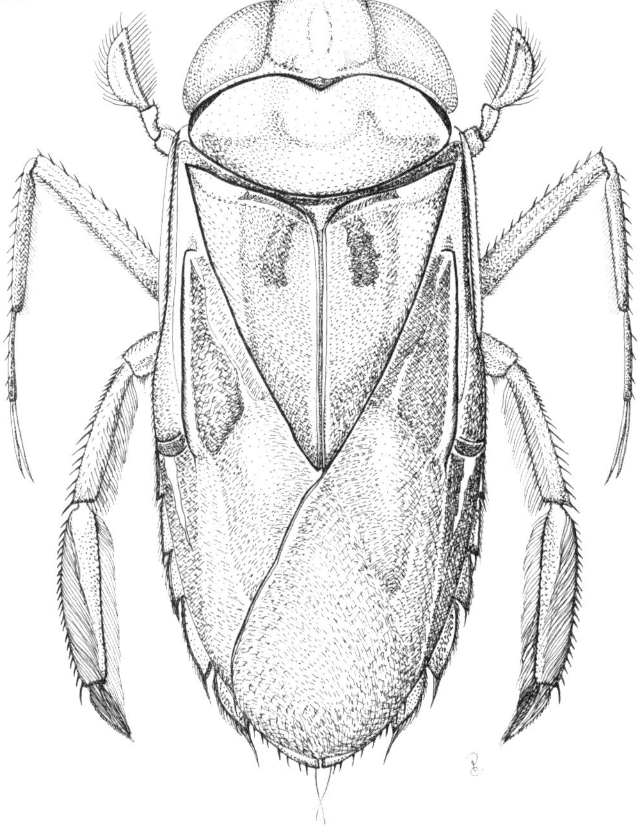

FIGURE 133
Agraptacorixa hyalinipennis, male, length 7.7 mm.

134

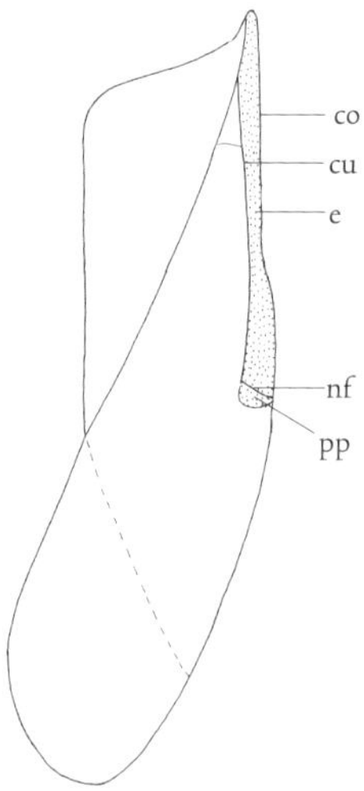

FIGURE 134
Right hemielytron of *Agraptacorixa hyalinipennis* in dorsal view; co = costa, cu = cubitus, e = embolium, nf = nodal furrow, pp = pruinose area posterior of nodal furrow, dotted = pruinose area.

Distribution in Malesia (fig. 572)
Three out of the ten species in this genus occur in New Guinea.

Biological notes
Agraptacorixa hyalinipennis occurs in ponds and sluggish parts of streams. This also seems to be the habitat of the other species. The egg has a long stalk.

Genus *Sigara*

Figure 135

135

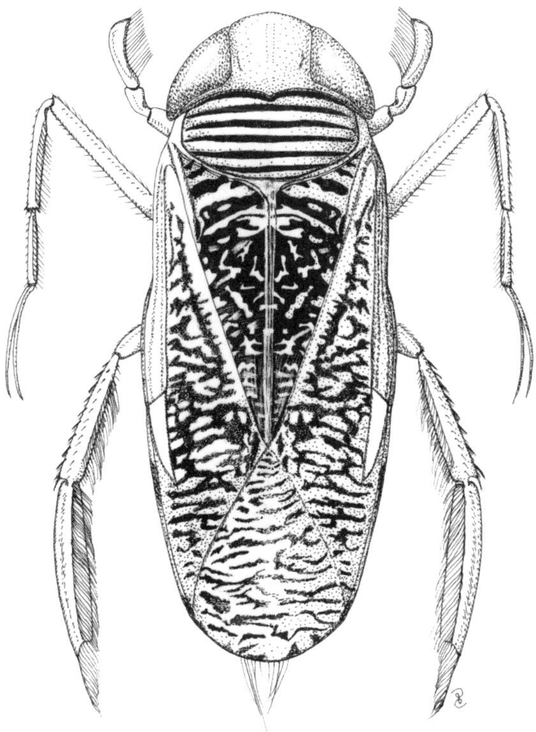

FIGURE 135
Sigara paivai,
male, length 5.0 mm.

Diagnosis
Small to medium-sized (4.5-8 mm). Pronotum brown with transverse yellow stripes; hemielytra brown with transverse interrupted yellow stripes. Media of hemielytra fused with cubitus at nodal furrow (fig. 130). Males have dextral abdominal asymmetry. Apices of clavi in females do not surpass line drawn through margins of hemielytra at nodal furrows. These features are diagnostic only for *Sigara* species occurring in Malesia.

Distribution in Malesia (fig. 572)

The genus has a worldwide distribution, but the number of tropical species is rather low. Of the about 50 species of subgenus *Tropocorixa* six occur in Malesia.

Biological notes

Habitat records indicate that some species prefer ponds and rice-fields, while others inhabit slow-flowing parts of streams. The egg has a short stalk (fig. 122).

REFERENCES – Corixinae. Species from Java and Sumatra (Lundblad, 1933a); notes on habitat and on larval instars of *Agraptacorixa hyalinipennis* (Fernando & Leong, 1963); revision of Australian *Sigara* (Lansbury, 1970); fossil Nepomorpha (Popov, 1971b); revision of Australian *Agraptacorixa* (Knowles, 1974); additions to the New Guinean fauna, redescription of *Cnethocymatia nigra* (Jansson, 1982); fossil Corixoidea (Nel & Paicheler, 1993a).

Family Micronectidae – pygmy water boatmen

Diagnosis

Small (0.8-5 mm), dorsoventrally flattened, dorsally brownish, usually with darker, sometimes indistinct longitudinal stripes. Eyes large; ocelli absent. Posterior margin of head covers anterior margin of pronotum. Rostrum broadly triangular, not segmented but usually with transverse grooves; antennae three-segmented. Scutellum exposed. Hemielytra of uniform texture, membrane without veins. Tarsal formula 1-1-2. Fore tarsus (pala) usually spoon-shaped with long bristles. Palar claw in males variously modified, in females bristle- or spine-like. Middle tarsus with pair of striking long claws. Male abdomen usually with strigil, distallly strongly asymmetrical. Female with fore tarsus and tibia fused; in *Synaptonecta* this is present in both sexes. Reliable identification is often possible for males only. Key structures are much the same as in Corixidae: bristles and spines on fore tibia and pala (figs. 136, 137), the dorsum of abdomen (fig. 141), notably the shape of the free lobe dorsally of the left part of segment 8 (fig. 139), and the genital capsule, especially both parameres (figs. 140, 141). Recently some Australian species were found to be very difficult to separate on preserved material, although they can be distinguished by their acoustic

FIGURES 136 – 137
Fore leg of *Micronecta*.
136 *Micronecta kymatista* male – **137** *Micronecta quadristrigata* female.

136

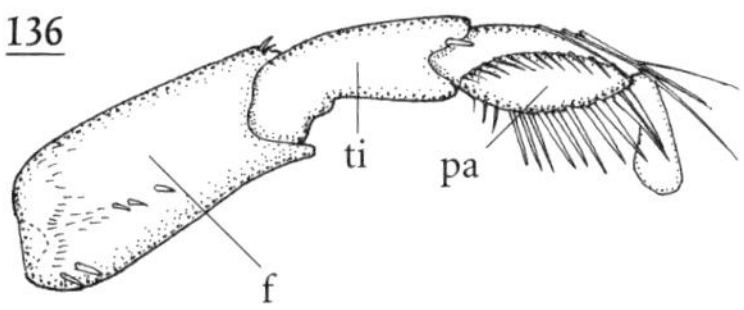

137

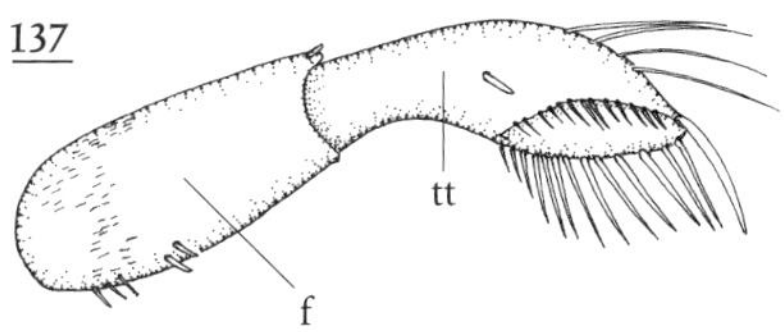

FIGURES 138 – 139
Abdomen of male *Micronecta quadristrigata* in dorsal view.
138 tergites 6-8 – **139** left part of segment 8; l = free lobe of left part of tergite 8, p = position of plectrum, s = strigil.

138

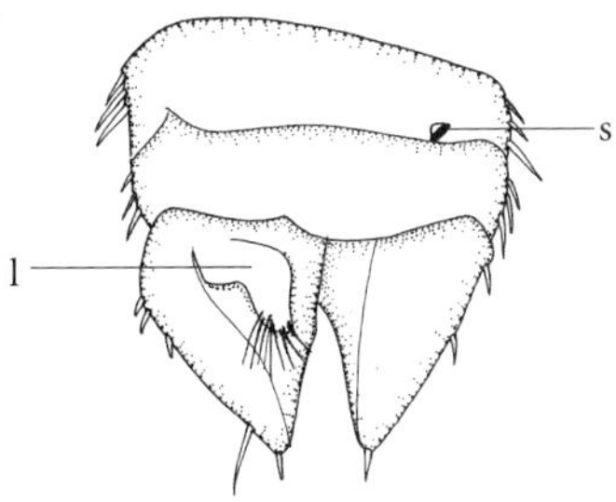

139

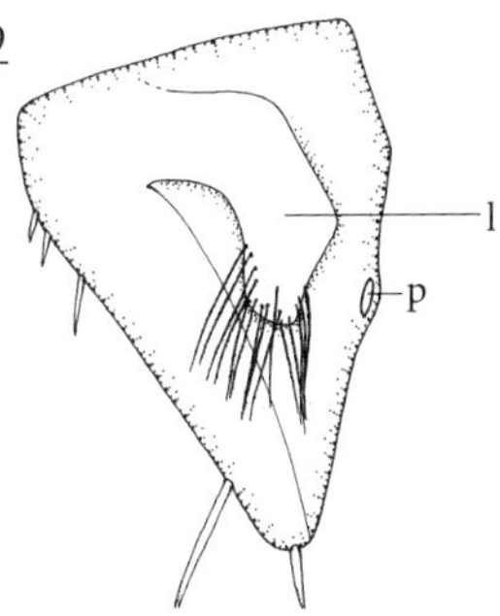

FIGURES 140 – 141
Micronecta kymatista, parameres.
140 right paramere – **141** left paramere; ps = pars stridens.

140

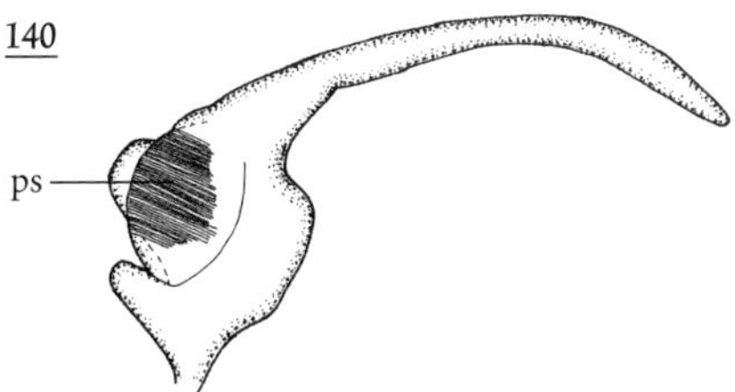

141

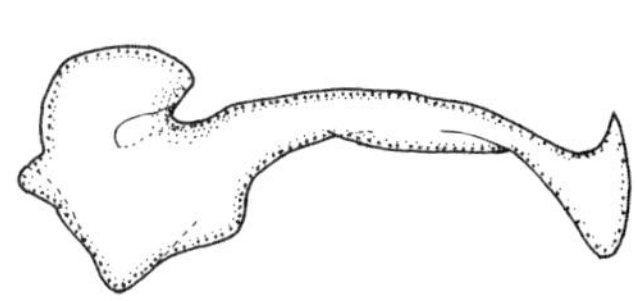

signals (King, 1997). This might also be the case in closely related species of the *M. quadristrigata*-group. Female *Micronecta* can be distinguished from males by having the pala and tibia of the fore leg fused (figs. 136, 137). In *Synaptonecta* the apex of the abdomen is asymmetrical in males and symmetrical in females.

Characteristics

Because of their size and relationship with Corixidae, morphology of Micronectidae has been little studied: it has been tacitly assumed that both are similar. However, Larsén (1938) points out that, although the structures are related, genitalia of both sexes in Micronectidae differ considerably from those of Corixidae. Moreover, *Micronecta* transmits its sperm in a spermatophore, which is not the case in Corixidae; the only other genus of Nepomorpha known to produce spermatophores is *Aphelocheirus*. Parsons (1976) demonstrated that respirational adaptations of the thoracic region are similar to Corixidae. The distribution of hydrofuge hairs also indicates that the air stores in Corixidae and Micronectidae are roughly the same.

Most *Micronecta* males possess a strigil. King (1976) supposed that in *Micronecta* the strigil was involved in stridulation (which is only performed by males). However, Jansson (1989) analyzed stridulation in Micronectidae more thoroughly and found that a ribbed process on the base of the right paramere (pars stridens, fig. 140 ps) is rubbed against one or two ridges near the median edge inside the pocket formed by the left lobe of segment 8 (plectrum, fig. 139 p). Wróblewski (1972a) considered an area of fine ridges near the median margin of the right lobe of segment 8 as the plectrum, and Nieser & Chen (1999) applied this term to the same structure. Jansson (1989) pointed out that since the muscles connect with the genital capsule, this structure thus lies on the 'wrong' place to be a stridulatory organ; the ridges are not developed enough to produce the rather strong sound. He suggested that this area is used to keep the ridges of the pars stridens clean. The stridulating apparatus in *Synaptonecta* is similar to that of *Micronecta* (Jansson & Meyer-Rochow, 1990). J. Polhemus (1994b) used a different terminology, which adds to the confusion: he called the stationary part of the stridulatory mechanism strigil, which is unfortunate, in view of the organ of the same name in Corixidae and Micronectidae, which not function in stridulation. Thus, J. Polhemus' (1994b) strigil is the plectrum of Jansson (1989). In addition J. Polhemus (1994b) considers plectrum and pars stridens synonymous.

Eggs of *Micronecta* differ from those of Corixidae in lacking the adhesive disc and stalk. They are glued with their flattened side on plants or hard substrates. The egg-shell may be beset with short wart-like protrusions (fig. 142). The long filaments reported by Poisson (1938) and figured by Hungerford (1948a) are apparently artefacts. Wróblewski (in Cobben, 1968) suggested they might be filaments of fungi. Eggs of the American vicariant of *Micronecta*, *Tenagobia* Bergroth, do have an adhesive disk and stalk (Hungerford, 1948b).

142

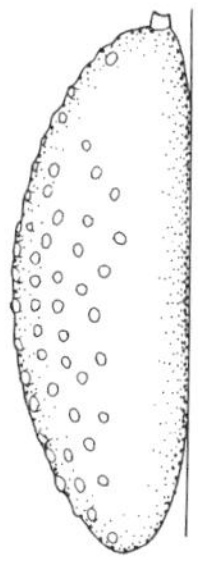

FIGURE 142
Egg of *Micronecta scutellaris*, length 0.5 mm.

Taxonomic note

Micronectidae are closely related to Corixidae and are united with Diaprepocoridae Lundblad (which occurs in Australia and New Zealand) into the superfamily Corixoidea. They are often included in Corixidae, with Micronectidae and Diaprepocoridae reduced to subfamily rank. However, Mahner (1993) has shown that Diaprepocoridae and Micronectidae are quite distinct from Corixidae. Nieser (2002b) re-instated Micronectinae as a family.

Micronecta was divided in six subgenera by Hutchinson (1940), and four more were added by Wróblewski (1962a, 1967). However, as some of the characters used to separate the subgenera are only visible on microscopic slides, and Wróblewski (1968, 1972a, b) did not continue using subgenera, we refrain from using them in this book.

Distribution worldwide

There are about 140 described species divided over three genera. *Synaptonecta* has two species in tropical Asia, and *Micronecta*, with about 110 species, is widespread in the Old World.

Distribution in Indo-Australia

There are 59 species in South-east Asia, Malesia and Australia, of which fifteen (thirteen endemic) occur in Australia including Tasmania. Many Asian species have a large range, though sometimes composed only of two or three localities which are wide apart.

Biological notes

Micronectidae are typically inhabitants of stagnant water, including potholes in streams. Some species prefer running waters but then usually stay in the slow flowing places such as bays or tributary mouths, or shelter behind e.g. boulders. Most prefer scarcely vegetated places and stay at the bottom except when refreshing the air supply. Micronectidae may be very abundant. As most species come regularly and sometimes in large numbers at light, most of the taxonomical work until now has been on specimens from light traps (Wróblewski, 1967) and field data are rather rare. Recently more intensive collecting in the field has been done in continental South-east Asia (Nieser, 2000, 2002a). It was found that some of the species preferring running waters are predominantly brachypterous.

Diet of most species is unknown, which makes it difficult to keep them in aquaria and restricts behavioural studies (Jansson, 1989). However, both Ahmad & Khan (1977) and Clark (1992) reared *M. scutellaris* in the laboratory by providing them with bottom material and decaying organic matter. Ahmad & Khan (1977) found about 24 °C to be the optimal temperature: development from egg to adult took on average 41 days. Gut contents of various Malaysian *Micronecta* species consisted mainly of various unicellular algae, remnants of filamentous algae and some setae of freshwater annelids. In addition bacteria were abundant (Fernando & Leong, 1963). At least one species, *M. ovivora* Westwood feeds on fish eggs. Fernando & Leong (1963) reared *M. quadristrigata* on a diet of *Pleurococcus*, although there was also bottom material from the habitat in the rearing vials. Development from egg to adult took on average 45 days at temperatures varying between 23.2-27.5 °C. The egg, like that of *M. scutellaris* (fig. 142), has short tubular projections on the convex side.

Key to Indo-Australian genera of Micronectidae

1 Vertex concave. Hemielytra of brachypterous morph with strongly reduced apically pointed membranes (fig. 146). Fore tibia and pala of males fused ▸ *Synaptonecta* p. 146

– Vertex convex (rarely flattened). Hemielytra of brachypterous morph moderately reduced, apically truncate (fig. 144). Fore tibia and pala of males separate (fig. 136) ▸ *Micronecta* p. 144

Genus *Micronecta*

Figures 143-145

Diagnosis

Small to large (0.8-4.0 mm). Posterior margin of pronotum straight or convex. Hemielytra with nodal furrow. Membrane of brachypterous morph broadly rounded. In males, pala and tibia of fore leg separate; strigil usually present.

Distribution in Malesia

Distributed throughout the Malesian region. Up to now 31 species recorded, but some more could occur. Eight are widespread, five are thus far only recorded from continental Malaysia, Java has three, Sulawesi two, and Borneo, Leyte and Palawan one possibly endemic species each. Continental Malaysia seems especially rich with sixteen species, although this may be due to recent special interest in the *Micronecta*-fauna of continental South-east Asia (Nieser, 2000, 2002ab, b).

FIGURE 143 – 145

143 *Micronecta haliploides*, macropterous female, length 3.0 mm – 144 *Micronecta haliploides*, brachypterous female, length 2.9 mm – 145 *Micronecta quadristrigata*, macropterous female, length 2.7 mm.

143

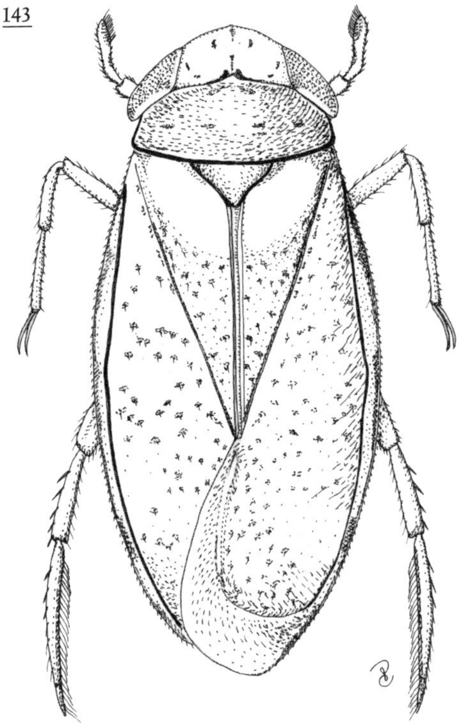

144

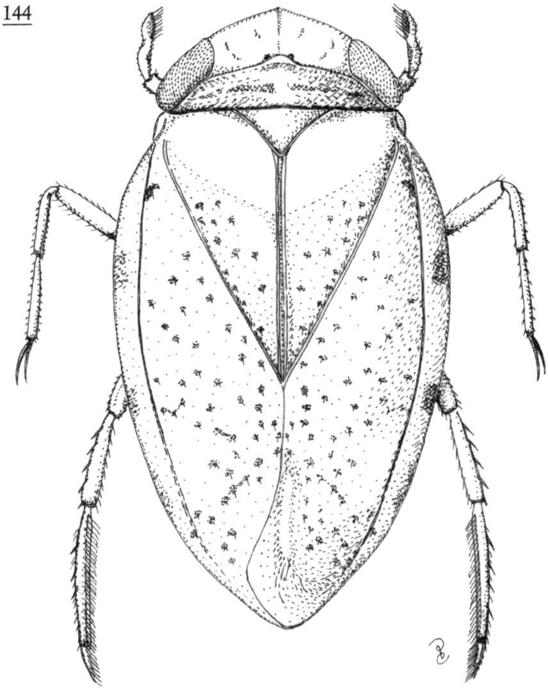

145

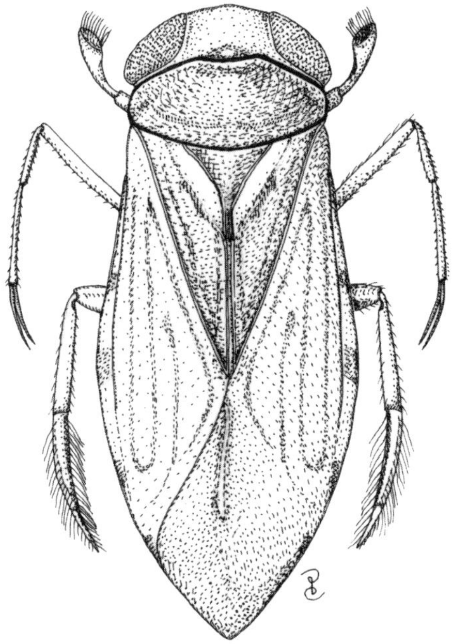

Biological notes

Micronecta species occur in stagnant water, but there are a few records from streams. Widespread species such as *M. haliploides* and *M. quadristrigata* have been reported from rice-fields and other agricultural areas.

Genus *Synaptonecta*

Figure 146

146

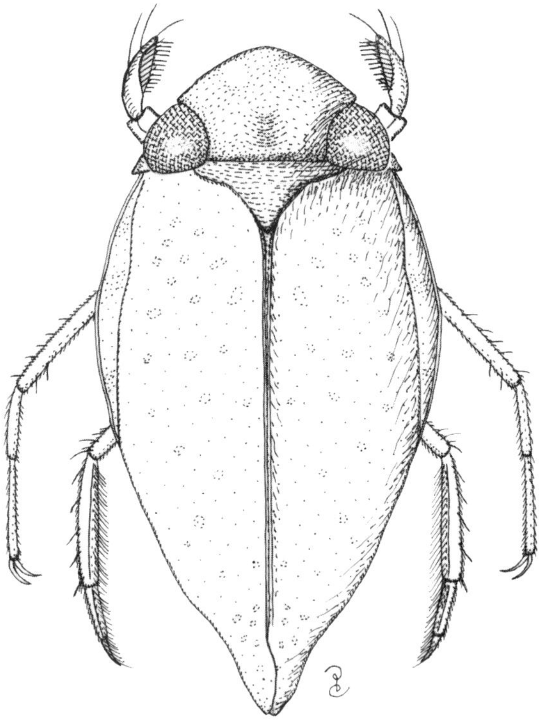

FIGURE 146
Synaptonecta issa, brachypterous female, length 1.7 mm.

Diagnosis

Small to medium-sized (1.9-2.6 mm). Posterior margin of pronotum straight or convex. Hemielytra with nodal furrow. Membrane of brachypterous morph pointed. In males, pala and tibia of fore leg fused; strigil present.

Distribution in Malesia (fig. 573)

Synaptonecta issa, native to India and through continental South-east Asia to Java, turned up recently in an aquarium in New Zealand (Jansson & Meyer-Rochow, 1990) and became established in Florida, U.S.A. It was probably imported with plants from Signapore, where it is common (J. Polhemus & Rutter, 1997). (*Synaptonecta capillata* Wróblewski is endemic to Sri Lanka.)

Biological notes

Synaptonecta issa lives in stagnant water. In Singapore it is common in swamps (Nieser, 2002b). Although Jansson & Meyer-Rochow (1990) report it to be negatively phototactic, and specimens are predominantly brachypterous when collected in the field, Fernando & Cheng (1974) found it abundantly in light trap catches.

REFERENCES – Micronectidae. Life history of *M. quadristrigata* (Fernando & Leong, 1963); checklist of Oriental species (Wróblewski, 1962b, 1968); key to continental South-east Asian species (Nieser, 2000); key to Sulawesian species (Nieser & Chen, 1999); new species from continental Malaysia (Nieser, 2002a); key to continental Malaysian species (Nieser, 2002b); species of the Philippines (Nieser & Chen, 2003).

Family Notonectidae – backswimmers

Diagnosis

Small to medium-sized (3.4–18.0 mm), elongate, boat shaped. Antennae two- to four-segmented, hidden beneath eyes. Eyes large to very large, occupying most of the head, providing almost 360° vision. Rostrum four-segmented, short. Fore leg not raptorial. Tarsal formula usually 2-2-2, but fore tarsus sometimes one-segmented. Hind legs flattened with a well-developed fringe of swimming hairs. Macropterous morph predominant in most species, membrane of hemielytra without veins, in macropterous morph with a fold, thus fitting like a tent over abdominal apex. Abdomen dorsally convex, venter concave with median keel fringed with long setae forming air chambers. Abdominal spiracles 2-8, located latero-ventrally on abdomen within the air chambers, thoracic and first abdominal spiracles slit-like. Abdominal apex symmetrical in both sexes, genital capsule symmetrical, except for parameres in some genera.

Characteristics

The boat-shaped body has apparently to do with the habit of back-swimming, as it is shared with other families swimming on their back: Helotrephidae and Pleidae. Until recently these three families were combined in superfamily Notonectoidea, but nowadays the Helotrephidae and Pleidae are combined in a separate superfamily Pleoidea. All Notonectidae share the structure of the abdominal venter, which is armed with a median keel which, combined with the connexiva, forms a pair of channels. The keel and ridges are beset with dense fringes of setae to close the channels, in which an air store is kept (fig. 147).

147

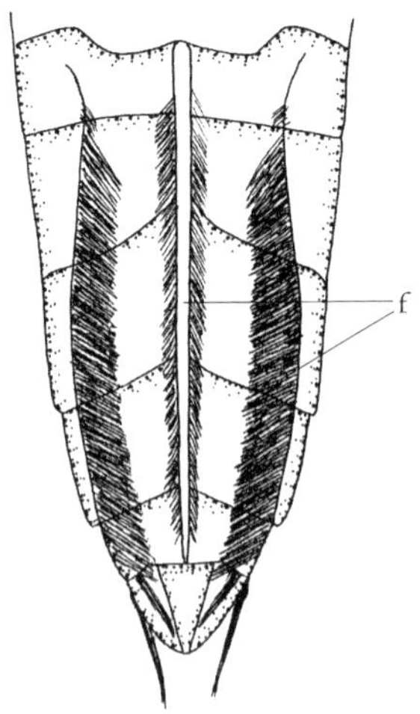

FIGURE 147
Venter of *Anisops*; f = ridges with fringes of setae.

Notonectidae are further distinguished from Helotrephidae and Pleidae by longer hind legs and larger eyes. The very large eyes occupying nearly the entire side of the head give them a very wide field of vision (fig. 5) and set them apart from all other Nepomorpha. Corixidae, which also have large eyes (although relatively smaller than Notonectidae) have a flat, not boat-shaped, body. The abdomen is symmetrical in both sexes. Within subfamily Notonectinae the external difference between the sexes is small. Males of *Enithares* are smaller, as usual in Nepomorpha, and have a more projecting sternite 7 (figs. 148, 149). However, the genital capsule of *Enithares* is large and usually recognizable from the outside. Males of *Nychia* have two-segmented foretarsi (fig. 165), females only one. In *Anisops* the reverse is true (figs. 154, 155). *Anisops* males have a lateral extension on the rostrum (rostral prong; fig. 5) and a stridulatory comb on the fore tibia. The genital capsule is symmetrical in *Enithares* and its shape provides clues to specific identity (fig. 151). The structure will be discussed under Notonectinae. The parameres of *Enithares* are small or very small and only occasionally useful in identification. In *Nychia* the parameres and capsule are diagnostic. In *Anisops* the shape, number and position of bristles and spines of the fore leg of the male, number, size and shape of teeth in the stridulatory comb of the male's fore tibia, and the relative length of the rostral prong are diagnostic. Genital capsule and parameres (in spite of them being relatively large; fig. 150) do not help in identification. Identification keys to species in both *Anisops* and

148

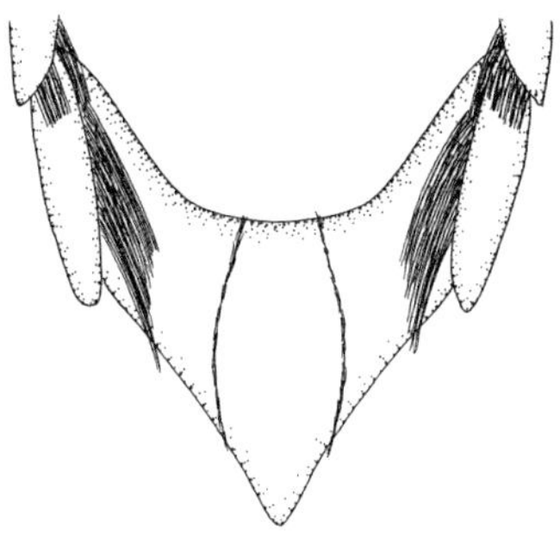

149

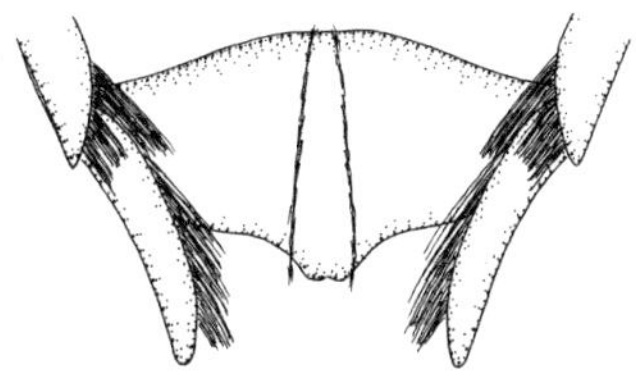

FIGURES 148 – 149
Sternite 7 of *Enithares ektakta*.
148 male – **149** female.

150

151

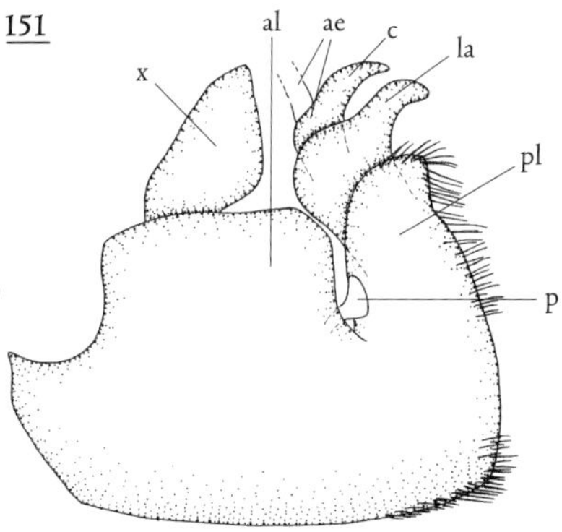

FIGURES 150 – 151
Genital capsules of Notonectidae.
150 *Anisops kuroiwae* –
151 *Enithares charakia*; ae = aedeagus, al = anterior lobe, c = conjunctival appendages, la = lateral arm of basal plate, lp = left paramere, p = paramere, pl = posterior lobe, x = tenth segment.

Buenoa usually refer to males only. Isolated females are often difficult, sometimes impossible, to identify. Other characters used in identification are width of the synthlipsis compared to the anterior width of the vertex, body size, shape of the pronotum, and measurements of the head compared to the pronotum.

Fossil record

There are many fossils of Notonectidae. Most of them are from the Cenozoic period and are related to recent genera, notably *Anisops* and *Notonecta*. Fossil *Anisops* go back to the Oligocene (50 million years ago). A large number of other fossils are either only identifiable as Notonectidae, or their identification as Notonectidae is even doubtful. This is partly because a considerable number are larvae. The only known mesozoic fossils of Notonectidae which have been identified with certain accuracy belong to the extinct subfamily Clypostemmatinae Popov. Clypostemmatinae was synonymized with the tribe Notonectini, and *Clematina* Popov with *Notonecta* L. by Papáček & Štys (1985). Subsequently, Nel & Paicheler (1992) synonymized *Clypostemma* Popov with *Notonecta*. As *Clypostemma* was based on fossils from the Lower Cretacous, *Notonecta* should be the oldest recent genus of Nepomorpha, dating back at least 120 million years.

Distribution worldwide

There are about 350 recent species worldwide, placed in two subfamilies and ten genera. Most species occur in the warmer regions. *Notonecta*, with 83 species and one of the two large genera of Notonectinae, does not occur in Malesia; it is well represented in temperate regions and poorly in warmer areas. About two-thirds of the species in this genus occur in the Americas. The other large genus is *Enithares* with 73 species, of which two in South America, fourteen in Africa and 56 in Asia and Australia. *Enithares* is very similar to *Notonecta* and seems to replace it in the warmer parts of the Old World. *Nychia*, with three species, is distributed in the Old World tropics, and *Aphelonecta* has six South-east Asian species. In Anisopinae, *Anisops* has about 130 species in the warmer parts of the Old World. *Paranisops* Hale has four species, two each in Australia and Thailand (Nieser & Zettel, 2001; Chen et al., 2002). *Walambianisops* Lansbury has one endemic species in Australia.

Distribution in Indo-Australia

Anisops has 81 Indo-Australian species, of which 35 in Malesia, which makes it one of the most speciose areas of Asia for this genus. India has 20 species and continental South-east Asia eleven (Nieser, 1998). *Enithares* has 49 species in Indo-Australia. All six species of *Aphelonecta* occur in Malesia, and only one extends into continental South-east Asia. For distribution *Paranisops* and *Walambianisops* see above.

Biological notes

Notonectidae species swim on their backs, as their vernacular name implies. Static organs, like in Nepidae, Aphelocheiridae and some Naucoridae, are lacking; the equilibrium is controlled by the Johnston's organ.

Notonectidae are generalized predators, hunting on sight or vibrations on the water surface. Prey is grabbed and handled with fore and middle legs, although the fore legs are not raptorial. Especially Notonectinae seem to have a preference for mosquito larvae and pupae (e.g. Papáček, 2000, 2001). Notonectidae also prey on larvae of various fish species living in stagnant waters. This was already noted by Bueno (1905). Both *Notonecta* and *Anisops* can be economically harmful in fish ponds (Papáček, 2000, 2001).

REFERENCES – Notonectidae. Fossil Nepomorpha (Popov, 1971; Nel & Paicheler, 1992); economic importance, with extensive references to the subject (Papáček, 2000, 2001).

Key to Indo-Australian subfamilies and genera of Notonectidae

1 Hemielytral commissure with well-defined hair-lined pit at anterior end (fig. 158) ▸ **Anisopinae 2**

– Hemielytral commissure without well-defined hair-lined pit at anterior end (figs. 165, 160-162) ▸ **Notonectinae**
▸ **Notonectinae 4**

2 Coxal plates of hind legs bare (fig. 152). Male fore tibia proximally with row of stridulatory pegs, usually placed on a protrusion (figs. 154, 156, 157) ▸ 3

– Coxal plates of hind legs covered with long black hairs (fig. 153). Males without stridulatory teeth or pegs on fore tibiae [Australia, Thailand] ▸ *Paranisops*

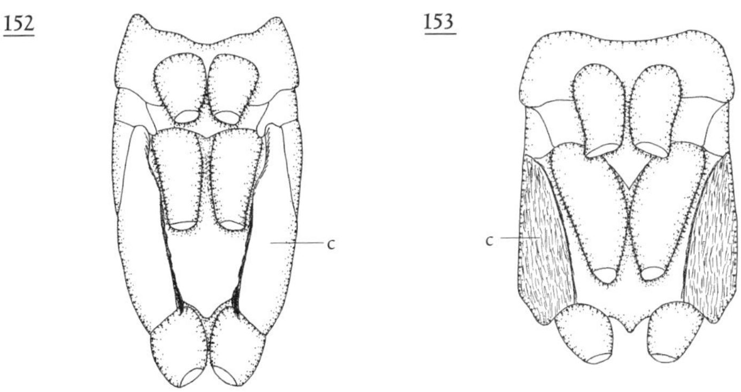

FIGURES 152 – 153
Thoracic venter of Anisopinae.
152 *Anisops* – 153 *Paranisops*; c = coxal plate.

3 Antennae three-segmented. Male rostrum with prominent lateral prong (fig. 5). Proximal stridulatory teeth or pegs on fore tibia of male packed close together and situated on a protuberance (figs. 154, 156) [Paleotropical and subtropical] ▸ *Anisops* p. 154

– Antennae two-segmented. Male rostrum without lateral prong. Proximal stridulatory pegs on fore tibia clearly separate, not situated on a protuberance (fig. 157) [Australia] ▸ *Walambianisops*

4 Anterolateral margins of prothorax foveate (figs. 161, 165) ▸ 5

– Anterolateral margins of prothorax not foveate. Middle femur with subapical pointed protuberance (fig. 14) [cosmopolitan, predominantly in temperate and subtropical zones, not recorded from Malesia] ▸ *Notonecta*

5 Middle femur with subapical pointed protuberance (fig. 14) [pantropical] ▸ *Enithares* p. 160

– Middle femur without subapical pointed protuberance ▸ 6

154

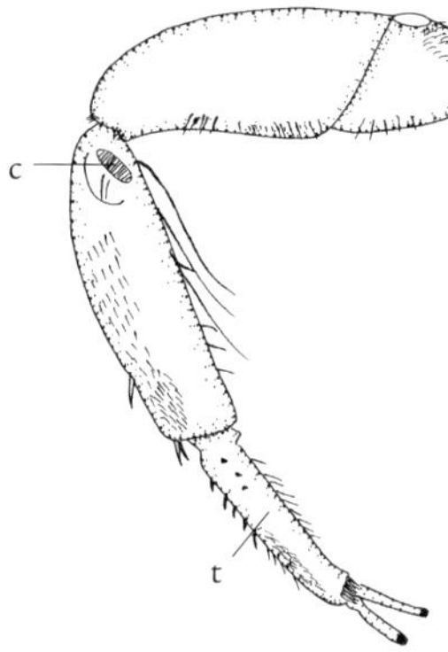

155

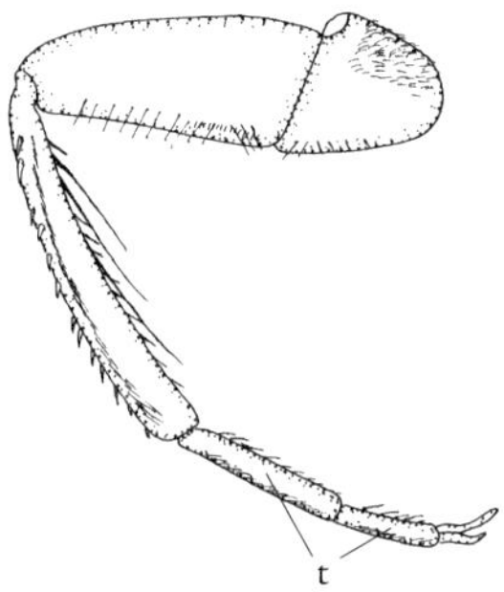

FIGURES 154 – 155
Fore leg of *Anisops kuroiwae*. 154 male – 155 female; c = stridulatory comb, t = tarsus.

156

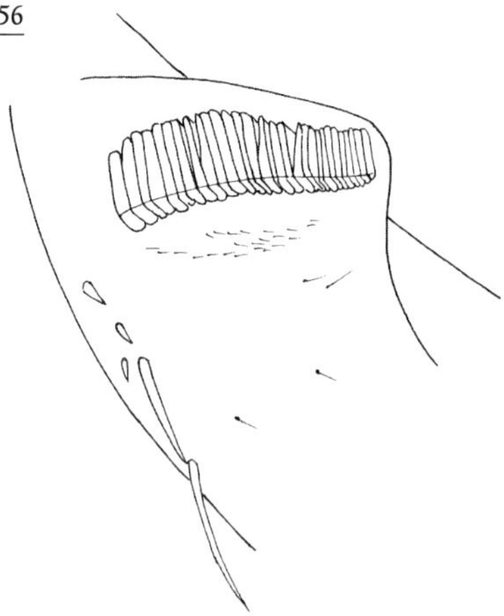

157

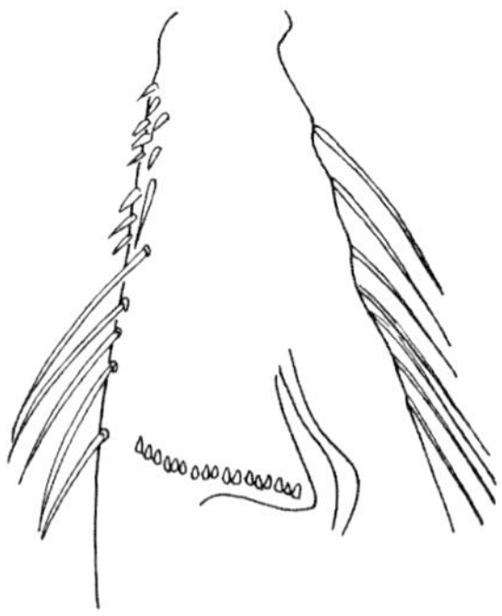

FIGURES 156 – 157
Detail of fore tibia of male Anisopinae.
156 *Anisops rhomboides* – 157 *Walambiopsis wandjina* Lansbury (157 after Lansbury 1984).

6 Eyes basally contiguous, forming an ocular commissure (fig. 165) [Paleotropical]
‣ *Nychia* p. 163

– Eyes basally widely spaced (fig. 160) [South-east and insular Asia]
‣ *Aphelonecta* p. 158

SUBFAMILY ANISOPINAE

Diagnosis

Small to large (3.4-10.6 mm). Antenna two- or three-segmented. Male rostrum usually with lateral prong. Claval commissure anteriorly with a large sensory pit lined with setae. Fore tibia of male usually with stridulatory pegs. Segments 3-7 usually with haemoglobin-containing cells. Parameres usually asymmetrical.

Characteristics

Specific identification of *Anisops* is usually only possible for males. Contrary to *Paranisops* the genital capsule and parameres in *Anisops* are very uniform (fig. 150) and of no use in identification. Useful identification characters are found in the anterior projection of the vertex (fig. 158), shape and size of the rostral prong (fig. 5), number and shape of the teeth of the stridulatory comb (fig. 154), and the shape of the fore femur. In addition, various measurements, notably the ratio of the anterior width of vertex/synthlipsis, are used in identification.

Biological notes

Anatomy of *Anisops sardeus* Herrich-Schäffer was studied by Poisson (1926). His main discovery were cells associated with tracheoles ventrally in the proximal part of the abdomen. These cells contain a pigment which, according to Poisson, is very close to haemoglobin. Poisson (1926) overlooked a paper by Hungerford (1922), who had already noted the presence of haemoglobin in *Buenoa margaritacea* de la Torre Bueno. Independent of Poisson, Bare (1928) made an extensive study of the morphology of *B. margaritacea* with emphasis on these haemoglobin cells. Miller (1964) found that the cells contain the oxygen store consumed during a dive. As *Anisops* and *Buenoa* are structurally very similar, these studies complement each other.

The external gas store is relatively smaller in *Anisops* than in *Enithares* and *Notonecta*. This smaller gas store enables *Anisops* to have a more or less neutral buoyancy at a specific depth. This depth where they float seems to depend on the species, and the neutral buoyancy is not perfect. At the start of a dive there is a tendency to float, which is counteracted by active swimming, as in Notonectinae. Later on there is a stage of neutral buoyancy, during which leg movement is used only for horizontal movement. At the end of the dive the bug starts to sink gradually, which is counteracted by occasional strokes. Apparently, the bugs do not have active control over the gas store, and gas

diffuses out just as in reservoir breathers. The near-neutral buoyancy makes them one of two insect types which are planktonic (the other one being larvae of Chaoboridae [Diptera: Nematocera]).
In Anisopinae, tibiae and tarsi of fore and middle legs are on both sides beset with long spine-like bristles and form a 'planctonic crib' (Hungerford, 1920). This structure enables them to catch and handle planctonic Crustacea, apparently their preferred food.

Genus *Anisops*

Figures 158-159

158

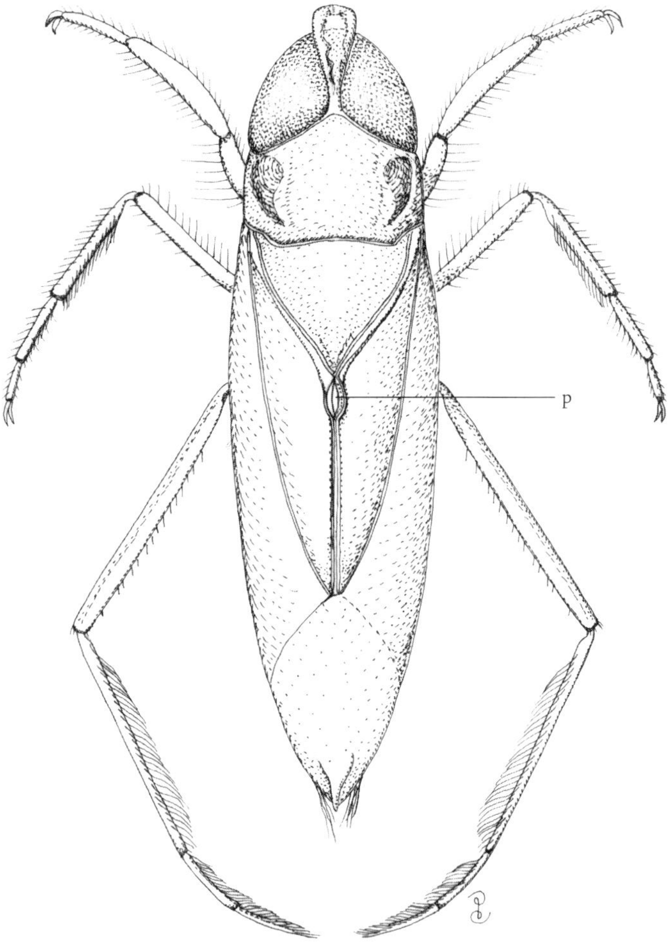

FIGURE 158
Anisops nasutus,
male, length 7.2 mm;
p = hair lined pit at anterior end of hemielytral commissure.

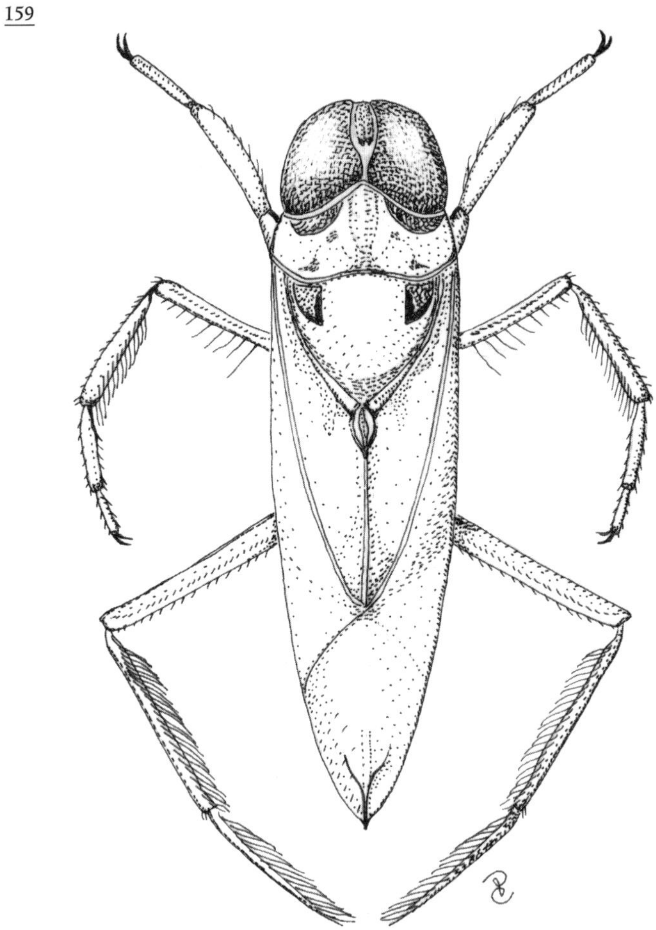

FIGURE 159
Anisops paracrinitus,
male, length 4.8 mm.

Diagnosis

Small to large (4.2–10.6 mm). Dorsally pale yellowish to blackish, scutellum in several species orange to reddish, specimens killed and stored in alcohol tend to become unicolorous sordid whitish. Usually macropterous, but some species with brachypterous morph. Antennae three-segmented. Coxal plates of hind legs bare. Ventral abdominal keel extending onto last abdominal segment. Male rostral prong present on third rostral segment (except in one African species); stridulatory pegs on fore tibia packed closely together and placed on a ridge; fore tarsi one-segmented; parameres asymmetrical.

Distribution in Malesia

Distributed throughout Malesia. Of the 35 Malesian species 20 are widespread. Five are endemic to New Guinea, and five to the Bismarck and Solomon Archipelagos.

Biological notes

Anisops species inhabit stagnant or virtually stagnant water. Some species prefer rock pools or pools along streams, but several, especially various widespread, species occur in ponds in villages and in rice fields.

Males of several *Anisops* have been recorded to stridulate and probably all except one African species are able to do so. To stridulate they rub the tibial comb against the rostral prong. An extensive analysis of sound production in *Buenoa* is given by Wilcox (1975). However, in many *Buenoa* the sound producing organs are more complicated than in *Anisops*. The sound of a single specimen of at least larger species is audible when one sits close to the aquarium. Stridulation is used in courtship and has been described for *A. thienemanni* (misidentified as *A. hyperion* Kirkaldy) by Hale (1923). A male stridulating rapidly during courtship positions itself beneath and slightly behind a female. In this position it follows every movement of her for a while until it tries to grasp her. If a female is not responsive the male will follow another one nearby. Copulation takes an hour or more and they swim together with the male below and slightly to the right of the female. The finger-like anterior claws help the male to maintain its embrace.

Eggs are elongate oval with a thin, transparent and smooth wall, except for an area at the anterior end which shows weak hexagonal sculpturing. This area forms a lid and is positioned at the opening of the slit in the plant; it opens at hatching. Eggs are deposited singly in plants. A female saws a hole in a stem and deposits one egg, leaving a small portion of the anterior part exposed. As all species have similar well developed ovipositors this probably holds for all species. However, Lahr (2000) and Lahr et al. (2000) found several species of *Anisops* in Senegal developing in temporary ponds without aquatic or emergent vegetation; where the eggs were deposited remained unknown. Larval development in South Australia of *A. thienemanni*, a medium-sized species (6-7 mm), takes about two months. Leong (1962) reared *A. breddini*, which is of similar size, in a laboratory in Singapore and found a development time of two to three months. He suggests that this slightly longer development may be due to lower temperatures with greater fluctuations compared to field conditions.

Food consists mainly of Entomostraca and mosquito larvae, but when hungry anything is taken. Young larvae also prey upon Rotifera, and are in turn preyed upon by older larvae and adults of their own species (Gilbert & Burns, 1999). The long setae along the posterior margin of the fore femur, tibia and middle femur, and the anterior margin of the middle tibia form a sieve with which Entomostraca are caught.

Although they do eat mosquito larvae their way of life makes them less suitable for mosquito control than Notonectinae. In addition, some species of *Anisops* have been used for evaluating toxic effects of pesticides used against locusts in Senegal (Lahr, 2000). On the other hand, some species can be harmful in fish ponds by eating fish larvae and small fish, but especially by competing for food (Gorai & Chaudhuri, 1962; Leong, 1962).

REFERENCES – Anisopinae. Biological notes on *A. thienemanni* (Hale, 1923); anatomy of *A. sardeus* (Poisson, 1926); study of haemoglobin cells in *Buenoa* (Bare, 1929); revision of *Anisops* (Brooks, 1951); additional records of *Anisops* (Brooks, 1952, 1954; Lansbury 1962, 1963, 1964a, 1965a, 1967, 1968a); feeding habits of *Anisops bouvieri* (Gorai & Chauduri, 1962); life history of *Anisops breddini* (Leong, 1962); function of haemoglobin cells in *Anisops* (Miller, 1964); key to males of Sulawesian *Anisops* (Nieser & Chen, 1999); key to males of Philippine *Anisops* (Zettel, 2003b).

SUBFAMILY NOTONECTINAE

Diagnosis

Small to large (4.0-18.0 mm). Antennae three or four-segmented. Claval commissure simple without anterior hair-lined sensory pit. Abdomen without haemoglobin containing cells. Male parameres usually symmetrical.

Characteristics

Although the European *Notonecta glauca* L. has been used in a large number of physiological and several biological and detailed morphological studies, there is apparently no extensive morphological study of other representatives of Notonectinae. Lansbury (1968b) gives some information on *Enithares*, but this is restricted to external and genital features useful in identification. *Nychia* was discussed by Lansbury (1985).

Specific identification in *Enithares* and *Nychia* is most reliable when based on male genitalia. In *Enithares* the shape of the genital capsule (fig. 151) is diagnostic. The capsule has a deep lateral cleft dividing it in an anterior and a posterior lobe. The parameres (p) are placed in the cleft; they are small and rarely diagnostic. The shape of the posterior lobe (pl) is also specific. In addition, the shape of the conjunctival appendages (c) at the apex of the phallosoma of the aedeagus may be diagnostic. In the apical part of the posterior lobe the aedeagus lies in the lateral arms of the basal plate (la). The pilosity of the fore and middle legs of the males is usually also diagnostic. Females may be identified on size, colour and relative measurements of body parts, and specific recognition is hence less reliable. Basic structure of *Aphelonecta* genitalia is similar to that of *Enithares*.

Biological notes

Respiration in Notonectinae is a combination of reservoir and air bubble breathing. The reservoir is situated mainly on the abdominal venter in two longitudinal bristle-covered grooves. Gas exchange takes place mainly in a gas film covering the hemielytra. Because they carry a large amount of air, Notonectinae will float unless they swim actively or hold on to something. They use two main ways of waiting for prey: some species hold on to a perch, other species float against the surface. Prey living under water is located by sight, prey on the surface by ripples made by the prey. The ripples are detected by sensory organs in the legs. Once detected, the bug slowly swims towards the source until it sees it, after which it strikes. Prey lying in front or to the side – up to five or six times the bug's length – is detected sooner than prey behind it (Giller & McNeil, 1981; Markl & Wiese, 1969). For *Enithares* this has not been verified but since structure and, at least for some species, floating behaviour is very similar it can be assumed that they hunt in the same way.

In stagnant pools inhabited by *Enithares* and *Notonecta*, ripples are likely to be caused by potential prey. Data on *Nychia* are not available but its New World vicariant *Martarega* detects prey on the surface by sight (Gittelman 1975). This is supported by opaque prey (e.g. ants, mosquito pupae) on or against the surface being more readily selected as prey than translucent or transparent prey (mosquito larvae). However, *Nychia* is found more often in isolated ponds than *Martarega*, so verification is needed.

Although Notonectinae will accept any suitable prey, their favourite prey is mosquito larvae and pupae. Actual observations on effectiveness in biocontrol are restricted to *Notonecta*. Especially populations of mosquito larvae living in open water can be virtually wiped out; apparently those associated with vegetation are less vulnerable (Blaustein et al., 1995). These authors studied *N. maculata* Fabricius, a species less effective near vegetation than other *Notonecta* species (Giller & McNeil, 1981). Mosquitoes are not only directly reduced by *Notonecta*. Chesson (1984) showed that *Culex* mosquitoes oviposit less in containers holding *Notonecta*. This type of biological control has some restrictions: firstly, Notonectinae fly easily, so one has to regularly restock the population. Secondly, alternative prey may considerably reduce predation on mosquito larvae (Chesson, 1989). Notonectidae also prey on fish larvae living in stagnant water, as already noted by De la Torre Bueno (1905). Both *Notonecta* and *Anisops* can be harmful in ponds for fish culture (Papáček, 2000, 2001).

Genus *Aphelonecta*

Figure 160

Diagnosis

Rather small to large (5.7-14.7 mm). Dorsally varying from predominantly pale yellow to anteriorly yellowish and posteriorly (scutellum, hemielytra) blackish. In some species pale and dark morphs occur. Antennae four-segmented. Eyes large, reniform, inner margins in dorsal view converging but widely spaced posteriorly. Pronotum broader than long with anterolateral margins distinctly foveate. Corium with thin cleft just anterior of membrane (nodal furrow), only known in macropterous morph. Middle femur without pointed subapical

160

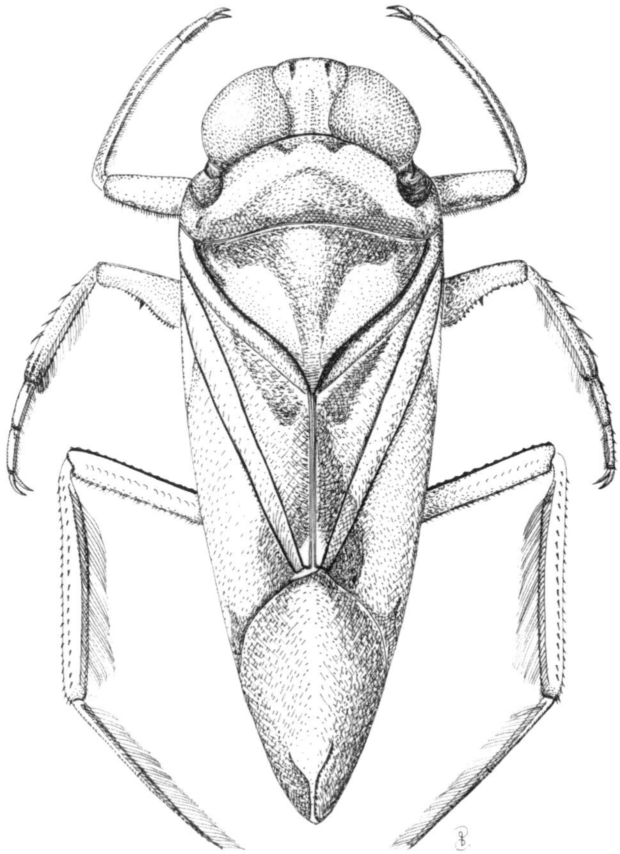

FIGURE 160
Aphelonecta alexis,
male, length 8.9 mm.

protuberance. Male genital capsule divided in anterior and posterior lobe. Parameres symmetrical, well developed, situated in cleft, dorsally reaching apex of posterior lobe. Female with short gonapophyses.

Distribution in Malesia (fig. 573)

There are six species and all are reported from Malesia (Malaysia, Brunei and Palawan).

Biological notes

The few data on habitat suggest that *Aphelonecta* species occur in similar habitats as most *Enithares*: virtually stagnant parts of streams. This, combined with morphological similarity to *Enithares*, suggests that biology is similar.

Genus *Enithares*

Figures 161-164

161

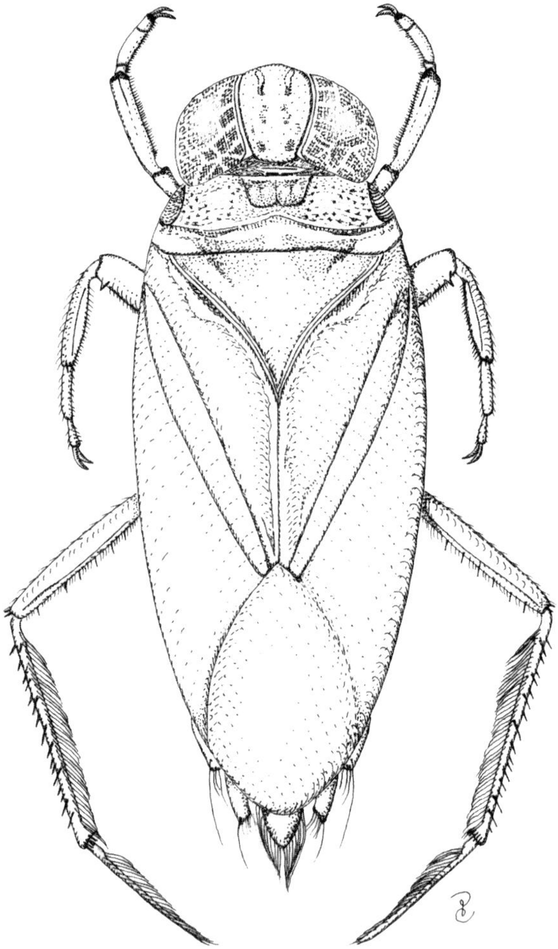

FIGURE 161
Enithares bakeri,
male, length 8.2 mm.

Diagnosis

Medium-sized to large (6-17 mm). Dorsal coloration varying from pale yellowish to blackish. Most species with light and dark morph, the latter often uniform blackish, but sometimes patterened with blackish marks. Antennae four-segmented. Eyes large, reniform, inner margins in dorsal view converging but widely spaced posteriorly. Pronotum broader than long with anterolateral margins distinctly foveate. Corium with thin cleft just anterior of membrane (nodal furrow; fig. 16). Middle femur with pointed subapical protuberance (fig. 14). Male genital capsule divided in anterior and posterior lobe (fig. 151). Parameres symmetrical, usually small, situated in cleft. Female with short gonapophyses.

162

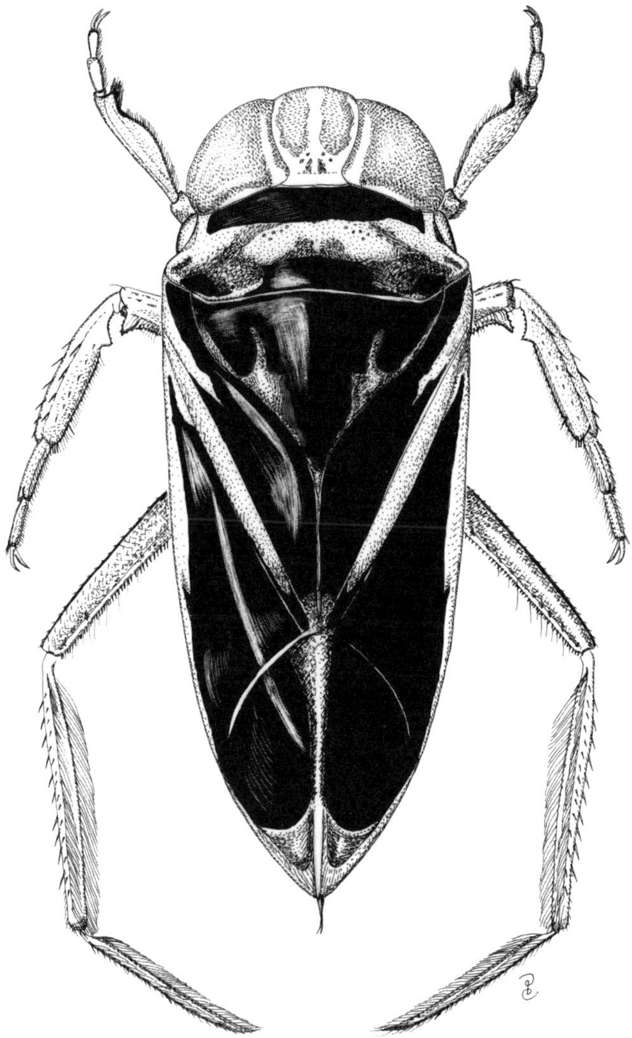

FIGURE 162
Enithares charakia,
male, length 12.5 mm.

Distribution in Malesia

The genus is distributed throughout the Malesian region. Of the 49 Indo-Australian species 44 occur in Malesia, of which nine in New Guinea (eight endemic) and twelve in Sulawesi (ten endemic). In Malesia the ranges tend to be limited, except for a few, usually rather small, species such as *E. bakeri* and *E. paramegalops*. The large number of species on Sulawesi, apparently with very similar habitat preferences, is striking (fig. 163).

Biological notes

Little is known of biology and ecology of *Enithares* species. At least two Malesian species, *E. nigra* and an undescribed species, have a stridulatory mechanism in

163

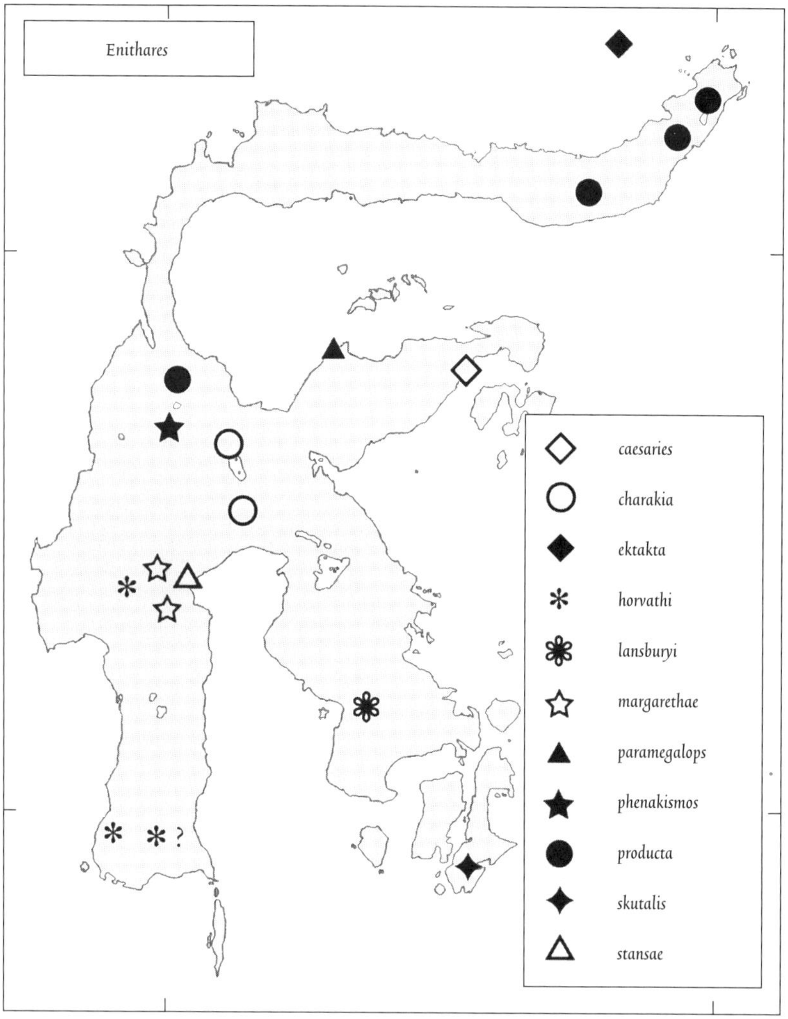

FIGURE 163
Distribution of *Enithares* in Sulawesi.

both sexes. It consists of ridges ventrally in the basal part of the connexiva against which a basal posterolateral ridge on the hind femur is rubbed (J. Polhemus, 1994b).

Apparently, two types of species with different ecological strategies occur in Sulawesi. Some smaller species (7.5-8.5 mm; fig. 161) are found mainly along roadsides, in village ponds and in pools without fish. Several larger species (9.4-12.8 mm; fig. 162) live in virtually stagnant parts of small streams or in pools associated with forest streams. One usually encounters a few late instar larvae or one or two adults in a pool. This led Nieser & Chen (1991) to the following hypothesis on the life history of stream-associated species: gravid *Enithares* females seek out suitable pools to deposit a number of eggs. The larvae resort to cannibalism when food supply becomes exhausted, so at least a few adults are produced. This strategy is also known from *Notonecta* (Fox, 1975a, b). Some intermediate-sized species, like *E. malayana* and species of the *E. producta*-group, also seem to be intermediate in habitat selection (Nieser & Chen, 1996; Nieser, 2004).

There is significant variation in the number of eggs per female. This variation may represent different reproductive strategies. Eggs were observed to be glued on submerged plants (Lansbury, 1968). Since all *Enithares* have rather short gonapophyses this is probably the rule. The eggs, which vary in size, have a simple elongate shape with rounded ends. Larvae of *Enithares*, as far as known, are pale with a dark transverse band (fig. 164).

164

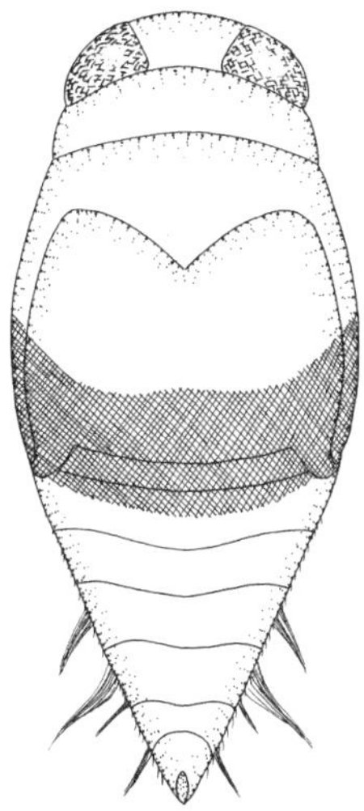

FIGURE 164
Larva of *Enithares*, appendages removed.

Genus *Nychia*

Figures 165-166

Diagnosis

Small (4.0-6.0 mm). Dorsally whitish. Most often brachypterous. Antennae three-segmented. Eyes large, reniform to semicircular, inner margins in dorsal view converging and touching each other in posterior third. Pronotum wider than long with anterolateral margins distinctly foveate. Corium without nodal furrow. Middle femur without pointed subapical protuberance but with pair of strongly developed bristles ventrally in basal half (fig. 166). Male genital capsule not divided in anterior and posterior lobe; parameres asymmetrical, small, situated on dorsal margin of capsule. Female with short gonapophyses.

165

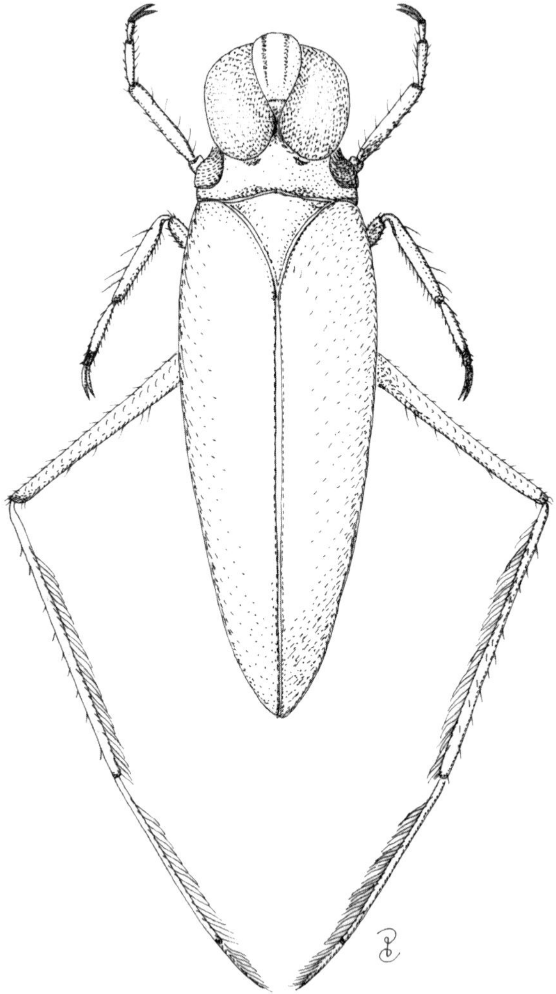

FIGURE 165
Nychia sappho, brachypterous male, length 4.0 mm.

166

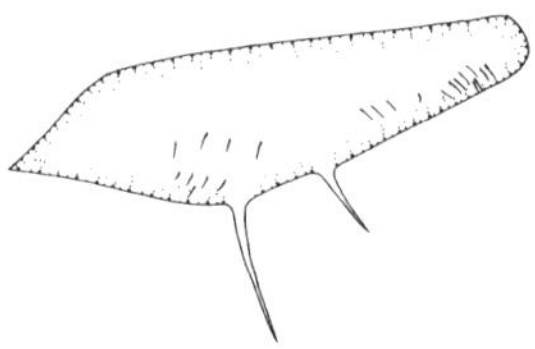

FIGURE 166
Middle femur of *Nychia sappho*, female, in ventral view.

Taxonomic note

Apart from the single Malesian species four more have been described from tropical Asia, but two of these have been synonymized by Lansbury (1985). The status of *N. limpida* Stål from China (Canton area) remains unresolved. *Nychia infuscata* Paiva is probably a synonym of *N. sappho*, as extensive material from Chiang Mai, Thailand, close to the type locality of *N. infuscata* (Inle Lake, Myanmar) contained only *N. sappho* (Nieser, unpublished). *Nychia sappho* is white with part of the hemielytra translucent, especially in the brachypterous morph. Females may show one or two dark spots on the median margin of the hemielytra.

Distribution in Malesia (fig. 573)

One species, *N. sappho*, occurs in Malesia.

Biological notes

Biology virtually unknown. The usual habitat is nearly stagnant stretches of streams or, in the dry season, potholes in the streambed. However, *Nychia* species have also been collected in ponds not connected with a stream. They rest against the surface film or swim slowly just below it. When disturbed they dive.

REFERENCES – Notonectinae. Revision of *Enithares* in tropical Asia and Australia (Lansbury, 1968); identity of Oriental *Nychia* (Lansbury, 1985); key to *Aphelonecta* (Zettel, 1995a); key to Sulawesi males of *Enithares* (Nieser & Chen, 1996).

Family Helotrephidae – hemispherical backswimmers

Diagnosis

Small (0.9-3.7 mm), roundish, semiglobular or flattened. Antenna 1- or 2-segmented (fig. 167), but lacking in brachypterous *Fischerotrephes*. Eyes relatively small, flat. Head and prothorax fused into a cephalonotum (figs. 228-234), with w-shaped suture usually present on dorsal surface (figs. 228, 233, 234). Labrum reduced; labium 4-segmented, with apical labellum. Thorax and anterior abdominal sternites with ventromedian laminate carina (figs. 223-225), secondarily reduced in benthic genera. Mesoscutellum usually large, but small in Fischerotrephinae; hemielytron entirely coriaceous, lacking distinct veins, apically overlapping (figs. 204, 229, 234; but relatively short in Fischerotrephinae, fig. 228). Macropterous morph with claval suture (fig. 205). Hind wing brachypterous morph without claval suture (fig. 204) with long hemielytra and with very reduced alae, in most species more common than macropterous morph. Tarsal formula 3-3-3 (basic), 1-1-2 (in most genera) or 2-2-3 (rare). Pretarsi with two claws, a pair of short parempodia, and at least with ventral arolium. Male genitalia (figs. 168-194) strongly asymmetrical, with rotated genital capsule; right paramere situated dorsally. Aedeagus variously modified, often asymmetrical. Parameres of different shape: right paramere shorter and narrower than left one. Female genitalia symmetrical or asymmetrical (to varying degree), usually with modified sternite 7 (subgenital plate) (figs. 211-222); ovipositor present but reduced.

167

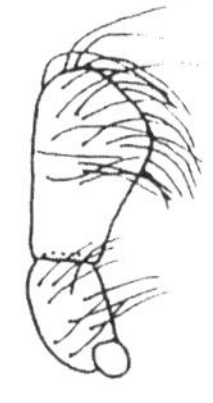

FIGURE 167
Hydrotrephes corporaali, antenna (adapted from China, 1930).

Characteristics

Helotrephidae are easily recognizable by the unique fusion of the head with the prothorax, forming the so-called cephalonotum (figs. 228-234). Mahner (1993) listed the following autapomorphies of the family within his superfamily Notonectoidea (= Notonectoidea + Pleoidea in this book): 1) cephalonotum with w-shaped cephalo-pronotal suture; 2) reduced ovipositor; 3) antennae reduced, with a maximum of two segments; 4) labrum reduced, membranous, more or less transparent; 5) propleuron lacking coxal fissure; 6) genital capsule

FIGURES 168 – 194
Male genital structures in Helotrephidae.

168, 174, 180, 186, 192 aedeagus; 169, 175, 181, 187, 193 right paramere; 170, 176, 182, 188, 194 left paramere; 171, 177, 183, 189 aedeagus; 172, 178, 184, 190 right paramere; 173, 179, 185, 191 left paramere

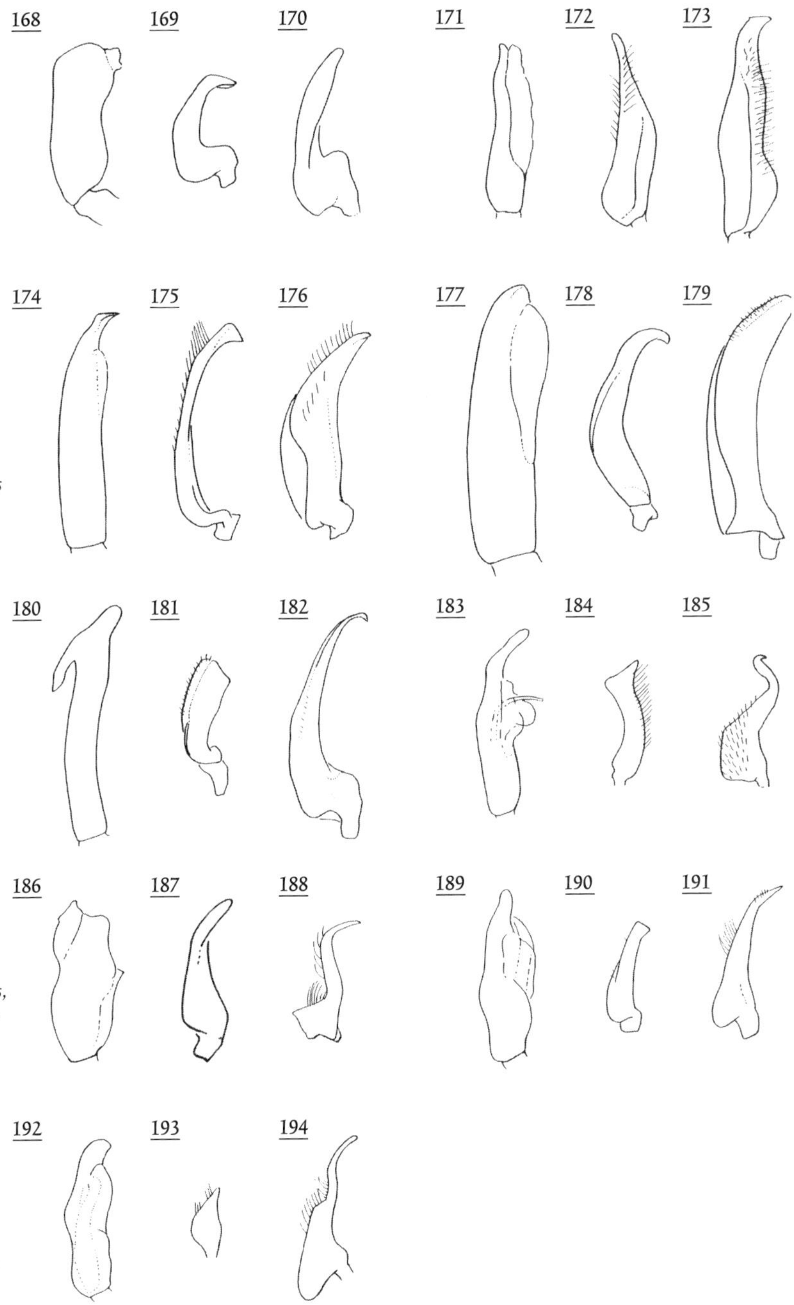

168 – 170 *Fischerotrephes depressus*, male; **171 – 173** *Platytrephes insulanus*, male.

174 – 176 *Helotrephes otoeis*, male; **177 – 179** *Hydrotrephes yangae*, male.

180 – 182 *Hydrotrephes jani*, male; **183 – 185** *Idiotrephes chinai*, male.

186 – 188 *Tiphotrephes indicus*, male; **189 – 191** *Distotrephes* (s.str.) *stysi*, male.

192 – 194 *Distotrephes* (*Polhemotrephes*) *sarawakensis*, male.

(of male) 90° rotated to the left (left paramere turned to a ventral position); 7) (?) first abdominal tympanum enlarged. Pleidae have been proved to be the sister taxon of the Helotrephidae by 26 characters (Mahner, 1993).
Some special characters in the classification of Helotrephidae are found in the hemielytra which completely cover the abdomen in all wing morphs. Caudally on a hemielytron there is a thickened ridge which keeps the hemielytra together, called the pseudomembrane (figs. 204, 205, pm). In most genera macropterous morphs are recognized by a claval suture in the hemielytra, which is absent in (hind wing-)brachypterous morphs. In addition macropterous morphs have a small suture at the base of the hemielytra near the ventroposterior angle of the cephalonotum. This suture delimits a small area, called clavulus (fig. 205), which is not or poorly developed in brachypterous morphs. In Trephotomasinae there is a suture in the hemielytra of both brachypterous and macropterous morphs which delimits the so-called pseudendocorium (fig. 204, pe). The pseudendocorial suture lies in about the same place as the claval suture, but as macropterous specimens of Trephotomasinae also have a true claval suture, the two are clearly not homologous.
Main characters for generic and suprageneric classification are: the tarsal formula (= number of tarsal segments of fore, middle and hind leg (figs. 198-203)); the lateral margin of the cephalonotum sometimes running through the eyes (figs. 206-210); the structure of the thoracic and abdominal venter (especially median carinae; e.g. figs. 223-225), which are highly diagnostic; the female subgenital plate (= sternite 7; figs. 211-222). Size, colour pattern, characters of ventral carinae and pronotal plate (figs. 195-197), and microstructures (punctation, micropunctation, reticulation) of the dorsum help to distinguish species, or to associate males and females of the same species. However, size, body shape, eye development, and colouration differ considerably between brachypterous and macropterous morphs in some species.

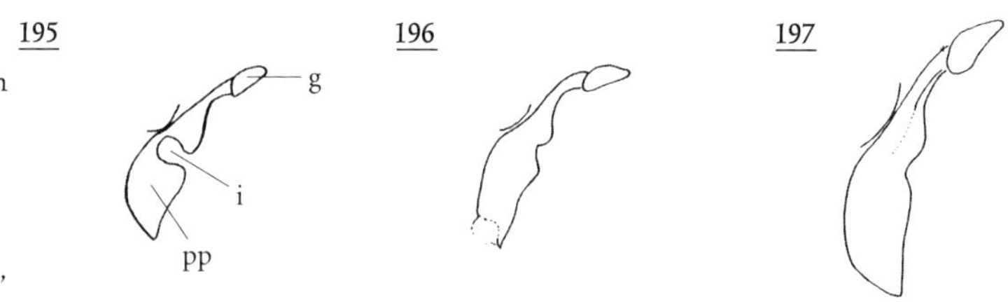

FIGURES 195 – 197 Pronotal and genal plate in Helotrephidae.
195 *Helotrephes australis* –
196 *Helotrephes otoeis* –
197 *Hydrotrephes martini*;
g = genal plate, i = incision, pp = pronotal plate.

Taxonomic notes and historical review

This chapter follows the phylogenetic concept by Papáček et al. (1988), J. Polhemus (1990a), and Mahner (1993), as the main problems pointed out by Mahner (1993) and Zettel (1997a) do not heavily concern the Malesian fauna, except for the position of the Limnotrephini (see below).

During the last fifteen years no other family of Nepomorpha has gone through such considerable progress in morphology, taxonomy and systematics than the small family Helotrephidae. However, the first classification of the family was published by Esaki & China (1927, 1928). Then, after detailed studies by W.E. China in the 1930's and early 1940's (e.g., China 1930, 1940), no thorough research was carried out until the 1980's, except for some descriptions of African taxa by R. Poisson and R. Linnavuori. The revival of Helotrephidae systematics started with the innovative papers by Papáček et al. (1988, 1989, 1990) which also included the description of the new subfamily Trephotomasinae. At the same time J.T. Polhemus (1990a) wrote an important paper on the Helotrephinae which contains the establishment of the new tribe Limnotrephini and a complete bibliography of the family. Mahner (1993) took up and elaborated the ideas presented in those papers, and created a phylogenetic concept of the Helotrephidae. Zettel (1994a) described a fifth subfamily, the Fischerotrephinae, from Malaysia. Further, he discussed the relationships of the tribes Helotrephini and Limnotrephini of the subfamily Helotrephinae with the African Idiocorinae (Zettel, 1997a). All Asian genera have been recently revised (Zettel, 1996a, 1999d, 2005b; Papáček & Zettel, 2000, 2001, 2003; Zettel & Polhemus, 1998; J. Polhemus & D. Polhemus, 2003). Nine species groups within the largest genus *Hydrotrephes*, were erected by Zettel (1998b, 2000a, 2003a, 2004f). In the last ten years numerous Oriental Helotrephidae were described by various authors (see references of Helotrephidae). Nowadays, the understanding of Helotrephidae has changed dramatically: it is, at least, in South-east Asia, an unexpectedly diverse family in morphological and biological regards, and offers interesting data for zoogeographical analysis of the area.

Distribution worldwide

Helotrephidae contains over 170 described species, mainly in tropical areas of the Old and New World. Highest species diversity is found in the Oriental region (J. Polhemus, 1990a). The family is not known to occur east of Weber's Line. Five subfamilies are recognized, of which Neotrephinae are restricted to the Neotropics and Idiocorinae to East Africa and South India.

Distribution in Indo-Australia

From India to the Philippines and Sulawesi four subfamilies with fourteen genera occur. Ten genera and about 120 species belong to Helotrephinae; Fischerotrephinae (one genus, six described species), Trephotomasinae (three genera, seven species) and Idiocorinae (one genus, two species) are small taxa. *Mixotrephes* (Limnotrephini) is distributed from Iran to Thailand, *Trephotomas* (Trephotomasinae) occurs in southern China and Indochina. Nine genera are reported from Malesia. Only *Hydrotrephes* extends its distribution east of Wallace's Line.

Biological notes

Compared with taxonomic knowledge, very little is known on the biology of Helotrephidae. Only two species have been studied more extensively: *Helotrephes semiglobosus formosanus* (Esaki & Miyamoto) by Miyamoto (1952), and *Idiotrephes asiaticus* by Papáček (1993). Fragmental notes are available for some other species, notably *Hydrotrephes balnearius* (Takahashi, 1941). Data on habitat preference of species can be found in several revisionary papers (see references below).

Helotrephidae are strictly subaquatic, inhabiting mainly lentic parts of running waters but also stagnant waters (rock pools, even small artificial water bodies such as ceramic bonsai bowls). Several genera are known to live benthic in unpolluted streams (*Fischerotrephes*, Trephotomasinae, *Distotrephes*) or lakes (*Idiocoris, Paskia*; e.g. Miyamoto 1952; Papáček et al., 1988; Zettel, 1994a, 1997c). Species of *Idiotrephes* and *Tiphotrephes* tolerate poor water quality. *Idiotrephes chinai* has been reported from hot springs in Sumatra and peninsular Malaysia (Lundblad, 1933a; Miyamoto, 1952), *Mixotrephes thermophilus* from similar habitats in Thailand (Papáček & Kovac, 2001b).

Larger distances are covered by swimming venter up (as in Pleidae and Notonectidae) or venter down, rowing with the hind legs (Miyamoto 1952). Benthic species are probably mainly creeping on the substrate, other species are found on vertical surfaces (banks) or between floating macrophytes. Flight has not been observed and there are no records of Helotrephidae collected at light.

Oxygen is usually taken directly from the air. A film of air working as a plastron covers the venter of *Helotrephes semiglobosus formosanus* (Miyamoto, 1952). At least in some benthic species oxygen is taken exclusively from a thin plastron from the water as in Aphelocheiridae. This is suspected for *Idiocoris, Paskia*, *Trephotomas* (Papáček et al., 1988), *Fischerotrephes* (Zettel, 1997c), and *Distotrephes*.

Helotrephidae feed on small invertebrates. Under laboratory conditions a large variety of aquatic organisms is accepted by *Idiotrephes asiaticus*.

Key to South-east Asian subfamilies, tribes, genera and subgenera of Helotrephidae

1 Tarsal formula 3-3-3 (first segment very short and difficult to see; figs. 198, 201). Very small (up to 1.6 mm), body depressed, ovate (fig. 228) (subfamily Fischerotrephinae) ▸ *Fischerotrephes* p. 175

– Tarsal formula 2-2-3 (first segment very short) or 1-2-2. Usually globular, rarely flat ▸ 2

2 Hemielytron with 'pseudendocorium' reaching pseudomembrane (fig. 204) in all morphs, and additionally with claval suture in macropterous specimens; tarsal formula 2-2-3 (figs. 199, 202), cephalonotum very large, much larger than hemielytra in dorsal and lateral view (fig. 229). (subfamily Trephotomasinae) ▸ 3

– Hemielytron without 'pseudendocorium', but macropterous morph with claval suture not reaching pseudomembrane but crossing lateral side of mesoscutellum (fig. 205); tarsal formula 1-1-2 (figs. 200, 203); cephalonotum usually much smaller (figs. 230-234) (subfamily Helotrephinae) ▸ 4

198

199

200

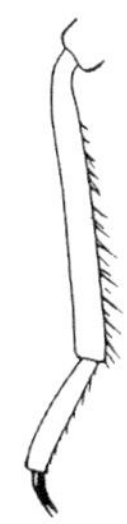

FIGURES 198 – 200
Fore tibia and tarsus of Helotrephidae.
198 *Fischerotrephes depressus* – **199** *Platytrephes insulanus* – **200** *Helotrephes otoeis*.

201

202

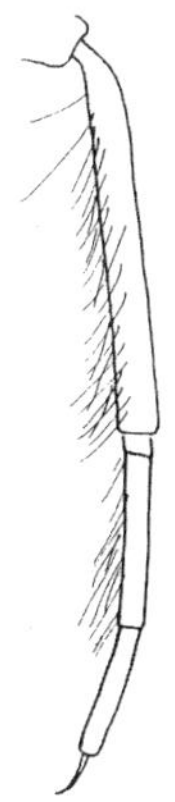

203

FIGURES 201 – 203
Hind tibia and tarsus of Helotrephidae.
201 *Fischerotrephes depressus* – **202** *Platytrephes insulanus* – **203** *Helotrephes otoeis*.

204

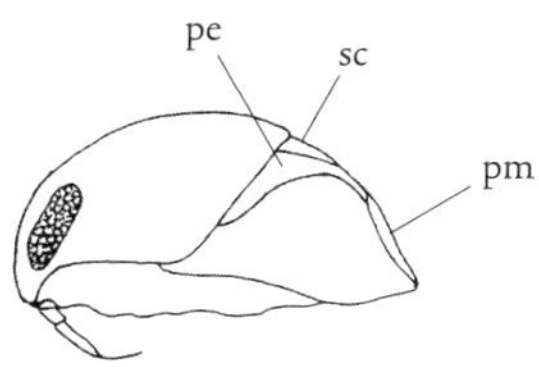

205

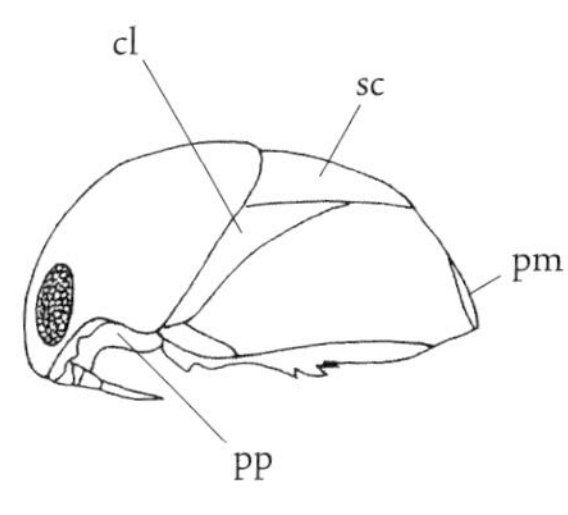

FIGURES 204 – 205
Helotrephidae in lateral view.
204 *Trephotomas compactus*, brachypterous specimen – **205** *Hydrotrephes* sp., macropterous specimen; cl = clavus, pe = pseudendocorium, pm = pseudomembrane, pp = pronotal plate, sc = scutellum.

3 Length 1.5 mm; dorsum brown; eyes extremely small ▸ *Papacekia* p. 177
– Length more than 2.2 mm; dorsum with yellow pattern; eyes relatively large; distance between hind coxae larger than width of coxa ▸ *Platytrephes* p. 177
– Length 1.9-2.0 mm; dorsum with yellow pattern; eyes relatively large; distance between hind coxae larger than width of coxa ▸ *Trephotomas* p. 178

4 Large (body at least 1.95 mm, in most species over 2.2 mm), cephalonotum usually densely punctated and therefore rather dull. Postero-lateral margin of cephalonotum continued ventrally of the eye, not dividing or indenting it (figs. 204, 206). Female subgenital plate usually symmetrical (figs. 213-218) (tribe Helotrephini) ▸ 5
– Small (body less than 2.1 mm, in Malesian species less than 1.9 mm); cephalonotum usually with small scarce punctations and appearing shiny. Postero-lateral margin of cephalonotum not continuing ventrally of the eye, either dividing or indenting it or pointed to the centre of the eye (figs. 207-210). Female subgenital plate usually more or less asymmetrical (figs. 219-222) (tribe Limnotrephini) ▸ 7

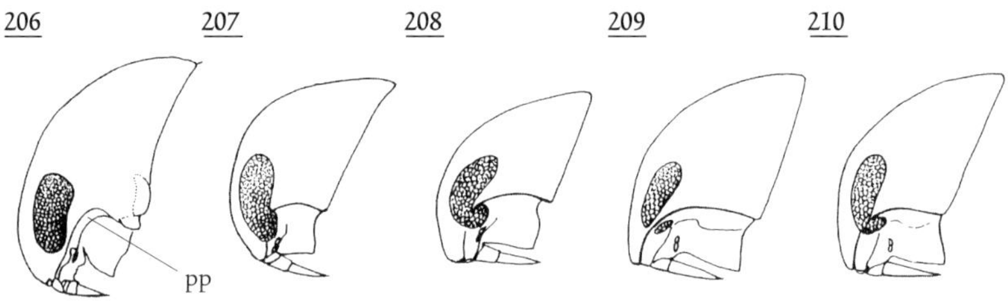

FIGURES 206 – 210 Cephalonotum of Helotrephidae in lateral view. **206** *Helotrephes otoeis* – **207** *Tiphotrephes indicus* – **208** *Distotrephes (Polhemotrephes) heveli*, macropterous morph – **209** *Distotrephes (Polhemotrephes) heveli*, brachypterous morph – **210** *Idiotrephes chinai*; pp = pronotal plate.

5 Abdominal sternites 4-5 (or 4-6) with median carina (fig. 223) ▸ *Helotrephes* p. 181
– Abdominal sternites 5-6 without median carina, sternite 4 usually without median structure (fig. 225), rarely with a small tooth (fig. 224) ▸ 6

6 Medial carina of mesosternum terminating in sharp apex. In frontal view anterior margin of head broadly rounded (fig. 227). Female subgenital plate asymmetrical, often with slender medial lobe (fig. 218) ▸ *Ascetotrephes* p. 180
– Medial carina of mesosternum either low or (more frequently) with laminate distal zone. In frontal view anterior margin of head usually narrowly rounded, never as broadly as in fig. 227. Female subgenital plate (sub-)symmetrical, variously modified, but without slender lobe (figs. 215-217) ▸ *Hydrotrephes* p. 183

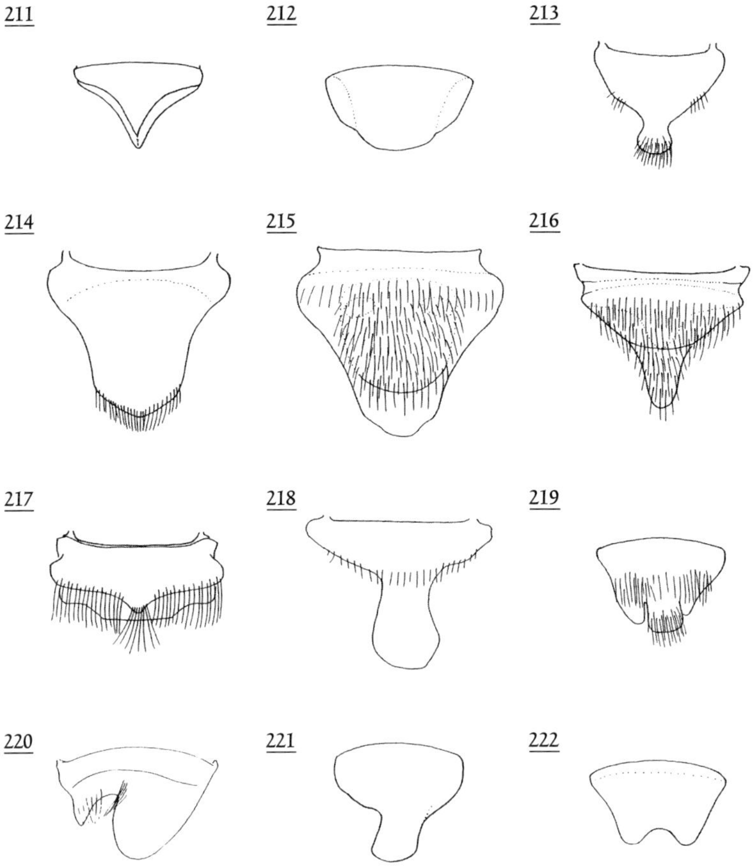

FIGURES 211 – 222
Female subgenital plate in Helotrephidae.
211 *Fischerotrephes depressus* – **212** *Platytrephes kodadai* – **213** *Helotrephes otoeis* – **214** *Helotrephes australis* – **215** *Hydrotrephes yangae* – **216** *Hydrotrephes corporaali* – **217** *Hydrotrephes martini* – **218** *Ascetotrephes mesilau* – **219** *Tiphotrephes indicus* – **220** *Idiotrephes chinai* – **221** *Distotrephes* (s.str.) *asymmetricus* – **222** *Distotrephes* (*Polhemotrephes*) *sarawakensis*.

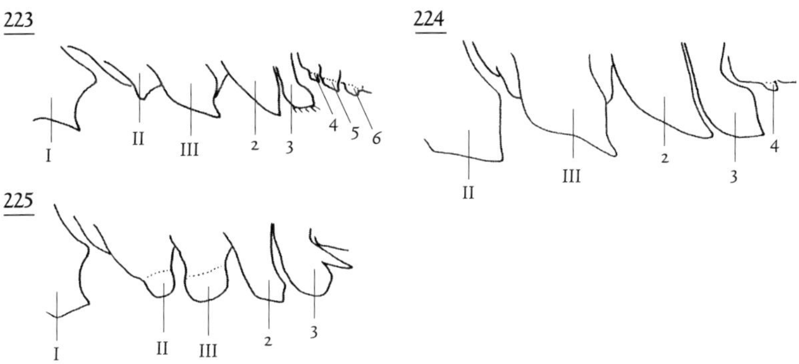

FIGURES 223 – 225
Ventromedian carinae in Helotrephidae.
223 *Helotrephes australis* – **224** *Hydrotrephes yangae* – **225** *Hydrotrephes jani*; I, II, III = prosternal, mesosternal and metasternal carina, 2 - 6 = carinae on abdominal sternites 2 - 6.

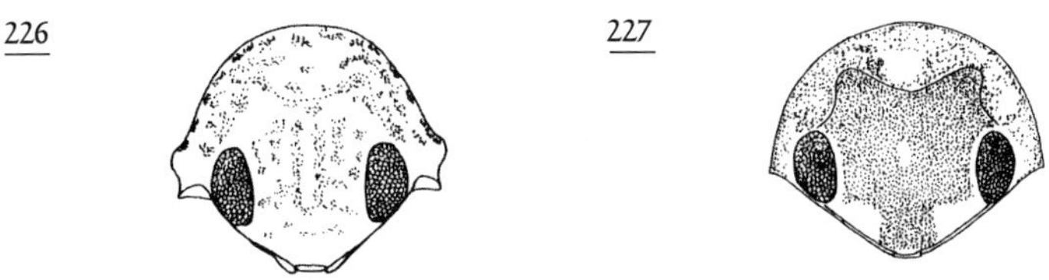

FIGURES 226 – 227
Head in frontal view.
226 *Helotrephes otoeis* – **227** *Ascetotrephes mesilau*.

7 Eye not divided by lateral margin of cephalonotum, but indented, sometimes deeply, in posterior half (figs. 208, 210). Metasternal carina simple ▸ 9

– In brachypterous morph eye completely divided in a dorsal and ventral part by the lateral margin of cephalonotum (fig. 209); in rare macropterous morph these parts are connected in anterior fourth (fig. 208). Metasternal carina in posterior half with two branches forming a diamond-shaped structure ▸ *Distotrephes* p. 185 8

8 Female subgenital plate with a more or less asymmetrical middle lobe (fig. 221). Aedeagus with narrow apex (fig. 189), left paramere with acuminate apex (fig. 191); right paramere with rounded, truncate, or acuminate apex (fig. 190) ▸ **subgenus** *Distotrephes* **p. 185**

– Female subgenital plate symmetrical, with hind margin medially concave (fig. 222). Aedeagus with broadly rounded apex (fig. 192), left paramere with slender distal part (fig. 194), right paramere strongly reduced, with acuminate apex (fig. 193) ▸ **subgenus** *Polhemotrephes* **p. 186**

9 Lateral margin of cephalonotum not extending onto eye surface, only indistinctly indenting the eye at posterior margin (fig. 207). Male left paramere distally with distinct bent apex (fig. 188). Subgenital plate of female nearly symmetrical, with two narrow incisions laterally of a median lobe (fig. 219). Predominately macropterous; body in both morphs highly domed ▸ *Tiphotrephes* **p. 188**

– Lateral margin of cephalonotum clearly extending onto eye surface, deeply indenting the eye (figs. 210). Male genitalia and female subgenital plate very different (e.g. figs. 183-185, 220). Brachypterous morph less domed ▸ 10

10 Left paramere similar to figure 185; female subgenital plate (usually strongly) asymmetrical, with more or less deep dextrocaudal incision or impression (fig. 220) ▸ *Idiotrephes* **p. 187**

– Left paramere relatively simple, triangular. Female subgenital plate subsymmetrical, without dextrocaudal incision, impression or break, with more or less distinct posteromedial lobe ▸ *Mixotrephes*

This genus has thus far not been recorded from Malesia. Nearest occurrence in North Thailand and Myanmar.

REFERENCES – Helotrephidae. Erection of family Helotrephidae (Esaki & China, 1927); revision of Helotrephidae (Esaki & China, 1928); description of some Helotrephinae (China, 1930); description of *Idiotrephes* (Lundblad, 1933a); biological notes on *Hydrotrephes balnearius* (Usinger, 1937); survey of South American Helotrephidae (China, 1940); biological notes on *Hydrotrephes* (Takahashi, 1941); biology of *Helotrephes formosanus* (Miyamoto, 1952); erection of subfamily Trephotomasinae (Papáček et al., 1988); morphology of Helotrephidae (Papáček et al., 1988); description of *Distotrephes* and world checklist of Helotrephidae (J. Polhemus, 1990a); cladistic analysis of phylogeny of Nepomorpha (Mahner, 1993); morphological notes on *Idiotrephes* (Papáček, 1994); description of *Fischerotrephes* (Zettel, 1994a); new species of *Hydrotrephes* from Sulawesi (J. Polhemus, 1997); notes on Helotrephidae from Borneo (Zettel, 1997c); revision of *Helotrephes* (Zettel & Polhemus, 1998); *Hydrotrephes*, new species and key to Sulawesi species (Nieser & Chen, 1999); survey with checklist and key to species of Borneo (Zettel, 2000a); survey and key to South American Helotrephidae (Nieser & Chen, 2002); survey of Helotrephidae of the Philippines (Zettel, 2003a); review of Trephotomasinae (Zettel, 2005b).

SUBFAMILY FISCHEROTREPHINAE

This subfamily contains only one Oriental genus. Diagnosis of the subfamily is therefore identical with the generic diagnosis. Despite numerous apomorphic characters connected with benthic life, Fischerotrephinae is a phylogenetically basic group of Helotrephidae.

Genus *Fischerotrephes*

Figure 228

228

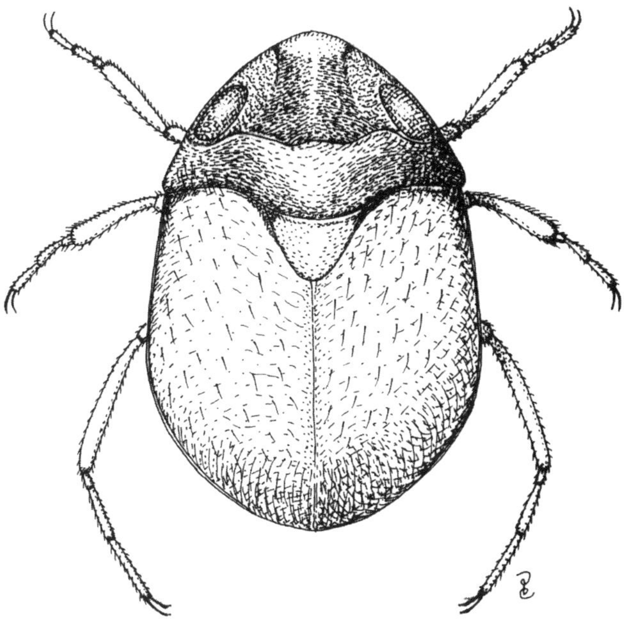

FIGURE 228
Fischerotrephes maryatiae, paratype, brachypterous male, length 1.4 mm.

Diagnosis

Small (1.3-1.6 mm), flattened. Dorsally varying from yellowish to dark grey. Antennae absent in brachypterous morph, one-segmented in macropterous morph. Cephalonotum short, about 1/3 of body length, very short behind eye (fig. 228); cephalonotal suture absent. Eye very small, divided by lateral margin of cephalonotum (brachypterous morph) or small and deeply incised posteriorly (macropterous morph). Rostral segment 4 more than twice as long as segment 3. Mesoscutellum small, triangular (fig. 228). Median carinae of thoracic sterna low, on mesosternum and metasternum with rhombic foveae. Hemielytron densely sculptured and matt, with scattered, long, appressed hairs. Legs without pectinate bristles; tarsal formula 3-3-3; segment 1 of all tarsi very short, segments 2 and 3 subequal in length (figs. 198, 201). Mesofemur of male with group of modified bristles. On abdomen only sternite 2 with carina. Aedeagus weakly sclerotized (fig. 168); parameres diagnostic (figs. 169, 170); female subgenital plate small, triangular, with very acute apex (fig. 211). Larvae flat.

Distribution in Malesia (fig. 574)

Seven species, of which six described, are known from tropical Asia, of which three have been recorded from Malesia. More species are expected (Zettel 1994a, 1997a, d, and unpublished material). Because of their small size and cryptic life, *Fischerotrephes* are rarely collected; three species are known only from single specimens, and five only from a single locality.

Biological notes

Fischerotrephes depressus and *F. maryatiae* inhabit fine gravel sediments of brooks and streams, in Sarawak and Sabah respectively. Several generic characters are adaptations to benthic life in streams, convergently developed, e.g., in *Trephotomas*, *Distotrephes*, Aphelocheiridae and certain Naucoridae. All known *Fischerotrephes* specimens, with the exception of one (unpublished) female from Laos, are brachypterous, which is probably an adaptation to stable habitats. The body is strongly flattened to reduce resistance of the body against water flow. Plastron respiration is suspected which enables specimens to stay at the stream bed. Biology is unknown.

REFERENCES – Fischerotrephinae. Description of *Fischerotrephes* (Zettel, 1994a); comparison of Bornean species (Zettel, 1997c).

SUBFAMILY TREPHOTOMASINAE

Diagnosis

Usually strongly flattened, very small to medium-sized (1.5-2.8 mm). W-shaped suture of cephalonotum present. Antennae two-segmented. Femora and tibiae with pectinate bristles; tarsal formula 2-2-3 (figs. 199, 202). Cephalonotum large, nearly circular (fig. 229), with continuous lateral margin ventral of eye (fig. 204); propleuron concave, in lateral view covered by pronotum, not divided into lateral pronotal plate and propleural plate. Hemielytron with pseudendocorium (unique characteristic) separated from corium, pseudendocorium reaching apex of hemielytron and there connected with pseudomembrane (fig. 204). Ventral carinae absent or low; metaxiphus with or without median carina, H-shaped or trapezoidal, postero-laterally enclosing anterior part of the sternite 2; abdominal sternites 3-6 without or with very low median carina. Ventral laterotergite 8 of female large, flap-like, with postero-lateral margin of segment 8 articulated. Larvae flat.

Taxonomic note

Trephotomasinae are a well-defined clade by tarsal formula 2-2-3, structures of thoracic and abdominal sterna and presence of pseudendocorium.

Distribution worldwide

Trephotomasinae are restricted to South-east Asia and Borneo.

Genus *Papacekia*

Diagnosis

Small (1.5 mm), moderately broad, moderately convex. Dorsum virtually uniform brown. Eyes very small and narrow. Mesoscutellum small. Pseudomembrane fused with pseudendocorium. Prosternum and mesosternum with narrow low carinae; metasternum flat; distance between middle coxae smaller than width of rostral segment 3; distance between hind coxae smaller than their diameter. Sternum 7 of female with transverse edge.

Distribution in Malesia

The genus, only known from the type species, *Papacekia microphthalma*, occurs in East Kalimantan.

Biological notes

Biology is probably similar with that of *Trephotomas*.

Genus *Platytrephes*

Figure 229

Diagnosis

Medium-sized (2.3-2.8 mm), very broad and weakly convex. Dorsum with conspicuous yellow marks (fig. 229). Eyes small, but much larger than in *Papacekia*. Pseudomembrane distinct from pseudendocorium. In ventral view prosternum triangular; mesosternum without narrow carina, transverse, relatively

229

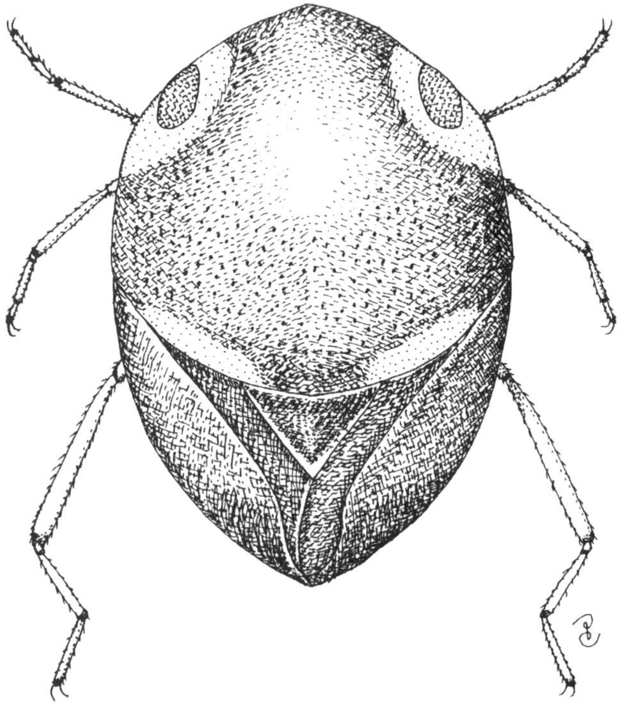

FIGURE 229
Platytrephes kodadai, paratype, brachypterous female, length 2.5 mm.

broad, and with diverse structures; metasternum flat; distance between middle coxae larger than width of rostral segment 3; distance between hind coxae much larger than their diameter. Male genitalia strongly modified (figs. 171-173). Subgenital plate of female simple (fig. 212).

Distribution in Malesia (fig. 574)

Two species inhabit Borneo; three other species are known from southern China, northern Laos and Thailand.

Biological notes

Biology is probably similar to that of *Trephotomas*.

Genus *Trephotomas*

Diagnosis

Medium-sized (1.9-2.0 mm), broad and weakly convex. Dorsum with conspicuous yellow marks. Eyes small, but much larger than in *Papacekia*. Pseudomembrane distinct from pseudendocorium. In ventral view prosternum triangular; mesosternum narrow, but without carina; metasternum posteriorly flat but narrow; distance between middle coxae smaller than width of rostral segment 3; distance between hind coxae smaller than their diameter. Male genitalia strongly modified. Subgenital plate of female simple.

Distribution in Malesia

Thus far, this genus has not been encountered in Malesia. The single known species is widely distributed in South-east Asia, with nearest occurrence in northern Thailand.

Biological notes

Trephotomas compactus inhabits the substrate of streams in forested areas (Papáček et al., 1988). Adaptations to benthic life are similar to *Fischerotrephes*, although to a lesser degree. Considering the preferred habitats, plastron respiration is suspected (Papáček et al., 1988).

REFERENCES – Trephotomasinae. Description of *Trephotomas* (Papáček et al., 1988); comparison of Bornean species (Zettel, 2000a); generic classification with description of *Papacekia* and *Platytrephes* (Zettel, 2005a).

SUBFAMILY HELOTREPHINAE

Diagnosis

Small to large (0.9-4.0 mm), usually distinctly convex, never strongly flattened. Antenna two-segmented. W-shaped suture of cephalonotum present. Mesoscutellum enlarged. Hemielytron without pseudendocorium. Claval suture of macropterous morph pointing at posterior third of mesoscutellum (fig. 205). Legs with pectinate bristles; tarsal formula 1-1-2 (figs. 200, 203).

Taxonomic note

Within Helotrephinae only Helotrephini are well defined by several synapomorphies. Although there are no proven apomorphies for either Helotrephinae (compared with Idiocorinae) nor Limnotrephini in their present range (Zettel, 1997a), the provisional arrangement used nowadays will be followed, but the two tribes are treated separately.

Distribution worldwide

Following the traditional system, the subfamily contains two tribes, Helotrephini and Limnotrephini. Helotrephinae inhabit tropical Africa and Madagascar and, with ten genera, the Oriental or South Palearctic regions.

– TRIBE HELOTREPHINI

Diagnosis

Usually relatively large (1.9-4.0 mm), always strongly convex. Cephalonotum usually densely punctated and dull, suture usually weakly developed; posterolateral margin of cephalonotum continued ventrally of the eye (fig. 206). Propleuron separated into 'pronotal plate' and 'pleural plate', situated in different levels to each other. Costal margin of hemielytron with a row of numerous fine ridges (presumed strigil for stridulation by J. Polhemus, 1990a). Hind femur with distal ridge (presumed plectrum for stridulation). Female subgenital plate usually symmetrical, except in *Ascetotrephes* (figs. 213-218).

Distribution in Indo-Australia

Three genera are known from the area: *Ascetotrephes* is restricted to continental Malaysia, Borneo and Sumatra. *Helotrephes* is widely distributed in South-east Asia, including Malesia west of Wallace's Line, and in China. Centre of distribution lies in continental South-east Asia. *Hydrotrephes* is the most widespread genus in the family: it occurs from Sri Lanka through continental South-east Asia to Sulawesi and the Philippines, and is thus far the only helotrephid genus found east of Wallace's Line. *Heterotrephes* Esaki & Miyamoto, with one species (which may be a species-group of *Hydrotrephes*), occurs in southern Japan.

Genus *Ascetotrephes*

Diagnosis

Rather large (2.7-3.6 mm). Brown to dark brown with yellowish or light brown markings. Dorsum moderately convex. Anterior margin of head weakly convex in frontal view; eyes small (fig. 227). Genal plate very narrow; pronotal plate anteriorly of eyes narrow, with usually shallow notch at eye level. Ventral carinae reaching onto abdominal sternite 3. Mesosternal carina usually high, without lamella; metasternal carina long, without lamella. Male genitalia: aedeagus with apical lamella; both parameres long. Female subgenital plate at least slightly asymmetrical (fig. 218).

Taxonomic note

This genus was recently described by J. Polhemus & D. Polhemus (2003) and more species were included by Zettel (2004f). It comprises a group of relatively closely related species which differ from *Helotrephes* by lacking carinae on abdominal sterna 4 and 5. Females can be distinguished from *Hydrotrephes* by the asymmetrical subgenital plate. Most species of *Hydrotrephes* differ in lamellate parts of meso- and metasternal carinae ('*Hydrotrephes*-type'), a medial tooth on sternite 4, relatively simple male genitalia, or strongly convex body. See also taxonomic note for *Hydrotrephes*.

Distribution in Malesia

This genus contains seven species, so far recorded only from Malesia. One species occurs in peninsular Malaysia, one in Sumatra, and five in Borneo.

Biological notes

Ascetotrephes species are apparently restricted to hilly and mountainous regions. Two North Bornean species were found in small rocky streams in the mountains at shaded places.

Genus *Helotrephes*

Figure 230

230

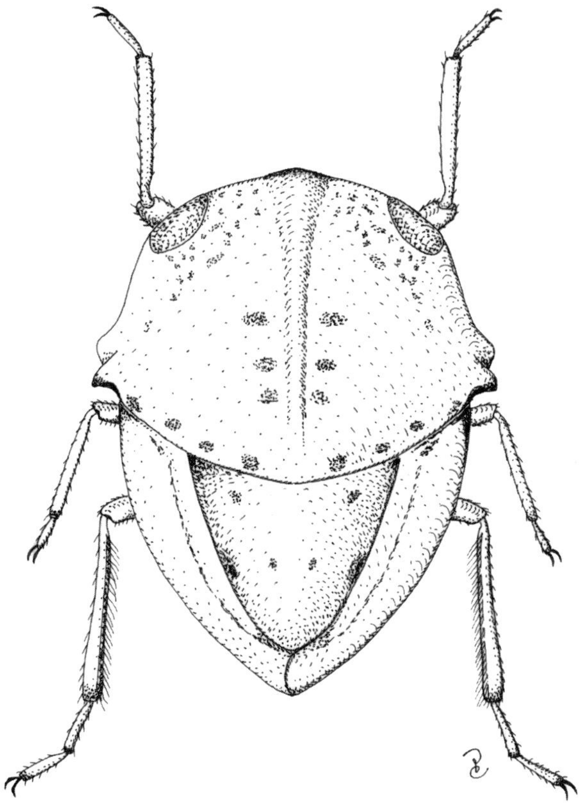

FIGURE 230
Helotrephes otoeis, paratype, brachypterous female, length 2.3 mm.

Diagnosis

Relatively large (2.2-4.0 mm), strongly convex (fig. 230). Dorsally usually yellowish to light brown with darker brown markings. Eye in macropterous morph slightly larger than in brachypterous morph. Posterior corner of cephalonotum in macropterous morph not modified. Rostral segment 4 1.6-2.6 times as long as segment 3. Pronotal plate with (usually deep) incision at eye level (figs. 195, 196). Mesosternal carina always obtuse posteriorly; metasternal carina with ventral margin straight, without lamella, posteriorly usually forming a distinct angle (fig. 223). Distal locking tab (pseudomembrane) on right hemielytron. Abdominal carinae reaching onto abdominal segment 5 (some females) or segment 6 (males and some females, fig. 223). Male genitalia variously modified: aedeagus (fig. 174) long, cylindrical, usually with modified apex, rarely unmodified and with simple tip. Right paramere always long, slender, more or less curved, usually twisted in distal half, with tapered, enlarged, hooked, or otherwise modified apex (fig. 175); left paramere rather uniform: long, tapering toward apex, on basal half with more or less developed mediad directed lobe which surrounding aedeagus posteriorly (fig. 176), in some species with apical hook. Female: abdominal segments symmetrical; sternites 5 and 6 more or less fused; sternite 6 often with convex or angulate hind margin, in ventral view more or less covering base of sternite 7; subgenital plate distally sculptured, of varying shape, in most species with distinct median lobe at hind margin (figs. 213, 214).

Taxonomic note

See under *Hydrotrephes* for relationship between the two genera. *Helotrephes* is defined by the presence of a median carina on abdominal sternite 5 and often on sternite 6. *Helotrephes* was revised and five species-groups were defined by Zettel & J. Polhemus (1998). Zettel (2001d) added a sixth species-group. At present there are approximately 30 species and subspecies described (Kovac & Papáček, 2000; Zettel, 2000a, c; 2001c, d, 2004f, h).

Distribution in Malesia (fig. 575)

Ten species occur in Malesia west of Wallace's Line: two species of the *H. australis*-group, and eight species (and one subspecies) of the *H. otoeis*-group.

Biological notes

Helotrephes species inhabit lentic or slow-lotic areas of running waters. They are usually found along edges of small and medium-sized streams, preferably with shaded banks and roots of trees hanging in the water. Some species (e.g. *H. otoeis*) can be collected in inlets with a large floating accumulation of plant debris (including twigs on which they probably rest; Zettel, 1997c). Some species live together with *Hydrotrephes*, which usually have similar habitat requirements (Miyamoto, 1952; Zettel & J. Polhemus, 1998).

Biology of Taiwanese *H. semiglobosus formosanus* was described in detail by Miyamoto (1952; partly noted above). During copulation (observed only from March to May) the male sits on the posterior-right side of the female. Eggs are laid separately on submerged vegetation. Larvae are observed from February to early June. Imagines from the preceding year could not be obtained after the end of

May, while newly emerged imagines have been found from the end of March until June. *Helotrephes s. formosanus* thus has only one generation per year. During the cold season they hibernate in mud at the banks. Tropical species do not hibernate and probably have no fixed life cycle.

In the laboratory *H. s. formosanus* can be reared on a diet of various aquatic insects. Adults also consume their own larvae. Prey is seized and held with fore and middle legs and pierced with the rostrum.

Genus *Hydrotrephes*

Figure 231

231

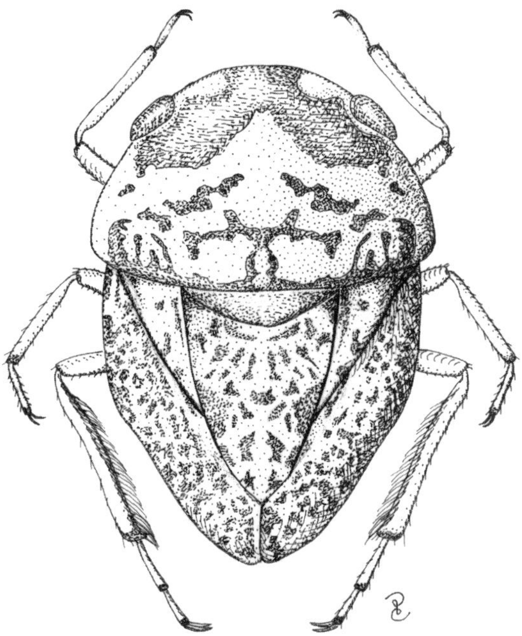

FIGURE 231
Hydrotrephes pardalos, paratype, macropterous male, length 2.2 mm.

Diagnosis

Small (2.1-3.8 mm), strongly convex. Dorsally usually yellowish to medium brown with, usually extensive, darker brown markings. Differs from *Helotrephes* mainly by lacking median carinae on sternites 5 and 6 (figs. 224, 225). Other differences not present in all species. Lateral margin of cephalonotum in macropterous morph simple or with angulation close to posterior corner. Rostral segment 4 2.0-2.9 times as long as segment 3. Pronotal plate with, often shallow, incision at eye level (fig. 197). Ventral carinae usually reaching onto abdominal sternite 3 or, rarely, sternite 4, with small median tooth (figs. 224, 225). Mesosternal carina either high and with ventral lamella or, rarely, low and without lamella; metasternal carina roundish and with thin distal lamella or, rarely, with ventral margin straight and without lamella (*'Hydrotrephes*-type': fig. 225; *'Heterotrephes*-type': fig. 224). Male genitalia variously modified: aedeagus long, cylindrical, rarely unmodified, with simple tip (*'Heterotrephes*-type'; fig. 177), but usually with modified apex (lamina or carina) and/or posterodistal surface (spur or denticle; *'Hydrotrephes*-type'; fig. 180). Right paramere usually distinctly shorter than left (fig. 181), rarely long, slender, and undulate (fig. 178), with apex variously modified; left paramere rather uniform: long, more or less tapering toward apex, on basal half without distinct mediad directed lobe (figs. 179, 182). Female: abdominal segments symmetrical; sternites 5 and 6 more or less fused; subgenital plate strongly varying, with different modifications (e.g., with medial tumescence, tooth, distal laminae, ridge, distomedial triangular process, or hair tufts), sometimes distinctly shortened (figs. 215-217).

Taxonomic note

The present generic definition is mainly based on the absence of a median carina on abdominal segments 5 and 6 (in comparison with *Helotrephes*). *Hydrotrephes* is possibly polyphyletic. There are several rather different species-groups. Nine species-groups have been recently defined (Zettel, 1998b, 2000a, 2003a), containing most but not all described Malesian species. One of these groups is identical with *Heterotrephes* (described from a single Japanese species). Phylogenetic relationships between some species groups of *Hydrotrephes, Heterotrephes, Ascetotrephes* and *Esakiella* are presently unclear.

Distribution in Malesia (fig. 576)

Hydrotrephes has 47 species in Malesia. Sulawesi with twelve species (Nieser & Chen, 1999; Zettel, 2004f), Borneo with eight and Luzon with six (Zettel, 2000a, 2003a) are especially rich. It is a widespread Oriental genus, occurring from Sri Lanka to Sulawesi and the Philippines, and is so far the only helotrephid genus found east of Wallace's Line. At present it contains more than fifty described species and subspecies. Many more undescribed species occur in the eastern Oriental and Wallacean realms.

Biological notes

The habitat is comparable to that of *Helotrephes*. Most *Hydrotrephes* species creep or swim in the lentic parts of running waters, usually at the waters' edge either on steep rocky banks of streams, where water passes evenly and usually rather slowly, between roots and plant debris, preferably at places with undercut river banks, or in large aggregations of plant debris in quiet bays of streams (Zettel, 2003a). *Hydrotrephes bouvieri* occurs in lakes and ponds (J. Polhemus, 1997; Nieser & Chen, 1999). Biology of *Hydrotrephes* is mostly unknown, but probably similar to that of *Helotrephes*. In the Philippines, imagines and larvae can be found together at least from November to July, so probably the year round. Some information on a *Hydrotrephes* species from Luzon (as *H. balnearius*) has been published by Usinger (1937) and Takahashi (1941).

TRIBE LIMNOTREPHINI

Diagnosis

Small (0.9-2.0 mm), usually strongly convex (except in *Distotrephes*). Cephalonotum smooth and shining, sparsely punctated, suture usually distinct. Posterolateral margin of cephalonotum pointed at centre of eye, indenting or dividing it (figs. 207-210). Propleuron not separated into pronotal plate and pleural plate. Costal margin of hemielytron without row of ridges. Hind femur without distal ridge. General shape of female subgenital plate strongly varying, often more or less asymmetrical.

Distribution in Indo-Australia

Six genera occur in Indo-Australia. *Limnotrephes* occurs in India, Sri Lanka (and Africa), *Nanotrephes* Papáček & Zettel is endemic to South India and *Mixotrephes* Papáček, Štys & Tonner is distributed from Iran through Afghanistan and India to North-west Thailand. The latter genus might extend into Malesia. Three genera occur in Malesia.

Genus *Distotrephes*

Figure 232

Diagnosis

Small (1.1-1.6 mm), slightly flattened (not strongly domed). Dorsally yellowish with extended dark marks (fig. 232). Macropterous morph rare. Lateral carina of cephalonotum continuing across eye, in brachypterous morph dividing it into dorsal and ventral part (fig. 209), in macropterous morph widely extended onto eye surface but not continuous (fig. 208). Cephalonotum in dorsal view longer than posterior part of body (fig. 232). Hemielytron with sparse but quite long hairs; propleural plate truncate at inner corner. Metasternal carina low, bifid, diamond-shaped; abdominal carinae extended onto sternite 4; carina of sternite 2 anteriorly inserted in diamond-shaped structure of metasternal carina. Aedeagus externally quite simple (figs. 189, 192), left paramere distally slender (figs. 191, 194). Larvae flattened.

Diagnosis of *Distotrephes* sensu stricto

Diamond-shaped structure of metasternum broad. Cephalonotum yellow with varying brownish or blackish marks. Male genitalia: pygophore usually with sinistral process; aedeagus with narrow apex (fig. 189); left paramere with acuminate apex (fig. 191); right paramere apically acuminate, rounded, or truncate (fig. 190). Female subgenital plate either slightly to strongly asymmetrical,

232

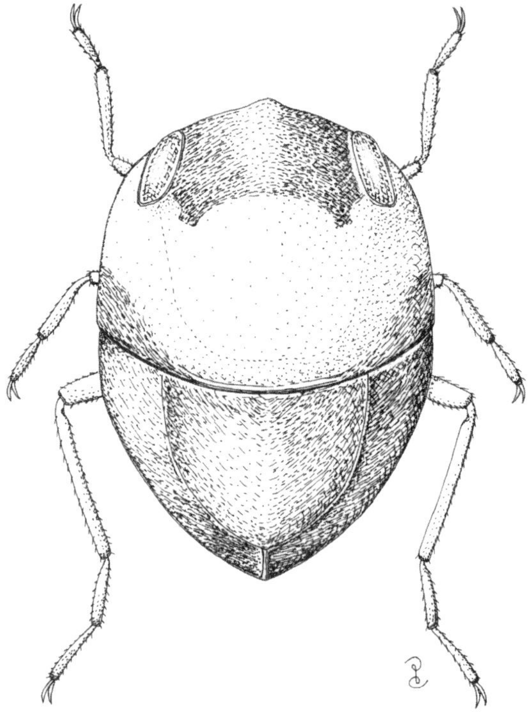

FIGURE 232
Distotrephes sarawakensis, paratype, brachypterous female, length 1.3 mm.

at posterior margin simple or (usually) with median process (fig. 221), or symmetrical and with sharp posteromedial tip.

Diagnosis of *Distotrephes* subgenus *Polhemotrephes*

Head usually uniformly blackish (fig. 232), rarely with five large confluent blackish marks. Pronotum completely yellowish, or, rarely, with small brownish marks; diamond-shaped structure of metasternum narrow. Male genitalia: pygophore without sinistral process; aedeagus apex broadly rounded (fig. 192); left paramere with distal part very slender, with apex narrowly rounded (fig. 194); right paramere strongly reduced and acuminate (fig. 193). Female subgenital plate symmetrical, at posterior margin emarginated (fig. 222). Because of numerous structural differences and the clear geographical separation, it might be considered to give *Polhemotrephes* generic rank.

Distribution in Malesia (fig. 575)

Thirteen species in two subgenera are described (Zettel, 1999d, 2000a, 2004h; Papáček & Kovac, 2001b), of which five occur in Malesia. The subgenus *Distotrephes* (nine described species) occurs on the South-east Asian mainland from southern China and Hainan to continental Malaysia. The subgenus *Polhemotrephes* (four species) is endemic to Borneo.

Biological notes

Biology is unknown. *Distotrephes* species are benthic, living on stream beds, but they are less rheophilous than *Fischerotrephes*, which sometimes inhabit the same stream. In contrast to other Limnotrephini in the area there are no records from stagnant waters. Currently, plastron respiration is assumed.

Genus *Idiotrephes*

Figure 233

233

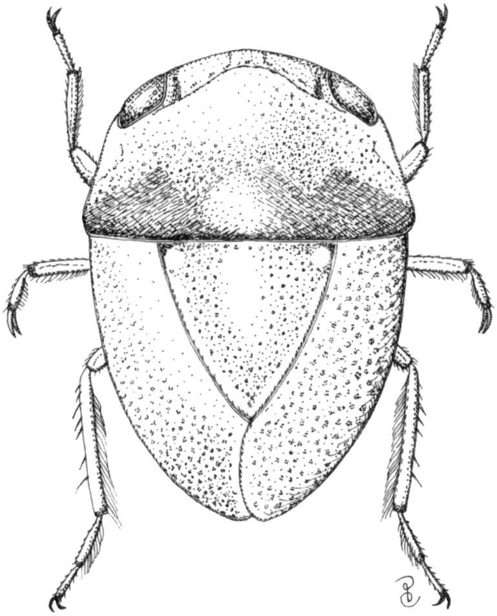

FIGURE 233
Idiotrephes chinai, brachypterous male, length 1.4 mm.

Diagnosis

Small (1.3-1.8 mm), domed, in macropterous morph more so than in brachypterous morph. Dorsally yellow with numerous brownish or blackish dots. Macropterous morph in some species common.

Cephalonotum distinctly shorter than posterior part of body (fig. 233); lateral margin of cephalonotum strongly indenting eye at posterior margin (fig. 210). Propleural plate at inner corner acuminate. Abdominal carinae extending onto sternite 4; sternites 4-6 medially fused, with sutures more or less reduced. Male genitalia: pygophore with sinistral process; aedeagus with distal part elongate and narrow in most species (fig. 183); left paramere of very characteristic shape, with more or less squared base and S-shaped, slender distal part (fig. 185); right paramere slightly shorter than left paramere, elongate, distally obliquely truncate (in Malesian species; fig. 184), tapered, or obtuse. Subgenital plate of female asymmetrical, with large (fig. 220) or small incision of its posterodextral margin (except in two species from Thailand), often with distinct hook-shaped ridge on right side of ventral surface. Larvae not strongly flattened. The diagnostic shape of the male left paramere, and the asymmetrical female sternite 7 with a dextrocaudal incision or impression and a dextrocaudal break, are diagnostic.

Distribution in Malesia (fig. 576)

Two species, belonging to the *I. chinai* species-group, have been recorded from Malesia west of Wallace's Line. A total of nine species in three species groups (Papáček & Zettel, 2000), and a tenth species apparently not belonging to any of these (Papáček & Kovac, 2001b), have their centre of diversity in continental South-east Asia.

Biological notes

Idiotrephes species inhabit stagnant waters, including small and artificial water bodies. The biology of *I. asiaticus* (under the name *I. chinai*) has been described by Papáček (1993), based on laboratory observations under artificial 'Central European conditions', and partly confirmed in the field in Vietnam later on; an English summary of this study is given in Papáček & Zettel (2000): *Idiotrephes asiaticus* is probably bivoltine. It can stand high water temperatures and is tolerant to temporary habitat changes (e.g. dry and wet season, variation in availability of prey type). Oviposition starts immediately after mating and lasts about two months. Eggs are deposited in small groups (2-4 eggs) on the surface of solid objects (e.g. stones, gravel, plants, artificial objects). Instar phases of larvae last 10-17 days. No preference of prey has been found and a wide variety of organisms is attacked (e.g. Annelida, Cladocera, Copepoda, small Diptera, Ephemeroptera, Heteroptera larvae, including younger instars of the own species, Protozoa, Rotifera). This species survives even in flower pots provided they contain water permanently.

Genus *Tiphotrephes*

Figure 234

Diagnosis

Small (1.4-1.6 mm), highly domed in both morphs. Macropterous morph common. Dorsally yellow with numerous brownish or blackish dots. Cephalonotum shorter than posterior part of body; lateral margin of cephalonotum not extending onto eye surface, only indistinctly indenting the eye at posterior margin (fig. 207). Propleural plate at inner corner acuminate. Metasternal carina simple; abdominal carinae extending onto sternite 4. Aedeagus broad (fig. 186); left paramere with basal lobe, distally with distinct bent (fig. 188); right paramere slightly reduced (fig. 187). Subgenital plate of female virtually symmetrical, with two narrow incisions laterally of a median lobe (fig. 219), which is most characteristic for the genus.

234

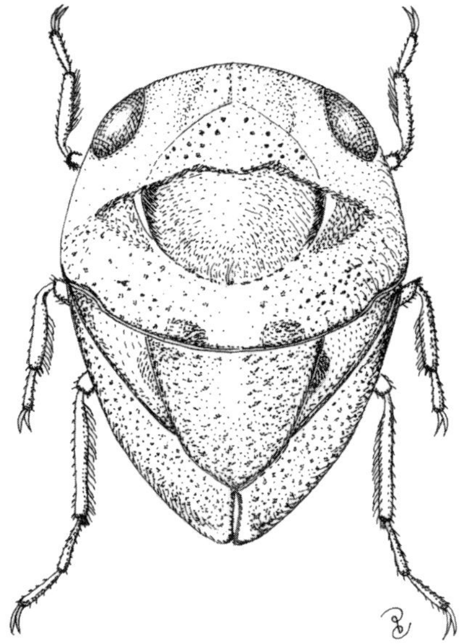

FIGURE 234
Tiphotrephes indicus, macropterous female, length 1.25 mm.

Distribution in Malesia (fig. 576)

Tiphotrephes indicus occurs in Borneo, peninsular Malaysia and elsewhere on the Asian mainland. A second species is known only from northern Vietnam (Esaki & China 1928; Zettel 1995, 2001c; Papáček & Zettel, 2003, and in press).

Biological notes

Tiphotrephes indicus inhabits stagnant waters, including small and artificial water bodies. Its biology is undescribed.

REFERENCES – Helotrephinae. Biology of *Helotrephes semiglobosus formosanus* (Miyamoto, 1952); classification and checklist of Helotrephidae (J. Polhemus, 1990a); key to *Helotrephes* (Zettel & J. Polhemus, 1998); key to Sulawesian *Hydrotrephes* (Nieser & Chen, 1999); description of subgenus *Polhemotrephes* (Zettel, 1999b); revision of *Idiotrephes*, biology of *Idiotrephes asiaticus* (Papáček & Zettel, 2000); key to Bornean species of Helotrephidae (Zettel, 2000a); revision of *Limnotrephes*, *Mixotrephes* and *Tiphotrephes* (Papáček & Zettel, 2003); survey of Philippine Helotrephidae (Zettel, 2003a); new species of Helotrephini from Malesia (Zettel, 2004f).

Family Pleidae – pygmy backswimmers

Diagnosis

Small (1.5-3.3 mm), yellowish with variable brownish marks, adults with dark longitudinal stripe on frons, boat-shaped to semi-globular, heavily punctated. Antennae three-segmented, rostrum four-segmented. Head broad and short, distinctly separated from thorax but relatively immobile. Eyes small, fronto-clypeus with distinctive sensory organ. Scutellum relatively large. Hemielytra coriaceous, membrane not recognizable. Legs apparently cursorial, although hind legs longest with some swimming hairs. Fore (fig. 235) and middle tarsi two- or three-segmented, hind tarsi three-segmented; all tarsi with two claws. Thoracic venter and abdominal venter on segments 2-5 or 2-6 with laminate keel. Male genitalia not strongly rotated, parameres asymmetrical (fig. 236-238). Ovipositor well developed, with large apical teeth (fig. 239).

235

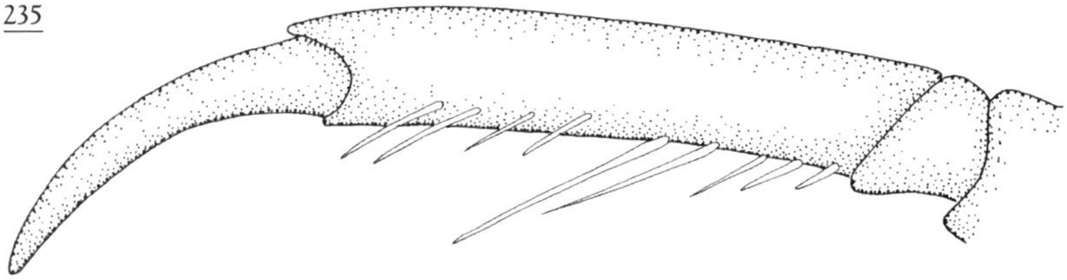

FIGURE 235
Fore tarsus of *Paraplea* sp.: two segments and claw.

236

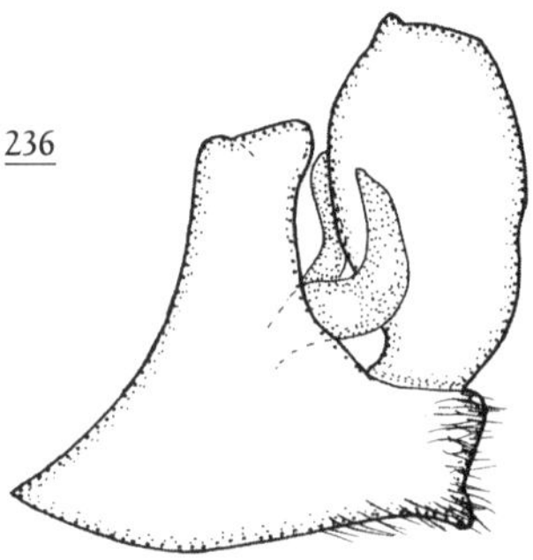

237

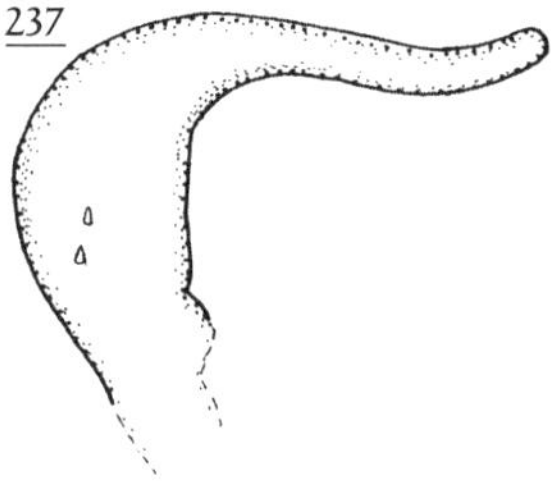

238

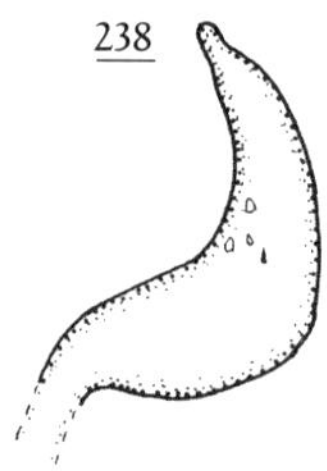

FIGURES 236 – 238
Male genitalia of *Paraplea frontalis*.
236 genital capsule – 237 dorsolateral view of right paramere – 238 dorsolateral view of left paramere.

239

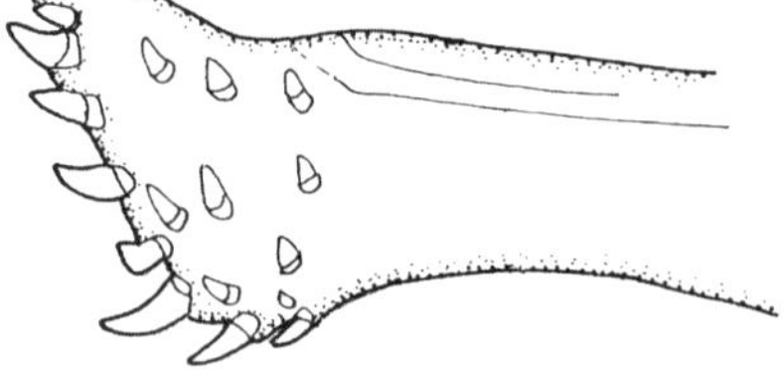

FIGURE 239
Apex of ovipositor of *Paraplea frontalis*.

Characteristics

Pleidae look like small stoutly built Notonectidae. The main difference, apart from the small eyes, are the shorter, 'normal' (not flattened) hind legs. The abdomen is symmetrical. Sexes can be separated only by small differences in the genital operculum (on sternite 7). As a rule the female operculum has a small acute tip, whereas that of the male is more obtuse or has a small notch (figs. 240, 241). In alcohol, specimens can be sexed by lifting the genital operculum.
The dark longitudinal stripe on the frons in adults consists of numerous fine channels with a small bristle on a cap on top.
Pleidae are of uniform structure. Differences between the genera are restricted to the number of tarsal segments. Characters used in species identification are ratios, size, and colour pattern. The parameres are useful in some, but in Malesian species they are of little diagnostic value.
The boat-shaped body is apparently related to the habit of back-swimming, as in Notonectidae and Helotrephidae.

240

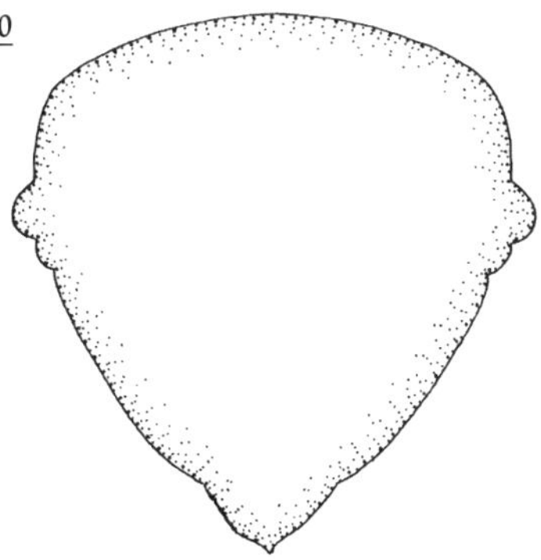

241

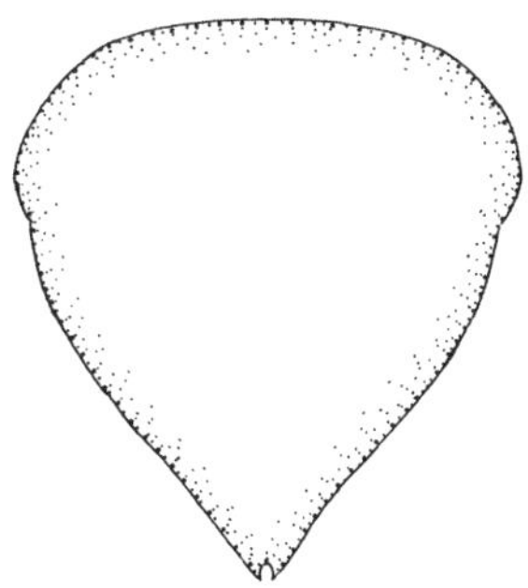

FIGURES 240 – 241
Outline of genital operculum of *Paraplea frontalis*. 240 female – 241 male.

Distribution worldwide

There are about 40 species divided over three genera. *Plea* Leach contains only the Palearctic species *P. minutissima*. *Paraplea* with about 25 species occurs in warmer areas of the world. However, lack of proper identification characters makes the status of many described species more or less dubious. Five species are reported to occur in Malesia, two are endemic to Australia.

Biological notes

Pleids are typical inhabitants of stagnant water with rich vegetation in which they hide. Collecting is done by putting plants from the net in a tray with water, in which they will start swimming after a while. In a wet mass of plants they hardly move and are easily overlooked. They swim on their back and are good swimmers, in spite of little hind leg adaptations.
The abdominal venter has lateral fringes of hairs. Together with the median keel these hold a large ventral air bubble. In addition there are air stores under

the pronotum and hemielytra. During summer pleids can stay alive in normal water (stagnant water with the regular amount of oxygen) about half a day without replenishing air stores, in well-aerated water about six days. In temperate regions the ability to stay under water increases considerably in autumn and winter, because the ventral air store becomes smaller and functions as a plastron (Kovac, 1982).

Copulation starts by the male grabbing the female dorsally. Soon it lowers itself to one side (usually the right), curves its abdominal tip and inserts the aedeagus. Eggs are elongate-oval and ventrally flattened. They are deposited in submerged plants, preferably in leaves, and glued with a secretion. The longitudinal axis makes a slight angle with the plant surface, leaving about $^3/_4$ of its length and about half of its width exposed. Egg and larval development take about four and six weeks respectively at a temperate summer for *P. minutissima* (Wefelscheid, 1912).

Like Notonectidae they are generalized predators taking small invertebrate prey, including younger instars of their own species. Specifically mentioned are mosquito larvae, ostracods and cladocerans (Papáček, 2001). They usually stay submerged between the vegetation and hunt on sight and vibratory signals.

Genus *Paraplea*

Figure 242

242

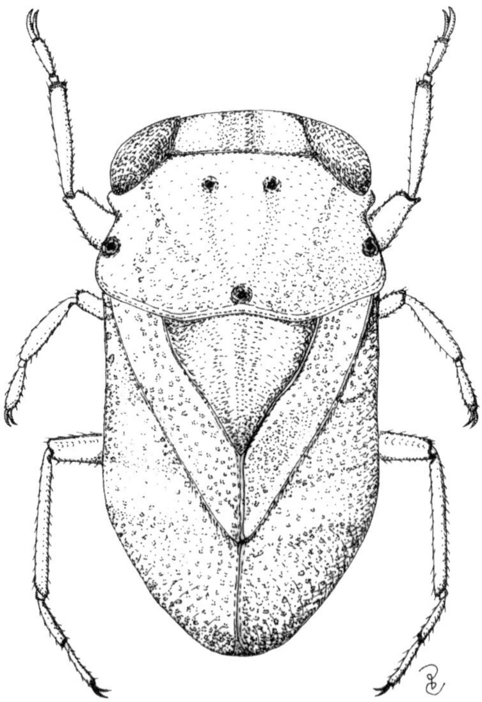

FIGURE 242
Paraplea liturata, male, length 1.6 mm.

Taxonomic note

Benzie (1989) found that characters used to separate species are more variable than previously thought. Several of the tropical Asian species, notably *P. quinquemaculata*, may turn out to be *P. frontalis*.

Diagnosis

Distinguished from other Pleidae by having fore and middle tarsi two-segmented and hind tarsi three-segmented. Ventral abdominal carina extending onto sternite 6.

Distribution in Malesia (fig. 574)

Distributed throughout, except for the Solomon Islands. Five species inhabit Malesia, two of which occur from India to the Moluccas, Australia and New Britain. The other three are endemic to Sumatra, Luzon and New Guinea respectively.

Biological notes

Paraplea species usually live in small stagnant waters, notably village ponds, with dense vegetation and often covered with floating plants such as Lemnacea, *Azolla* or *Salvinia*. Pleidae belong to the few Nepomorpha preferring waters covered with such floating vegetation.

REFERENCES – Pleidae. Morphology and biology of *Plea minutissima* (Wefelscheid, 1912); description and redescription of Indo-Australian species (Lundblad, 1933a); redescription of adult *Paraplea frontalis*, description of larvae, discussion of specific characters in genus (Benzie, 1989); key to Sulawesi species (Nieser & Chen, 1999); economic importance (Papáček. 2001).

Family Mesoveliidae – water treaders

Diagnosis

(This diagnosis refers exclusively to Malesian species. Subfamily Madeoveliinae has not been considered.) Small (1.2-4.2 mm). Layer of macrotrichia and microtrichia cover most of head, pleurosternal parts of thorax and ventral part of abdominal segment 2. Macropterous morph with ocelli. Base of labium in lateral view not obscured by bucculae. Pronotum truncate posteriorly, scutellum in macropterous morph exposed, in apterous morph absent. Hemielytra with reduced venation (fig. 252), usually not extending onto apical half. Tarsi three-segmented, claws inserted apically.

Characteristics

Mesoveliidae is a morphologically diverse family. Characters mentioned in the generic keys are easily observed. Differences between *Mesovelia* species are found in body size, setae on antennae and legs (figs. 248, 249), tufts of setae on the ventral side of segment 8 in males (fig. 243), and shape of the parameres (figs. 244, 245). Eggs of *Mesovelia* have a cap; they are elongate with the anterior end obliquely truncate (fig. 246). They are similar to eggs of various terrestrial Heteroptera (e.g. Nabidae, various Miridae, some Reduviidae), which is an additional argument to consider Mesoveliidae related to the ancestral Gerromorpha.

Fossil record

There is only one known fossil mesoveliid (Andersen, 1998): *Duncanovelia*, which is one of the oldest gerromorphan fossils. It originates from the Lower Cretaceous of Australia, about 120 million years ago (Jell & Duncan, 1986).

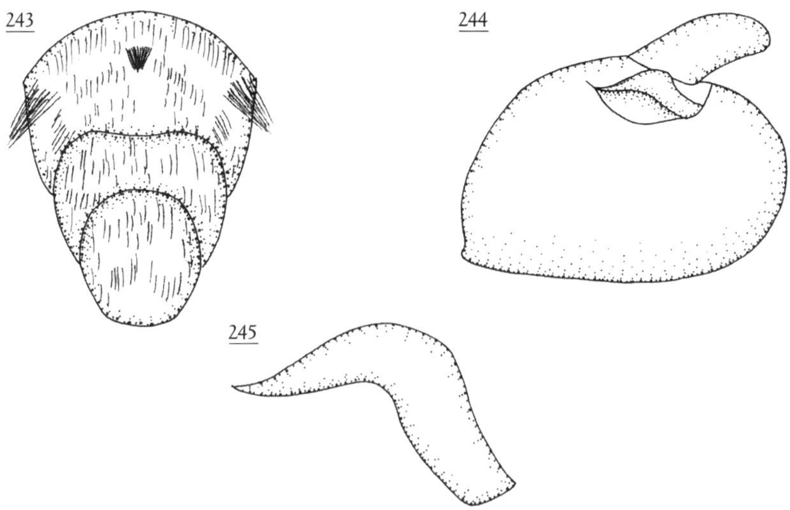

FIGURES 243 – 245
Mesovelia vittigera male. 243 apex of abdomen, ventral view – 244 genital capsule – 245 paramere.

246

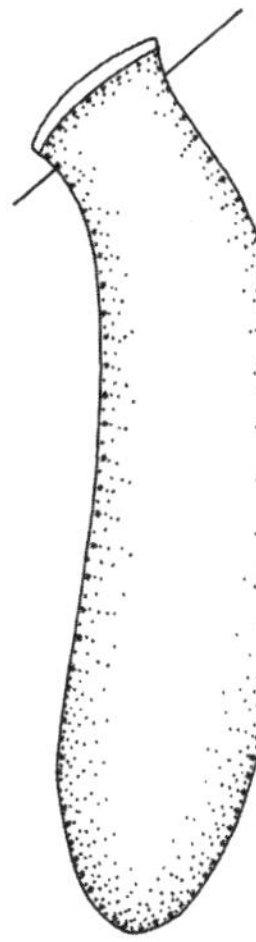

FIGURE 246
Mesovelia sp., egg.

Distribution worldwide

The family contains 43 recent species divided over two subfamilies and eleven recent genera. Ten of these contain only one to three species. *Mesovelia*, occurring worldwide, contains 25 species. A few *Mesovelia* species have a very wide distribution, e.g. *M. vittigera* occurs from the Mediterranean and the whole of Africa through the warmer parts of the Palearctic to Japan and through tropical Asia to the Philippines, New Guinea and Australia (J. Polhemus & D. Polhemus, 2000b). Several of the small genera occur on isolated islands or archipelagos. Because of the cryptic habits of various species and their small size, our knowledge of the distribution of Mesoveliidae is still incomplete (Andersen, 1999a).

Distribution in Indo-Australia

There are five genera, all belonging to Mesoveliinae, of which *Austrovelia* Malipatil & Monteith with two species is restricted to Australia and New Caledonia. The other four genera are discussed below.

Biological notes

Biology of *Mesovelia vittigera* is probably similar to that of the well-studied temperate Palearctic species *M. furcata* Mulsant & Rey (excepting for the cold season adaptations). At least the common and widespread species of *Mesovelia* are found in stagnant water with floating vegetation such as duckweed (Lemnaceae) or nymphoids. They prefer to stay on the floating plants but run quickly over open water when disturbed.

During copulation the male grabs the female with his fore legs around her mesothorax and middle legs either around the base of the abdomen or resting on the water. The male's hind legs are held upward. Copulation is short: from a few to fifteen minutes (Ekblom, 1930).

Eggs are deposited in plants with the anterior part protruding. The female saws a hole in a withered stem with the strongly developed ovipositor (fig. 247) and deposits an egg, which takes 30-60 seconds. Hatching is different from other Gerromorpha (which have a sharply pointed sclerotized egg buster to cut the egg shell): the embryo of *Mesovelia* develops a bladder-like outgrowth of the frontal cuticle with which it pushes off the egg cap. Incubation and hatching can take place both under water and in a moist atmosphere. If hatched under water the larva swims up and breaks through the surface.

247

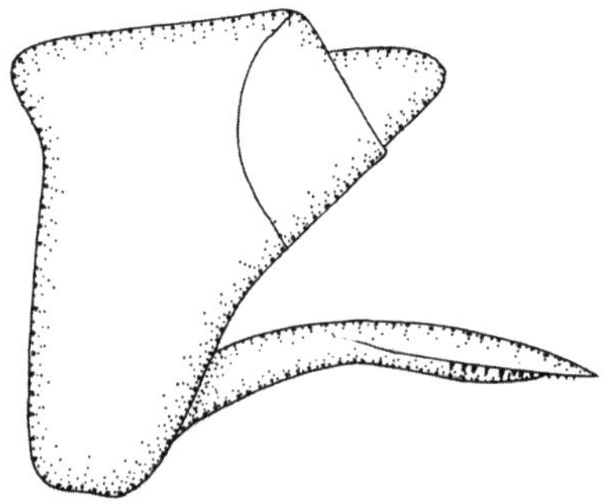

FIGURE 247
Mesovelia vittigera, female, apex of abdomen in lateral view, ovipositor slightly expanded.

Food consists mainly of dead or nearly immobile insects, notably larvae and pupae of Chironomidae and Culicidae, but also dying adults. Although larvae of Diptera, including Culicidae, are an important part of their diet, Mesoveliidae are apparently not of economic importance: they are generalized predators and take damaged or dying prey.

Key to Indo-Australian genera of Mesoveliidae

1 Middle and/or hind femora with strong spinous hairs (fig. 248). Apterous or macropterous; apterous morph with mesonotum prolonged along median line, distinctly longer than pronotum (fig. 251) ‣ ***Mesovelia* p. 198**

– Middle and hind femoras without or with weakly developed spinous hairs (fig. 249). Only apterous morph. Mesonotum not prolonged, as long as or shorter than pronotum (figs. 250, 253, 254) ‣ 2

248

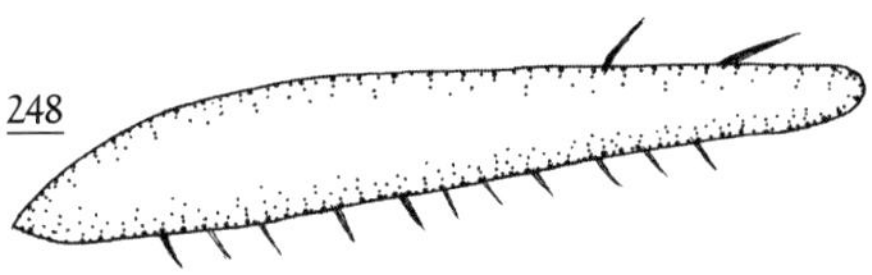

249

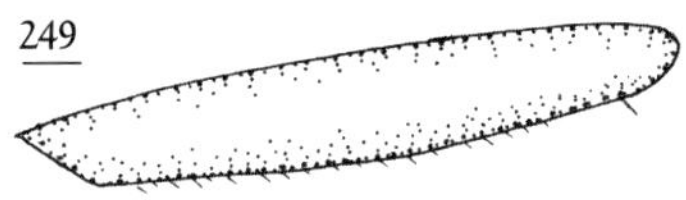

FIGURES 248 – 249
Middle femura of Mesoveliidae.
248 *Mesovelia vittigera* –
249 *Phrynovelia papua*.

2 Eyes vestigial, consisting of three large ommatidia (fig. 250). Terrestrial, living in leaf litter in rainforests ‣ *Cryptovelia* **p. 197**

– Eyes well developed with over 20 ommatidia ‣ 3

3 Apex of rostrum reaching base of abdomen. Femoral spinous hairs absent ‣ *Nereivelia* **p. 201**

– Apex of rostrum not reaching beyond the hind coxa ‣ 4

4 Femoral spinous hairs present. Posterior margin of metanotum straight (fig. 254) ‣ *Phrynovelia* **p. 202**

– Femoral spinous hairs absent. Posterior margin of metanotum sinuate [Australia and New Caledonia] ‣ *Austrovelia*

Genus *Cryptovelia*

Figure 250

250

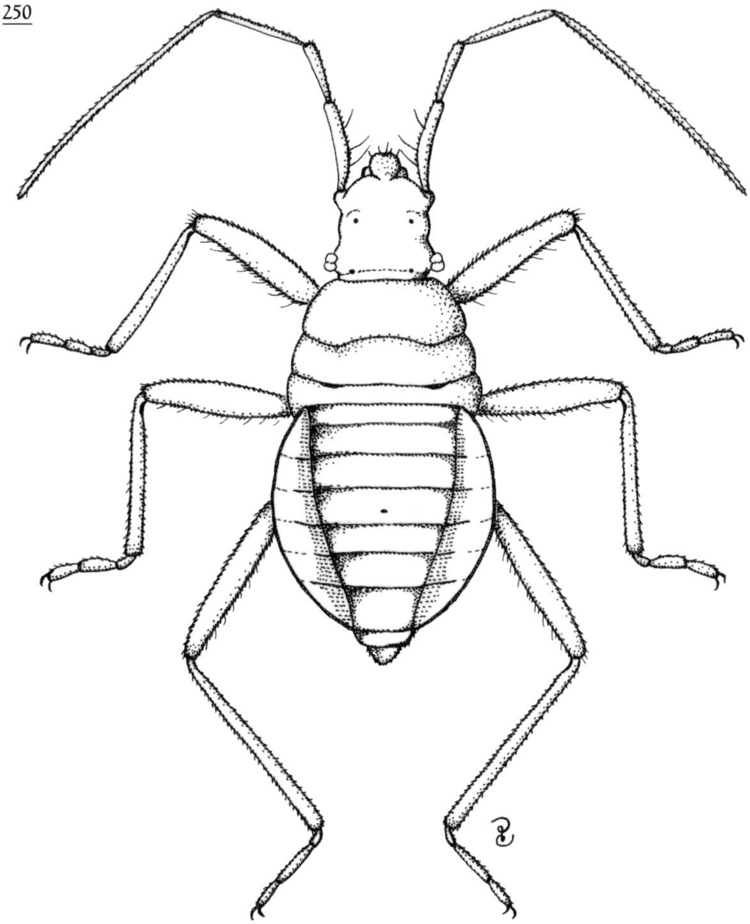

FIGURE 250
Cryptovelia stysi, paratype, apterous male, length 1.1 mm (after Andersen, 1999a).

Diagnosis

Small (1.0-1.3 mm). Ground color yellowish or pale brownish. Monomorphic apterous. Head elongate with subparallel sides. Antenna long and thin, subequal to slightly shorter than body length. Eyes vestigial, composed of 3-4 large ommatidia. Rostrum long and slender, reaching at least basal third of abdominal venter. Mesonotum on median line slightly shorter than pronotum; hind margin of metanotum virtually straight. Abdomen broad, length subequal to or slightly more than greatest width. Connexiva broad. Male genitalia small. Ovipositor well developed.

Distribution in Malesia (fig. 577)

Cryptovelia stysi is known from Sarawak. There are probably more species in this genus.

Biological notes

Both representatives of this genus live in leaf litter of rainforests (*C. terrestris* Andersen & J. Polhemus in the Amazon area, Brazil), but as the habitat is permanently wet they can be compared to hygropetric species. The reduction of eyes may be related to living in a dark habitat. The well-developed ovipositor suggests that eggs are deposited in plant tissue, as in *Mesovelia*.

Genus *Mesovelia*

Figures 251-252

Diagnosis

Small to large (2.0-3.5 mm). Ground colour pale brown or olive green. Both apterous and macropterous morphs occur. Macropterous morph with ocelli, apterous morph without ocelli. Head with subparallel sides, variable, either rather short and wider than long, or elongate and distinctly longer than wide. Antenna of variable length. Eyes well developed with 50 or more ommatidia. Rostrum reaching at most to hind coxa. Mesonotum wider than pronotum, hind margin of metanotum virtually straight. Abdomen elongate, distinctly longer than wide; connexiva of medium width. Male genitalia prominent. Ovipositor well developed. Anterior end of egg obliquely truncate, with egg cap (at least in the common species).

Distribution in Malesia

Six species (out of 25) occur in Malesia. *Mesovelia vittigera* is distributed throughout Malesia and well beyond (see Distribution worldwide), *M. horvathi* is also widespread in Malesia and tropical Asia. Two species are known only from New Guinea, one from New Guinea and the Moluccas (J. Polhemus & D. Polhemus, 2000b), and one from Sumatra and Luzon.

251

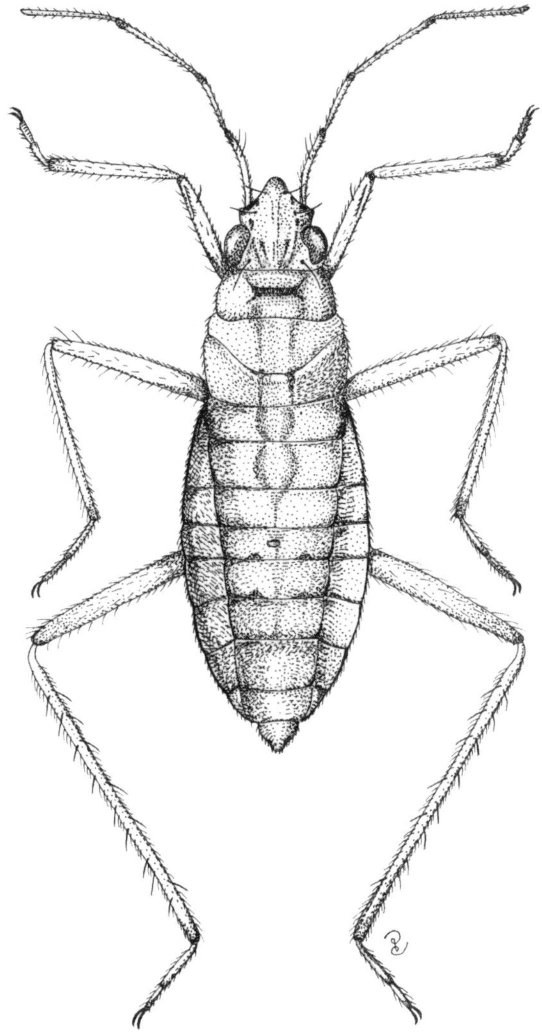

FIGURES 251
Mesovelia vittigera, apterous female, length 3.3 mm.

Biological notes

As far as known there are two main habitat preferences: some species, such as *M. vittigera*, live rather exposed on floating water plants, others more cryptically at the edge of various water bodies. *Mesovelia horvathi* may fall in this latter category. *Mesovelia* species in New Guinea have been frequently found flying around (Lansbury pers. comm.). The most widespread species, *M. subvittata*, was collected from an array of habitats (e.g. margins of streams, wet rock faces, seeping clay banks, damp depressions on the forest floor). It thus appears to be capable of adopting a semi-terrestrial mode of life, particularly during the dry season (J. Polhemus & D. Polhemus, 2000b). *M. horvathi* and *M. vittigera* have been found regularly in light traps in Thailand.

252

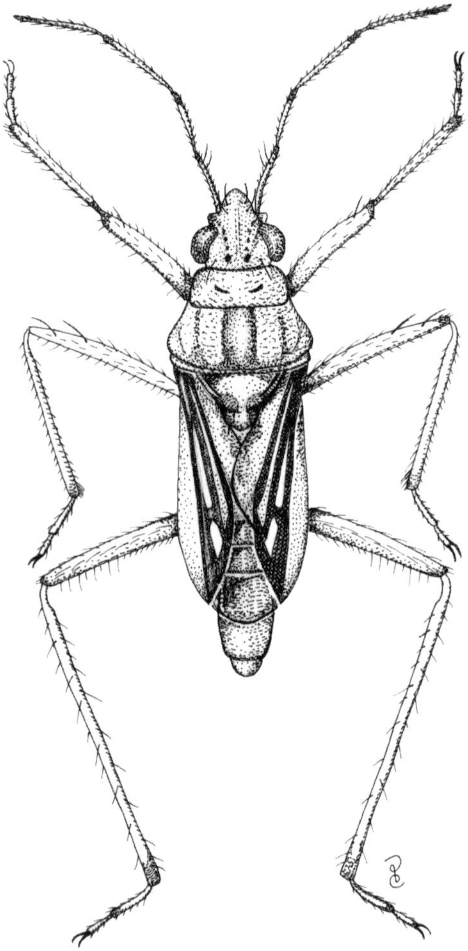

FIGURES 252
Mesovelia vittigera,
dealate male,
length 2.7 mm.

The eggs of *Mesovelia* (fig. 246), notably *M. furcata* and the American *M. mulsanti* White, possess symbionts (Cobben, 1965, 1968), which is not only unique within Gerromorpha but also for predacious insects. The precise nature of these symbionts is still unknown. They are micro-organisms with an individual size of 1-2 μ. During the embryological development they unite to a red ball. This ball is initially situated at the place where the thorax develops, but by the time of eclosion it lies in the intestine in the first abdominal segment. After eclosion the ball of symbionts breaks up at first, but then recombines to a single body in the hind part of the mid-gut. In the adult bug the symbionts swim freely in the digestive fluids of the gut. Shortly before oviposition they assemble to irregularly-shaped clumps and lie between the cells of the gut wall. Apparently the female delivers a drop of fluid containing symbionts from the anus on an egg. The symbionts are supposed to enter the egg through the micropyle (Cobben, 1968).

Genus *Nereivelia*

Figure 253

253

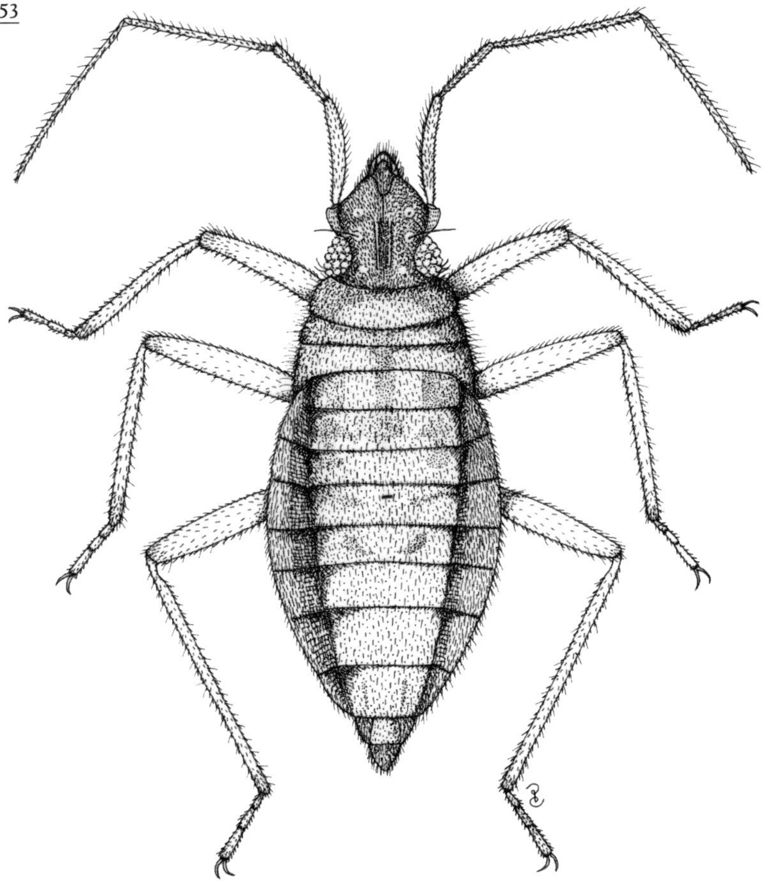

FIGURE 253
Nereivelia murphyi, paratype, apterous female, length 2.1 mm.

Diagnosis

Small to medium-sized (1.7-2.2 mm). Ground colour yellowish to pale brownish. Monomorphic apterous. Head slightly broader than long with subparallel sides. Antennae long and thin, subequal to slightly shorter than body length. Eyes well developed, composed of about 30 large ommatidia. Rostrum long and slender, reaching to base of abdominal venter. Pronotum slightly wider than mesonotum; hind margin of metanotum virtually straight to slightly concave. Abdomen rather broad, about 1.5 times as long as wide; connexiva narrow. Male genitalia prominent. Ovipositor well developed.

Distribution in Malesia (fig. 577)
One species, N. *murphyi*, has been described and so far only three specimens have been collected. They were found at the base of logs in mangrove swamps on the western coast of continental Malaysia (J. Polhemus & D. Polhemus, 1989).

Genus *Phrynovelia*

Figure 254

254

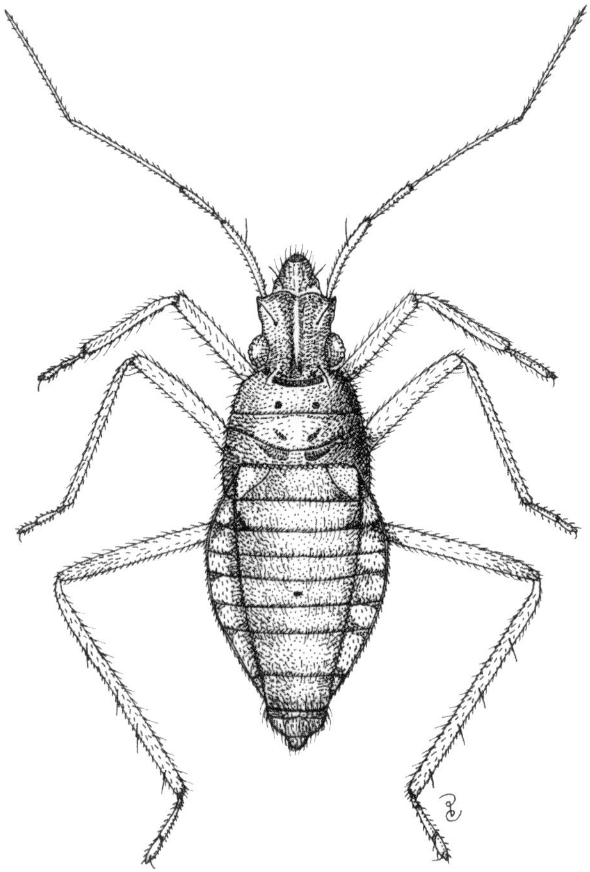

FIGURE 254
Phrynovelia papua, female, length 2.4 mm.

Diagnosis

Small (1.5-1.8 mm). Ground colour yellowish to pale brownish. Monomorphic apterous. Head about 1.3 times as long as wide, with subparallel sides. Antennae about 3/4 of body length. Eyes rather small, composed of about 20 large ommatidia. Rostrum reaching to hind coxa. Pronotum slightly wider than mesonotum; hind margin of metanotum straight. Abdomen broad, about 1.7

times as long as wide; connexiva well developed. Male genitalia prominent. Ovipositor well developed.

Distribution in Malesia (fig. 577)

Until 1983 one species, *P. papua*, was known, occurring in New Guinea. (Malipatil & Monteith (1983) described two new species from New Caledonia.)

Biological notes

All three species are found in forest leaf litter. Judging from the eye development they may live higher up in the litter than *Cryptovelia*.

REFERENCES – Mesoveliidae. Biology of *Mesovelia furcata* (Ekblom, 1930); synonymy and distribution of *M. vittigera* (Andersen, 1967); embryogenesis (Cobben, 1968); key to genera, checklist of species (Andersen & J. Polhemus, 1980); phylogeny, adaptations, biogegraphy and classification of Gerromorpha (Andersen, 1982a); fossil Mesoveliidae (Jell & Duncan, 1986); description of *Nereivelia*, comparison with related genera (J. Polhemus & D. Polhemus, 1989); cladistic analysis of the family on genus level, biogeography (Andersen, 1999a); descriptions and key to species of *Mesovelia* occurring in New Guinea (J. Polhemus & D. Polhemus, 2000b).

Family Hebridae – velvet water bugs

Diagnosis

Small (1.3-3.7 mm), usually stout (figs. 283, 286). Head usually elongate, with bucculae plate-like expanded (figs. 272-276). Ocelli usually present. Rostrum long and slender, segment 4 prolonged, about half as long as segment 3. Mesoscutellum reduced to narrow transverse plate; metanotum usually with medial 'scutellum-like' elevation (fig. 255). Fore wing venation apically reduced, with one or two closed cells in basal half (figs. 283, 285). Legs of moderate length, not modified for rowing. Coxae separated medially by rostral groove (figs. 281, 282). Tarsus two-segmented, segment 1 very short, with long ventral bristle, segment 2 long; claws inserted apically (figs. 277-280). Genital segments of both sexes inserted proximal of abdominal apex. Gonocoxae simple, plate-like.

Characteristics

The bucculae (figs. 272-276) are a pair of ventral plates of the head capsule forming a (usually deep) rostral groove, in genera of Hebrini bearing two or three circular or ovate impressions (figs. 275, 276). The rostral groove extends to the thorax and is demarcated by ridges (figs. 281, 282). The antennae are of varying structure according to habits: long in terrestrial genera (figs. 270, 271), short in aquatic and semi-aquatic genera (figs. 267-269). The metanotal elevation or pseudoscutellum (formerly referred to as scutellum) is part of the metanotum (fig. 255); it is lacking in the apterous morph of some species. The male genitalia (figs. 259-266) are usually small, symmetrical and simple, often with strongly reduced parameres (e.g., in *Nieserius*, *Merragata* and most *Hebrus*; figs. 261-263), but parameres and proctiger are complex and asymmetrical in some *Timasius* (e.g. fig. 266).

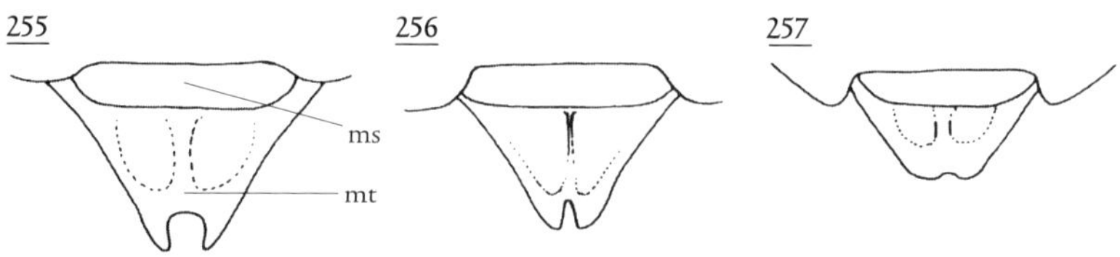

FIGURES 255 – 257 Hebridae, mesoscutellum and metanotal elevation in dorsal view. 255 *Hebrus cruciatus* – 256 *Hebrus harrisi* – 257 *Hebrus bergrothi*; ms = mesoscutellum, mt = metanotal elevation.

Main characters for generic and suprageneric classification are size, shape, and position of the eye; general structure of bucculae; length of individual antennal segments and total length; pilosity of antennal segment 4 (preapical cluster of hairs, preapical pit); structure of antennal segment 4 (showing a constriction or false joint, making the antennae appear five-segmented in *Hebrus* and *Timasius*);

258

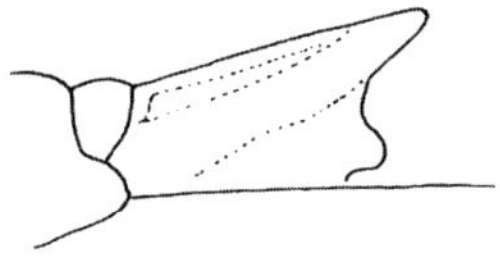

FIGURE 258
Timasius chinai, meso-scutellum and metanotal elevation in lateral view.

259

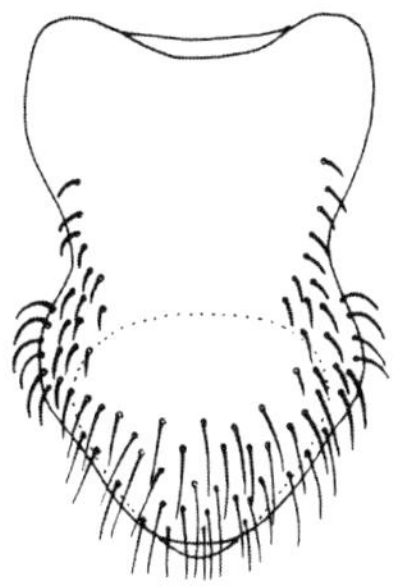

FIGURE 259
Hyrcanus varicolor, proctiger.

260

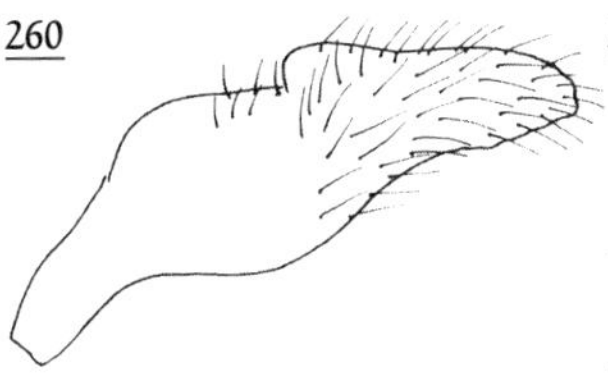

261

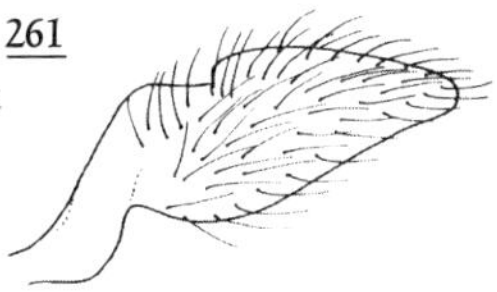

262

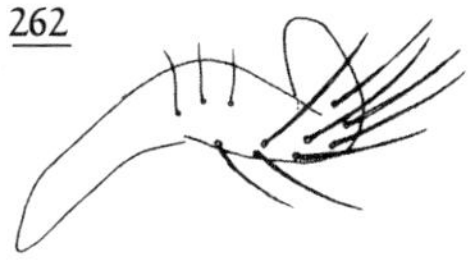

263

264

FIGURES 260 – 264
Hebridae, paramere.
260 *Hyrcanus varicolor* – **261** *Nieserius subaquaticus* – **262** *Merragata pallescens* var. *whitei* – **263** *Hebrus cruciatus* – **264** *Hebrus harrisi*.

265

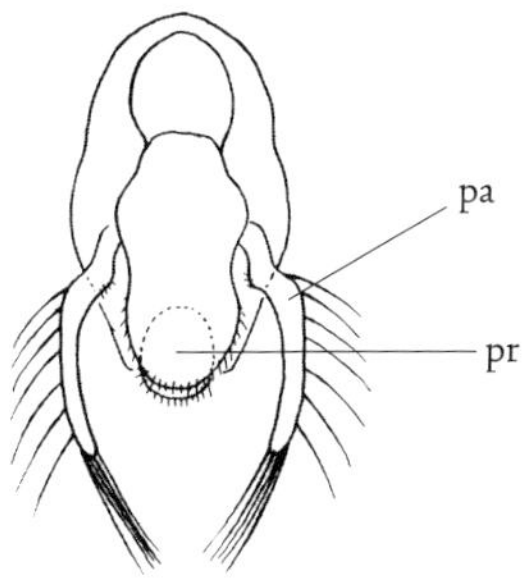

266

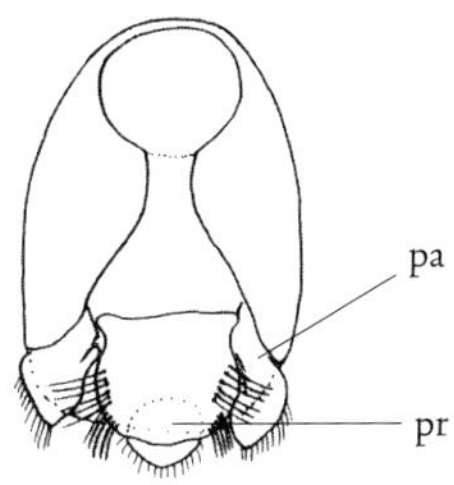

FIGURES 265 – 266
Hebridae, male genitalia in dorsal view.
265 *Hebrus harrisi* – **266** *Timasius chinai*; pa = paramere, pr = proctiger.

length of rostrum; height and position of ventral thoracic carinae; and general structure of male and female genitalia (Andersen, 1981a, 1982a; Zettel, 1999e). Specific characters lie mainly in the male genitalia, especially of proctiger and parameres. However, in most species genitalia are small, and therefore high magnification is needed (scanning electron microscope, see e.g. Zettel & Sehnal, 1995). Body pilosity (e.g. hemielytron, male legs), colour, shape and structure of bucculae, and certain microstructures are also helpful to distinguish species and other infrageneric units. The structure of the metanotal elevation has been widely used for distinction, but it differs between macropterous and meiopterous specimens in a single species. Females of some species are not identifiable with certainty.

Taxonomic note

Here we follow the phylogenetic analysis and concept by Andersen (1981a, 1982a).

Distribution worldwide

The family is distributed in all zoogeographical regions and from temperate to tropical zones. Because of their small size and generally cryptic habits in the land-water zone, tropical Hebridae are still very poorly known. Two subfamilies are recognized: Hebrinae (seven genera, approximately 180 described species) and Hyrcaninae (two genera, thirteen species). Most Hebrinae belong to the cosmopolitan genus *Hebrus* (with several subgenera and about 130 described species) or to the Oriental genus *Timasius* (31 described species). *Merragata* is also cosmopolitan, with eight described species, most of them in the New World. The other four genera are confined to single zoogeographical regions. Hyrcaninae is an Oriental subfamily, with most species occurring from India to South China and North-east Thailand.

Distribution in Indo-Australia

Hebrinae has five genera in Indo-Australia. *Neotimasius* Andersen is restricted to the Indian subcontinent, *Austrohebrus* Andersen & Weir to Australia. *Merragata* is only represented by two or three species (Lundblad, 1933a; Andersen & Weir 2004). *Hebrus* and *Timasius* are represented by a good number of species (many still undescribed), *Hebrus* all over the region, *Timasius* mainly west of Wallace's Line and Dickerson's Line. Ten *Hyrcanus* species are known from the southern part of the Asian continent, and from Sumatra and Java. Highest species diversity is found in North Thailand (four species) (Zettel, 1998d, 2000b). *Hyrcanus varicolor* is widely distributed in Indochina, Sumatra and Java and is the only species of the subfamily Hyrcaninae reported from Malesia.

Biological notes

Hebridae are primarily terrestrial, inhabiting banks of waters, and so represent the ancestral type of Gerromorpha (Andersen 1982a, 1995a). Especially the two largest genera, *Hebrus* and *Timasius*, live on land, although *Hebrus* may occasionally run on the water surface. A few species of *Hebrus* live in the intertidal zone in mangroves and stay submerged during high tide. Different habitats are used by *Merragata* (water surface and water plants, sometimes entering the water), *Hyrcanus* (land-water transition zone and accumulated plant material in streams, frequently entering the water) and *Nieserius* (obligatory subaquatic)

(Drake, 1917; Andersen, 1982a; J. Polhemus & D. Polhemus, 1989; Zettel, 1998d, 1999e, 2004b). For more details see information on genera.

Little is known of hebrid biology. Locomotion on ground and on water is done by walking and running (Andersen, 1982a). Rowing and jumping is not developed. Swimming in subaquatic Hyrcaninae has not been observed and is unlikely, as no swimming device on the legs is developed. Judging from field observations they crawl over the substrate when under water. Most species have well-developed wings (characteristic for most insects living in the land-water transition zone). The subaquatic species of *Nieserius* and a few species of *Hebrus* living in constantly moist habitats are commonly brachypterous or micropterous, rarely apterous. *Timasius* and the Ethiopian *Hebrometra* Cobben frequently take flight when disturbed, as do some species of *Hebrus* (Andersen 1982a; pers. obs.). Respiration of Hebridae living under water has not yet been studied. See also notes under *Hyrcanus* and *Nieserius*.

Structure, function, and evolution of reproductive organs were studied in European *Hebrus* species by Heming-van Battum & Heming (1986, 1989). Female *Hebrus* deposit eggs free or partly exposed in a humid place, preferable at secluded places. In *H. ruficeps* an egg is lengthwise attached to the substrate by a gelatinous substance (Hungerford, 1920; Andersen, 1982a). Eggs of *Hebrus* have an entirely solid shell and a single micropyle at the anterior pole (Cobben, 1968). In *T. major* and *Neotimasius orientalis* eggs have a dorso-ventral differentiation. Prolarvae of *Hebrus* have a paired clypeal ruptor to burst the egg shell. As far as known, all Hebridae have five nymphal instars (Andersen, 1982a). The duration of nymphal life of the West Palearctic *Hebrus pusillus* is approximately 21 days at 19 °C (Lebrun, 1960).

Although sucking on *Sphagnum* leaves in the European *Hebrus ruficeps* Thomson has been observed (Jordan, 1935, 1952), any kind of phytophagy in Hebridae was questioned by Cobben (1978), as all other observations (e.g. Hungerford, 1920; Heming-van Battum & Heming, 1986) indicate that Hebridae are predators.

Key to South-east Asian subfamilies and genera of Hebridae

1 Antennae distinctly shorter than greatest width of pronotum; antennal segments stout, segment 4 without desclerotized zone, antennae thus clearly four-segmented (figs. 267-269) ▸ 2

– Antennae at least subequal in length to greatest width of pronotum; antennal segments slender; last (4th) segment more or less clearly divided by a desclerotized zone, antennae thus appearing five-segmented (figs. 270, 271) (Hebrinae, partim) ▸ 4

2 Medium-sized to large (more than 2.1 mm). Head long, narrow and pointed (figs. 272, 273, 286, 287). Corium of fore wing dark. Antenna relatively stout (figs. 267, 268). Femora stout (figs. 277, 278) (Hyrcaninae) ▸ 3

– Very small (less than 1.8 mm). Head short and broad, in lateral view anteriorly truncate (figs. 274, 284). Antenna comparatively slender (fig. 269). Corium of fore wing with white patch. Femora slender (fig. 279) (Hebrinae, partim) ▸ *Merragata* **p. 212**

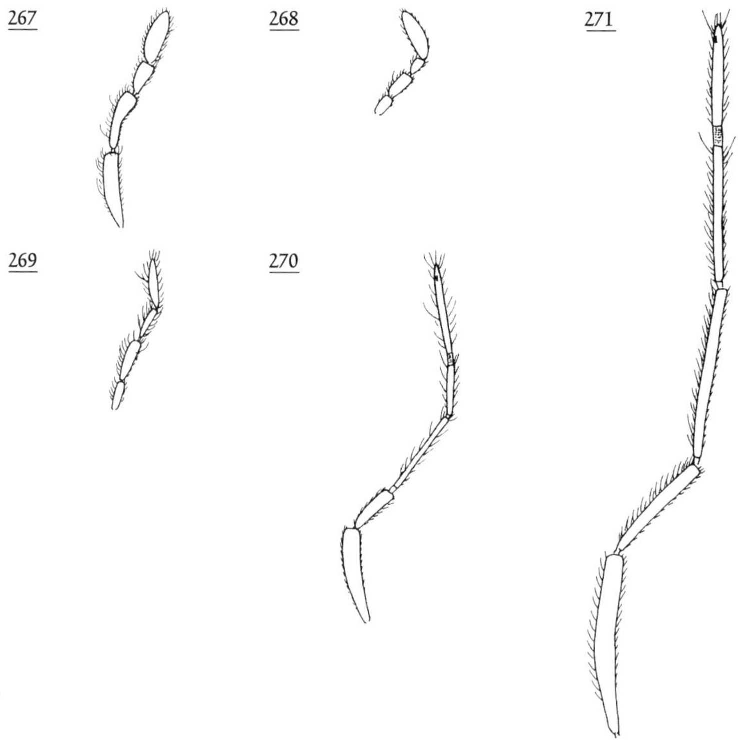

FIGURES 267 – 271
Hebridae, antenna.
267 *Hyrcanus varicolor* – **268** *Nieserius subaquaticus* – **269** *Merragata pallescens* var. *whitei* – **270** *Hebrus cruciatus* – **271** *Timasius chinai*.

3 Antennal segment 1 shorter than segments 2 and 4 (fig. 268). Distance between eye and anterolateral margin of pronotum more than 0.4 times eye width (fig. 273). Head below eyes without conspicuous long bristles, only with some relatively short hairs. Legs slender, mesofemur less than 3.7 times as long as broad (fig. 278). Brachypterous or macropterous ▸ *Nieserius* **p.217**
(This genus is so far not recorded from Malesia. Nearest occurrence lies in North Thailand.)

– Antennal segment 1 longer than segment 2 and subequal to segment 4 (fig. 267). Distance between eyes and anterolateral margin of pronotum clearly less than 0.4 times width of eye. Head below eyes with tuft of conspicuous long bristles (fig. 272). Legs stout, mesofemur more than 3.5 times as long as broad (fig. 277). Monomorphic macropterous ▸ *Hyrcanus* **p. 215**

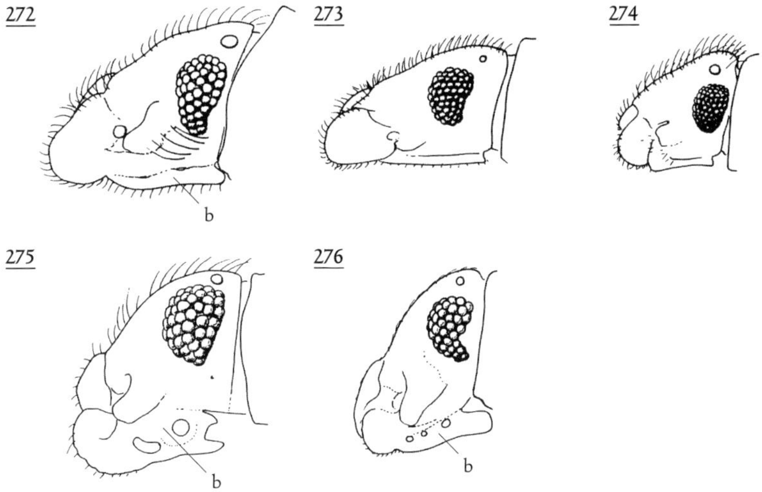

FIGURES 272 – 276 Hebridae, head in lateral view.
272 *Hyrcanus varicolor* – 273 *Nieserius subaquaticus* – 274 *Merragata pallescens* var. *whitei* – 275 *Timasius chinai* – 276 *Hebrus cruciatus*; b = buccula.

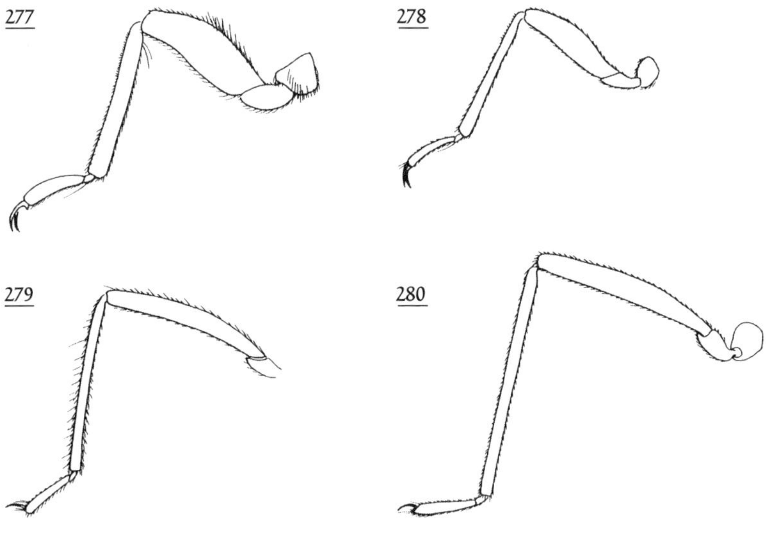

FIGURE 277 – 280 Legs of Hebridae.
277 *Hyrcanus varicolor*, middle leg – 278 *Nieserius subaquaticus*, hind leg – 279 *Merragata pallescens* var. *whitei*, hind leg – 280 *Timasius chinai*, middle leg.

4 Paired longitudinal carinae of thoracic venter converging and meeting before hind margin of metasternum (fig. 281). Usually longer than 2.5 mm (range 2.45-3.70 mm) ▸ *Timasius* p. 214

– Paired longitudinal carinae of thoracic venter parallel throughout and continuing separately onto base of abdomen (fig. 282). Usually shorter than 2.5 mm (range 1.30-2.54 mm) ▸ *Hebrus* p. 210

SUBFAMILY HEBRINAE

Diagnosis

Antenna usually long and appearing five-segmented, because segment 4 has a membranous zone or false joint (except in *Merragata*); segment 4 subapically

FIGURES 281 – 282
Hebridae, schematic view of mesosternum and metasternum and base of abdomen in ventral view, showing position of carinae.
281 *Timasius* – **282** *Hebrus*;
a = base of abdomen,
g = rostral groove,
ms = mesosternum,
mt = metasternum.

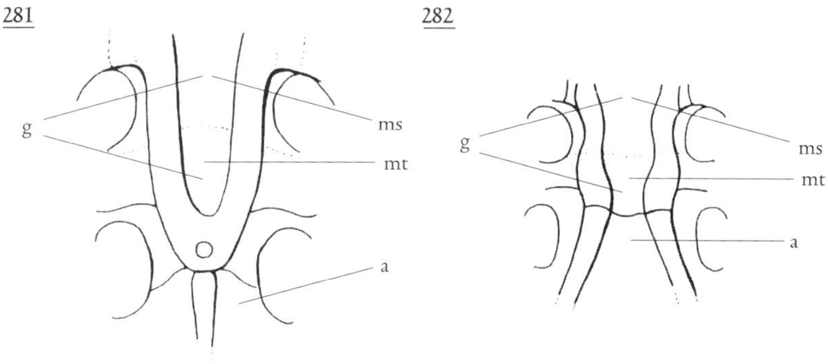

with group of sensorial hairs (figs. 269-271). Eyes at or very close to anterolateral margin of pronotum (figs. 274-276). Buccula wide, usually with two or three impressions (figs. 275-276). Rostral segment 3 not surpassing posterior margin of head. Fore wing at base and/or on membrane often with whitish or bluish-frosted patches. Legs slender (figs. 279, 280). Dorsal and ventral arolium subequal in length.

Genus *Hebrus*

Figure 283

Note

Because of the numerous species and their strong morphological differences, there is no 'typical character' defining *Hebrus*. Even Andersen (1981a, 1982a) could not indicate any autapomorphy for the genus. Within Malesia however, *Hebrus* can be easily recognized by a combination of characters of antennae and sternal carinae. Species discrimination is mainly based on male genitalia and structure of the bucculae. Additional reliable characters are found in colour, pilosity, and structure of legs. In the past, too much emphasis has been put to the shape of the metanotal elevation (pseudoscutellum). This character varies depending on the development of wings and is therefore only useful when the same morphs are compared. Although most *Hebrus* are macropterous, some degree of brachyptery is observed in several species. A distinct group of species, with long, externally visible parameres (fig. 265), is presently known from the Philippines and Australia.

Diagnosis

Relatively small (1.30-2.54 mm), usually stout. Dorsally variable in colour, but many species with a black or reddish body and blackish hemielytra with pale stripes proximally. Predominately macropterous, short winged morphs only known in some species, wingless morph extremely rare. Antenna long, segment 4 subdivided by membranous zone (fig. 270). Buccula strongly developed (fig. 276), with two or three impressions. Paired longitudinal carinae on thoracic

283

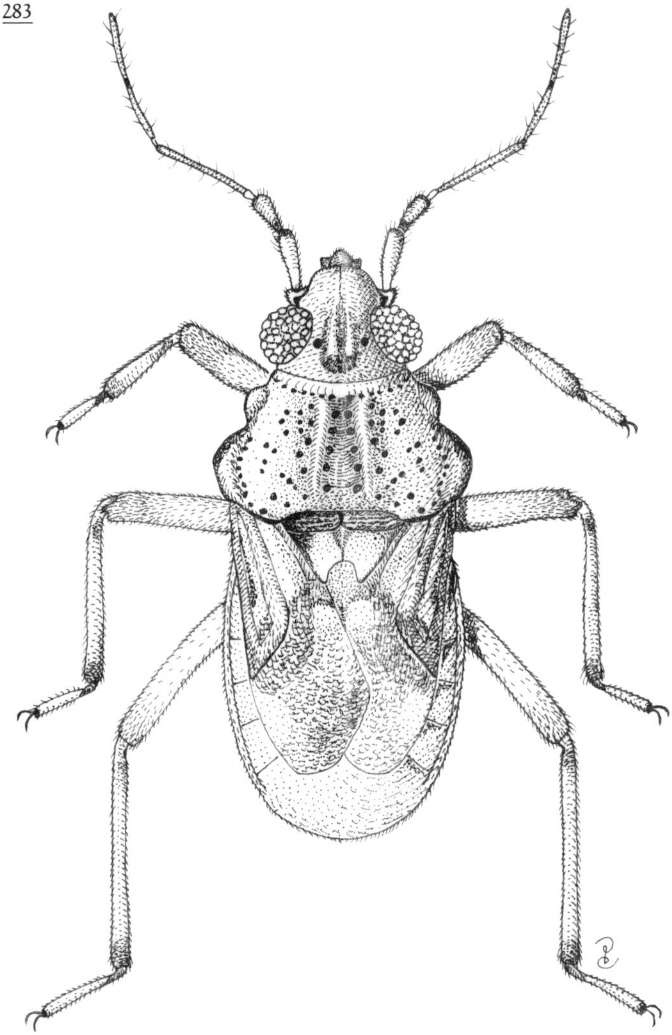

FIGURE 283
Hebrus hoberlandti, brachypterous male, length 1.6 mm.

venter parallel throughout and continuing separately onto the base of abdomen (fig. 282). Legs rather long and slender, claws relatively strong. Male genitalia usually small, always symmetrical, parameres usually short and hooked (fig. 263), in a few species elongate and surpassing end of abdomen (figs. 264, 265).

Distribution in Malesia (fig. 577)

Until now twenty described species are known from Malesia. Based on recent collecting in this area however, species diversity must be considerably larger. On the Philippines about 20 species are currently recognized (half of which undescribed), and other areas probably have similar high species numbers. There are apparently some widely distributed species, like *H. cruciatus*, described from the Philippines and recently recorded from Thailand (Zettel, 1999h). Other species are probably more restricted, but only rarely to single islands like in several Philippine endemics. As identification of species is extremely difficult and should be based on examination of male genitalia at high magnification, confirmed distribution patterns of species are presently not available.

Biological notes

Microhabitats of *Hebrus* seem to be more diverse than those of any other hebrid genus, due to the large number of species. Most oriental species can be collected on banks of running and stagnant fresh and brackish waters (Andersen, 1982a; J. Polhemus & D. Polhemus, 1989). Some species have been collected in light traps. Many species are hygropetric. *Hebrus nereis*, *H. mangrovensis*, and *H. murphyi* have exceptional life forms: they inhabit mangroves, where they hide during high tide in air bubbles kept by small holes in submerged, dead wood (Polhemus & Polhemus, 1989; Zettel, 2004b). Some Australian species have been collected in pitfalls or on twigs in rainforests (Lansbury, pers. comm.). Personal observations indicate that habitat requirements of single species are rather specific and influenced by sediment (rock, gravel, sand), vegetation (lichens, mosses), inclination, humidity, light conditions, et cetera. Some species flee to the surface of stagnant waters, especially after disturbance, eventually also for hunting. Specimens of *H. cruciatus* were found under stones on a river bank in Mindanao, Philippines, where they took shelter from the heat at noon (Zettel, 1999h). For biology of *Hebrus*, see the biological notes under the family text.

Genus *Merragata*

Figure 284

Diagnosis

(Indo-Australian species only) Very small (up to 2 mm, but up to 1.8 mm in Malesian species) and stout. Malesian species dorsally blackish brown, with clavus and membrane white. Old World species monomorphic macropterous. Antenna short and slender, segment 4 without membranous zone (fig. 269). Eyes relatively large and close to anterolateral margin of pronotum (fig. 274). Buccula rather narrow. Claws relatively small. Male genitalia minute. In Malesia this genus is immediately recognized by its antennal structure. *Merragata pallescens* var. *whitei* has a high body, relatively coarse sculpture of pronotum (within Hebrinae), and dense semi-erect pilosity on head, pronotum, and basal areas of fore wing.

Taxonomic note

Conspecificity of *M. pallescens* var. *whitei* with *M. pallescens* from India was recently discussed and doubted by Zettel (1999f). The Oriental species and the sibling

284

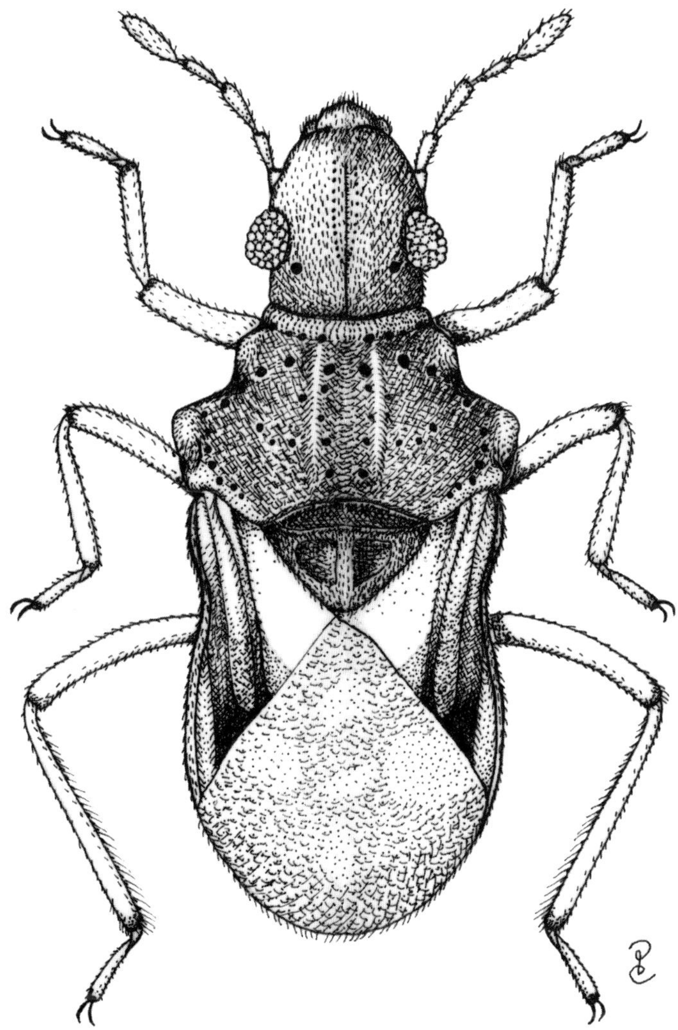

FIGURE 284
Merragata pallescens var. *whitei*, macropterous female, length 1.65 mm.

M. hackeri Hungerford from Australia form an aberrant group which may deserve subgeneric rank.

Distribution in Malesia (fig. 577)

Only one species occurs in Malesia, *M. pallescens* var. *whitei*, described from Java (Lundblad, 1933a).

Biological notes

Merragata species are reported to live on plant-covered water surfaces and rarely enter the damp ground at the shore (Drake, 1917; Andersen, 1982a). In Northeast Thailand *M. pallescens* var. *whitei* was found under the same conditions (Zettel, 1999f). A few observations on North American *Merragata* were reported by Drake (1917): specimens stand, walk or run on the water surface, or creep on submerged plants below the surface, their body enclosed in an air layer held by the velvety hairs. Nothing is known about food habits and life history.

Genus *Timasius*

Figure 285

285

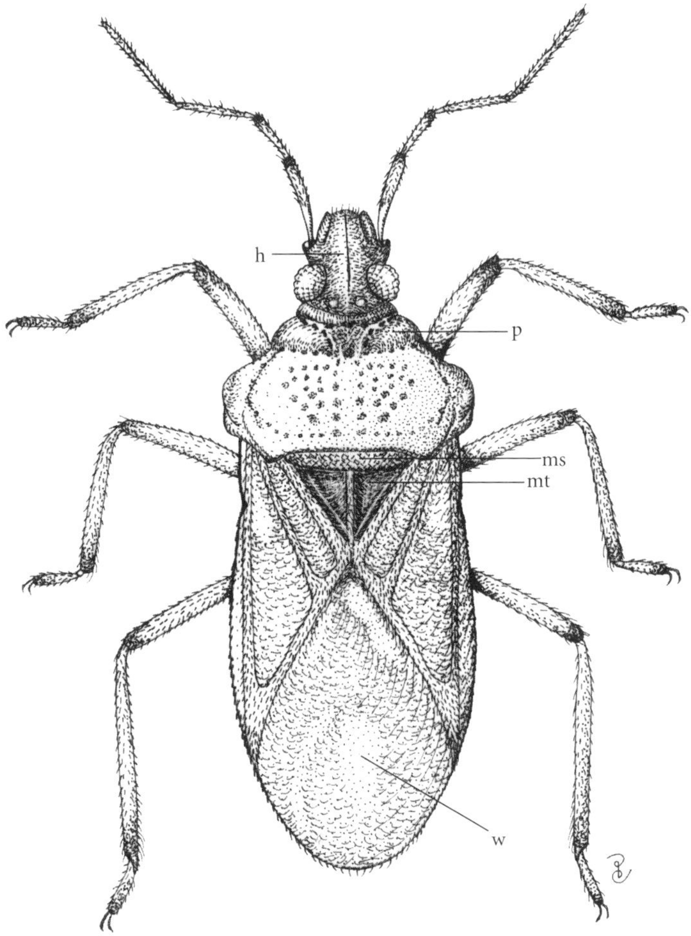

FIGURE 285
Timasius jaechi, paratype, male, length 2.9 mm; h = head, ms = mesoscutellum, mt = metanotal elevation or pseudoscutellum, p = pronotum, w = wings.

Diagnosis

Small to relatively large (2.5-3.7 mm). Dorsally dark brown to black, usually with golden pubescence; hemielytra proximally usually with pair of elongate whitish or bluish-grey pruinose spots. Monomorphic macropterous. Antenna very long, segment 4 subdivided by a membranous zone (fig. 271). Buccula strongly developed with two or three impressions (fig. 275). Paired longitudinal carinae of thoracic venter converging and meeting before hind margin of metasternum (fig. 281), not lace-like. Legs long and slender, claws relatively strong (fig. 280). Male genitalia usually large and asymmetrical, in some species strongly modified. *Timasius* is easily distinguished from te more common occurring *Hebrus* by structure of ventral carinae (compare figs. 281 and 282). Species groups distinguished e.g. by structure of bucculae, general build of male genitalia, and dorsal pilosity (see Andersen, 1981a; Zettel, 2004d). Identification bases mainly on male genitalia, which consist of modified pygophore, paramere, and proctiger; genitalia often strongly asymmetrical and sometimes of bizarre shape (Zettel, 1998e). Only some females can be specifically identified with certainty by special characters.

Distribution in Malesia (fig. 577)

This genus is strictly Oriental and distributed from Sri Lanka and South India to Taiwan, Java, Borneo and Flores. Except for Taiwan, which is inhabited by an endemic species-group, and a record of an undescribed species from Flores (J.T. Polhemus, pers. comm.), it seems to be restricted to areas with Pleistocene connections to the Asian mainland, as *Timasius* is absent from the Philippines and Sulawesi. Six described species occur in Malesia: two species, belonging to the *T. jaechi*-group, on Borneo (Zettel & Chen, 2000b), and four species, belonging to the *T. chinai*-group, on continental Malaysia and Java (Zettel, 2004d).

Biological notes

Timasius species typically inhabits steep surfaces of wet, shaded rocks along streams and rivers, especially under overhanging rocks and under the stones along streams. All species are agile runners and quickly take to flight when disturbed. Biology is unknown.

SUBFAMILY HYRCANINAE

Diagnosis

Fore wing uniformly dark. Head elongate with dorsal median furrow. Antenna short, clearly four-segmented, segment 4 thicker than preceding segments, without membranous zone and without preapical group of sensorial hairs. Eye distinctly removed from anterolateral margin of pronotum, with two curved ocular setae. Bucculae narrow, without impressions. Rostral segment 3 surpassing posterior margin of head. Legs stout and relatively short. Dorsal arolium distinctly shorter than ventral arolium.

Genus *Hyrcanus*

Figure 286

Diagnosis

Relatively large (2.5-3.2 mm) and usually stout, rarely more slender. Monomorphic macropterous. Dorsally blackish with scattered minute golden hairs;

286

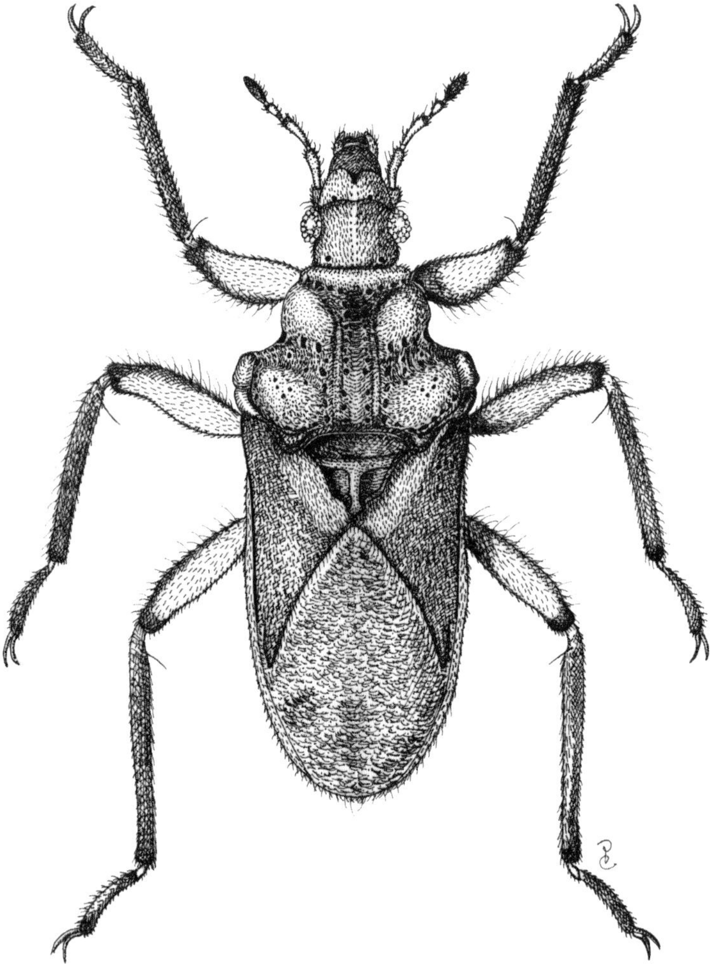

FIGURE 286
Hyrcanus varicolor, macropterous male, length 2.9 mm.

many species with brownish pronotum. Antenna short, as long as or shorter than head, segments 2 and 4 subequal in length, slightly shorter than segment 1, segment 3 shortest (fig. 267). Eye small, slightly removed from anterolateral margin of pronotum. Head below eye with conspicuous tuft of long hairs (fig. 272). Ventral lobe of head large, protruded anteriad. Pronotum (fig. 286) anteriorly with median furrow and paired lateral tumescences, posteriorly with pair of longitudinal carinae. Metanotal elevation triangular, laterally with pair of small subapical tubercles. Thorax ventrally with pair of widely separated longitudinal carinae reaching base of abdomen. Mesosternum with deep median rostral furrow; metasternum shorter than mesosternum along midline; orifice of metasternal glands at hind margin of metasternum. Fore wing with strongly thickened veins bearing rows of bristles, with one closed cell. In resting position, costal margin close to connexivum. Femora strongly enlarged, mesofemur about three times as long as wide. Tarsi and claws long and stout (fig. 277). Sutures between abdominal sternites medially bent anteriad. Genital segments of male and female small. Genitalia of male symmetrical, with relatively simple proctiger (fig. 259) and long parameres (compared with most other Hebridae, fig. 260). *Hyrcanus* can be distinguished from the highly specialized genus *Nieserius* mostly by plesiomorphic characters (e.g. relatively large eyes, strictly macropterous). Apomorphic characters are the tuft of hairs ventral of eye, modifications of antennal segment 2, and the strongly thickened femora. Species discrimination is mainly based on characters of male genitalia (paramere, proctiger). *Hyrcanus varicolor*, the only Malesian species, has unique peg-like bristles on the proctiger (fig. 259). Additional specific characters are found in pilosity and shape of head and pronotum. Females of some species are not distinguishable with certainty.

Distribution in Malesia (fig. 578)

Hyrcanus varicolor is widely distributed in Myanmar, Thailand, Vietnam, Sumatra and Java, and is so far the only species of this genus recorded from Malesia.

Biological notes

There are few notes on habits of *Hyrcanus*. Those given by Lundblad (1933a) on *H. varicolor* (as *H. capitatus* Distant) in Sumatra and Java indicate a terrestrial (hygropetric) life. Andersen (1981a) reports similar field notes by M. Holmen on *H. saxatilis* Andersen in North Thailand. The authors collected *H. draculus* Zettel, 1998 and *H. varicolor* always at the edge or between accumulated plant material in running waters (small streams or small rivers). Apparently *Hyrcanus* temporarily walks into the water and lives there at least facultatively. However, habitats are very different from the strictly benthic living genus *Nieserius*. A plastron-like pubescence on the abdominal sternites may support the supposedly partially aquatic habits of *Hyrcanus*. Modifications of antennal segment 2 and the long bristles below the eyes may eventually be part of a breathing mechanism as convergently developed in Hydrophilidae (Coleoptera) (see Zettel, 1998d for more details).

Genus *Nieserius*

Figure 287

Diagnosis

Relatively large (2.3-3.3 mm). Dorsally brownish-black with short, fine, apressed golden hairs. Structurally similar to *Hyrcanus*, but differing mainly in the following characters: antennal segment 1 shorter than segment 2 and segment 4 respectively (fig. 268); eye small, distance from pronotum at least 0.6 times eye width; posterior end of buccula behind hind margin of eye; head below eye

287

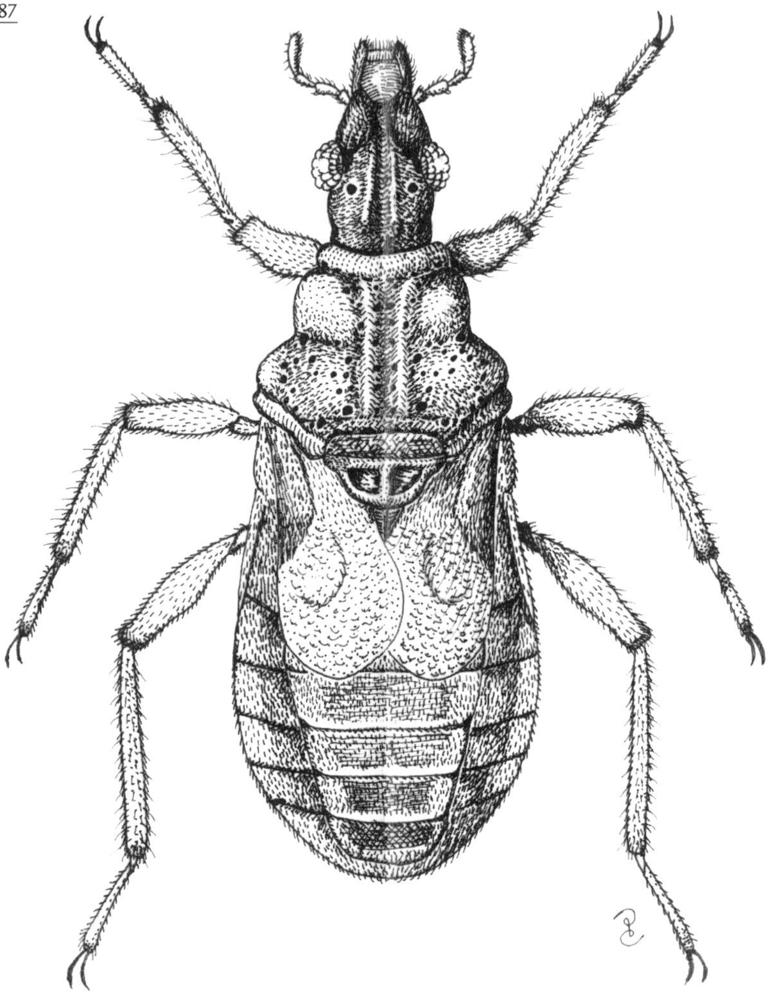

FIGURE 287
Nieserius subaquaticus, paratype, brachypterous female, length 2.7 mm.

without tuft of long bristles (fig. 273); ventral groove of head shallow; legs not very stout (fig. 278); mesofemur 3.8-4.5 times as long as wide; tarsi slender; proctiger with sharp constriction; paramere short (fig. 261). Brachypterous morph common. *Nieserius* species are the only permanently subaquatic Gerromorpha.

Distribution in Malesia

Thus far, this genus has not been encountered in Malesia. Three species have been described from Nepal, North Thailand and North Laos (Zettel, 1999e). Another representative of this genus was found in Myanmar (J. Polhemus, pers. comm.).

Biological notes

Field observations have only been published for one species. *Nieserius subaquaticus* Zettel inhabits the fine gravel beds in the current of streams, similar to *Aphelocheirus* or certain Elmidae (Coleoptera) (Zettel, 1999e). In Myanmar specimens were found while collecting elmid beetles under stones in rapids (J. Polhemus pers. comm.). The benthic habits of *Nieserius* are unique among Gerromorpha. *Nieserius subaquaticus* creeps on the bed of streams, where it probably feeds on small insect larvae and other animals. Specific apomorphic characters are small eyes, short antennae, and the fact that it is usually brachypterous. According to the preferred habitat, Zettel (1999e) supposes that the bugs get oxygen from a plastron: the differently developed body hair layers seem to have this function, enabling the animal to stay permanently submerged. Hence *N. subaquaticus* is probably confined to waters with a high oxygen content. It is unknown from where it gets the primary gas bubble, but it is probably built from small bubbles collected from the substrate, as suspected for *Aphelocheirus* and elmid beetles.

REFERENCES – Hebridae. Revision of *Timasius* and *Hyrcanus* (Andersen, 1981a); morphology, phylogeny, adaptations, biogeography and classification of Gerromorpha (Andersen, 1982a); revision of *Hyrcanus* (Zettel, 1998d); description of *Nieserius* (Zettel, 1999e); notes on Oriental *Merragata* (Zettel, 1999f); key to genera of Hebridae in Thailand (Zettel, 1999h); new species of Hebridae from southern India (Zettel, 2000); *Timasius* from Borneo (Zettel & Chen, 2000b); description of apterous *Hebrus* (Zettel, 2004b); description of *Hebrus* species from Malesia (Zettel, 2004c); revision of *Timasius chinai* group (Zettel, 2004d).

Family Hydrometridae – marsh treaders or water measurers

Diagnosis

Small to large (2.7-22 mm), usually very slender. Extent of micro- and macrotrichia variable. Head elongated in front of and behind the compound eyes (figs. 292, 293). Ocelli absent except in Heterocleptinae. Venation of hemielytra somewhat reduced, extending onto apical part except in Heterocleptinae. Tarsi three-segmented, claws inserted apically (ventrally in the Neotropical genus *Limnobatodes*). Ovipositor strongly reduced: only a pair of valves remains at abdominal tip, consisting mainly of first gonocoxae. First larval instars proportionally much broader than adults, recognizable by head already prolonged behind eyes.

Characteristics

Specific identification in *Hydrometra* involves the form of the anteclypeus (figs. 288, 289), ratio between length of head anterior of the eyes to length posterior of the eyes, and structure of abdominal segments 8-9, especially in males (fig. 290). Single females are often difficult to identify; the available keys mostly refer to male characters.

288

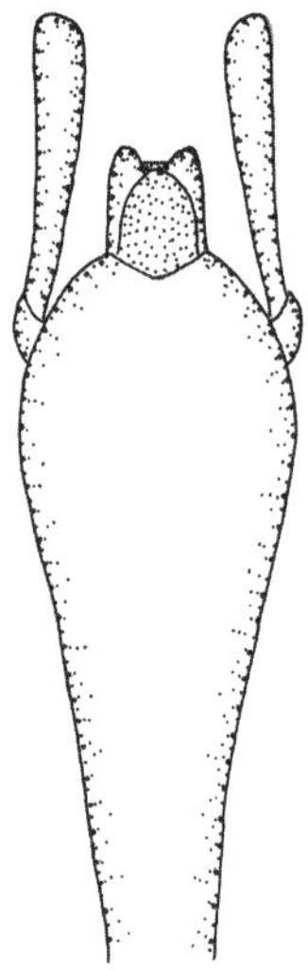

289

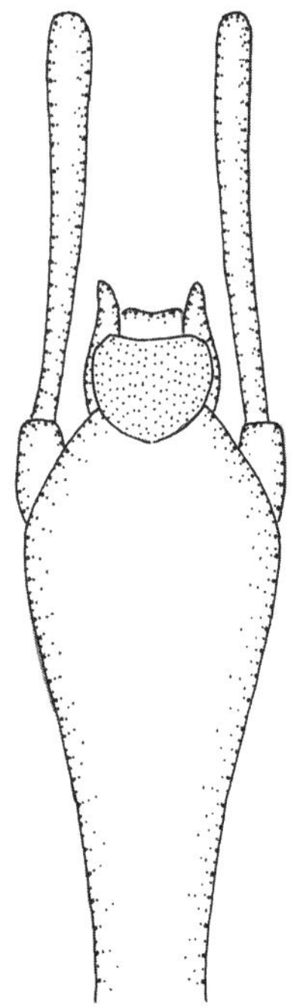

FIGURES 288 – 289
Hydrometridae, apex of head, anteclypeus densely punctate.
288 *Hydrometra orientalis* –
289 *Hydrometra papuana*.

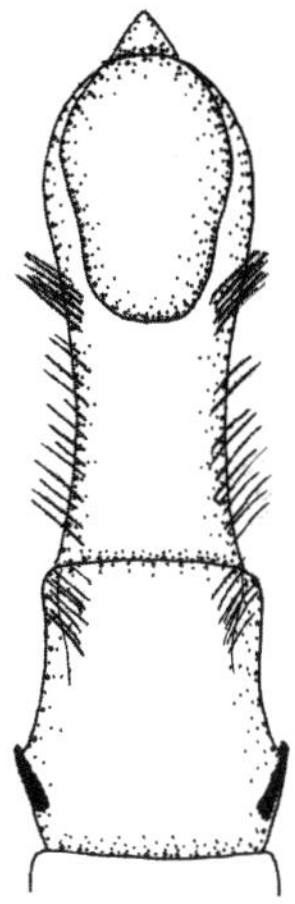

290

FIGURE 290
Hydrometra mindoroensis, male, apex of abdomen in ventral view.

Fossil record

Hydrometridae is the gerromorphan family with the most extensive fossil record. There are seven fossil genera, among which *Cretaceometra* Neil & Popov from the Lower Cretaceous of Brazil, and *Carinametra* Andersen & Grimaldi from the Middle Cretaceous of Myanmar. Andersen & Grimaldi (2001) present a complete phylogenetic tree of both fossil and recent genera of Hydrometridae.

Distribution worldwide

Although there are seven extant genera, 115 of the about 125 species belong to the cosmopolitan genus *Hydrometra*. The number of species is rather low in temperate and Arctic areas compared to warmer areas, e.g. only two species occur in Europe and in the northern half of North America.

Distribution in Indo-Australia

Indo-Australia has one species of *Heterocleptes* and 29 of *Hydrometra*. The fauna of Malesia contains nineteen species of *Hydrometra* and one of *Heterocleptes*. Australia is comparatively rich with nine species of *Hydrometra*, of which six endemics. Three species are restricted to Queensland (Andersen, 1982b; J. Polhemus & Lansbury, 1997; J. Polhemus & D. Polhemus, 1995a).

Biological notes

Hydrometra species live along the edge of stagnant and slowly running waters. They usually stay on the wet margins but in order to catch prey or when disturbed they also walk on the water. The long legs are used for walking, not for rowing. Individuals usually walk slowly, but they can run quickly when disturbed.

In order to copulate the male approaches the female, usually from the side, swings his body parallel to hers and mounts, grasping her pronotum with his fore legs. Abdominal segment 8 is then protracted in both sexes to free the genital capsule, and the venter of abdominal segment 7 (which in males often has small teeth of hairy tubercles) rests on her abdominal tergite 8. The male's segment 8 is bent downward and the genital capsule (segment 9) is directed under the female's segment 9. The phallobase is inserted between the gonocoxae and the aedeagus is inserted in the genital tract of the female. Copulation takes only a few minutes but the male may stay on the female for about fifteen minutes (Sprague 1956).

Eggs are very elongate and have an attachment disc. From the disc a stalk leads to the main egg body, which shows distinct longitudinal ribs. They are whitish at laying but darken very soon to brown or almost black, depending on species. Eggs are glued one by one to a hard substrate above the water surface (fig. 291), usually objects on the bank, but also emerging tips of floating water plants like *Ceratophyllum.* Before the egg is laid a drop of fluid comes out of the abdominal apex, apparently the glue with which the disc is fastened. After a few minutes the female moves away, thus pulling the egg from her abdomen. The whole process takes 5-15 minutes. In captivity eggs are often glued to aquarium walls.

291

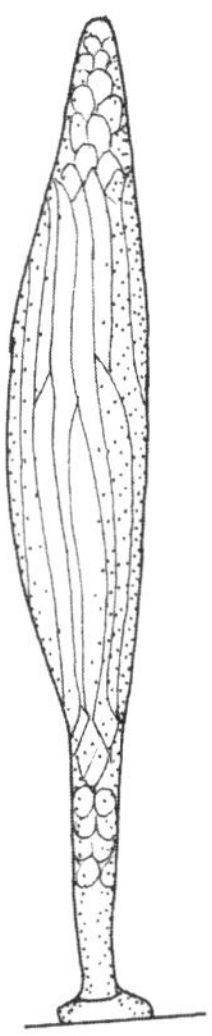

FIGURE 291
Hydrometra sp., egg.

Food consists of dead or small living insects which are pierced and held by the rostrum only. The maxillary stylets are, as in most Heteroptera, barbed and these barbs help to keep the prey fixed; these stylets are very long and are moved inside the prey, quickly sucking it clean. Prey includes adult Chironomidae, Cladocera, Collembola, Culicidae larvae and adults, Ephemeroptera, and Ostracoda. Some authors mention aerial insects trapped in the surface film. In addition, cannibalism is frequent, at least in captivity. First instar larvae are able to kill maggots of *Musca*-flies considerably larger than themselves. Some preys, e.g.

Leptinotarsa larvae and eggs, are rejected immediately after piercing (Cobben, 1978). Although they prey on culicid larvae and adults this is done too infrequently to have any impact on mosquito populations.

Key to Indo-Australian subfamilies and genera of Hydrometridae

1 Longer than 6 mm. Body very elongated, needle-like (fig. 293, Hydrometrinae)
▸ *Hydrometra* p. 225

– Length about 3 mm. Body not needle-like (fig. 292, Heterocleptinae)
▸ *Heterocleptes* p. 223

SUBFAMILY HETEROCLEPTINAE

Diagnosis

Small (2.7-3.2 mm), rather broad (length 3-4 times maximal width). Layer of micro- and macrotrichia covering head, thorax and base of abdomen. Posterior pair of trichobothria very long, positioned on prominent round elevations. Ocelli present.

Genus *Heterocleptes*

Figure 292

Diagnosis

Malesian species dorsally yellowish, hemielytra pale brownish with whitish marks. Antennal segment 4 with membranous ring, giving the antenna a five-segmented appearance. Diagnosis further as in subfamily.

Distribution in Malesia (fig. 578)

Heterocleptes spinosus has been described from Sabah.

Biological notes

The habits of *H. spinosus* are unknown. The very few records of the African species indicate a semi-terrestrial habitat (wet moss and soil in gallery forest; Andersen, 1982a).

292

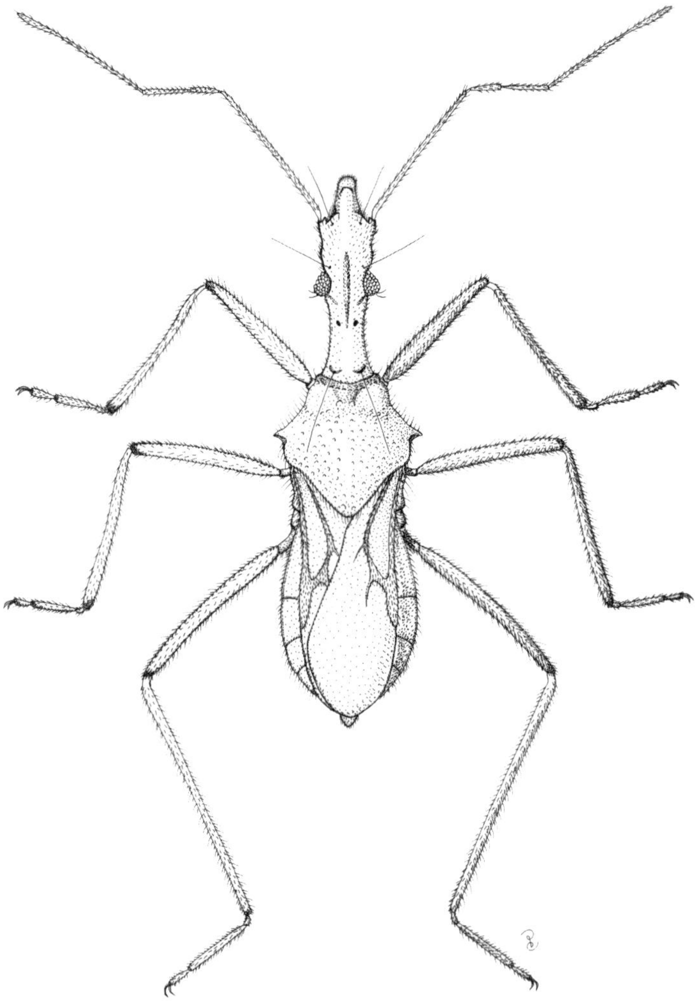

FIGURE 292
Heterocleptes spinosus,
macropterous female,
length 3.0 mm.

SUBFAMILY HYDROMETRINAE

Diagnosis

Medium to large (6.5-22 mm), elongate, needle-like (length six times or more maximal width). Layers of micro- and macrotrichia cover the entire body. Posterior pair of trichobothria rather short and usually on low elevations. Ocelli absent.

Genus *Hydrometra*

Figure 293

Diagnosis

Dorsally usually brown to blackish, hemielytra with whitish marks. Diagnosis further as in subfamily. In addition, body length ten times or more its maximal width. Head in front of eyes distinctly longer than behind. Posterior pair of

293

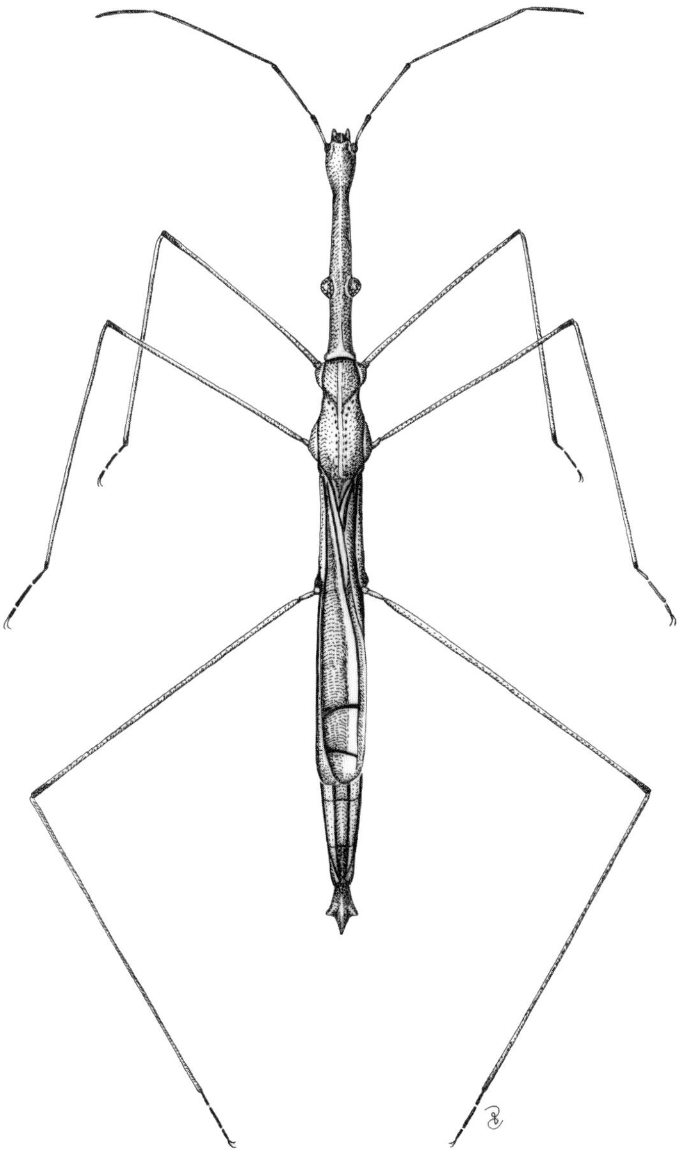

FIGURE 293
Hydrometra lineata, macropterous male, length 11.1 mm.

trichobothria not on elevations. Middle coxa distinctly closer to fore coxa than to hind coxa. Mesosternum and metasternum without sulci. Abdominal tergites longer than wide. Brachypterous morph with vestigial wings. There are however species with a morph with extremely short wing rudiments; in this family they constitute the true micropterous specimens (Andersen, 1992a).

Distribution in Malesia

The genus is distributed throughout the Malesian area. There are nineteen species in Malesia, many of which have wide ranges. There are apparently no areas with an exceptionally high number of species. New Guinea and Nusa Tenggara each have two endemic species.

Biological notes

Most *Hydrometra* species are found between vegetation of stagnant or slow flowing waters. Several species are found in rice fields or village ponds. Some have varied habitat requirements, e.g. *H. longicapitis* occurs both near muddy banks of lowland swamp forest streams and at mountain streams, especially near waterfalls (J. Polhemus & D. Polhemus, 1995a).

REFERENCES – Hydrometridae. Biology and morphology of *Hydrometra stagnorum* (Ekblom, 1926); biology and morphology of *Hydrometra martini* (Sprague, 1956); cladistic analysis, key to genera, remarks on biogeography (Andersen, 1977b); phylogeny, morphology, biogeography and classification of Gerromorpha (Andersen 1982a); description of *Heterocleptes spinosus* (Andersen, 1982b); key to species of West and Central Malesia (J. Polhemus & D. Polhemus, 1995a); key to species of East Malesia and Australia (J. Polhemus & Lansbury, 1997); Lower Cretaceous fossil hydrometrid (Nel & Popov, 2000); phylogeny of Hydrometridae, including both recent and fossil genera (Andersen & Grimaldi, 2001); key to Philippine species (Gapud et al., 2003).

Family Hermatobatidae – coral treaders

Diagnosis

Rather small (2.7-4.0 mm). More or less uniformly dark brown to blackish with dense greyish pilosity; head between eyes with yellowish marks. Monomorphic apterous. Layer of macrotrichia and microtrichia cover entire body. Head very short, at least three times as wide as long; ocelli absent. Pronotum short; mesonotum and metanotum indistinguishably fused in dorsal view. Tarsi three-segmented; middle and hind claws apical.

Characteristics

Systematics of this small family has recently been improved considerably by Andersen & Weir (2000). They present a revised key to males of all species and also present distributional data. Reliable characters are the structure of the male metasternum and fore femur.

Taxonomic note

Andersen & Weir (2000) synonymized *Hermatobates weddi* with *H. marchei*. We here follow the opinion of J. T. Polhemus (pers. comm.) and consider *H. marchei* and *H. weddi* as separate species.

Distribution worldwide

There is only one genus with nine described species. One species is known from the Caribbean, another one from the Red Sea area and the western part of the Indian Ocean (J. Polhemus pers. comm.). The other seven occur in the Pacific Ocean Basin, six of which occur in the Indo-Australian region.

Biological notes

Until recently *Hermatobates* species were considered to be very rare. The first to collect and study ample material was Cheng (1977). She concluded that specimens of most species collected up to several 100 meters off-shore were strays and that both adults and larvae live on coral reefs and shores. A possible exception is *H. hawaiiensis* China, which was regularly and in one instance in fair numbers collected in off-shore tows with a plankton net. This species becomes active during low tide, skating over pools or walking on the rocks. During high tide it hides under stones or coral debris containing air bubbles.

Foster (1989) extensively studied *H. weddi* in several locations in the South-west Pacific. He found that, when looked in appropriate places, *Hermatobates* appears to constitute a regular component of the intertidal fauna of tropical rocky and coral shores. The species hides during high tide under stones or coral debris and is restricted to a narrow zone of about 20 cm around the mean low tide line. The activity period is roughly from one hour before to one hour after low tide. The bugs emerge from holes in the rocks as the tide passes them, skate on the remaining puddles and retreat into the holes when the water returns. There are four larval instars; the younger instars stay close to the water edge and are very difficult to collect amongst the lapping waves. The last instar and adults, usually comprising 95% of the total catch, skate somewhat farther out on the open water. These observations were done during daytime, so the suggestion of previous authors that *Hermatobates* is mainly nocturnal is at least invalid for *H. weddi*. Some adults stay active after their hiding places are flooded and it is suggested that these constitute the bulk of nocturnal catches, which were mostly done during plankton surveys. However, the activity pattern at night on the shore is unknown. I. Lansbury (pers. comm.) found *Hermatobates* in large numbers off-shore, mainly at night, at Madang, New Guinea, where there is very little difference between high and low tide and tidal pools on the beach are absent. Both adults and larvae are often taken at light. Under stones younger instars constitute about 90% of the total number. Moult apparently takes place in air bubbles under water.

Copulation has been observed twice on a tidal pool: the male mounts the female when both are on the water surface. They then skate swiftly to the edge of the pool, where they climb a few millimetres out of the water and remain in copulation for about three minutes. They then separate and the female skates away (Foster, 1989). Feeding has been observed once: the prey consisted of recently dead adult marine Chironomidae (Foster, 1989).

J. Polhemus (1982) collected *H. haddoni* at Darwin, Northern Territories, Australia. He found most specimens at shallow pools on extensive sand flats with coral and rock debris. They ran rapidly over the rocks and were difficult to catch. When they entered a pool they skated extremely fast. The only abundant animal which could be potential prey was an unidentified collembolan.

Genus *Hermatobates*

Figure 294

Diagnosis

As in family.

Distribution in Malesia (fig. 578)

Three species are known from Malesia. A fourth species, *H. haddoni*, is endemic to Australia but also occurs on Mabuiag Island, just south of New Guinea, and may thus be found in Malesia.

294

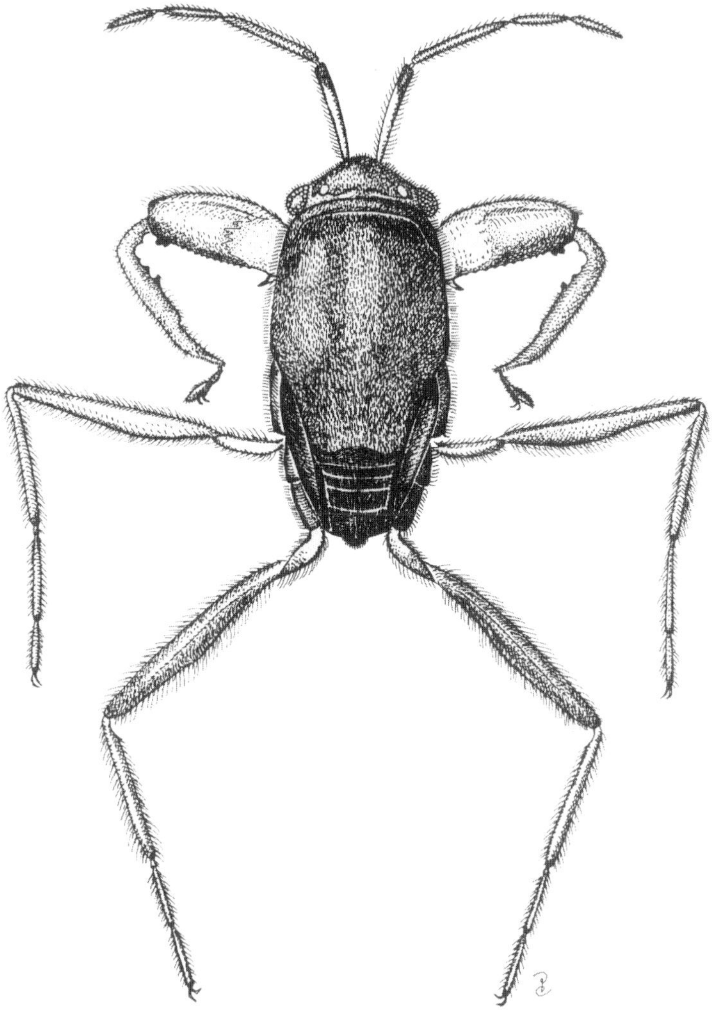

FIGURE 294
Hermatobates sp.,
male, length 3.5 mm.

REFERENCES – Hermatobatidae. Popular account of *Hermatobates* (Cheng, 1966a); description of *H. singaporensis* (Cheng, 1976); distribution and biology of *Hermatobates* (Cheng, 1977); phylogeny, adaptations, biogeography and classification of Gerromorpha (Andersen, 1982a); biology of *H. marchei* (Foster, 1989); key to species of *Hermatobates* (Andersen & Weir, 2000).

Family Veliidae – small waterstriders or watercrickets

Diagnosis

Small to medium-sized (1-10 mm). Entire body covered with layer of macrohairs and microhair. Head rather short and broad, in lateral view truncate, dorsally with median impressed line and pair of deep pits near hind corners of eyes. Three pairs of cephalic trichobothria. Antennal sockets usually ventrally of eyes and not visible in dorsal view. Compound eyes usually large. Ocelli absent (except in Ocelloveliinae, not in Malesia). First rostral segment small, may be up to twice as long as second segment. Pronotum often with collar and/or punctuated. Middle and hind coxae widely separated. Tarsi one- to three-segmented, tarsal claws nearly always placed subapically in distinct cleft, apical tarsal segment of middle legs often with modified ventral arolium and claws. hind femur often distinctly thicker than middle femur. Middle and hind leg often of subequal length, when of different length either one can be longest. Fore tibia of male apically usually with grasping comb of short stout spines on medio-posterior margin. Dorsal abdominal scent gland orifice absent.

Characteristics

The pits dorsally on the head near the inner posterior angles of the eyes are the points of attachment of an internal sclerite on which the antennal musculature is attached. The specialisation of the middle tarsus varies from laminar claws and ventral arolium in *Xiphovelia* and *Papuavelia* (figs. 305, 308) to an elaborate swimming fan in *Rhagovelia* (fig. 297). Characters for generic and specific identification are varied. Apart from general characters such as size, colour, pilosity, development and dentation of leg segments (notably the hind femur), there are many structures which, depending on taxon, may be diagnostic. In males the parameres (figs. 298, 299), and often the length of the grasping comb on the fore tibia (fig. 13), are usually diagnostic. Unfortunately the parameres may be rather small to apparently absent and difficult to evaluate. In *Pseudovelia* it is not the parameres but modifications of abdominal segment 8 in males which give specific clues (figs. 302, 303), whereas in *Neusterensifer* the shape of the process of the proctiger (fig. 314) is species specific. In females the shape of the abdomen, notably the connexiva, may be diagnostic. Apart from normal pilosity, especially in *Rhagovelia* there are tiny black denticles on the body, the extension of which is important in separating species (fig. 363). The number and development of closed cells in the hemielytra of macropterous specimens (figs. 300, 301, 295) and the extension of the pronotum in apterous ones (figs. 319, 321, 366, 362) often give clues to genera or species groups.

Taxonomic note

Recently, different opnions have been expressed as to the systematic position of the genera *Tetraripis* and *Chenevelia*. D. Polhemus (1997), J. Polhemus & D. Polhemus (1999) and D. Polhemus & J. Polhemus (2003) argue that *Tetraripis* and *Chenevelia* Zettel (the latter with one species from Thailand and Vietnam (Zettel, 1996b; D. Polhemus & J. Polhemus, 2003)) belong to the Veliinae. Andersen (2000c) argues that *Tetraripis* and *Chenevelia* belong to the Rhagoveliinae, based on the synapomorphy of the unique mesotarsal structures shared with *Rhagovelia* (figs. 296, 297). This book follows the latter opinion. Veliinae in the present sense may remain as a paraphyletic group (Andersen, 2000c).

Fossil record

The fossil record of Veliidae is rather poor: fossils of five taxa have been found so far. There are three extinct genera from the Eocene/Miocene, about 40 million years ago (Andersen, 1998). In addition, one of the oldest gerromorphan fossils known is an unnamed veliid described by Jell & Duncan (1986) from the Lower Cretaceous of Australia, about 120 million years ago (see also the paragraph on gerromorphan phylogeny in the chapter on taxonomy). A fossil *Halovelia* was found in amber from Dominica (see below under *Halovelia*).

Distribution worldwide

Veliidae presently contains more than 850 described species and is the largest family within Gerromorpha (Andersen & Weir, 2001, 2003a). They are distributed over all zoogeographical regions and from temperate to tropical zones. Most species are found in the tropics. Six subfamilies are recognized, of which Microveliinae (approximately 37 genera), Rhagoveliinae (three genera) and Veliinae (nine genera) are widely distributed in both Old and New World. Haloveliinae (five genera) occurs from East Africa (including Madagascar) through the Oriental and Australian areas into the Pacific (but there is one fossil record from the Dominican amber; see Andersen & Poinar, 1998), Perittopinae (one genus) occurs in the Oriental region (Andersen, 1982a). Only subfamilies Veliinae and Microveliinae, as well as a very few species of Rhagoveliinae, occur also in temperate areas. The genera *Microvelia* and *Rhagovelia* each hold several hundred species (among which many undescribed).

Distribution in Indo-Australia

Five subfamilies and 27 genera occur in Indo-Australia. Most species belong to the Microveliinae (fifteen genera, about 110 described species) and Rhagoveliinae (three genera, about 125 described species). Perittopinae contains one genus with eleven described species and Veliinae three genera with eight described species. Haloveliinae contains two limnic genera (sixteen described species) and three marine genera (41 described species). Microveliinae, Haloveliinae and Rhagoveliinae are distributed throughout the area. The distribution of Perittopinae is limited to the east by Wallace's Line and Dickerson's Line. Veliinae transgresses this line with only one species of *Angilovelia*, which was recently found in South-east Sulawesi.

Biological notes

Veliidae is a large and diversified family; representives are adapted to a large variety of habitats, from wet soil and rock to fast flowing streams.
An interesting method of locomotion so far only observed in various Veliidae (*Microvelia*, *Velia* Latreille and *Rhagovelia* larvae) is expansion skating or 'skimming' (Andersen, 1976). This locomotion is based on the tension of the water surface. The veliids release a drop of salivary liquid to move, using the tension in the surface film. Direction of movement is quite variable, depending where the drop of saliva is deposited. In *Microvelia* the process can be easily observed when the animal moves on a surface with dust or bacterial film. In the field one can often see a *Velia* skating upstream suddenly dash forward, seemingly effortlessly. Expansion skating is believed to have evolved as a way to avoid predators.
Copulation starts in many species by the male jumping on the female, grasping her thorax firmly with the fortibial grasping comb. Several species exhibit phoresy: the male rides on the female more or less continuously. Females of *Halovelia*, Haloveloides and various species-groups of *Microvelia* have modifications of the thorax and abdomen to accomodate males during their ride (Kellen, 1959; J. Polhemus, 1974). Wilson (1958) reported that in the American *Platyvelia brachialis* (Stål) the female mounts the male and stimulates him to copulate by rubbing his abdomen with her hind legs. As in several genera of the Veliinae, both males and females have grasping combs on the fore tibia. It would be interesting to check wether this reversed behaviour is more widespread among Veliinae.
Veliidae is the gerromorphan family with the largest number of genera possessing a stridulatory device. In South America two species of *Paravelia* and all species of *Stridulivelia* sensu stricto have stridulatory mechanisms. In Malesia these mechanisms are restricted to *Rhagovelia lugubris* and *Angilovelia y-alba*. In addition, *Chenevelia stridulans* Zettel from North Thailand and Vietnam has stridulatory devices. Stridulation is much more widespread among Nepomorpha. Possibly, visual clues are much less effective under water, which could have led to the use of stridulation for intraspecific communication. This is weakly supported by the fact that both *Angilovelia* and *Chenevelia* have been collected in dark localities, and the terrestrial stridulating nepomorphan genus *Nerthra* is apparently active at night.
Eggs are glued lengthwise to a substrate usually just in contact with or slightly above the water, often plants but also stone or wood. They are attached by their less convex side and embedded in a gelatinous substance.
Of all Veliidae, especially *Microvelia* preys upon larvae and pupae of Culicidae. There are two detailed reports of *Microvelia* attacking pests (Spence & Andersen,

2000). The first deals with M. *douglasi atrolineata* predating upon the brown planthopper *Nilaparvata lugens* (Stål), a major rice pest in rice. The second report is by Laird (1956) who noted an unidentified species of *Microvelia* attacking egg rafts of the mosquito *Culex femineus* on Espiritu Santo in Vanuatu (New Hebrides). Some species of *Microvelia* are easily reared in large quantities. Moreover, many species have a considerable fraction of apterous specimens in the population. The combination of these qualities would make further investigation to use *Microvelia* in biological control in very small waterbodies (such as pots with marsh plants) worthwhile. See also biological notes under Gerridae.

Key to Indo-Australian subfamilies of Veliidae[1]

1 Middle tarsus deeply cleft with a fan of plumose or hairy swimming fans arising from its base (figs. 296, 297) ▸ **Rhagoveliinae p. 278**

– Middle tarsus less deeply cleft, lacking plumose or hairy swimming fans ▸ 2

2 Middle tarsus three-segmented, first segment may be short (figs. 362, 370) ▸ 3

– Middle tarsus two-segmented ▸ 4

3 Fore tarsus two-segmented; first segment of middle tarsus more than half as long as second segment (fig. 362). Hemielytron divided in a proximally thickened coriaceous portion with two closed cells, and a membranous distal portion (fig. 295) ▸ **Perittopinae (only one genus)** *Perittopus* **p. 276**

– All tarsi three-segmented, first segments less than half as long as second (fig. 370) ▸ **Veliinae p. 286**

295

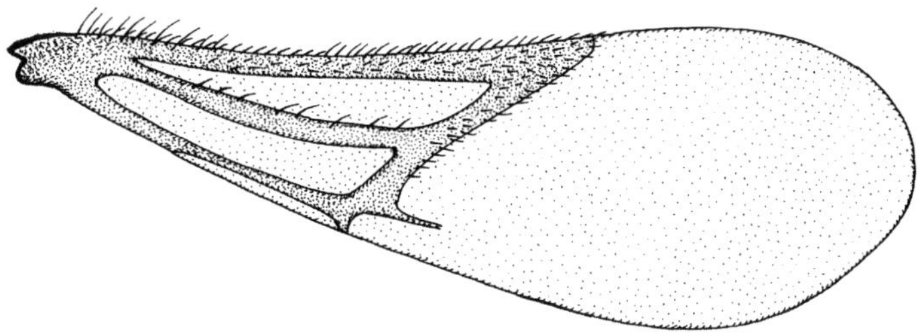

FIGURE 295
Perittopus sp., right hemielytron.

1 Adapted from Andersen (1982a).

296

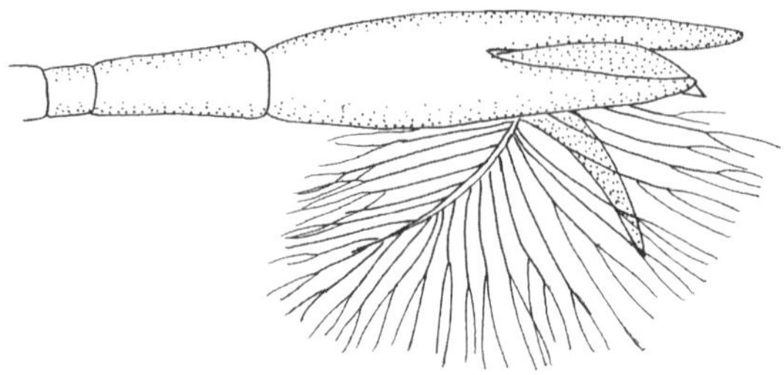

297

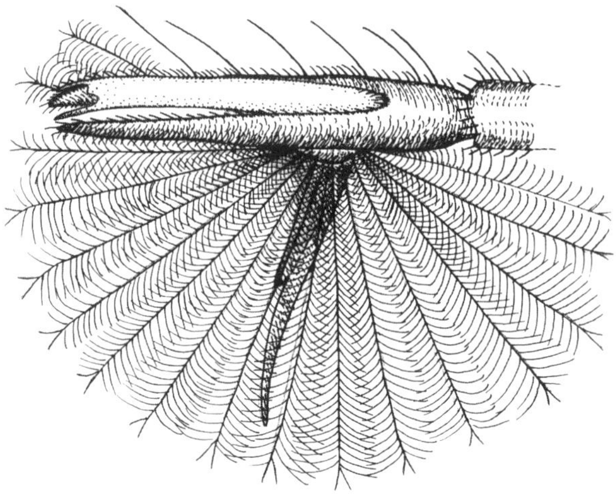

FIGURES 296 – 297
Rhagoveliinae, swimming fan.
296 *Tetraripis borneensis*, hind tarsus – **297** *Rhagovelia* sp., apical segment of middle tarsus (adapted from Andersen, 1982a).

4 Fore tarsus one-segmented (fig. 13). Middle tarsus rarely more than twice as long as hind tarsus ▸ **Microveliinae p. 234**

– Fore tarsus two-segmented (fig. 340). Middle tarsus three or more times the length of hind tarsus ▸ **Haloveliinae p. 264**

REFERENCES – Veliidae general. Malesian species (Lundblad, 1933a); locomotion (Andersen, 1976); morphology, phylogeny, adaptations, biogeography and classification of Gerromorpha (Andersen, 1982a); fossil veliid from Lower Cretaceous (Jell & Duncan, 1986); assignment of *Tetraripis* to Veliinae (D. Polhemus, 1997); key to genera of Thailand (Hecher, 1998); fossil Gerromorpha (Andersen, 1998, 1999c); classification of subfamilies (Andersen, 2000c); key to Indo-Australian Veliinae (D. Polhemus & J. Polhemus, 2003).

SUBFAMILY MICROVELIINAE

Diagnosis

Small to medium-sized (1.5-6 mm). Anterior margin of pronotum nearly always with transverse yellow or orange band. Apart from apterous specimens, individuals usually with fully developed or rarely with reduced wings. Hemielytra at most with five closed cells (figs. 300, 301). Tarsal formula unique: fore tarsi one-segmented, middle and hind tarsi two-segmented, middle tarsi not deeply cleft and at most twice as long as hind tarsi. Pretarsal appendages sometimes leaf-like (figs. 305, 308).

Characteristics

This is the most ecologically diverse subfamily of Veliidae, which is reflected in morphological specialisations mentioned in the generic diagnoses. Important characters for both generic and specific identification are the structure of the thorax, especially in apterous specimens, and the morphology of the legs and antennae. The modification of pretarsal appendages into leaf-like structures apparently developed independently in various taxa.

Distribution worldwide

There are 37 genera with about 350 described species. Of these, the cosmopolitan *Microvelia* holds about 150 described species (Andersen & Weir, 2003a), more or less evenly divided over the Old and the New World. *Pseudovelia*, mainly in the Old World tropics, has about 50 described species. Of both genera numerous undescribed species have been collected. Of the remaining 35 genera 19 are Indo-Australian.

Distribution in Indo-Australia

Of the 21 genera in the Indo-Australian region the terrestrial genus *Geovelia* Zimmermann occurs in the Himalayas while six genera are Australian endemics. The remaining fourteen genera hold about 90 species, of which 75 described (belonging to all fourteen genera). Eighteen species belong to *Pseudovelia*. At least 21 undescribed species of *Pseudovelia* from the Philippines have been recognized (Sehnal, 1999). *Microvelia* has fourteen described and a considerable number of undescribed species in Malesia. Australia, with 40 species of *Microvelia* (most of them endemic; Andersen & Weir, 2003a), is extremely rich, or perhaps better surveyed, for this genus. Apart from Australia, the greatest variability is found in New Guinea and adjacent islands with six genera, of which five endemic, and eighteen species, only two of which belong to *Microvelia*. Also here, several undescribed species, at least for the moment assigned to *Microvelia*, have been collected. So far no specialized genera have been recorded from the Bismarck and Solomon Archipelagos.

Biological notes

Most Microveliinae species, notably *Microvelia*, live at the edge of stagnant water between vegetation. However many genera have invaded special habitats. There are at least two genera living on wet vertical rock faces: *Aegilipsovelia* J. Polhemus from Mexico and *Aegilipsicola* from New Guinea. Various *Microvelia* live in small confined habitats such as Bromeliaceae, crab holes or tree holes. Species of *Lathriovelia* live in internodes of bamboo partly filled with water. Several species, in Malesia notably *Velohebria* from New Guinea, are terrestrial whereas quite a number of species, including most *Pseudovelia*, are associated with running water.

Microveliinae move about by walking, not rowing as in Gerridae or Rhagoveliinae. The American *M. longipes* Uhler, which lives on open but shaded water, is the only veliid thus far reported to produce ripple signals. Apparently these are warning signals to keep other males out (J. Polhemus, 1990c). Ripple signals are observed more often in various Gerridae.

In *Phoreticovelia* and several species of *Microvelia* males are reported to stay on the back of the female for a long time. This is called riding piggy-back or phoresy. In Malesia this behaviour is most pronounced in *M. leveillei*. As far as observed, eggs of *Microvelia* are glued with the flattened side on floating objects on the bank of the water. However, egg laying habits of most genera and species are unknown.

Food consists of all kinds of small soft-bodied invertebrates. Preys specifically mentioned are anopheline larvae (Frick, 1949), newly laid mosquito eggs (Laird, 1956) and larvae of their own species (by adults; Miyamoto, 1953). Adults also frequently attack adult conspecifics, but usually without success (Andersen, 1982a). Young larvae seem to feed especially on micro-crustaceans. Large dead prey is located by sight, small prey such as fruit flies apparently not (Jackson & Walls, 1998). Chemical clues are apparently not involved in prey detection.

Some species of *Microvelia* have been observed to feed in groups, which means that several individuals suck the same prey, a behaviour also known from various Gerridae. Initially the function of this behaviour was thought to enable individuals to utilize a larger prey: the saliva of one bug would not be sufficient to histolyze its contents. Travers (1993) analyzed group foraging in the North American *M. austrina* de la Torre Bueno using dead fruit flies as prey. He found that the time to locate a prey was lower in groups than in isolated individuals, in spite of the fact that individuals in groups moved less on average. When a *M. austrina* found a prey it increased tarsal movement. As foraging microveliids quickly responded to conspecifics which located food by moving in the direction of the feeding animal(s) it is supposed that these rapid tarsal movements function as signals. Similar results were found by Jackson & Walls (1998) for the New Zealand *M. macgregori* (Kirkaldy). In addition they found that struggling prey was found quickly both during daylight and at night, but that as long as they struggled vigorously, especially larger prey, they were not pierced.

Key to Indo-Australian genera of Microveliinae[1]

1	Specimens from New Guinea ▸	13
–	Specimens from other parts of Malesia ▸	2

1 Adapted from Andersen (1982a) and J. Polhemus & D. Polhemus (1994). For Australian fauna see Andersen & Weir (2001, 2003a).

2 Species at least 3.5 mm. First antennal segment usually slender and long, at least 3/4 of distance between the eyes ▸ 3

– Species less than 3.5 mm. First antennal segment usually incrassate and mostly shorter and stouter ▸ 7

3 Third antennal segment 1.1 times as long as fourth, length of second antennal segment 2/3 of third, total length of antenna half of body length. Length 3.6-4.5 mm ▸ *Microvelia albolineata* p. 246

– Third antennal segment usually shorter than or equal to fourth, occasionally slightly longer but then second antennal segment about half as long as third and length of antenna 3/4 or more than body length ▸ 4

4 Posterior surface of fore femora, at least in males, modified (with hair tufts, spines, tubercles, apical swellings, etcetera) on posterior surface ▸ 5

– Posterior surface of femora not modified ▸ 6

5 Eyes distinctly separated from lateroanterior angles of pronotum (fig. 317) [peninsular Malaysia] ▸ *Lathriovelia* p. 244

– Eyes touching or very close to anterolateral angles of pronotum (fig. 316) [Oriental] ▸ *Baptista* p. 242

6 Second antennal segment about half as long as third (fig. 327) [Borneo] ▸ *Polhemovelia setosa* p. 255

– Second and third antennal segments of subequal length (fig. 322) [Oriental] ▸ *Neoalardus* p. 248

7 First antennal segment slender, its length subequal to the interocular space ▸ 8

– First antennal segment incrassate or, if slender, distinctly shorter than the interocular width ▸ 10

8 Second antennal segment half as long as third ▸ 9

– Second antennal segment 3/4 as long as third; third and fourth antennal segments not very thin, only slightly narrower than second segment. Length of female 3.2 mm [Bali] ▸ *Microvelia magnifica* p. 246

9 Antennae only slightly over half of body length. Third and fourth antennal segments thin, half as wide as second segment (fig. 316). Male parameres falciform (fig. 298). Length of Malesian species 2.2-3.5 mm [peninsular Malaysia] ▸ *Baptista* p. 242

– Length of antennae 3/4 to equal the body length. Third and fourth antennal segments thin but wider than half of width of third segment. Male parameres with a widened round tip (fig. 299). Length 2.4-4.1 mm [Borneo] ▸ *Polhemovelia* p. 255

FIGURES 298 – 299
Microveliinae, right parameres.
298 *Baptista femoralis* –
299 *Polhemovelia setosa*.

298

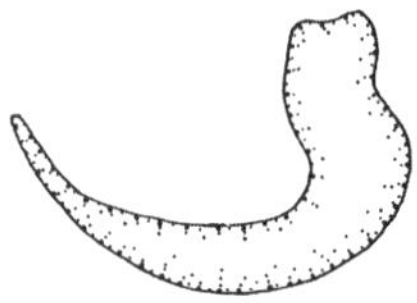

299

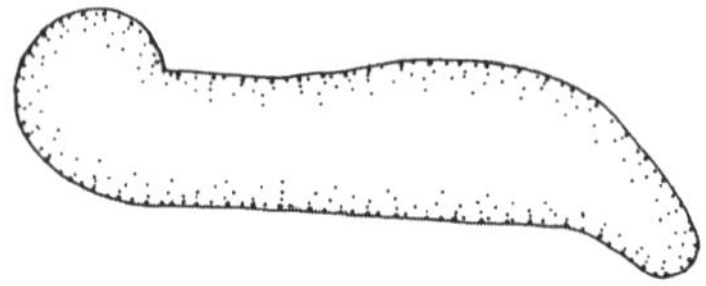

10 First antennal segment incrassate, extending more than $^2/_3$ of its length beyond apex of head. Apical cells of fore wings reduced, distal $^1/_3$ of hemielytron with two longitudinal weak veins, often difficult to detect (fig. 300). Pronotal lobe of apterous morph covers mesonotum and most of metanotum ‣ 12
– First antennal segment more slender and usually much shorter; if not, then venation of fore wing normal with closed cells extending into distal third of hemielytron (fig. 301), or with shortened pronotal lobe (apterous morph) ‣ 11

FIGURES 300 – 301
Microveliinae, right hemielytron.
300 *Pseudovelia sangihe* –
301 *Microvelia douglasi*.

300

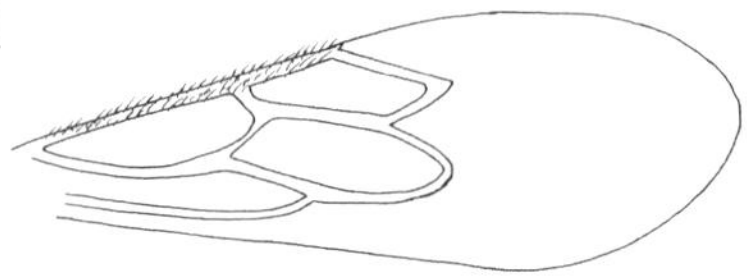

301

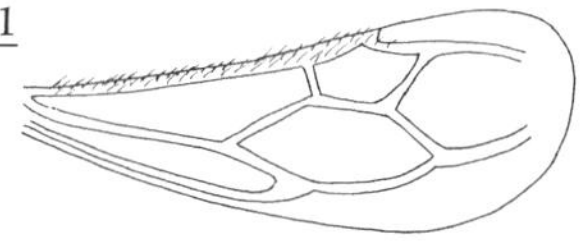

11 Middle tarsi with three leaf-like subapical structures (claws and ventral arolium; fig. 305) [Oriental] ‣ ***Xiphovelia* p. 262**
– Middle tarsi not modified (fig. 304) [cosmopolitan] ‣ ***Microvelia* p. 246**

12 Parameres very small, segment 8 of male ventrally strongly modified (fig. 302) [Paleotropical and Far Eastern] ‣ ***Pseudovelia* p. 256**
– Parameres large, segment 8 of male not strongly modified [Himalayas] ‣ *Geovelia*

302

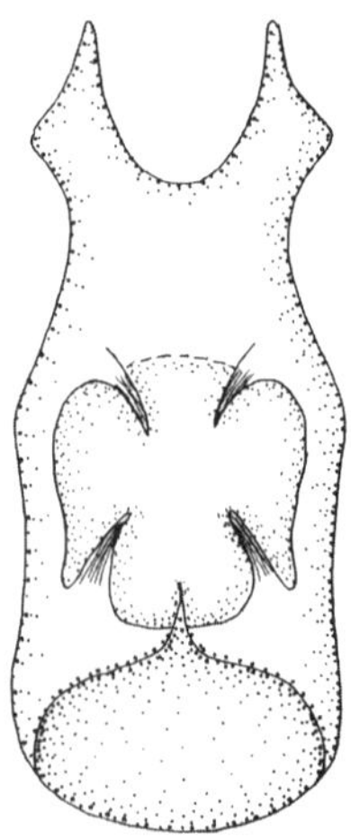

303

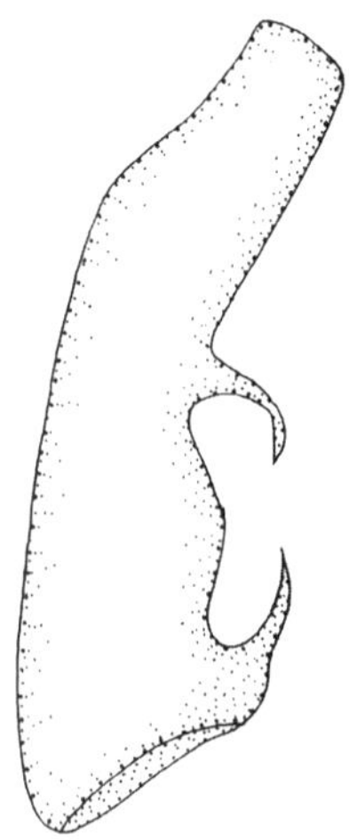

FIGURES 302 – 303
Pseudovelia epimekta male, abdominal segment 8.
302 ventral view –
303 lateral view.

Key to New Guinean genera of Microveliinae

13 Claws all subterminal, inserted very close to apex of tarsus (fig. 306) ▸ *Velohebria* p. 261

– Claws all preterminal, inserted distinctly before apex of tarsus (fig. 304) ▸ 14

14 Males up to 1.3 mm in apterous morph, up to 1.6 mm in macropterous morph. Female with large and deep depression on mesothorax and metathorax and base of abdomen (fig. 309, 326) ▸ *Phoreticovelia* p. 252

– Males larger, 2 mm or more. Females without depression on mesothorax and metathorax (figs. 310, 320) ▸ 15

15 Tarsi with four leaf-like structures (claws and arolia; fig. 308). Apex of male abdomen with broad and deep groove, segment 8 ventrally with large anteriorly directed hook (fig. 311). Large: males 3.3-3.5, females 3.8-4.1 mm ▸ *Papuavelia* p. 251

– Tarsi with pair of slender claws and pair of thin arolia (fig. 304). Smaller: males 3.0 mm or less, females less than 3.7 mm ▸ 16

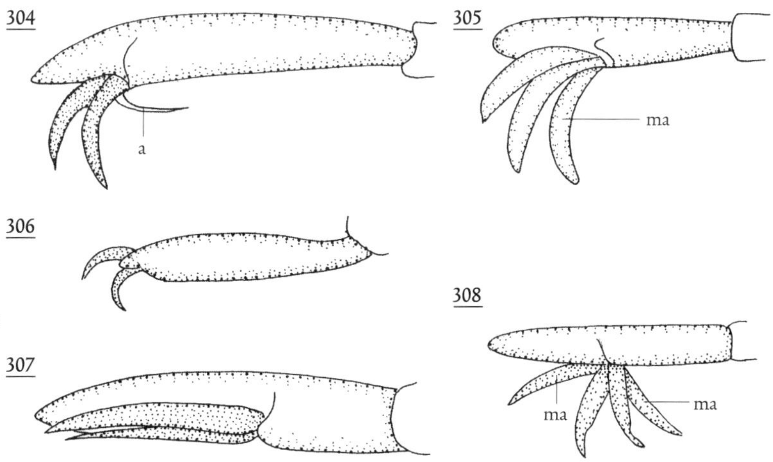

FIGURES 304 – 308 Microveliinae, apical tarsal segment in lateral view. **304** *Microvelia* sp., fore tarsus – **305** *Xiphovelia skoteina*, middle tarsus – **306** *Velohebria antennalis*, fore tarsus – **307** *Aegilipsicola rapida*, fore tarsus – **308** *Papuavelia*, hind tarsus; a = unmodified arolium, ma = modified arolium.

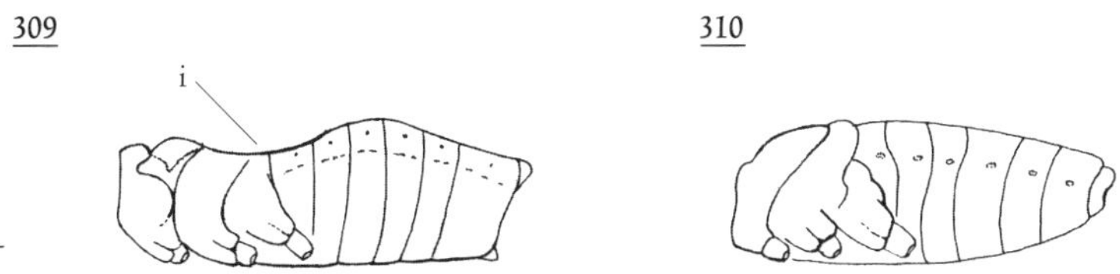

FIGURES 309 – 310 Microveliinae, females, body in lateral view, appendages removed. **309** *Phoreticovelia notophora* – **310** *Microvelia douglasi*; i = depressed area on dorsum.

311

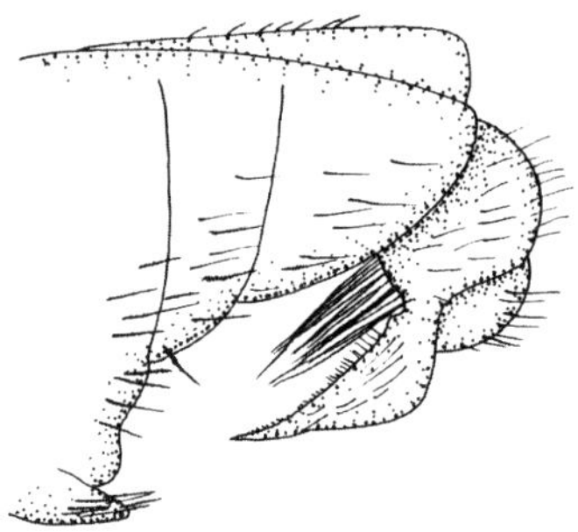

FIGURE 311
Papuavelia siculifera, male, apex of abdomen in lateral view.

16 Middle tarsi $^3/_4$ or more of length of middle tibia (fig. 330) ▸ *Tarsovelia* p. 259
– Middle tarsi half or less of length of middle tibia (fig. 318) ▸ 17

17 Claws very long (fig. 307), gula long (fig. 312). Living on steep wet rock surfaces ▸ *Aegilipsicola* p. 241
– Claws of normal length (fig. 304), gula shorter (fig. 313). Living on or at the margin of water bodies ▸ 18

312

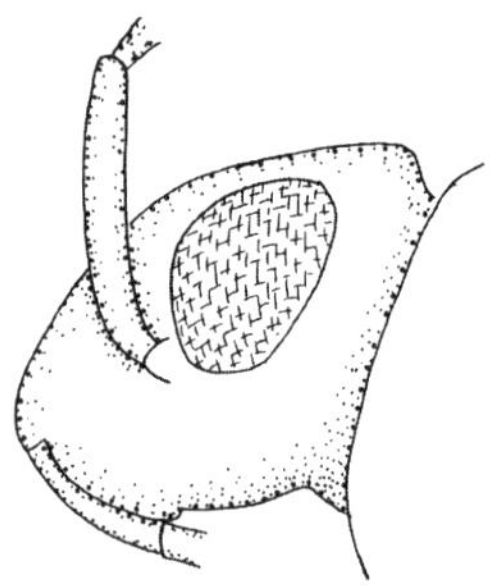

313

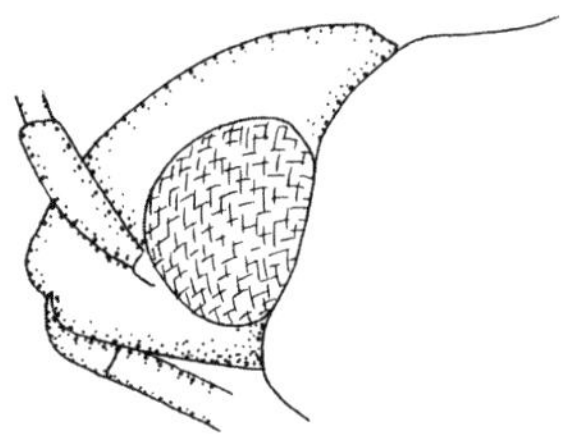

FIGURES 312 – 313
Microveliinae, head in lateral view.
312 *Aegilipsicola rapida* – 313 *Microvelia douglasi atrolineata*.

18 Antennae long, $^3/_4$ or more of body length (fig. 329). Monomorphic micropterous ▸ *Tanyvelia* p. 258
– Antennae shorter, $^2/_3$ or less of body length. Only apterous and macropterous morphs known ▸ 19

19 Male genitalia with long anterior ventral projection (fig. 314). tergite 7 of female folded, basal part horizontal, caudal part vertical, tergite 8 also vertically folded, together forming anal plate covering apex of abdomen in caudal view ▸ *Neusterensifer* p. 249
– Male genitalia without long anterior ventral projection. Female tergites 7 and 8 horizontal, not forming anal plate ▸ *Microvelia* p. 246

314

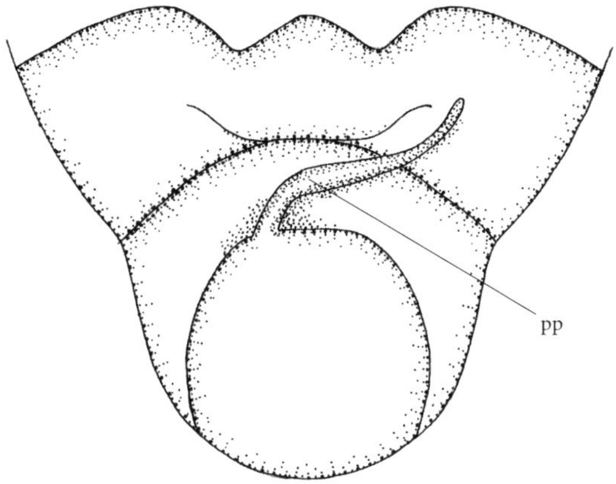

FIGURE 314
Neusterensifer compactus, male, apex of abdomen in ventral view; pp = process of proctiger.

Genus *Aegilipsicola*

Figure 315

Diagnosis

Small to medium-sized (2.0-3.7 mm), elongate and slender. Macropterous or micropterous. Dorsal colour dark brown to black, variably marked with yellowish-brown to brown; hemielytra black with six elongate bluish-grey pruinose streaks, two confined within basal cells, four others in distal third. Head not produced posteriorly. Antennae very long, about $^2/_3$ of body length. Eyes globose, laterally reaching beyond anterior pronotal angles and removed from anterior pronotal margin. Rostrum reaching to middle of mesosternum; rostral cavity posteriorly closed by strong carinate edge. Gular region long, plainly visible. Macropterous hemielytra with five closed cells, basal part not thickened; rows of rather long setae on veins in basal half. Legs slender, long, hind legs very long, claws subapical, very long (fig. 307). Fore femur in male with a small ventral tubercle on $^1/_3$ from tip. Grasping comb on fore tibia long, $^3/_4$ of of tibial length. Male genital segments moderately large, protruding; sternite of abdominal segment 7 apically and sternite 8 basally with distinct impression. Parameres small, short and broad, symmetrical. Female tergite 8 on level with tergite 7, protruding posteriorly over tergite 9. Gonocoxae somewhat retracted in apex of abdomen.

Distribution in Malesia (fig. 578)

Three species are endemic to the mountainous areas of Central New Guinea. One species is described from Batanta Island.

Biological notes

All four species have been taken from wet vertical rock faces, where they ran quickly across the surface (D. Polhemus & J. Polhemus, 2000a).

315

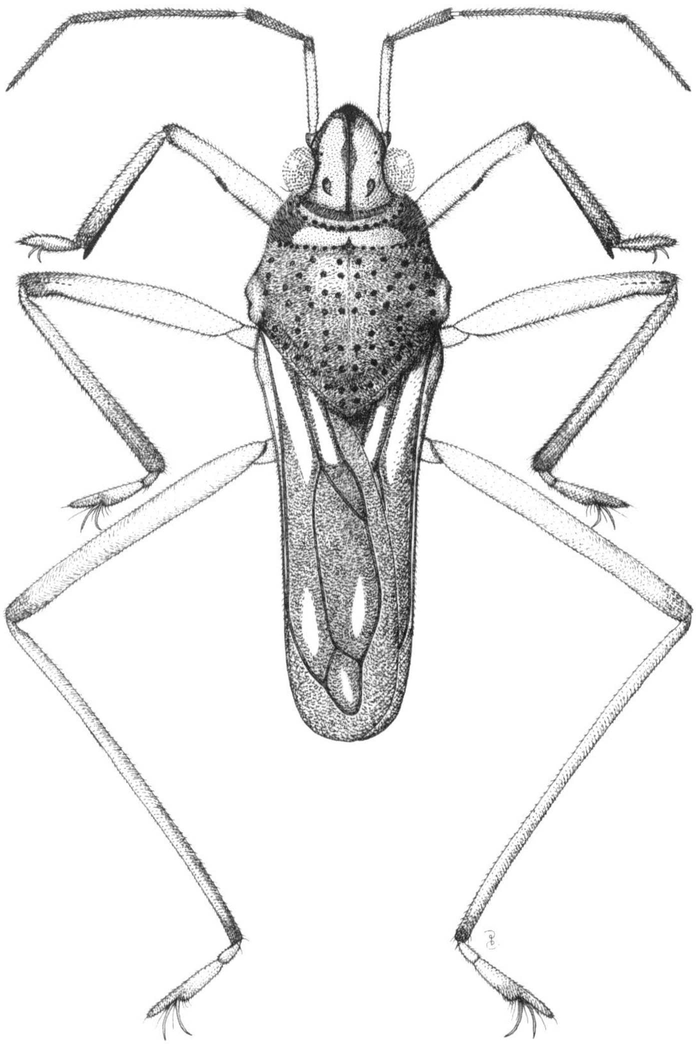

FIGURE 315
Aegilipsicola rapida, paratype, macropterous male, length 2.8 mm.

Genus *Baptista*

Figure 316

Diagnosis

Medium-sized (2.4-4.1 mm), elongate. Apterous or macropterous. Dorsally yellowish to dark reddish-brown. Silvery pubescence on head, first mediotergite and laterotergites. Head produced posteriorly, extending well behind distal margin of eyes. Antennae slightly over half of body length; first segment long and slightly curved, extending more than $^2/_3$ of its length beyond tip of head; third segment twice as long as second segment, third and fourth segment very slender, almost filiform, setose. Eyes globose, laterally reaching, or extending beyond, anterior pronotal angles. Rostrum reaching to middle of mesosternum.

316

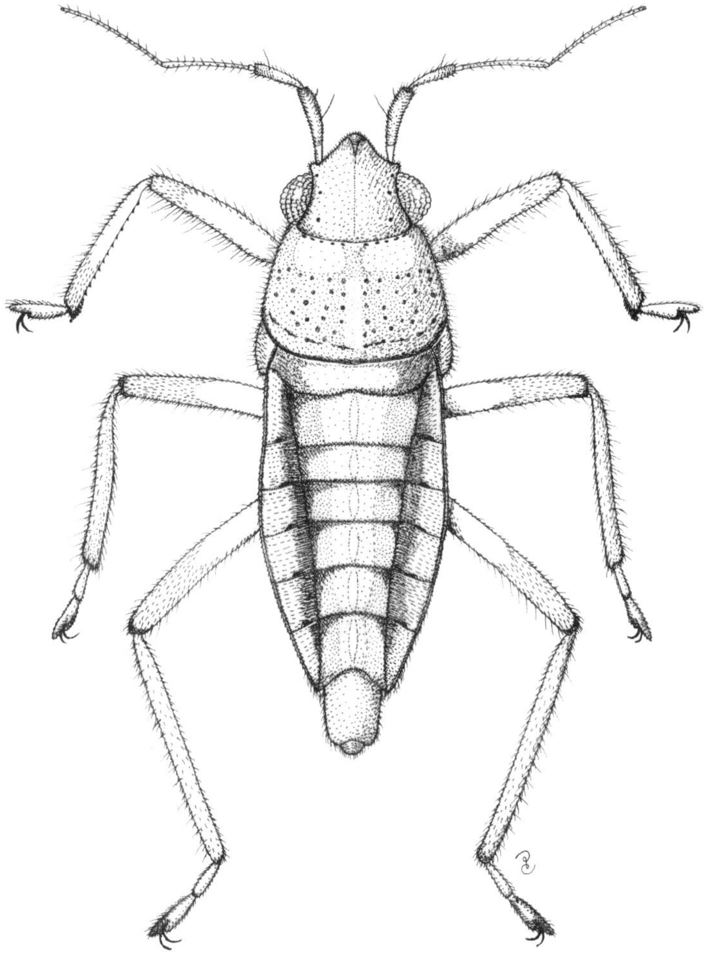

FIGURE 316
Baptista collaris, apterous male, length 2.2 mm.

Gular region moderately developed, clearly visible. Pronotal lobe of apterous morph large (fig. 316) with broadly rounded posterior margin, almost completely covering mesonotum. Macropterous hemielytra with four closed cells, basal third of hemielytra thickened and covered with dense short hairs. Wings distinctly surpassing tip of abdomen. Claws subapical. Male fore leg usually modified (femur and/or tibia curved with tumescences and specialised pilosity), except in *B. collaris*. Grasping comb on fore tibia short, $^{1}/_{4}$ or less of tibial length. Genital segments small to medium-sized, distinctly protruding. Abdominal segments 6 and/or 7 usually strongly modified with finger-like extensions (not in *B. collaris*). Parameres large, falciform, symmetrical. Female tergite 8 horizontal, level to tergite 7.

Distribution in Malesia (fig. 579)

There are six described species, all from India and South-east Asia (Andersen, 1989c; Zettel, 2004g). Two species have been recorded from Malesia: *B. femoralis* from North Thailand and continental Malaysia (Perak), and *B. collaris* from continental Malaysia. A third species, *B. digitata* Andersen was described from North Thailand and may occur in continental Malaysia too.

Biological notes

Collecting records show that species of this genus live in dark water-filled holes on rocky banks, small water-filled cavities under turf at the foot of rocks where water seeps out, and among coarse litter at a seepage on a river bank. The silvery pubescence makes them visible in the dark (Andersen, 1989c).

Genus *Lathriovelia*

Figure 317

Diagnosis

Large (4.4-6.0 mm), elongate. Monomorphic macropterous. Dorsally dark reddish brown to dark grey. Head posteriorly produced, extending well behind hind margin of eyes. Antennae from slightly less to slightly more than half of body length; first segment long and slightly curved, extending half to more than $^2/_3$ its length beyond tip of head, third segment $1^1/_2$ to twice as long as second segment, third and fourth segment very slender, almost filiform, setose. Eyes globose, laterally reaching to or just beyond anterior pronotal angles, situated well in front of anterior margin of pronotum. Rostrum reaching middle of mesosternum. Gular region moderately developed, plainly visible. Hemielytra with four closed cells, basal third of hemielytra thickened and covered with dense short pilosity. Wings leave last abdominal tergite uncovered or reach just to tip of abdomen. Claws subapical. Abdomen relatively long and elongate, especially in males. Male fore leg usually modified (femur and/or tibia apically somewhat swollen); grasping comb on fore tibia short, $^1/_4$ or less of tibial length. Genital segments small to medium-sized, distinctly protruding. Abdominal segments 6 and/or 7 modified. Parameres large, falciform, symmetrical. Female tergite 8 horizontal, level to tergite 7. Genital segments small but clearly visible behind abdominal sternite 7.

Distribution in Malesia (fig. 579)

The two described species are known from Malaysia, Sumatra and Thailand.

Biological notes

Both species are active at night. They are found mainly in partly water-filled internodes of large bamboo species such as *Gigantochloa*. They enter the internodes through holes made by other animals (e.g. beetles, caterpillars, wood-

317

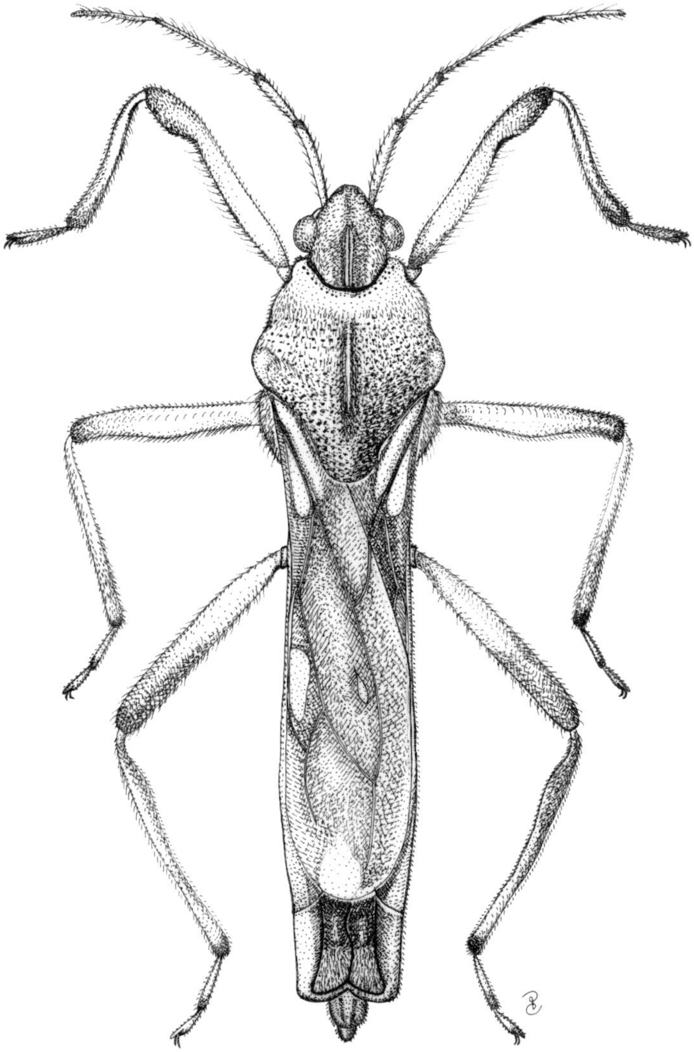

FIGURE 317
Lathriovelia rickmersi,
male, length 5.7 mm.

peckers), but are also occasionally found in water-filled bamboo stumps. Larvae were found only once in another plant than bamboo (in water-filled leaf axils of *Colocasia esculenta* (L.)) (Kovac & Yang, 2000). In their specialized habitat they seem to be rather common (Kovac & Yang, 2000). Food consists of Collembola and other invertebrates.

Genus *Microvelia*

Figures 318-321

Diagnosis

As this genus is very variable (and apparently not monophyletic) a consistent description cannot be given. Microveliinae without distinct apomorphies are provisionally placed in *Microvelia*.

Distribution in Malesia

This genus has a worldwide distribution and is recorded throughout Malesia. There are fourteen described species in Malesia but, as in *Pseudovelia*, many more species have been collected and are awaiting description. Two species are widespread: *M. douglasi* (figs. 318, 319) and *M. leveillei* (figs. 320, 321).

FIGURE 318 – 319
Microvelia douglasi.
318 macropterous male, length 1.6 mm – **319** apterous male, length 1.4 mm.

318

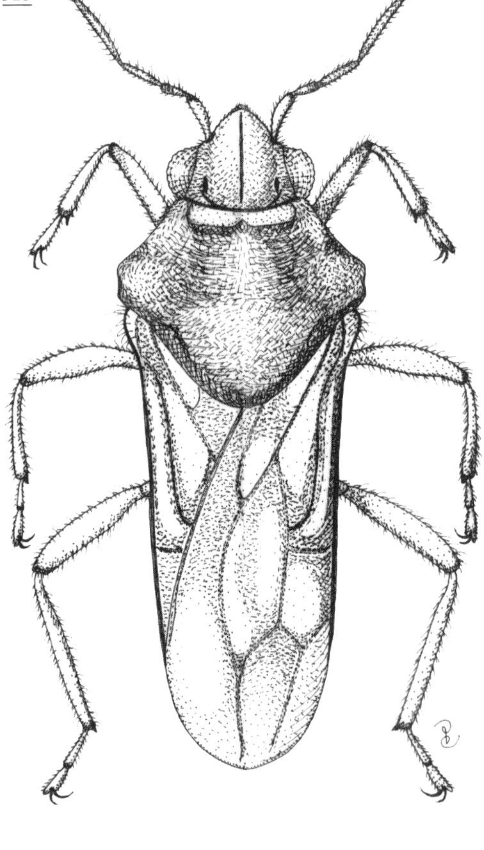

319

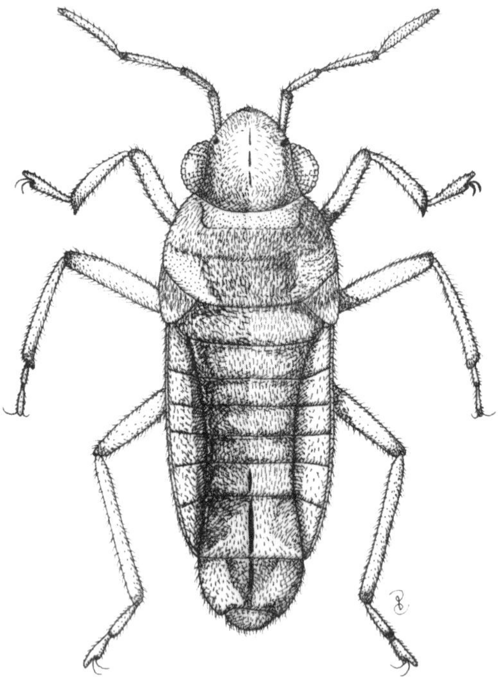

FIGURE 320 – 321
Microvelia leveillei.
320 apterous female, length 2.5 mm – **321** apterous male, length 1.5 mm.

320

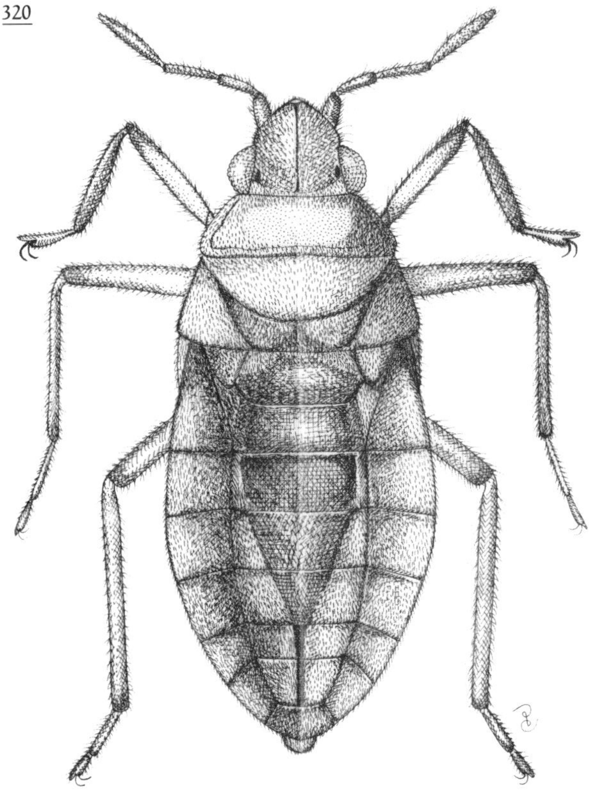

321

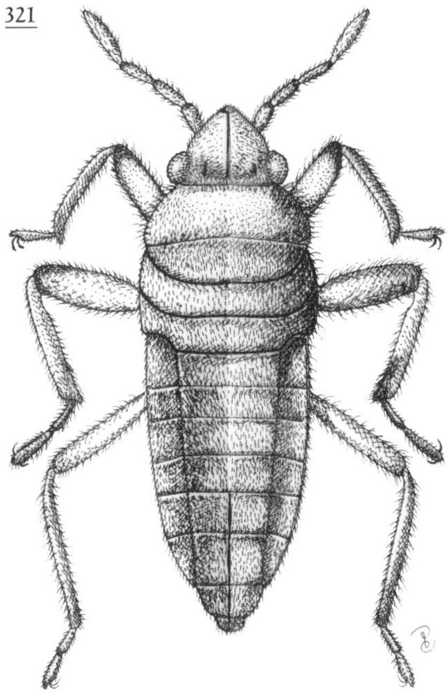

Biological notes

Many species, including the two widespread ones, occur at the edges of stagnant water, including small and artificial ponds. Other species live at edges of streams or wells. At least some macropterous *Microvelia* are attracted by light at night; among Malesian species this has been recorded for *M. douglasi*.

Genus *Neoalardus*

Figure 322

Diagnosis

Rather large (4.4-4.9 mm), elongate. Apterous or macropterous. Dorsally reddish to dull brown. Head produced posteriorly, extending well behind hind margin of eyes. Antennae setose, slightly more than half the body length, first segment long and slightly curved, extending more than $^2/_3$ of its length beyond tip of head, second and third segment of subequal length, second to fourth segment of equal width, distinctly thinner than first but not filiform. Eyes

322

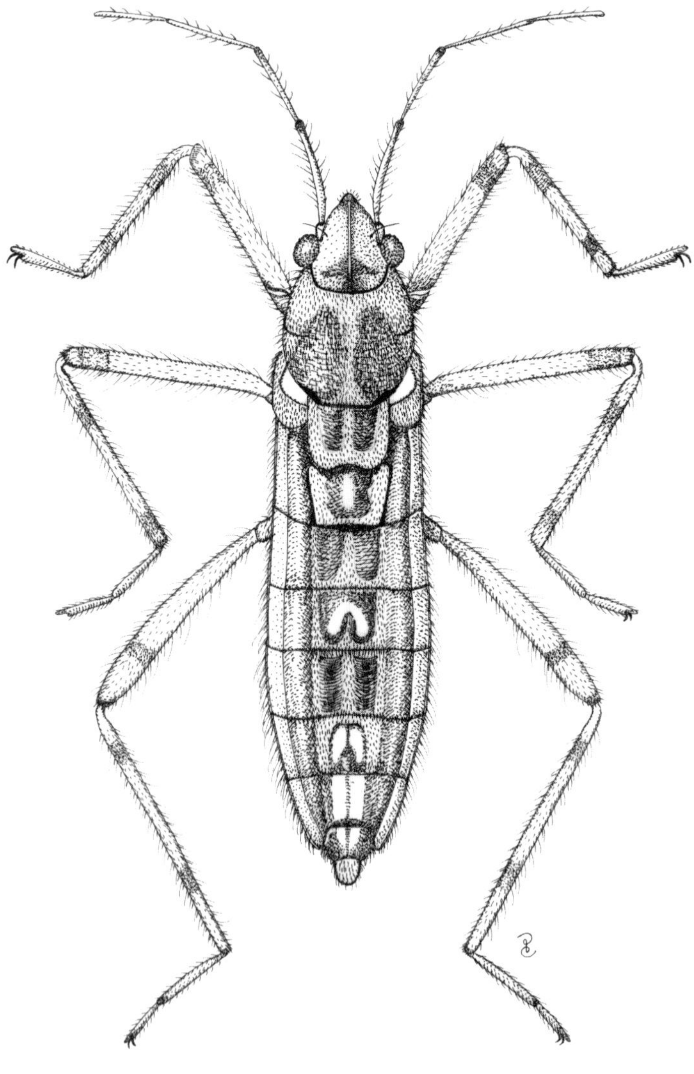

FIGURE 322
Neoalardus typicus, apterous female, length 4.6 mm.

globose, laterally reaching to anterior pronotal angles and situated at base of head, adjacent to anterior margin of pronotum. Rostrum reaching middle of mesosternum. Gular region moderately developed, clearly visible. Pronotum in apterous morph elongate, covering mesonotum and most metanotum. Macropterous hemielytra with four closed cells, almost reaching apex of abdomen. Basal third of hemielytra not thickened. Claws subapical. Abdomen very long and elongate. Male fore leg not modified; grasping comb on fore tibia absent. Genital segments medium-sized, distinctly protruding and well developed; abdominal segments not modified; parameres large, stout with median tooth on dorsal margin, symmetrical. Female tergite 8 horizontal, level to tergite 7.

Distribution in Malesia (fig. 580)

There is only one described species, N. *typicus*, occurring from India to Hainan. It has been recorded also from continental Malaysia and Sumatra (Zettel, 1998g).

Biological notes

The preferred habitat is supposed to be moist sand and mud along small streams, usually in deep shade from rocks (Andersen, 1982a). However, in Thailand N. *typicus* has been usually found between emerging vegetation at the edge of ponds, less frequently at the edge of sluggish streams. When the vegetation is pressed down onder water, the insects run swiftly over the water and try to climb up some emergent object, notably the collector's boots.

Genus *Neusterensifer*

Figure 323

Diagnosis

Small to medium-sized (2.0-3.2 mm), stout. Apterous or macropterous. Sexually dimorphic: females larger and with broader abdomen than males. Dorsally black, tinged with orange, heavily marked with silvery pubescence in apterous morph. Posterior head margin nearly straight, not produced behind hind margin of eye. Antennae long and slender, about $^2/_3$ of body length, first segment medium-sized and slightly curved, extending about $^2/_3$ beyond head; second segment short, about $^2/_3$ to $^4/_5$ the length of first and third segment (which are of subequal length); third and fourth segment thinner than first and second but not filiform. Eyes globose, laterally reaching to anterior pronotal angles and situated at base of head, appressed to anterior margin of pronotum. Rostral cavity posteriorly open, rostrum reaching middle of mesosternum; gular region short, not visible. Pronotum in apterous morph wider than long, covering mesonotum; metanotum about as long as pronotum. Macropterous hemielytra with four closed cells, occasionally five, almost reaching to abdominal tip. Basal third of hemielytra not thickened. Claws subapical. Abdomen

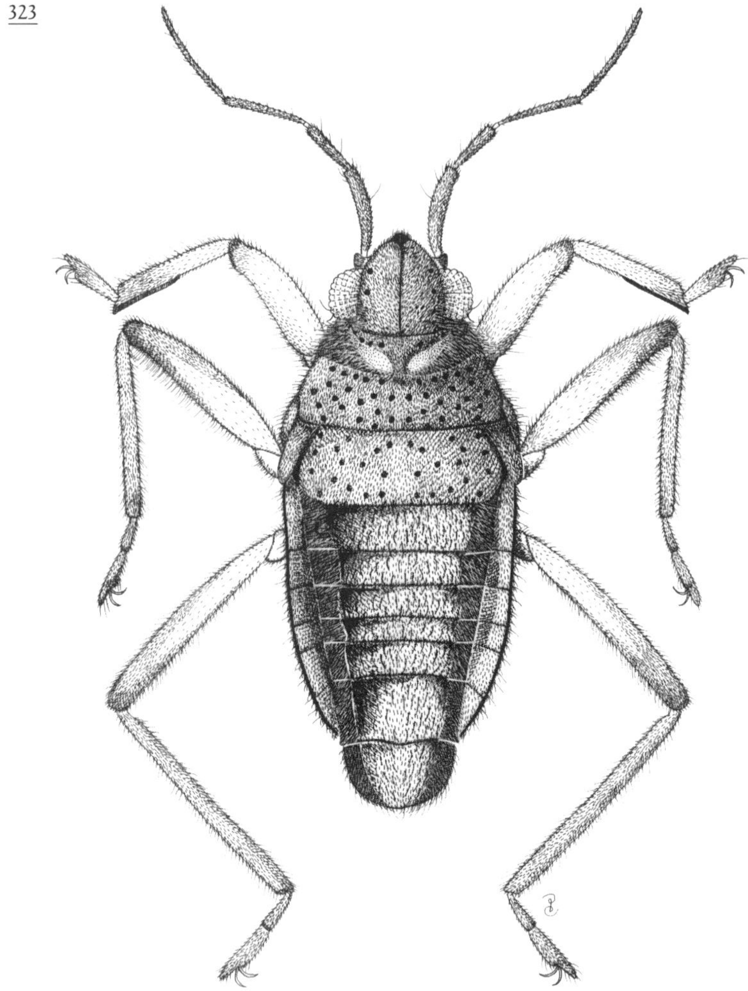

FIGURE 323
Neusterensifer compactus,
paratype,
apterous male,
length 2.4 mm.

broad and compact and extensively beset with silvery setae, more distinctly so in females. Male fore leg modified or not, according to species; grasping comb on fore tibia $^1/_2$ to $^3/_5$ of tibial length. Abdominal sternite 5 or 6 caudally with pair of dense tufts of dark erect setae on either side of midline. Median part of sternite 7 broadly impressed. Genital segments large, strongly modified, distinctly protruding; parameres vestigial or absent. In ventral view proctiger bearing large elongate appendage on anterior margin, directed anterolaterally beneath abdomen (fig. 314). Apical part of female tergite 8 curved downward, covering most of the genital segments.

Distribution in Malesia (fig. 579)
There are fourteen described species, all endemic to New Guinea and adjacent islands.

Biological notes

Neusterensifer species are most commonly encountered in sheltered areas along the margins of stream pools or on the surface of water-filled potholes in bedrock adjacent to flowing sections of streams (J. Polhemus & D. Polhemus, 1994a).

Genus *Papuavelia*

Figure 324

324

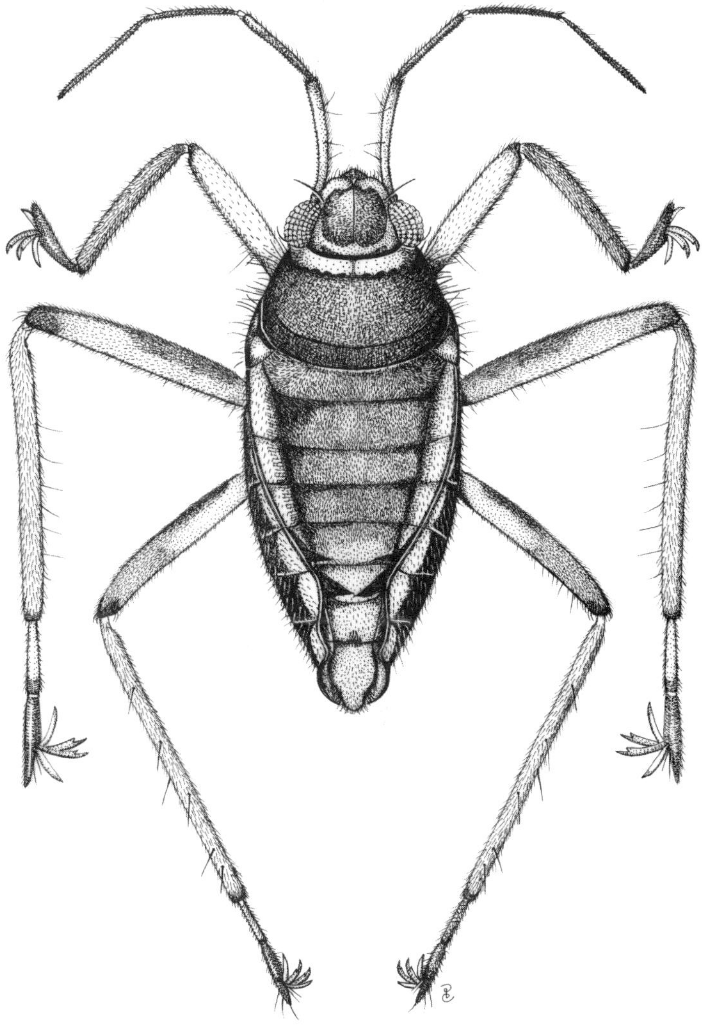

FIGURE 324
Papuavelia siculifera, apterous female, length 3.7 mm.

Diagnosis

Medium-sized to rather large (male 3.3-3.5, female 3.8-4.1 mm). Apterous or macropterous. Blackish-brown tinged with orange, some silvery pubescent markings on pronotum and first abdominal tergite, transverse yellowish band anteriorly on pronotum sharply contrasting. Head posteriorly nearly straight. Antennae long, about 0.8 times body length; first antennal segment longest, about $^{1}/_{3}$ of total antennal length and distinctly stouter than the three apical segments; fourth antennal segment only slightly shorter than first. Eyes appressed to anterior margin of pronotum, globose, in apterous morph laterally just reaching beyond anteriolateral angles of pronotum. Rostrum reaching middle of mesosternum. Gular region short, plainly visible. Macropterous hemielytra reaching beyond abdominal tip, with five closed cells, uniformly dark brown without pale markings, basal third of hemielytra not thickened. Legs long and stout, middle and hind legs of about equal length. Tarsi long, claws and both arolia modified into leaf-like structures (fig. 308). Abdomen relatively long, sides almost parallel, in apterous male elongate ovate, in apterous female posteriorly convergent (fig. 324). Male fore legs not modified; grasping comb on fore tibia long, about half the tibial length. Abdominal sternites 2-5 with broad median carina, sternite 6-7 broadly impressed, sternite 8 modified into large anteriorly directed sharply pointed projection (fig. 311). Genital segments small, retracted into pregenital abdomen; parameres small, symmetrical. Female tergite 8 level to tergite 7, gonocoxae large, exposed. *Papuavelia* is recognizable among New Guinean Microveliinae by its relatively large size and the blade-like claws and arolia.

Distribution in Malesia (fig. 581)

So far one species has been described from mountainous areas in Papua (Irian Jaya).

Biological notes

Papuavelia siculifera is found in dark sheltered places in streams with fair to swift current. It is able to skate quickly and erratically and superficially resembles *Rhagovelia* (D. Polhemus & J. Polhemus, 2000a).

Genus *Phoreticovelia*

Figures 325-326

Diagnosis

Small (male 0.9-1.6, female 1.6-2.2). Strongly sexually dimorphic. Apterous or macropterous. Within a species female $1^{1}/_{2}$-2 times longer than male. Black to orange brown, anterior transverse band on pronotum brown to orange-brown. The two species from South-east New Guinea and North-east Australia dorsally

325

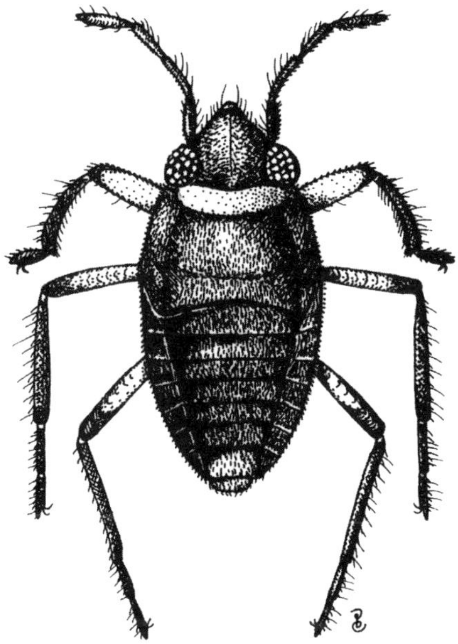

FIGURE 325
Phoreticovelia notophora, apterous male, length 1.0 mm.

with patches of silvery setae. Head posteriorly slightly convex. Antennae moderately stout and short, about half of body length; first antennal segment somewhat stouter than three apical segments; relative length of antennal segments variable but fourth segment longest. Eyes close to or appressed to anterior margin of pronotum, globose, in apterous morph laterally just reaching anterolateral angles of pronotum. Rostrum reaching middle of mesosternum. Gular region short but visible. Macropterous hemielytra reaching beyond apex of abdomen, colour uniformly dark or with light areas proximally and distally, with four closed cells of which, however, only the proximal two are distinct; basal third of hemielytra not thickened. Legs short and moderately stout, middle and hind legs about equal in length; tarsi moderately long, claws and ventral arolium originating at about middle of apical tarsal segment; middle tarsal segments long, together almost as long as tibia; claws and ventral arolium modified into long slender blade-like structures (apparently for rowing). Pronotum short, medially about half the length of an eye, laterally subequal to eye length. Apterous female with mesonotum, metanotum and anterior abdominal tergites modified, strongly depressed medially; sutures between mesonotum and metanotum medially absent; mesonotum in male normal. Abdomen broad in females, more slender in males (figs. 325, 326). Fore tibia in male slightly bent, flattened and expanded distally, tibial comb $^1/_5$ to $^1/_3$ of tibial length. Abdominal sternites not modified; genital segments small (parameres not visible at 80× magnification), not modified. Female tergite 8 level to tergite 7. Gonocoxae large, exposed, visible in lateral and ventral view.

326

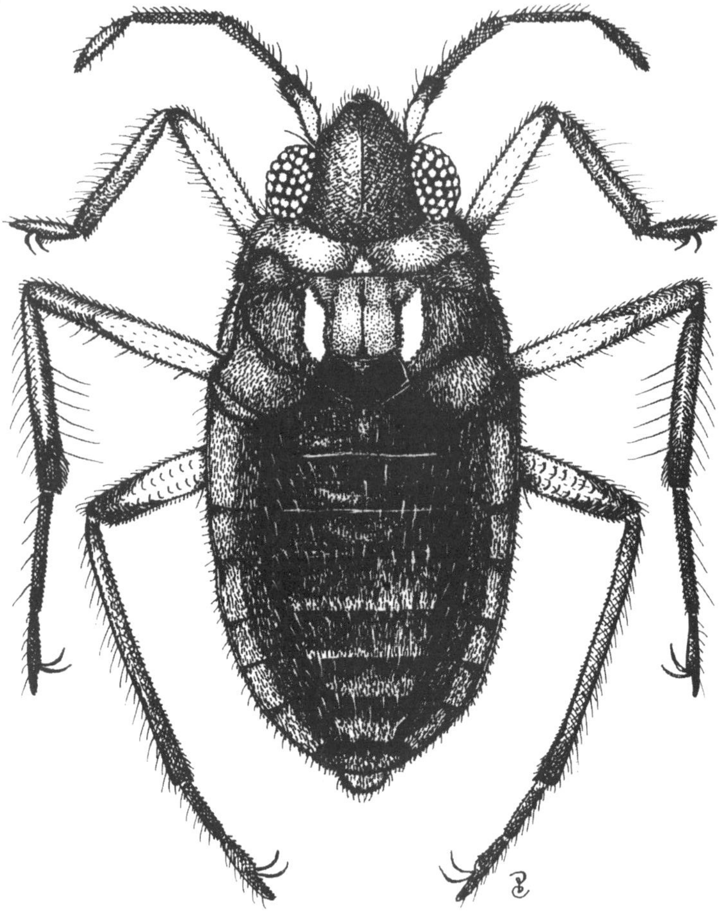

FIGURE 326
Phoreticovelia notophora,
apterous female,
length 2.0 mm.

Distribution in Malesia (fig. 580)

There are four described species, two of which occur on New Guinea and adjacent islands, and on Cape York, Australia. The other two occur on the Palau Islands, Micronesia, and in North Queensland.

Biological notes

Phoreticovelia species are found on quiet or stagnant parts of various types of streams. During copulation males ride on the back of the females. The latter have pronounced dorsal adaptations to accomodate the males.

Genus *Polhemovelia*

Figure 327

Diagnosis

Small to rather large (2.4-4.1 mm), elongate. Monomorphic macropterous. Dorsally dark brown to blackish. Head posteriorly produced behind eyes, more or less distinctly set in an anterior emargination. Antennae long and slender, length $^3/_4$ to equal to body length; first segment medium-sized and slightly curved, extending about $^2/_3$ of its length beyond apex of head; second short, about $^2/_3$ to $^4/_5$ of the combined length of first and third segment, which are of subequal length. Third and fourth segment thinner than first and second but not filiform. Eyes globose, laterally reaching to slightly beyond anterior pronotal angles; eyes adjacent to anterior border of pronotum or somewhat removed from it. Rostrum reaching middle of mesosternum. Gular region well developed, clearly visible. Hemielytra with four closed cells, reaching tip of

327

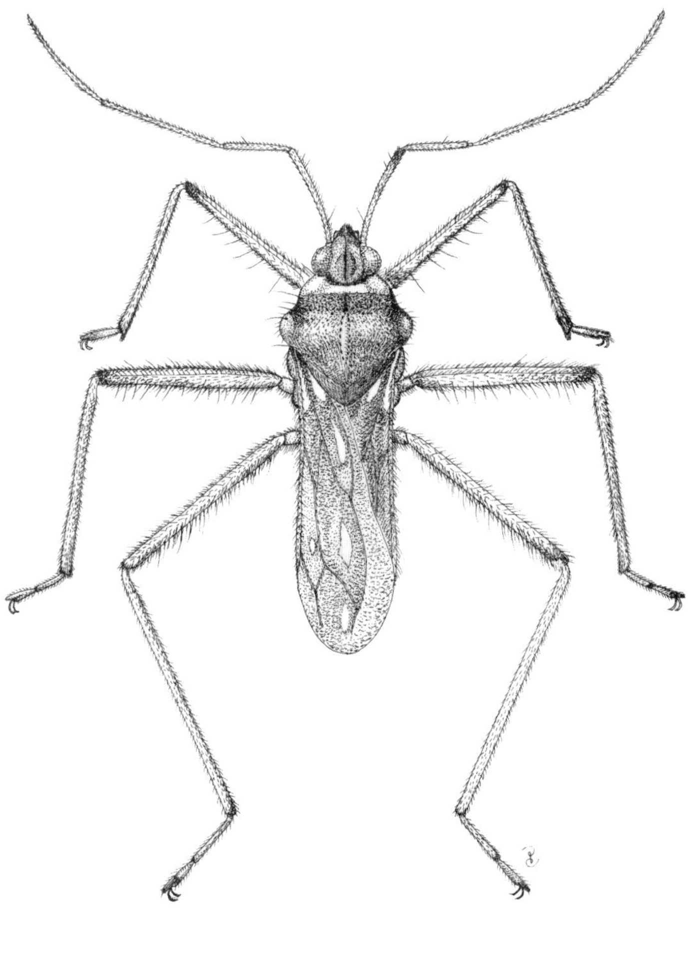

FIGURE 327
Polhemovelia setosa, paratype, macropterous male, length 3.5 mm.

abdomen. Basal third of hemielytra not thickened. Legs long and slender, claws subapical. Fore legs of male not modified; grasping comb on fore tibia short, 1/5 to 1/4 of tibial length. Abdomen elongate and slender. Median part of sternite 7 and 8 broadly impressed. Genital segments well developed, distinctly protruding but in dorsal view covered by hemielytra. Parameres small with widened and rounded apex (fig. 299). Tergite 8 of female level to tergite 7. Gonocoxae well developed, clearly visible in lateral and ventral view (Zettel & Sehnal, 2000).

Distribution in Malesia (fig. 580)

There are three described species, all known only from Sabah, North Borneo.

Biological notes

Polhemovelia species are associated with streams in rainforests. They were found to be hygropetric on shady banks, or collected at light (Zettel & Sehnal, 2000).

Genus *Pseudovelia*

Figure 328

Diagnosis

Small (2.0-3.0), rather stout. Apterous or macropterous. Dorsally yellowish brown to blackish.

Head posteriorly nearly straight or produced behind eyes, depending on species. Antennae of medium length and not obviously slender, 1/2 to 3/4 of body length; first segment medium-sized and slightly curved, extending about 2/3 beyond apex of head; second and third segment more slender than first, fourth segment stout, first and fourth segment usually of subequal length, like second and third segment, which are about 2/3 to 3/4 of length of first and fourth. Eyes globose, in apterous morph laterally reaching beyond anterior pronotal angles; eyes adjacent to anterior border of pronotum or at some distance depending on species. Rostrum reaching middle of mesosternum. Gular region small, usually visible but sometimes virtually hidden by ventral lobes. Macropterous hemielytra reaching beyond apex of abdomen, with four closed cells; distal third of hemielytra with two weak longitudinal veins, often difficult to detect (fig. 300; this part of wing is often broken off by autotomy); basal third of hemielytra not thickened. Legs not elongate, hind legs distinctly longer than middle legs; claws normal, subapical, ventral arolium usually bristle-like, rarely leaf-like. Fore legs in male of most species not modified; grasping comb on fore tibia long, in most species 1/2 to 3/4 of tibial length, sometimes 1/2 to 4/5 of tibial length. Abdomen relatively long, sides in apterous male almost parallel, in apterous female posteriorly convergent. Median part of sternite 7 and 8 broadly impressed; genital segments of male small and retracted into the pregenital abdomen, segment 8 ventrally modified (figs. 302, 303). Parameres small, symmetrical.

328

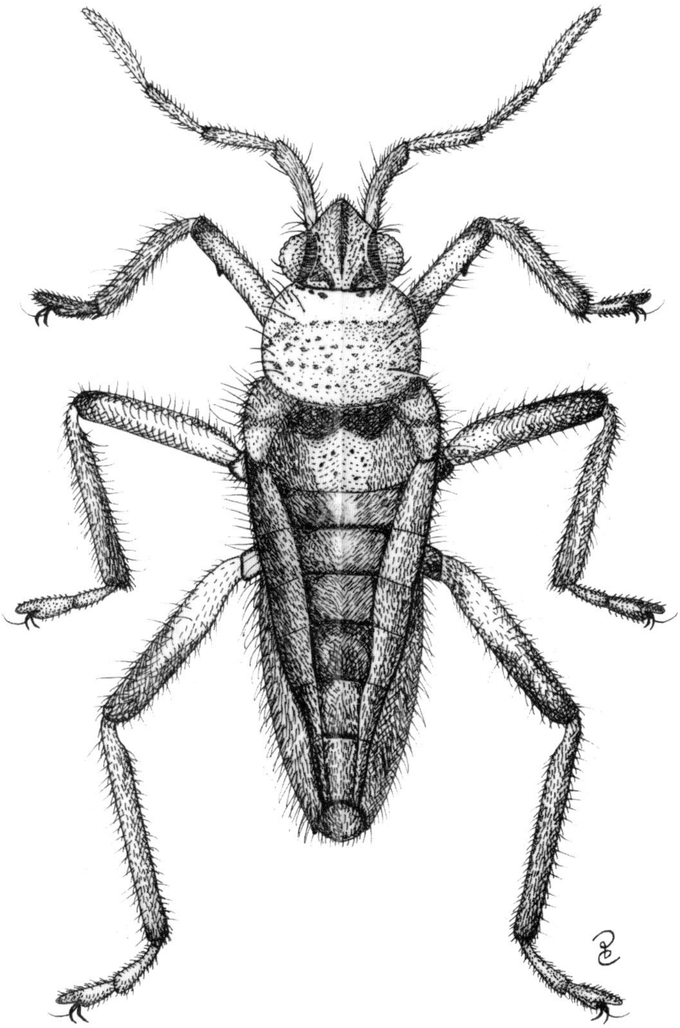

FIGURE 328
Pseudovelia pyrokrene, paratype, apterous female, length 2.7 mm.

Tergite 8 of female level to tergite 7. Gonocoxae elongate, well developed, usually clearly visible in lateral and ventral view (adopted from Andersen, 1983).

– *Pseudovelia* **sensu stricto**

Diagnosis

Body hairs of variable length. Female connexiva more or less raised but rarely covering abdominal tergites. Gonocoxae moderately elongate. Male hind tibia usually modified only at apex, hindtarsus usually with ventral row of bristle-like hairs. All Malesian species belong to this subgenus.

Distribution in Malesia (fig. 580)

There are eighteen described species of *Pseudovelia* in the western part of Malesia. At least 21 as yet undescribed species have been recognized from the Philippines (Sehnal, 1999).

Biological notes

Pseudovelia species are associated with streams where they are found on side pools or at quiet places at the edge or behind boulders. Several species have been recorded from patches of foam or under overhanging banks. They often occur in large numbers.

Genus *Tanyvelia*

Figure 329

329

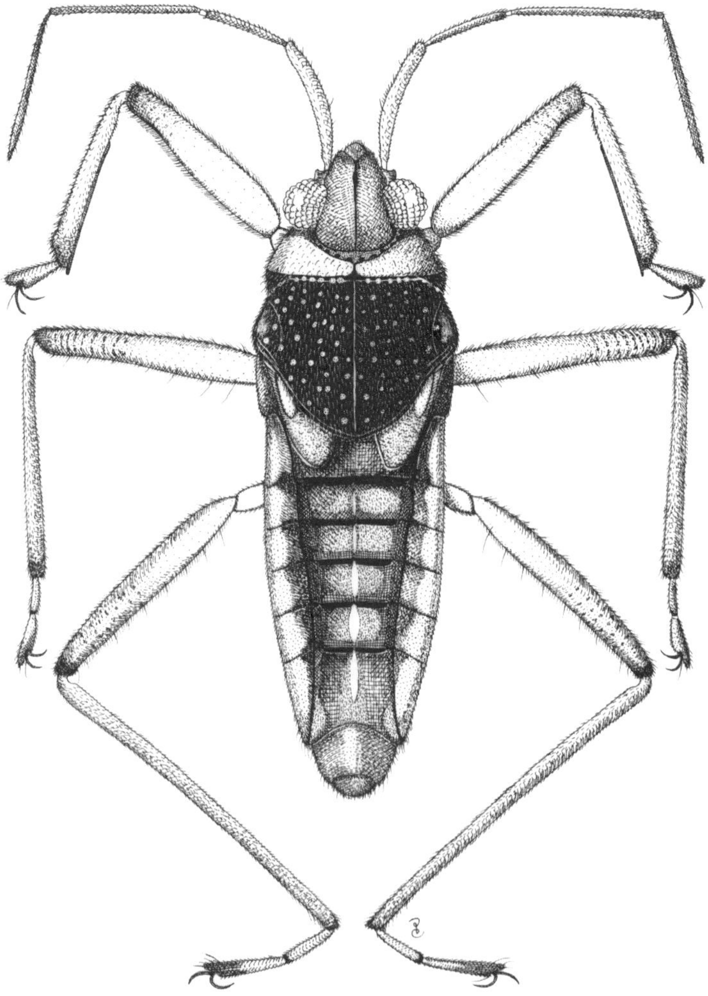

FIGURE 329
Tanyvelia missim, paratype, male, length 2.8 mm.

Diagnosis

Medium-sized (2.7-3.1 mm), elongate and slender. Monomorphic micropterous. Dorsally blackish-brown tinged orange. Wing pads brown with anterolateral light streaks. Head not produced posteriorly. Antennae very long and slender, about 3/4 of body length. Eyes globose, laterally not reaching beyond anterior pronotal angles and appressed to anterior margin of pronotum. Rostrum reaching to middle of mesosternum, rostral cavity closed posteriorly; gular region long, clearly visible. Legs slender, long, hind legs very long, all tarsi long, claws subapical, moderately long. Male fore femur unmodified; grasping comb on fore tibia long, 3/4 of tibial length. Male genital segments moderately large, protruding; parameres rather small, with narrow elongate tip, symmetrical. Female tergite 8 horizontal, in the same plane as tergite 7; gonocoxae well developed and well visible posteriorly.

Distribution in Malesia (fig. 581)

There are two described species, both occurring on eastern New Guinea.

Biological notes

Both species have been taken from small rocky streams with a boulder-strewn bed heavily shaded by primary rainforest (J. Polhemus & D. Polhemus, 1994a, D. Polhemus & J. Polhemus, 2000a).

Genus *Tarsovelia*

Figure 330

Diagnosis

Rather small to medium-sized (2.4-3.5 mm), robust. Macropterous or micropterous. Dorsally black or, sometimes in micropterous specimens, blackish-brown; micropterous wing pads white. Head not produced posteriorly. Antennae slender, about 1/2 of body. Eyes globose, laterally not reaching beyond the anterior pronotal angles and appressed to anterior margin of pronotum. Rostrum reaching almost to metasternum, rostral cavity closed posteriorly; gular region moderately long, plainly visible. Pronotum of micropterous morph only slightly reduced compared to macropterous morph, covering mesonotum and metanotum, reaching halfway second abdominal tergite. Macropterous hemielytra with four closed cells, not reaching the apex of abdomen, leaving part of the seventh tergite uncovered, or reaching beyond apex of abdomen, depending on species. Legs slender, middle legs longer than hind legs; middle tarsi very long, over 3/4 to longer than middle tibia; claws subapical, short. Fore femur of male slightly modified or unmodified, depending on species. Grasping comb on fore tibia short, 1/4 or slightly less of tibial length. Male genital segments moderately large, protruding; parameres rather small, with a narrow

330

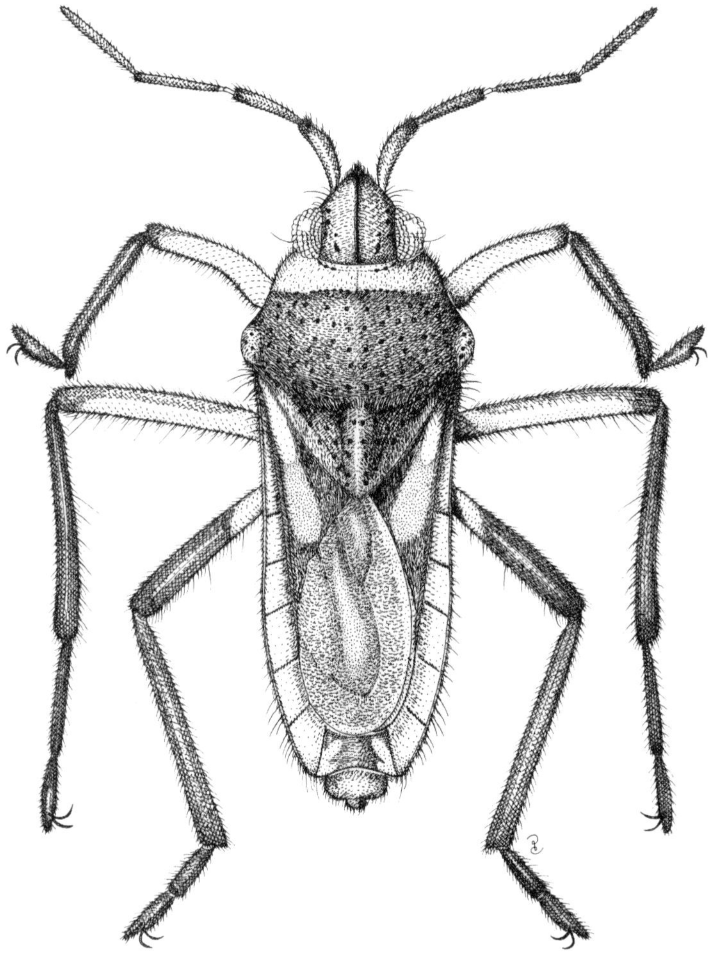

FIGURE 330
Tarsovelia dani, paratype, macropterous male, length 2.5 mm.

elongate tip, symmetrical. Female tergite 8 horizontal, level to tergite 7. Gonocoxae small, just visible in lateral and ventral view (J. Polhemus & D. Polhemus, 1994a).

Distribution in Malesia (fig. 581)

There are eight species, all from mountainous areas in New Guinea and Batanta Island, but not necessarily from high altitudes.

Biological notes

Tarsovelia species have been encountered in pools associated with streams, including pools under waterfalls. They usually stay at the edge of the pools and sometimes hide in crevices, but skate rapidly when disturbed. Macropterous individuals fly readily from a net (J. Polhemus & D. Polhemus, 1994a).

Genus *Velohebria*

Figure 331

Diagnosis

Small (2.1-2.2 mm), not very robust. Monomorphic macropterous. Dorsally pale to dark brown, head yellow to orange. Head not produced posteriorly. Antennae slender, about $^3/_4$ of body length, first two segments short and stout, third and fourth segments flagelliform and very long. Eyes globose, laterally reaching to or just beyond anterior pronotal angles and adjacent to the anterior margin of the pronotum. Rostrum long, reaching to between middle coxae. Hemielytra with four closed cells, reaching apex of abdomen, leaving part of the

331

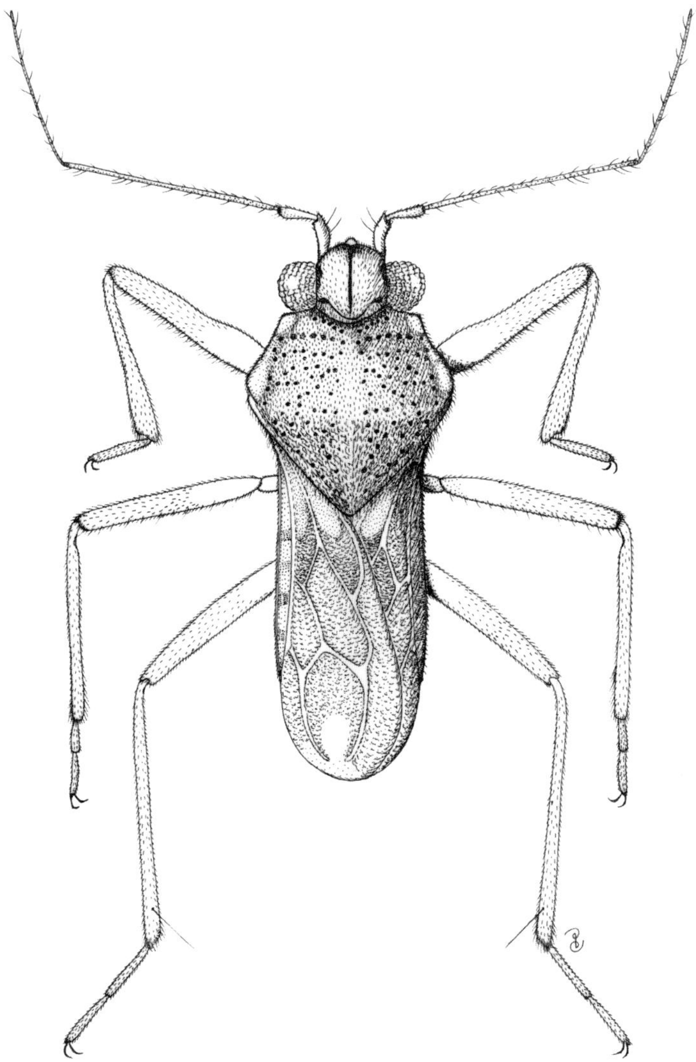

FIGURE 331
Velohebria antennalis, holotype, macropterous male, length 2.15 mm.

connexiva uncovered. Legs slender; claws virtually apical, short. Fore femur in male modified or unmodified, depending on species. Fore tibia somewhat widened apically, grasping comb on fore tibia very short, slightly over $^1/_{10}$ of tibial length. Male genital segments large, protruding; parameres large, distinctly protruding caudally, blade-like with a tapering but rounded tip, symmetrical. Female tergite 8 horizontal, level to tergite 7; gonocoxae well developed, clearly exposed in lateral and ventral view.

Distribution in Malesia (fig. 580)

The single described species, *V. antennalis*, is endemic to New Guinea.

Biological notes

Nothing is known about its habitat. Considering the tarsal structure it is probably terrestrial.

Genus *Xiphovelia*

Figure 332

332

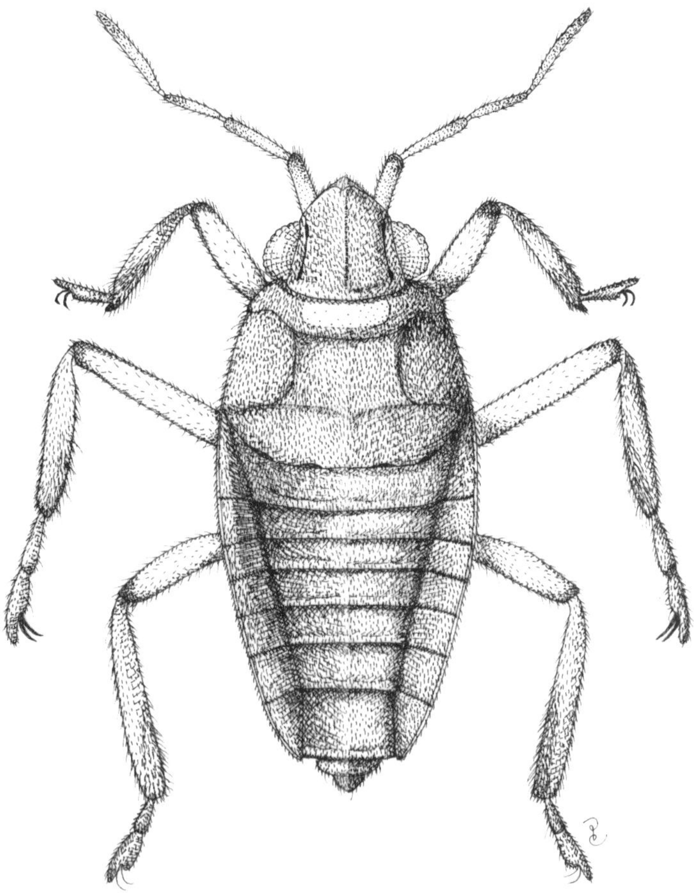

FIGURE 332
Xiphovelia skoteina, paratype, apterous male, length 1.8 mm.

Diagnosis

Small (1.5-2.4 mm). Apterous or macropterous. Sexual dimorphism rather distinct: in apterous morph with elongate (males) to broadly oval or globular (females) body. Dorsal colour dark grey to black, in some species with distinct silvery pubescence. Head not produced posteriorly. Antennae about 1/2 of body length, first segment stout, apical segments thinner but not flagelliform. Eyes globose, laterally reaching beyond anterior pronotal angles and adjacent to anterior margin of pronotum. Rostrum reaching to almost middle of mesosternum; rostral cavity posteriorly closed. Gular region short but clearly visible. Pronotum of apterous morph short, less than 1/3 of length of mesonotum and metanotum together. Macropterous hemielytra with four or five closed cells, reaching abdominal apex. Legs of 'normal' thickness, middle legs slightly longer than hind legs; middle tibia short, second segment of middle tarsi about half as long as middle tibia. Claws on fore and hind tarsi subapical, short; claws and ventral arolium of middle tarsi leaf-like (fig. 305). Fore femur of male and tibia slightly thickened or unmodified, depending on species. Grasping comb on fore tibia well developed, 1/4 to 1/2 of tibial length. Male genital segments moderately large, only partly protruding. Parameres vestigial, but there are two symmetrical projections on the dorsal margin of the pygophore, which in caudal view are round or ear-shaped. Female tergite 8 horizontal, in the same plane as tergite 7; gonocoxae elongate but retracted into apex of abdomen.

Distribution in Malesia (fig. 580)

Two described and several undescribed species are found in Malesia. Distribution of the genus is limited to the east by Weber's Line. In total there are seven described species from the Oriental region. Some also occur in the Palearctic (Japan).

Biological notes

Xiphovelia species are usually found at the edges of streams or springs.

REFERENCES – Microveliinae. Indo-Australian Veliidae (Lundblad, 1933a); description of *Velohebria antennalis* (Štys, 1976); morphology, phylogeny, adaptations, biogeography and classification of Gerromorpha (Andersen, 1982a); revision of Oriental *Pseudovelia* (Andersen, 1983); key to genera (excluding New Guinea; Nieser, 1995); revision of *Baptista*, description of *Lathriovelia* (Andersen, 1989c); descriptions of Microveliinae from New Guinea (J. Polhemus & D. Polhemus, 1994a); key to Oriental *Pseudovelia*, Sulawesi *Xiphovelia* (Nieser, 1995); redescription of *Neoalardus typicus* (Zettel, 1998g); survey of Philippine *Pseudovelia* and description of two species from Palawan and Busuanga (Sehnal, 1999); key to species of *Microvelia somnokrene*-group (Zettel & Gapud, 1999); key and species description of Microveliinae of New Guinea (D. Polhemus & J. Polhemus, 2000a, c); revision of *Lathriovelia* (Kovac & Yang, 2000); description of genus and key to species of *Polhemovelia*, colour plate of *P. septuaginta* (Zettel & Sehnal, 2000).

SUBFAMILY HALOVELIINAE – marine waterstriders

Diagnosis

Small to very small (1.0-2.8 mm), blackish. Females larger than males. In apterous morph posterior margin of pronotum reduced; intersegmental sutures between mesonotum, metanotum and basal abdominal segments indistinct. In macropterous morph – only known in freshwater genera – fore wing venation strongly reduced to basal thickening and a pair of obsolete longitudinal veins in proximal part of hemielytron (figs. 333, 334). Middle legs conspicuously prolonged. Dorsal vesical sclerites (in males) proximally shortened (figs. 336-339, except in *Entomovelia*, fig. 347), lateral sclerites well developed (except in at least some species of *Strongylovelia*, fig. 350).

333

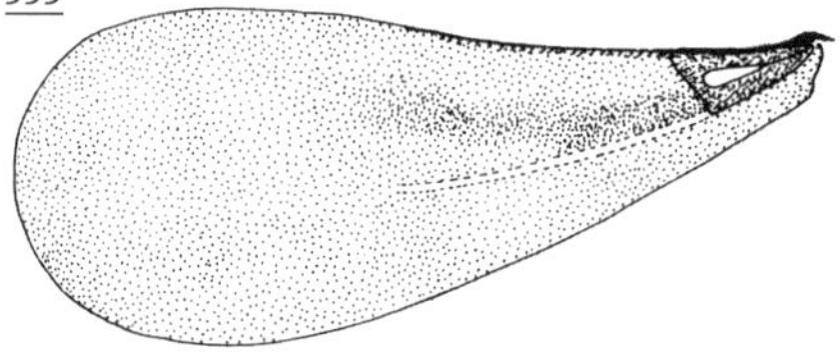

334

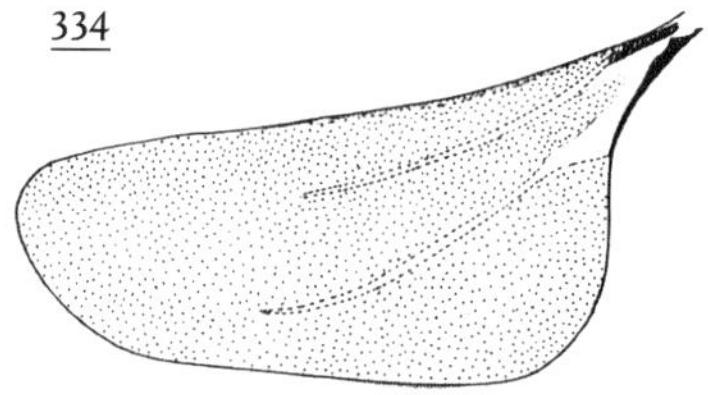

FIGURES 333 – 334
Wings of *Entomovelia doveri*. 333 hemielytron – 334 hind wing.

335

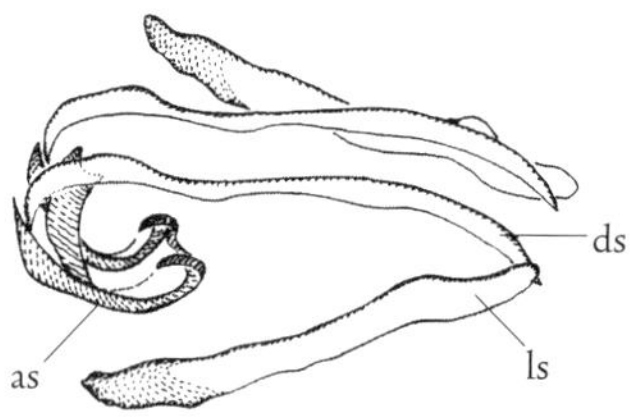

336

337

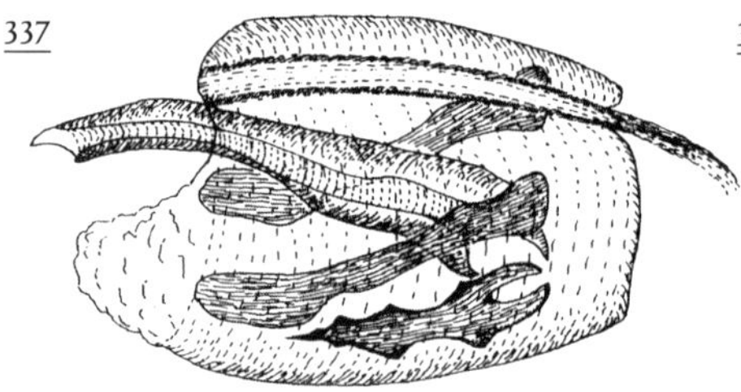

338

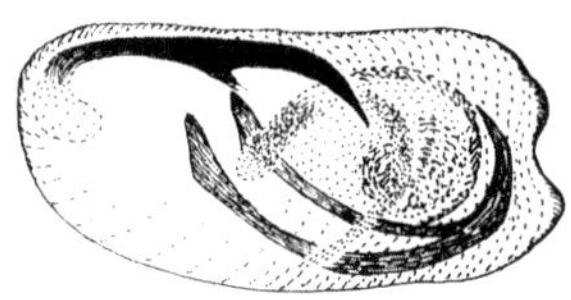

339

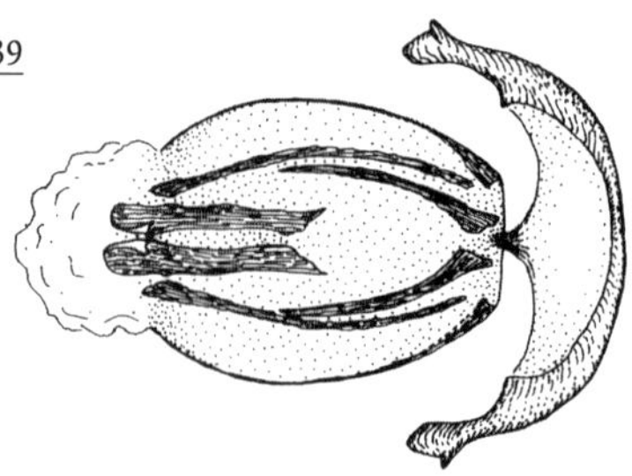

FIGURES 335 – 339
Haloveliinae, vesical sclerites.
335 *Entomovelia doveri*, dorsolateral view – 336 *Haloveloides browni*, dorsolateral view of phallotheca – 337 *Halovelia annemariae*, ventral view of phallotheca – 338 *Strongylovelia philippinensis*, lateral view of phallotheca – 339 *Xenobates* sp., ventral view of phallotheca; as = apical sclerites, ds = dorsal sclerites, ls = lateral sclerites.

Fossil record

Although the subfamily nowadays only occurs throughout the Old World tropics, the fossil *Halovelia electrodominica* Andersen & Poinar (1998) was found in Dominica, The Antilles. The specimen is 15-45 million years old, indicating that the genus used to have a much wider distribution.

Distribution worldwide

Subfamily Haloveliinae contains at present two groups, divided on basis of habitat preference. The marine species (sometimes called coral bugs) belong to the genera *Halovelia* (Indian & West Pacific Oceans), *Haloveloides* (Malesia) and *Xenobates* (from South India through Malesia to New Caledonia). At least in Papua New Guinea specimens of these three genera are frequently collected together (Lansbury, pers. comm.). The freshwater species belong to *Strongylovelia* (from Malesia until Sri Lanka) and *Entomovelia* (Continental Malaysia, North Borneo, North Vietnam, South China).

Distribution in Indo-Australia

All genera of Haloveliinae occur in the Indo-Australian region and most species are found in the Papuan subregion. *Halovelia* is the most widespread genus with an Indo-West Pacific distribution. The other haloveliine genera are distributed throughout Malesia.

Key to the genera of Haloveliinae[1]

1 Head anteriorly strongly deflected. Eyes large, half or more as wide as interocular width of head (figs. 348, 359). First antennal segment shorter than third segment. Male fore tibia without grasping comb (fig. 342). Freshwater species ▸ 2

– Head anteriorly moderately deflected. Eyes less than half as wide as interocular width of head (figs. 357, 358). First antennal segment usually longer than third segment. Male fore tibia with (*Halovelia* fig. 340) or without (*Haloveloides*, *Xenobates* figs. 341, 343) grasping comb. Mainly marine species ▸ 3

2 Second antennal segment longer than first segment (fig. 345). Tarsal segments of hind legs subequal in length (fig. 347) ▸ ***Strongylovelia* p. 272**

– Second antennal segment shorter than first segment (fig. 344). First tarsal segment of hind leg about 1/2 of second segment (fig. 346) ▸ ***Entomovelia* p. 267**

1 Adapted from Andersen (1992b).

FIGURES 340 – 343
Haloveliinae, fore leg of male.
340 *Halovelia annemariae* – **341** *Haloveloides browni* – **342** *Strongylovelia philippinensis* – **343** *Xenobates* sp.

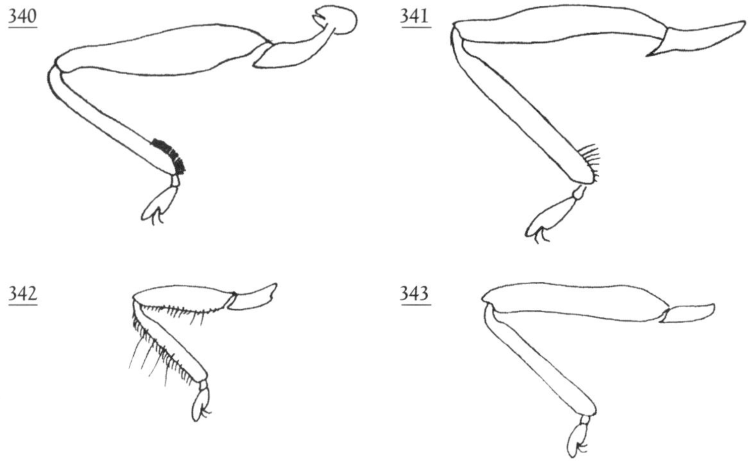

FIGURES 344 – 345
Haloveliinae, antenna.

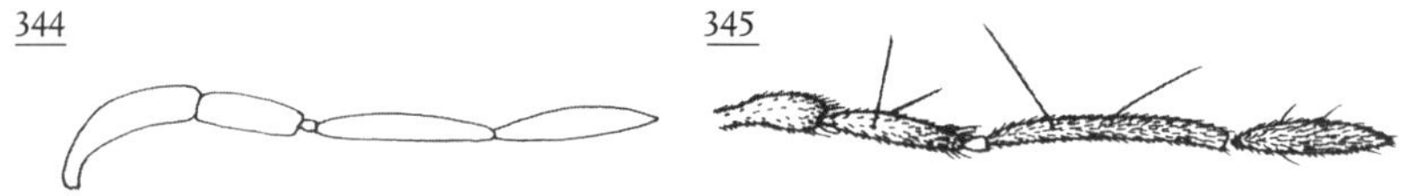

FIGURES 346 – 347
Haloveliinae, hind tarsus.
346 *Entomovelia* – **347** *Strongylovelia.*

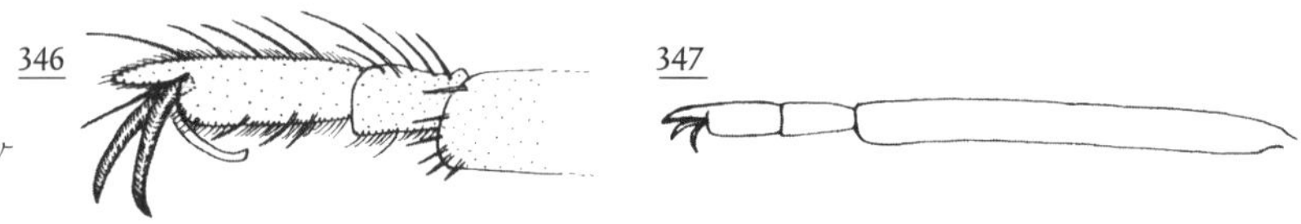

3 Pronotum usually totally dark (fig. 357). Eye width less than 0.3 times the interocular width of head. Fore tibia of male with grasping comb (fig. 340). Abdominal venter of male rarely modified. Male genital segments withdrawn into pregenital abdomen and only slightly protruding from abdominal end ▸ *Halovelia* p. 269

– Pronotum with pale marking or spots (figs. 358, 360). Eye width more than 0.4 times the interocular width of head. Fore tibia of male without grasping comb (figs. 341, 343). Abdominal venter of male usually with basal segments distinctly higher than apical segments; genital segments distinctly protruding from abdominal end ▸ 4

4 Usually dark and pale; thoracic dorsum with well-defined spots of silvery hairs. Fore trochanter of male not modified (fig. 343). Middle femur less than 0.9 times body length, with or without a row of long, bristle-like hairs along anterior margin; ratio of middle tarsus and tibia less than 0.7:1 ▸ *Xenobates* p. 274

– Chiefly coloured dark; thoracic dorsum at most with scattered silvery hairs. Fore trochanter of male usually with a tubercle or spine before or at apex (fig. 341). Middle femur more than 0.9 times body length, with short pubescence along anterior margin; ratio of middle tarsus and tibia about 0.8:1 ▸ *Haloveloides* p. 271

Genus *Entomovelia*

Figures 348-350, 352

348

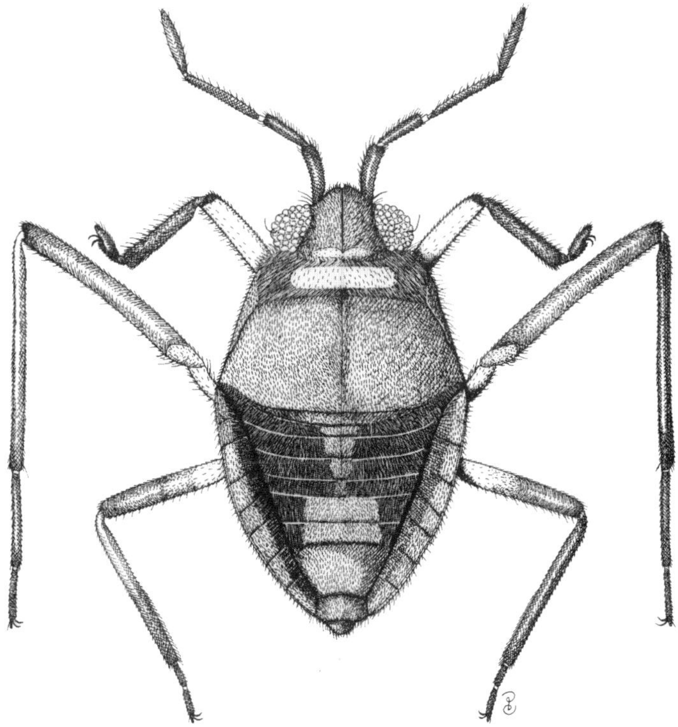

FIGURE 348
Entomovelia doveri, apterous female, length 1.9 mm.

Diagnosis

Small (males 1.1-2.1, females 1.8-2.2 mm) and hairy. In some species sexes dimorphic. Body more or less roundish, black. Diagnostic characters are: (1) antenna shorter than body; (2) first antennal segment longer than second (fig. 344); (3) propleura confluent with pronotum, distinctly shorter than median length of pronotum (fig. 350); (4) mesopleura wide and large, distinctly separated from mesonotum; (5) first tarsal segment of hind leg about half as long as second (fig. 346). First antennal segment and femora of all legs partly pale in some species. Venation of hind wing virtually absent, only two pale longitudinal folding lines remaining (fig. 334). Mesonotum and metanotum with deep median longitudinal groove. For specific identification male parameres (fig. 352) and vesical sclerites (fig. 335) are important.

Distribution in Malesia (fig. 582)

Only one described species has so far been reported from Malesia (Continental Malaysia). A number of undescribed species of this genus have recently been recognized from southern China, northern Vietnam and Borneo. Some areas with as yet undescribed species have been incorporated in figure 582.

349

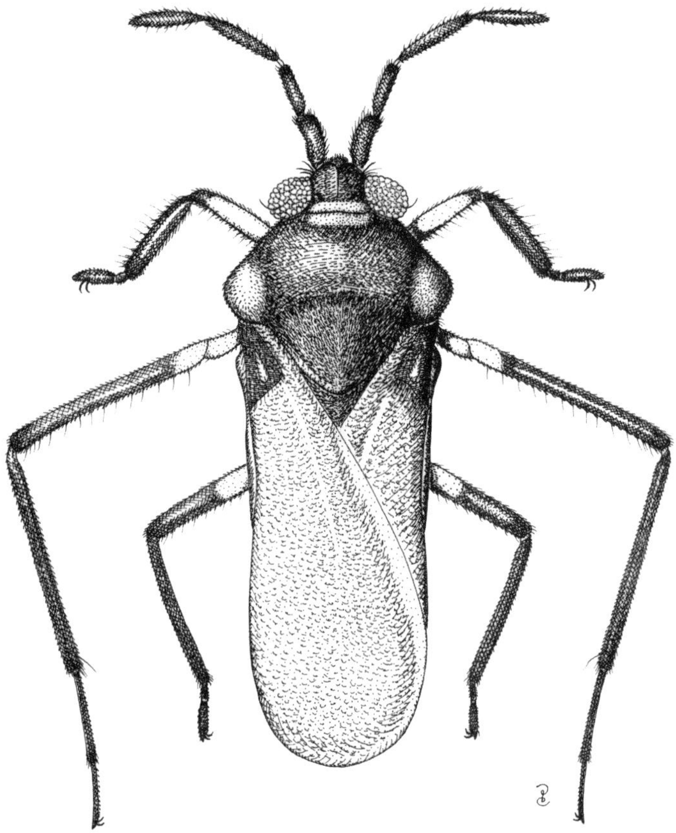

FIGURE 349
Entomovelia doveri,
macropterous male,
length 2.3 mm.

350

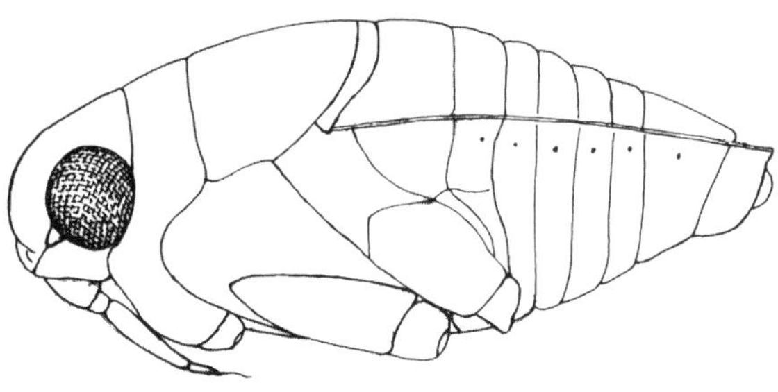

351

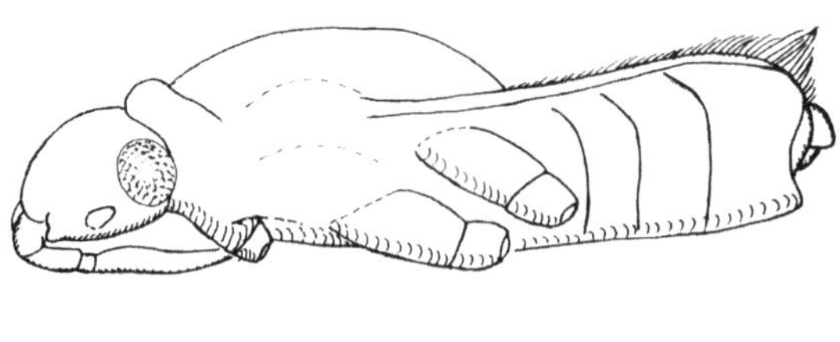

FIGURES 350 – 351
Haloveliinae in lateral view.
350 *Entomovelia* sp. –
351 *Haloveloides* sp.

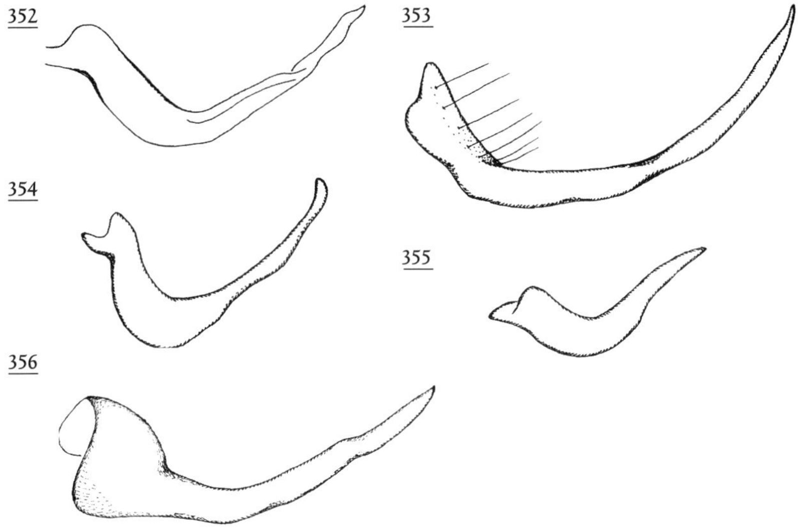

FIGURES 352 – 356
Haloveliinae, parameres.
352 *Entomovelia doveri* –
353 *Halovelia annemariae* –
354 *Haloveloides browni* –
355 *Strongylovelia philippinensis* – 356 *Xenobates* sp.

Biological notes

Entomovelia species have been collected mostly from well-vegetated mountain streams and near waterfalls. *Rhyacobates* and *Metrocoris* (Gerridae) have sometimes been found in the same stream.

Genus *Halovelia* – coral bugs

Figure 357

Diagnosis

Small (males 1.28-1.95, females 1.75-2.60 mm), subovate. Mainly dark, covered by greyish thick pilosity, especially on abdomen; thoracic and abdominal dorsum without spots of silvery hairs; head and pronotum without extensive pale markings. Head short, dorsal surface at most with median groove. Antenna long and slender, 0.6-0.8 times body length; segment 1 as long as head, subequal to or slightly shorter than segment 4, segment 3 longer than segment 2. Eyes globular. Pronotum very short; suture between pronotum and mesonotum usually laterally obscured, rarely distinct throughout, reaching lateral margins of thorax. Dorsal boundaries between mesothorax and metathorax and between metathorax and abdominal terga indistinct. Ventral sutures of thorax and abdomen distinct. Fore tibia of male with a grasping comb (fig. 340) composed of a compact row of short spines along the inner margin. Mesotrochanter prolonged, middle femur very long, middle tibia and tarsus very slender and long; claws very small. Hind leg short, second tarsal segment much longer than first segment. Abdomen short with broadly rounded sides in male, longer and usually with more straight sides in female. Abdominal terga 4-7 distinctly separated by intersegmental sutures. Male genital segments relatively

357

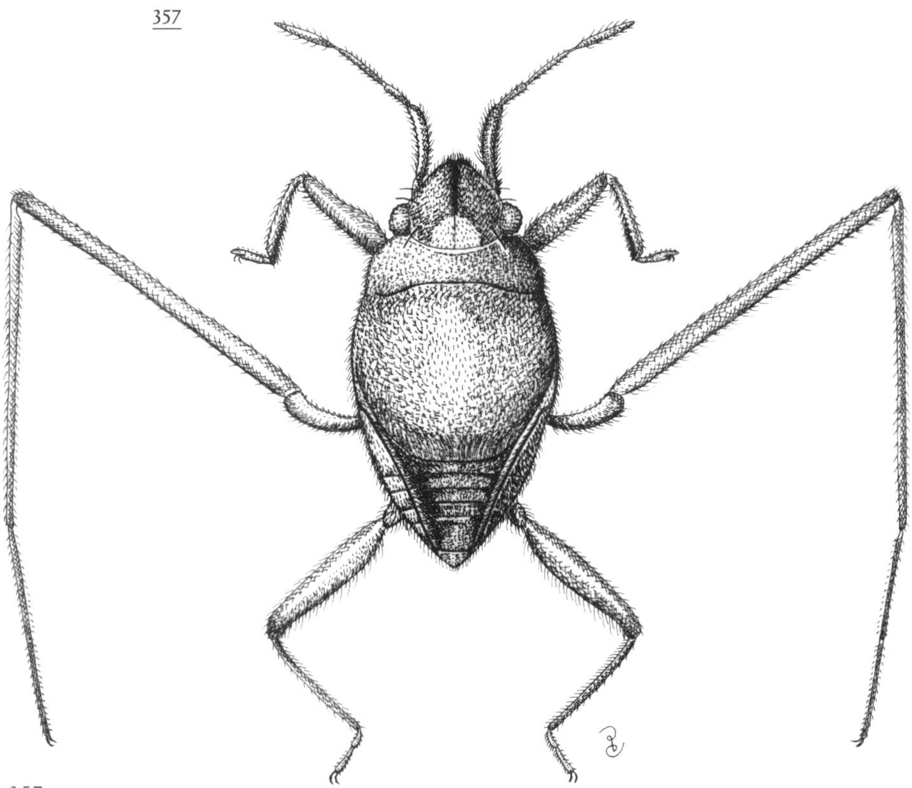

FIGURE 357
Halovelia novoguineensis, apterous male, length 1.9 mm.

large but withdrawn into pregenital segments, only slightly protruding from abdominal end. Parameres large (fig. 353); vesica with three asymmetrical dark sclerites (one dorsal and two laterals, fig. 337). Female genital segments clearly visible behind tergum 7; proctiger concealed beneath tergum 8.

Distribution in Malesia (fig. 582)

Sixteen species have been reported from Malesia. *Halovelia* is reported from the entire area, with the exception of Sumatra. Areas relatively rich in species are New Guinea (six species), Sulawesi (four), Mindanao (three) and continental Malaysia (three).

Biological notes

Halovelia species are inhabitants of intertidal reefs, where they live on the surface of tidal pools among stands of *Porites* and *Acropora* corals on the mid-reef flat. They rarely enter the inner-reef flat or the outer reef margin. They have also been recorded from rocky coasts without corals and mangroves, especially where these border coasts are fringed by coral reefs. They usually appear when the tide recedes. When the tide rises, both adults and nymphs retreat into cavities and holes and stay submerged during high tide, surrounded by an air bubble.

Although *Halovelia* have long middle legs and skate more like Gerridae than Veliidae, they are able to move very fast on solid substrates such as rock or coral. Usually they use all six legs but sometimes, especially when moving short distances, they fold the middle legs. With the use of the middle legs the bugs are also able to jump several centimetres, like Gerridae. Jumping has never been observed in other veliids (Andersen, 1989b).

Halovelia probably deposit their eggs on blocks of dead coral; ecdysis of nymphs takes place in the cavities.

Genus *Haloveloides*

Figure 358

Diagnosis

Small (males 1.4-1.9, females 1.7-2.2 mm), fusiform coral bugs. Always wingless. Chiefly dark, covered by thick, usually greyish pubescence. Head largely

358

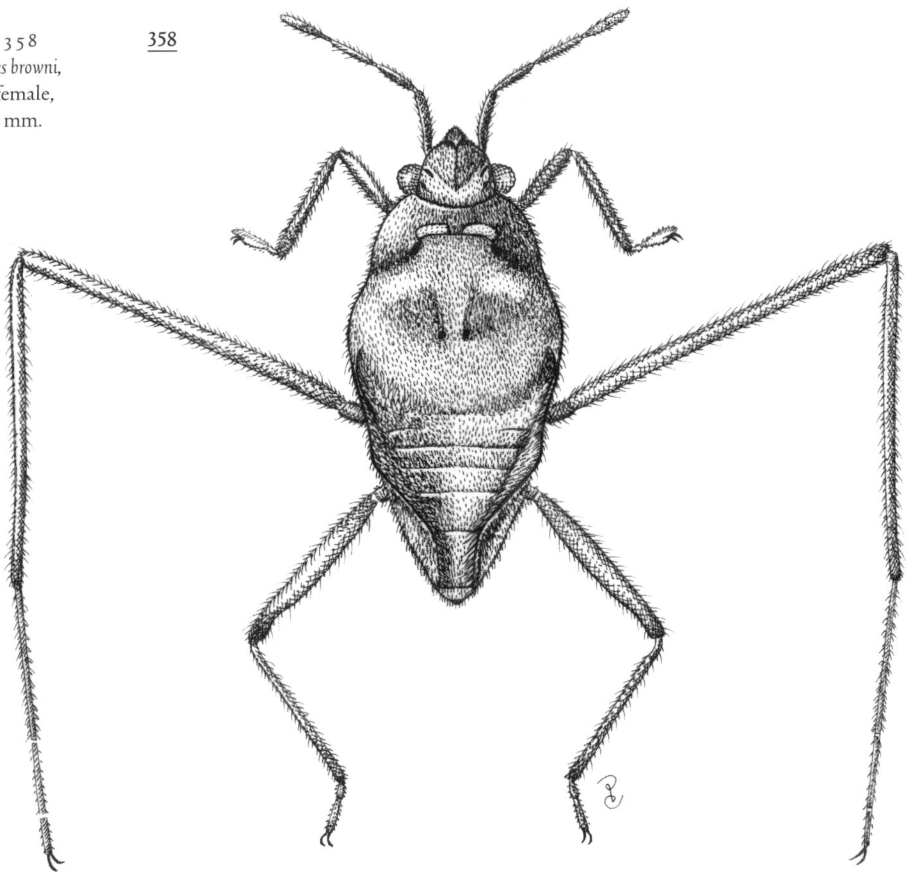

FIGURE 358
Haloveloides browni, apterous female, length 2.1 mm.

pale between eyes; pronotum with transverse pale marking. Thoracic dorsum usually without well-defined spots of silvery hairs. Pronotum short, suture between prothorax and mesothorax laterally turned obliquely backwards, terminating far from lateral margins of thorax; suture between propleura and mesopleura lost (fig. 351). Ventral sutures of thorax and abdomen distinct. Fore trochanter of male usually with tubercle or spine (fig. 341). Fore tibia of male without well-defined grasping comb, but with scattered distal spinose hairs along inner surface (fig. 341). Middle femur very long; tibia and tarsi prolonged. Abdominal venter usually with basal tumescence and median keel. Male genital segments distinctly protruding at abdominal end; pygophore usually modified distally; parameres large (fig. 354. See also fig. 336 for vesical sclerites).

Distribution in Malesia (fig. 583)

So far *Haloveloides* is known only from Malesia. Thirteen species have been described; nine of them are endemic to the Philippine Islands.

Biological notes

Adults and nymphs of *Haloveloides* have been collected in an estuary and on a (freshwater) spring near the sea, at the edge of dead coral reefs, on intertidal pools connected with open sea, and a mile or more off-shore. It has apparently a somewhat broader ecological range than *Halovelia*. In one sample, *Haloveloides* were collected at light together with *Halovelia*, *Halobates* and *Hermatobates* (Andersen, 1992b).

Genus *Strongylovelia*

Figure 359

Diagnosis

Small (males 1.0-1.4, females 1.3-1.7 mm). Globular, dark, thorax and often also abdomen dorsally with distinct yellowish marks. Head wide, coarsely punctuate, with dense pilosity. Antennae (fig. 345) long and slender; first segment slightly thickened, second segment longer than or subequal to first, third segment longest, almost twice as long as first; fourth segment subequal to first. Rostrum reaching fore coxae. Pronotum short, mesonotum and metanotum together very much rounded and convex, almost hemispherical. Fore legs short, femur thicker and subequal in length to fore tibia; first tarsal segment very short, second segment three times as long as first; claws situated preapically. Middle legs very long, tibia slightly shorter and thinner than femur; first tarsal segment about twice as long as second. Hind legs much shorter than middle legs; femur thickened and produced beyond abdominal apex; tibia subequal in length to femur; first tarsal segment of hind leg subequal to second segment (fig. 347). Parameres (fig. 355) of related species differ only slightly from each other, hence not diagnostic. Vesical sclerites, which show species-

359

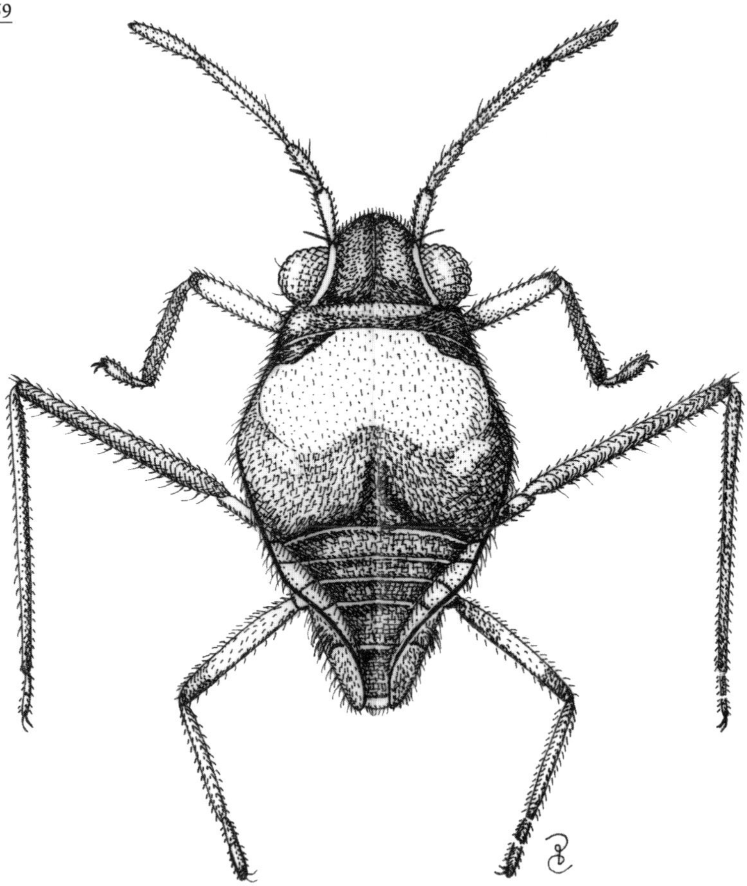

FIGURE 359
Strongylovelia priori, apterous female, length 1.7 mm.

specific variation, have hardly been studied due to small size. Therefore, females, which vary considerably in shape, size, relative measurements, pilosity and coloration, are the important sex for identification.

Distribution in Malesia (fig. 584)

Fifteen of the sixteen described species of *Strongylovelia* have been described from Malesia: four from Borneo, eight from the Philippines, one from Sulawesi, one from New Guinea, and one from the Bismarck Archipelago. Unidentified records are from peninsular Malaysia, Singapore, and Sumatra. All species are regional endemics. As the genus is only collected by experienced collectors it is probably more widespread in the area.

Biological notes

Strongylovelia is one of the two described limnic genera of Haloveliinae. The species are found only on running water, preferably small brooks, creeks and channels, occasionally also in more extended habitats, where they frequent shaded, slowly flowing or lentic areas. *Strongylovelia* species tend to congregate in groups; adults are found together with nymphs in very low current running waters. During copulation the female carries the male on her dorsum. Macropterous specimens are rarely caught with intact wings: these are mostly broken off at the base. Adult *Strongylovelia* are agile considering their size, and move very quickly in straight lines.

Genus *Xenobates* – mangrove bugs

Figure 360

Diagnosis

Small (males 1.3-2.3, females 1.5-2.8 mm), ovate. Monomorphic apterous. Mainly dark, covered by thick layer of hairs. Head largely pale between eyes; pronotum with transverse pale marking(s) in middle; thoracic and abdominal dorsum usually with well defined spots of greyish or silvery hairs. Head shorter than wide. Antennae slender, segment 1 always shorter than head. Pronotum very

360

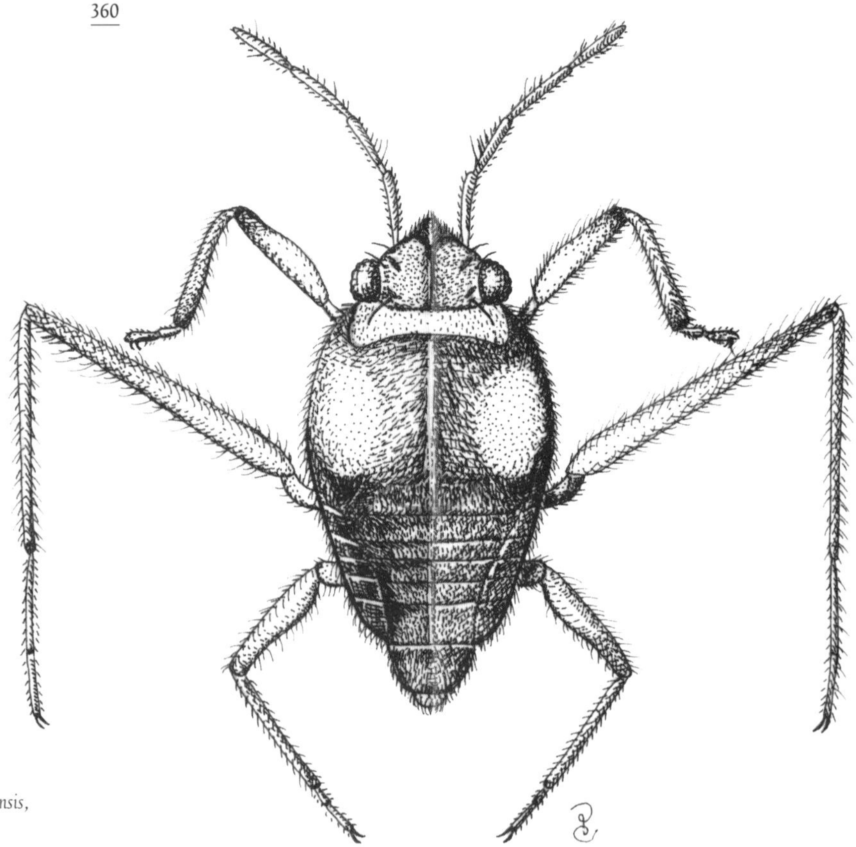

FIGURE 360
Xenobates solomonensis, apterous male, length 1.5 mm.

short, suture between pronotum and mesonotum distinct in middle, terminating far from lateral margins of thorax; suture between propleura and mesopleura absent. Dorsal suture between mesothorax and metathorax and suture between metathorax and abdominal terga indistinct. Ventral sutures of thorax and abdomen distinct. Fore tibia of male without grasping comb (fig. 343) but usually with row of scattered, spine-like hairs along inner surface before apex. Mesotrochanters prolonged (fig. 361); middle femur very long, distinctly thickened at basal part. Hind leg short, second tarsal segment much longer than first. Abdomen relatively short, terga 4-7 distinctly separated by intersegmental sutures. Abdominal venter of male modified with basal tumescence and median ridge. Male genital segments conspicuous, distinctly protruding from pregenital abdomen; parameres large and usually falciform (fig. 356); vesica with 3-4 asymmetrical dark sclerites. Female proctiger variable in shape, protruding from segment 8.

361

FIGURE 361
Xenobates sp., mesotrochanter.

Distribution in Malesia (fig. 585)

Xenobates species occur from India through the Malesian area to Australia and New Caledonia. So far twenty species have been described. Eight of these occur in the Northern Territories and Queensland and eleven are found in Malesia (Andersen & Weir, 1999; Andersen, 2000a).

Biological notes

Xenobates species are typical inhabitants of near-shore marine habitats with a preference for mangroves, but are also found in estuaries, intertidally and on coral reefs. There are probably five nymphal instars. Very little is known about the biology. The present records for Australian *Xenobates* indicate that species may breed during most of the year. Adults and nymphs belonging to various instars are usually collected together, and samples often contain more than one species. In Papua New Guinea mercury-vapour light has been used to attract *Xenobates*; these were often found in places where saline water and fresh water mingle (Andersen & Weir, 1999).

REFERENCES – Haloveliinae. Description of *Entomovelia doveri* (Esaki, 1930a); morphology, phylogeny, adaptations biogeography and classification of Gerromorpha (Andersen, 1982a); revision of *Halovelia* (Andersen, 1989a, b); notes on Haloveliinae from Australia, Papua New Guinea and the Solomon Islands (Lansbury, 1989a, 1993a, 1996a); description of *Haloveloides* and key to species and key to Haloveliinae (Andersen, 1992b); *Strongylovelia* from Borneo and the Philippines, key to species (Lansbury & Zettel, 1997, Zettel, 2003d); notes on Haloveliinae from the Philippines (Zettel, 1998f); Haloveliinae of North Australia and South-east Papuasia, descriptions and key to *Xenobates* of Australia, southern New Guinea and New Caledonia (Andersen & Weir, 1999); marine Haloveliinae of Singapore, Malaysia and Thailand (Andersen, 2000a); marine Haloveliinae of Philippines (Zettel, 2003c).

SUBFAMILY PERITTOPINAE

Diagnosis

See diagnosis of *Perittopus*.

Genus *Perittopus*

Figure 362

Diagnosis

Medium-sized (2.8-4.6 mm), relatively short and stout. Usually both apterous and macropterous morphs present, apterous morph usually more common.

362

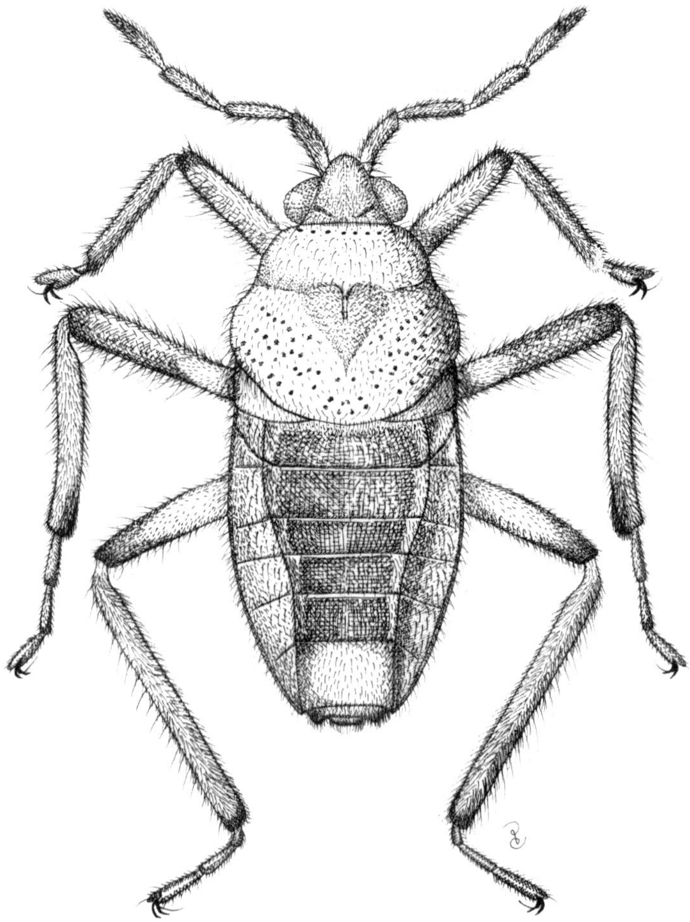

FIGURE 362
Perittopus breddini, apterous female, length 3.4 mm.

Bright red or orange, often with black marks; hemielytron of macropterous morph blackish, without white marks. Head relatively small, deflected. Antenna relatively short and stout, segment 1 longest. Pronotum of apterous morph long, covering mesonotum and metanotum at least medially (fig. 362). Legs relatively stout, but male hind femur not obviously thickened; fore tibia of male with long grasping comb (lacking in female); tarsal formula 2-3-3, tarsi of middle and hind legs long and slender; apical segment of middle tarsi without swimming fan. Hemielytron with thick corium with two closed cells, well separated from membrane (fig. 295). Abdominal tergites not fused; tergite 1 laterally covering laterotergite 2; abdominal sternites without modifications. Male segment 8 with dorsal hind margin deeply concave. Pygophore and proctiger simple; parameres symmetrical, long, in all Malesian species distally twisted. Female tergite 8 directed posteriad; female gonocoxa simple, plate-like, in Malesian species exposed, at most weakly inserted within abdominal segment 7; proctiger of female small, in Malesian species directed posteriad.

Distribution in Indo-Australia

Eleven species have been described so far, but the true number of species is much higher. Highest species diversity is observed in southern India and Indochina (Zettel, 2001a, e).

Distribution in Malesia (fig. 585)

In Malesia seven, mostly regionally distributed, species occur in continental Malaysia, Sumatra, Java, Bali, and Borneo.

Biological notes

Information on habitat preference of *Perittopus* is scarce. Lundblad (1933a) cited some (partly doubtful) localities from Java and Bali (spring, waterfall, stream, pond in a cave, between leaf debris). Personal observations on species from Thailand and Borneo indicate different habitat requirements: most species prefer quiet bays in streams, where some may gather in groups of hundreds. Other species have been found in remnant pools of small streams, but *P. breddini* was found by us under an overhanging wall at the edge of a rushing stream on top of a waterfall, supporting Lundblad's (1933a) observations on this species. The prolongation of the middle leg probably supports a more effective way of rowing on running water (Andersen, 1982a). The biology of *Perittopus* is so far unknown.

REFERENCES – Perittopinae. Redescription of Javanese species (Lundblad, 1933a); morphology, phylogeny, adaptations and classificatiion of Gerromorpha (Andersen, 1982a); new species from South-east Asia and Sri Lanka (Zettel, 2001a, e).

SUBFAMILY RHAGOVELIINAE – riffle bugs

Diagnosis

Small to medium-sized (2.2-5.1 mm). Ocelli absent. Tarsal formula 3-3-3 (rarely 2-2-2); apical segment of middle tarsus (sometimes also of hind tarsus) with deep cleft and with swimming fan arising from this cleft (diagnostic for this subfamily; figs. 296, 297). Fore wing without thick corium, with 2-4 closed cells.

Characteristics

The presence or absence of a cleft in the hindtarsi and stridulatory organs are used to separate the genera. Apart from size and colour, body shape, especially of the abdomen, presence and extension of small black denticles, notably on the ventral part of the body (fig. 363), and shape and denticulation of the hind femur and tibia are characters for specific identification in both sexes, although the hind leg characters are often more pronounced in males. When checking for the black denticles, one should start at the prothorax on both sides of the rostrum (pro-episterna); if present there, check other areas such as around the base of the rostrum, thoracic and abdominal venter, sides and dorsum, to ascertain the extension of the denticles. In addition the parameres usually give good clues to specific identity.

363

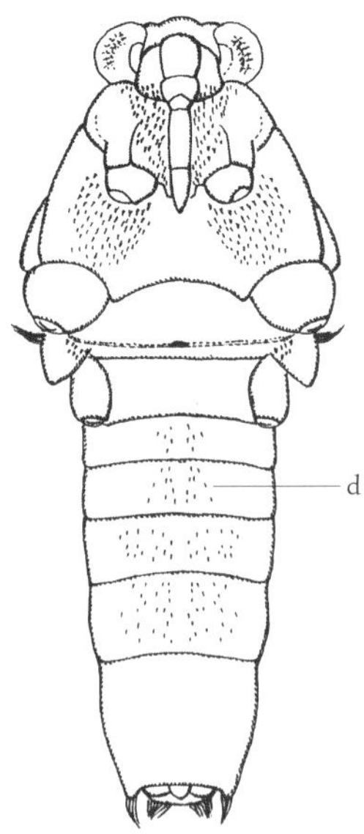

FIGURE 363
Rhagovelia kalami, in ventral view, legs and antennae removed, showing extension of small denticles on body; d = denticles.

In one species of *Rhagovelia*, *R. lugubris* from Sumatra and Java, both sexes possess a stridulatory device; this consists of fine ridges on the hind trochanter, which are rubbed against a field of stridulatory pegs laterally on sternite 3 (J. Polhemus & D. Polhemus, 1988). Also *Chenevelia stridulans* has such a stridulatory device (see key below).

Distribution worldwide

There are three or four genera (depending on the rank assigned to *Trochopus* Carpenter): *Chenevelia* with one species and *Tetraripis* with seven species are both restricted to tropical Asia. *Rhagovelia* has over 270 species, which have a pantropical distribution, with a few species extending into temperate regions. About 115 species of Rhagoveliinae are known from Asia, 112 of which occur in Malesia.

Key to Indo-Australian genera of Rhagoveliinae

1 Apical segment of middle and hind tarsi deeply cleft with a swimming fan, composed of a basal stem with dichotomously branching hairs arising from both sides (fig. 296) ▸ 2

– Only apical segment of middle tarsus deeply cleft with a swimming fan, composed of about twenty plumose structures branching off one by one from a common stem (fig. 297) [warmer regions of the World] ▸ *Rhagovelia* p. 279

2 Hind femur and connexiva without stridulatory devices [tropical Asia] ▸ *Tetraripis* p. 284

– Hind femur in both sexes proximally with a small field of stridulatory pegs on its dorsal surface; numerous fine vertical ridges on outer side of the connexiva of abdominal segments 2 and 3 [North Thailand] ▸ *Chenevelia*

Genus *Rhagovelia*

Figures 364-368

Diagnosis

(Modified after Andersen, 2000c). Small to medium-sized (2.2-5.1 mm). Extremely variable in habitus and colour, but most species either blackish or orange to reddish. Eyes large. Pronotum of apterous morph variable, usually either long and covering mesonotum completely (fig. 364), or shortened and then usually not longer than length of eye (fig. 366, 367). Fore tibia of male with short grasping comb, lacking in female, sometimes strongly curved. Middle legs obviously prolonged. Tarsal formula 3-3-3. Apical segment of middle tarsus with cleft over more than $^{2}/_{3}$, bearing swimming fan at base (dorsal extension of ventral arolium). Swimming fan composed of about twenty plumose structures branching off one by one, forming common stem. Claws long, asymmetrical, flattend (fig. 297). Hind femur often strongly enlarged, thickened (fig. 364) to strongly incrassate (fig. 365), teeth and denticles on femur and tibia especially in male; hind tarsus simple. Fore wing with two to four closed cells, without distal white spots (sometimes with one whitish streak at base). Stridulatory devices absent except in *R. lugubris*. Male segment 8 usually without

364

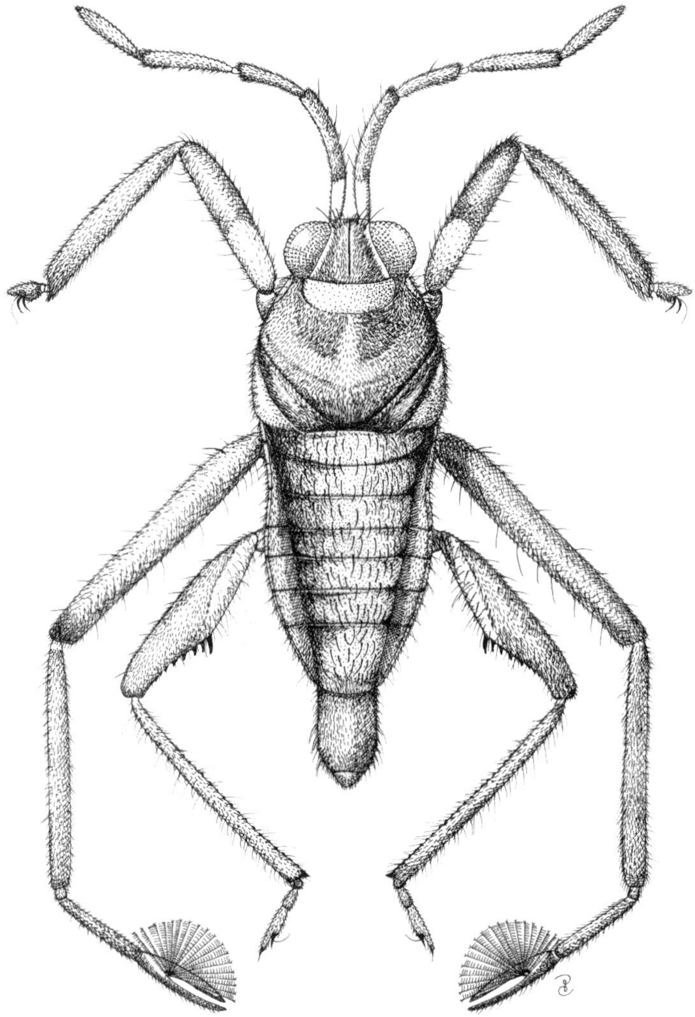

FIGURE 364
Rhagovelia chrysomalla,
paratype, apterous male,
length 3.8 mm.

365

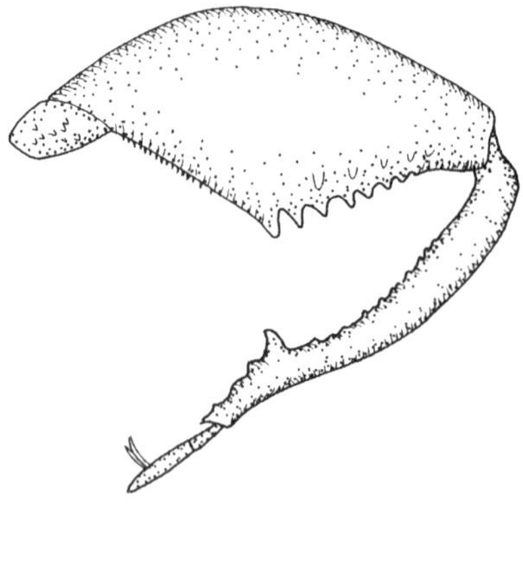

FIGURE 365
Rhagovelia horaia, right hind
leg of male.

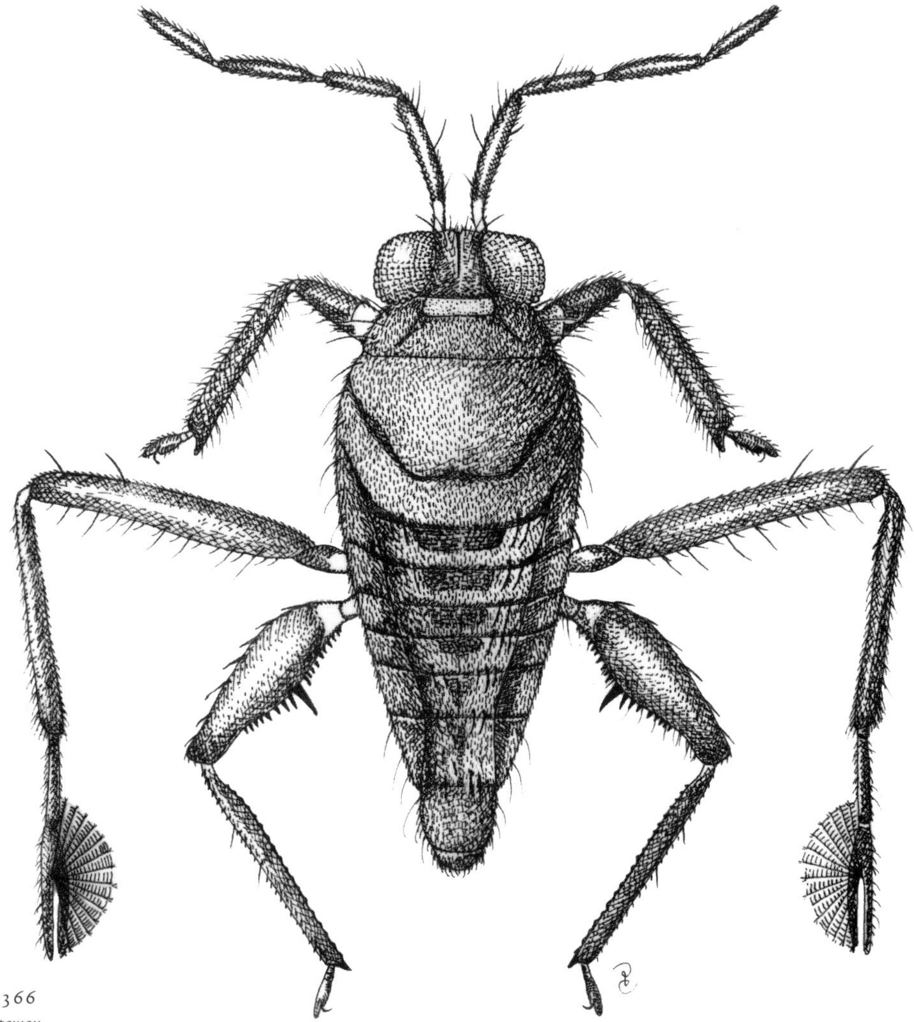

FIGURE 366
Rhagovelia tawau, apterous male, length 2.3 mm.

modifications. Parameres of variable shape. Female gonocoxa plate-like and usually exposed, or partly covered by proctiger.

Taxonomic note

The genus is presently arranged in species groups of various size and distributional range (J. Polhemus & D. Polhemus, 1988; Zettel, 1994c; Nieser et al., 1997).

Distribution in Malesia

In Malesia *Rhagovelia* holds 108 described species, which are distributed over the entire Malesian area. Numerous species are still undescribed. Most are restricted to single islands or even smaller areas. Centres of diversity are found in

367

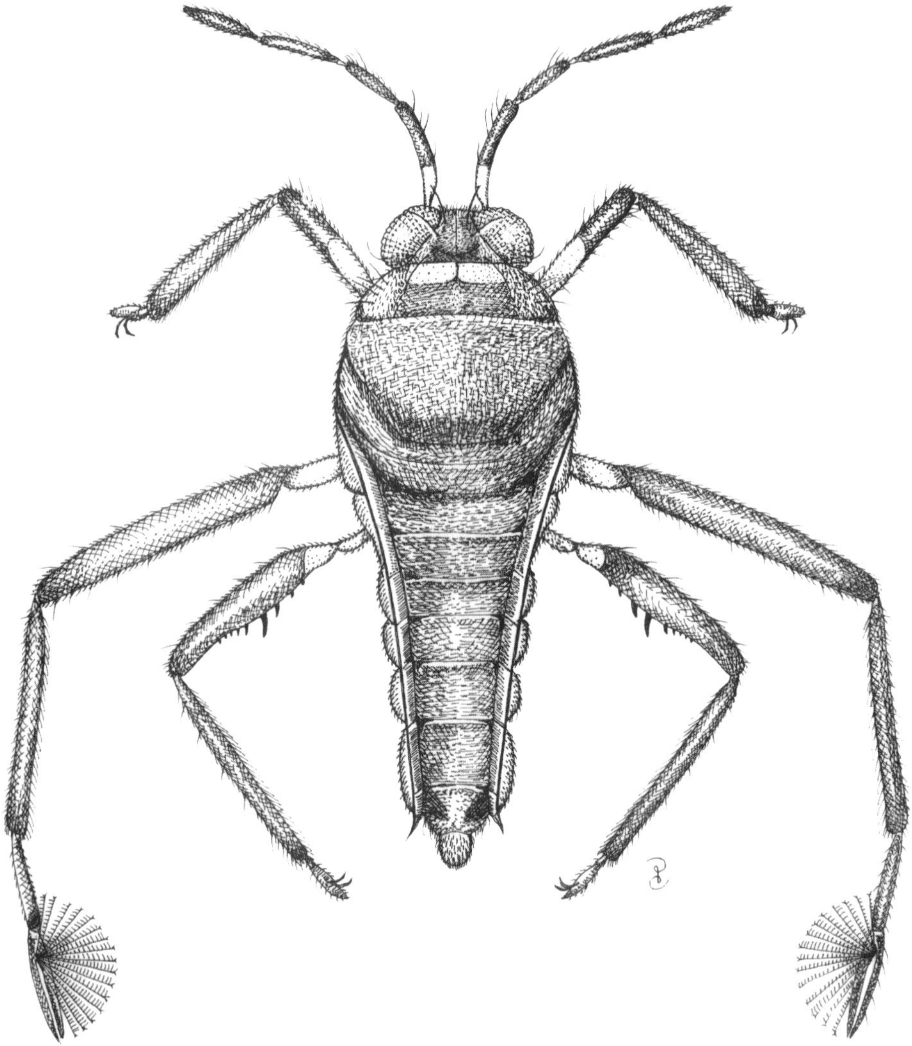

FIGURE 367
Rhagovelia sumatrensis, apterous female, length 2.8 mm.

Borneo (fourteen described species), Philippines (38), Sulawesi (30) and New Guinea (nine).

Biological notes

Rhagovelia is the most numerous life form on the surface of tropical streams. Different species inhabit various types of running waters, from minute streamlets to the edge of large rivers. The occurrence on stagnant water is probably accidental. Often several species inhabit the same section of a stream (up to five species observed in Palawan). They seem to select a niche mainly by differences in velocity and waves. J. Polhemus & D. Polhemus (1988) showed that occurrence of some species depended on altitude on different islands. Many

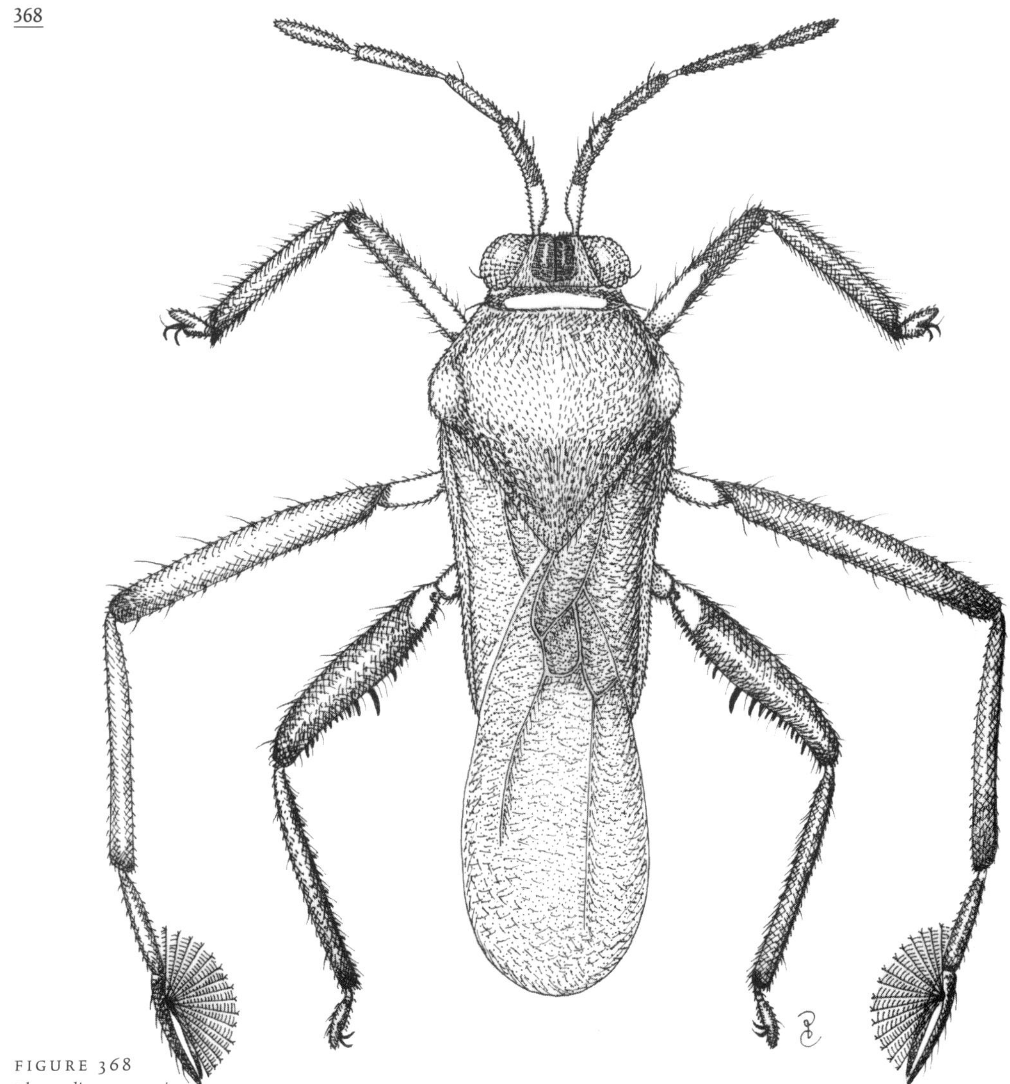

FIGURE 368
Rhagovelia sumatrensis, macropterous male, length 3.1 mm.

Rhagovelia live in swarms, which disperse when disturbed and eventually re-associate (Cheng & Fernando, 1971). Apart from finding mates easily and perhaps sharing food, this behaviour apparently impedes predation. Deshefey (1980) experimented with bird and fish models. He found that when a model was approaching slowly the swarm maintained tight formation and avoided the shadow with coordinated movements, moving forward at the same speed of subsurface models. Both moderate and rapid introduction of overhead models resulted in immediate dispersal. The same reaction occurred when gentle irregular wave action was generated by throwing a piece of wood within a few feet of the swarm. The dispersal was supposed to distort the search image of approaching predators. The segregation of swarms of different developmental stages is probably caused by a lower tolerance to water velocity of the small nymphs (Hynes, 1955).

Rhagovelia species are highly adapted to fast rowing on moving water (Andersen, 1976). As most species live in stable habitats, the apterous morph is usually much more common than the macropterous one. There are no observations of flying *Rhagovelia*. In some infrageneric units (e.g., *R. orientalis*-group, *R. sarawakensis*-group), macropterous specimens show autotomy, probably after a 'compensation flight' (upstream against drift).

During mating the female is usually grasped and held by the male with the protibial grasping comb and the hind leg, which is modified in many species. Phoresy is found in some species from New Guinea and the Philippines: the much smaller male holds the female with the strongly curved fore tibiae. Secondary sexual characters on the abdomen of both sexes (dorsally in female, ventrally in male) are probably important for species recognition and to keep the male in position. Nymphal development (of the North American species *R. obesa* Uhler) takes 48 days at a temperature of 19 °C (Cheng & Fernando 1971).

Rhagovelia feed on small insects and crustaceans trapped on the surface (de la Torre-Bueno, 1907; Bacon, 1956). The habitat, where water flow accumulates floating food, allows relatively high population densities. Food is probably recognized by sight.

Genus *Tetraripis*

Figure 369

Diagnosis

(modified after Andersen, 2000c). Relatively large (3.3-6.0 mm). Yellowish to brownish, dorsum with silvery shining areas, legs striped dark. Pronotum of apterous morph long, completely covering mesonotum. Fore tibia in both sexes with grasping comb. Tarsal formula 3-3-3. Apical segment of middle tarsus and hind tarsus with cleft, bearing swimming fan, developed asymmetrically; blade-like claws at base. Swimming fan composed of basal stem with dichotomously branching hairs arising from both sides (fig. 296). Cleft and swimming fan on hind leg smaller than on middle leg. Hind femur strongly incrassate, especially in male. Fore wing with four closed cells and three white spots. Stridulatory devices absent. Paramere of male usually (in all Malesian species) simple and curved upwards. Female gonocoxa plate-like and exposed.

Distribution in Malesia (fig. 584)

Ten species have been described, of which five occur in Malesia. All species are regional endemics.

369

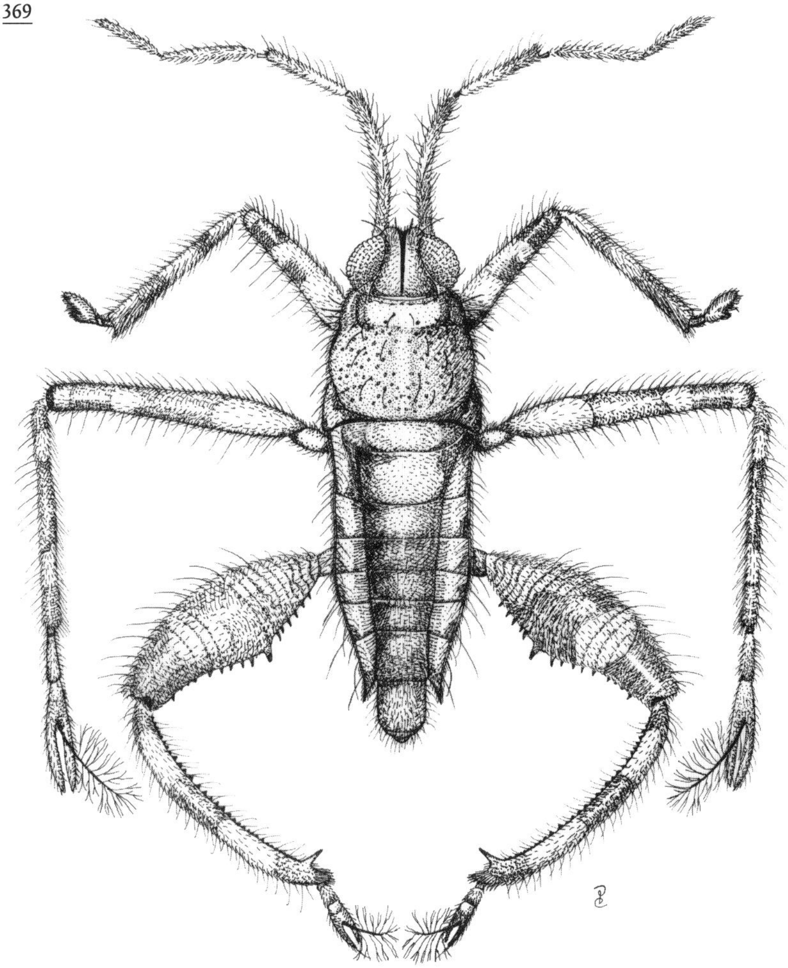

FIGURE 369
Tetraripis borneensis, paratype, apterous male, length 4.0 mm.

Biological notes

Habitat preferences seem to be similar to those of *Angilia* (see below). In South India *Tetraripis* species have been collected in a small ditch and nymphs have been found by turning stones on a stream bank (Andersen, 1982a). In Myanmar *T. chinthe* J. Polhemus & D. Polhemus has been collected in dark sheltered sites under steep stream banks with tangles of roots, but once also under detritus or small logs along the sloping shore of a swift clear stream (J. Polhemus & D. Polhemus, 1999). When disturbed individuals skate to shady spots near the shore. A nocturnal life is suspected (J. Polhemus & D. Polhemus, 1999). *Tetraripis asymmetricus* Polhemus & Karunaratne from Sri Lanka has two 'forms' of apterous females, one of which has well-developed abdominal carinae on tergites 1-4 as in the macropterous morph (Zettel & Yang, 2002); this may indicate an environmentally influenced mode of wing development of females.

REFERENCES – Rhagoveliinae. Revision of Old World Rhagoveliinae (Lundblad, 1936); morphology, phylogeny, adaptations and classification of Gerromorpha (Andersen, 1982a); key to *Rhagovelia* of Borneo, Moluccas and Sulawesi (J. Polhemus & D. Polhemus, 1988); key to South-east Asian *Rhagovelia* (J. Polhemus, 1990b); key to *Rhagovelia* of Australia, Papua New Guinea and Solomon Islands (Lansbury, 1993b); key to *Rhagovelia* of Singapore (Yang & D. Polhemus, 1994); description of new species of *Rhagovelia* from Philippines (Zettel, 1994b, 1991, 2003e); revision of Philippine *Rhagovelia*, part 1, 2 and 3 (Zettel, 1994c, 1995d, 1996c); description of new species of *Tetraripis* from Borneo (Zettel, 1995e); key to Sulawesi species of *Rhagovelia* (Nieser et al., 1997); key to *Tetraripis* (J. Polhemus & D. Polhemus, 1999); critical assessment of generic classification of subfamily Rhagovelia, key to species, and description of new species (Andersen, 2000c); study of South-east Asian species of *Tetraripis* and *Angilia* (Zettel & Yang, 2002).

SUBFAMILY VELIINAE

Diagnosis

Medium-sized to large (4.0-10 mm). Malesian species yellowish with brown marks. Hemielytra proximally not coriaceous, with four closed cells. All tarsi three-segmented, middle tarsus not deeply cleft, lacking plumose or hairy swimming fans.

Characteristics

Characters separating both genera and species are the antennal structure, tumescences on the thoracic venter, teeth on the male hind femur, flattened tarsal claws and/or arolia and relative length of tarsal segments, presence of stridulatory devices, and the structure of the female proctiger. However, South-east Asia has a rather poor fauna of Veliinae and identification of the genera is simple in this area. Apart from size and colour and characters mentioned above, characters to identify species are structure of the humeral angles of the pronotum, shape of connexiva (especially the caudal apices), shape of the parameres, endosomal sclerites and female proctiger. In *Angilovelia* both sexes have a stridulatory device, consisting of a patch of sharp pegs proximodorsally on the hind femur (fig. 371), which are rubbed against fine ridges laterally on the connexival margin (Andersen, 1981b).

Biological notes

Although most species are associated with running water, the microhabitats are found usually on virtually stagnant or very slow flowing parts of streams. Many species (especially from South America) live at the edge of streams under overhanging banks, between marginal vegetation, or on small mountain streams consisting of small pools connected by trickles of water for a part of the year. However, several species live on ponds, lakes or marshes, usually hidden between vegetation along the edges. A few South American species live in water held by Bromeliaceae and tree holes. In spite of the fact that the middle legs are not strongly lengthened, all species studied in this respect can walk on land. On water they skate by simultaneous strokes of the middle legs. Expansion skating in Veliinae has only been observed in *Velia*.

Eggs of *Velia* are glued to plants at or slightly above the water surface. Considering their habitats most other veliine genera probably deposit their eggs in the same way.

Distribution worldwide

There are nine recent genera, three of them divided into two or three subgenera, with a total of about 130 described species. Of the three Old World genera, *Velia* has 28 species, most of which are Palearctic. *Angilia* has nineteen species of which seven, all belonging to the subgenus *Adriennella*, occur in South-east Asia. *Angilovelia* has one described species in South-east Asia.

Distribution in Indo-Australia

See under *Angilia* and *Angilovelia*.

Key to Indo-Australian genera of Veliinae

1 Hind femur in both sexes with a group of small black pegs proximally on the dorsal surface (fig. 371) ‣ *Angilovelia* p. 289

– Hind femur without such small black pegs ‣ 2

2 Macropterous (occasionally brachypterous), hemielytron distally with a large white spot surrounded by 2-4 smaller spots (fig. 370). Antennal segments 3-4 distinctly more slender than segment 2 ‣ *Angilia* s.l. p. 287

– Predominantly apterous, hemielytron of macropterous morph distally with a different arrangement of white spots. Antennal segments 3-4 not noticeably thinner than segment 2 [Palearctic, mountains of South-east Asia; not in Malesia] ‣ *Velia*

Genus *Angilia*

Figure 370

Diagnosis

Medium-sized to large (5.2-7.2 mm), macropterous or (occasionally) brachypterous. Medium to light brown, variegated white. Hemielytra apically with large white spot surrounded by 2-4 small spots. Antennal tubercles in dorsal

370

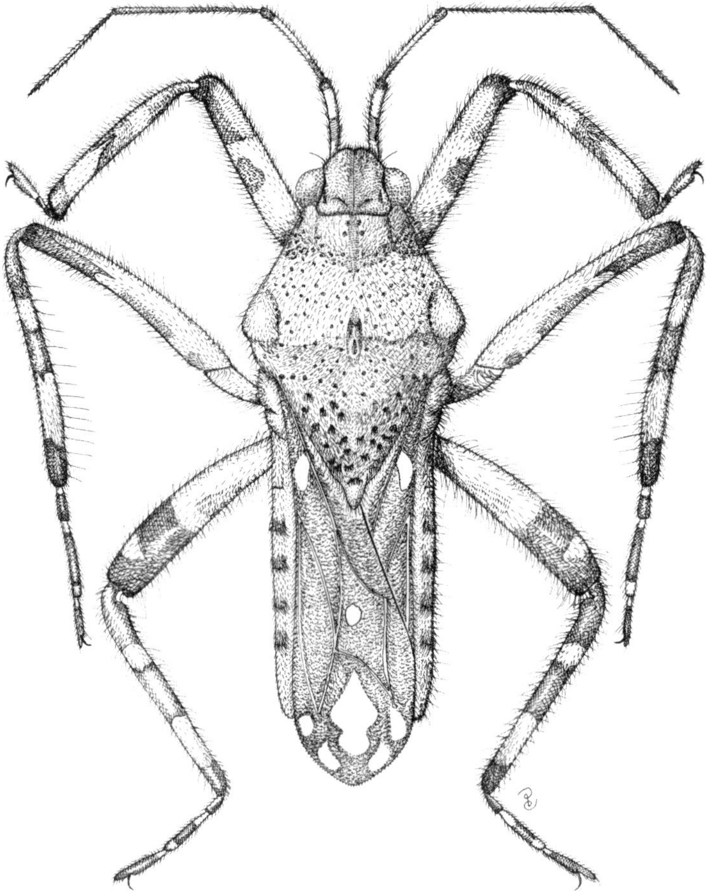

FIGURE 370
Angilia orientalis, macropterous male, length 5.6 mm.

view not visible; apex of rostrum not surpassing middle of mesosternum. Pronotum large, strongly prolonged posteriorly, with distinctly converging sides and more or less pointed apex. Fore leg with grasping comb in both sexes, in males longer than half of tibial length. Hind femur ventrally with numerous, sometimes very small, dark teeth or pegs, but lacking field of minute pegs proximally on dorsal surface. Female proctiger cone-shaped, not covering the gonocoxae.

– *Angilia* **subgenus** *Adriennella*

Diagnosis

(after Andersen, 1981a; Zettel & Hecher, 1998). Macropterous morph common. Head strongly deflected; ventral lobes of head prominent. Antennal segment 1 shorter or subequal to head width, segments 3 and 4 longer and more slender than segment 1 and 2. Pronotum of macropterous morph strongly elevated between humeri, with finger-like projection in middle of hind margin, often with spine-like structures at humeri (and additionally in centre). Both sexes with protibial grasping comb. Middle tarsus more than 0.65 times as long as middle tibia and distinctly longer than hind tarsus; hind femur more or less incrassate, with teeth or denticles in both sexes; hind femur and connexiva without stridulatory devices. Fore wing with four closed cells and white spots; segment 8 of male often with ventral modifications; parameres symmetrical, long, slender falciform or short and broad. Gonocoxa of female exposed, proctiger conical.

Distribution in Malesia (fig. 586)

Of the eleven species in this subgenus, seven occur in the Oriental region and six in Malesia. The genus is unknown from areas east of Wallace's Line (in its original concept), but *A. philippiensis* transgresses Dickerson's Line and reaches Luzon and Mindoro in the Philippines.

Biological notes

Andersen (1981b) gives notes on two species collected in Thailand. *Angilia bispinosa* Andersen has been found in a very small (0.5 m) water-filled, probably spring-fed and partly overgrown hole in the shade in a secondary forest at 500 m altitude. They do not run on open water but stay under a stone or at the muddy shores. *Angilia orientalis* has been swept from vegetation overhanging a small stream. This species and *A. philippiensis* have been collected in the Philippines at the rocky edges of streams and rivers in small sheltered places (F. Seyfert, pers. comm.). Although *Angilia* are associated with streams, they appear to occur mainly on stagnant or virtually stagnant water like springs or spring-fed pools (Andersen, 1981b; J. Polhemus & D. Polhemus, 1999).

Genus *Angilovelia*

Figure 371

Diagnosis[1]

Medium-sized to large (5.2-7.2 mm). Macropterous or (occasionally) brachypterous, apterous morph unknown. Medium to light brown, variegated white. Hemielytra apically with large white spot surrounded by 2-4 small spots. Head moderately deflected; ventral lobes of head rather prominent. Antennal segment 1 subequal to head width, segments 3 and 4 longer and more slender than segment 1 and 2. Pronotum of macropterous morph large, strongly prolonged posteriorly, with distinctly converging sides and more or less pointed apex. Fore leg of male only with grasping comb. Middle tarsus less than 0.65 times

1 After Andersen (1981b).

371

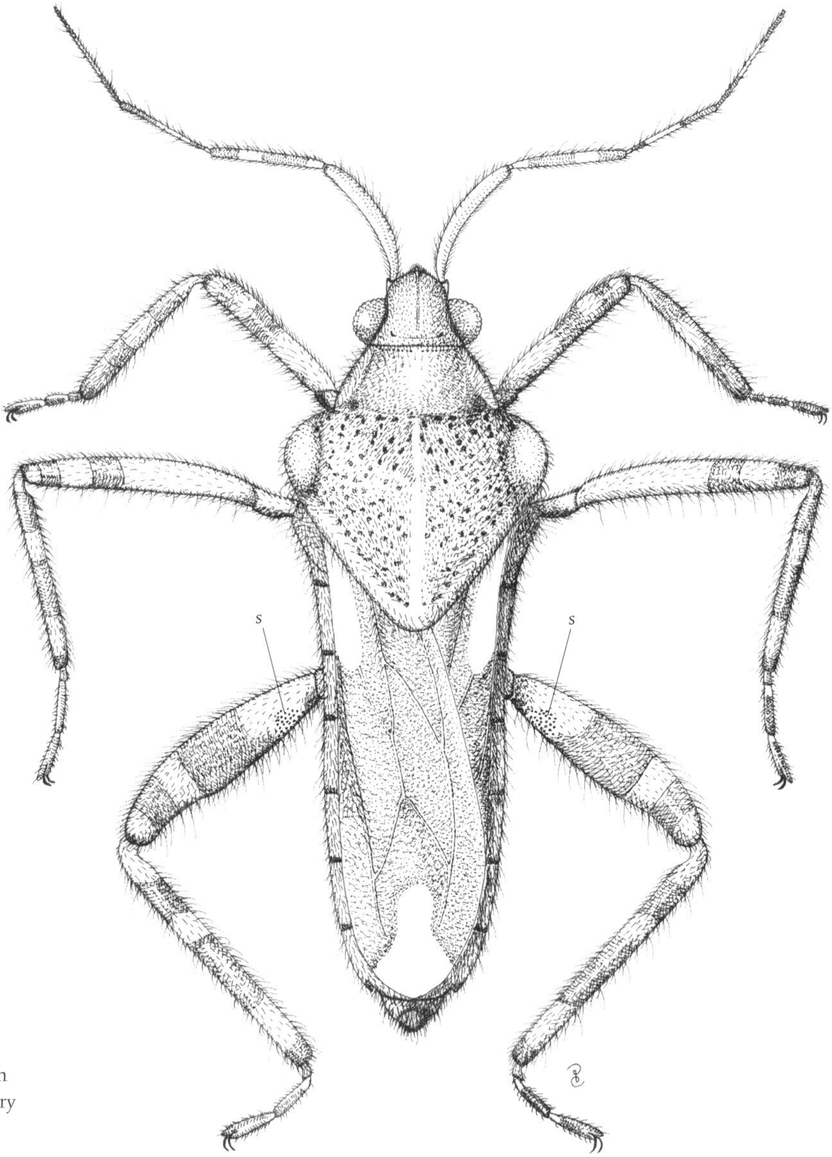

FIGURE 371
Angilovelia sp., macropterous female, length 5.4 mm; s = stridulatory pegs on hind femur.

as long as middle tibia and equally long as hind tarsus; hind femur weakly incrasssate, without teeth or denticles, but with a field of minute pegs proximally on the dorsal surface, which is supposed to be a plectrum for stridulation. Connexiva 2-3 with fine ridges (supposed strigil for stridulation). Segment 8 of male with ventral modifications; parameres symmetrical, long, broadly falciform. Gonocoxa of female exposed, proctiger conical.

Distribution in Malesia (fig. 586)

There is only one described species, *A. y-alba*, which is rarely encountered. The records are far apart. The species extends from North-east India and South China to continental Malaysia (Andersen, 1981b, J. Polhemus & D. Polhemus, 1999). Nieser (unpubl.) collected this or a closely related species in South-east Sulawesi.

Biological notes

Angilovelia species probably prefer stagnant water: the Sulawesi specimens (two females) were collected at the edge of a marsh between vegetation of dense rushes, and on a pool at the banks of a stream (cf. J. Polhemus & D. Polhemus, 1999).

REFERENCES – Veliinae. Description of *Angilovelia*, review of Oriental *Angilia* (Andersen, 1981b); morphology, phylogeny, biogeography, adaptations and classification of Gerromorpha (Andersen, 1982a); key to Oriental *Angilia* species (Zettel & Hecher, 1998).

Family Gerridae – pond skaters or waterstriders

Diagnosis

Small to large (1.6-36 mm), elongate to globular with long legs. Wing polymorphism common, apterous and macropterous morphs predominant. Dorsal face of head without median impressed line and without a pair of deep pits near inner posterior angles of eyes. Antennal sockets usually in front of eyes, visible from above. First antennal segment usually longer and somewhat stouter than remaining three segments. Pronotum without collar, never punctuated. Mesothorax elongate compared to other Gerromorpha. Middle and hind legs slender, greatly elongated; middle and hind coxae widely separated, middle coxa large and usually greatly extended along lateral margin of body. All tarsi two-segmented; claws almost always inserted preapically. Venation of hemielytra variable but never with more than four cells; two apical cells often terminating in relatively long free vein. Abdominal length variable. Parameres symmetrical, usually present, but very small in some groups. First gonocoxae plate-shaped, usually covering short ovipositor.

Characteristics

Gerridae is a very diverse family. Most important apomorphic characters are: four pairs of cephalic trichobothria (fig. 8; but secondarily reduced to three pairs in a few taxa), the strongly prolonged mesothorax, and two-segmented tarsi. Characters separating subfamilies are mainly the position of thoracic sutures dorsally in apterous morphs (figs. 378-381), and development of the metasternum and first abdominal sternum (figs. 376, 377). Main characters to differentiate genera are the relative length of the appendages and their segments, structure of the metasternum including the area around the scent gland orifice, and the structures of the pretarsus. In addition, in some genera the position of the metathoracic spiracle is diagnostic. Characters to identify species are variable among subfamilies and genera. In addition to colour and relative length of the appendages, the male fore leg is modified in some genera. The parameres may also be useful in a number of genera. Vesical sclerites are often diagnostic as well and are used especially in taxa with reduced parameres (figs. 372-374). In females the shape of the abdominal apex, notably the last abdominal sternite, the gonocoxae, and the length of the connexival spines may give clues to specific identity. Additional features are mentioned under the subfamilies.

Fossil record

There are quite a number of fossil Gerridae (Andersen, 1998). All are from the Cenozoicum, up to 60 million years ago. With the exception of two species they

FIGURES 372 – 374
Development of vesical sclerites in Gerridae (adapted from Andersen, 1982a); as = apical sclerites, bs = basal sclerites, ds = dorsal sclerites, ls = lateral sclerites, vs = ventral sclerites.

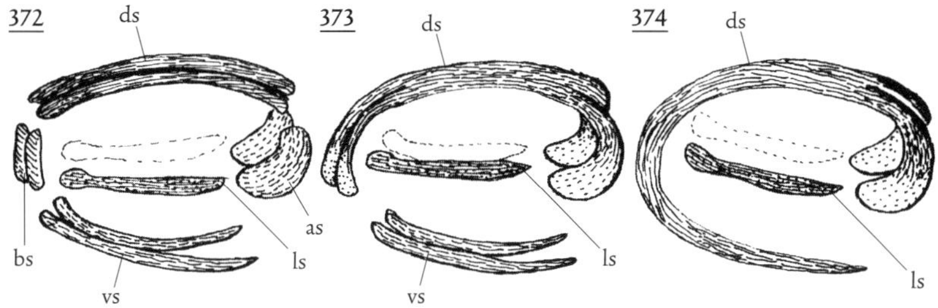

belong to extinct and extant Gerrinae. These fossils prove that some of the present genera existed already over 50 million years ago. Fossils of two extant gerrine genera *Aquarius* and *Limnoporus* (Stål) have been found in deposits from Denmark (55 million years ago) and British Columbia, Canada (50 million years ago; Andersen et al., 1993). Andersen (1998) presents a phylogenetic tree showing the relationship between fossil and recent genera. Of the two non-gerrine genera *Electrobates* Andersen & Poinar belongs to an extinct subfamily *Electrobatinae* (see chapter 3). The other non-gerrine fossil belongs to *Halobates* and is discussed there.

Distribution worldwide

There are eight subfamilies with 80 genera (of which six are extinct) and about 700 described species. Rhagadotarsinae has two genera, one of which has five widespread species in the Oriental and Australasian regions. Eotrechinae with five genera and 60 species is restricted to Asia; only a few species occur outside tropical areas. Ptilomerinae with twelve genera and about 100 species is also mainly Asian. Halobatinae has ten genera and about 150 species, of which a few live on open oceans. One species, *Halobates micans*, occurs in the Pacific and the Atlantic Oceans. The remaining species are confined to the Old World, with the vast majority in tropical areas. The other subfamilies are more evenly distributed, or Neotropical.

Distribution in Indo-Australia

Details are given under subfamilies or tribes.

Biological notes

Although a few gerrids (Eotrechinae) are hygropetric or live at the edge of the water, and especially larvae of some species live amongst emergent vegetation, most Gerridae live on open water. A few even occur on the open ocean. Gerridae are usually the first water bugs observed when inspecting a water body.

Most gerrids are dimorphic, either macropterous or apterous, although polymorphic and monomorphic species also occur. Monomorphic apterous species live in stable permanent habitats (e.g. all *Halobates*). Monomorphic macropterous species tend to live in unstable temporary habitats, which holds for all waterbugs (see chapter on Morphology). Autotomy in macropterous morphs occurs in the Metrocorini of the Halobatinae, Ptilomerinae and Trepobatinae. In Ptilomerinae wings are torn off at the base; in Trepobatinae wings are torn off in a rather straight line distally of the basal venation, at about $^1/_3$ from the base; in Metrocorini wings are torn off rather irregularly near the abdominal apex.

Ripples are used in interspecific communication in several gerrine genera, including several species of *Rhagadotarsus* (apart from its function in locating prey). This behaviour was analyzed most extensively in *R. anomalus* (Wilcox, 1972, 1980). Males send out ripples especially when holding a soft free-floating or fixed object. Some ripple patterns are aggressive, in order to keep competing males away and defend the floating object which may become an oviposition site. There are calling signals to attract females, and courtship signals when a female is close enough. Ripple signals are sent during copulation and during oviposition (Wilcox, 1972, 1980. The species studied was supposed to be *R. kraepelini* but was in fact *R. anomalus*; J. Polhemus & Karunaratne, 1993). Ripple communication has also been observed in *R. kraepelini* (J. Polhemus, 1990c). In *Tenagogonus* and other gerrine taxa the main function of ripple signals appears to be protection of a territory, although other functions have been observed as well, e.g. repelling males from trying to grasp females which already carry a mate (Wilcox & Di Stefano, 1991). More details will be given under genera.

In most species, with males somewhat smaller than females, copulation is apparently straightforward: the male sits on the back of the female and body axes are kept parallel. Usually the male stays with the female for several hours after copulation has taken place. In Gerridae phoresy is quite rare and in Malesian species it seems to be restricted to *Rheumatogonus* and some Trepobatinae. In some genera males are larger than females and/or have longer legs. Cheng (1967b) reports that in *Metrocoris tenuicornis* the male carries the passive female under his body. Andersen (1982a) reports the same for a species of *Ptilomera*. A third genus occurring in Malesia with males larger than females, to which the same kind of sexual behaviour may also apply, is *Limnometra*. In most species eggs are glued to floating plants, in some species just above, in others just below the surface. Some species glue their eggs to more solid substrates. *Rhagadotarsus* lay their eggs in plants (see below).

Prey consists of soft invertebrates, mainly insects. An important aspect of searching for prey is the ability to locate the origin of ripples in the surface film (as in Notonectinae and many Veliidae). Orientation receptors are located either at the extreme tip of the tibia or on the first tarsal segment and is probably associated with the tibio-tarsal joints. When ripples reach the animal it will turn in the direction of the source. Only when the source is behind the bug and the angle between the source of the ripples and the longitudinal axis is about 180° they may turn to the wrong side; the next turn however will be in the right direction. This system works only on virtually undisturbed water and only when the bug is not using its legs for propulsion (Murphey, 1971). When close to the prey visual stimuli take over. How gerrids locate their prey on turbulent streams has not been investigated. They possibly react on visual stimuli. Smell does not play a role in prey detection: reaction to small empty boxes or boxes containing a dead fly are identical. *Velia* can select the box with a dead fly, but when searching for prey *Velia* apparently does not use smell. However, *Hydrometra* do make use of smell (Rensing, 1962).

The economic importance of Gerridae and Gerromorpha in general seems to be small but is poorly documented (Spence & Andersen, 2000). Some species have been observed to consume mosquito eggs and larvae (Gupta, 1985), but most mosquitoes consumed by Gerridae are specimens trapped in the surface film and are bound to die anyway. Nummelin (1988b) studied the pest mangement potential of two African Gerrinae on newly emerged adults of *Aedes aegypti* mosquitoes. He concluded that their predation efficiency was 'too low to encourage experiments for the use of these species of water striders as a controlling agent for mosquitoes'. Two species of *Limnogonus* are reported to be of economic importance in rice fields. Gupta et al (1990) state that *L. nitidus* (as *Gerris nitida*) causes significant damage to rice in North-east India by sucking grains and stems. Spence & Andersen (2000) state that these observations need further investigation as they are the only published report of phytophagy among Gerromorpha. *Limnogonus fossarum* and *L. nitidus* are reported to be important predators of the brown plant hopper *Nilaparvata lugens* (Stål) which is one of the most important pests in rice (Selvanayagam & Rao, 1986). So far no quantitative studies of their impact on *N. lugens* have been published (Spence & Andersen, 2000).

REFERENCES – Gerridae general. Checklist Indo-Australian species, description and redescription of species (Lundblad, 1933a); structural analysis and classification of Gerridae (Matsuda, 1960); description of various species from Brunei (Miyamoto, 1967); keys to peninsular Malaysian species (Cheng & Fernando, 1969); orientation in *Aquarius* (Murphey, 1971); morphology, phylogeny, adaptations, biogeography and classification of Gerromorpha (Andersen, 1975, 1982a); ripple communication (Wilcox, 1972, 1980); bionomics of gerrids (Gupta, 1985); feeding behaviour (Selvanayagam & Rao, 1986); poor potential for mosquito control by Gerridae (Nummelin, 1988b); new record of bugs infesting paddy fields in Orissa (Gupta et al., 1990); description of extinct genus *Electrobates* (Andersen & Poinar, 1992); key to genera from Indonesia excepting Papua (Irian Jaya), survey of Sulawesi species (Chen & Nieser, 1992); fossil Gerromorpha (Andersen, 1998); simplified key to genera from Thailand (Chen & Zettel, 1998); economic importance (Spence & Andersen, 2002).

Key to Indo-Australian subfamilies of Gerridae

Based on apterous specimens (adapted from Anderson, 1982a). Samples consisting of only macropterous specimens usually belong to Gerrinae.

1 First abdominal sternite present (the abdomen counts seven unmodifidied sternites). Ventral lobes of head produced (fig. 375). Ovipositor well-developed (fig. 383) ▸ **Rhagadotarsinae p. 298**

– First abdominal sternite absent or fused with metasternum (the abdomen counts six unmodifidied sternites). Ventral lobes of head not produced (fig. 8). Ovipositor short ▸ 2

375

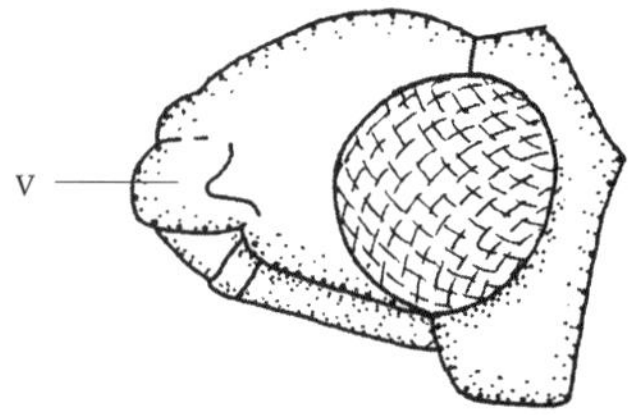

FIGURE 375
Head of apterous *Rhagadotarsus kraepelini*; v = ventral lobe.

2 Middle femur stout, distinctly shorter than middle tibia and usually shorter than hind femur (figs. 412, 422). Metasternal scent gland orifice absent in most genera (present in the marine Stenobatini) ▸ **Trepobatinae p. 301**

– Middle femur slender, usually distinctly longer than middle tibia. Metasternal scent gland orifice present ▸ 3

3 Metasternum extremely reduced, usually only represented by a very short subtriangular plate enclosing the scent gland orifice (fig. 377). Claws of hind legs modified, straight or S-shaped ▸ **Halobatinae p. 384**

– Metasternum well-developed, clearly reaching metacetabular region laterally (fig. 376). Claws of hind legs, if present, falcate ▸ 4

376

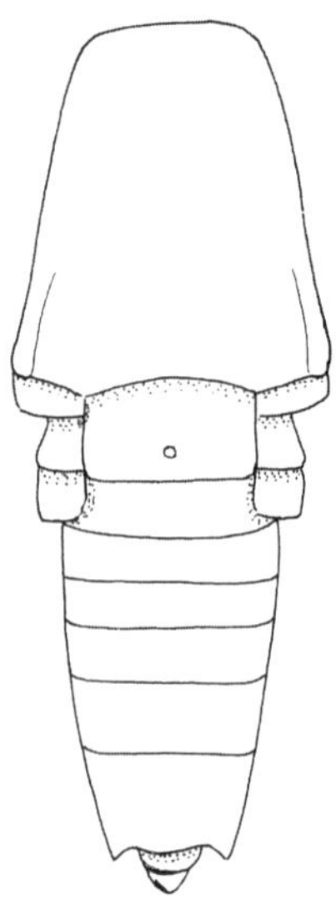

377

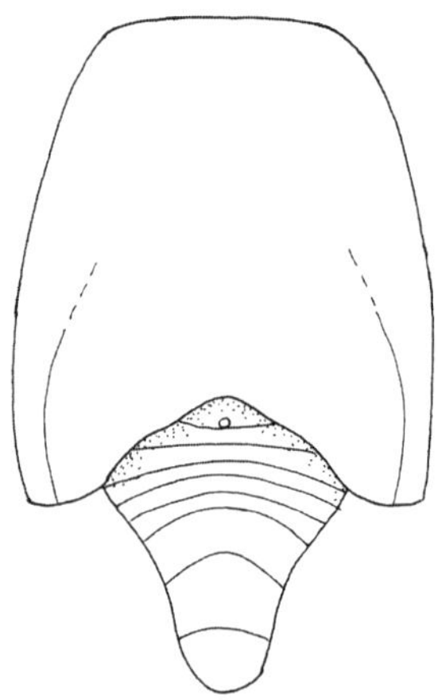

FIGURES 376 – 377
Semidiagrammatic representation of mesosternum and metasternum and second abdominal sternite. **376** Gerrinae – **377** Halobatinae.

4 Metacetabular groove dorsally reaching anterior end of first abdominal tergum (fig. 380). Fore tarsus very long, at least $^{1}/_{2}$ of tibial length ▸ **Ptilomerinae p. 368**

– Metacetabular groove not reaching anterior end of first abdominal tergum (fig. 381). Fore tarsi shorter ▸ 5

5 Metacetabular groove dorsally connected with hind margin of mesopostnotum (fig. 378). Fourth antennal segment short, curved. Rostrum short, not reaching beyond posterior margin of prosternum. First segment of fore tarsus less than $^{1}/_{2}$ as long as second segment ▸ **Cylindrostethinae p. 365**

– Metacetabular groove dorsally indistinct, not connected with hind margin of mesopostnotum (fig. 379). Fourth antennal segment straight. Rostrum reaching beyond posterior margin of prosternum. First segment of fore tarsus usually more than $^{1}/_{2}$ as long as second segment ▸ 6

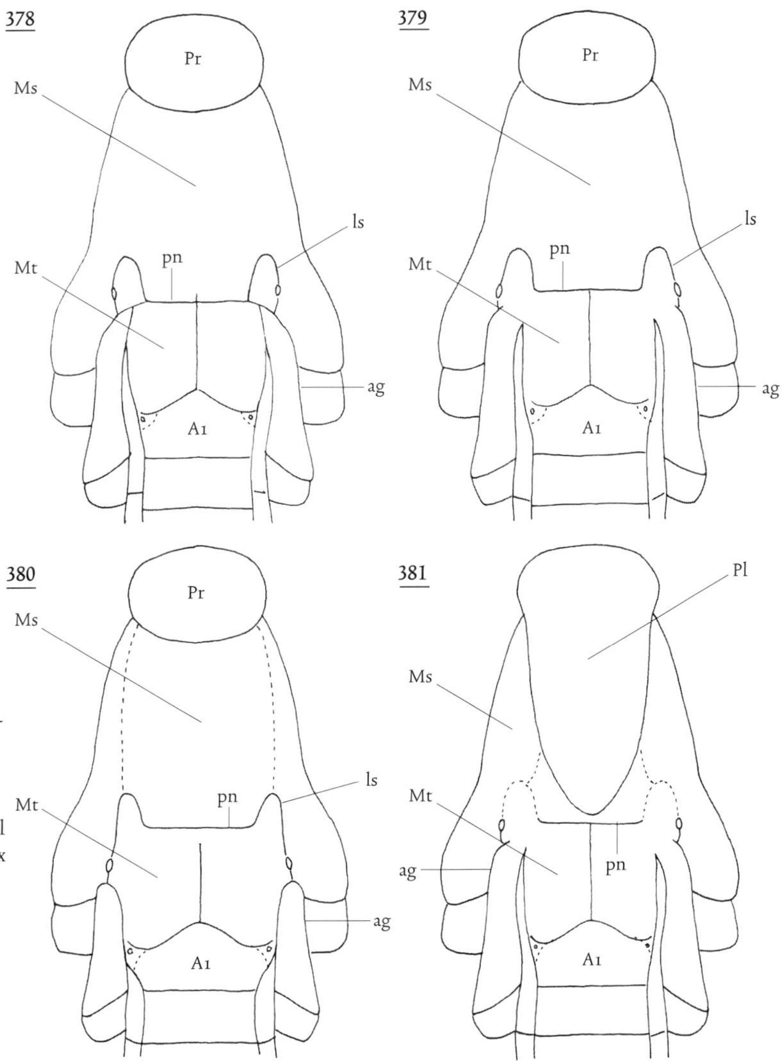

FIGURES 378 – 381 Gerridae, semidiagrammatic representation of dorsal sutures of thorax (adapted from Andersen, 1982a). **378** Cylindrostethinae – **379** Eotrechinae – **380** Ptilomerinae – **381** Gerrinae; A1 = first abdominal (medio)tergite, ag = metacetabular groove, ls = lateral suture between mesothorax and metathorax distinct, Ms = mesonotum, Mt = metanotum, pn = hind margin of the mesopostnotum, Pl = pronotal lobe, Pr = pronotum.

6 Pronotal lobe in apterous specimens always reduced. Pretarsal arolia present. Lateral suture between mesothorax and metathorax distinct (fig. 379)
▸ **Eotrechinae p. 356**

– Pronotal lobe in apterous specimens usually not reduced; if so, then pretarsal arolia absent or lateral suture between mesothorax and metathorax indistinct (fig. 381) ▸ **Gerrinae p. 340**

SUBFAMILY RHAGADOTARSINAE

Diagnosis

Small (1.6-6.0 mm), elongate, moderately flattened. Antenna slender, relatively short. Ventral lobes of head large, produced in front of rostral base (fig. 375). Mesonotum laterally defined by longitudinal suture. Paired longitudinal suture separating mesosternum from mesopleural regions lost. Metathoracic scent apparatus absent. Abdomen with distinct first laterotergites and sternum, segments 2-6 shorter than segment 8. Genital segments conspicuously enlarged. Parameres absent; vesical sclerites completely fused, forming ring. Female genitalia distinctly projecting posteriorly (fig. 383), forming serrate ovipositor. Rhagadotarsinae is morphologically the most deviating subfamily of Gerridae.

Genus *Rhagadotarsus*

Figures 382-385

Diagnosis

Medium-sized to large (males 3.0-5.3, females 3.3-6.0 mm). Both apterous and macropterous. Black, covered with grey or bluish-grey pubescence on dorsum. Antennal segments of male not modified; first antennal segment longest. Rostrum slightly surpassing anterior margin of mesosternum. Pronotum short in apterous morph, with median yellowish mark obscured by grey pubescence. Mesothorax much larger than metathorax; latter fused with first abdominal tergite. Fore tibia short and flattened. Middle leg much longer than hindleg. Hind leg of male not modified. Wings when folded reaching abdominal apex (male), or leaving tip of genital segments exposed (female; fig. 383). Abdomen long. Male genital segments very long and slender (fig. 382), segment 8 dorsoventrally depressed, strongly modified ventrally. Parameres absent, sclerites of vesica forming loop-like structure (fig. 385). Female gonocoxae very large and elongate.

Distribution worldwide

Rhagadotarsus has two subgenera with five species in total. The nominate subgenus contains four species which are widespread in Indo-Australia. The subgenus *Caprivia* China, 1931 has one species in Africa south of the Sahara.

382

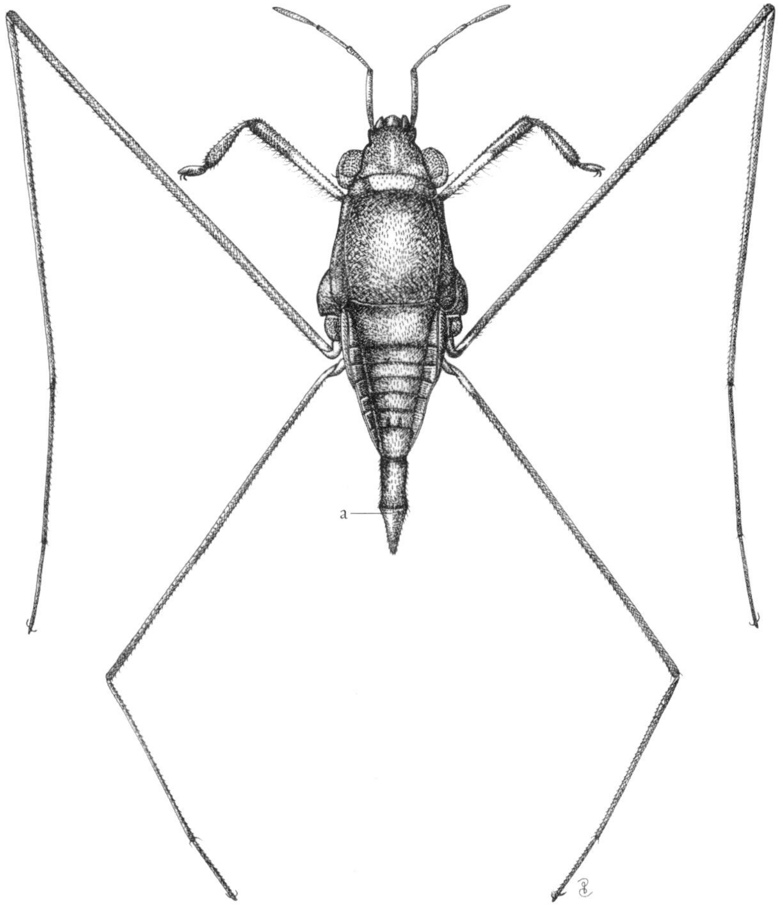

FIGURE 382
Rhagadotarsus kraepelini, apterous male, length 3.4 mm; a = apex of genital segments.

– *Rhagadotarsus* sensu stricto

Note

Differences between species are inconspicuous. Main diagnostic characters include length of hind femur, intensity of pilosity, relative length of antennal segments and – in males – modification of fore femur, form of the sulcus on sternite 7 and the depression on sternite 8.

Distribution in Malesia (fig. 587)

Three species of *Rhagadotarsus* occur in Malesia. The most widely distributed species, *R. kraepelini*, occurs from eastern India and Sri Lanka through the Malay Archipelago to Micronesia.

Biological notes

The habitat of *Rhagadotarsus* species is diverse: lakes, ponds, ephemeral pools, ditches, small rivers, mangrove streams, and brackish freshwater. Individuals

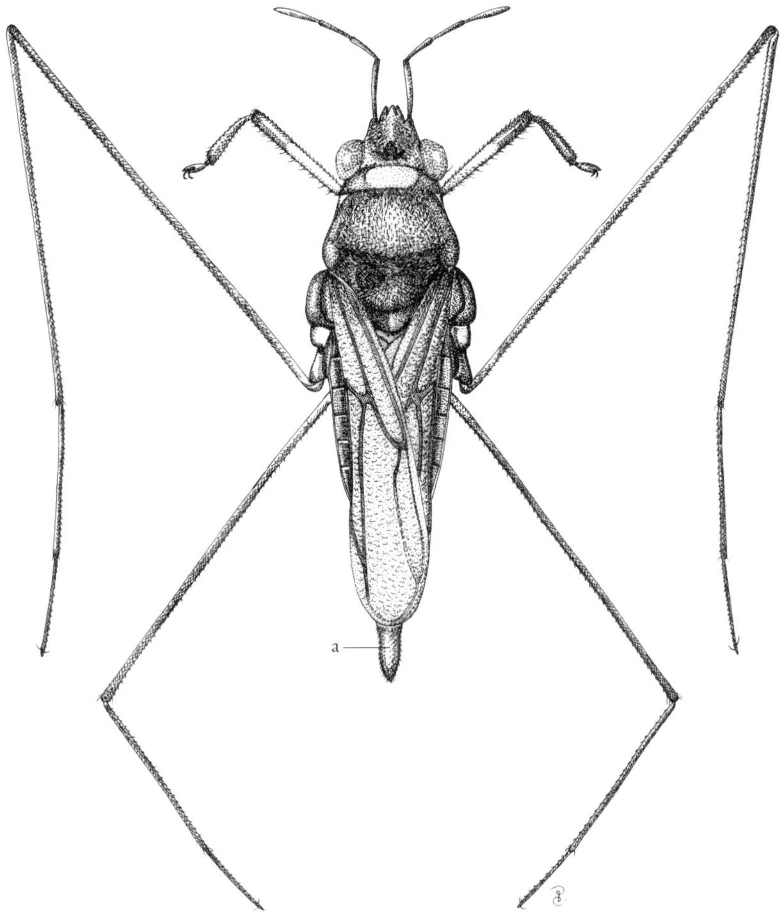

FIGURE 383
Rhagadotarsus kraepelini, macropterous female, length 4.4 mm; a = apex of genital segments.

are usually found on a clear water surface; if there is floating debris they may become entangled and drown.

Populations of *Rhagadotarsus* species are predominantly apterous, but in most species macropterous individuals are known. Males produce different wave patterns for communication: some for calling and courtship, others for competition. The life history of *R. kraepelini* and *R. anomalus* was studied in China and Australia respectively (J. Polhemus & Karunaratne, 1993). Some majore differences between these species were found: *R. kraepelini* has five or six generations per year, breeds during most of the year, and adults overwinter, while *R. anomalus* stays active and mates all year round.

The sulcus of the male venter of all species and the pilosity are considered to ensure that the long slender female genital segments fit closely during mating. The ventral hairs on the male genital segments are sensory hairs and may assist in positioning before copulation. *Rhagadotarsus* males have the peculiar habit of pushing or grasping small bits of soft floating material, which are eventually used as oviposition sites.

384

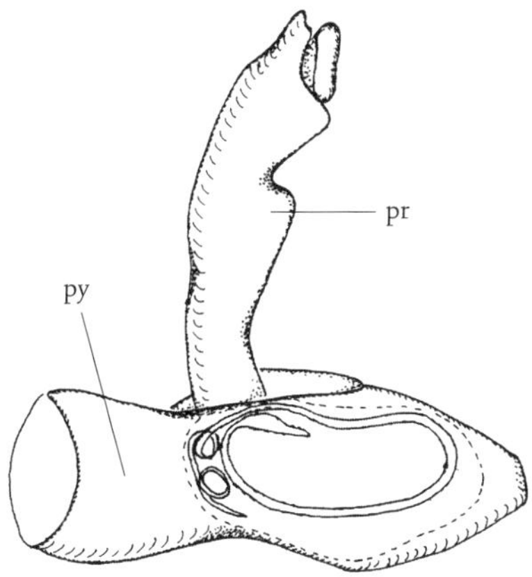

385

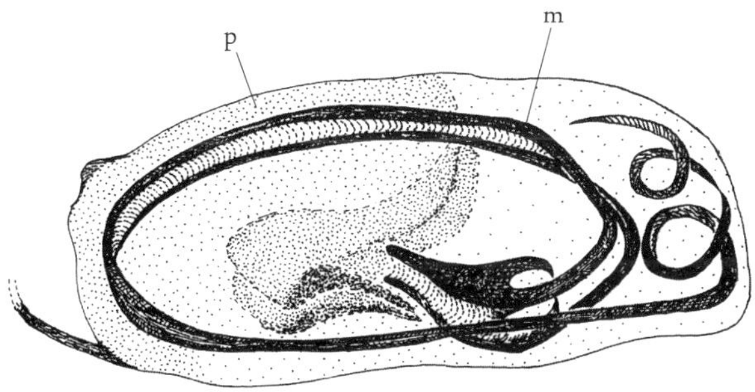

FIGURES 384 – 385
Rhagadotarsus kraepelini male. **384** pygophore and proctiger folded – **385** vesical sclerite; m = median vesical sclerite, p = sclerotized part of wall of vesica, pr = proctiger, py = pygophore.

REFERENCES – *Rhagadotarsinae.* Ripple communication (Wilcox, 1972, 1980); morphology, phylogeny, adaptations, biogeography and classification of Gerromorpha (Andersen, 1982a); revision of *Rhagadotarsus* (J. Polhemus & Karunaratne, 1993); ripple communication (Gogala, 1996); Australian Rhagadotarsinae (Andersen & Weir, 1998).

SUBFAMILY TREPOBATINAE[1]

Diagnosis

Small (1.27-5.70 mm), females usually larger than males, macropterous morph larger than apterous morph. Usually dark with pale markings and grey pruinose areas. Intersegmental suture between mesonotum and metanotum weakly sinuated, almost straight; metathoracic scent apparatus absent in most trepobatine species, but retained in a few Oriental genera. Middle femur short, slightly thickened, always shorter than middle tibia and usually shorter than hind femur. Hind tarsi often fused into one segment (except in Stenobatini). Claws of hind legs long and slender, arising almost from middle of second tarsal segment, or far from apex if tarsi fused. Some Australian and Oriental genera

1 Based on J. Polhemus & D. Polhemus (1993, 1994b, 1995b, 1996, 2002).

with distinct process (= spur) on fore tibia. In some genera, male proctiger with anterolaterally directed lateral processes. Parameres reduced except in Trepobatini; plates of male phallotheca in some groups strongly sclerotized. Ovipositor short, non-serrate except in Stenobatini.

Distribution worldwide

Trepobatinae, divided in four tribes, presently contains 26 genera (J. Polhemus & D. Polhemus, 1993, 2002). Of these, fifteen occur in Malesia. The tribe Metrobatini, the dominant tribe in Malesia, holds nine freshwater genera which have their main distribution in the Papuan subregion of Australia. The tribe Naboandelini contains three freshwater genera, ranging from Africa through southern Asia to Australia. Trepobatini holds nine genera, of which one (*Cryptobates*) occurs disjunct in South-east Asia. Stenobatini is composed of five entirely marine genera which are widespread in the Malay Archipelago. The four tribes and fourteen genera recognized in Malesia present 56% of the world fauna of trepobatids. In the Paleotropical region the greatest species diversity occurs in Australasia, especially in New Guinea.

Biological notes

Freshwater Trepobatinae live in stagnant to moderately flowing waters, notably streams and rivers virtually free of vegetation. They avoid the swiftly flowing parts, except during feeding. Marine species occur in mangrove swamps. Adults and nymphs occur together in large compact aggregations.

Key to tribes and genera of Trepobatinae in Indo-Australia[1]

1 Raised scent gland orifice on metasternum well-developed in both sexes (figs. 387, 390), in males more than in females. Male proctiger with long, stout, anterolaterally directed lateral processes (figs. 386, 389). Tarsi of fore leg broad, blade-like. Mesosternum of female medially membranous. Head long, antennal socket removed from eye by width of socket. Marine [Australasia]
▸ **Stenobatini 13**

– Raised scent gland orifice on metasternum poorly developed or absent in both sexes. Male proctiger unmodified, or with smaller anterolaterally directed or ventrolaterally directed lateral processes. Tarsi of fore leg cylindrical or flattened, not blade-like. Female mesosternum medially sclerotized. Head short, antennal socket closer to eye than width of socket. Freshwater [in New World also brackish] ▸ 2

1 Adapted from J. Polhemus & D. Polhemus (1993, 1994b, 1995b, 1996, 2000a, 2002).

FIGURES 386 – 388
Pseudohalobates inobonto, male.
386 apex of abdomen, lateral view – **387** metasternum with scent gland orifice – **388** right fore leg. (after J. Polhemus & D. Polhemus, 1996)

386 387 388

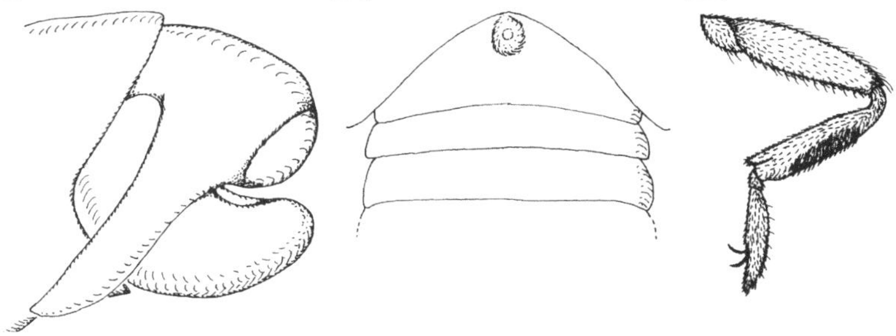

389 390 391

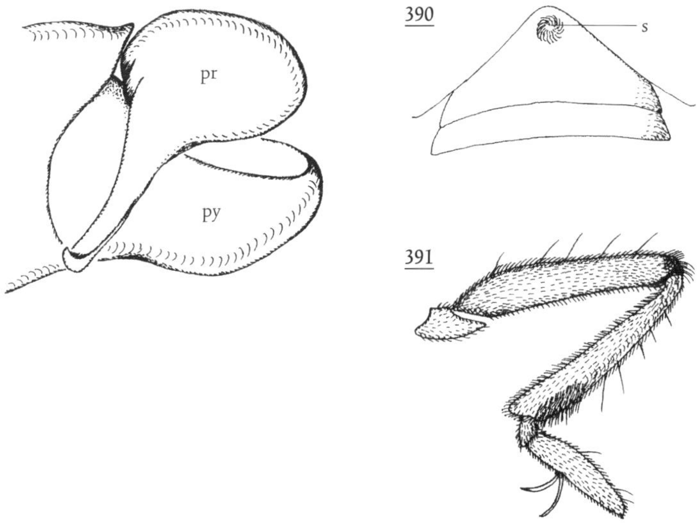

FIGURES 389 – 391
Rheumatometroides sele, male.
389 apex of abdomen, lateral view – **390** metasternum with scent gland orifice – **391** right fore leg; pr = proctiger, py = pygophore, s = metasternal scent gland orifice.
(after J. Polhemus & D. Polhemus, 1996)

2 First tarsal segment of fore leg with long erect setae (except in genus *Rheumatometra*). Distal comb of fore tibia with a differentiated pad of closely packed short or medium-long setae, or, if absent, antennal segments 1-3 with distal spines (*Metrobates* Uhler). Antennae with stiff spine-like setae on distal $^1/_4$ or distal extreme of at least segments 1-2 [Australia and Papua New Guinea, with one genus, *Metrobates*, in New World] ▸ **Metrobatini 4**

– First tarsal segment of fore leg without long erect setae. Distal comb of fore tibia without differentiated pad of closely packed setae. Antennal segments 1-3 without distal spines; antennae without stiff spine-like setae or, if present, scattered and not restricted to distal $^1/_4$ or distal extreme ▸ 3

3 Male proctiger with long slender anterolaterally directed lateral processes; male parameres reduced, small, vestigial or absent. Ground colour grey to black with pale markings. Head with gular region not tumid, base of rostrum arising anteriorly, not enclosed; rostral cavity at or above gular region arising at about ventral eye level, not directed posteriorly [Australia, Papua New Guinea, Asia, Ethiopian region] ▸ **Naboandelini 11**

– Male proctiger without long slender anterolaterally directed lateral processes; male parameres well-developed, long, blade-like, falcate or corkscrew-shaped. Ground colour pale with dark markings. Head with gular region usually tumid; base of rostrum enclosed in a ventrally projecting differentiated rostral cavity, arising below ventral eye level, directed slightly posteriorly [New World and Asia] ▸ **Trepobatini 12**

4 (Metrobatini) Middle tarsi with swimming plume (long hair fringe, figs. 407, 409). Entire median part of mesonotum with well defined quadrate grey pruinose area, often depressed ▸ 5

– Middle tarsi without swimming plume (figs. 408, 410-412). Median part of mesonotum with or without grey pruinose area, but if present small, not well defined, not quadrate, and not depressed ▸ 6

5 Body broad, subparallel, strongly flattened. Male genital segments with stout lateral projections on both proctiger and pygophore (fig. 409) ▸ *Metrobatoides* **p. 314**

– Body elongate (fig. 407), moderately flattened. Male genital segments without lateral projections ▸ *Ciliometra* **p. 310**

6 Eyes small: width about half of interocular space, widely divergent posteriorly ▸ 7

– Eyes large: width more than half of interocular space, not widely divergent posteriorly ▸ 8

7 In lateral view, eyes covering most (in female) to all (in male) of lateral margin of pronotum; mesonotum without a median sulcus ▸ *Talaudia* **p. 320**

– In lateral view, eyes covering about half the lateral margin of pronotum; mesonotum with median sulcus, especially prominent in female [Australia] ▸ *Rheumatometra*

8 Black, but mesonotum of apterous morph with pair of large ovate yellowish markings. Male (1.4-1.9 mm) about half as long as female (2.7-3.4 mm) ▸ *Andersenella* **p. 309**

– Black or grey, but mesonotum of apterous morph with longitudinal lighter stripes and irregular or ovate grey spots on midline; never with pair of large yellowish markings. Length of male $^{3}/_{4}$ or more the length of female ▸ 9

9 Male abdominal segment 8 greatly elongated (fig. 410); pygophore with prominent ventrally or posteriorly directed spine (fig. 393). Proctiger with lateral finger-like processes. Female tergite 6 prominent, usually projecting posteriorly beyond proctiger and anal cone (fig. 392); female sternite 7 folded caudolaterally ▸ *Metrobatopsis* **p. 316**

392

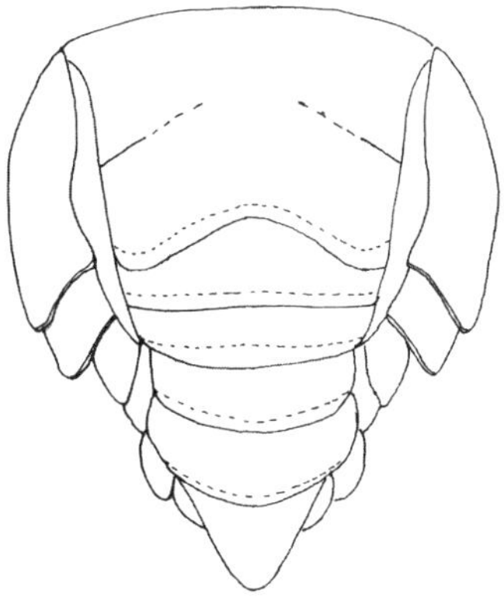

393

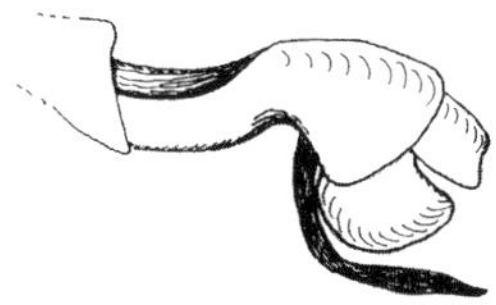

FIGURES 392 – 393
Metrobatopsis flavonotatus
392 female, dorsal view of abdomen – **393** male, genital segments, lateral view.

– Male abdominal segment 8 not greatly elongate; pygophore without prominent ventrally or posteriorly directed spine; proctiger without lateral finger-like processes. Female tergite 6 not prominent; female sternite 7 may be indented or notched, but not folded ▸ 10

10 Mesonotum with pair of ovate silvery grey pruinose areas. Hind femur distinctly shorter than total body length; middle tibia slightly longer than middle femur. Male phallotheca lightly sclerotized; vesica with only lateral sclerites present. Female tergite 8 not plate-like, posterior margin sinuate, anal cone directed posterodorsally (fig. 408) ▸ *Iobates* p. 312

– Mesonotum without pair of ovate silvery grey pruinose areas. Hind femur equal to or longer than total body length. Middle tibia distinctly longer than middle femur. Male phallotheca heavily sclerotized both dorsally and ventrally, acuminate dorsodistally; vesica with dorsal and lateral sclerites present. Female tergite 8 plate-like, posterior margin sinuate or lateral margins produced posteriorly, anal cone directed posteriorly ▸ *Stygiobates* p. 318

11 (Naboandelini) Body with bright shiny pale marks, at least in females. Mesonotum medially without pruinose grey markings. Venter pale with varied dark markings in Malesian species. Fore femur in male without basal protuberance ▸ *Calyptobates* p. 323

– Body not brightly marked shiny pale. Mesonotum medially with pruinose grey markings. Venter not extensively pale. Fore femur in male with or without a basal protuberance ▸ *Naboandelus* p. 326

12 (Trepobatini) Thorax soft brown to yellow-brown without highly contrasting dark markings, or yellowish with contrasting markings (as in fig. 420). Antennae almost as long as body. Not sexually dimorphic. Fore femur in male straight. Parameres spiralled or strongly twisted with basal projection (fig. 394) ▸ *Cryptobates* p. 328

394

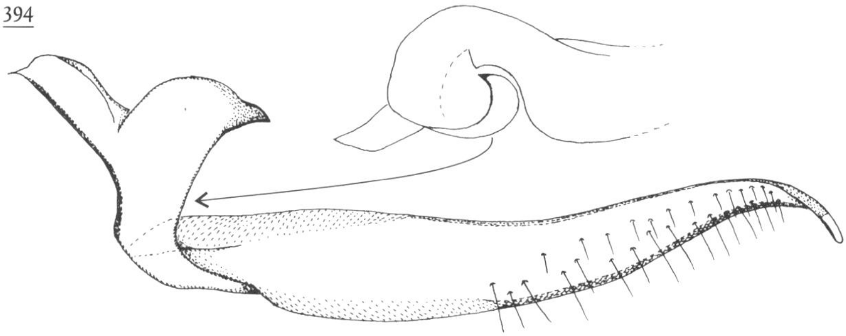

FIGURE 394
Cryptobates johorensis,
paramere.

395

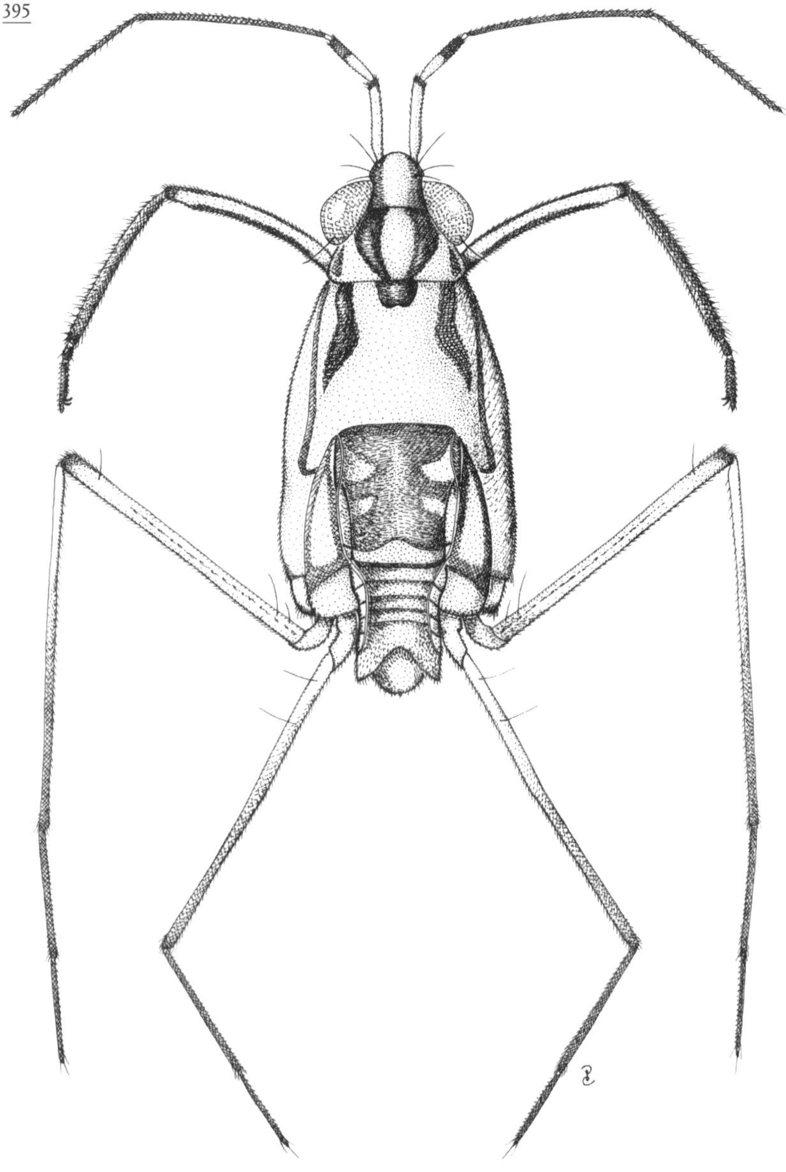

FIGURE 395
Gnomobates kuiterti,
apterous male,
length 2.1 mm.

– Thorax of dorsum as in fig. 395. Antennae clearly shorter than body. Sexually dimorphic: male 1.91-2.27 mm, female 2.38-2.92 mm. Fore femur of male curved (fig. 395). Parameres straight without basal projection [India to North Thailand] ▸ *Gnomobates*

13 (Stenobatini) Body strongly flattened. No contrasting lateral marks at mesonotal midline (figs. 421, 424). Sides of mesothorax in male parallel. Lateral arms of proctiger narrowly rounded or distally truncate (figs. 386, 396). Anterodistal pad of stiff setae on fore tibia oval (fig. 388) ▸ 14

– Body not flattened. Contrasting lateral marks at mesonotal midline (figs. 422, 425). Sides of mesothorax in male not typically parallel, divergent posteriorly. Lateral processes of proctiger distally not truncate or rounded (figs. 402, 405, 399). Anterodistal pad of stiff setae on fore tibia elongate and more or less rectangular (fig. 391) ▸ 15

14 First antennal segment in male incrassate and relatively short (fig. 424). Vertex of head always with one large dark median mark (fig. 424). Fore femur less hirsute (fig. 398). Middle tibia of male without dorsal row of erect setae. Male metasternum anteriorly slightly extended, with small scent gland orifice (fig. 397) ▸ *Stenobatopsis* p. 336

– First antennal segment in male not incrassate (fig. 421). Vertex of head always with several pairs of dark marks (fig. 421). Fore femur heavily hirsute (fig. 388). Middle tibia of male with dorsal row of erect setae. Male metasternum anteriorly extended, with large scent gland orifice (fig. 387) ▸ *Pseudohalobates* p. 330

15 Pronotum always with median pale spot. Mesonotum sublaterally with pair of roundish pale marks (fig. 425). Male metasternum anteriorly weakly extended, with small scent gland orifice (fig. 400). Fore tibia of male with patch of dark stiff setae restricted to distal $^1/_7$, with strong distal spur (fig. 401). Eyes globose; basal abdominal tergite of female with well defined sutures. Tergite 7 not completely sclerotized ▸ *Thetibates* p. 338

– Pronotum either with median or with paired pale marks. Mesonotum sublaterally with longitudinal pale marks (figs. 422, 423). Eyes elongate. Male metasternum anteriorly weakly to strongly extended, with small to large scent gland orifice (figs. 390, 403). Fore tibia of male with patch of dark stiff setae occupying distal $^1/_5$ – $^1/_2$, without strong distal spur (figs. 391, 404). Basal abdominal tergite of female without well defined sutures. Tergite 7 completely sclerotized ▸ 16

16 Pronotal marks variable, along midline of mesonotum with longitudinal pale stripes (fig. 423). Male metasternum anteriorly usually strongly extended, with scent gland orifice moderately large and raised (fig. 403). First antennal segment of male in most species distally thickened (fig. 423). Male antennae ventrally with dense short setae ▸ *Stenobates* p. 334

FIGURES 396 – 398
Stenobatopsis stygius, male.
396 apex of abdomen, lateral view – **397** metasternum with scent gland orifice – **398** right fore leg.
(figs. 397 – 398 after J. Polhemus & D. Polhemus, 1996)

396

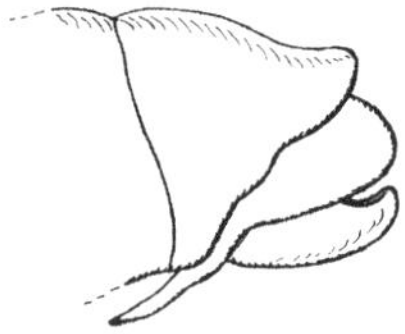

397

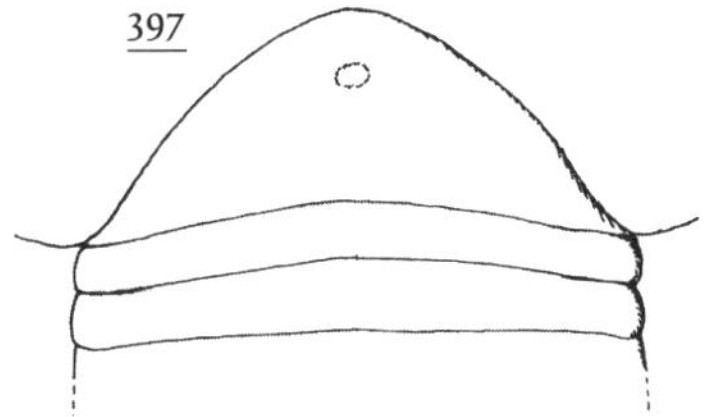

398

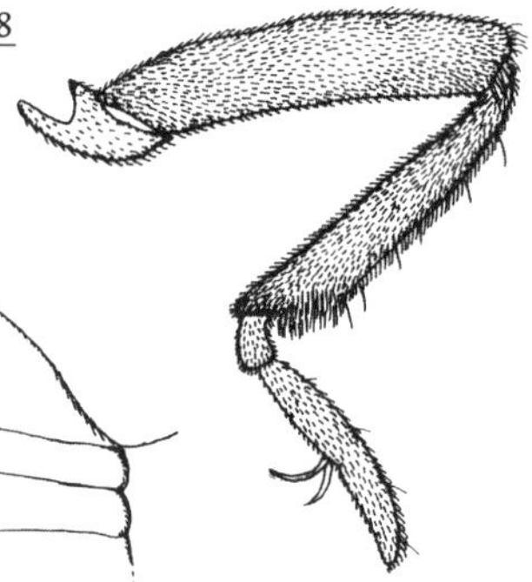

FIGURES 399 – 401
Thetibates serena, male.
399 apex of abdomen, lateral view – **400** metasternum with scent gland orifice – **401** right fore leg.
(after J. Polhemus & D. Polhemus, 1996)

399

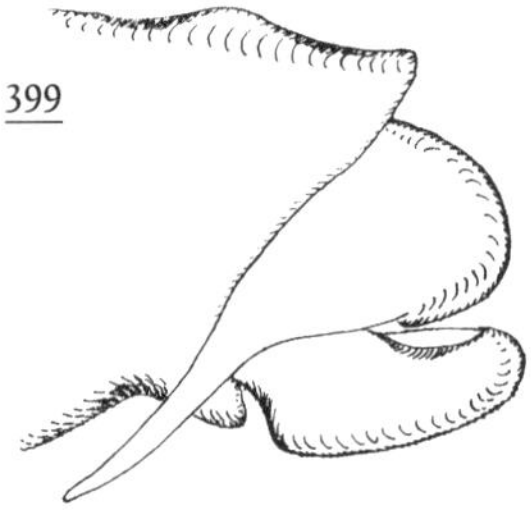

400

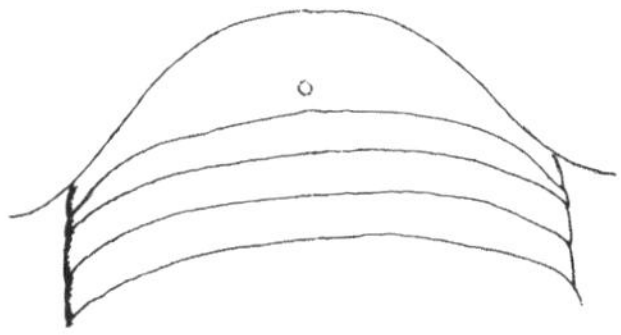

401

402

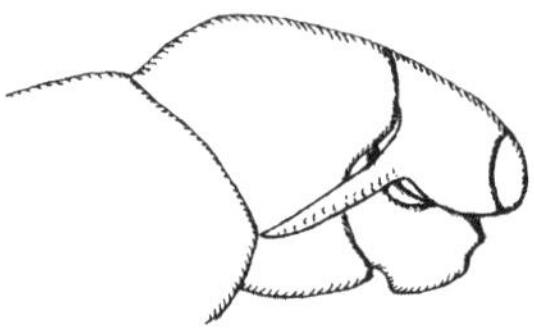

403

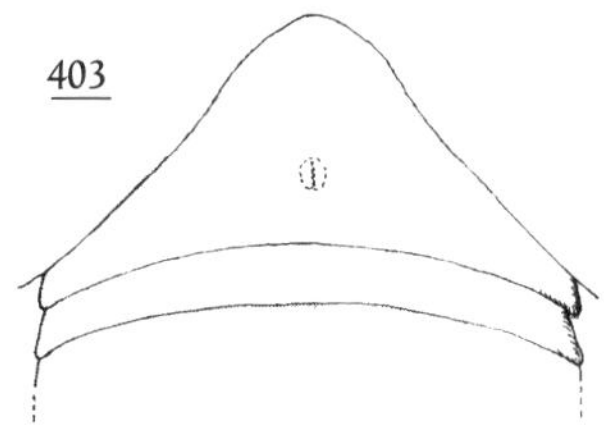

404

405

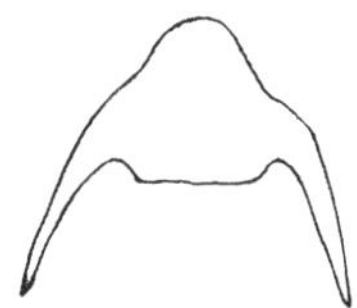

FIGURES 402 – 406
Stenobates makraitos, male.
402 apex of abdomen, lateral view – **403** metasternum with scent gland orifice – **404** right fore leg – **405** proctiger, dorsal view – **406** vesica with vesical sclerites, ventrolateral view; ds = dorsal sclerite, ls = lateral sclerite.
(figs. 402-405 after J. Polhemus & D. Polhemus, 1996)

406

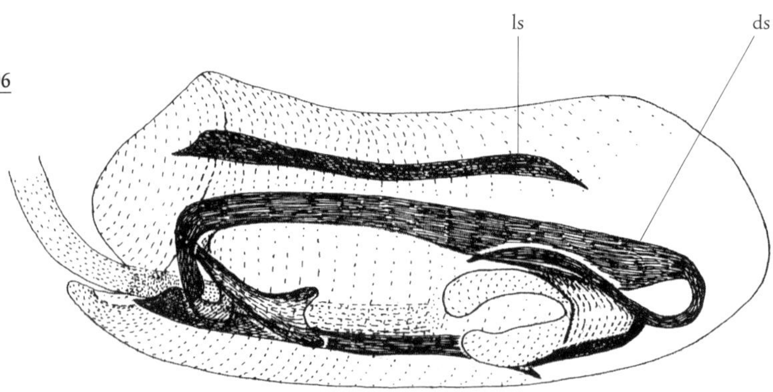

– Pronotal marks variable, pale, along midline of mesonotum longitudinal or quadrate (fig. 422). Male metasternum anteriorly weakly to moderately extended, with straight anterior margin, forming regular triangle with small scent gland orifice (fig. 390). First antennal segment of male not distally thickened (fig. 422). Male antennae ventrally without dense short setae
▸ *Rheumatometroides* p. 332

– TRIBE METROBATINI[1]

Diagnosis

See key above. Small to medium-sized (males 1.7-3.3 mm, females 1.7-3.7 mm). Indo-Australian species dimorphic (apterous or brachypterous), with fused tarsal segments of hind legs, and small or vestigial parameres in all Old World genera.

Distribution in Indo-Australia

Except for *Metrobates*, Metrobatini are distributed exclusively in the Australasian region. Seven genera are found in this area. Except for *Rheumatometra*, which seems to be endemic to Australia, Metrobatini are found in New Guinea, with additional records from Sulawesi, the Moluccas, Solomon Islands and Bismarck Archipelago.

Biological notes

Metrobatini occupy moderately to swiftly flowing streams (J. Polhemus & D. Polhemus, 1993). The tuft of long setae at the base of the first tarsal segment of the fore legs possibly acts as a sensor to surface currents, aiding in prey detection. In Metrobatini certain genera are predominantly phoretic, e.g. *Andersenella* and *Talaudia*. This also explains why species in these genera prefer quieter water than most other members of this tribe.

Genus *Andersenella*

Diagnosis

Small, strongly sexually dimorph (males 1.67-1.92, females 3.13-3.43); macropterous morph unknown. Dorsally black. Head with yellowish marking on posterior margin and yellowish quadrate spot anterior to antennal sockets. Pronotum with median quadrate yellowish region. Mesonotum with pair of large, posteriorly diverging, elongate oval yellowish spots. Thoracic venter yellowish. Appendages blackish brown, bases of first antennal segments and fore femora, fore- and middle coxae and trochanters at least ventrally yellowish.

1 Based on J. Polhemus & D. Polhemus (1993).

Head short, anterior margin rounded. Eyes nearly reaching anterior margin of mesonotum, their width $^2/_3$ the width of interoculus. Antennae moderately long, relatively stout and slightly shorter and thicker in males. Segment 1 longest, about 1.5 times length of segment 2, segments 3 and 4 about equal in length, distinctly shorter than segment 2 in females, only slightly shorter than segment 2 in males. Pronotum short, width less than head measured across eyes. Mesonotum long, much wider than head, suture between mesonotum and metanotum only visible laterally. Metanotum fused with abdominal tergites 1 and 2. Femur of fore leg slightly thickened, in male flattened ventrally, in female cylindrical; inner side of femur and tibia covered with short stiff setae; first tarsal segment short and cylindrical, with tuft of long setae at base, second tarsal segment longer; claws small, distinct, arising beyond middle of segment 2. Middle leg long and slender, femur stouter and only slightly shorter than tibia; tarsi long, first segment about twice as long as second, claws long and slender, subapical. Hind femur slender, more than twice as long as tibia, tarsal claws reduced to bristles. Abdominal segments subequal in length. Pregenital abdomen and genital segments in male not modified. Vesical sclerites: dorsal sclerites long and recurved proximally, lateral sclerites short.

Distribution in Malesia (fig. 587)

The single species, *A. binotata*, is so far only collected in eastern New Guinea.

Biological notes

Andersenella binotata is a weak skater. It prefers quiet places along the edges of shaded streams.

Genus *Ciliometra*

Figure 407

Diagnosis

Medium-sized (males 2.22-3.22, females 2.37-3.44 mm), ovate. Dark with pale markings and extensive pruinose grey areas. Head dull, sometimes with indistinct lighter markings at anterior and posterior margin. Pronotum with yellowish median marking. Mesonotum tinged black, pruinose, with medially depressed area pruinose grey, laterally distinctly demarcated. Metanotum and abdominal tergite 1 brown, medially with pruinose grey. Thoracic venter dull grey with weak lateral brown markings. Abdominal sterna brown in males, grey in females, in both with median and lateral dark marks. Appendages usually dark but sometimes with pale bases. Head short, anterior margin rounded, sides subparallel with number of long black stiff setae in addition to normal three pairs of facial trichobothria. Antennae moderately long, in males slightly thicker; segment 1 longest, about two times the length of 2, segment 3 shorter than 2 but distinctly longer than 4. Eyes almost reaching anterior margin of

407

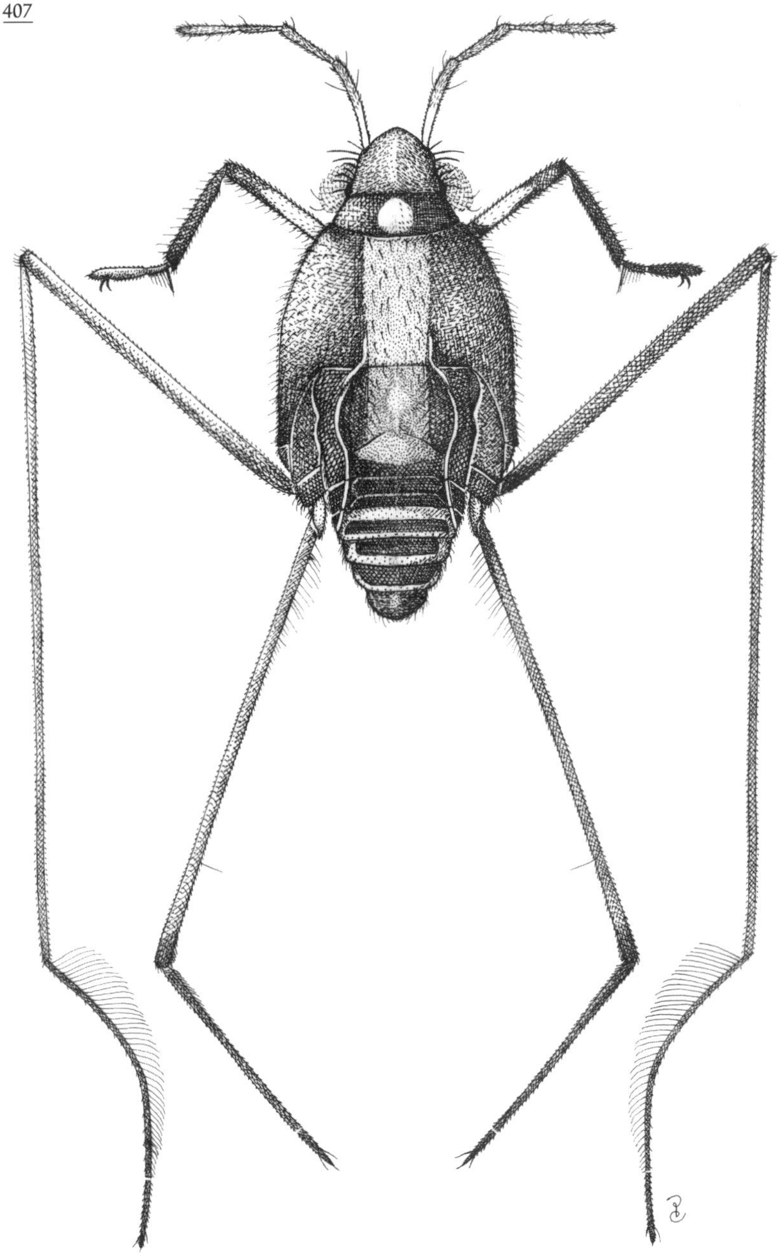

FIGURE 407
Ciliometra femorata,
paratype, apterous female,
length 2.7 mm.

mesonotum. Rostrum large. Dorsum of thorax in some species clothed with both fine pubescence and long stiff setae. Pronotum short, 0.4 times length of mesonotum, width less than head measured across eyes. Mesonotum long, medially broadly depressed and differentiated, suture between mesonotum and metanotum nearly straight. Metanotum indicated by lateral suture, fused with abdominal tergites 1 and 2. Femur of fore leg in male slightly thickened, basal $^2/_3$ to $^4/_5$ of inner face covered with long stiff setae; tibia distally slightly broadened, with small apical spur bearing oval dark patch of short setae; inner side of fore femur and tibia covered with short stiff setae; first tarsal segment short, cylindrical, with tuft of long setae at base; second tarsal segment two times longer than first; claws small, short and slender, arising preapically. Female fore leg not modified. Middle leg long and slender; tibia very long, more than 1.5 times longer than femur; tarsi very long, first segment about 4 times as long as second, inner side with fringe of long hairs; claws very small, arising near apex. Hind femur long and slender, more than three times as long as hind tibia; claws long and very slender, arising from distal third. Hind tibia of female with long erect dark setae on inner side and basal third. Abdominal tergites 3-4 subequal in length, segment 7 longer. Male genitalia prominent, not modified, with large pygophore; proctiger laterally extended but without lateral processes. Phallotheca distally truncate and sclerotized. Vesical sclerites: dorsal sclerites long and proximally recurved, lateral sclerites short. Female tergite 8 short, posterior margin notched or rounded; sternite 8 truncate.

Distribution in Malesia (fig. 588)

Ciliometra is apparently restricted to Papua New Guinea (Lansbury, 1996b). Five species have been described (but there are some more undescribed species, J. Polhemus pers. comm.).

Biological notes

Ciliometra species live in relatively wide rivers in rainforest, preferring fast currents and the heads of rapids. They skate rapidly on open water and are difficult to see and collect due to their size and agility (J. Polhemus & D. Polhemus, 1993).

Genus *Iobates*

Figure 408

Diagnosis

Medium-sized (males 2.44-2.89, females 2.83-3.39 mm), ovate. Dark with pale and grey pruinose markings. Head dark with pale margins. Pronotum dark with a yellowish median square mark. Mesonotum dark, tinged grey pruinose, two grey pruinose ovate marks at either side of midline. Metanotum and abdominal tergites medially grey pruinose. Posterior abdominal tergites 2-6 in female often brownish, tergite 8 pale. Thoracic venter grey, weakly tinged brown on acetabulae. Abdominal sterna usually pale but dark in some species. Appendages mostly dark but sometimes with pale base. Head short, anterior margin rounded, sides subparallel. Antennae moderately long, slightly thicker in males; segment 1 longest, about 1.5 times the length of 2, segment 3 and 4 subequal in length and slightly shorter than 2. Eyes posteriorly extending

408

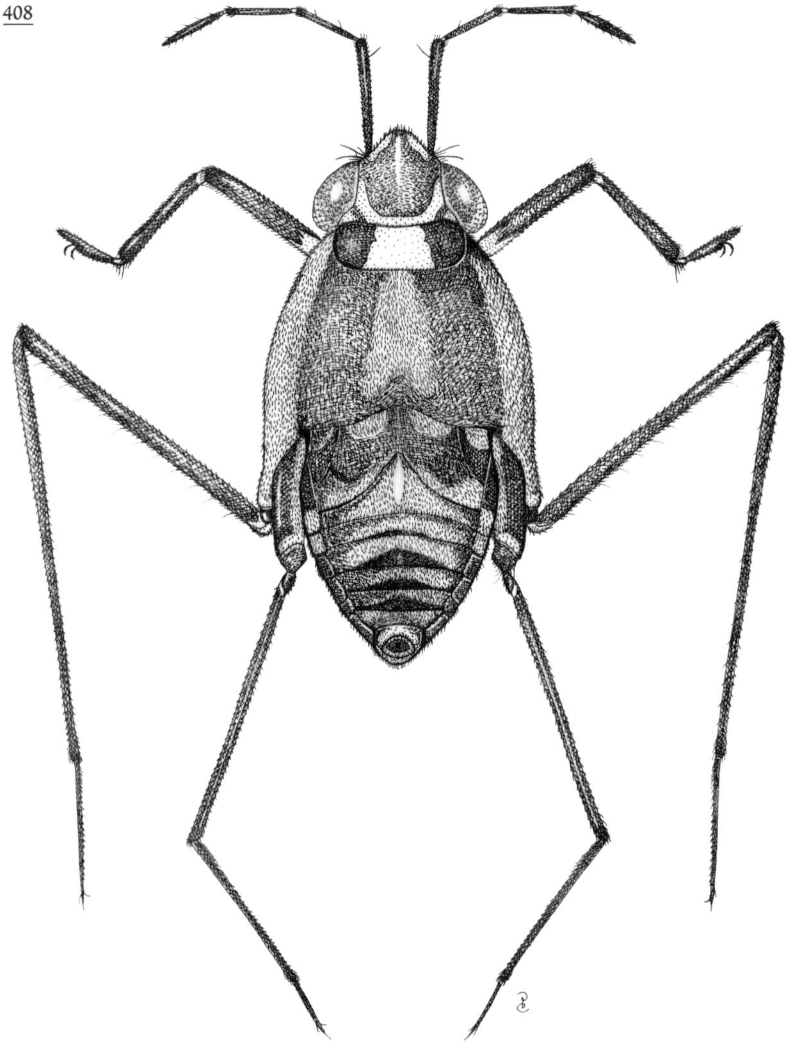

FIGURE 408
Iobates sp.,
apterous female,
length 3.0 mm.

almost to mesonotum. Rostrum large. Pronotum short, about 0.25–0.33 times of mesonotal length, anterolateral angles of mesonotum embracing lateral margins of pronotum. Mesonotum with convex lateral margin, anterolateral angle almost reaching anterior margin of pronotum, suture between mesonotum and metanotum extending forward. Metanotum indicated by a lateral suture, fused with abdominal tergites 1 and 2. Male fore leg with basally thickened femur, covered over entire length of inner side with short stiff setae; tibia distally slightly widened and with thick brush of short setae and, in addition, a pad of short stiff setae on the small apical spur. First tarsal segment short and cylindrical, with longer setae at base, second tarsal segment three to six times longer. Claws small, short and slender, arising from middle of second segment. Segments of female fore leg of similar proportions, but cylindrical and without specialized hairs and bristles. Middle leg long and slender, femur shorter than tibia or hind femur; tibia long; tarsi long, first segment about 2-3 times as long as second; claws long, very slender, arising near apex. Hind femur long and slender, at least two times as long as hind tibia; claws long and slender, arising preapically. Abdominal tergites 3-6 subequal in length, segment 7 longer. Male genital segments with large pygophore, proctiger with broad internal phragma; distal part of phallotheca laterally expanded. Vesical sclerites strongly reduced: only lateral sclerites present; other sclerites absent or vestigial. Anal cone in female directed posterodorsally.

Distribution in Malesia (fig. 589)

The genus *Iobates* holds four species, all from New Guinea and Salawati.

Biological notes

Iobates species are abundantly in moderate-sized streams near the coast or in primary rainforest, where they prefer slow moving open water.

Genus *Metrobatoides*

Figure 409

Diagnosis

Medium-sized (males 3.03-3.33, females 3.38-3.69 mm), ovate. Macropterous morph unknown. Head black, posterior margin with brown markings; pronotum dark with median pale marking; mesonotum dark with pruinose grey median depressed area, in females sharply demarcated; metanotum and abdominal tergites dark, tinged pruinose grey; thoracic venter dull grey, males with posterior part of median mesosternum with long golden setae; acetabulae always lighter; pleural regions grey pruinose. Abdominal sternites brown in males, grey marked with brown medially and laterally in females. Appendages mostly dark, but fore femora basally, and coxae and trochanter wholly pale. Head short, anteriorly truncate, sides subparallel with longitudinal sulcus widening posteriorly. Antennae moderately long and stout, thicker in males; segment 1 longest, thickened in males, in length subequal to segment 2, segment 4 longer than 3 but shorter than 2. Eyes posteriorly (in males) extending almost to or (in females) reaching mesonotum. Pronotum short; sides of mesonotum subparallel (male) or slightly diverging posteriorly (female), suture between mesonotum and metanotum nearly straight. Metanotum indicated by lateral suture, fused with abdominal tergites 1 and 2. Fore femur constricted at distal $^{1}/_{3}$, inner side covered with short stiff setae; tibia distally slightly widened, with large apical spur; tarsi subequal to or slightly longer than tibia, first tarsal segment short and cylindrical, basally with tuft of long setae, second tarsal segment two times longer; claws small, short and slender, arising preapically. Middle leg long and slender, femur distinctly shorter than tibia or hind

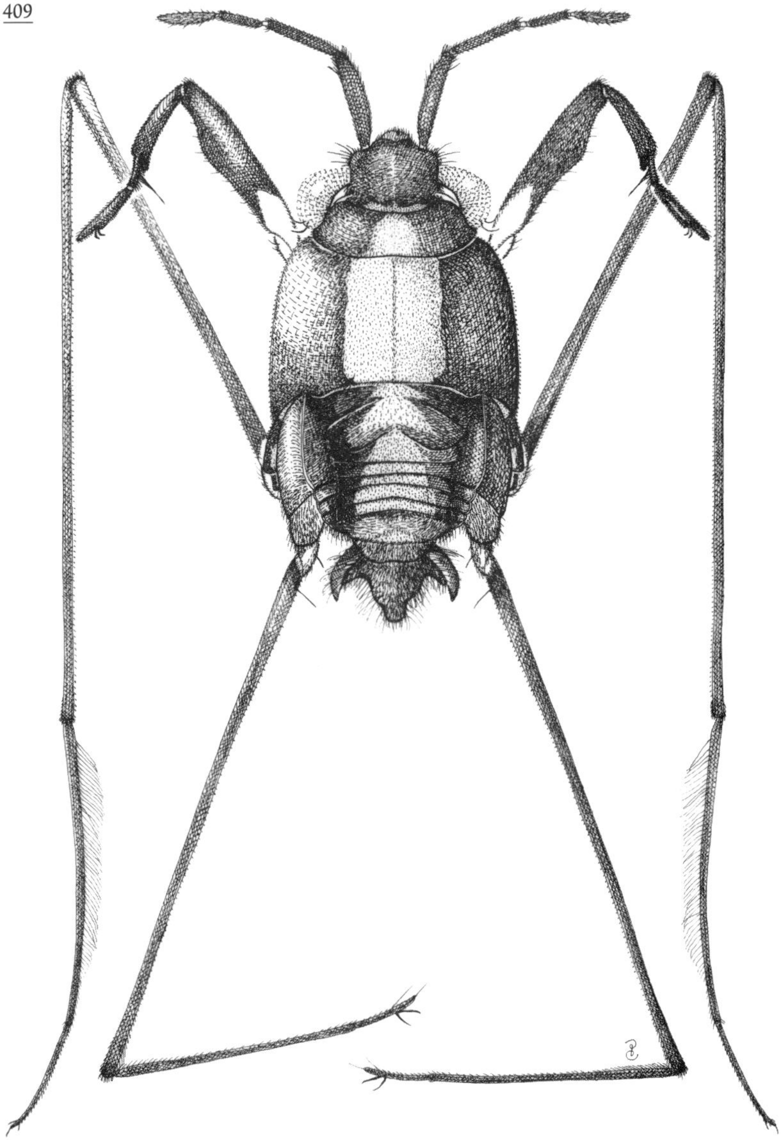

409

FIGURE 409
Metrobatoides genitalis, paratype, apterous male, length 3.1 mm.

femur; tibia about $^{1}/_{5}$ longer than femur; first tarsal segment about three times as long as second, inner side with fringe of long hairs (swimming plume); claws very small, arising preapically. Hind femur long and slender, at least $2^{1}/_{2}$ times as long as hindtibia; claws long and slender, arising preapically. Abdominal tergites 3-6 subequal in length, 7 much longer. Male pregenital abdominal sternites short and compressed, except sternite 7; male sternite 8 long and broad; genital segments modified, both pygophore and proctiger (fig. 409) with large lateral projections; phallotheca distally truncate. Vesical sclerites contain distinctly long dorsal sclerites and short lateral sclerites. Female sternite 8 triangular with median longitudinal carina, first gonocoxae folded, hollow, closed, extending posteriorly. Although *Metrobatoides* is more closely related to *Ciliometra*, its appearance is similar to *Metrobates* (the unique metrobatine genus occurring in the Western Hemisphere). The striking genital segments of *Metrobatoides* easily separate it from other metrobatine genera.

Distribution in Malesia (fig. 587)

Two species have been described. They occur in the northern part of New Guinea.

Biological notes

Both species have been found on open midstream waters of streams. *Metrobatoides genitalis* is a strong skater, preferring the midstream currents. *Metrobatoides bifurcatus* however, although scattered individuals were found midstream, apparently prefers the edges and side pools of streams (J. Polhemus & D. Polhemus, 1993, 2002).

Genus *Metrobatopsis*

Figure 410

Diagnosis

Small (males 2.29-2.61, females 1.76-3.00 mm), ovate. Head dark prominently, margins pale. Pronotum dark with a median pale marking. Mesonotum dark, with two pale markings at either side of mid line, sometimes coalescing into a single median elongate spot. Metanotum and abdominal tergites dark. Female abdominal tergites 2-9 often brownish posteriorly, connexiva pale. Dorsum tinged with grey pruinose. Thoracic venter usually tinged with pale markings, acetabulae always brown. Abdominal sterna usually mostly pale. Appendages mostly dark, basal parts of first antennal segment, fore femora, and coxae pale. Head short, anterior margin rounded, sides subparallel. Antennae moderately long, thicker in males; segment 1 longest, in some species distinctly thickened, about 1.3-1.5 times length of 2, segment 3 and 4 subequal in length and slightly shorter than 2. Eyes posteriorly extending almost to mesonotum. Pronotum short, lateral margin embraced by inner margins of eyes. Mesonotum with convex lateral margin, suture between mesonotum and metanotum extended forward, less so in females. Metanotum indicated by lateral suture, fused with abdominal tergites 1 and 2. Fore leg with thickened femur, entire length of inner side covered with short stiff setae; tibia distally slightly broadened, with small apical spur bearing pad of very short stiff setae; first tarsal segment short and cylindrical, second tarsal segment eight times longer; claws small, short and slender, arising from middle of second segment. Middle leg long and slender, femur distinctly shorter than tibia or hind femur; tibia long; first tarsal segment about 2-3 times as long as second; claws long, very slender, arising near

410

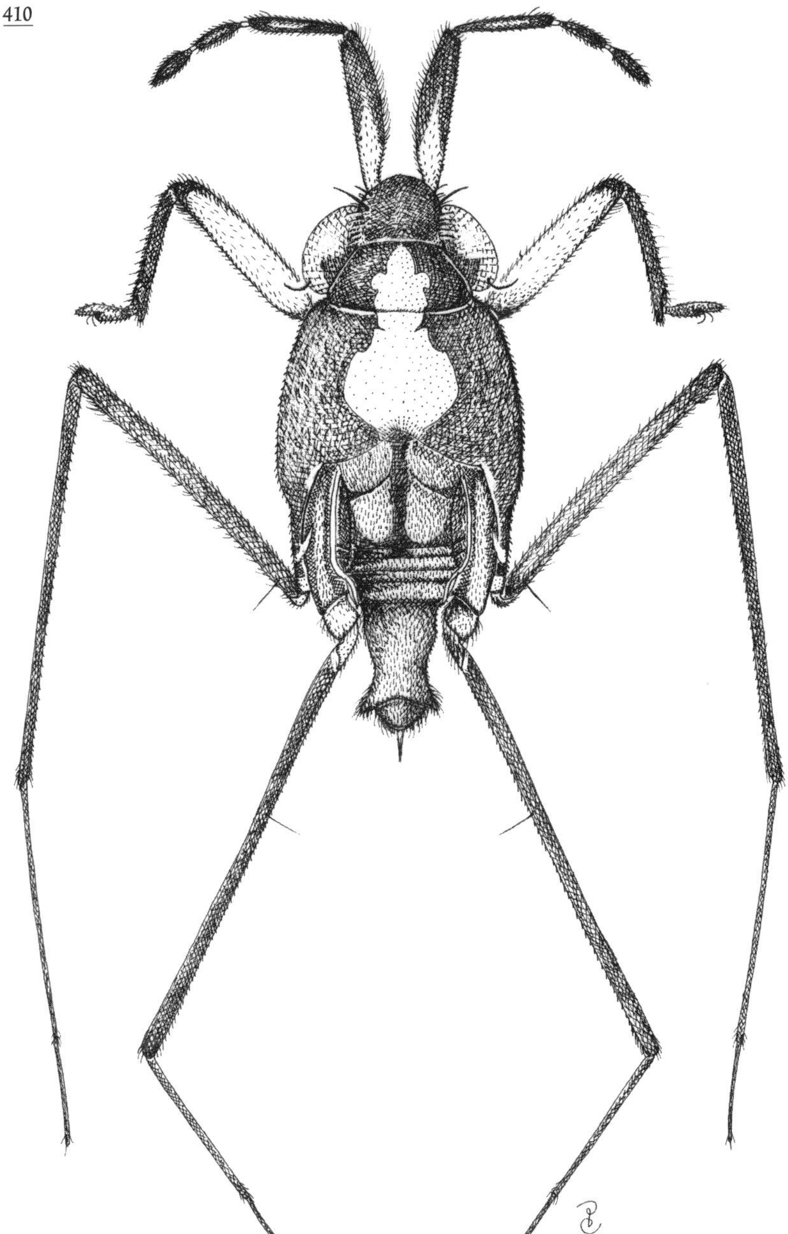

FIGURE 410
Metrobatopsis insularis, paratype, apterous male, length 2.4 mm.

apex. hind femur long and slender, at least two times as long as hind tibia; tarsal segments fused; claws long and slender, arising preapically. Abdominal tergites 3-7 subequal in length; in some females tergite 5-6 with hind margin projected posteriorly. Segment 8 in male long, sometimes ventrodistally carinate, angulate posterolaterally. Pygophore with distinct ventral projection; proctiger with broad internal phragma and lateral long projections; phallotheca long and slender, dorso-distally sclerotized and acuminate. Vesical sclerites reduced: dorsal sclerites short, lateral sclerites long. Female gonapophyses plate-like, partly sclerotized, distally membranous, not serrate. *Metrobatopsis* is similar to *Iobates* and *Stygiobates*; apart from characters given in the key, it is characterized by a long first genital segment (segment 8) in males, and the rounded or angulate hind margins of tergite 5-6 in females of most species.

Distribution in Malesia (fig. 590)

Six species are recognized in this genus. It is restricted to eastern Malesia, occurring in New Guinea, Bismarck Archipelago and Solomon Islands.

Biological notes

Metrobatopsis species are abundant in moderate-sized streams near the coast with primary or secondary rainforest, where they prefer slowly moving open water.

Genus *Stygiobates*

Figure 411

Diagnosis

Medium-sized (males 2.69-2.94, females 2.77-3.27 mm), ovate. Head medially dark, posterior and lateral margins pale. Pronotum dark with median pale marking. Mesonotum dark, grey pruinose at least medially, occasionally with pair of elongate light stripes at sides of mid line. Metanotum and abdominal tergites dark, tinged grey pruinose, but in female abdominal tergite 8, connexiva and dorsum of mesocetabulae and metacetabulae in both sexes paler. Thoracic venter usually dull grey tinged with paler markings. Abdominal sterna usually dark grey. Appendages mostly dark, basal parts of first antennal segment, fore femora, coxae and trochanter pale. Wings black, vaguely marked brown. Macropterous morph mainly coloured like apterous morph. Head short, anterior margin rounded, sides subparallel. Antennae moderately long, thicker in males; segment 1 longest, more than 1.5 times length of 2, segment 3 shorter than 4, about $^1/_3$ shorter than 2. Eyes extending posteriorly almost to mesonotum. Pronotum short, lateral margin embraced by inner margins of eyes. Mesonotum with convex lateral margin, suture between mesonotum and metanotum extended forward, less so in females. Metanotum indicated by lateral suture, fused with abdominal tergites 1-2. Trochanter of fore leg of male distally with tuft of dark setae; femur thickened, entire length of inner side covered with short stiff setae; tibia distally slightly broadened, with apical spur bearing pad of very short stiff setae; first tarsal segment short and cylindrical, distally with tuft of long setae, second segment three times longer; claws small, short and slender, arising from middle of second segment. Middle leg long and

411

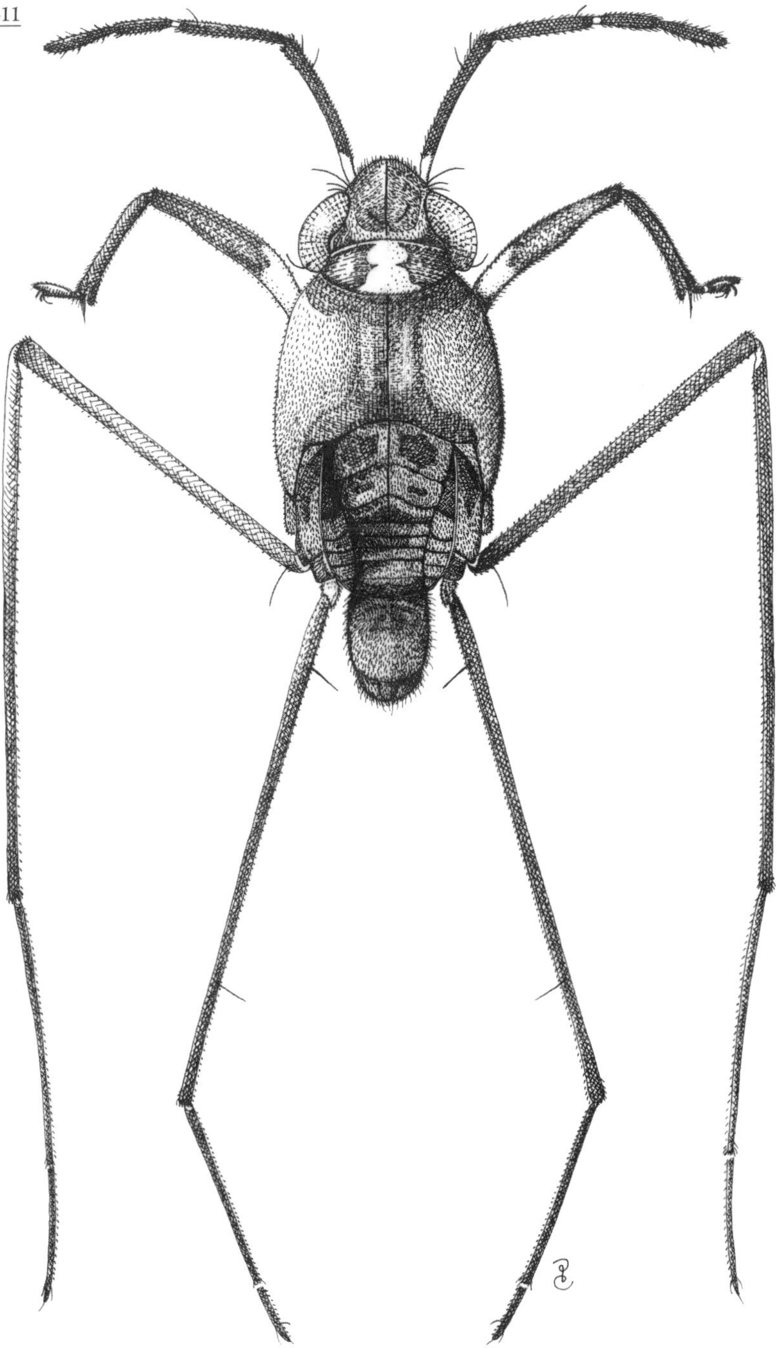

FIGURE 411
Stygiobates iriana,
paratype, apterous male,
length 2.6 mm.

slender, femur distinctly shorter than tibia or hind femur; tibia long; tarsal segment 1 at least two times as long as 2; claws long, very slender, arising near apex. Hind femur long and slender, at least $2^{1}/_{2}$-3 times as long as hind tibia; claws long and slender, arising preapically. Abdominal tergites 3-6 subequal in length, male segment 8 unmodified or ventrally with tooth-like projection. Pygophore large; proctiger with rounded lateral lobes and broad internal phragma; phallotheca acuminate, strongly sclerotized except ventrodistally. Vesical sclerites reduced, apicolateral sclerites short; only lateral sclerites distinct and long. Features of male sternite 7 and female abdominal tergite 8 are used for specific identification.

Distribution in Malesia (fig. 590)

The five species of *Stygiobates* are reported from New Guinea and the Moluccas.

Biological notes

Stygiobates species live in moderate-sized streams in rainforest, preferring slow moving open waters. They were collected in Papua (Irian Jaya) only between sea level and 200 m (J. Polhemus & D. Polhemus 1993).

Genus *Talaudia*

Figures 412-414

Diagnosis

Both apterous and macropterous morphs known. Apterous morph: small, strongly sexually dimorph (males 1.35-1.50, females 2.79-3.05). Dorsally black. Head with indistinct brownish transverse stripe on posterior margin, extending forward along inner eye margin. Pronotum with median yellowish spot covered by pruinosity. Mesonotum with posterior part indistinctly pruinose in male, and lateral quarters pruinose in female, but without yellowish marks. Inner face of connexival segments and, especially in females, sutures of abdominal tergites yellow. Metanotum in female with irregular median yellowish to fuscous median stripe extending onto the first two abdominal tergites, outer faces of connexival segments yellow. Venter in male light to dark brown, in female yellowish except for mesosternum, which is black with bluish pruinose bloom. Appendages blackish-brown, base of first antennal segment, fore acetabulum and fore femur, fore and middle coxae and all trochanters yellowish. Head short, anterior margin rounded. Antennae moderately long, relatively stout and distinctly shorter and relatively thicker in males. Segment 1 longest, about 1.5 times the length of segment 2, segments 3-4 about equal in length, in females distinctly shorter than 2, in males only slightly shorter than or subequal to segment 2. Eyes reaching or nearly reaching anterior margin of mesonotum, their width $^{1}/_{2}$ (male) to $^{2}/_{3}$ (female) the width of interoculus. Pronotum short, about 0.11 times the length of body in males, to 0.08 times the length of

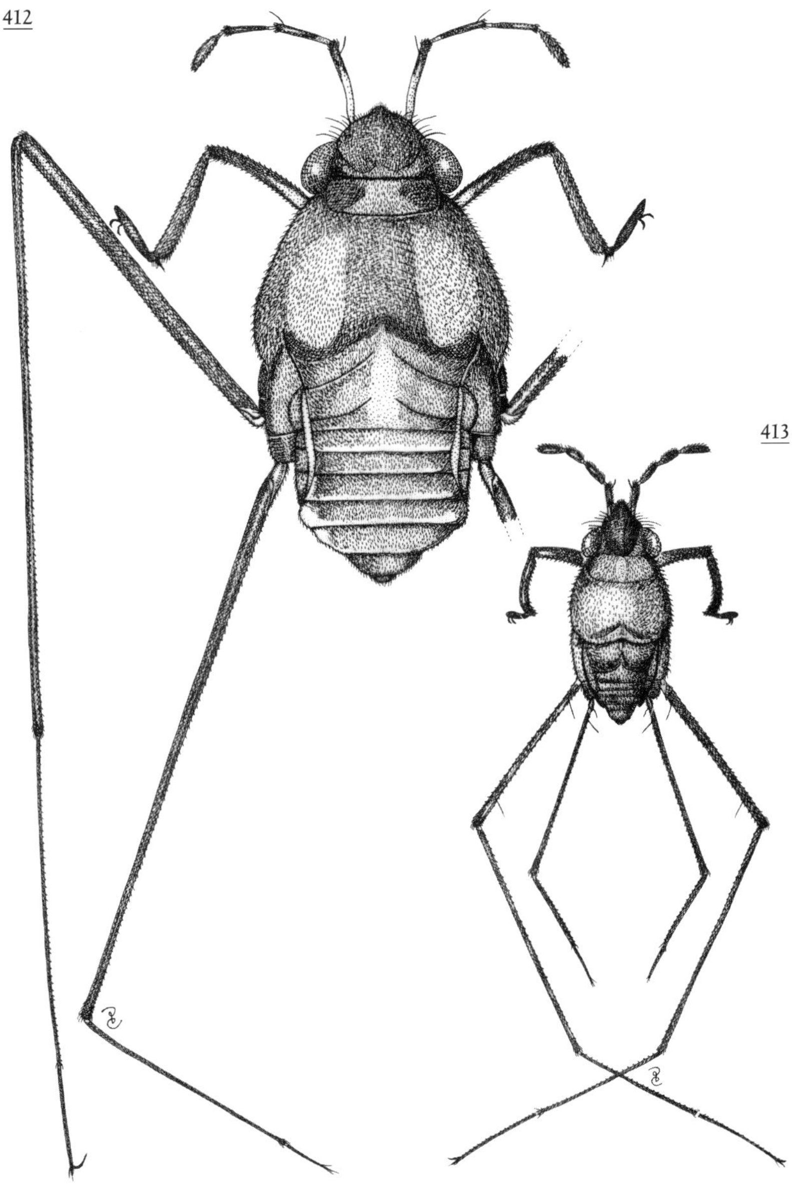

FIGURES 412 – 413
Talaudia nilsi
412 holotype, apterous female, length 2.7 mm – **413** paratype, apterous male, length 1.35 mm.

body in females; width less than head measured across eyes. Mesonotum long, much wider than head, suture between mesonotum and metanotum only visible laterally. Metanotum fused with abdominal tergites 1 and 2. Fore leg: in male femur thickened and slightly curved and flattened, tibia distinctly curved distally; in female femur and tibia long, slender and unmodified; in both sexes first tarsal segment short and cylindrical, with tuft of long setae at base in female, not evident in male, second tarsal segment longer; claws small, distinct, arising beyond middle of segment 2. Middle leg long and slender, femur

414

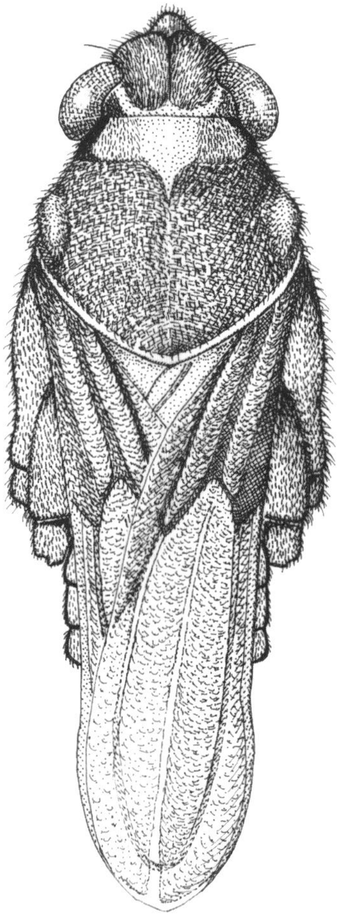

FIGURE 414
Talaudia nilsi,
paratype, macropterous female, legs an antennae removed,
length 3.8 mm.

stouter and distinctly shorter than tibia and hind femur; tarsi long, first segment about 2.5 times as long as second segment in females, about 1.5 times as long as second in males, claws long and slender, subapical. Femur of hind leg slender, three times as long as tibia in female, about twice as long as tibia in male, tarsal claws reduced to bristles. Abdominal segments subequal in length. Mesonotum and metanotum, pregenital abdomen and genital segments in male not modified; in female the median part of mesonotum, metanotum and abdominal tergites 1 and 2 depressed, forming a space for the male in phoresy. Vesical sclerites: dorsal sclerites comparatively short and not recurved proximally, lateral sclerites short. Macropterous morph (fig. 414): length of dealate male as apterous morph, length of female 3.80 mm. Pronotum and wings well developed, otherwise structurally as apterous morph. Pronotum black, anteriorly with median yellowish spot and transverse pruinose band, posterior margin narrowly and indistinctly brown. Median length of pronotum one half (male) to one third (female) of body length.

Distribution in Malesia (fig. 587)

This genus contains only one species, *T. nilsi*, known from the Talaud Islands in Indonesia.

Biological notes

T. nilsi was collected on side pools of a stream in the shade of forest.

– TRIBE NABOANDELINI[1]

Diagnosis

See above (Key to tribes and genera of Trepobatinae in Indo-Australia). Species are dimorphic, with apterous and macropterous morphs, but the latter is rare in most species. Rostrum short and robust; anterior margin of mesonotum slightly wider than head; metasternal scent gland orifice is vestigial.

Distribution in Indo-Australia

The tribe contains three genera and 30 species and is distributed across the Paleotropical region. Greatest diversity is found in Africa; secondary areas of diversity are Australia and Sri Lanka. In the Malay Archipelago records are relatively scarce. *Naboandelus* (twenty species and subspecies) is paleotropical, *Calyptobates* (nine species) occurs from Sri Lanka and the Andaman Islands across Malesia to northern Australia.

Biological notes

These small freshwater-striders are weak skaters, occupying sheltered, vegetation-free habitats along margins of ponds and streams.

Genus *Calyptobates*

Figures 415-416

Diagnosis

Small (males 1.77-2.17, females 1.97-2.32 mm). Head posterolaterally with pale markings. Pronotum and mesonotum dorsally without pruinose silvery-grey pubescence. Pronotum dark with single median pale mark of variable shape. Mesonotum dark, in some species with lateral or posterior pale stripes or bands. Metanotum and abdominal tergites dark. Sometimes pale spots on metacetabulae, tergites 2-5 and connexiva. Pleura with contrasting longitudinal stripe. Prosternum pale. Mesosternum dark tinged with pale markings (fig. 416). Appendages mostly dark, but pale at basis of antennal segment 1, basis of fore and middle femora, coxae and trochanters. Head short, anteriorly

---

1 Based on J. Polhemus & D. Polhemus (1994b).

415

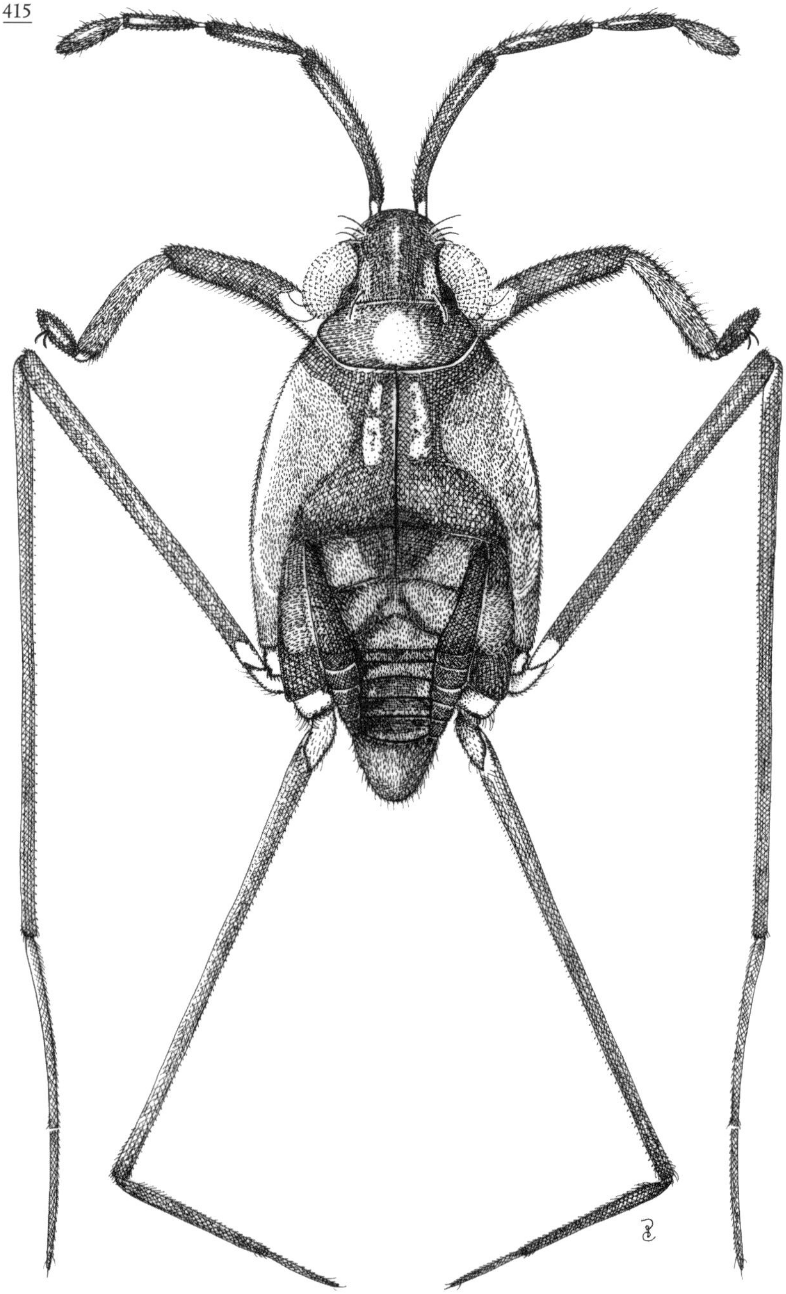

FIGURE 415
Calyptobates jourama, paratype, apterous male, length 2.05 mm.

rounded, sides subparallel (excluding eyes). Antennae moderately long and stout, segment 1 longest, about 1.7 times length of 2, segments 3-4 about equal in length and both shorter than 2. First antennal segment distinctly thickened in males. Anteclypeus not prominent. Eyes extending posteriorly onto anterior part of pronotum. Pronotum short, anterior margin almost straight, convex

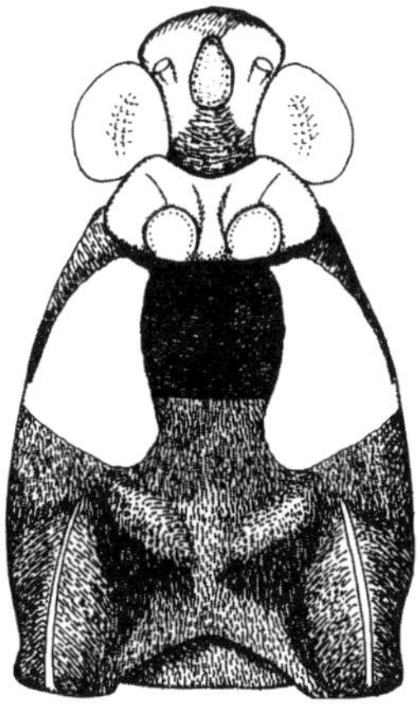

FIGURE 416
Calyptobates jourama, female, ventral view of head and thorax (appendages removed).

posteriorly; in males length about 1/2 of length of mesonotum, in females about 1/3. Posterior margin of mesonotum in females produced, carinate. Metanotum laterally indicated by weak suture, invisible medianly, fused with abdominal tergite 1. Female metacetabula with dorsal tuft of stiff erect black setae. Legs similar in both sexes except for male fore femur, which has slight swelling at base and is slightly arched. Inner side of fore femur covered with short stiff spines; fore tibia with a flattened inner surface. Anterior tarsal segment 1 short, segment 2 long; claws of fore legs small but distinct. Middle femur slightly thickened, shorter than tibia, tarsi long, first segment 1.3 times length of second; claws reduced to bristles. Hind femur about three times as long as tibia, tarsus long, fused into one segment; claws long and slender, arising before middle of tarsus. Abdominal tergites 3-7 subequal in length, not modified. Male genital segments large; proctiger with long narrow lateral processes directed forward. Vesical sclerites contain dorsal sclerite, long lateral sclerite, short ventral sclerite and thin apical sclerite. Parameres small, distally weakly bifurcate.

Distribution in Malesia (fig. 591)

Calyptobates is distributed from Sri Lanka (an undescribed species) and the Andaman Islands through Malesia to northern Australia. Of the nine described species three occur in Australia and three in New Guinea. Borneo (Kalimantan Timur), Moluccas and Andaman Islands have one species each.

Biological notes

Calyptobates species are weak skaters and behave secretly. They prefer quiet places near streams and rivers and can be found on virtually stagnant pools and slow moving margins of shaded rivers in rainforest or on broad exposed rivers. Macropterous morphs tear the distal 2/3 of their wings off, just behind the apices of the basal thickened fore wing veins.

Genus *Naboandelus*

Figures 417–419

417

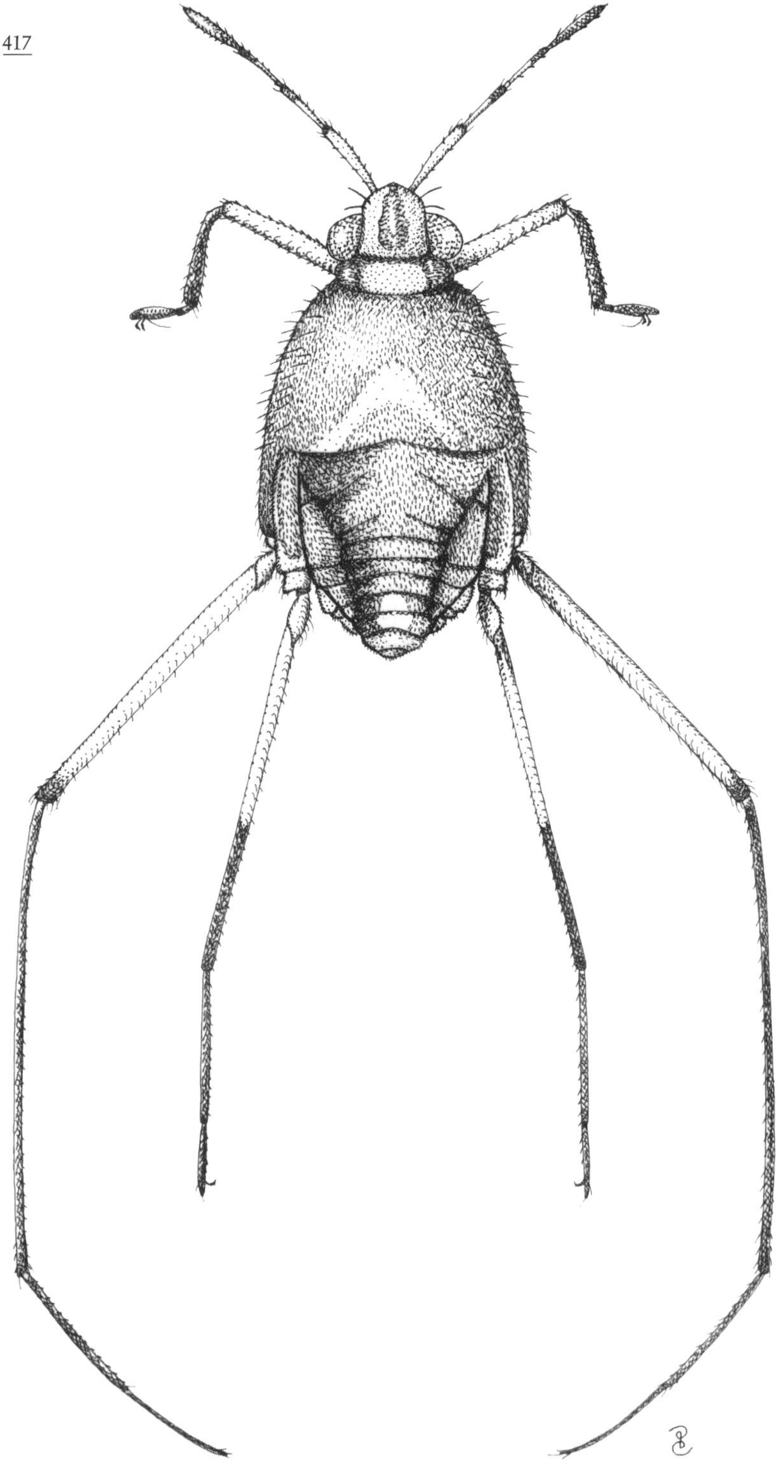

FIGURE 417
Naboandelus signatus, apterous female, length 2.3 mm.

Diagnosis

Small to very small (males 1.46-2.50, females 1.57-2.89 mm). Never with longitudinal pale stripe on mesopleura. Head striped dark; clypeal region with pale mark. Pronotum dark with single median pale mark of various shape. Mesonotum dark, in some species with lateral or posterior pale stripes or bands. Metanotum and abdominal tergites dark with pale spots on some tergites. Pale marks sometimes also present on metacetabulae and connexiva. Prosternum pale. Mesosternum and mesopleura always grey. Appendages mostly dark, but pale in at base of antennal segment 1, bases of fore and middle femora, coxae and trochanters. Head short, about half as long as wide across eyes, anteriorly rounded. Antennae relatively long and slender, segment 1 longest, thicker and longer in male than in female; segment 2-4 subequal in length. Anteclypeus not prominent. Eyes extending posteriorly onto anterior part of pronotum. Pronotum short, length in males about half of mesonotum, in females about $^1/_3$. Posterior margin of mesonotum in both sexes straight or slightly sinuated. Metanotum fused with tergite 1, laterally indicated by weak suture. Legs similar in both sexes. Fore leg slightly thickened and arched; inner side covered with short stiff spines; tarsal segment 1 short, segment 2 long; claws small but distinct. Middle femur slightly thickened, shorter than tibia, tarsi long, first segment 1.3 times length of second; claws reduced to bristles. Hind femur about three times as long as tibia, tarsus short, fused into one segment; claws reduced to bristles in middle legs, long and arising from middle of the segment in hind legs. Abdominal tergites 3-6 subequal in length, segment 7 slightly longer, not modified. Male proctiger with long narrow lateral processes directed forward (fig. 418). Vesical sclerites (fig. 419) contain dorsal sclerite, long lateral sclerite, short ventral sclerite and apical accessory sclerite.

418

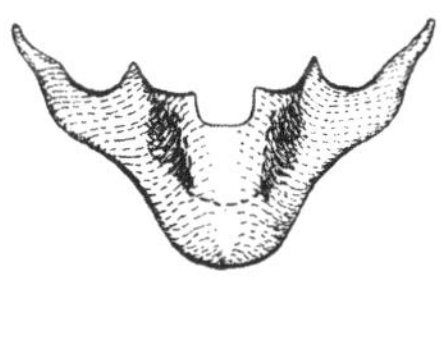

419

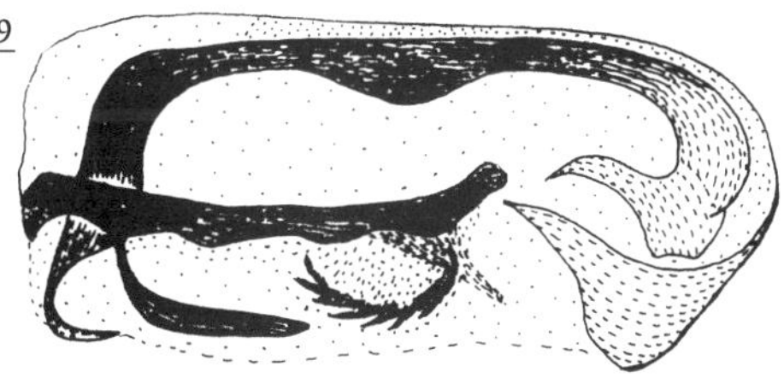

FIGURES 418 – 419
Naboandelus signatus, male.
418 proctiger, dorsal view – **419** vesical sclerites, lateral view.

Distribution in Malesia (fig. 591)

Of *Naboandelus* four species are recorded from the Oriental region and two in South-west Asia (India, Sri Lanka, Nepal and Thailand). In Malesia two species are found: in continental Malaysia (Johor) and Borneo (Kalimantan Timur) respectively.

Biological notes

The tiny *Naboandelus* species are weak skaters. They prefer sheltered habitats in freshwater ponds and along the margin of streams, but also occur in groups on sunny ponds or rivers.

– TRIBE TREPOBATINI[1]

Diagnosis

See above (Key to tribes and genera of Trepobatinae in Indo-Australia).

Distribution worldwide

Nine genera belong to the tribe Trepobatini, of which two occur in the Oriental region: *Cryptobates* and *Gnomobates* J. Polhemus & D. Polhemus. The species of the latter genus are found outside Malesia only: India, Myanmar and northern Thailand. The tribe shows a prominent distribution in Western Hemisphere, where seven genera are found.

Genus *Cryptobates*

Figure 420

Diagnosis

Medium-sized (males 2.70-3.46, females 2.66-3.67 mm). Pale to brown with dark markings. Thorax with dark median longitudinal stripe, often also with sublateral dark stripes. Metanotum and abdominal tergites 1-2 with dark median stripe. Thoracic venter pale with dark lateral stripes in some species. Appendages mostly pale, variously tinged with darker brown. Head elongate, anteriorly rounded, sides subparallel (excluding eyes). Anteclypeus not prominent. Antennae long and slender, segment 1 about 1.5 times the length of 2, segments 3 and 4 longest, about equal in length and about 1.5 times the length of 1. Eyes posteriorly extending to middle of pronotum. Rostrum relatively long. Pronotum short, anterior and posterior margins almost straight, length about $^1/_3$ of mesonotum length. Anterior margin of mesonotum about as wide as head. All *Cryptobates* have an elongate pruinose patch or stripe on middle of mesopleura. Metanotum laterally indicated by transverse indistinct suture (invisible in median part), and fused with abdominal tergite 1. Some species with metasternal scent gland orifice. Legs similar in both sexes. Fore femur slightly thickened, not modified; tibia distally broadened, covered with short stiff spines; tarsal segment 1 short, segment 2 long; claws small but distinct. Middle femur slender, shorter than tibia, tarsi long, first segment longer than second; claws reduced. Hind femur slender, more than twice as long as hind

1 Based on J. Polhemus & D. Polhemus (1995b).

420

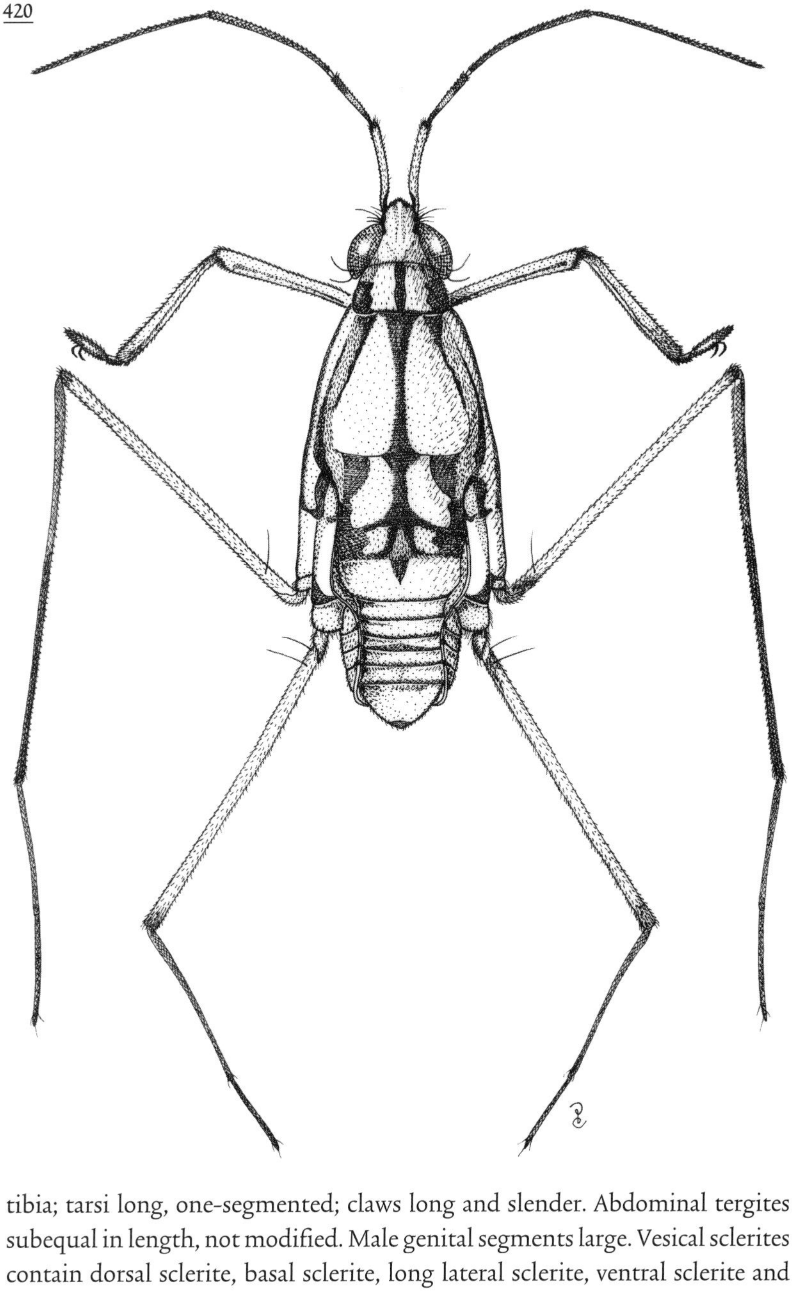

FIGURE 420
Cryptobates johorensis,
paratype, apterous female,
length 3.2 mm.

tibia; tarsi long, one-segmented; claws long and slender. Abdominal tergites subequal in length, not modified. Male genital segments large. Vesical sclerites contain dorsal sclerite, basal sclerite, long lateral sclerite, ventral sclerite and apical sclerite. Parameres (fig. 394) strongly twisted or spiralled in all species, with basal structure unique within Gerridae.

Distribution in Malesia (fig. 592)
Cryptobates contains four described species, which are distributed from southern India and Sri Lanka eastward through continental Malaysia to Borneo. It is

apparently restricted to the western side of Wallace's Line. Three species are recorded from peninsular Malaysia, Singapore, and Borneo.

Biological notes

Cryptobates species are weak skaters, preferring quiet places along the banks of the stream, together with *Rhagadotarsus* and *Ventidius*. They have also been collected together with a marine *Stenobates* species, indicating that they may live close to a mixohaline biotope (J. Polhemus & D. Polhemus, 1995b).

– TRIBE STENOBATINI[1]

Diagnosis

See above (Key to tribes and genera of Trepobatinae in Indo-Australia). Monomorphic apterous. Brown to blackish with lighter markings, head light with dark markings or dark with light markings. Rostrum relatively stout, apex surpassing anterior margin of mesonotum; pronotum shorter than head, truncate with posterior margin straight to slightly convex; median length of mesonotum 3-3.5 times median length of pronotum. Hind femur two times length of hind tibia, tarsal segments of hind legs not fused. Parameres small but distinct. Mesonotum in female divided into two lateral parts connected by membranous region, diverging through metanotum, running along side of abdominal tergites. Second gonapophyses serrated.

Distribution worldwide

Stenobatini is restricted to Malesia and Australia, ranging from the Philippines and southern continental Malaysia to northern Australia and Solomon Islands. It contains five genera, which have all been found in Malesia.

Genus *Pseudohalobates*

Figure 421

Diagnosis

Medium-sized to large (males 4.9-5.7, females 4.6-5.3 mm), strongly dorsoventrally compressed. Pronotum dark with pair of pale sublateral markings. Mesonotum of male dark or with indistinct paired pale sublateral stripes. Metanotum light. Abdominal tergites posteriorly lighter. Appendages generally dark, but first two antennal segments, fore femora and fore tibia, all coxae and trochanters pale. Head relatively long, angulate anteriorly, dorsoventrally compressed; on ventral side with broad weak sulcus; anteclypeus prominent. Antennae relatively long and slender, segment 1 longest, about twice as long as

---

1 Based on J. Polhemus & D. Polhemus (1996).

421

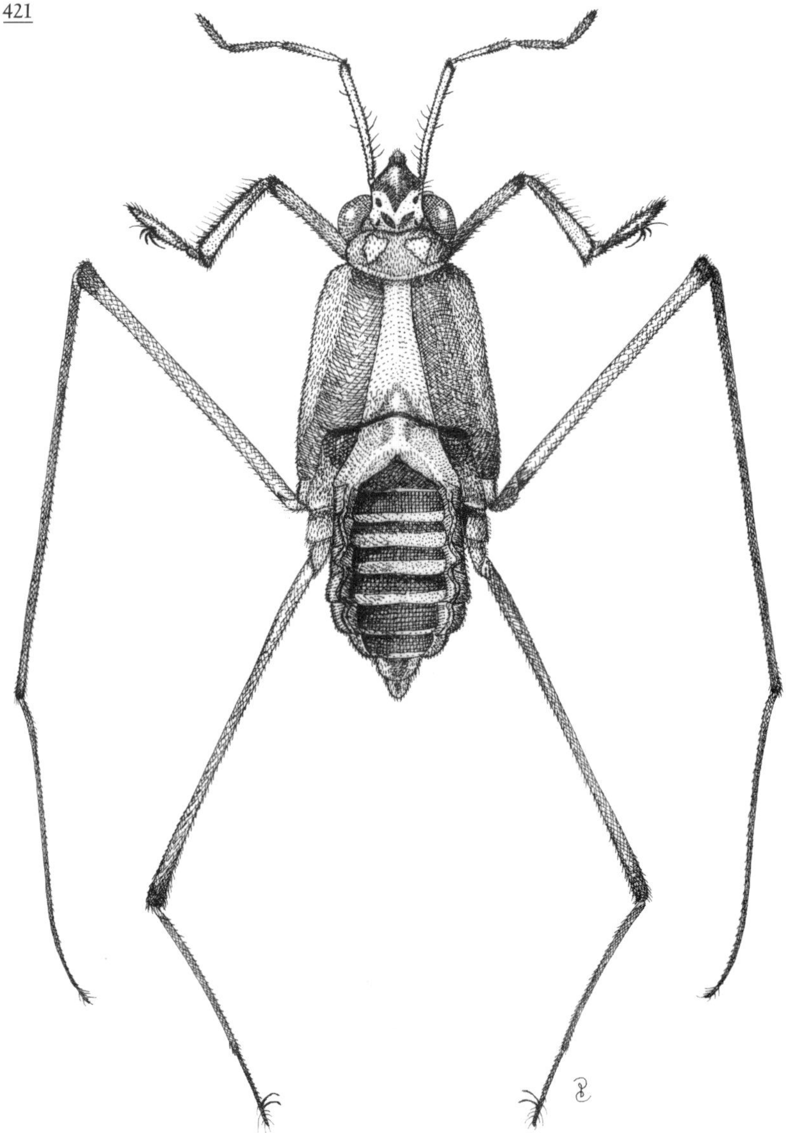

FIGURE 421
Pseudohalobates inobonto, paratype, apterous female, length 5.7 mm.

segment 2, distally slightly widening, inner margin with fine hair fringe and scattered longer stiff setae; segments 3 and 4 about equal in length, slightly shorter than 2 in both sexes. Eyes extending posteriorly nearly to middle of pronotum. Metanotum posteriorly indented, defined by weak suture with abdominal tergite 1. Metasternum of male strongly produced anteriorly (fig. 387); metasternal scent gland orifice large and distinct. Metasternum in females anteriorly much more weakly curved, scent orifice small. Legs similar in both sexes; fore leg (fig. 388) slightly thickened; fore tibia distally broadened, with apical spur, on both sides covered with short stiff spines and scarce long setae;

distally with patch of stout black setae; tarsal segment 1 short, segment 2 about 6-7 times longer; claws of fore leg long and slender, arising from middle of the segment. Middle tibia with row of erect setae, much shorter in females; tarsus long, segment 1 longer than 2; claws reduced to bristles. First segment of hind tarsus shorter than 2; claws long and very slender, arising from distal $^1/_3$ of segment. Abdominal tergites 1-2 longest, 3-7 subequal in length. Male genital segments large; proctiger with long, broad lateral processes, directed forward (fig. 386). Vesical sclerites contain long dorsal sclerite, long lateral sclerite, short ventral sclerite, and apical accessory sclerite. Second gonapophyses of female long and narrow, with serrate distal lobes.

Distribution in Malesia (fig. 592)

The single species of *Pseudohalobates* is reported only east of Wallace's Line. Records come from Sulawesi, Talaud Archipelago, Bacan, Halmahera, Irian Jaya, Biak, and Yapen. It has also been recorded from Mindanao.

Biological notes

Contrary to the other genera in Stenobatini, *Pseudohalobates* species are strong skaters, often occurring on coastal waters but sometimes away from the coast, in company with strong swimming species of *Halobates.* They usually live gregariously in a stony environment, preferring shadow, and are also found on open and shallow pools around corals and boulders which are exposed at low tide.

Genus *Rheumatometroides*

Figure 422

Diagnosis

Small to medium-sized (males 2.7-4.0, females 2.8-4.0 mm), elongate ovate, not dorsoventrally compressed. Head usually with several pairs of lateral and one median markings. Pronotum dark with pair of sublateral pale markings, or with single median light region. Mesonotum always with paired contrasting light marks; marks basally quadrate but with variable anterior dark indentation which may divide it in two longitudinal stripes. Metanotum light brown with dark markings. Abdominal tergites posteriorly paler than anteriorly. Appendages usually dark with contrasting pale markings. Head not dorsoventrally compressed; on ventral side without broad sulcus. Antennae relatively long and slender, segment 1 longest, about twice as long as segment 2, distally not distinctly widening, inner margin with fine hair fringe and scattered longer stiff setae; segment 3 and 4 about equal in length and slightly shorter than 2. In females antennal segment 2-4 subequal in length. Eyes posteriorly extending nearly to middle of pronotum. Metanotum posteriorly indented, defined by weak suture with tergite 1. Metasternum of male anteriorly

422

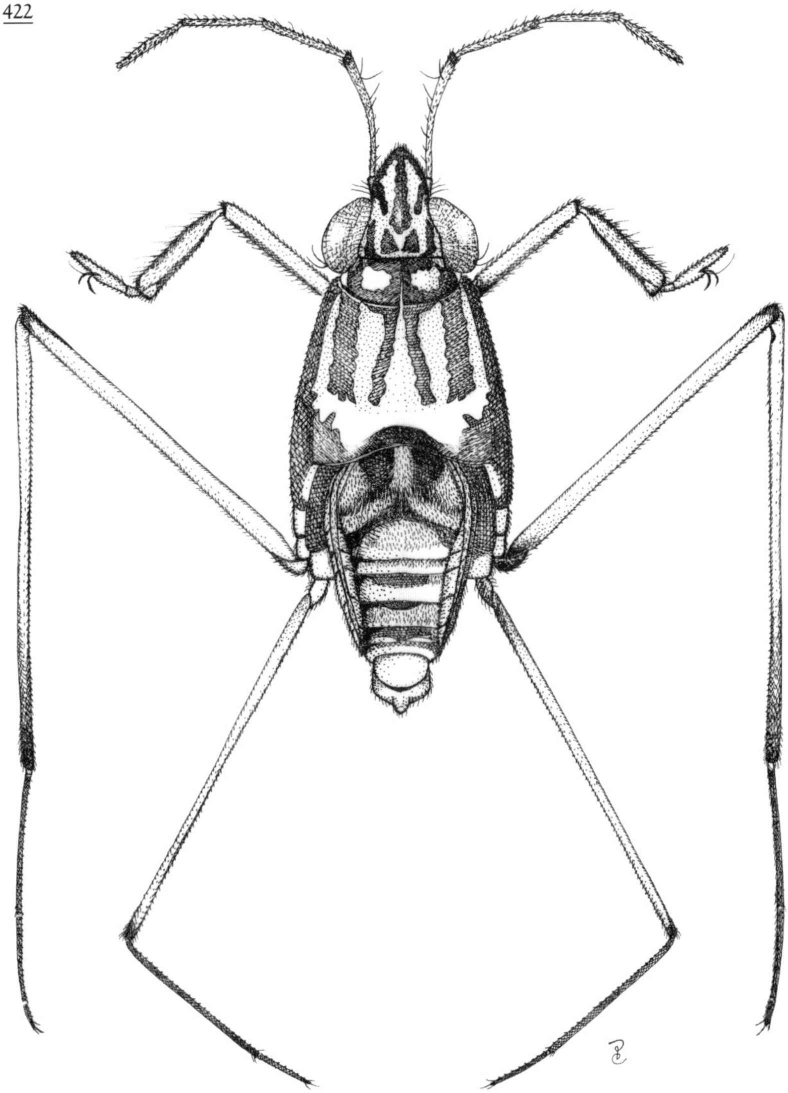

FIGURE 422
Rheumatometroides drepanephoros,
paratype, apterous male,
length 3.1 mm.

strongly produced (fig. 390); metasternal scent orifice not prominent, except in *R. kikori* and *R. sele*. Metasternum in females anteriorly only weakly curved, scent orifice small. Legs similar in both sexes; fore leg slightly thickened (fig. 391); fore tibia distally broadened, without apical spur, inner side covered with short stiff spines and near distal end with patch of stout black setae; tarsal segment 1 short, segment 2 about six times longer; claws long and slender, arising from middle of the segment. Middle tarsi long, segment 1 and 2 subequal in length; claws reduced to bristles. Hind tarsal segment 1 shorter than 2; claws long and slender, arising preapically. Tergites 1-2 longest, 3-7 subequal in length. Male genital segments large; proctiger with long lateral processes, directed forward (fig. 389). Vesical sclerites contain long dorsal sclerite, long lateral sclerite, short ventral sclerite, and apical accessory sclerite. Female gonocoxae large, ovipositor well developed, first gonapophyses with membranous apices, distal lobes on second gonapophyses on outer margin serrated. For specific identification number and form of pale markings on pronotum and mesonotum, colour of first antennal segment, and shape of metasternum are important.

Distribution in Malesia (fig. 593)

Rheumatometroides is one of two stenobatine genera with an Oriental-Australian distribution, ranging from continental Malaysia through Borneo, Sulawesi, New Guinea, Solomon Islands to northern Australia. Seven species belong to this genus, six of which occur in Malesia. The greatest species diversity occurs in New Guinea, where five species are reported. The other two species have a disjunct distribution.

Biological notes

Rheumatometroides species live in marine habitats. Some species are found on deep and open water, but most live in shallow near-shore water like sheltered tidal creeks, edges of mangrove estuaries, spring-fed rivers in lowland rainforest, interfaces of fresh and salt water, etcetera. Various species of *Rheumatometroides* are often found in company with species of *Asclepios*, *Halobates*, *Stenobates*, and marine veliids.

Genus *Stenobates*

Figure 423

Diagnosis

Medium-sized to large (3.1-5.1 mm), elongate, not dorsoventrally flattened. Head with lateral pale markings. Pronotum dark, with pair of pale sublateral markings. Mesonotum dark, always with contrasting pale stripes. Most possess elongate rather than quadrate pale markings on mesonotum. Metanotum and abdominal tergites dark. Head short, not dorsoventrally compressed; clypeal region not prominent. Antennae moderately long and slender, segment 1 longest, about twice as long as segment 2, inner margin with fine hair fringe; segments 3-4 about equal in length and slightly longer than 2. Eyes posteriorly extending to anterior part of pronotum. Metanotum laterally indicated by weak suture, fused with abdominal tergite 1. Metasternum of male anteriorly strongly produced, about $^1/_3$ of mesosternal length (fig. 403); metasternal scent gland orifice large. Metasternum in females much shorter, scent orifice small. Legs similar in both sexes; fore leg slightly thickened (fig. 404), fore tibia distally broadened, with apical spur; inner side covered with short stiff spines; tarsal

423

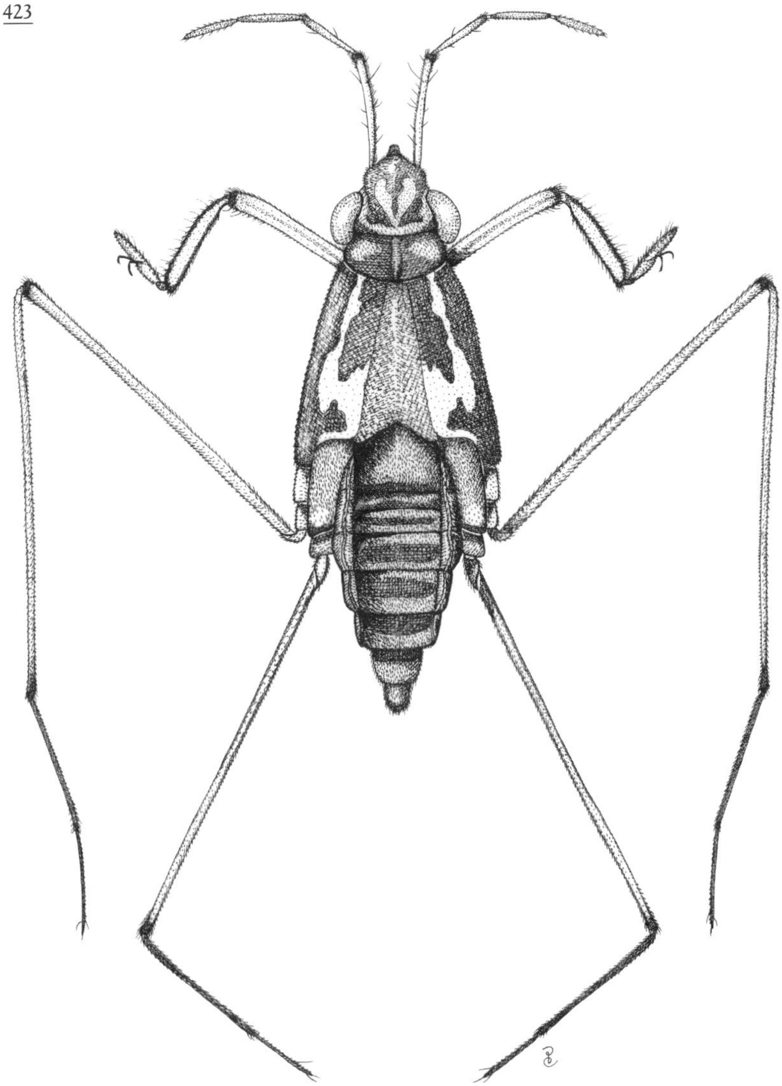

FIGURE 423
Stenobates sangihe,
paratype, apterous female,
length 3.8 mm.

segment 1 short, segment 2 three times longer; claws of fore leg long and slender. Middle tarsi long, first segment 1.3 times the length of second; claws reduced to bristles. Hindtarsal segment 1 shorter than 2; claws long and slender. Abdominal tergites 1-2 fused, 3-5 subequal in length, 6-7 longer, not modified. Male genital segments large; proctiger with long lateral processes, directed forward (fig. 402, 405). Vesical sclerites (fig. 406) contain dorsal sclerite, long lateral sclerite, short ventral sclerite and apical sclerite. Female gonapophyses blunt, second gonapophyses with distal lobes serrated on outer margin. For specific identification first antennal segment, pilosity of male middle tibia, colour and pigmentation of body, and shape of metasternum in male are important.

Distribution in Malesia (fig. 594)

Stenobates contains ten species, which are distributed from Singapore to Northeast Australia. All species are reported from Malesia. The number of species found in Papua New Guinea is four, in the Moluccas and Sulawesi three each, in the Philippines two and in Singapore one.

Biological notes

As usual in marine waterstriders *Stenobates* adults are always apterous. They are weak skaters, occupying sheltered mangroves or reefs along the coast, tidal creeks of interior deltas, small inlets and pools on the shore, etcetera. The habitat is often shared with other species of Trepobatinae or with *Halobates*. Females have the second gonapophysis distally serrated, but whether eggs are laid in or on the substrate it is still uncertain. Life history is largely unknown.

Genus *Stenobatopsis*

Figure 424

Diagnosis

Small to medium-sized (males 3.9-4.3, females 3.2-3.8 mm), elongate, strongly dorsoventrally compressed. Head with lateral pale stripes. Pronotum dark with large quadratic paired spots. Mesonotum in males completely dark, in females membranous part brownish. Metanotum and abdominal tergites dark, posterior margins brownish. Appendages mostly dark. Head moderately long, dorsoventrally compressed; clypeal region prominent. Interocular space about two times eye width. Antennae relatively short, segment 1 longest, about 1.4 times segment 2, incrassate in males, inner margin with fine hair fringe; males with segments 2, 4 and 3, respectively, progressively shorter, in females segment 3 shortest. Eyes extending to anterior part of pronotum. Suture between mesonotum and metanotum distinct. Metanotum of male indented posteriorly, fused with abdominal tergite 1. Metasternum of male anteriorly weakly produced (fig. 397); metasternal scent orifice small, in females reduced. Male fore leg (fig. 398) thickened, fore tibia distally broadened, with apical spur, anterodistally with dark patch of stiff setae; inner face clothed with long and short setae; tarsal segment 1 short, segment 2 longer and flattened; claws long and slender. Middle tarsi long, segment 1 slightly longer than 2; claws reduced to bristles. Hind tarsal segment 1 about half of length of 2; claws long and slender. Abdominal tergites 1-2 fused and longest, 3-7 subequal in length; female tergite 7 short. Male sternite 8 caudally produced (fig. 396). Proctiger produced laterally into long slender processes, tapering (fig. 396). Vesical sclerites: long dorsal and lateral sclerites, short and broad ventral sclerite and apical sclerite. Female gonapophyses blunt, second gonapophyses lanceolate, serrate on outer margin. The genus resembles *Pseudohalobates*. Features of antennae are diagnostic for specific identification.

424

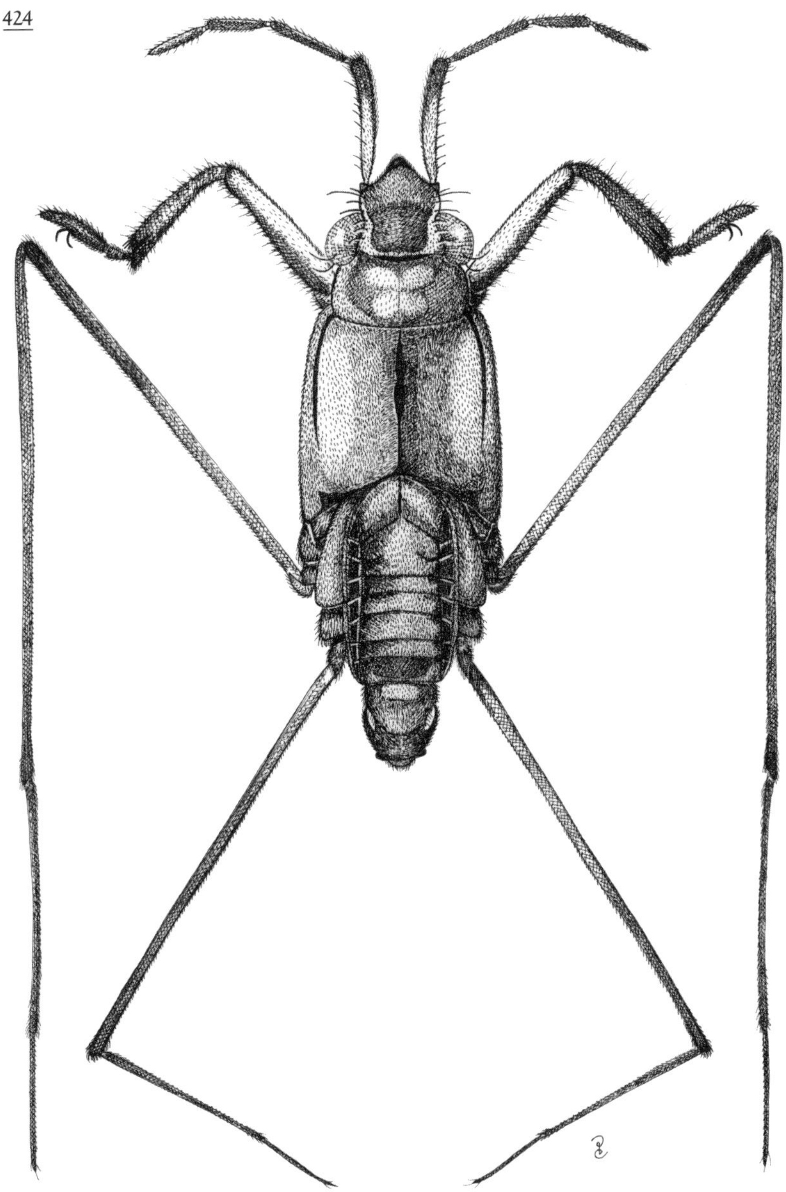

FIGURE 424
Stenobatopsis sp., apterous male, length 4.4 mm.

Distribution in Malesia (fig. 594)
The single described species of *Stenobatopsis* is reported from the northern Moluccas. It also occurs in the Palawan province of the Philippines (H. Zettel, unpubl.).

Biological notes

Little is known about the biology of *Stenobatopsis* species, except that they share the same marine habitats with *Pseudohalobates inobonto* and *Stenobates labuha* in the northern Moluccas.

Genus *Thetibates*

Figure 425

425

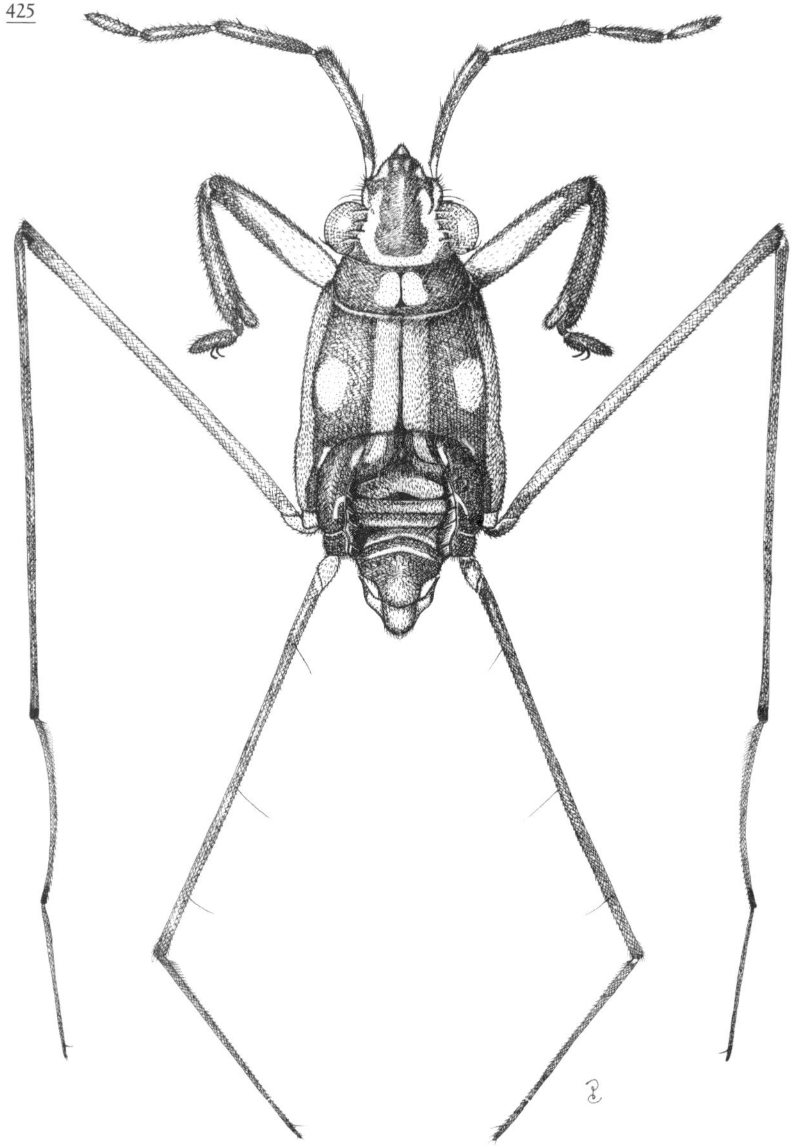

FIGURE 425
Thetibates serena, apterous male, length, 2.8 mm.

Diagnosis

Small to medium-sized (males 2.7-3.1, females 2.7-3.0 mm), elongate oval, not dorsoventrally flattened. Head pale with large dark median mark. Pronotum dark with central pale markings separated by dark median line. Mesonotum dark with paler area along median line, in middle sublaterally with pair of pale roundish markings. Metanotum with pale markings along median line. Abdominal tergites dark with posterior lighter margins. Antennae dark, first segment with pale base. Legs dark but ventrally lighter. Head short, not dorsoventrally compressed, without broad longitudinal sulcus; clypeal region not prominent. Interocular space about 1-1.5 times eye width. Antennae relatively long, stouter in male; inner surface with dense hair fringe and scattered longer setae; segment 1 longest, about 1.5 times as long as segment 2, distally widening in males; segments 3 and 4 subequal in length and slightly shorter than segment 2; in females segments 2, 3 and 4 subequal in length. Eyes posteriorly extending to anterior margin of pronotum. Suture between mesonotum and metanotum distinct. Male metanotum posteriorly indented, demarcated by weak suture. Metasternum of male anteriorly weakly produced (fig. 400). Metasternal scent gland orifice in males distinct (fig. 400), in females more reduced. Male fore leg (fig. 401) thickened, distally narrower, fore tibia distally broadened with large apical spur, anterodistally with dark patch of stiff setae; inner side covered with short setae; tarsal segment 1 short, segment 2 longer; claws slender, arising from middle of second tarsal segment. Middle tarsi long, segment 1 slightly longer than 2; claws reduced to bristles. Hindtarsal segment 1 slightly shorter than 2; claws long and slender, arising from middle of second tarsal segment. Abdominal tergites 1-2 longest, 3-7 subequal in length; female tergite 7 medially more or less sclerotized. Male sternite 8 produced caudad. Proctiger laterally produced into long slender processes, tapering toward distal end (fig. 399). Vesical sclerites: dorsal and lateral sclerites long, short and thick ventral sclerite apically broadened, apical sclerite short.
The genus resembles *Stenobates* and *Rheumatometroides*. The pale marks on mesonotum and lateral processes of male proctiger are important for specific identification.

Distribution in Malesia (fig. 593)

Thetibates contains two species. Both occur in Malesia, east of Weber's Line, in East New Guinea and on Bismarck and Solomon Islands.

Biological notes

Thetibates serena was collected from sheltered branching inlets in Papua New Guinea, together with some *Halobates* species. The inlets were protected from the open sea by reefs and offshore islands.

REFERENCES – Trepobatinae. Structural analysis and classification of Gerridae (Matsuda, 1960); morphology, phylogeny, adaptations, biogeography and classification of Gerromorpha (Andersen, 1982a); general information, keys, descriptions to genera and species, and biological notes of Metrobatini (J. Polhemus & D. Polhemus, 1993); key to genera, description of species, and biological notes (J. Polhemus & D. Polhemus, 1994b); keys, descriptions to genera and species, and biological notes of Trepobatini (J. Polhemus & D. Polhemus, 1995b); additional species of *Ciliometra* (Lansbury, 1996b); keys, descriptions to genera and species, and biological notes of Stenobatini (J. Polhemus & D. Polhemus, 1996); key and description to Australian species of Trepobatinae (Andersen & Weir, 1998); additional species of genus *Andersenella* (Chen & Nieser, 2000); additional species and description of *Talaudia* (J. Polhemus & D. Polhemus, 2002a).

SUBFAMILY GERRINAE

Diagnosis

Small to large (4-36 mm). Usually dark with pale markings. Apterous individuals with long pronotal lobe (figs. 431, 447), but this in some genera secondarily reduced (e.g. *Neogerris*). Intersegmental suture between mesonotum and metanotum either distinct or laterally obliterated. Lateral longitudinal suture (fig. 379) separating mesonotum and mesopleura absent in most genera (fig. 381). Metathoracic scent apparatus well-developed, omphalium always present. Fore wing with veins M+Cu connected with Sc+R by anterior crossvein; Sc+R and M+Cu basally fused (fig. 18). Claws of hind legs in adults of some genera disappeared (*Limnogonus*, *Limnometra*, *Tenagogerris* Hungerford & Matsuda and *Tenagometrella*). First abdominal segment ventrally lost. Dorsal plate of male vesica distinctly sclerotized (fig. 435); parameres reduced. Second gonapophyses of females usually membranous.

Characteristics

Apart from colour, characters to separate genera are the structure of the metathoracic scent gland orifice and the omphalium, position of the metathoracic spiracle, and the development of the abdomen. For identification the following characters are important: body size, relative length and width of appendages and their segments including the rostrum, development of the connexival spines (fig. 427), presence or absence of cilia or spines on the legs, shape of the abdominal apex in both sexes, notably the variation in the V-shaped emargination on the hind margin of male sternite 7 (fig. 432) and the shape of the genital segments, shape of the distal part of male phallic vesica (fig. 430) and vesical sclerites (figs. 429, 435), and shape of male proctiger (figs. 428, 433). The tribe Tachygerrini are recognized by the presence of the retained pretarsal arolia, the tribe Gerrini by the reduced pretarsal arolia in the adult stage (arolia present in nymphs).

Distribution worldwide

In the present sense the cosmopolitan subfamily Gerrinae contains two tribes, Tachygerrini and Gerrini, fourteen recent genera and 168 recent species (Andersen, 1995b; Andersen & Weir, 1997).

Distribution in Indo-Australia

Six genera belonging to the subfamily Gerrinae have been recognized from Indo-Australia: *Limnometra* with 22 species in the area is by far the most speciose, followed by *Tenagogonus* with seven, *Limnogonus* with six, and *Aquarius* with five species. *Tenagogerris* with three species is restricted to Australia.

According to Andersen (1975) 'the zoogeographical pattern of Gerridae in general within the Indo-Australian area shows a progressive reduction in number of species, but an increasing dominance of one particular taxon (Gerrinae) eastward from the Asian continent. This "filter-bridge" effect is a well-known general phenomenon in the zoogeography of this area. Although *Limnometra* is richest in species, the genera *Tenagogonus* and *Limnogonus* are the most widespread. They are the only genera of freshwater gerrids which extend into the Pacific area. Only species of *Limnogonus* go as far east as Samoa and the Society Islands. In the Pacific area freshwater gerrids are only found on volcanic islands and raised atolls. Each of the larger islands show a similar pattern: one, two or three recent immigrants belonging to widespread species and one endemic species. The population of the immigrant species, which differ from the continental populations only in minute characters, have retained their wide habitat tolerance and are mainly found in the lowland. Most of the endemic species are now confined to the mountainous regions of the islands, where they live in quiet parts of streams or in rock pools and pot-holes beside streams. In some way these endemic species occupy the same habitats as do the non-gerrine waterstriders in the continental areas'. In general, this still holds, especially when only freshwater species are considered. However, the strong speciation of Metrobatini and Stenobatini (J. Polhemus & D. Polhemus, 1993, 1996) around New Guinea clashes with this general pattern.

Key to genera and subgenera in Indo-Australia[1]

1 Pronotum with a central pale spot (fig. 442). Claws absent on middle legs ▸ *Neogerris* p. 350

– Pronotum dark with a median yellowish line, or a pair of yellow spots, or both (fig. 431), or pronotum light with a dark median line (fig. 437). Middle legs with small but distinct claws ▸ 2

1 Adapted from Andersen (1995b).

2 First antennal segment distinctly longer than second and third segment together (fig. 426). Metathoracic spiracle located only a little more than its own length from the fore wing base in macropterous morphs, or a little more than its own length from the lateral margin of the pronotal lobe in apterous mohps. Length 9.1-15.2 mm for Malesian species ▸ *Aquarius* **p. 342**

– First antennal segment distinctly shorter than second and third segments together (fig. 431). Metathoracic spiracle located much more than its own length from the fore wing base in macropterous morphs, or much more than its own length from the lateral margin of the pronotal lobe in apterous morphs ▸ 3

3 Metathoracic scent gland orifice circular, located on small, conical tubercle lying in a depressed area of metasternum. Pronotal lobe predominantly blackish, but frequently with distinct yellow or orange marks (middle line, spot, hind margin, etcetera; fig. 431) ▸ *Limnogonus* **p. 345** 4

– Metathoracic scent orifice transversely ovate, enclosed between two lip-shaped tubercles in posterior parts of metasternum. Pronotal lobe predominantly yellowish or brownish with blackish colour patterns, especially with blackish median line (fig. 437, 447) ▸ 5

4 Anterior part of pronotum with a pair of yellowish spots (fig. 431) ▸ **subgenus** *Limnogonus* **p. 345**

– Anterior part of pronotum without such spots ▸ **subgenus** *Limnogonoides*

5 Rostrum relatively short, third segment not reaching mesosternum from ventral view [Australia] ▸ *Tenagogerris*

– Rostrum relatively long, third segment reaching mesosternum from ventral view ▸ 6

6 Abdominal connexiva caudally produced into long connexival spines (fig. 437). Abdomen relatively long ▸ *Limnometra* **p. 348**

– Abdominal connexiva caudally not produced into long connexival spines (fig. 447). Abdomen relatively short ▸ *Tenagogonus* **p. 353**

Genus *Aquarius*

Figures 426-430

Diagnosis

Large to very large (males 8.3-26.5, females 11-26 mm for all *Aquarius*; 9.1-15.2 mm for Malesian species). Dark. No marked sexual dimorphism, except in the extralimital *A. najas*-group, although males usually smaller. Pronotum usually dark, anteriorly with median pale stripe. Fore femora mostly dark, light markings restricted to basal and distal end. First antennal segment long, subequal to or longer than segments 2 and 3 together. Anterior pronotal lobe separated

426

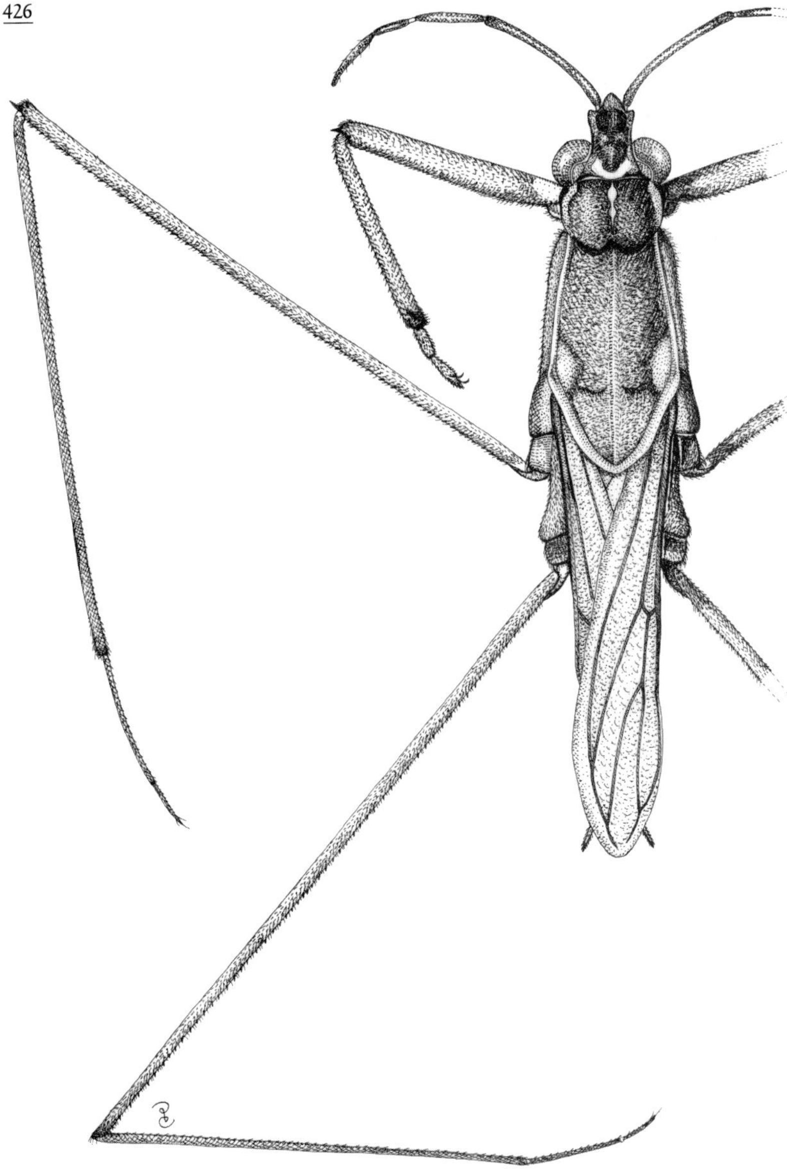

FIGURE 426
Aquarius adelaidis,
macropterous male,
length 11.5 mm.

from posterior lobe by obscure transverse depression; posterior lobe entirely covering mesonotum, which is about 2.5 times as long as metasternum. Omphalium close to posterior margin of metasternum. Fore leg with straight femur and tibia; second tarsal segment maximally 1.5 times as long as first segment. Middle leg with femur slightly longer than tibia; tibia about 3.5 times as long as first tarsal segment, first tarsal segment three to eight times as long as second. Hind tibia at least four times as long as first hindtarsal segment. Tip of abdomen usually with well-developed connexival spines (fig. 427), surpassing posterior margin of segment 8. Tergite of each abdominal segment relatively long. Ventral longitudinal margin of connexiva distinct. Sternite 8 anteriorly emarginated, with variable but conspicuous elevation. Proctiger cone-shaped (fig. 427). Parameres in Malesian species reduced but visible (fig. 428). Endosoma (fig. 429) with prolonged apical margin more strongly sclerotized; dorsal plate angularly produced (fig. 430); dorsal sclerite long, proximally recurved, fused with robust bifurcate apical sclerite; ventral sclerite short with two pairs of accessory sclerites (fig. 429). Gonocoxa relatively short, each with oblique lateral impression and dorsal tubercle. First gonapophyses with bipartite pointed apex (external part pointed, internal part rounded).

427 428

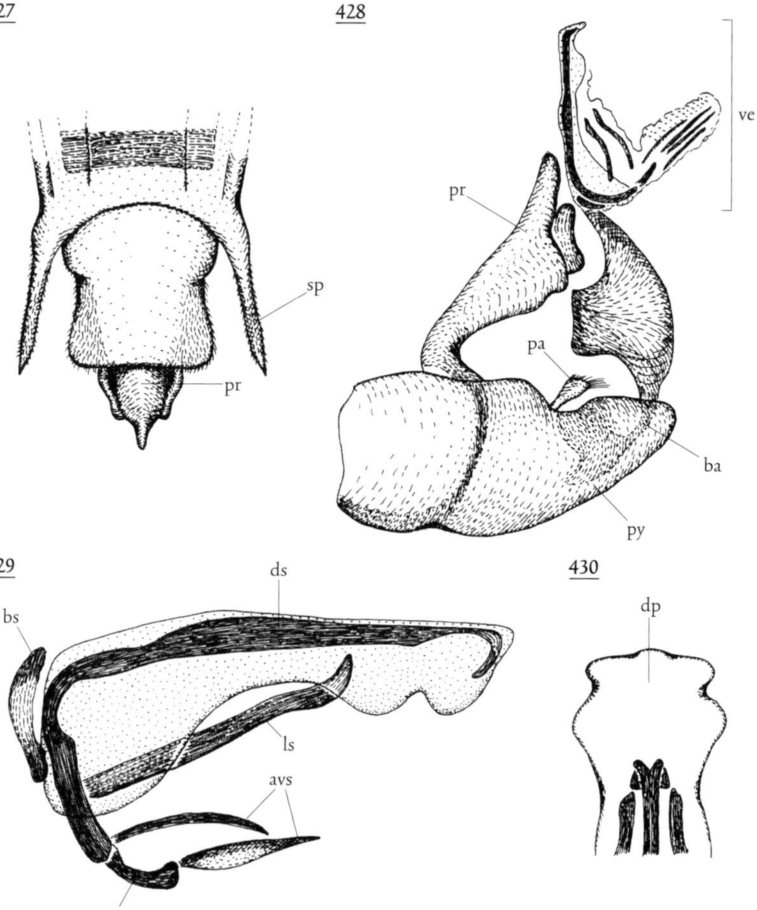

FIGURES 427 – 430
Aquarius adelaidis, male. **427** apex of abdomen, dorsal view – **428** genital segments, lateral view – **429** vesical sclerites, lateral view – **430** apex of vesica; avs = acessory ventral sclerites, ba = basal articulatory apparatus of aedeagus, bs = basal sclerite, dp = dorsal plate, ds = dorsal sclerite, ls = lateral sclerite, pa = paramere, pr = proctiger, py = pygophore, sp = connexival spine, ve = vesica, vs = ventral sclerite.

Distribution in Malesia (fig. 595)

Aquarius, with sixteen species, is chiefly Holarctic, but six widely disjunct species are known from the Southern Hemisphere. Three species occur in Malesia (Andersen, 1990; D. Polhemus & J. Polhemus, 1994; Zettel & Ruiz, 2003).

Biological notes

Aquarius species inhabit both lotic and lentic habitats. *Aquarius adelaidis*, confined to lentic habitats, has often been recorded from larger stagnant waters such as fish ponds, canals, lakes, temporary pools and flooded fields. (Cheng & Fernando, 1969; Andersen, 1990). *Aquarius lili* has been found on quiet pools in streams (D. Polhemus & J. Polhemus, 1994).

Genus *Limnogonus*

Figures 431-436

431

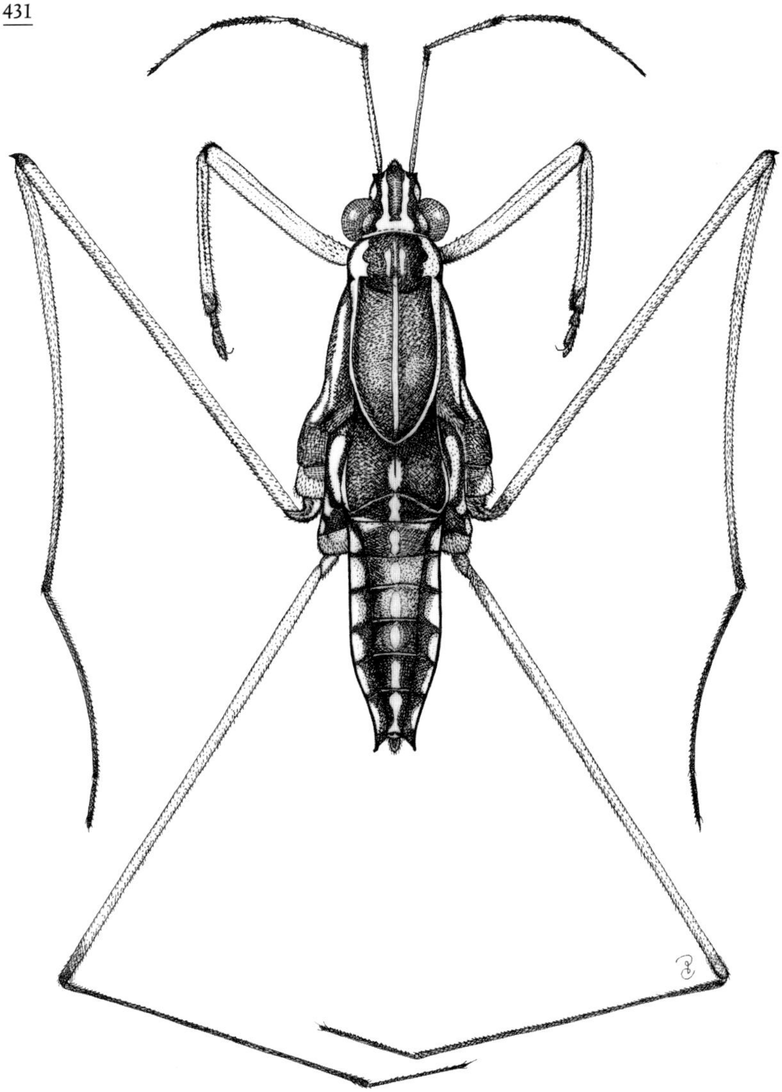

FIGURE 431
Limnogonus fossarum gilguy, apterous female, length 8.9 mm.

Diagnosis

Small to medium-sized (males 4.2-9.6, females 5.5-10.6 mm). Elongate, macropterous morph larger than apterous morph. Dorsally predominantly dark, blackish in most species; venter yellowish. Head typically with pair of yellow sublateral stripes. Pronotum in most species with longitudinal median yellow stripe and yellow margins. Pleura dark and yellowish, covered by silvery pubescence. Antennae slender, much shorter than body. Rostrum slender, fairly long, clearly surpassing hind margin of prosternum. Pronotal lobe prolonged, in apterous morph covering mesonotum. Prosternum with distinct median pit. Suture between mesothorax and metathorax in lateral view indistinct. Fore wings with two closed apical cells, without any distinct vein segment in common. Metasternal scent gland well-developed. Fore femur longer than tibia. Middle leg longer than hind leg. Hind claws in adults totally reduced. Abdomen in most species only moderately shortened. Connexiva usually not terminated by spines (fig. 432). Hind margin of sternite 7 simply concave or with median projection of varying shape and length. Phallotheca (fig. 434) boat-shaped, with dorsal bridge. Proctiger (fig. 433) narrow, without lateral projections. Paramere (fig. 433) symmetrical and very small, usually with dense short hairs. Vesical armature very complicated, with well developed conjectivum and sclerites (fig. 435): basal sclerite distinct; dorsal sclerite separated at apical half and proximal part; pair of subtriangular-shaped ventral sclerites partly fused with basal plate, occasionally also with pair of thin accessory lateral sclerites; in some species with pair of basal lobes connected with both conjectivum and vesica (fig. 435, bl), which are apically darkened and have sculptured pegs. Dorsal plate (figs. 435, 436) of vesica well-developed. Hind margin of sternite 7 in females simply concave or medially produced. Both pairs of gonapophyses basally partly sclerotized.

Limnogonus in Malesia is divided into *Limnogonus* s.str. and *Limnogonus* (*Limnogonoides*). In *Limnogonus* s.str. the anterior pronotal part usually has a pair of yellowish marks, whereas in subgenus *Limnogonoides* the pronotal lobe is usually dull and the anterior part of the pronotum lacks the yellowish marks.

Distribution in Malesia (fig. 595)

Limnogonus is a pantropical genus. Six species are found in Malesia. *Limnogonus* and *Tenagogonus* are the only freshwater gerrids extending into the Pacific area. *Limnogonus* s.str. is distributed throughout the Malesian area, with three species in Indonesia and four in New Guinea. Subgenus *Limnogonoides* (eight species) occurs in the Ethiopian region, with a single species (*L. pectoralis*) in India and Sri Lanka. This species has also been found in Malesia (Phuket, Thailand; Andersen, 1975).

432

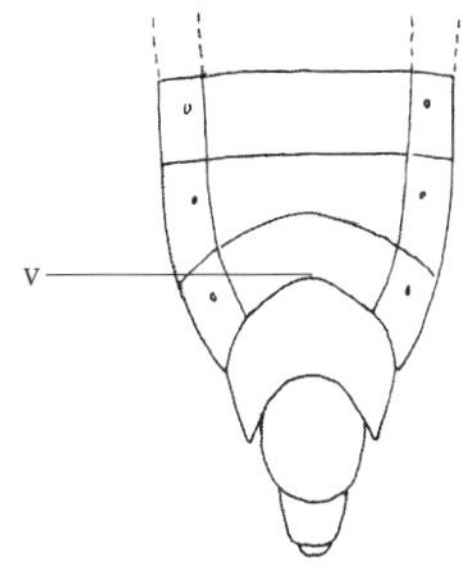

433

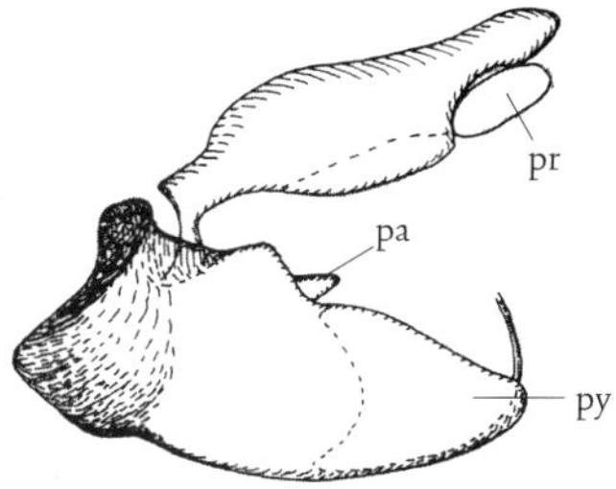

434

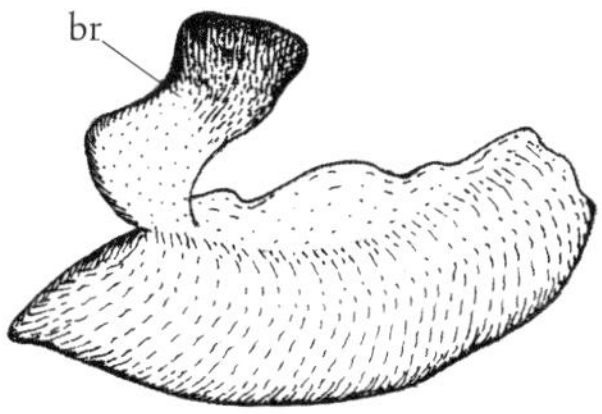

435

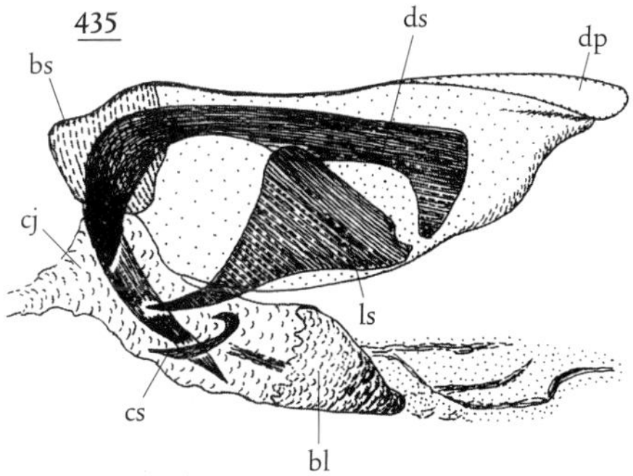

436

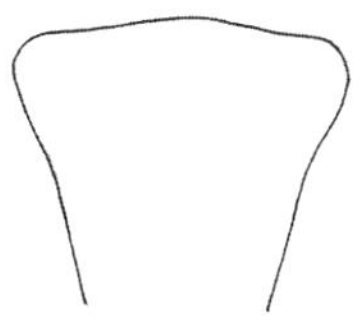

FIGURES 432 – 436 *Limnogonus fossarum gilguy*, male. 432 apex of abdomen, ventral view – 433 genital segments, lateral view – 434 phallotheca, lateral view – 435 vesical sclerites, lateral view – 436 dorsal plate of vesica, dorsal view; bl = basal lobe of vesica, br = dorsal bridge of phallotheca, bs = basal sclerite, cj = conjectivum, cs = conjunctival sclerite, dp = dorsal plate, ds = dorsal sclerite, ls = lateral sclerite, pa = paramere, pr = proctiger, py = pygophore, v = v-shaped emargination of hind margin of sternite 7.

Limnogonus s.l. includes many important colonizers of temporary habitats, and the frequency of winged individuals is generally higher than in other groups of waterstriders. Some species may disperse by air currents to isolated islands.

Biological notes

Limnogonus species are somewhat gregarious. Most species prefer relatively sheltered places on standing water. Widespread species like *L. fossarum* and *L. nitidus* colonize a great variety of habitats, including lakes, water reservoirs, rice fields, temporary pools, slow flowing streams, sulphur pools, hot springs and brackish water. Among tropical Gerridae these two species are the most successful colonizers of temporary habitats. When disturbed they readily fly off. Individuals are frequently attracted to light. Spontaneous flights are probably associated with rains.

In the laboratory *L. f. fossarum* deposited eggs in batches of about 10 eggs under leaves and other floating objects. A single female could lay about 100 eggs in 1-2 months. The relative frequency of winged and wingless individuals varies between species and may also show geographical variation within a single species (Andersen, 1975).

Food consists of various animal organisms but insects are the main prey. Emerging mosquitoes, midges, mayflies, damselflies, as well as terrestrial insects are caught on the water surface (Andersen, 1975). The feeding habits and morphology of the alimentary canal of *L. f. fossarum* were studied by Cheng (1967a). Prey was grabbed with one or both fore legs, after which the rostrum was inserted. When feeding commenced the prey was usually held with the rostrum only. Cheng confirmed that a strong histolytic enzyme is inserted and that the prey is nearly entirely cleaned by the maxillae. See also the biological notes under Gerridae.

Genus *Limnometra*

Figures 437-441

437

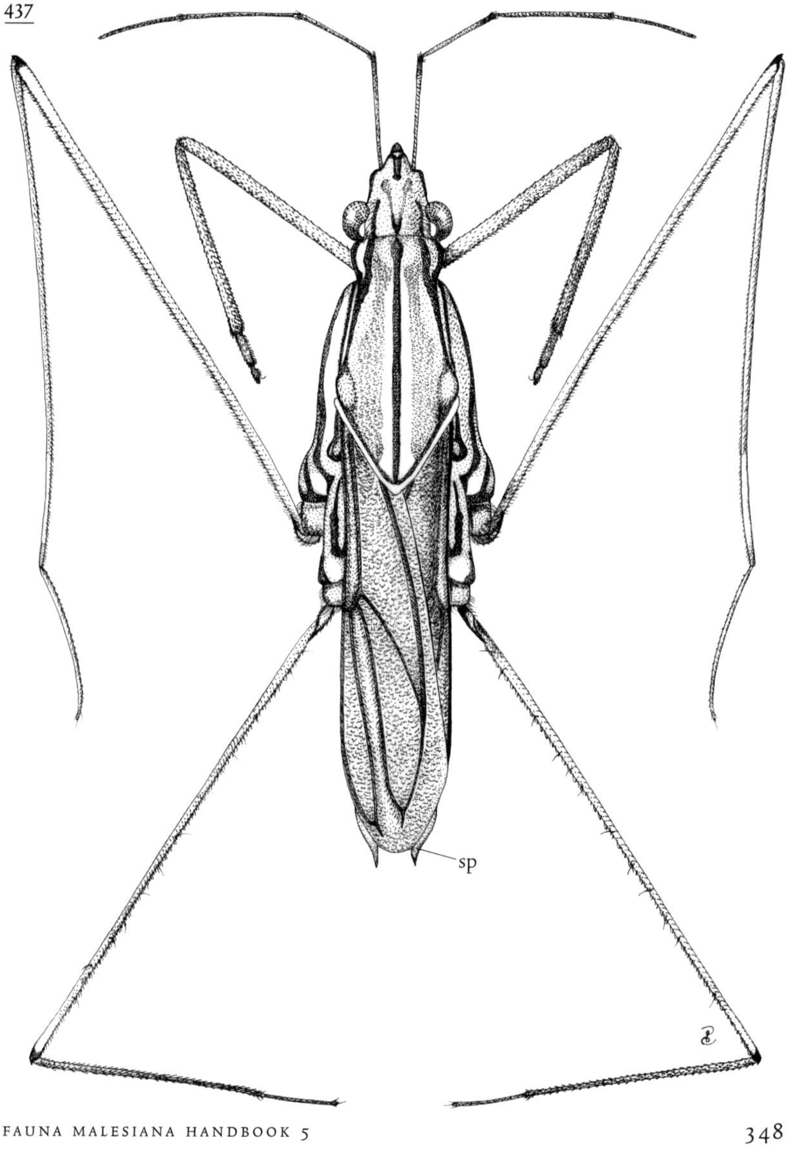

FIGURE 437
Limnometra aploa, paratype, macropterous female, length 11.4 mm; sp = connexival spine.

Diagnosis

Medium-sized to large (males 6.9-23, females 7.6-24.2 mm), in most species males larger than females. Some species monomorphic macropterous, other species dimorphic with macropterous and apterous or brachypterous morphs. Reddish-brown to yellowish-brown. Head with V-shaped dark median markings, always with pair of dark lateral and median markings. Pronotum with longitudinal dark medial stripe over entire length; anterior lobe of pronotum sometimes with pair of sublateral dark markings. Fore femora usually pale without longitudinal dark stripes. Ventral surface pale. Antennae long and very slender. First antennal segment longest, but much shorter than second and third segments together; fourth segment longer than third. Rostrum slender, reaching anterior part of mesosternum. Metathoracic spiracle located more than its own length from wing basis. Mesosternum 2.9-3.4 times as long as metasternum. Metasternal scent orifice transversely elongated, located near posterior margin of metasternum, enclosed between pair of low transverse tubercles; omphalial groove absent. Fore femur of male slender, about 1.5 times as thick as middle femur; first tarsal segment subequal to or longer than second segment. Middle femur as long as (female) or slightly longer (male) than body; tibia slightly shorter than femur; tarsus about 0.3-0.4 times as long as tibia; claws small but distinct. Hind femur subequal to or longer than middle femur; tarsus 0.2-0.3 times the tibial length; claws reduced. Fore wings of macropterous morph reaching abdominal end in both sexes; second bifurcation of M+Cu originating from point of origin of anterior crossvein. Male abdomen long, almost parallel-sided; posterior corners of connexiva produced, forming distinct connexival spines. Abdominal venter more or less carinated along longitudinal midline; hind margin of seventh segment usually concave. Male genital segments simple (fig. 438). Parameres very small (fig. 438). Phallotheca with dorsally sclerotized bridge (figs. 439, 440). Conjunctivum with lateral groups of pegs (fig. 441). Vesica in some species with highly variable armature, usually composed of single dorsal sclerite, fused and widened in middle but separated distally and proximally, two lateral sclerites and pair of ventral sclerites (fig. 441). Female abdomen long, parallel-sided or posteriorly slightly tapering. Gonocoxae small, plate-shaped, ventrally exposed. First gonapophyses relatively long, well sclerotized, apices membranous; second gonapophyses membranous and lobate, connected to first by sclerotized bridge. See also diagnosis of *Tenagogonus*.

Distribution in Malesia

So far 28 species are listed for *Limnometra*. Representatives are restricted to the Oriental and Australian regions as far eastward as the Fiji Islands. Malesia holds 25 species. Speciose regions within Malesia are Sulawesi with eight species, peninsular Malaysia with six and New Guinea with five.

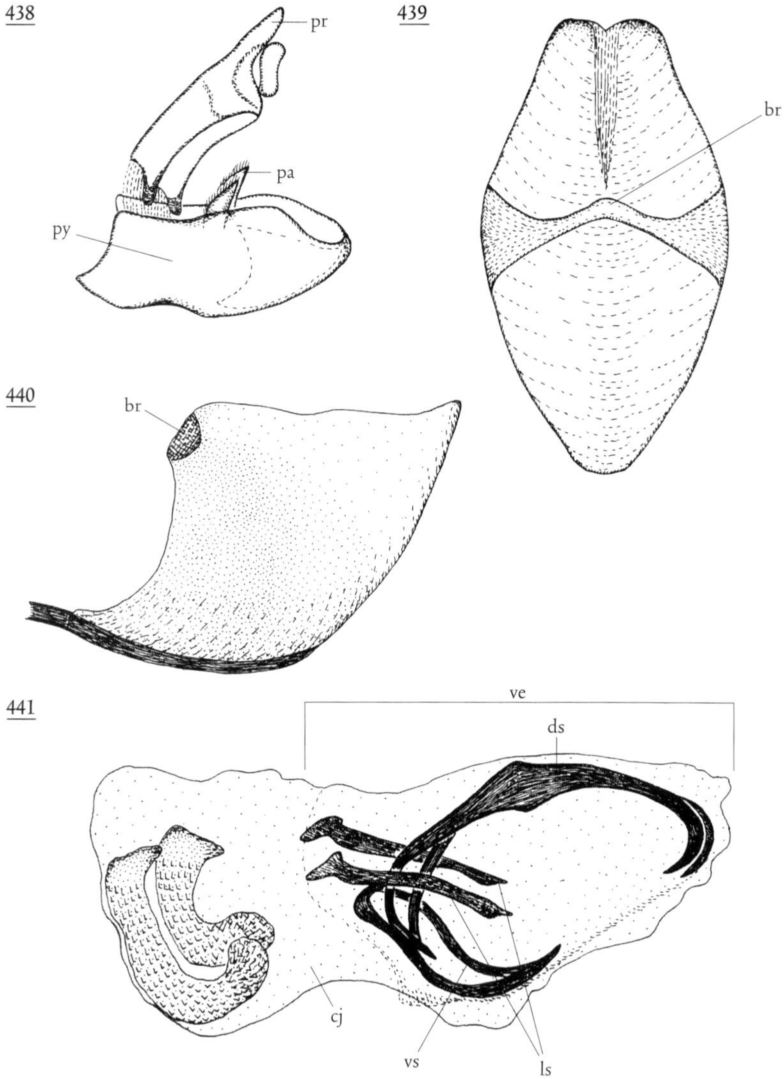

FIGURES 438 – 441
Limnometra aploa, male. **438** genitalia, lateral view – **439** phallotheca, dorsal view – **440** phallotheca, lateral view – **441** conjunctivum and vesica, lateral view; br = bridge, cj = conjunctivum, ds = dorsal sclerites, ls = lateral sclerites, pa = paramere, pr = proctiger, py = pygophore, ve = vesica, vs = ventral sclerites.

Biological notes

Limnometra species are commonly seen on the surface of streams in South-east Asia. Most species occur on sheltered pools of small shaded forest streams. At various locations two species of *Limnometra* were found on the same stream.

Genus *Neogerris*

Figures 442-446

Diagnosis

Small (males 3.6-5.2, females 4.2-6.3 mm) elongate, macropterous or apterous. Macropterous morph larger than apterous morph. Dorsally predominantly

442

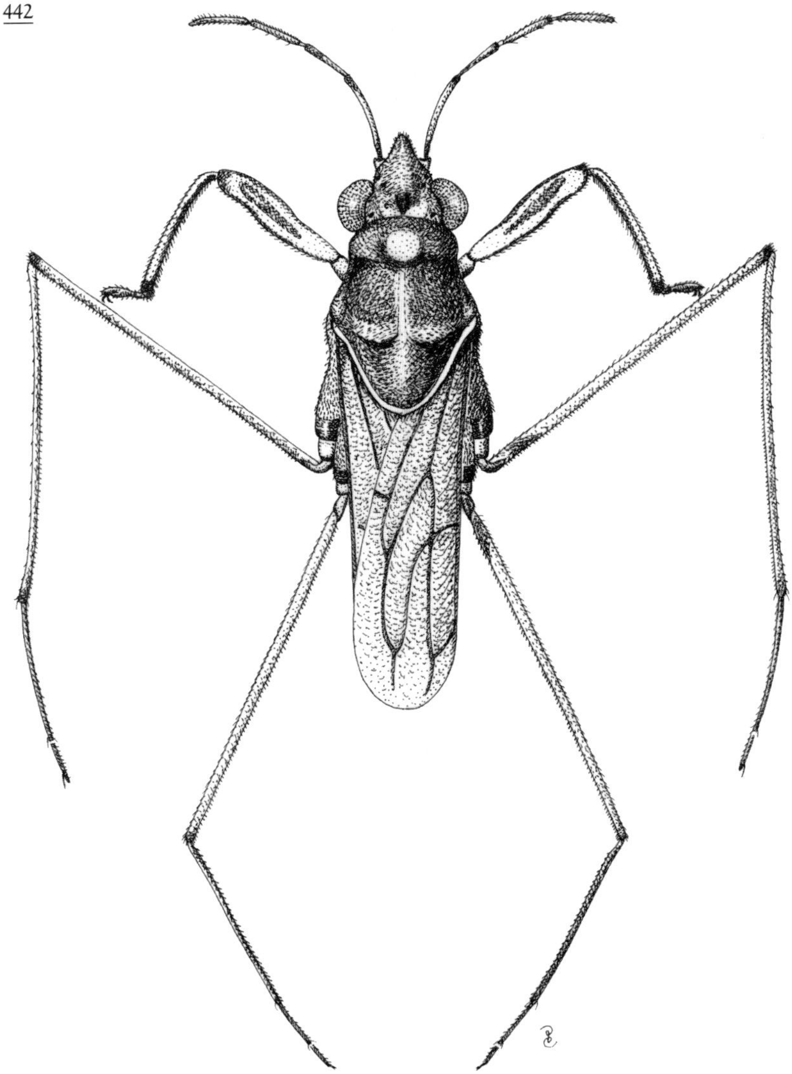

FIGURE 442
Neogerris parvulus, macropterous male, length 5.8 mm.

dark, blackish in most species. Venter usually pale, omphalium dark. Head typically with yellowish U-shaped band. Pronotum with oval or roundish spot in front; pronotal lobe dark with yellowish margins. Pleuron uniformly dark with silvery bands. Head directed forward, eyes large and globular. Antennae relatively short and robust, first segment shorter than or subequal to second and third segments together; fourth segment only slightly longer than third. Rostrum slender, just surpassing hind margin of prosternum. Prosternal pit missing. Mesothoracic and metathoracic suture laterally lost. Second spiracle removed more than twice its own length from wing base. Fore wings with two closed cells apically, without any distinct vein segment in common. Wing rudiments in flightless individuals only occasionally present. Metasternal scent gland apparatus well developed. Ventral pubescence of fore femur consisting of short, dense hairs and several longer dark bristles. Middle leg longer than hind leg. First pretarsal claw well-developed; middle and hind claws in adults totally reduced, distinct in nymphs. Pretarsal arolia of all legs absent, present in nymphs. Abdomen moderately shortened in most species. Connexiva never terminating in spines. Male: hind margin of sternites 7 and 8 not modified; pygophore subrectangular (fig. 443); proctiger narrow (fig. 445), usually without lateral projections; parameres symmetrical and very small (fig. 443); basal articulatory apparatus with additional piece (fig. 444); vesica (fig. 446) without

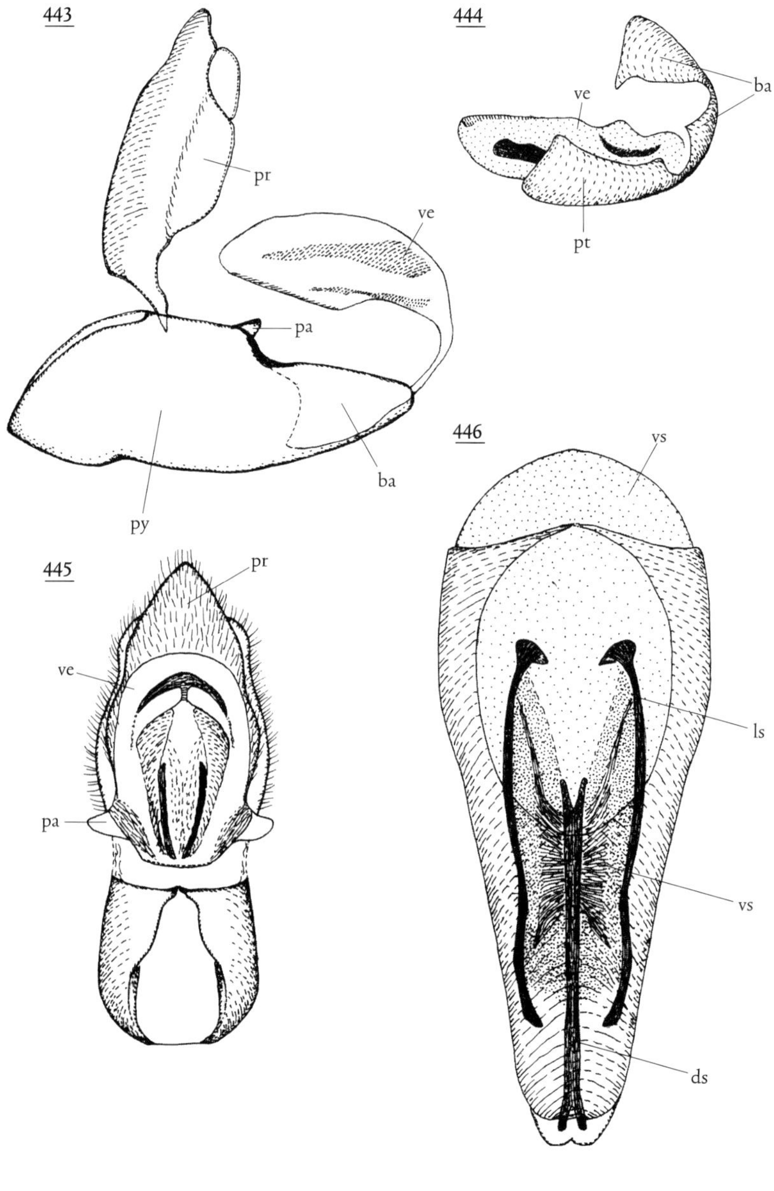

FIGURES 443 – 446
Neogerris parvulus, male.
443 genital segments, lateral view – 444 phallus, lateral view – 445 genital segments, dorsal view – 446 phallotheca and vesica, dorsal view; ba = basal articulatory apparatus, ds = dorsal sclerite, ls = lateral sclerite, pa = paramere, pr = proctiger, pt = phallotheca, py = pygophore, ve = vesica, vs = ventral sclerite.

basal sclerite, dorsal sclerite seldom prolonged proximally, ventral sclerites plate-shaped, lateral sclerites well developed and elongate. Female: hind margin of sternum 7 in middle produced to varying degree; both pairs of gonapophyses basally relatively well sclerotized (Andersen, 1975).

Distribution in Malesia

The genus is chiefly pantropical, with greatest diversity in the Neotropical region (eight species; Nieser, 1994). The Oriental region has three species, of which two are found in Malesia.

Biological notes

Neogerris species are gregarious, like *Limnogonus*. Most species prefer relatively sheltered places on standing water. The widespread N. *parvulus* colonizes a great variety of habitats. It is probably one of the most successful colonizers of temporary habitats among tropical Gerridae (Andersen, 1975).

Genus *Tenagogonus*

Figures 447-452

Diagnosis

Small to medium-sized (males 5.2-8.5, females 6.7-9.9 mm), macropterous or apterous. Reddish-brown to yellowish-brown. Head with median, V-shaped dark markings. Median longitudinal dark stripe over entire length of pronotum; anterior lobe sometimes with pair of sublateral dark markings. Fore femora usually pale, with or without longitudinal dark stripes. Ventral surface pale. Antennae long and very slender. First antennal segment longest but much shorter than second and third segments together; second segment shorter than third segment; fourth segment subequal to first segment. Rostrum slender, reaching anterior part of mesosternum. Metathoracic spiracle located more than its own length from wing basis. Metasternal scent orifice transversely elongated, located near posterior margin of metasternum, enclosed between pair of low, transverse tubercles; omphalial groove absent. Fore wings surpassing abdominal end in both sexes; second bifurcation of M+Cu approaching point of origin of anterior cross-vein. Fore leg longer than length of body; fore femur of male moderately thickened, about 1.5 times as thick as middle femur; first tarsal segment shorter than second. Middle femur as long as body; tibia slightly shorter than femur; tarsus about 0.4 times as long as tibia; claws small but distinct. Hind femur subequal to or longer than middle femur; tarsus 0.25 times tibia length; claws reduced. Male abdomen relatively short. Posterior corners of connexiva obtuse, never forming distinct connexival spines (fig. 447). Abdominal venter more or less carinate along longitudinal midline; hind margin of seventh sternum simple and concave. Male genital segments ventrally in some species more or less modified (fig. 451); parameres strongly reduced

447

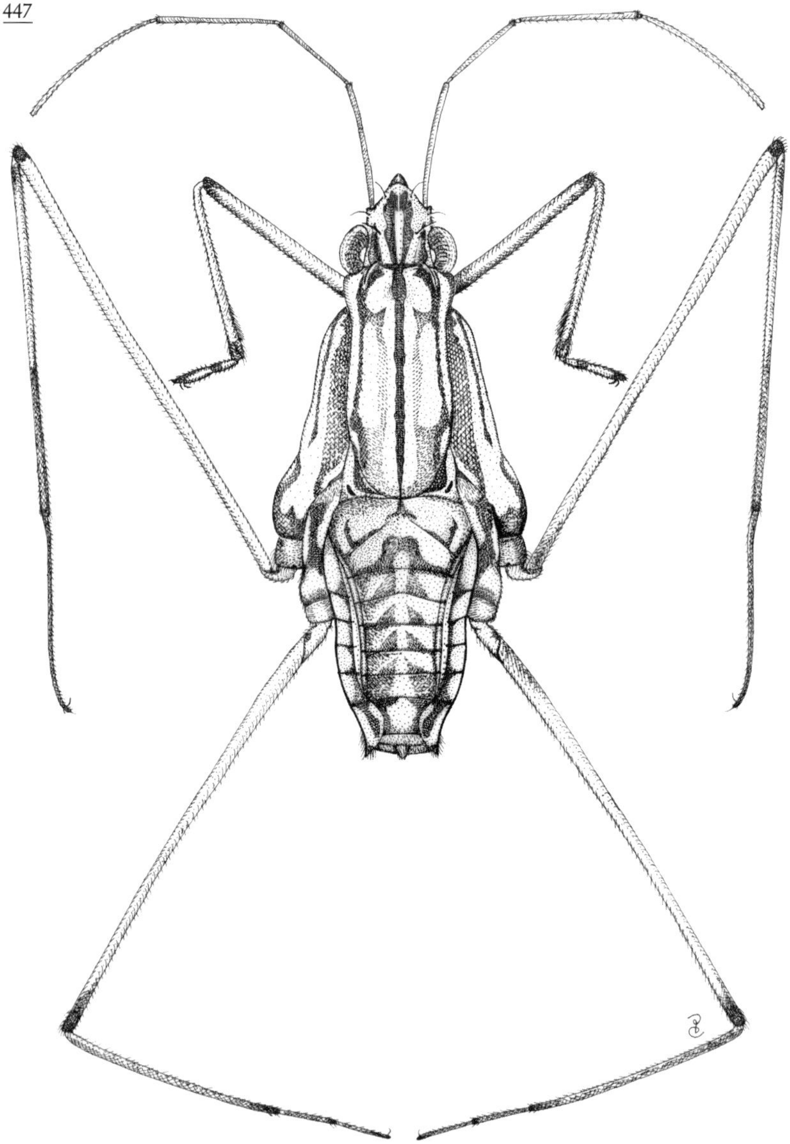

FIGURE 447
Tenagogonus akanthinos, paratype, apterous female, length, 7.0 mm.

(fig. 448); phallotheca with dorsal sclerotized narrow bridge (fig. 452); conjunctivum (fig. 449) with lateral groups of pegs; phallus with basal plate and dorsal plate posteriorly distinctly prolonged; vesica (fig. 449): dorsal sclerite well developed, apical accessory sclerites plate-shaped, lateral sclerites membranous, ventral sclerite small and membranous. Female abdomen longer than in male, in some species posterior corners of connexiva obtuse or produced into short spinose processes; seventh sternum with straight or slightly produced hind margin; gonocoxae small, plate-shaped, ventrally exposed; first gonapophyses relatively short and weakly sclerotized, apices membranous; second

448

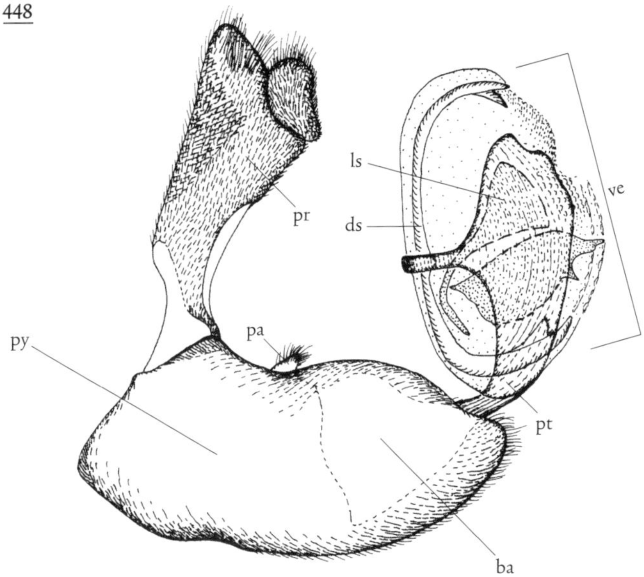

449

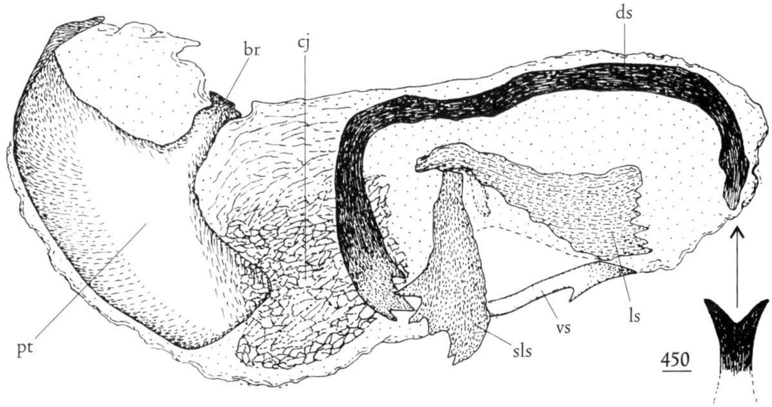

450

451

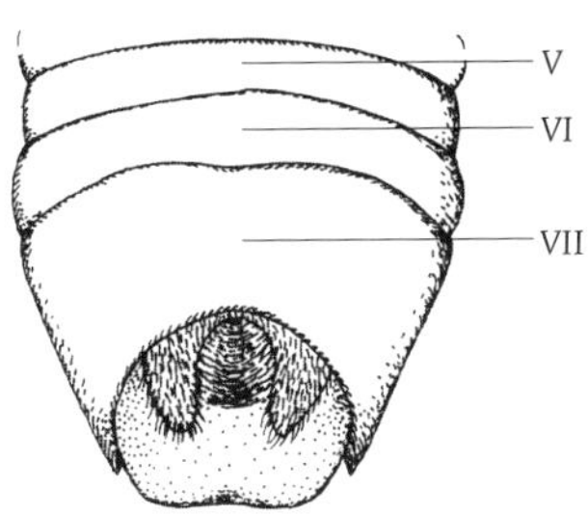

452

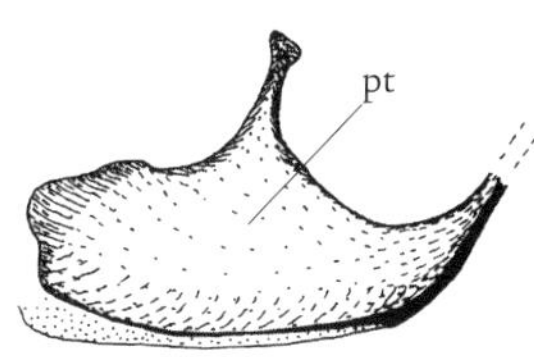

FIGURES 448 – 452 *Tenagogonus akanthinos*, male. **448** genital segments, lateral view – **449** endosoma, lateral view – **450** detail of apex of dorsal sclerite, frontal view – **451** apex of abdomen, ventral view – **452** phallotheca, lateral view; ba = basal apparatus of phallus; br = dorsal bridge of phallotheca, cj = conjunctivum, ds = dorsal sclerite, ls = lateral sclerite, pa = paramere, pr = proctiger, pt = phallotheca, py = pygophore, sls = second pair of lateral sclerites, ve = vesica.

gonapophyses membranous and lobate, sometimes connected by sclerotized bridge. *Limnometra* and *Tenagogonus* are closely related. They are morphologically very similar but differently shaped: elongated in *Limnometra*, short and robust in

Tenagogonus. *Tenagogonus* has a relatively short abdomen (and connexival spines), *Limnometra* has a long abdomen (and connexival spines).

Distribution in Malesia

So far fifteen species are recognized. The genus is distributed throughout the Afrotropical, Oriental and Australian regions, extending eastward to the Fiji Islands. Seven species are distributed throughout Malesia, but most areas have only two species.

Biological notes

Tenagogonus species prefer rather small streams in forests, as do many species of *Limnometra*.

REFERENCES – Gerrinae. Revision of *Limnometra* and *Tenagogonus* (Hungerford & Matsuda, 1958a); structural analysis and classification of Gerridae (Matsuda, 1960); feeding in *Limnogonus fossarum* (Cheng, 1967a); revision of Old World *Limnogonus* and *Neogerris* (Andersen, 1975); morphology, phylogeny, adaptations, biogeography and classification of Gerromorpha (Andersen, 1982a); revision of *Aquarius* (Andersen, 1990); revision of *Limnometra* in the Malay Archipelago (Nieser & Chen, 1992c); checklist, key to genera and distribution of Gerrinae (Andersen, 1995b); revision of Australian Gerrinae (Andersen & Weir, 1997); *Limnometra* of New Guinea (D. Polhemus & J. Polhemus, 1997); survey of Philippine *Limnometra* (Zettel & Chen, 2000a).

SUBFAMILY EOTRECHINAE

Diagnosis

Small to medium-sized (3.2-10.8 mm), apterous or macropterous. First antennal segment usually with preapical spine-like hairs. Pronotum of apterous morph short, posterior lobe completely reduced; intersegmental suture between mesonotum and metanotum laterally distinct. Metasternum well developed, scent gland ostiole present, lateral evaporative channel absent. Legs relatively long and slender. Middle and hind legs with rows of spinous hairs on tibia and tarsi. Claws inserted apically or close to apex of second tarsal segment. Female genital segments with distinctly bilobed first gonapohyses, inner lobe sclerotized and hairy, second gonapophyses membranous. Vesica of male with relatively large roundish apical sclerites and membranous ventral sclerites (Andersen, 1980).

Distribution worldwide

Subfamily Eotrechinae is exclusively Oriental and only recorded west of Wallace's Line, except for one species from Mindanao. Five genera, holding 60 species, are described in the subfamily, of which two occur in Malesia. The largest genus,

Amemboa with 29 species, is distributed from India eastward to the Philippines, and from Taiwan southward to Borneo, Sumatra and Java. Each described species has a very limited range, with a few exceptions. The species diversity is greatest in Thailand and adjacent countries (23 species). *Onychotrechus* has ten described species of which two occur in peninsular Malaysia.

Eotrechinae can be divided in two ecological groups: genera inhabiting hygropetric and other wet terrestrial habitats (e.g. *Onychotrechus*) and genera occupying the water surface (e.g. *Amemboa*).

Key to genera and subgenera of Eotrechinae in continental South-east Asia and Malesia

1 Hind leg distinctly shorter than middle leg. Hind tibia and tarsus together less than 3/4 of length of hind femur ▸ 2

– Hind leg about equal in length to middle leg. Hind tibia and tarsus together more than 3/4 of length of hind femur ▸ ***Onychotrechus* p. 362**

2 First segment of middle tarsus about twice as long as second segment. Male fore femur ventrally modified, bearing tufts or patches of black hairs and/or tubercles. Metasternal scent gland orifice usually not situated on elevated area (omphalium). Male pygophore usually apically modified, proctiger with very long lateral arms, parameres always rudimentary ▸ ***Amemboa* s.str. p. 357**

– First segment of middle tarsus distinctly less than twice as long as second segment. Male fore femur basally incrassate but not ventrally modified, except for rows of short spine-like hairs. Metasternal scent gland orifice situated on elevated area in both sexes. Male pygophore usually simple, proctiger without distinct lateral arms, parameres often large ▸ **subgenus *Amemboides* p. 361**

Genus *Amemboa*

Figures 453-466

Diagnosis

There are two subgenera; see below.

Distribution in Malesia (fig. 596)

Of *Amemboa* s.str. eight species are found in Malesia, of which six occur in continental Malaysia. Three species extend from continental Malaysia to Borneo, Java and Sumatra. Of the remaining two species, one is endemic on Borneo and one is endemic on Mindanao. The single species of subgenus *Amemboides* is recorded from continental Malaysia.

453

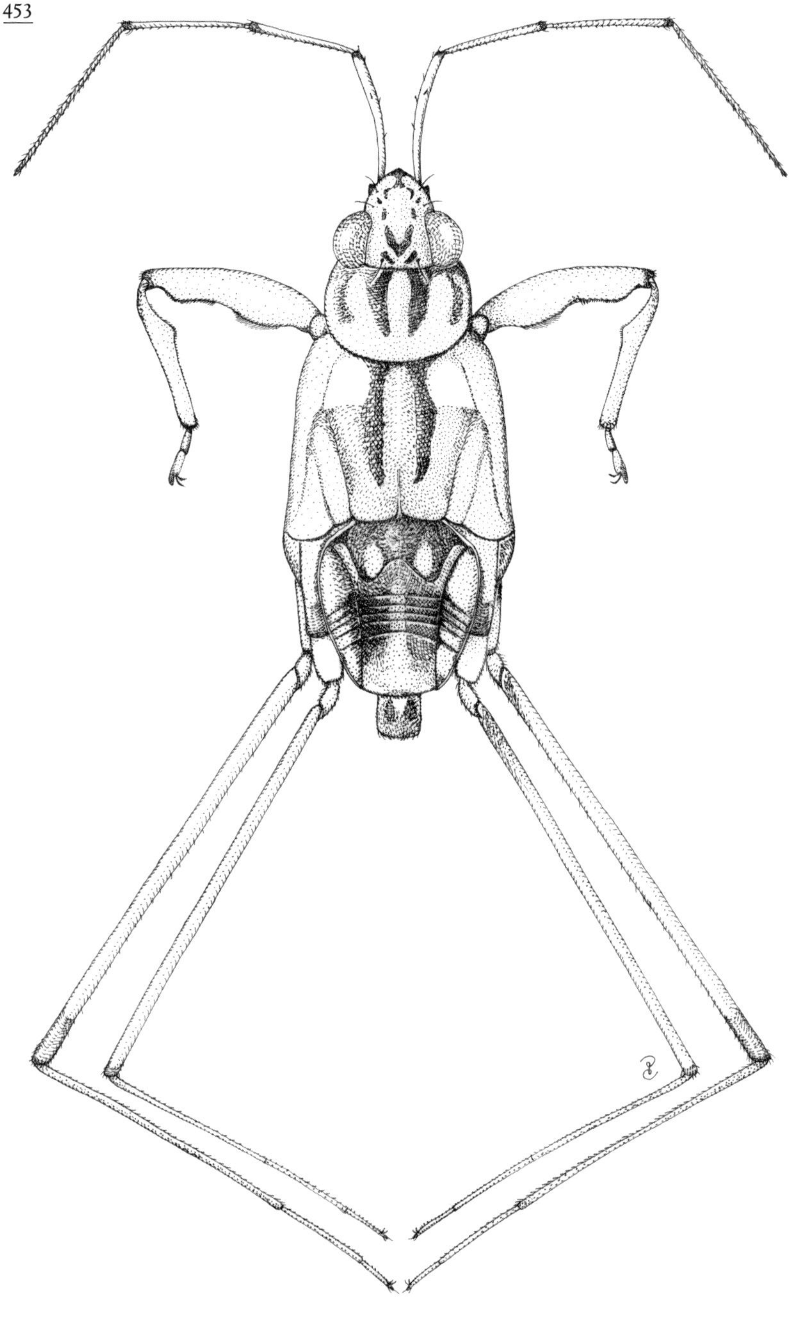

FIGURE 453
Amemboa cristata,
apterous male,
length 4.0 mm.

Biological notes

Many species belonging to the subgenus *Amemboa* live near springs, small pools or puddles in sandy or muddy stream beds or on river banks. Some are found on rockpools along torrents or below waterfalls. In their preferred habitats, species always stay close to the water edge and are usually found resting on the shore. Only when pursued they jump or skate on the water. To catch these bugs, a net is set at the edge of the pool after which the bugs are chased into it by waving a hand towards the opening. In the net *Amemboa* jump about with great agility. The single species of subgenus *Amemboides* prefers more secluded habitats and can be collected from humid or water-filled holes in rocks adjacent to streams or torrents. Individuals are rarely seen on the water and when pursued usually retreat to their holes. In the shade only the silvery markings are visible. Some *Amemboa* species can be found on seeping rock faces together with *Onychotrechus* (J. Polhemus & Andersen, 1984).

Amemboa sensu stricto

Diagnosis

Small (males 3.2-4.0, females 3.5-5.2 mm). Dorsal and pleural surface of body yellow to orange-yellow with brown or black marks, great variation in dorsal pattern. Pronotum of apterous morph has two median and two sublateral dark marks, macropterous morph with anterior lobe of pronotum entirely dark or with two median and two sublateral dark marks; posterior lobe light. Antennae subequal to body; first three segments subequal in length, fourth segment longest; first segment set with 8-11 short dark spines scattered over entire length. Rostrum long and slender (fig. 454), reaching to anterior $^1/_4$ of mesosternum; first and second labial segments small, together much shorter than fourth labial segment. Pronotum slightly wider than head (measured across eyes); intersegmental suture between mesothorax and metathorax laterally fairly distinct. Mesosternum 8-10 times as long as metasternum; scent gland orifice slit-shaped, situated on posterior part of metasternum, usually not on elevated area. Wings (figs. 455, 456) surpassing abdominal apex in both sexes. The two apical fore wing cells share common vein along basal part. Fore femur of male slender or more or less incrassate, on ventral side with tufts of black hairs, tubercles, and/or teeth; fore tibia with median tumescence or (at least) slightly thickened. Fore leg of female not modified. Middle leg distinctly longer than hind leg; middle tibia about $^2/_3$ of femur, more than twice as long as tarsus. Hind tibia less than half as long as femur, about 1.5 times length of tarsus. First segment of middle tarsus about twice as long as second. Tibiae and tarsi of two posterior legs with ventral row of spine-like hairs; claws well-developed but very small, dorsal arolium bristle-like, ventral arolium broader. Male abdominal venter strongly depressed but without median groove, except sometimes first genital segments. Genital segments very large, strongly modified, much longer than pregenital abdominal venter. Pygophore ventrally well exposed (fig. 457, 460), distally more or less modified; proctiger usually with very long, caudally directed lateral arms (fig. 458). Parameres small, club-shaped, usually hidden. Vesica (figs. 461, 462) with dorsal sclerites fused or separated from by

FIGURES 454 – 462

Amemboa cristata

454 rostrum, lateral view – **455** hemielytron – **456** hind wing – **457** male, pygophore, lateral view – **458** male, proctiger, lateral view – **459** female, gonocoxa, lateral view – **460** male, abdominal segment 8 and pygophore, dorsal view – **461** male, phallotheca and vesica, ventral view – **462** male, phallotheca and vesica, lateral view; as = apical sclerite, ds = dorsal sclerite, pt = phalloteca, py = pygophore.

454

455

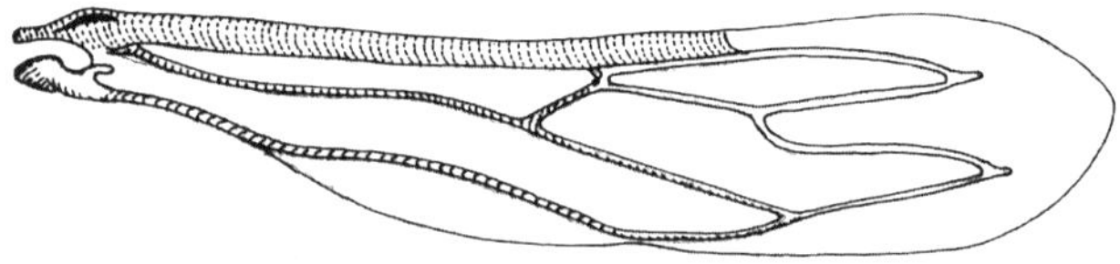

456

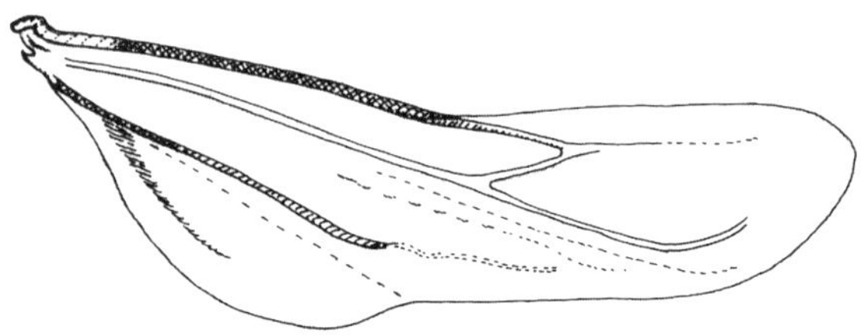

457

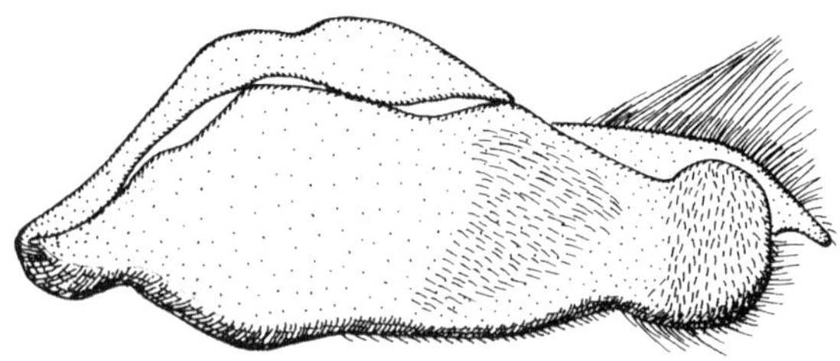

458

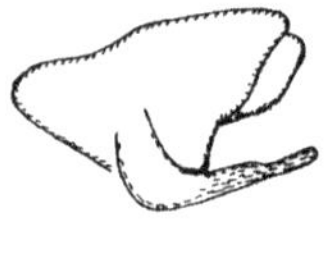

459

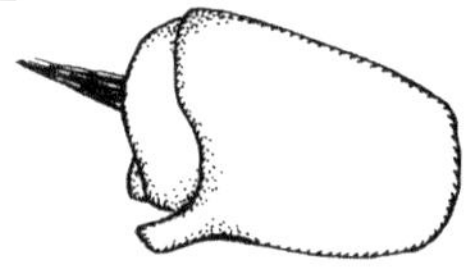

460

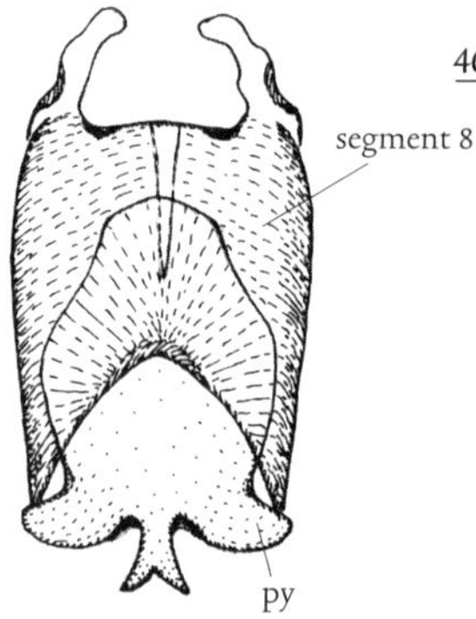

461

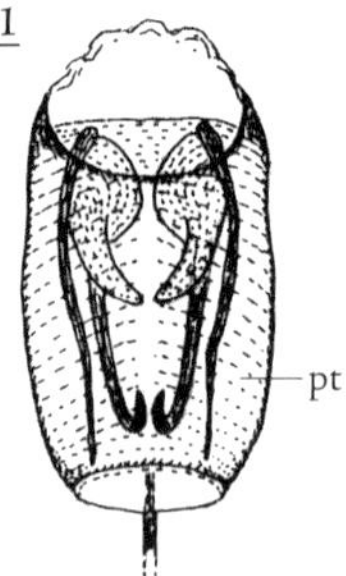

462

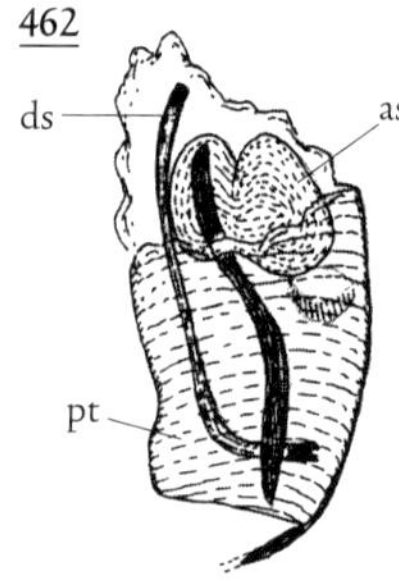

pair of round apical sclerites; lateral sclerites long, ventral lobe membranous. Female abdominal venter less than $^{1}/_{3}$ of body length. Sternum 7 longer than all preceding sterna together, with posterior margin in middle slightly produced. Genital segments lie hidden in pregenital abdomen (fig. 459) (J. Polhemus & Andersen, 1984).

Amemboa subgenus *Ameboides*

Diagnosis

Small (males 3.2-4.4, females 3.5-4.5 mm). Very similar to *Amemboa* s.str. except for the following characters: apterous morph typically with more or less continuous longitudinal light stripe, bordered brownish, medially on pronotum and mesonotum, continuing on metanotum and first abdominal tergite as silvery spots; anterior lobe of pronotum of winged morph with lighter median longitudinal stripe continuing on posterior lobe. Mesopleuron with slender brownish and silvery longitudinal stripe. All acetabulae with prominent silvery spot. First antennal segment has 3-5 dark spines in distal part. First and second labial segment longer than in *Amemboa*, together much longer than fourth labial segment. Scent gland orifice of both sexes situated on elevated area or tubercle in posterior part of metasternum. Two apical cells in fore wing (in winged morph) not sharing common vein segment along basal part. Fore femur basally incrassate, ventral side more or less extensively covered with close-set short black hairs; otherwise not modified. Fore tibia almost straight. Middle tibia about $^{3}/_{5}$ of femur length, more then twice as long as tarsus. First segment of middle tarsus less than twice as long as second segment. First segment of hind tarsus shorter than second. Male abdominal venter not medially depressed; sternum 7 only slightly longer than preceding two sternae together, posterior margin sinuated in middle. Genital segments of varying size; pygophore usually small and not modified; proctiger without lateral arms. Parameres distinct and diagnostic (fig. 463). Phallotheca sclerotized at basal half, apical half membranous (fig. 466); vesica with dorsal sclerites apically widely separated (fig. 464); ventral sclerites very small (fig. 466); lateral sclerites long; apical sclerite slightly sclerotized at apical edge (figs. 464, 465); pair of small round ventral plates at apical half (fig. 465). Female abdomen either with pregenital segments extended (as usual), or with posterior 4-5 segments shortened and folded in preceding segments and thorax.

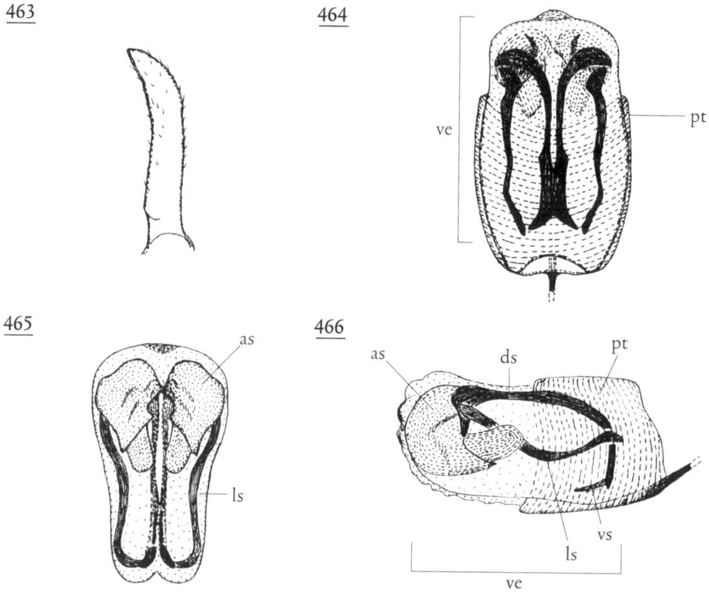

FIGURES 463 – 466
Amemboa (*Amemboides*) *velaris*, male.
463 left paramere – **464** phallotheca and vesica, dorsal view – **465** vesica, ventral view – **466** phallotheca and vesica, lateral view; as = apical sclerite, ds = dorsal sclerite, ls = lateral sclerite, pt = phallotheca, ve = vesica, vs = ventral sclerite.

Genus *Onychotrechus* – hygropetric waterstriders

Figures 467-469

Diagnosis

Small (males 3.7-6.5, females 4.8-7.4 mm), elongate. Apterous morph smaller than macropterous morph. Dorsal and pleural surface usually blackish with yellowish-brown stripes. Pronotum dark with pale median and two pale sublateral stripes; posterior lobe of winged morph blackish; hind margin between humeri with silvery pubescence. Antenna longer than body, first segment longer than second with 2-4 dark spine-like hairs in apical third (fig. 467); second and third segments subequal in length, fourth segment longer than third, subequal to first. Rostrum long and slender, in most species almost reaching middle of mesosternum. Mesosternum 6-7 times as long as metasternum. Scent ostiole orifice slit-shaped, situated in middle or posterior third of metasternum, not elevated. Fore femur slender, in male sometimes apically modified. Middle and hind legs subequal in length; hind tibia and tarsi together longer than hind femur; first segment of middle and hind tarsi shorter than second; preapically inserted claws relatively long. Venter of male more or less grooved along median line. Genital segments relatively small (fig. 468), pygophore ventrally well exposed. Paramere clearly visible (fig. 468), club-shaped, with numerous short hairs. Phallotheca well developed (fig. 469). Vesica with dorsal sclerites thick and roundish (fig. 469), lateral sclerite long, ventral sclerites membranous (Andersen, 1980).

467

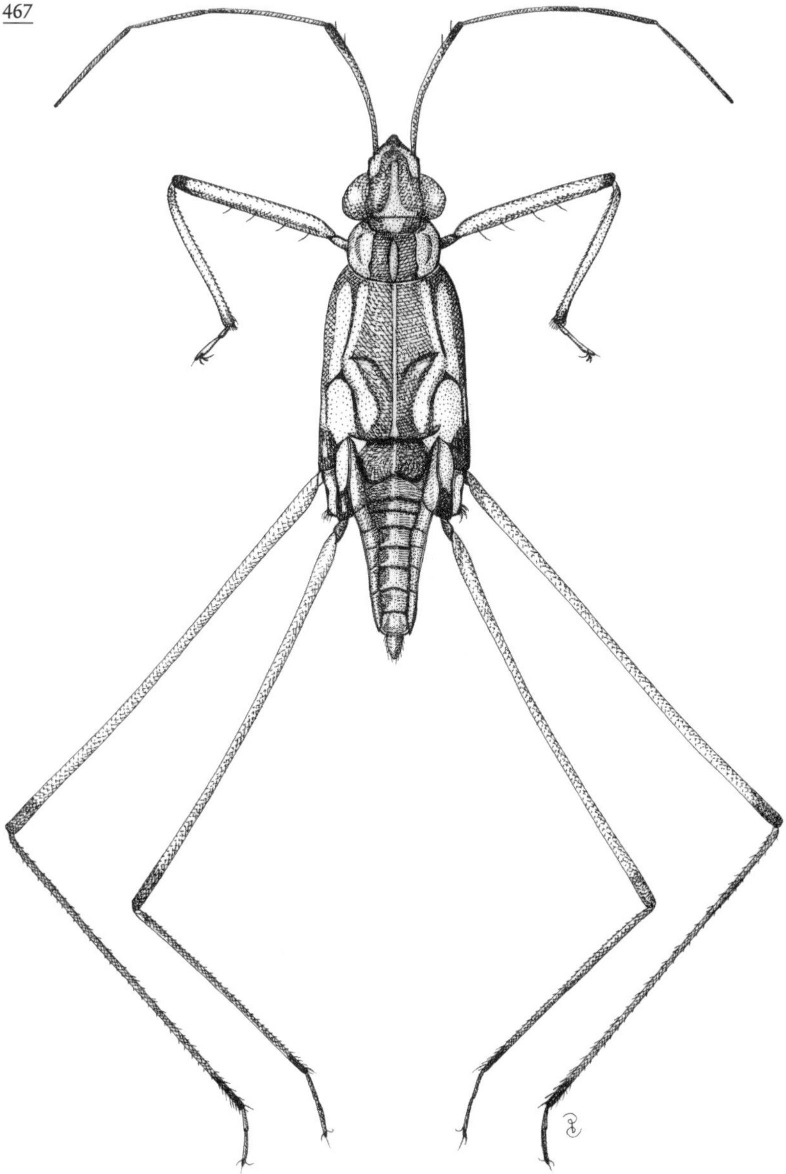

FIGURE 467
Onychotrechus esakii, apterous female, length 5.75 mm.

Distribution in Malesia (fig. 596)
Of the ten described species of *Onychotrechus* two have been found in the northern part of continental Malaysia. *Onychotrechus esakii* has a wide distribution, from Vietnam through Myanmar to the south of Thailand (Andersen, 1980). The other species, *O. pallidus*, was only known from the type series from Penang, Malaysia, but has also been collected in Perak (J. Polhemus, unpubl.).

468

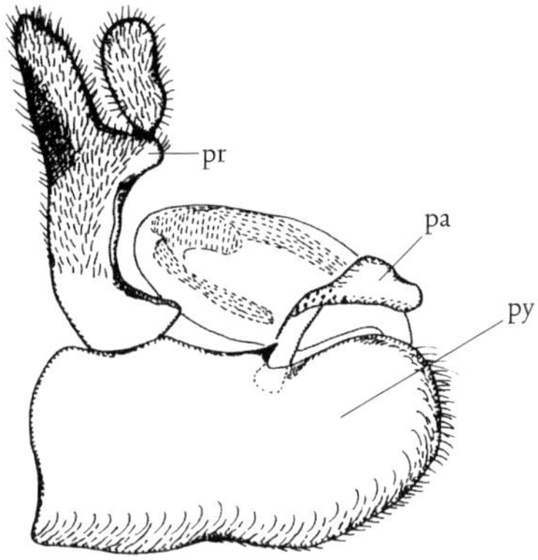

469

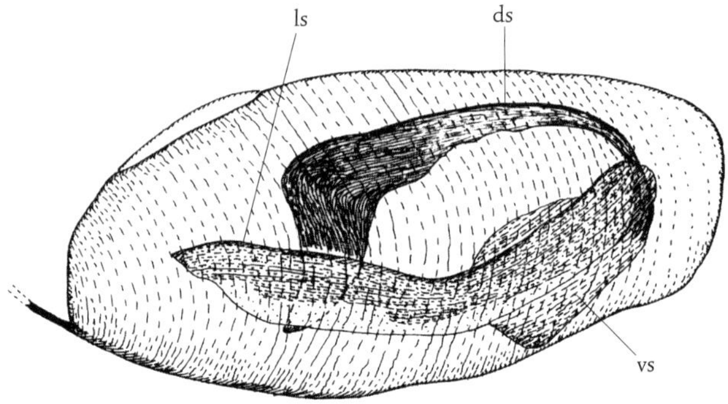

FIGURES 468 – 469
Onychotrechus esakii, male. **468** genital segments, lateral view – **469** phallotheca and vesica, lateral view; ds = dorsal sclerite, ls = lateral sclerite, pa = paramere, pr = proctiger, py = pygophore, vs = ventral sclerite.

Biological notes

Onychotrechus species are hygropetric. They live on seeping rock faces and on rocks splashed by waterfalls. They were found resting, running and jumping on rock faces covered by algae and a thin sheet of trickling water. When occurring together, different species occupy different microhabitats. The posture of resting *Onychotrechus* on a solid substrate is quite characteristic and different from that of other gerrids: the long and slender femora of the two posterior leg pairs point obliquely upwards and the tibiae are held almost perpendicular to the substrate, while the insects rests on the short tarsi. With the long middle and hind leg claws it is possible to grasp any irregularity of the substrate, thus enabling it to climb steep or even vertical slopes.

REFERENCES – Eotrechinae. Structural analysis and classification of Gerridae (Matsuda, 1960); revision of *Onychotrechus* (Andersen, 1980); morphology, phylogeny, adaptations, biogeography and classification of Gerromorpha (Andersen, 1982a); review of *Eotrechus* and *Chimarrhometra* (Andersen, 1982c); revision of *Amemboa*, phylogeny and ecology of subfamily Eotrechinae (Polhemus & Andersen, 1984); two new species of *Amemboa* from Borneo and Vietnam (Zettel, 1995f).

SUBFAMILY CYLINDROSTETHINAE

Diagnosis

Large (7.9-27.0 mm) with silvery longitudinal strips on lateral thorax. Posterior interoculus much broader than anterior part; clypeal region (fig. 477) anteriorly well produced, basal margin absent. Antennae much shorter than body. Rostrum short and thick, never reaching beyond prosternum. Pronotum in apterous morph short, posteriorly strongly widened. Intersegmental suture between mesonotum and metanotum faint, forming small subtriangular space anterior to metacabular groove. Metanotum with distinct median longitudinal sulcus; lateral longitudinal suture nearly reaching metacetabular groove (fig. 378); metacetabular groove joining posterior margin of mesonotum. Metasternum with distinct omphalium, omphalial groove leading to metacetabular region (fig. 471). Inner apical angle of fore tibia with inconspicuous process. Femur of middle leg distinctly longer than tibia, tarsus usually flatted and curved. Hind leg shorter than middle leg, femur 1-1.5 times as long as tibia. Lateral margins of proctiger in male asymmetrically modified, at posterior edge with number of short and thick spinules. Parameres reduced. Vesical sclerites: dorsal sclerite long, proximally recurved, apically fused and proximally separated; ventral sclerite long, weakly sclerotized; lateral sclerites short, in some species absent. Female second gonapophyses heavily sclerotized.

Distribution worldwide

The subfamily Cylindrostethinae contains three genera, of which *Cylindrostethus*, with fourteen species, is pantropical. The other two genera are Neotropical.

Genus *Cylindrostethus*

Figures 470-477

Diagnosis

Small to large (7.9-27.0 mm), elongate, cylindrical. Black to brown, often with yellowish stripes on pronotum, longitudinal band of silvery pubescence on mesopleuron; mesonotum often with paired longitudinal stripes. Vertex of head strongly constricted in middle towards median line; anterior interocular much narrower than posterior interocular, clypeal region produced (fig. 477). Antennae thick, relatively short; segment 1 longer than segment 2 and 3 together, segment 4 longer than segment 3. Rostrum reaching to posterior margin of head. Pronotum roughly hexagonal, shorter than head, posterior margin slightly concave, posterolateral margins oblique. Mesonotum long, posterior margin rounded and medially weakly indented, midline with incipient longitudinal sulcus. Length of mesosternum 2-3.5 times that of metasternum. Metanotum with distinct longitudinal sulcus along midline. Metasternum subequal in length to abdominal sternite 2. Fore femur incrassate, ventral side with some

470

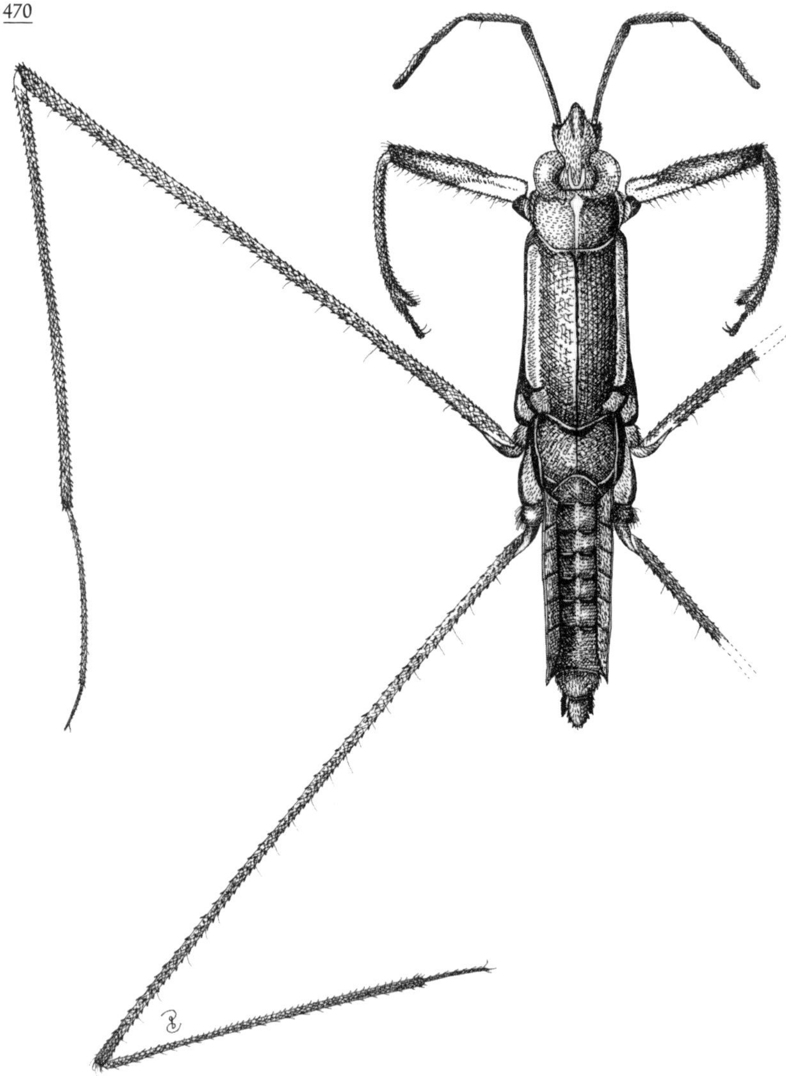

FIGURE 470
Cylindrostethus persephone,
apterous male,
length 11.8 mm.

small teeth; second segment of fore tarsus 2-4 times length of first segment; claws subapical. Middle leg longer than hind leg; middle femur subequal to or shorter than body, longer than middle tibia. Middle tibia 2-3 times longer than segment 1 of middle tarsus. Middle tarsus flattened, segment 1 at least three times as long as segment 2. Hind femur at least 1.2 times as long as hind tibia; hind tibia about half as long as hind tarsi; first tarsal segment of hind leg twice longer than segment two. Claws on middle and hind legs small and subapical (absent in Neotropical species). Abdomen long, parallel sided, connexival segment 7 ends in distinct, usually long, spine-like projection in both sexes. Male pygophore with apex rounded; shape of proctiger varied, often with asymmetrical lateral projections (fig. 474); parameres (fig. 473) small and symmetrical.

FIGURES 471 – 477
Cylindrostethus persephone
471 female, abdomen, ventral view – **472** female, head and thorax, dorsal view – **473** male, left paramere, lateral view – **474** male, proctiger – **475** male, genital segments – **476** male, vesica – **477** head dorsofrontal view; ds = dorsal sclerite, ls = lateral sclerite, og = omphalial groove, pr = proctiger, py = pygophore, vs = ventral sclerite.

471

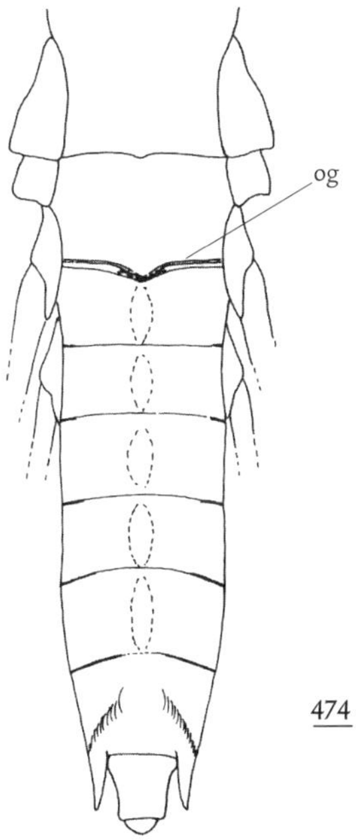

472

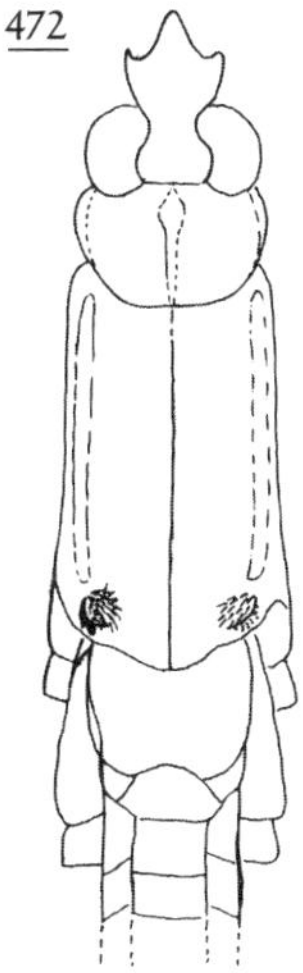

473

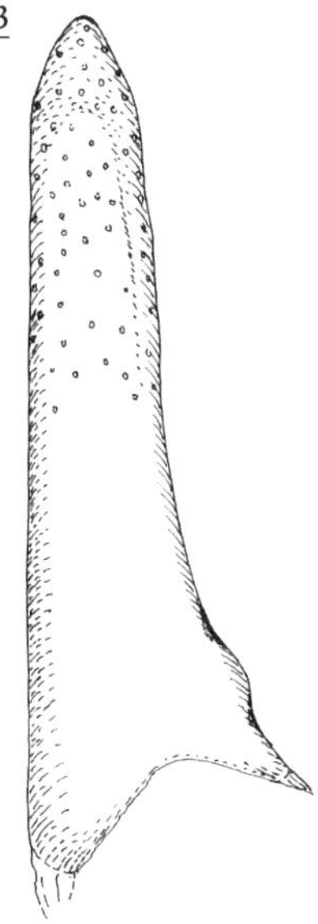

474

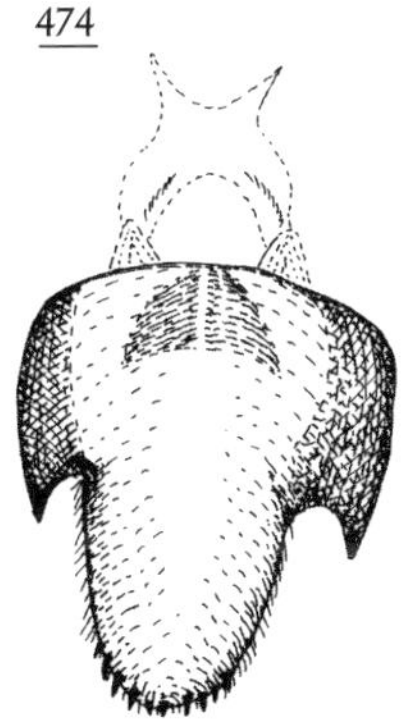

475

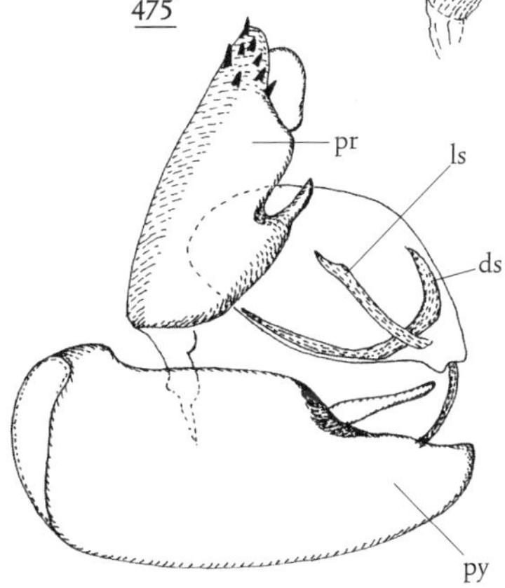

476

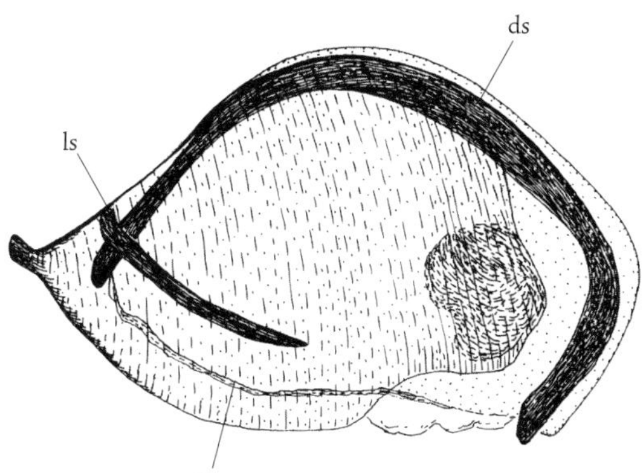

477

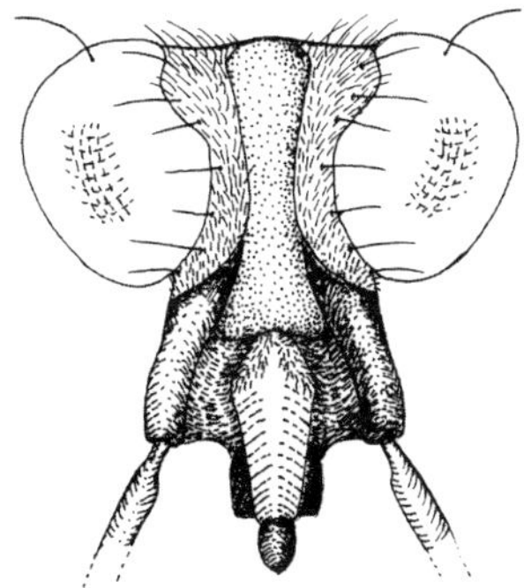

Vesica (fig. 476): dorsal sclerite long and proximally recurved, apically fused and separated proximally; ventral sclerite long and weakly sclerotized; lateral sclerites short, in some species absent.

Distribution in Malesia (fig. 597)

Cylindrostethus is present in Malesia with six species. One species is distributed throughout Sundaland except for Palawan, and one occurs in continental Malaysia and Sumatra. Sumatra, Borneo, the Philippines and Sulawesi each have one endemic species. Sumatra with three has the highest number of species, followed by continental Malaysia and Borneo with two each.

Biological notes

Cylindrostethus species tend to gregariousness and are usually encountered on the quiet parts of lowland rainforest streams. They may also occur in upland streams on smoothly flowing pools or behind rocks. Typical locomotion consists of a strong kick followed by a long glide. Populations of *Cylindrostethus* generally contain a mixture of winged and wingless morphs, with the latter usually dominating. Some species are found together with *Ptilomera* on turbulent rocky upland streams (D. Polhemus, 1994b).

REFERENCES – Cylindrostethinae. Structural analysis and classification of Gerridae (Matsuda, 1960); revision of Old World species (Hungerford & Matsuda, 1962b); morphology, phylogeny, adaptations, biogeography and classification of Gerromorpha (Andersen, 1982a); revision of Old World species, biogeography (D. Polhemus, 1994b).

SUBFAMILY PTILOMERINAE

Diagnosis

Small to large (4-20 mm), mostly elongated. Pronotum of apterous morph short, without pronotal lobe; metasternum well developed, laterally reaching metacetabular region, with metasternal scent orifice; metacetabular groove dorsally reaching anterior end of abdominal tergite 1 (fig. 380); abdominal sternite 1 absent. Length of protarsus at least half of protibial length (fig. 491). Mesofemur slender, distinctly longer than mesotibia. Paramere of male always well developed, usually curved and with pointed apex (figs. 486, 494, 501). Ovipositor short.

Distribution worldwide

Ptilomerinae are restricted to Central and tropical Asia and New Guinea, except for *Potamometroides* Hungerford from Madagascar. There are about 120 species divided over twelve genera. Of these, five genera are found in Malesia: *Ptilomera* (about 70 species), distributed from India to New Guinea and adjacent islands,

478

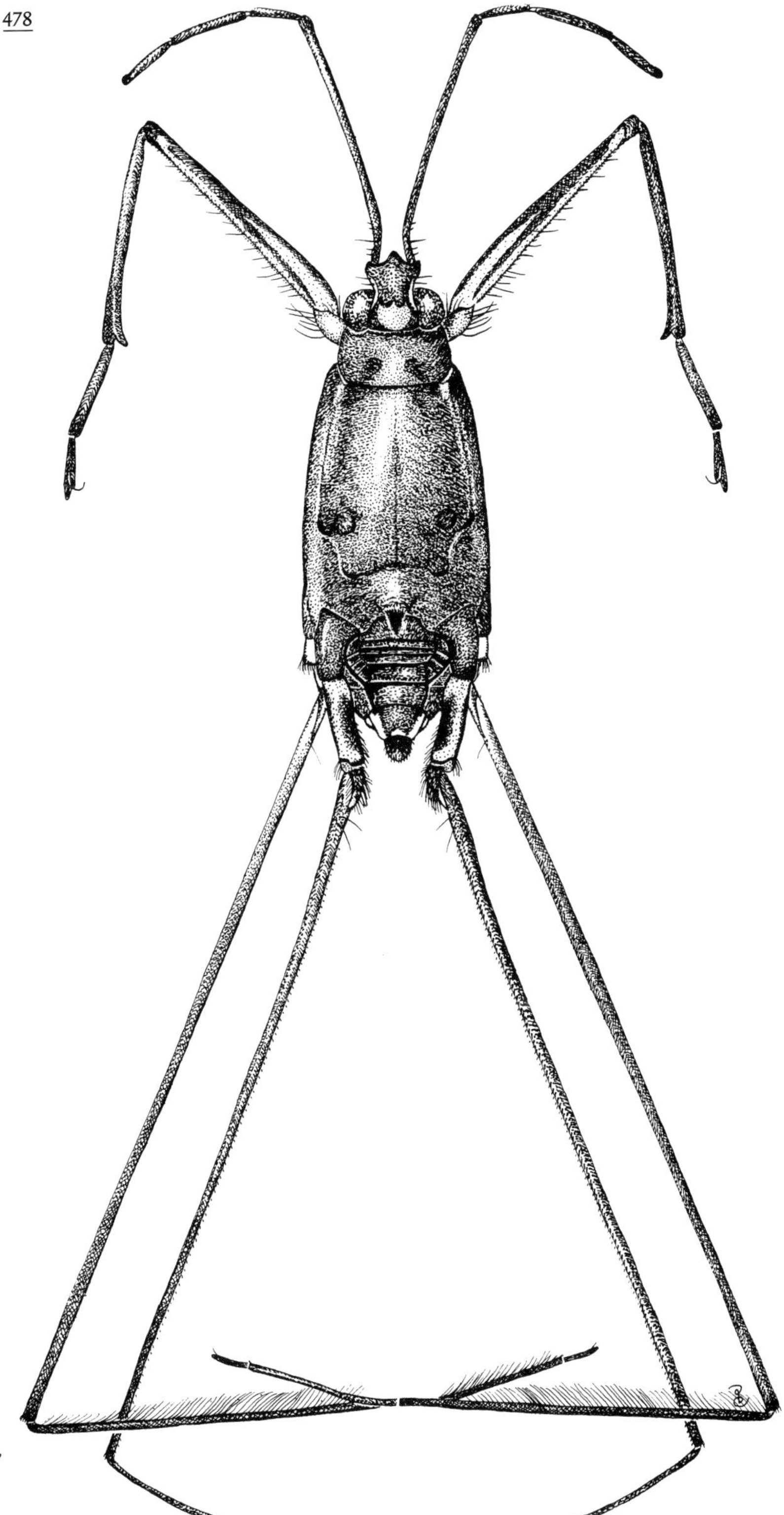

FIGURE 478
Andersenius dentifer,
paratype, apterous female,
length 6.8 mm.

is the largest genus in the subfamily, and also the only one extending to New Guinea, where the genus is abundant; *Potamometropsis* (twelve species) is restricted to Malesia; *Rheumatogonus* (nine species) is distributed from Sri Lanka through Borneo and continental South-east Asia to the Philippines; *Pleciobates* (five species) is mainly distributed in Indo-China, but one species reaches the continental Malaysia. *'Rhyacobates' imadatei* from Borneo belongs to an undescribed genus. Of the remaining genera *Andersenius* Zettel & Chen is found in Vietnam with two species; the monotypic *Pleciogonus* Chen, Nieser & Wattanachaiyingcharoen was recently described from northern Thailand; *Potamometra* Bianchi, with four described and a number of undescribed species, is endemic to China; *Heterobates* Bianchi ranges from Central Asia to India; and *Jucundus* Distant was found in India.

Key to genera of Ptilomerinae in South-east Asia and Malesia

1 Hind femur distinctly longer than middle femur (fig. 489); hind coxa with small spine on apical margin. Middle femur of male with ventral hair fringe ▸ ***Ptilomera* s.str. p. 375**

– Hind femur subequal to or shorter than middle femur; hind coxa never with spines. Middle femur of male without ventral hair fringe ▸ 2

2 Hind coxa longer than wide, especially in females (1.2-2 times as long as wide in male, 3.5-4 times in female (fig. 478) [Vietnam] ▸ ***Andersenius***

– Hind coxa shorter than wide ▸ 3

3 Anterior margin of head rounded or obtusely pointed. First antennal segment about half of length of the three following segments together (fig. 499) ▸ ***Rheumatogonus* p. 379**

– Anterior margin of head truncate. First antennal segment 3/4 or more of length of the three following segments together ▸ 4

4 First antennal segment subequal to or shorter than length of the three following together. Middle and hind tarsi with distinct claws (fig. 485) ▸ ***Potamometropsis* p. 373**

– First antennal segment longer than the three following together. Middle and hind tarsi without distinct claws ▸ 5

5 Connexiva 6 of female with a long caudal process [North Thailand] ▸ ***Pleciogonus***

– Connexiva 7 of female with a long caudal process ▸ 6

6 Hind femur about twice as long as tibia (fig. 479) ▸ ***Pleciobates* p. 371**

– Hind femur about three times as long as tibia ▸ 7

7 Mesonotum in apterous morph entirely black. Middle and hind tarsi with small but distinct claws (fig. 503) ▸ *'Rhyacobates' imadatei* p. 383

– Mesonotum in apterous morph with median longitudinal pale stripe. Middle and hind tarsi without claws [China, North Myanmar, North Thailand] ▸ *Rhyacobates* p. 382

Genus *Pleciobates*

Figures 479-484

479

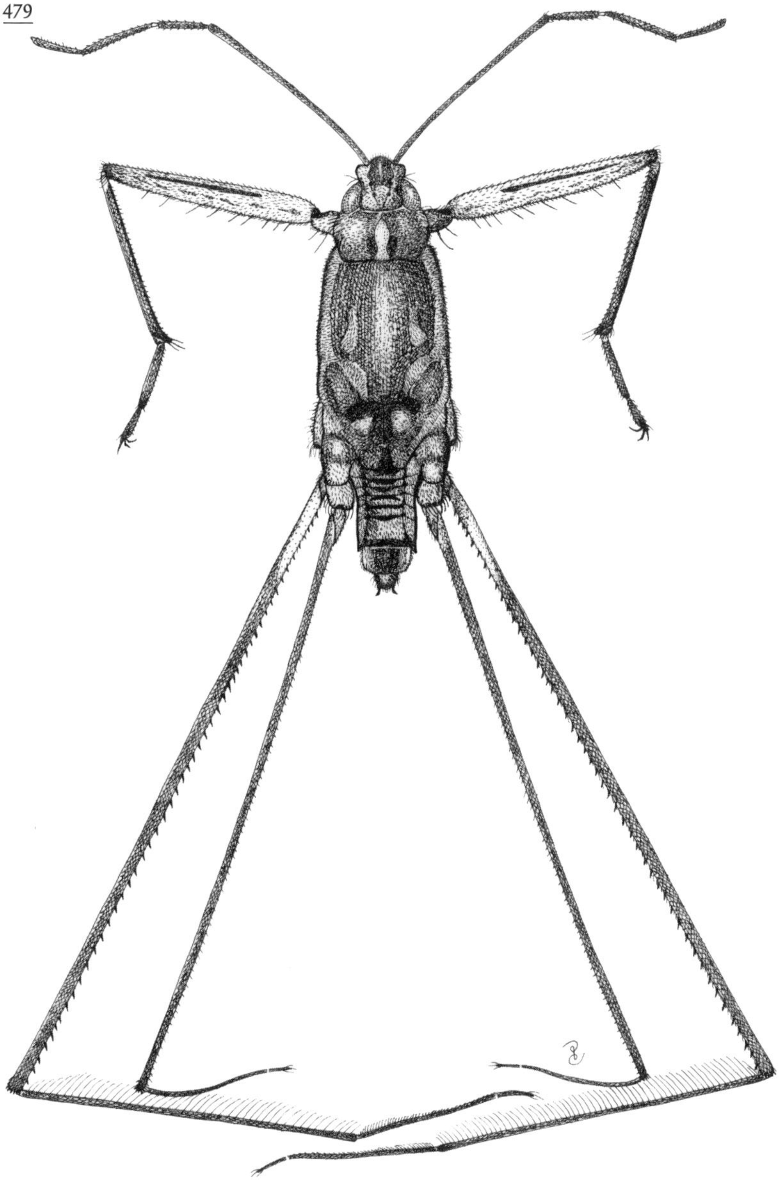

FIGURE 479
Pleciobates tuberculatus, apterous male, length 6.7 mm.

Diagnosis

Small (6.1-7.3 mm). Female subequal to male but distinctly stouter. Upperside blackish. Macropterous specimens are very rare. Antennal tubercles pronounced, angularly produced in dorsal view; antennal segment 1 longer than segments 2-4 together, segment 2 shorter than segment 3. Hind femur slightly shorter than middle femur; hind coxa without apical spine, not elongate; middle tarsus and hind tarsus without claws. Male: fore femur slender, subapically on ventral surface without tooth; mesofemur without fringe of long black hairs. Female: metanotum of apterous morph with more or less distinct tubercle; mesosternum of female without special features; abdomen shortened, apical segments withdrawn into segment 7, with process on connexivum 6 (fig. 480); abdominal segment 7 without connexival spine, sternite 7 medially with concave hind margin (fig. 481). Male genitalia: pygophore without lateral process, proctiger laterally angular or rounded, paramere (fig. 484) long and curved dorsad, distally with long pilosity.

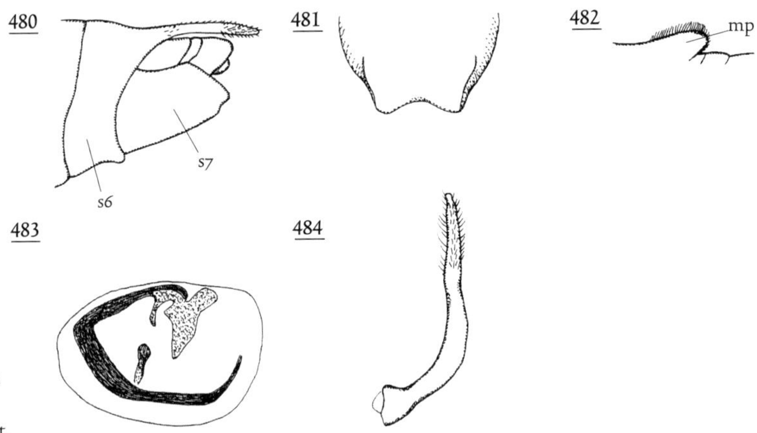

FIGURES 480 – 484 *Pleciobates tuberculatus* (from Zettel & Chen, 1996). **480** female, apex of abdomen, lateral view – **481** female, dorsal view of sternite 7 – **482** metanotal projection of apterous female, lateral view – **483** male, phallotheca and vesica, lateral view – **484** male, left paramere, external view; mp= metanotal projection, s6 = sixth abdominal segment, s7 = seventh abdominal segment.

Distribution in Malesia (fig. 597)

Pleciobates tuberculatus is endemic to continental Malaysia and the only species occurring in Malesia. Two other species have been described from continental South-east Asia.

Biological notes

Pleciobates tuberculatus has been found in the current of a middle-sized stream.

Genus *Potamometropsis*
Figures 485-488

485

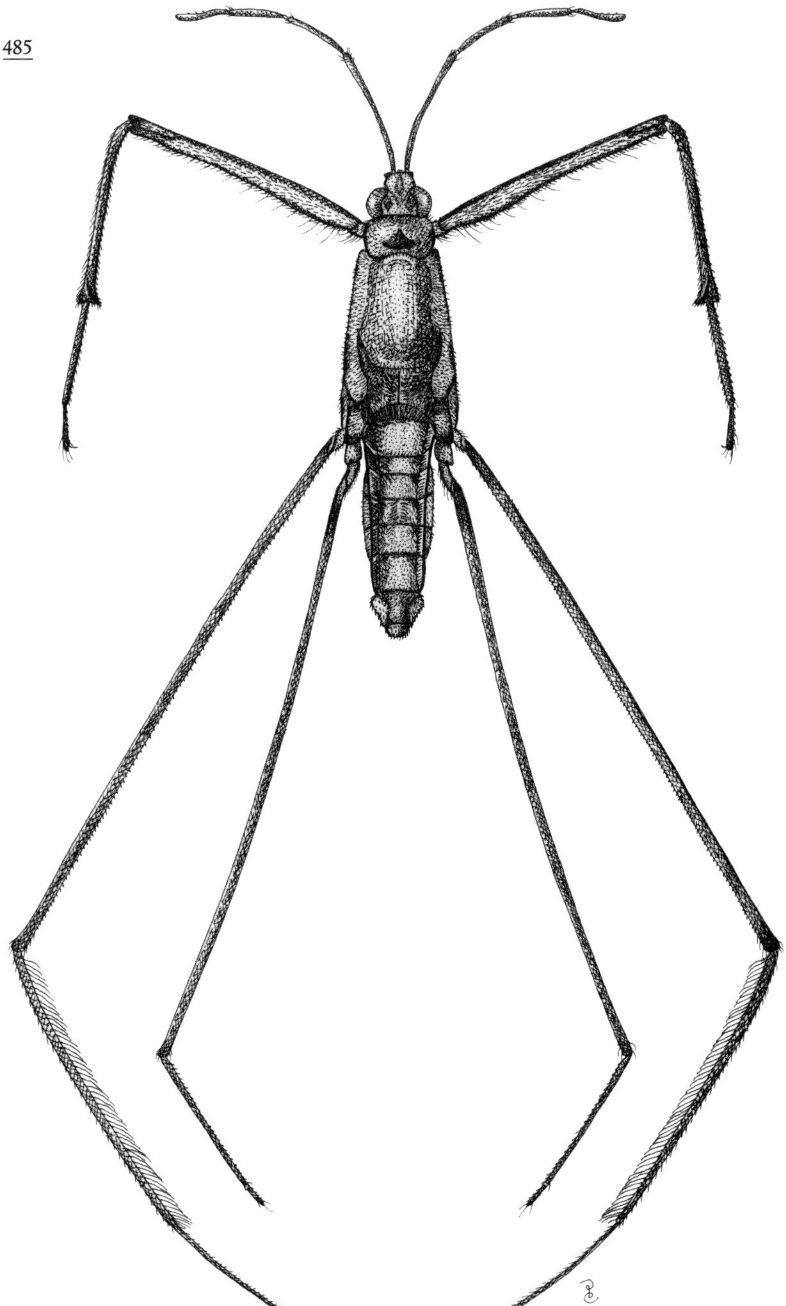

FIGURE 485
Potamometropsis anomalis,
paratype, apterous female,
length 9.2 mm.

Diagnosis

Medium-sized (5.6-10.9 mm), female on average larger than male. Dorsum varying from greenish (brownish in preserved specimens) to blackish. Antennal tubercles pronounced, angularly produced in dorsal view; antennal segment 1 subequal to segments 2-4 together, segment 2 usually shorter, rarely slightly longer than 3. Metanotum of apterous female usually without tubercle; mesosternum of female without special features. Profemur of male subapically on ventral surface at least with minute tooth; in some species profemur strongly dilated; mesofemur of male without fringe of long black hairs; hind femur longer than middle femur; hind coxa without apical spine, not elongate; hind tarsus with distinct claws. Female abdomen either elongate and straight or shortened and with apical segments more or less withdrawn into segment 7, without process on connexivum 6; abdominal segment 7 with modifications (with ventral lobe in most species) but always without connexival spine. Male genitalia: pygophore without lateral process, proctiger laterally angular or rounded, paramere (fig. 486) long and curved dorsad, with long pilosity; phallotheca modified, membranous distally, dorsally sclerotized in middle and thus forming 'bridge', sclerotized ventrally and with distal half membranous along mid-line; extra apical piece (fig. 487) loosely attached with ventral side distally directed. Vesica (fig. 488): dorsal sclerites fused but separated and curved proximally; ventral sclerites thin and membranous; pair of lateral sclerites slightly curved. Additional pair of lateral dark stripes very distinct, from lateral view looking very much like normal lateral sclerites, which might be described as second pair of lateral sclerites, but clearly attached with phallic plate basolaterally. Further study needed on status of this 'second pair of lateral sclerites' in Gerridae.

Distribution in Malesia (fig. 598)

Presently twelve species are described in the genus *Potamometropsis*, all from Malesia. Highest diversity is found in the Philippines (five species) and Borneo (five species). Single species inhabit Sumatra and Sulawesi.

Biological notes

Potamometropsis species inhabit fast flowing streams and rivers. Some (e.g., *P. luzonica*) live in and close to rushing, turbulent currents in small to middle-sized streams. Others (e.g. *P. werneri*) prefer less turbulent flowing, middle-sized to large streams where they sometimes aggregate behind boulders in large schools, with adults and immatures often mixed (Zettel, 1999g). Macropterous specimens are rare in most species and are nearly always collected de-alate. Complete macropterous specimens are only found when freshly moulted (Chen & Nieser, 1992).

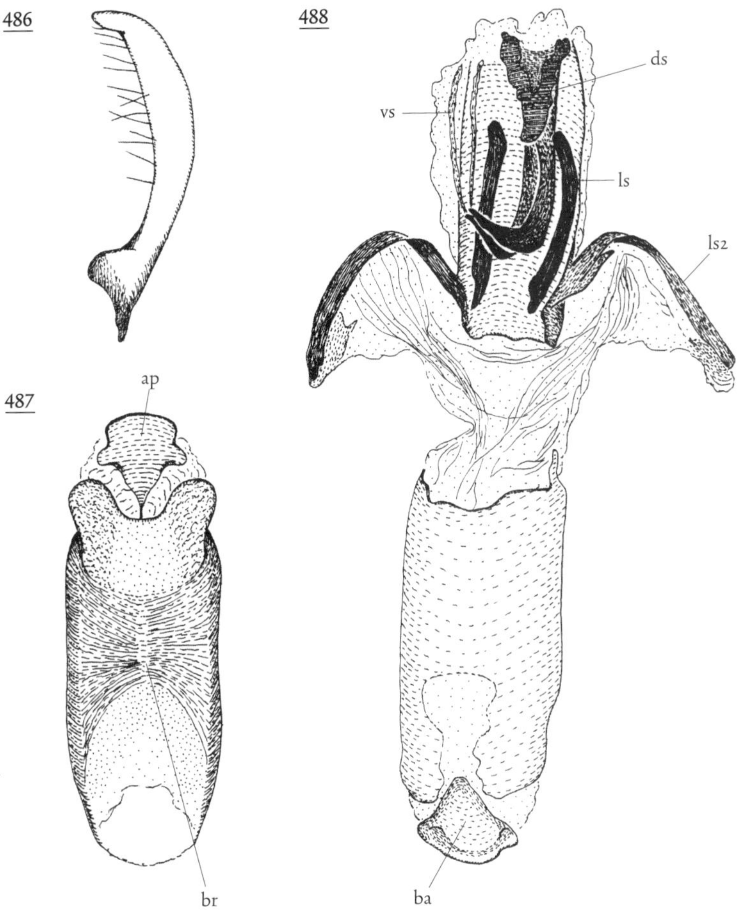

FIGURES 486 – 488
Potamometropsis anomalis, male.
486 left paramere, external view – **487** phallotheca, dorsal view – **488** phallotheca and vesica, ventral view; ap = ventral apical piece, ba = basal articulatory apparatus, br = bridge, ds = dorsal sclerites, ls = lateral sclerites, ls2 = second pair of lateral sclerites, vs = ventral sclerites.

Genus *Ptilomera* sensu stricto

Figures 489-498

Diagnosis

(This diagnosis refers to *Ptilomera* s.str. only.) Large (10-20 mm), elongate. Dorsum greenish (brownish in preserved specimens). Antennal tubercles pronounced, angularly produced in dorsal view; antennal segment 1 longer than segments 2-4 combined, segment 2 shorter than segment 3. Metanotum of apterous female without tubercle; mesosternum of female without special features. Ventral side of male fore femur with small teeth, which grip into basal teeth on ventral side of fore tibia (fig. 491); mesofemur of male with fringe of long black hairs; hind femur longer than middle femur; hind coxa with apical spine, not elongate (fig. 489); hind tarsus with distinct claws. Male abdominal segment 8 (fig. 493) with large median ventroapical carina, dorsal side with pair of longitudinal hairy sublateral patches, which suround spiracle 8 (fig. 493). Female abdomen elongate, straight, without process on connexivum 6 but with strong spines

489

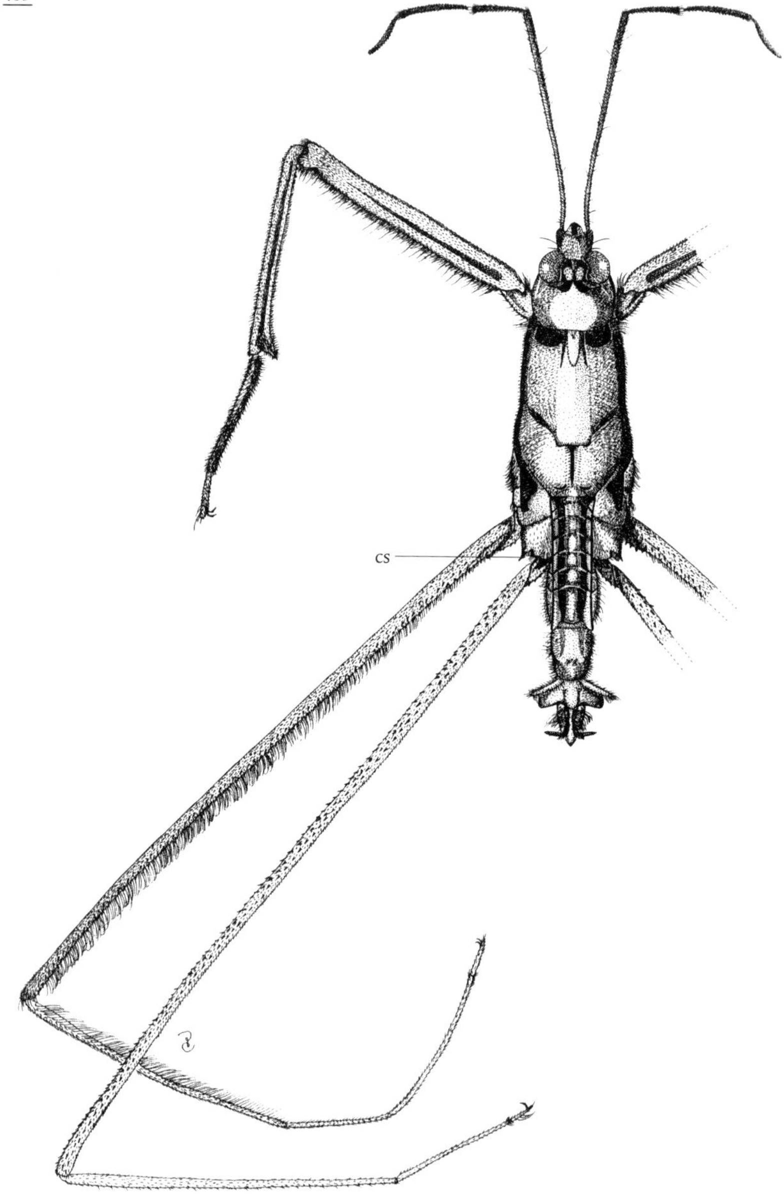

FIGURE 489
Ptilomera dromas, apterous male, length 14.9 mm; cs = metacoxal spine.

and lobes on sternite 7 and connexivum 7 (fig. 495). Male genitalia (fig. 490): pygophore with lateral process and pair of sclerotized plates which connect with distal wall of pygophore and phallic articulatory apparatus, suggesting strengthening of parameres (fig. 498); proctiger (fig. 490) laterally angularly produced ('lateral wings' *sensu* Hungerford & Matsuda, 1965). In general, *Ptilomera* species have a pair of strongly developed parameres but relatively small phallus. Phallic structure: basal articulatory apparatus with second diamond-shaped part connecting phallus (fig. 496, 497) (this part is also found in *Neogerris* (Gerrinae) (fig. 444)); phallotheca (fig. 496) ventrally and dorsodistally sclerotized; paramere (fig. 494) strongly developed, long and hooked, subapically with long black pilosity. Vesical sclerites (fig. 497): dorsal sclerite long, proximally curved, ventral sclerite membranous, with two pairs of lateral sclerites and basal sclerite.

FIGURES 490 – 493
Ptilomera dromas
490 male, apex of abdomen, dorsal view – **491** male, fore leg – **492** male, apex of fore tarsus – **493** male, abdominal segment 8, lateral view; lp = lateral process of pygophore, lw = lateral wing of proctiger, pa = paramere, pr = proctiger, sp = eighth spiracle, vc = ventral carina.

490

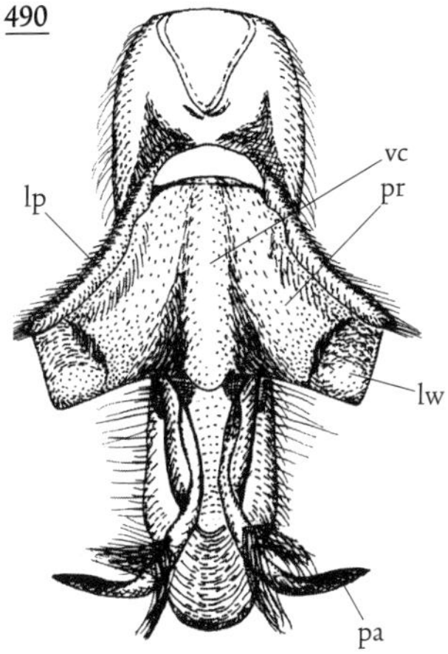

491

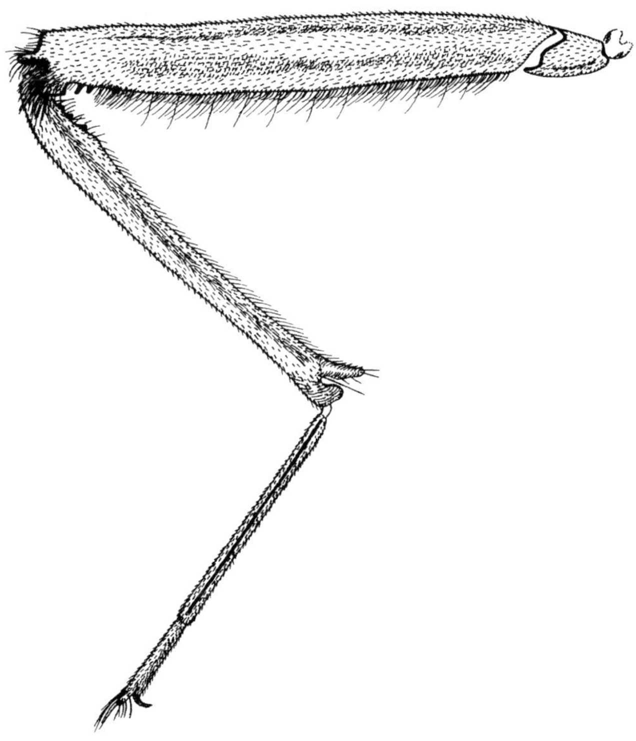

492

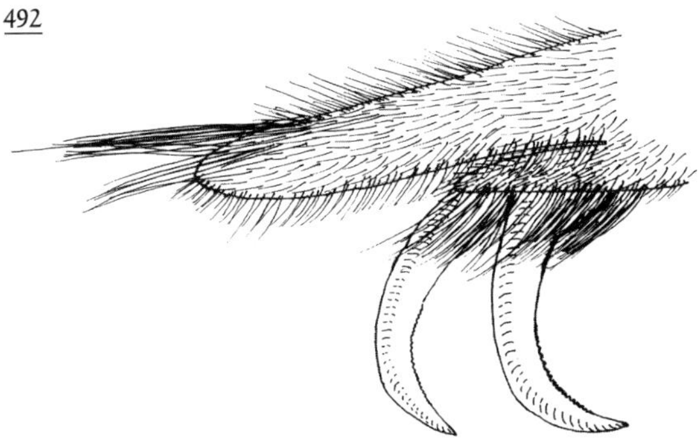

493

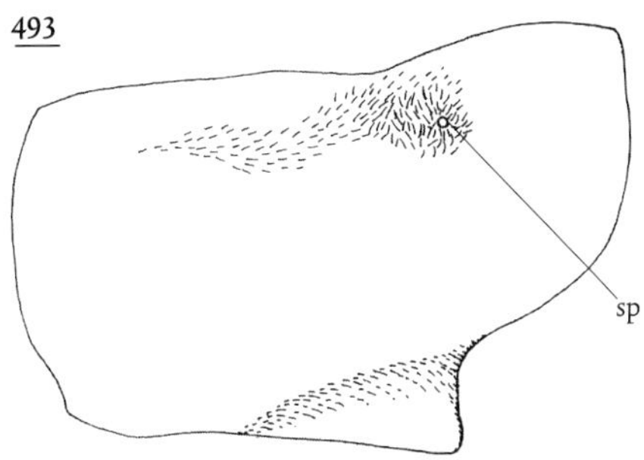

FIGURES 494 – 498
Ptilomera dromas
494 male, left paramere, dorsal view – **495** female, apex of abdomen, lateral view – **496** male, lateral view of phallotheca with vesica removed – **497** phallotheca and vesica, lateral view – **498** dissected genitalia, dorsolateral view; ba = basal articulatory apparatus, bs = basal sclerite, dp = diamond shaped piece of basal articulatory apparatus, ds = dorsal sclerite, ls = lateral sclerites, pt = phalloteca, vs = ventral sclerite.

494

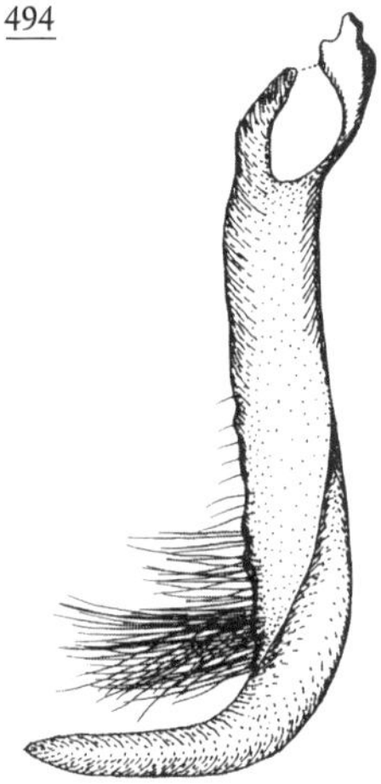

495

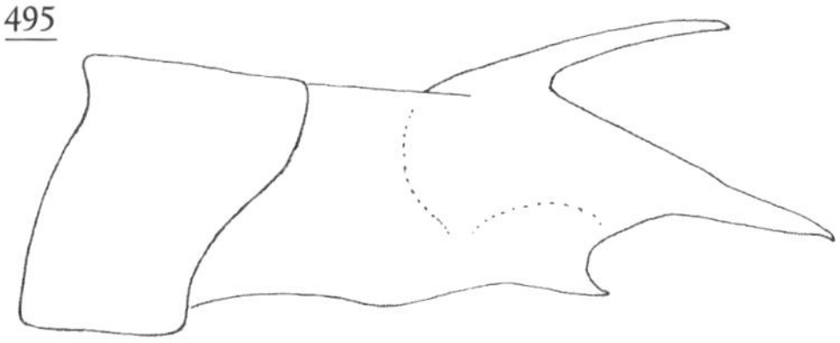

496

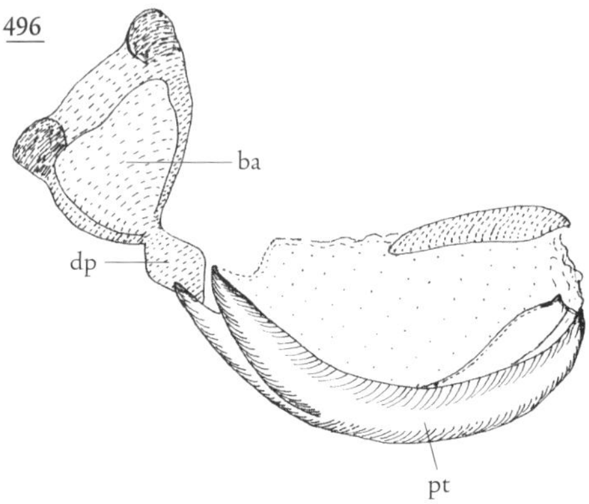

497

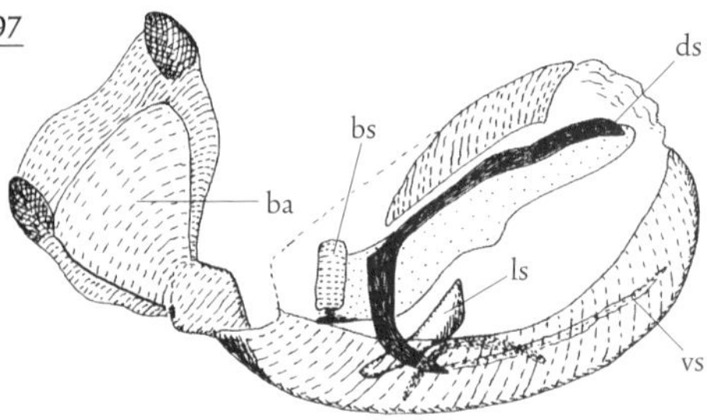

498

Distribution in Malesia (fig. 598)

Only the subgenus *Ptilomera* is present in Malesia. About 70 species have been described and 64 of them occur in Malesia. Most are regional endemics, but some have a wider distribution. The subgenus has a wider distribution and occurs from India and Sri Lanka to New Guinea and the Bismarck Archipelago;

on the Philippine Islands it is present only on Palawan. *Ptilomera* is the only genus of Ptilomerinae transgressing Weber's Line eastward. New Guinea and surrounding islands with nineteen species is by far the richest area. D. Polhemus & J. Polhemus (2001), who mention 22 species, include the Bismarck and Solomon islands.

Biological notes

Ptilomera inhabit rocky tropical streams, but usually stay in more or less lentic sections. They move back and forth on the water and can skate backwards. However, individuals do not continuously row against the current but wait more or less motionless in very slowly flowing or circulating water, often behind boulders. If prey is detected, they quickly jump out in the current, grasp it and return as fast as possible. Adults of different sexes occur on different parts of streams: males have a fringe of stiff setae on the middle femur (lacking in females), which allows them to row more effectively (D. Polhemus & J. Polhemus, 2001). Hence, males are found in fast midstream waters while females congregate in slower sections along the margins. On one occasion, thousands of *P. tigrina* were observed resting on a quiet and shaded section of Khek River, North Thailand (P. Chen pers. obs.). Macropterous individuals are rare (even unknown in many species) and often found dealate. This possibly limits gene flow among populations and has probably contributed to the development of the numerous geographically allopatric species presently seen in New Guinea and elswhere in the Malay Archipelago (D. Polhemus & J. Polhemus, 2001). Usually, only a single species is present in one stream, indicating that competitive exclusion may occur. However, at contact zones in New Guinea and other large islands of the Malay Achipelago, as well as in parts of Indochina, two or three species may occupy the same stream (D. Polhemus & J. Polhemus, 2001). Unlike most gerrids, *Ptilomera* emit a strong odour when captured by secreting a drop of yellowish liquid from the abdominal tip.

Genus *Rheumatogonus*

Figures 499-502

Diagnosis

Small (4.1-6.9 mm); female distinctly larger than male. Greenish (yellowish in preserved specimens) with some dark dorsal markings, especially along median line of thorax and abdomen. Antennal tubercles small, in dorsal view rounded; antennal segment 1 shorter than segments 2-4 combined, segment 2 slightly shorter or subequal to segment 3. Rostrum short, just reaching anterior margin of mesosternum. Metanotum and mesosternum of female without special features. Thorax almost parallel-sided in male, diverging posteriorly in female. Mesonotum longest, 2-3 times as long as pronotum. Metacetabular groove transverse and in dorsal view connected with very short longitudinal ridge.

499

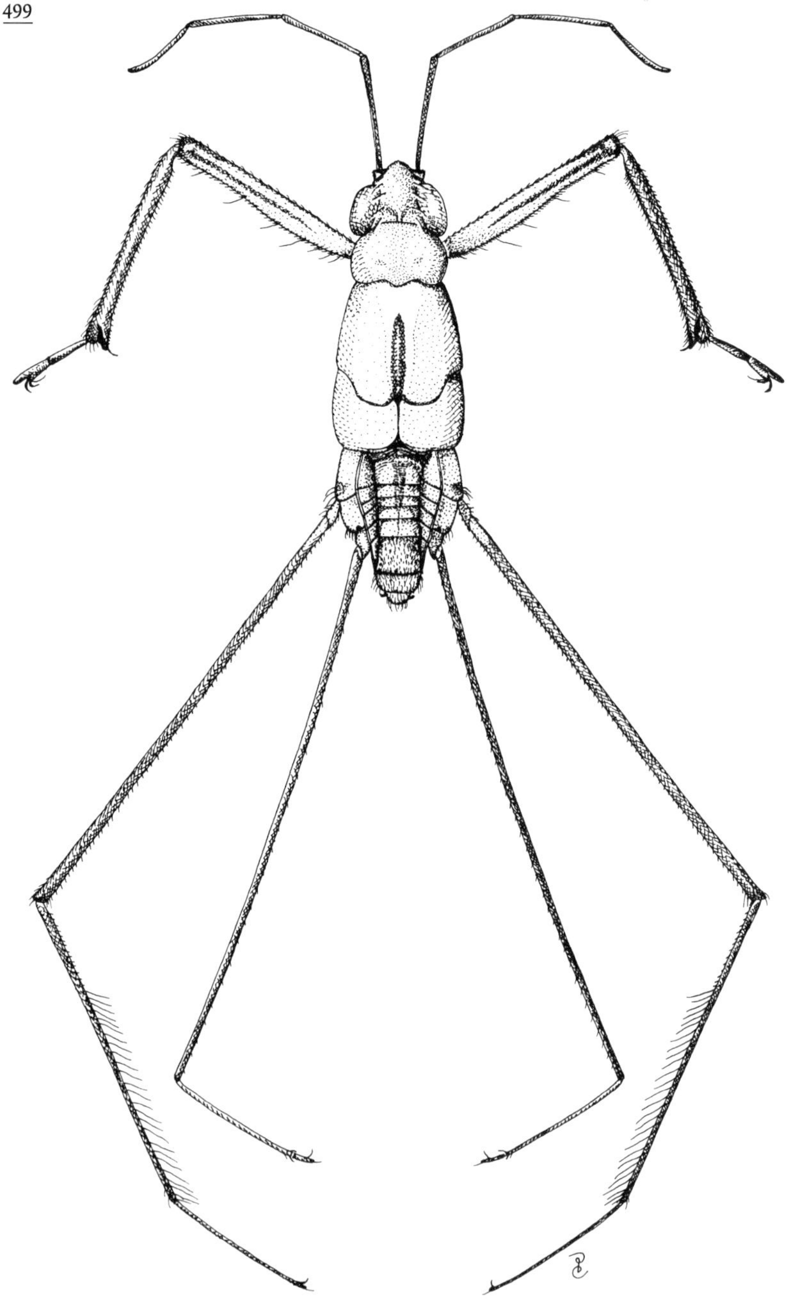

FIGURE 499
Rheumatogonus esakii,
apterous male,
length 4.2 mm.

Scent gland orifice near posterior margin of metanotum, not on tubercle or other elevation. Profemur of male moderately incrassate, apically slightly tapering, posteroventral surface with about 10 long bristals; fore tibia somewhat dorsoventrally flattened and with short distal spur at internal margin. Middle and hind legs long and slender, subequal in length, distal $^{2}/_{3}$ of middle tibia with long hair fringe. Basal 1/3 of hind femur broadened and with group of longer hairs on ventral surface; middle tarsus 2 without claws. Abdomen slightly shorter than thorax, sutures between abdominal tergite 1-7 distinct; sternite 1 fused with metasternum. Female without process on connexiva 6 or 7, sternite 7 on hind margin with broad medial lobe. In ventral view length of genital segments slightly longer than preceding abdominal sternites together. Male genitalia: pygophore without lateral process (fig. 500); proctiger laterally rounded (fig. 500), each side dorsally with group of thick setae, which can be modified as spinules in certain species; parameres symmetrical (fig. 501), falciform, curved dorsad, shaft broad, bearing sparse long pilosity. Vesica: phallotheca uniformly sclerotized, widely opened at apical half; vesical structure (fig. 502): dorsal sclerite asymmetrical, less sclerotized apically, lateral plate subtriangular, ventral sclerite modified according to species. Ovipositor: gonapophyses not serrate, first pair less sclerotized than second. Species are very similar in external appearance and structure, hence, for reliable identification careful examination of male genitalia is required.

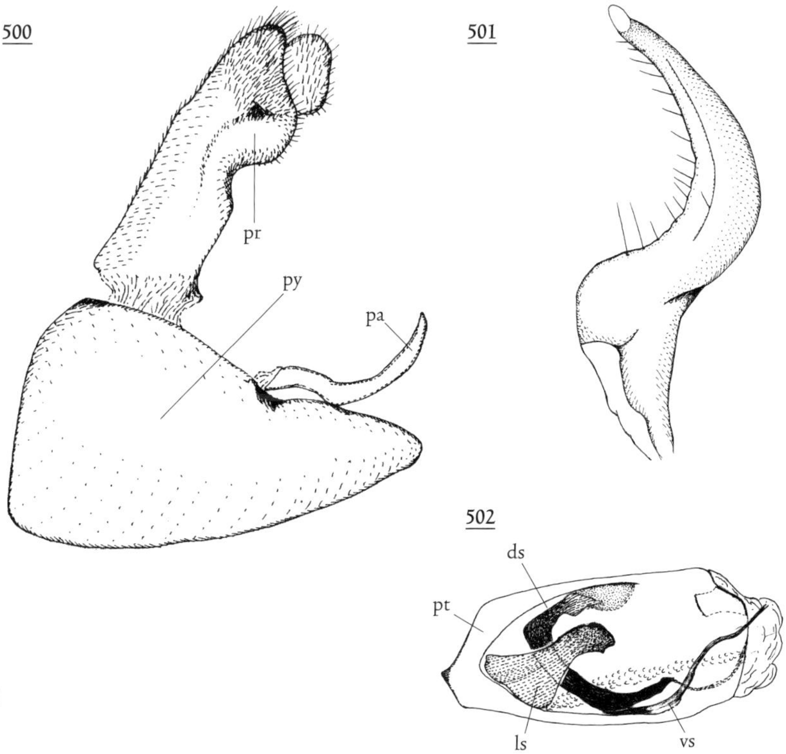

FIGURES 500 – 502 *Rheumatogonus esakii*, male. **500** pygophore and proctiger, lateral view – **501** left paramere, external view – **502** vesica, lateral view; ds = dorsal sclerite, ls = lateral sclerite, pa = paramere, pr = proctiger, pt = phalloteca, py = pygophore, vs = ventral sclerite.

Distribution in Malesia (fig. 599)

Rheumatogonus species are widely distributed in the Oriental realm, from Sri Lanka to Borneo and the Philippines. There are presently nine described species of which seven occur in Malesia: three in the Philippines, four in Borneo and one each in Sumatra, peninsular Malaysia and southern Thailand (Chen & Nieser, 2002).

Biological notes

Rheumatogonus species inhabit mountainous streams, creeks and waterfalls in tropical rainforests in the Oriental region. They share the habitat with other rheophilous gerrids and veliids, but generally prefer less lotic areas. In shaded, steady slow-lotic sections of streams *R. luzonicus* can aggregate into schools of adults and large nymphs. When disturbed, all individuals move rapidly back and forth (a behaviour also observed in other ptilomerine genera), presenting a confusing picture of jumping greenish, leggish dots. *Rheumatogonus esakii* from Borneo prefers slightly turbulent sections of small streams and is found only in relatively small groups. In northern Thailand, *R. vietnamensis* can be collected from broad rivers or narrow creeks, but only in quiet areas without turbulence. Macropterous individuals are rare and mostly found dealate. The distinctly smaller male rides on the back of the female during copulation.

Genus *Rhyacobates*

Figure 503

Diagnosis

Medium-sized (6.7-10.4 mm), female distinctly larger than male. Dorsum blackish, mesonotum entirely black. Antennal tubercles pronounced, angularly produced in dorsal view; antennal segment 1 longer than segments 2-4 combined, segment 2 shorter than segment 3; metanotum of apterous female without tubercle. Venter with narrow median line reaching from anterior $^1/_4$ of mesosternum to middle of sternite 7 and consists of appressed hairs distinctly longer than regular pubescence. Profemur of male slender, subapically without tooth on ventral surface; mesofemur of male with distinct row of black spines along proximal $^3/_4$ of ventral margin; metacoxa without apical spine, not elongate; mesotarsus and metatarsus with distinct claws. Female abdomen elongate and straight, without process on connexivum 6, with short connexival spines on segment 7. Female gonocoxa directed caudad, withdrawn into sternite 7. Male genitalia: pygophore without lateral process, proctiger with blunt lateral corners, paramere long and curved dorsad, without long pilosity.

Taxonomic notes

The only Malesian species *R. imadatei* was originally described in *Rhyacobates* after specimens from Borneo and Thailand (Miyamoto, 1967). In a phylogenetic analysis, Andersen & Chen (1995) have shown that this species does not belong to the monophylum comprising *Pleciobates*, *Heterobates* Bianchi and *Rhyacobates* (a clade now also containing *Andersenius* Zettel & Chen), but probably belongs to

503

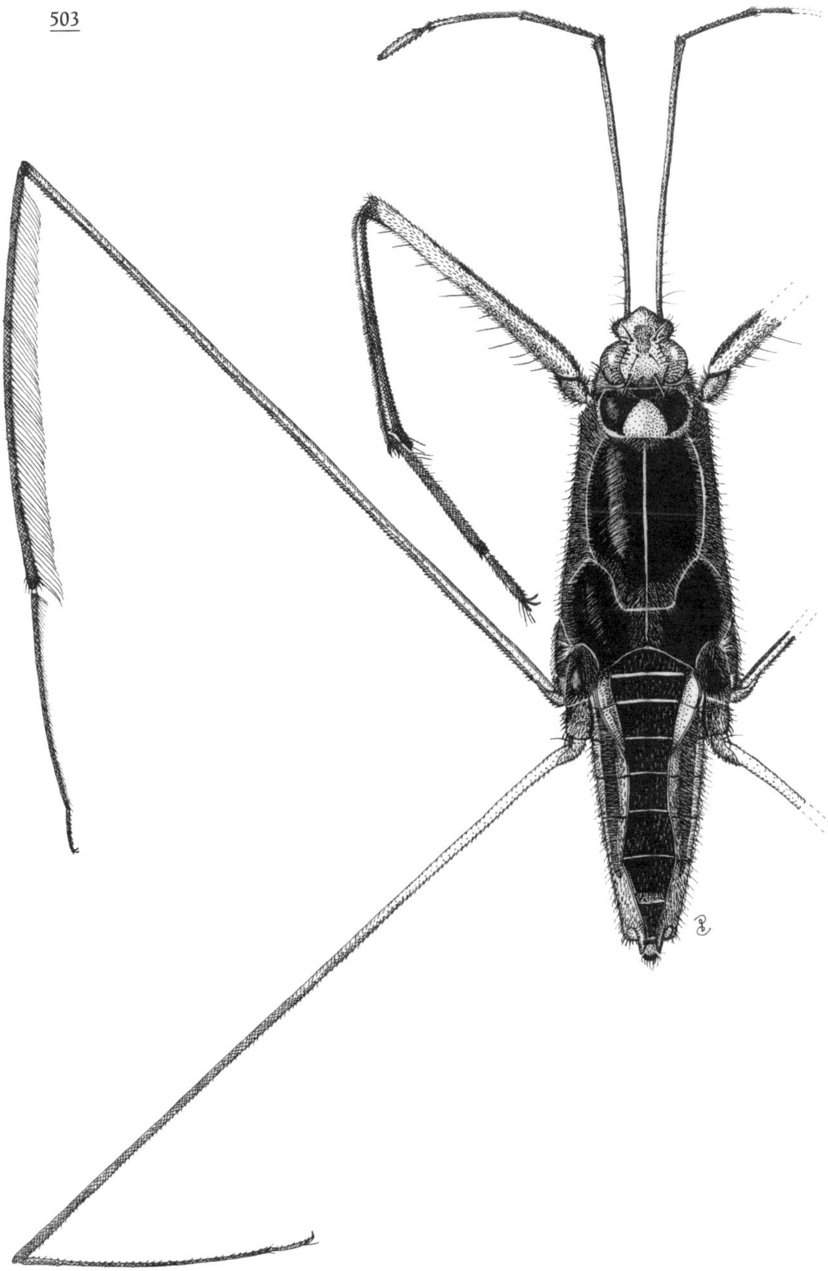

FIGURE 503
'Rhyacobates' imadatei, apterous female, length 9.0 mm.

an undescribed genus. *Rhyacobates imadatei* has been excluded from *Rhyacobates* without proposal of another generic name. Also other recent studies on Ptilomerinae have consistently stated that *'Rhyacobates' imadatei* does not comfortably settle into any described genus (Andersen & Chen, 1995; Zettel, 1994d; Zettel & Chen, 1996; Polhemus & Zettel, 1997). For the moment we will leave it in

Rhyacobates. So far, no close relative of *R. imadatei* is known. The diagnosis above is based on *R. imadatei* and not valid for other *Rhyacobates* occurring in China, Myanmar and North Thailand.

Distribution in Malesia (fig. 597)

The single species is probably endemic in Borneo. The record from Thailand is regarded as wrong (Andersen & Chen, 1995; Polhemus & Zettel, 1997).

Biological notes

This species inhabits large rivers, where it skates with great speed and agility in strong currents. Jumps up to 15 cm have been observed. Specimens rest on wet, shady, vertical surfaces of rock boulders, waiting to leap off onto the water to pursue prey. The rapidity of skating against the current is astonishing: even females in copula, with the smaller males on their backs, are able to skate and jump against a current of about 1.5 m/s. Winged specimens are unknown (Polhemus & Zettel, 1997).

REFERENCES – Ptilomerinae. Description of *Ptilomera (Proptilomera) himalayensis* (Hungerford & Matsuda, 1958e); structural analysis and classification of Gerridae (Matsuda, 1960); revision of *Ptilomera* (Hungerford & Matsuda, 1965); description of *Rhyacobates imadatei* (Miyamoto, 1967); *Ptilomera* species from Philippines and their genital structures (Andersen, 1976); morphology, phylogeny, adaptations, biogeography and classification of Gerromorpha (Andersen, 1982a); key to Sulawesi *Ptilomera* (Chen & Nieser, 1992); *Pleciobates tuberculatus* (Thirumalai, 1992); revision of *Potamometropsis* (Zettel, 1994d); revision of the genus *Rhyacobates* (Andersen & Chen, 1995); Gerridae of Vietnam, review of *Pleciobates* (Zettel & Chen, 1996); *Potamometropsis* from Borneo (J. Polhemus & Zettel, 1997); *Potamometropsis* of the Philippines (Zettel, 1999g); revision of *Ptilomera* of New Guinea and nearby islands (D. Polhemus & J. Polhemus, 2001); genus *Stridulobates* (Zettel & Thirumalai, 2001a); re-establishment of genus *Jucundus* (Zettel & Thirumalai, 2001b); male genital structure and review of *Rheumatogonus* Kirkaldy (Chen & Nieser, 2002); genus *Pleciogonus* (Chen et al., 2002).

SUBFAMILY HALOBATINAE

Diagnosis

Small to medium-sized (usually 3-8 mm). Head and thorax short and wide, metasternum strongly reduced, usually existing of very short subtriangular plate enclosing scent orifice. Intersegmental sutures between mesothorax and metathorax laterally indistinct, sutures between basal abdominal tergites obliterated (figs. 508, 538). Abdominal segments shortened. Middle femur slender, usually distinctly longer than middle tibia. Claws of hind tarsi modified, straight or S-shaped. Fore wing completely membranous. (Andersen, 1982a). Marine tribe Halobatini always apterous, with metasternum reaching metacetabular region laterally. Middle tibia clothed with conspicuous row of long hairs; hind coxae longer than wide. Male genital segments strongly modified: sternite of eighth abdominal segment with long styliform processes, with spatulate or spiny tips; proctiger is usually dilated; pygophore with narrow bridge connecting dorsal base; phallotheca connected to basal apparatus by short and broad median cord. Phallotheca ventrally and laterally distinctly sclerotized, but dorsally membranous. Lateral part of phallotheca often separated from ventral part by less sclerotized longitudinal areas. Vesical sclerites thick, with two pairs of lateral sclerites, and often with basal sclerite. Parameres reduced or absent (Matsuda, 1960; Andersen, 1982a; Andersen & Weir, 1994b). Freshwater tribe Metrocorini has apterous and macropterous morphs. Metasternum reduced to subtriangular plate, not reaching metacetabular region laterally, enclosing scent orifice. Male genital segments without processes on eighth abdominal sternite, proctiger not distinctly dilated. Vesical sclerites with one pair of lateral sclerites, although pair of small acessory sclerites may be present (figs. 531, 556). Parameres well developed to large (figs. 530, 536) (Matsuda, 1960; Andersen, 1982a).

Characteristics

Structure of fore leg and genitalia of male are of taxonomical importance at species level. This will be discussed under each genus. Some species bear modified pilosity or finger-like projections on tergites, legs or terminal segments, which can be useful for identification. The fore wing venation is relatively complete in *Metrocoris* and *Eurymetra*, and reduced in *Ventidius* and *Esakia*. Autotomy of wings occurs along a transverse line in the apical part of the wings. After autotomy the remaining part of the wings reaches more or less to the abdominal apex.

Biological notes

The species of Halobatini are, apart from a single exception, living along the sea coast, a few even on the open ocean. The species of the freshwater genus *Metrocoris* are typical inhabitants of heavily shaded mountainous streams in South and South-east Asia; they are especially abundant in areas of undisturbed rainforest. *Ventidius* and *Esakia* species usually prefer small or medium-sized streams and the shores of large rivers and mountainous lakes, where they inhabit open quiet areas.

Distribution worldwide

Ten genera are recognized in Halobatinae, and half of them occur in Malesia. The other genera occur mainly in Africa. The largest freshwater genus is *Metrocoris* with 60 species. It is distributed from Saudi Arabia to Japan, the Philippines, Sulawesi and the Lesser Sunda chain. The centre of diversification covers northern Thailand, Myanmar and South and South-west China, with 44% of the described species. Several subcentres of diversification have been noticed on both sides of Wallace's Line (25% of the species) and India (22%). *Ventidius* contains 24 species (Chen & Zettel, 1999), distributed from North-east India eastward to the Philippines, Sulawesi, and Nusa Tenggara Timur. Continental South-east Asia and Borneo together hold 70% of the species. *Esakia* has a similar distributional pattern as *Ventidius*, with over 60% of species occurring in continental South-east Asia and Borneo (Matsuda, 1960; Andersen, 1982a; Chen & Zettel, 1999). Six species have been described and a number of undescribed species exist in Malesia. A typical vicariance distributional pattern is observed among the freshwater genera.

Distribution in Indo-Australia

So far there are six genera recognised in the Indo-Australian area, with among them the typical Oriental genera *Esakia*, *Metrocoris* and *Ventidius*. The marine genera are more widely distributed. *Asclepios* occurs mainly in the eastern Palearctic, with one species in the south of continental Malaysia and in India. The highly specialized genus *Halobates* is pantropical. *Austrobates* is so far only known from North Queensland, Australia (Andersen & Weir, 1994a).

Key to the genera in Indo-Australia

1 Basal margin of clypeus well defined (fig. 515), anterior margin of head not smoothly rounded (marine, except *Austrobates*) ▸ 2

– Basal margin of clypeus not present, anterior margin of head broadly and evenly rounded (freshwater, rarely reaching mangroves) ▸ 4

2 Middle tibia and segment 1 of middle tarsus in most cases with hair fringe (fig. 514); proctiger strongly dilated laterally (fig. 516) ▸ *Halobates* p. 390

– In most cases, middle tibia with hair fringe, and segment 1 of middle tarsus without hair fringe ▸ 3

3 Proctiger cylindrical. Abdominal sternite 8 of male longer than wide, without spiracular processes; styliform processes short, straight, with blunt apices (fig. 509) [Asian continental shelf] ▸ *Asclepios* p. 387

– Proctiger (fig. 516) laterally moderately dilated. Abdominal sternite 8 of male wider than long, with slender styliform processes (fig. 505) [Australia] ▸ *Austrobates*

4 Eyes not overlapping anterolateral angles of mesonotum (fig. 506). Body yellowish with black pattern (fig. 532) ▸ *Metrocoris* p. 396

– Eyes overlapping anterolateral angles of mesonotum (fig. 507). Body greenish (after death becoming yellowish) with black (fig. 537) ▸ 5

504

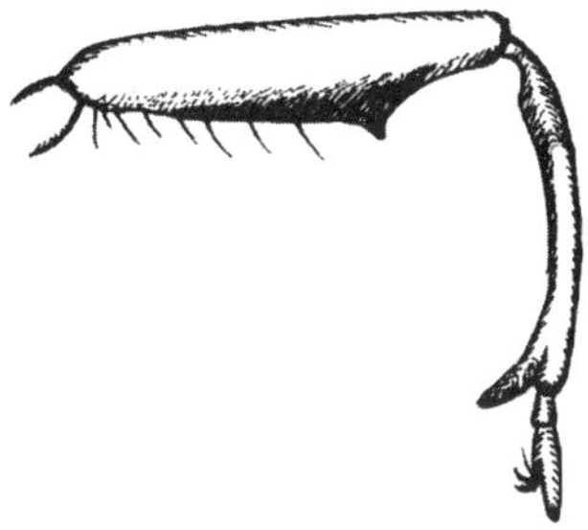

505

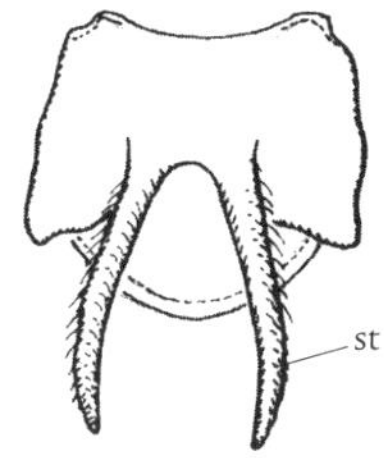

FIGURES 504 – 505
Austrobates rivularis Andersen & Weir, male (adapted from Andersen & Weir, 1994a). **504** fore leg – **505** eighth abdominal segment, ventral view; st = styliform process of abdominal segment 8.

506

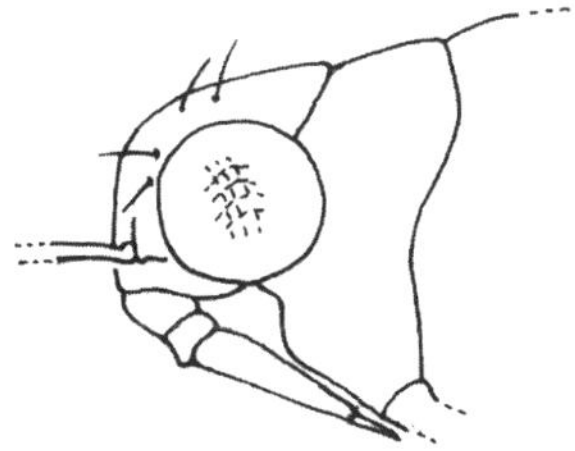

507

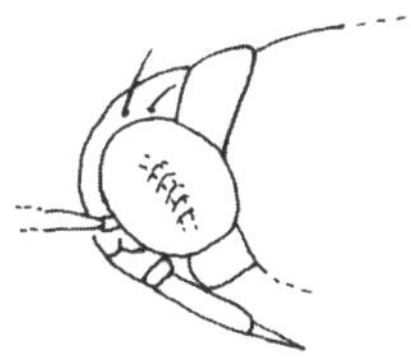

FIGURES 506 – 507
Metrocorini, head and prothorax in lateral view. **506** *Metrocoris* – **507** *Ventidius*.

5 Body usually not distinctly flattened (except in female *V. hungerfordi*), in anterior portion of thorax distinctly narrowed (figs. 537-540). Antennal segment 3 of male slender (fig. 541) ▸ *Ventidius* p. 399

– Body strongly flattened, in anterior portion of thorax less narrowed (fig. 524). Metanotum with lateral longitudinal elevation reaching intersegmental suture between mesonotum and metanotum (fig. 529). Antennal segment 3 of male enlarged and with a fringe of stiff hairs along margin (figs. 525, 524) ▸ *Esakia* p. 394

Genus *Asclepios*

Figures 508-513

Diagnosis

Small to medium-sized (3.0-4.0 mm), always apterous. Extensive yellowish markings on head, pronotum, thoracic pleura and abdominal dorsum; venter chiefly pale. Intersegmental suture between mesonotum and metanotum distinct laterally. Male fore femora always either ventrally dilated, tuberculate or spinous. First segment of fore tarsus very short. Male abdominal segment 8 (fig. 509) longer than wide, without spiracular processes, styliform processes straight, apically blunt and denticulate. Male proctiger (fig. 510) ovate; pygophore (fig. 511) with narrow bridge to connect ventral part dorsally; parameres absent; basal articulatory apparatus (fig. 512) of phallus distinctly fused with ventral part of phallotheca; vesica (fig. 513) weakly laterally sclerotized, with dorsal part membranous; vesical sclerites thick, consisting of dorsal sclerite, pair of long ventral sclerite, and two pairs of lateral sclerites. Ovipositor well developed, with longer and more hairy gonapophyses than in most *Halobates*

508

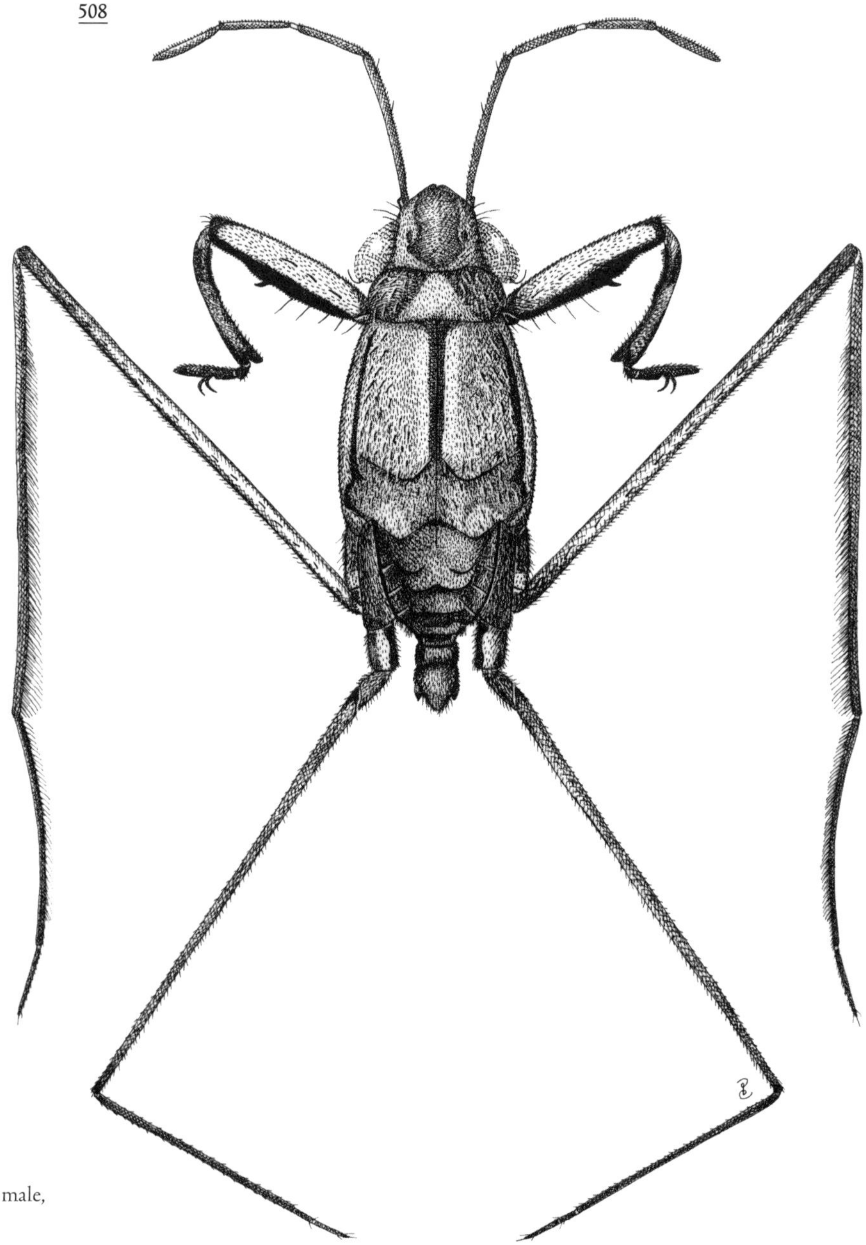

FIGURE 508
Asclepios annandalei, male, length 2.9 mm.

species; second gonapophysis with subapical filiform outgrowth; median lobes between two pairs of gonapophyses rounded.

Distribution in Malesia (fig. 599)
Of the three known species only *A. annandalei* occurs in Malesia (Phuket Island, Thailand and Singapore).

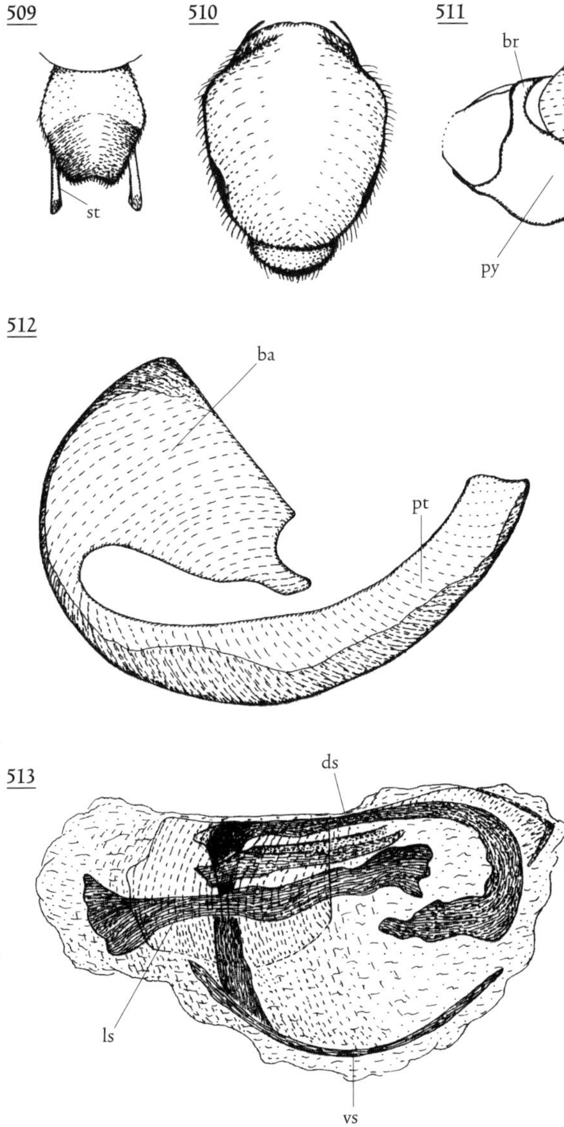

FIGURES 509 – 513 *Asclepios annandalei*, male. **509** eighth abdominal segment, ventral view – **510** proctiger, dorsal view – **511** pygophore, dorso-lateral view – **512** basal articulatory apparatus and ventral part of phallotheca, lateral view – **513** vesica, lateral view; ba = basal articulatory apparatus, br = bridge, ds = dorsal sclerite, ls = lateral sclerites, pt = phallotheca, py = pygophore, st = styliform process of abdominal segment 8, vs = ventral sclerite.

Biological notes

Asclepios species live along the coast (closer to the shore than *Halobates*), but also on small streams, ponds and irrigation ditches near estuaries. There is no biological information on the Malesian species, but *A. apicalis* (Esaki) from the eastern Palearctic coast has been found in large schools in a shaded area of a saline streamlet connecting a pond to the sea. More observations were made on *A. coreanus* Esaki in Korea. These swarmed on the sea surface close to the beach, with many pairs in copula, females carrying males on their back. More information on habitat, feeding behaviour, locomotion, copulation, and life history of

A. coreanus is given by Miyamoto (1937). Eggs of *A. coreanus* hatch after about ten days. There are five nymphal instars; throughout the nymphal stages colour and pilosity vary with instar stage. Nymphs are more inclined to stay in shaded places than adults. They feed on insects floating on the water. Cannibalism occurs when other prey is absent.

Genus *Halobates* – sea skaters, ocean striders

Figures 514-523

514

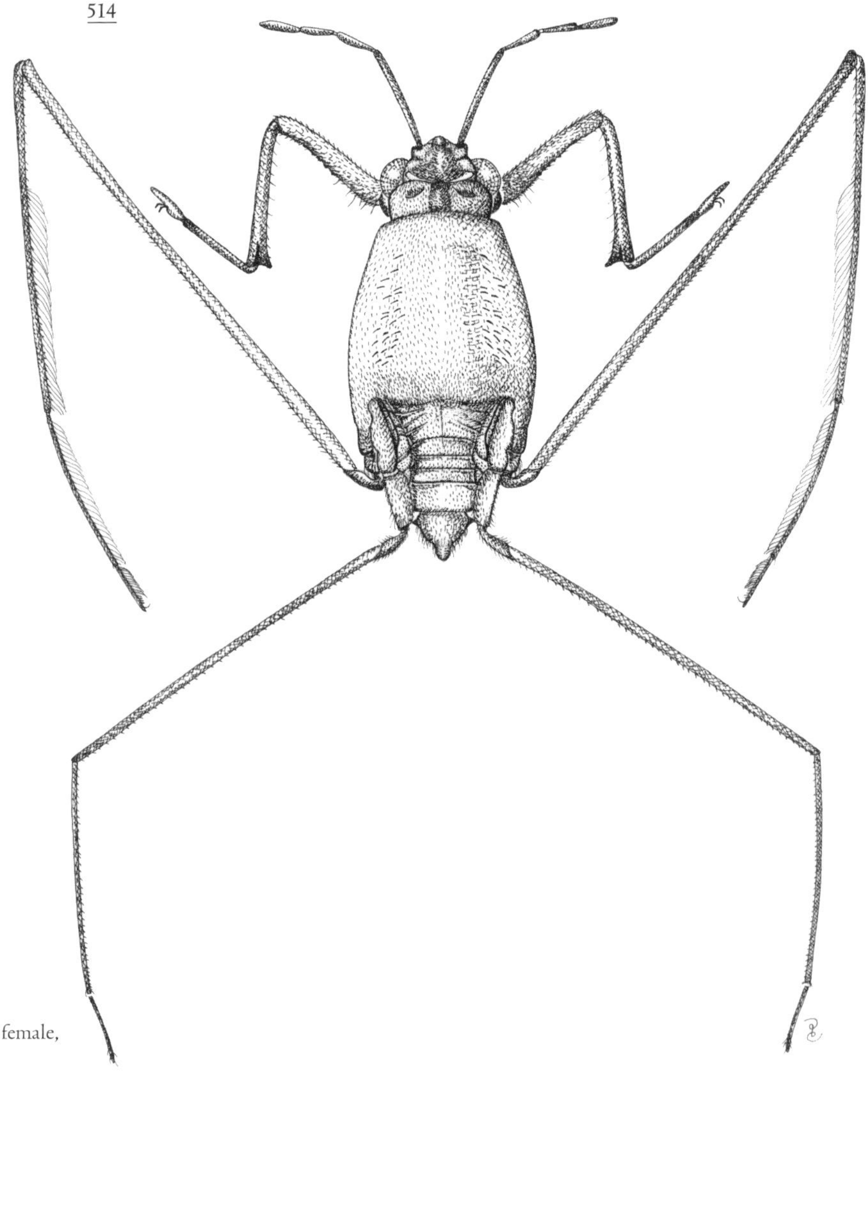

FIGURE 514
Halobates princeps, female, length 6.0 mm.

Diagnosis

Rather small (3.2-6.5 mm), apterous, typically short and broad. Males longer and thinner than females. Mostly dark or silver-grey without extensive pale markings except on venter and, in some species, on head. Antennae long, never modified. Intersegmental suture between mesonotum and metanotum reduced to pair of lateral V-shaped pits. Male fore femora rarely ventrally modified. First segment of fore tarsus of variable length. Flexible intersegmental membranes between segments 6-7 enable female abdomen to expand to nearly twice the normal length to accommodate large amounts of eggs. Male genital segments very prominent (fig. 516). Male abdominal segment 8 wider than long (figs. 517, 518) with rounded, tuberculate or finger-like spiracular processes; styliform processes present, usually long and slender. Male proctiger pentagonal in outline (fig. 516), usually distinctly laterally dilated. Pygophore boat-shaped (fig. 520, 521), tergal part reduced to narrow bridge. Parameres were considered absent in this genus except in *H. mjobergi* Hale (Andersen, 1991b), but we have found that the left paramere, although very much reduced, exists in *H. micans* (fig. 519): the left upper margin of pygophore has a shallow notch (fig. 521), fitting the position of the paramere. Basal articulatory apparatus of phallus strongly developed (fig. 522), reaching half way of phallotheca. Phallotheca ventrally and laterally sclerotized, but dorsally membranous. Vesical sclerites (fig. 523) thick and strongly darkened: dorsal sclerites distally and proximally fused, on dorsal side separated from each other; ventral sclerites fused with dorsal sclerite and very long; usually one pair of lateral sclerites, but two pairs in some species; basal sclerites small and less sclerotized. Ovipositor well-developed, generally shorter than in *Asclepios*; especially open-sea species have shorter gonapophyses. Nymphs are very similar to adults in external morphology, but there is no external sexual dimorphism until last nymphal instar (Herring, 1961; Andersen & Polhemus, 1976; Andersen, 1982a; Andersen & Weir, 1994b).

Fossil record

Although all recent species are restricted to the Indo-Pacific, with the exception of *H. micans*, there is a 45 million year old fossil from a marine deposit in the Mediterranean, named *H. ruffoi* Andersen et al. At the time, the Mediterranean was part of the Tethys Sea and had a wide connection with waters around India. This record gives a minimum age for the differentiation of this genus.

Distribution in Malesia

Halobates is distributed throughout Malesia, both in coastal water and on open sea, with fourteen species. The genus can be divided in two groups: 1) a 'strictly open-ocean group' (five species of tropical ocean striders), and 2) a 'coastal inshore group' (including the additional 38 species living in sheltered coastal waters throughout the tropical Indo-Pacific region) (Cheng, 1989). *Halobates* generally occurs between 40° N and 40° S. The majority of the species are found

FIGURES 515 – 523
Halobates micans, male.
515 head, frontal view – **516** apex of abdomen, dorsal view – **517** eighth abdominal segment, ventral view – **518** eighth abdominal segment, dorsal view – **519** left paramere, external view – **520** pygophore and proctiger, left lateral view – **521** pygophore, right lateral view – **522** basal articulatory apparatus and vesica, dorsal view – **523** vesical sclerites, dorsolateral view; ba = basal articulatory apparatus, bm = basal margin of clypeus, ds = dorsal sclerites, fp = finger-like spiracular process, i = indentation for attachment of paramere, ls = lateral sclerite, pr = proctiger, py = pygophore, st = styliform process of abdominal segment 8, vs = ventral sclerite.

515

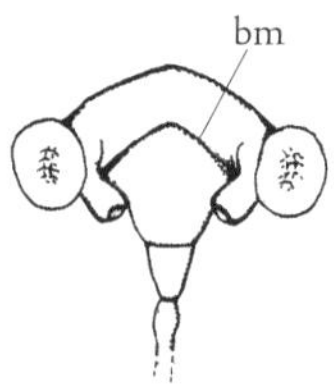

516

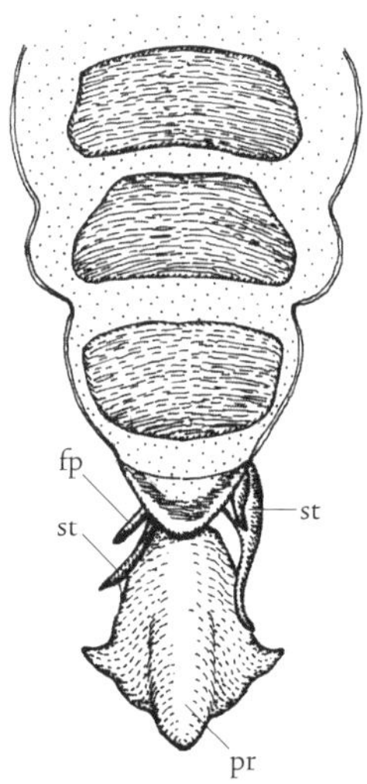

517

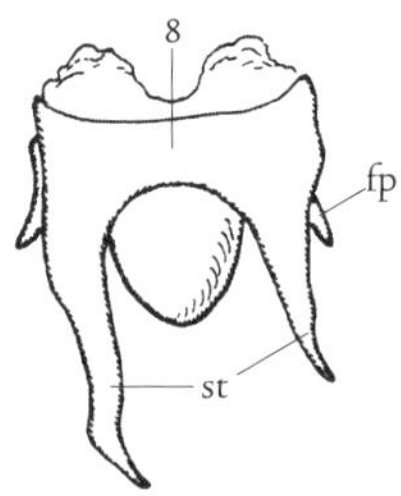

518

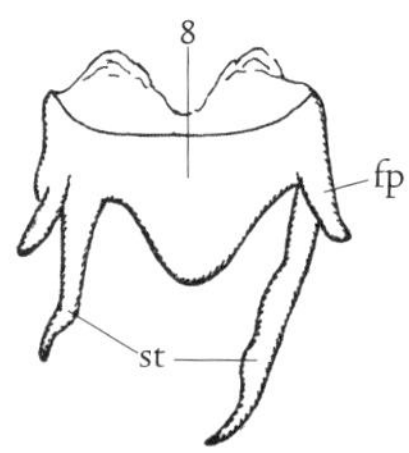

519

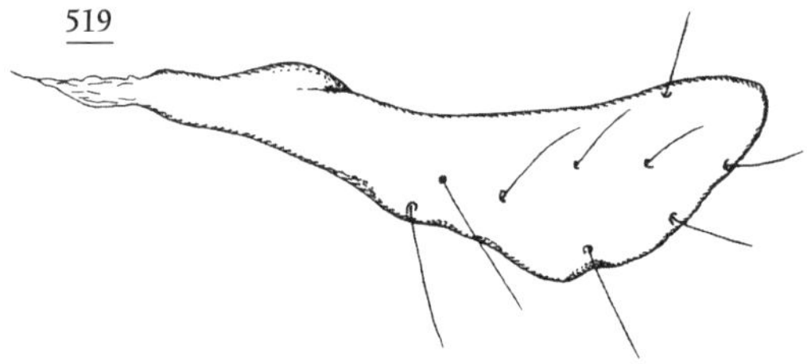

520

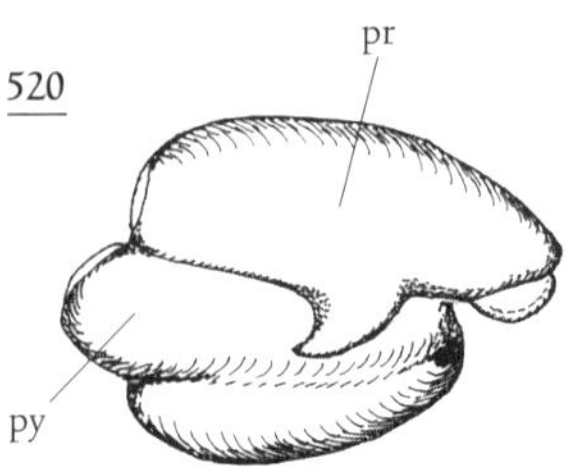

521

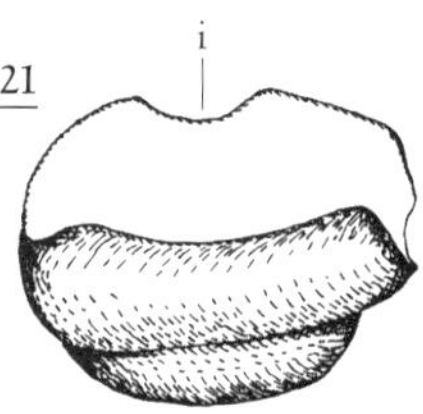

523

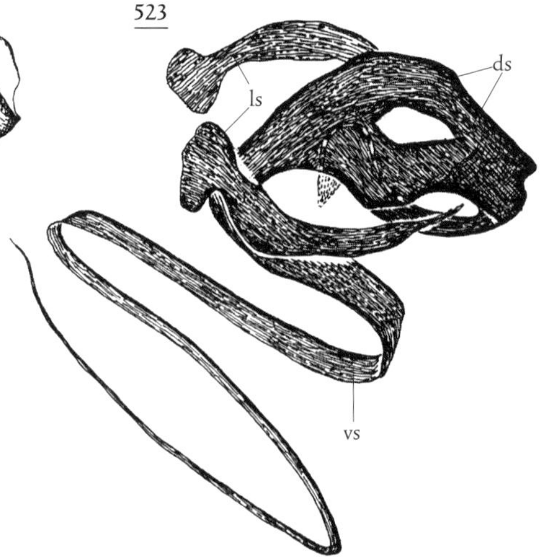

522

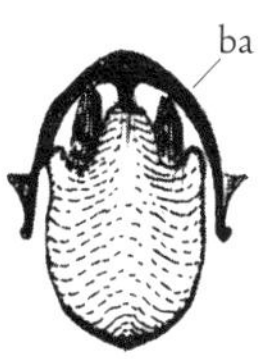

in the Indo-Malaysian and Australian region (Herring, 1961; Andersen & J. Polhemus, 1976; Polhemus & Cheng, 1982; Cheng, 1989; Andersen, 1991a; Andersen & Weir, 1994b).

Biological notes

Most *Halobates* species live near the coast, but a group of five species is pelagic. Individuals of this 'open-ocean group' are seldom found near the shore except after storms, when they can even be washed ashore. Near-shore species of *Halobates* seem to prefer coastal areas fringed with mangrove, where especially nymphs may be abundant. In coastal areas without mangroves young nymphs can be found close to overhanging branches or rocks. Adults usually frequent the more open areas away from the mangroves. Along exposed coasts *Halobates* are rarely found. Two species have been found on (freshwater) rivers, far from the sea (J. Polhemus, 1982; Andersen & Weir, 2003b).

Detailed biological observations on *Halobates* have been made on a few coastal and open-ocean species by Cheng (1989), who proved that there are several consistent differences between the two groups. The olive-green eggs of coastal *Halobates* are laid singly or in clusters of 2-16 together on substrates near the shore. In *H. fijiensis* Herring oviposition takes place almost entirely within one hour of low tide, when the substratum is exposed (Cheng & Frank, 1993). In an aquarium *H. flaviventris* glued the eggs to the wall of the aquarium 9-39 mm below the surface (Cheng, 1985). In the open ocean, eggs are laid on floating substrates (feathers, wood, seeds, tar lumps, etcetera). On one occasion fourth-instar nymphs were found carrying eggs on their back (Cheng, 1985). In the laboratory, development of off-shore living species took much longer than in coastal species (Cheng, 1985). Development from egg to adult may last 60-70 days. The most important factor for embryonic and nymphal development appears to be water temperature.

Halobates have a varied diet. Near-shore species feed on insects fallen in the water, which can be abundant, open-ocean species feed on various planktonic organisms as well as on floating insects. Cannibalism has been observed both in the field and in the laboratory, also among young nymphs. Seabirds are the most important predators of *Halobates*, but warblers and fishes feed on them as well. *Halobates* are strongly attracted to light, males more so than females.

Certain organic compounds and heavy metals are known to be concentrated on the sea-surface, but concentrations are difficult to measure. High levels of DDT and heavy metals such as cadmium (Cd), copper (Cu), iron (Fe), lead (Pb), nickel (Ni) and zinc (Zn) have been found in several pelagic water bugs, in which they can be 5-10 times higher than in near-shore water bugs. The cadmium concentration in open-ocean *Halobates* tends to be much higher than the highest values obtained for zooplankton samples. They obtain these metals either by accumulation through the food chain or by drinking. *Halobates* could conceivably be used as an indicator of heavy metal pollution in the open sea. Especially monitoring of cadmium concentrations is of great importance as it is toxic to humans and other animals (Herring, 1961; Andersen & J. Polhemus, 1976; Andersen, 1982a, 1999b; Cheng, 1981,1985, 1989).

Genus *Esakia*

Figures 524-531

524

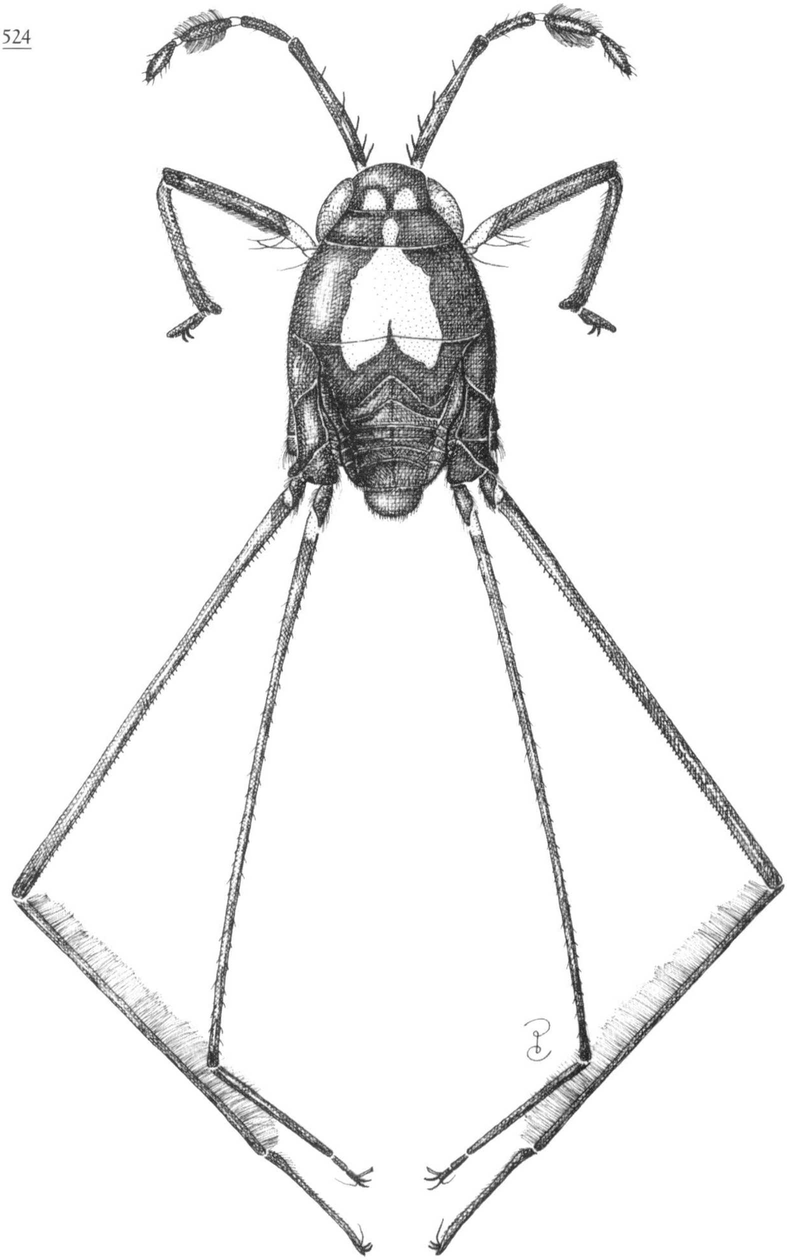

FIGURE 524
Esakia ventidioides, apterous male, length 2.2 mm.

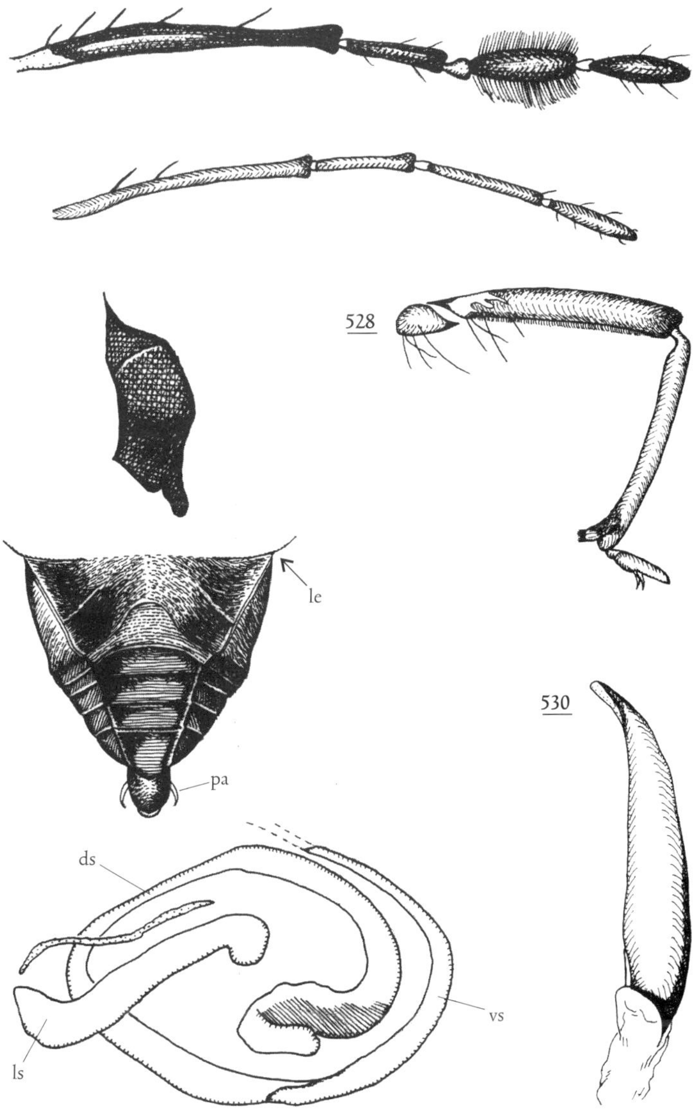

FIGURES 525 – 531
Esakia ventidioides
525 antenna of male – **526** antenna of female – **527** metacetabula, latero-dorsal view – **528** male, fore leg – **529** male, metanotum and abdomen, dorsal view – **530** male, paramere – **531** male, vesical sclerites, lateral view; ds = dorsal sclerite, le = lateral longitudinal elevation, ls = lateral sclerite, pa = paramere, vs = ventral sclerite.

Diagnosis

Dark and small (males 1.93-2.87 mm long, 1.30-1.70 mm wide, females 2.10-3.10 long, 1.76-1.90 wide). Eyes extend to anterolateral angle of mesonotum (fig. 507). First antennal segment of male longer than three following segments together (see *Ventidius*) and third antennal segment with stiff hairs on margin (fig. 525), female antennae without modification (fig. 526). Lateral longitudinal elevation on metanotum reaches intersegmental suture between mesonotum and metanotum (fig. 529). Metacetabula with bilobate posterolateral angle (fig. 527). Male fore tibia with subdistal patch of stiff setae, distal spur with thick stiff setae (fig. 528). Genital segments not prominent. Paramere small (fig. 530); vesical sclerites thick, with dorsal sclerite, pair of lateral sclerites and pair of long ventral sclerites (fig. 531).

Distribution in Malesia (fig. 600)

Six species of *Esakia* have so far been reported in Malesia. A vicariant distributional pattern has been noticed around continental Malaysia, Sumatra and the Philippines. Undescribed species extend the range to Vietnam, Thailand and Borneo.

Biological notes

Little information has been published on biology of *Esakia* species. They have been collected from irrigation ditches and mountain lakes, but less commonly than lentic mountainous species of *Metrocoris*. Observations from the Philippines and Borneo indicate that some species skate individually in lentic bays of streams, similar to, and sometimes mixed with, *Ventidius*, while other species live in small or large groups under lotic conditions in the middle of streams with low regular water velocity, smooth surface, and usually in deep water (Zettel pers. obs.). In the latter group adults are usually separated from nymphs.

Genus *Metrocoris*

Figures 532-536

Diagnosis

Medium-sized to large (3.3-7.6 mm, maximal body width 2.6-3.2 mm), males larger than females. Yellowish to light brown with complex black pattern. Body subtriangular, dorso-ventrally moderately flattened. Eyes not extending to anterolateral angles of mesonotum (fig. 506). In males fore femur and genital segments prominent, pygophore boat-shaped (fig. 534), parameres well developed. Female fore femur slender, not distinctly modified. Important diagnostic characters are male fore legs, especially amount of swelling and tooth-like armature on ventral surface of fore femur (fig. 532), fore tibia, and genitalia. Most reliable diagnostic character is the usually large and falciform paramere (fig. 536). Vesical sclerites (fig. 535) show distinct characters for dividing genus in species groups. Phallic articulatory apparatus (fig. 536) is sclerotized along edge, remaining part almost transparant (this type of articulatory apparatus also exist in *Ptilomera* (fig. 496)). Phallotheca dorsally open (fig. 536) with three longitudinal sclerotized strips. Basic model of vesical sclerites contains clearly separated dorsal sclerites, pair of ventral sclerites and pair of lateral sclerites; in addition, apical accessory sclerites and apical wishbone-shaped sclerites are found in some species groups. Except for shape of sternum 7 (fig. 533), which differs according to species groups, female terminalia are difficult to use for identification.

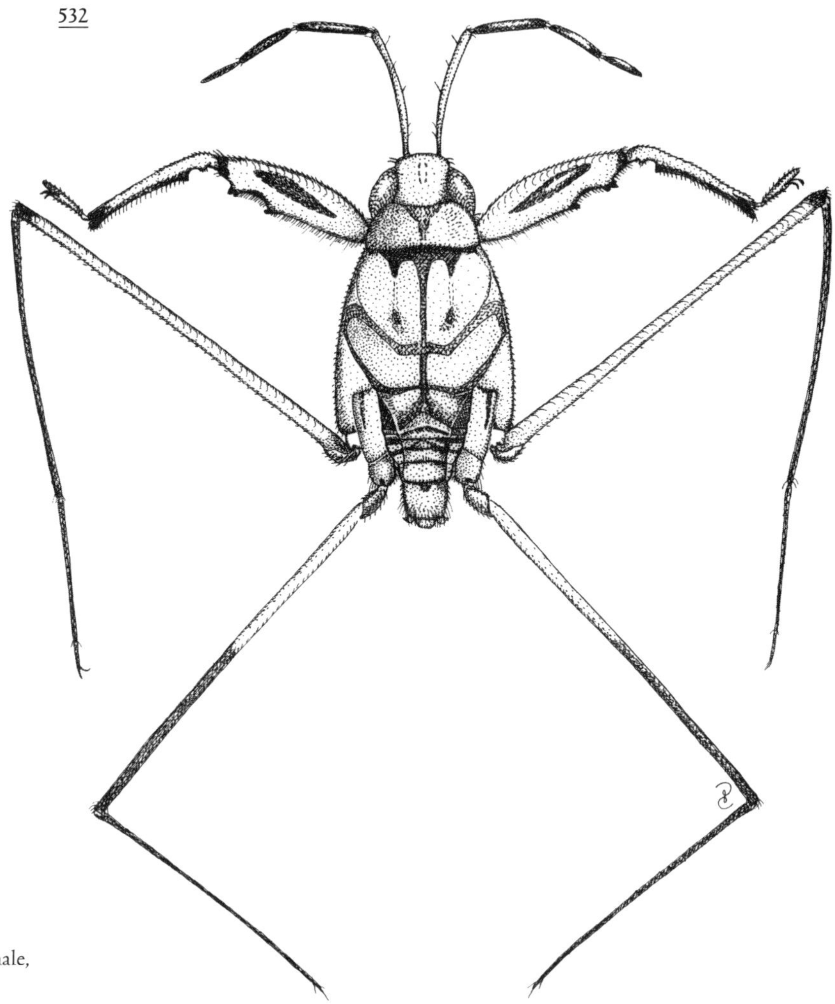

FIGURE 532
Metrocoris malayensis, paratype, apterous male, length 5.3 mm.

Distribution in Malesia (fig. 600)

Fourteen species have been reported in Malesia (Chen & Nieser, 1992, 1993, 1996; D. Polhemus 1990, 1998); a number of undescribed species have been collected from the Philippines. *Metrocoris* shows an active divergence at either side of Wallace's Line. Except for a few widely distributed species (e.g. *M. tenuicornis* and *M. nigrofasciatus*) most are endemic to areas such as continental Malaysia, each of the Greater Sunda Islands, the Philippines, or Sulawesi.

Biological notes

Metrocoris prefer quiet bays, smoothly flowing parts of streams, or the lower side of large rocks in the middle of the stream. Because of their preference for headwater streams and streamlets, they are usually encountered in upland habitats. However, they are not restricted to mountainous areas and some species are

533

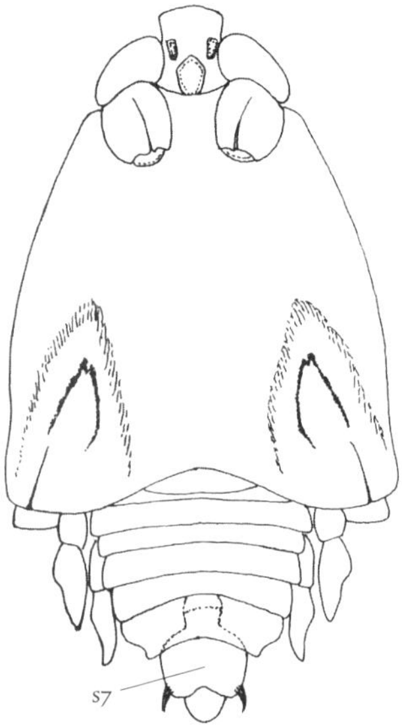

534

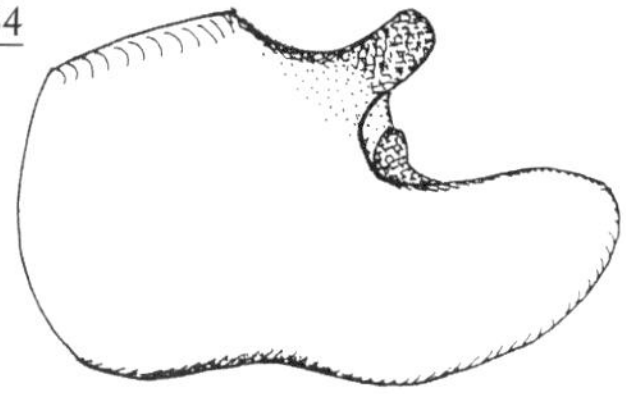

535

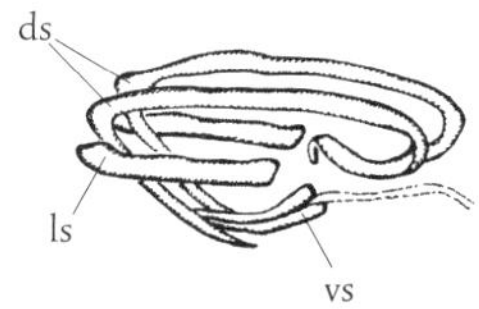

FIGURES 533 – 536
Metrocoris malayensis
533 female, ventral view, appendages removed – 534 pygophore, lateral view – 535 vesical sclerites – 536 paramere, basal articulatory apparatus and phallotheca, lateral view; ba = basal articulatory apparatus of phallus ds = dorsal sclerite, ls = lateral sclerite, pa = paramere, pt = phallotheca, s7 = seventh abdominal sternite, vs = ventral sclerite.

536

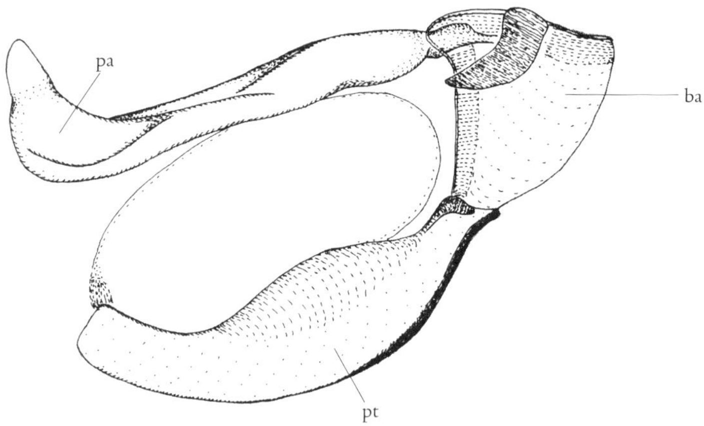

found almost at sea level. *Metrocoris* occurring at higher elevations tend to be endemic, whereas lowland species, such as M. *lituratus* (Stål) and M. *tenuicornis*, are widespread.

Eggs of M. *tenuicornis* hatch under laboratory conditions within 10 days; the nymphal stage takes approximately 40 days. Mating occurs soon after adulthood is reached, and eggs are laid 14-20 days later, giving a total generation time of about 60 days (Cheng, 1967b). *Metrocoris histrio* (White), studied in a headwater stream in Central Honshu, Japan, has two generations per year. Mating was observed only in December and overwintering of eggs was postulated. Nymphs were observed from May to October (Ban et al., 1988).

In *Metrocoris* intraspecific variation is observed in some species, such as M. *acutus*, M. *tenuicornis* and M. *nieseri*: individuals differ in size, length of appendages and/ or fore femur in males, but the genital structures remain the same.

Genus *Ventidius*

Figures 537-561

537

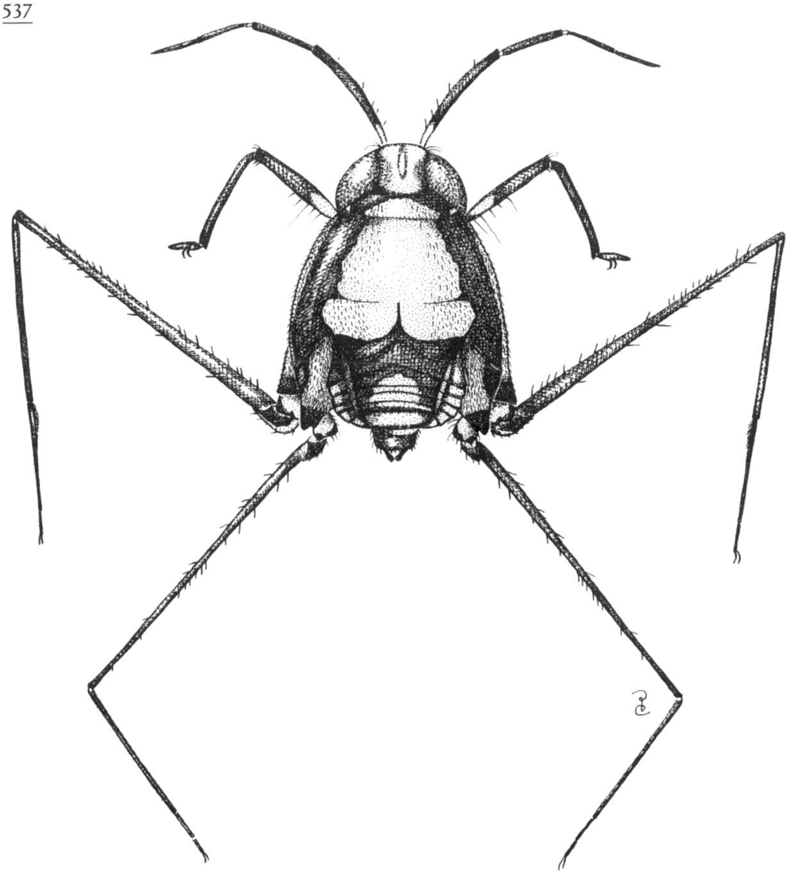

FIGURE 537
Ventidius werneri, apterous male, length 2.4 mm.

Diagnosis

Small to medium-sized (2.37-4.60 mm), males usually smaller than females, but some exceptions occur. Apterous or macropterous. Black and green (yellowish after death), more or less shiny. Body relatively broader than *Metrocoris*, dorsal surface convex. Eyes extend to anterolateral angles of mesonotum (fig. 507). First antennal segment of male always shorter than following segments together. Pronotum short in apterous morph, not longer than half of head length, pronotum subtriangular in macropterous morphs (figs. 544, 551). Metanotum slightly shorter than mesonotum, suture in-between posteriorly situated at middle, often very faint. Metanotum fused with first abdominal tergite. Posterolateral angle of metacetabula (figs. 545, 546, 552, 553, 558) in most species bilobed, except in *V. malayensis*-group. Fore leg of males slightly stouter; in some

538

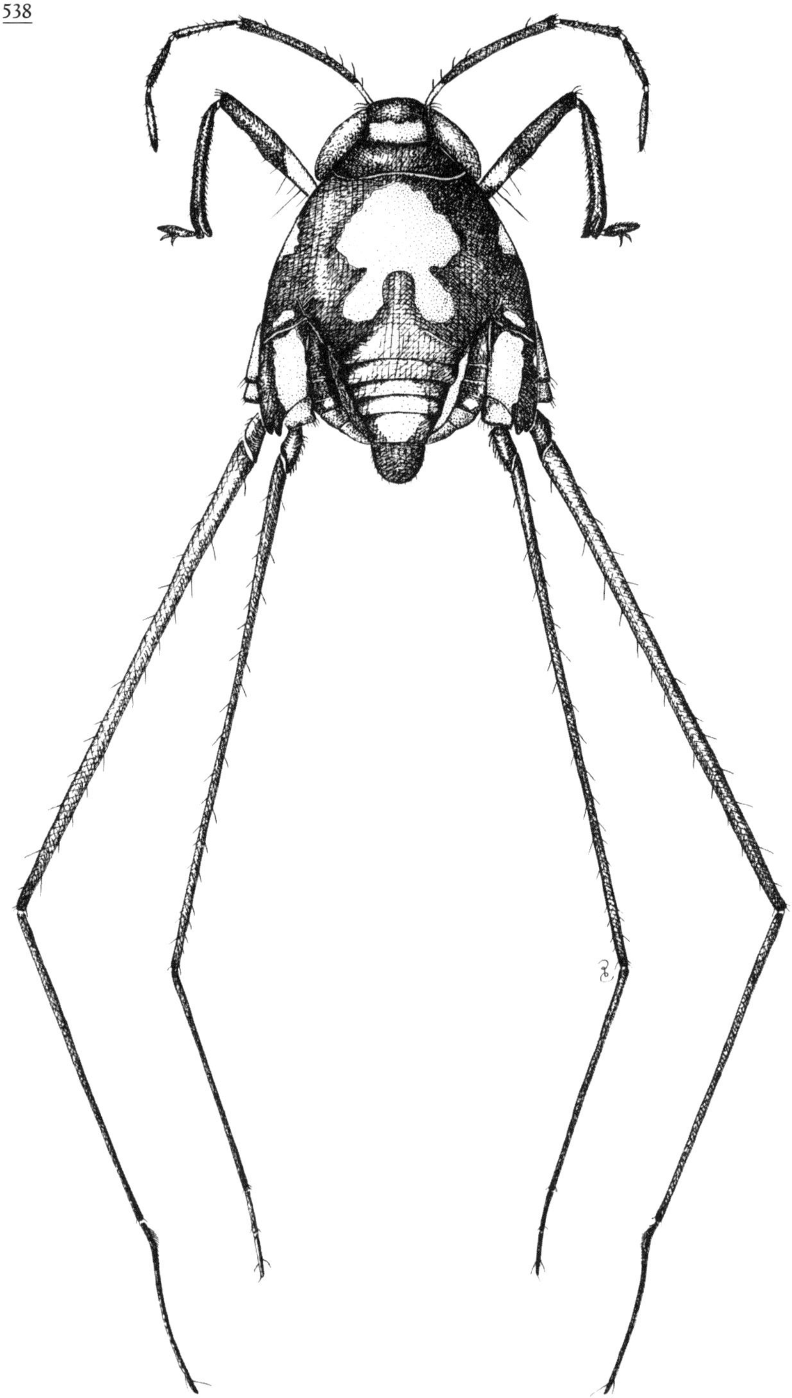

FIGURE 538
Ventidius (Ventidioides) xyele,
paratype, apterous male,
length 3.0 mm.

539

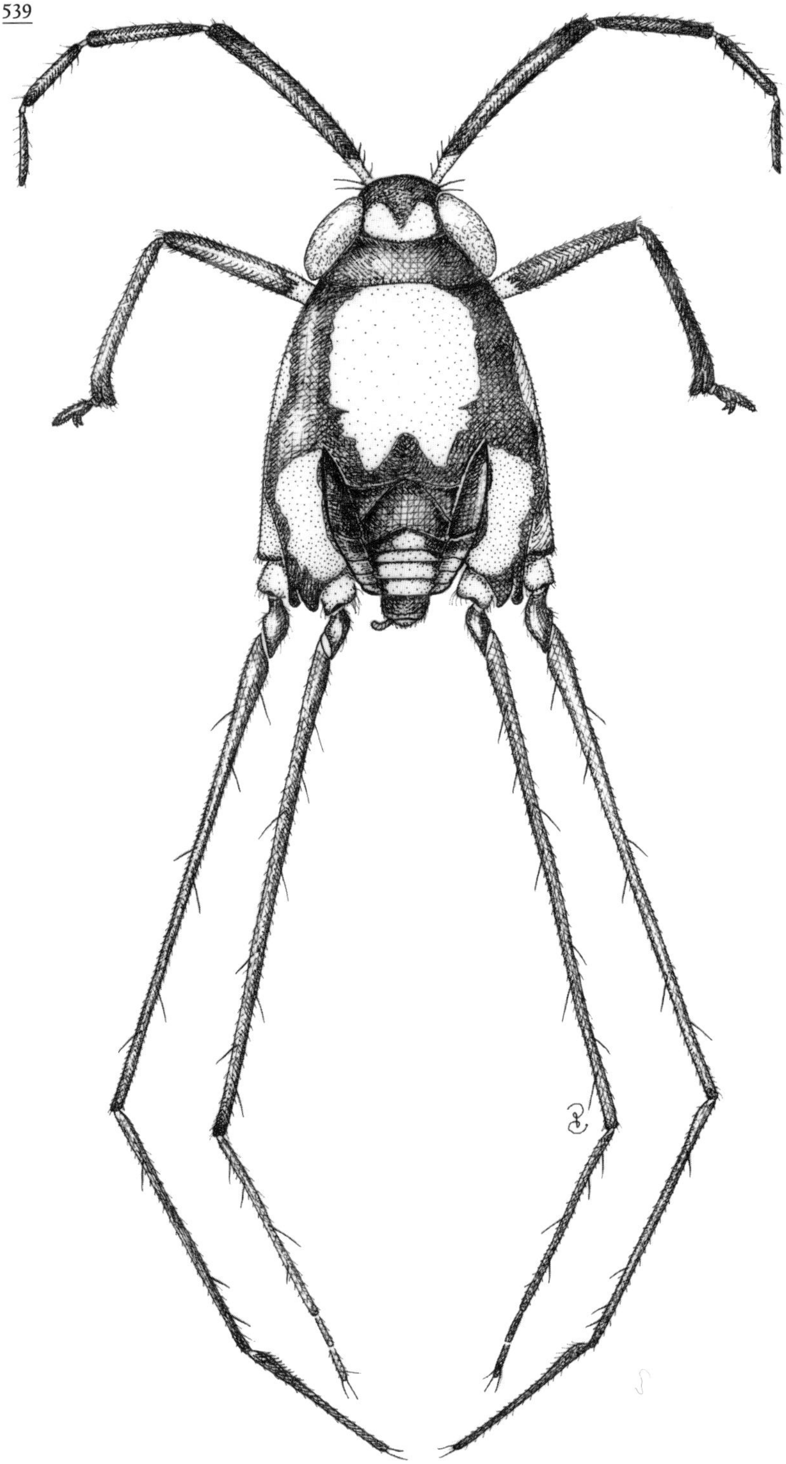

FIGURE 539
Ventidius (Ventidiopsis) yangae, paratype, apterous male, length 2.7 mm.

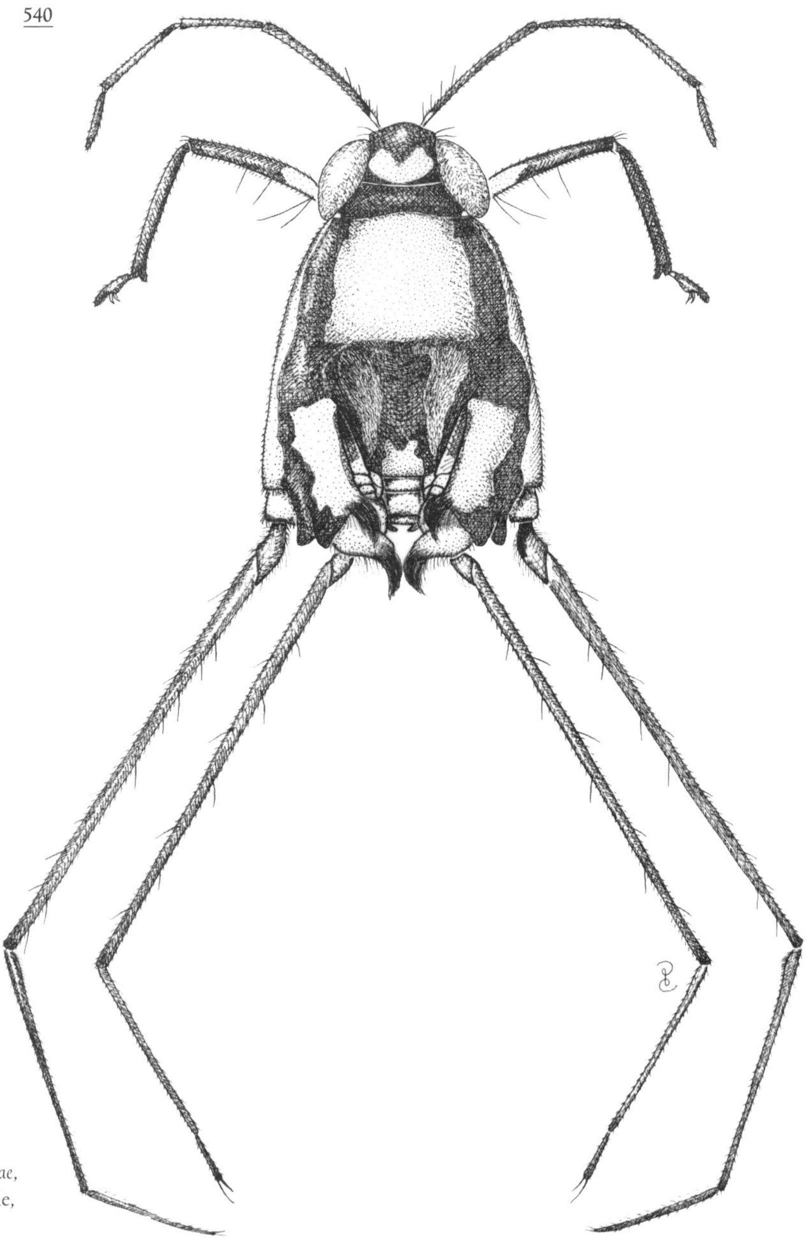

FIGURE 540
Ventidius (Ventidiopsis) yangae, holotype, apterous female, length 2.9 mm.

species male fore trochanter with hair tuft (fig. 550) and fore tibia with distal patch of spines (figs. 542, 549). Abdomen broad, male genital segments not prominent. Specific identification usually reliable in males only. Most important are parameres (which may be asymmetrical, figs. 547, 554, 555, 560) and vesical structures (figs. 548, 561). Basic structure of vesical sclerites consists of dorsal sclerite, pair of lateral sclerites and pair of ventral sclerites; some species have additional pair of long apical sclerites (fig. 556). Seventh sternite of female (fig. 559) modified in subgenus *Ventidiopsis*, with two pairs of lateral triangular lobes. Other important characters are length of antennae and fore tarsi, shape of posterolateral angle of metacetabula (truncate or bilobate), pilosity on mesosternum and eighth abdominal segment, and modification on ventral surface of male fore femur. *Ventidius* is divided in subgenera *Ventidius* s.str., *V.* (*Ventidioides*), and *V.* (*Ventidiopsis*) (Chen & Zettel, 1999).

Distribution in Malesia (fig. 601)

Ventidius have an Oriental distribution. So far 23 species have been described, seventeen of which are found in Malesia (Chen & Zettel, 1999). Centres of diversity are found in Borneo (nine species), Continental Malaysia (six), and continental South-east Asia (excluding continental Malaysia; seven).

Key to the subgenera of genus Ventidius

1 Posterolateral angle of metacetabula truncate or slightly bilobate in caudal view (figs. 545, 546); parameres symetrical; pronotum yellowish with dark lateral stripes and/or triangular dark mark in middle of hind margin (pronotum blackish in *V. henryi*) (fig. 537) ▸ *Ventidius* **s.str.**

– Posterolateral angle of metacetabula distinctly bilobate in dorsal view (figs. 552, 553); parameres asymmetrical; pronotum usually blackish ▸ 2

2 Abdominal tergites 3 and 4 in apterous morph fused; female with metaceabula and metacoxae each with a stout, medianly directed hairy processus (fig. 540); metanotum and anterior tergite concave; suture between mesonotum and metanotum with or without two hair tufts; male with very stout antennae, segment 1 thicker than fore tibia; segments 2 and 3 slightly flattened ▸ *Ventidius* (*Ventidiopsis*)

– Abdominal tergites 3 and 4 not fused; female without medianly directed processes on metacetabula and metacoxae; metanotum and anterior tergites slightly concave; suture between mesonotum and metanotum without tufts of hairs; male with slender antennae; segment 1 thinner than fore tibia, segments 2 and 3 not flattened (fig. 538) ▸ *Ventidius* (*Ventidioides*)

Biological notes

Ventidius species are typical inhabitants of slow-flowing forest streams. The preferred habitats appear to be lotic areas of brooks, streams and rivers. Because of this preference *Ventidius* are mainly encountered in hilly and lowland areas. The life history remains unknown.

541

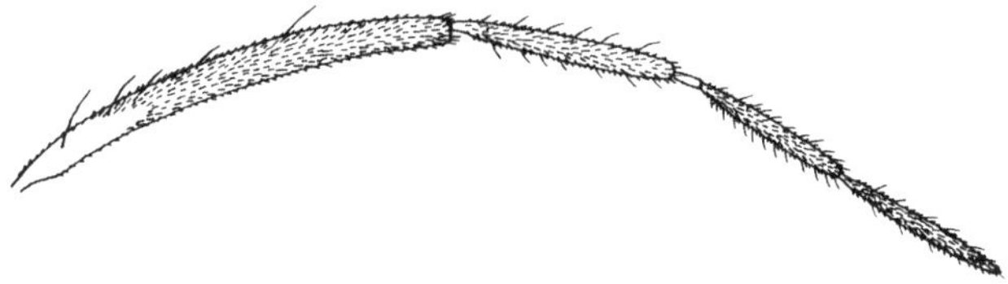

542

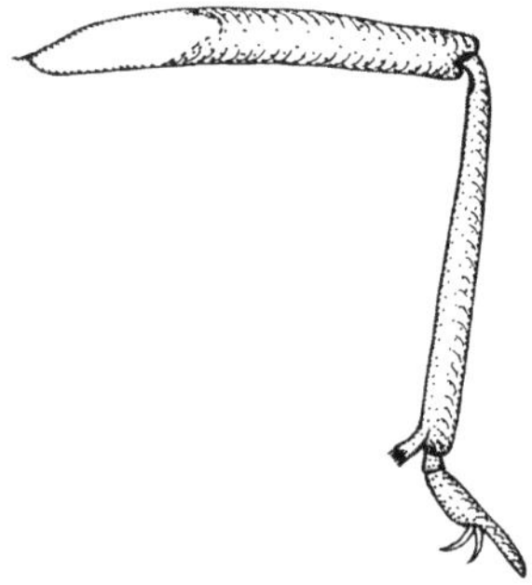

543

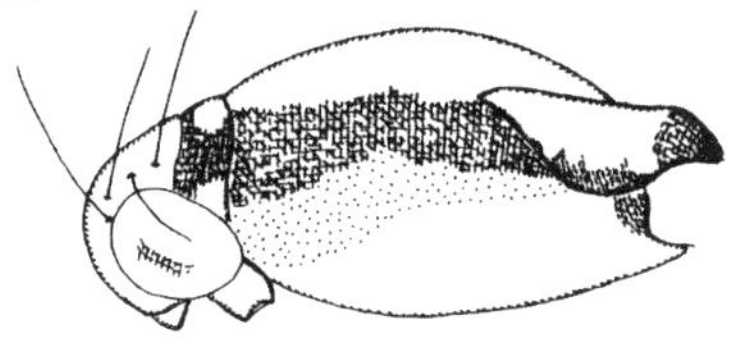

544

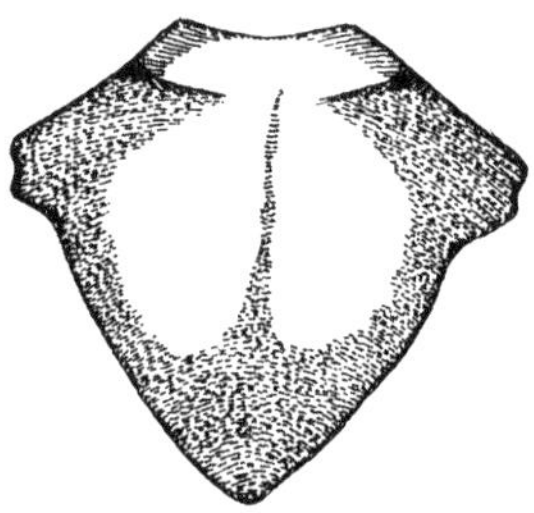

545

546

547

548

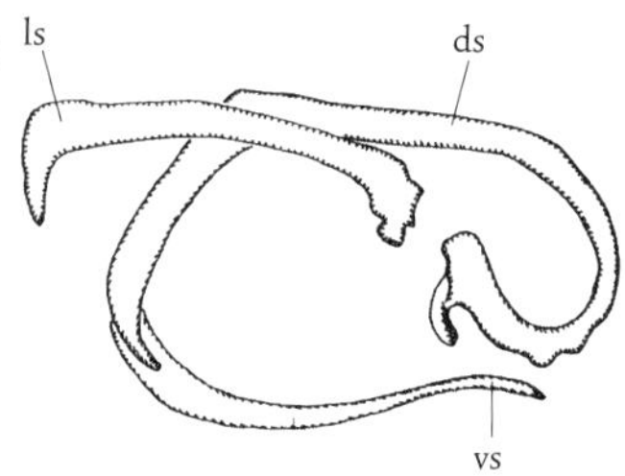

FIGURES 541 – 548
Ventidius werneri, male.
541 antenna – **542** fore leg – **543** head and thorax, lateral view, apterous morph – **544** pronotum of macropterous morph – **545** metacetabulum, dorsal view – **546** metacetabulum, dorsolateral view – **547** left paramere, external view – **548** vesical sclerites, lateral view; ds = dorsal sclerite, ls = lateral sclerite, vs = ventral sclerite.

Taxonomic notes

(On the status of *Ventidius modulatus* Lundblad, 1933.) Considering the large scale of distribution in Oriental region, and in order to try to avoid further confusion, the species *Ventidius* s.str. *modulatus* Lundblad, 1933 is synonymized here. This species was described from Java, Indonesia. It is widely distributed in the Oriental region (Sri Lanka, India, Malay Archipelago, Borneo, Indonesia). In their revision of the genus *Ventidius*, Chen & Zettel (1999) suspected that *V. modulatus* is *V. distanti* Paiva, but lacked the possibility to compare material with the type of *V. distanti*, so that *V. modulatus* was kept as a valid species in that

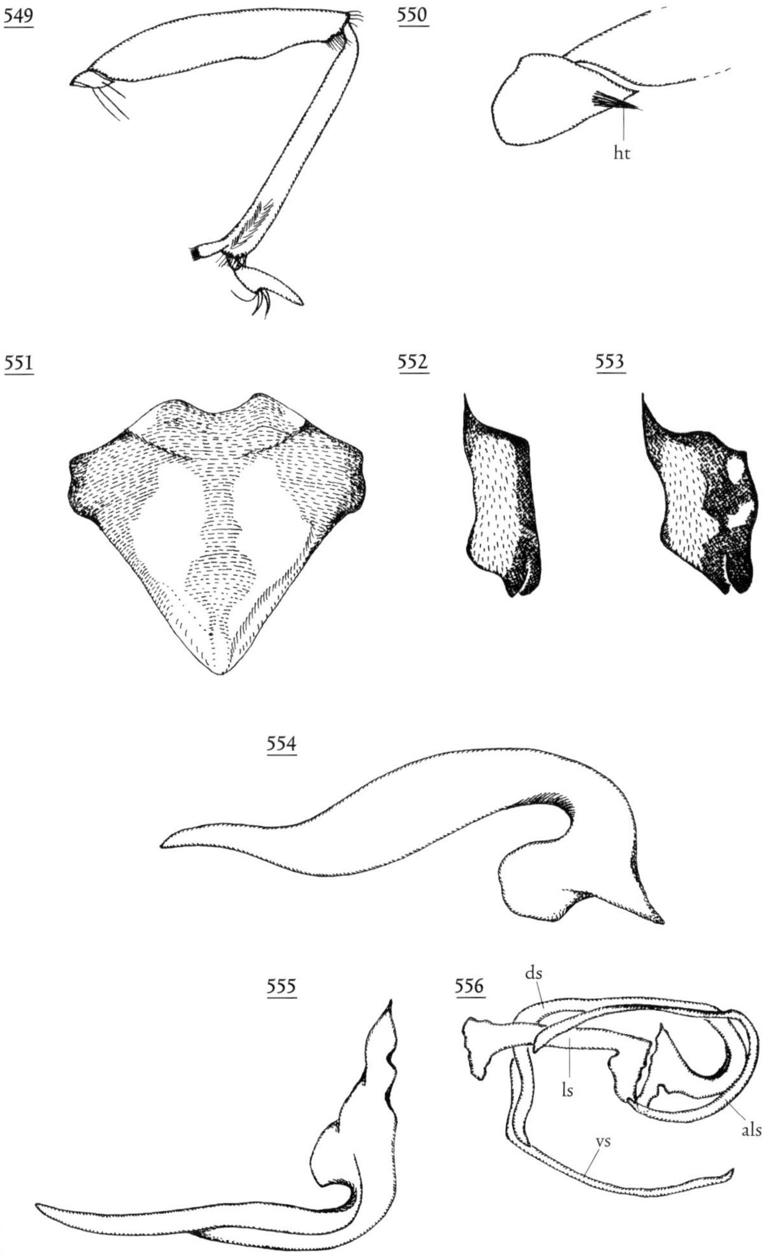

FIGURES 549 – 556
Ventidius (Ventidioides) xyele. All male, except 552 (female).
549 fore leg, dorsal view – **550** fore trochanter, ventral view – **551** pronotum of macropterous morph – **552** metacetabulum, dorsal view – **553** metacetabulum, dorsolateral view – **554** left paramere, external view – **555** right paramere, external view – **556** vesical sclerites, lateral view; als = additional lateral sclerite, ds = dorsal sclerite, ht = tuft of hairs, ls = lateral sclerites, vs = ventral sclerite.

paper. Since more samples were recently collected near the type locality of *V. distanti* (Yawnghwe State, 20°35′ N, 96°21′ E), and the specimens are inseparable from *V. modulatus*, we consider that *V. modulatus* is synonymous to *V. distanti*. As the type series of *V. distanti*, deposited in Zoological Survey of India, type No. 7152/H.i., is not traceable there, and no record is available even in the type-data sheet (Bal & Thirumalai, personal comm.), a neotype is designated here:

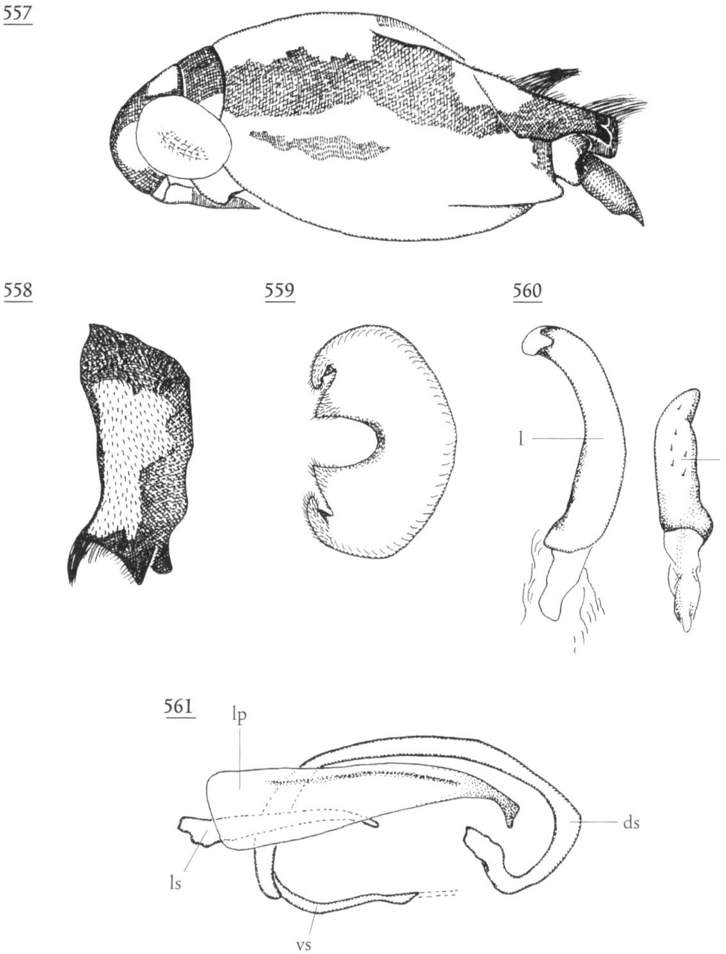

FIGURES 557 – 561 *Ventidius (Ventidiopsis) yangae* **557** female, head and thorax, lateral view – **558** metacetabulum, dorsolateral view – **559** female, sternite 7, frontal view after dissection – **560** male, parameres, external view; l = left paramere, r = right paramere – **561** vesical sclerites, lateral view; ds = dorsal sclerite, vs = ventral sclerite, ls = lateral sclerite, lp = lateral plate.

#36 in Natural History Museum Vienna, Austria, macropterous male, Myanmar: Bago Division, Bago Yoma (ca. 18° 47′ N, 96° 21′ E), Sein Ya Forest Camp, 1700 m, 29.10.1998, leg. H. Schillhammer.

REFERENCES – Halobatinae. Note on biology of *Asclepios*, differences between *Asclepios* and *Halobates* (Esaki, 1930b); description of Indonesian Halobatinae (Lundblad, 1933a); biology of *Asclepios coreanus* (Miyamoto, 1937); revision of *Esakia* (Hungerford & Matsuda, 1958c); structural analysis and classification of Gerridae (Matsuda, 1960); revision of *Halobates* (Herring, 1961); report of *Esakia* in peninsular Malaysia (Cheng, 1966b); life history of *M. tenuicornis* (Cheng, 1967b); general introduction to marine Gerromorpha (Andersen & Polhemus, 1976); review of *Halobates* in Micronesia (Cheng, 1981); rearing experiment of *H. mariannarum* (Cheng, 1981); morphology, phylogeny, adaptations, biogeography and classification of Gerromorpha (Andersen, 1982a); distributional notes on Halobatini (J. Polhemus & Cheng, 1982); biology of *Halobates* (Cheng, 1985); life history of *M. histrio* (White) (Ban et al., 1988); biogeography of *Halobates* (Cheng, 1989); revision of *Metrocoris* from the Malay Archipelago and the Philippines (D. Polhemus, 1990); biogeography of marine Gerromorpha (Andersen, 1991a); genitalia and phylogeny of Halobatini (Andersen, 1991b); key to Indian Ocean species of Halobatini (Andersen & Foster, 1992); key to Sulawesi *Metrocoris* (Chen & Nieser, 1992); revision of *Metrocoris* (Chen & Nieser, 1993); key, phylogeny and distribution of Australian *Halobates* (Andersen & Weir, 1994b); description of *Metrocoris medius* from Sangihe (Chen & Nieser, 1996); description of new species of *Metrocoris* (D. Polhemus, 1998); evolution of adaptation to marine habitats in Gerromorpha (Andersen, 1999b); revision of *Ventidius* (Chen & Zettel, 1999).

chapter 5

Checklist of Malesian species of Nepomorpha and Gerromorpha

In this checklist, families and subfamilies are arranged according to recent insight of phylogeny (Andersen, 1982a; Mahner, 1993), with some modifications (see chapter 3). Genera are alphabetically arranged within subfamilies. In a few cases tribes have been incorporated in the checklist; they also are arranged in alphabetical order within the subfamily. The exception are subspecies, in which the nominate subspecies are mentioned first. If only one subspecies is mentioned the other subspecies occur(s) outside Malesia.
The classification above species level of Gerromorpha is treated by Andersen (1982a), a few older names of categories above generic level can be found in Štys & Kerzhner (1975). Higher classification of Nepomorpha is treated in Štys & Jansson (1988). Taxonomic details such as author and year of establishing suprageneric categories can be found in these publications. The name of the genus in which a species was originally described is given between brackets. Synonyms are given only when the junior synonym has been widely used until recently.

There are many recognized but as yet undescribed species in various collections. These are ignored in this checklist. One of the consequences is that the distribution of some genera provides an incomplete picture.
Some unpublished records based on material in our private collections have been added (indicated with '!' after the abbreviation of the area). Doubtful records are indicated with a question mark (?).
Species described after the closing date of the manuscript, and thus not considered in the main text of this book, are indicated with an asterisk (*).
The distribution of each species is indicated according to the following abbreviations (fig. 562). The area within the lines is Malesia (see fig. 1).

We have tried to incorporate literature published up to and including 2002. Papers published in 2003 have been checked, but possibly not all of them were seen by us.

Abbreviations for geographical areas:

BO: Borneo (Kalimantan, Sabah, Sarawak and Brunei)
BS: Bismarck and Solomon Archipelagos
JA: Java (including Madura and Bali)
LU: Philippines, Luzon (including Mindoro, Marinduque & Catanduanes)
MA: southern part of Thailand, peninsular Malaysia and Singapore
MI: Philippines, Mindanao (including Bohol, Leyte, Samar & Sulu Islands)
ML: Moluccas
NG: New Guinea (including adjacent islands, notably the Raja Empat Islands)
NT: Nusa Tenggara (smaller Sunda Islands)
PA: Philippines, Palawan

PH: remaining Philippine Islands
PL: Philippines (unspecified locality)
SL: Sulawesi
SM: Sumatra (including Siberut & Nias)

Extralimital (EL):
AU: Australia
HT: China:
Hainan, Hong Kong, Macao and Taiwan
NC: New Caledonia
SA: continental South-east Asia – Myanmar, Thailand, Laos, Cambodia, Vietnam. In addition peninsular Malaysia and Singapore are usually considered to belong to this area. However, in this book they are not included when 'continental South-east Asia (SA)' is used.
India: India, Bangladesh and Sri Lanka

NEPOMORPHA Popov, 1971
NEPOIDEA Latreille, 1802
NEPIDAE Latreille, 1802
NEPINAE Latreille, 1802

Borborophyes Stål, 1871
Type species: *Borborophyes mayri* Stål, 1871
mayri Stål – Stål, 1871: 707 ‣ LU

Laccotrephes Stål, 1866
Type species: *Nepa atra* Linnaeus, 1758
archipelagi (Ferrari) – Ferrari, 1888: 183-184 (*Nepa*) ‣ SM, JA, BO; – ?EL: India
celebensis J. Polhemus & Keffer – J. Polhemus & Keffer, 1999: 4-9 ‣ SL
grossus (Fabricius) – Fabricius, 1794: 62 (*Nepa*)
‣ MA, SM; – EL: SA, HT, China, Japan
japonensis Scott – Scott, 1874: 450 ‣ MA, SM; – EL: HT, China, Korea, Japan
maculatus (Fabricius) – Fabricius, 1775: 692 (*Nepa*)
‣ MA, JA; – EL: HT, India, Japan
occultus Lundblad – Lundblad, 1933a: 23-26 ‣ SM, BO, SL, ML
papuus Montandon – Montandon, 1900: 539-541 ‣ NG
pfeiferae (Ferrari) – Ferrari, 1888: 187 (*Nepa*)
‣ MA, SM, JA, BO, SL, NT; – EL: SA, HT, India
robustus Stål – Stål, 1871: 706 ‣ LU, PL
simulatus Montandon – Montandon, 1913: 122-123 ‣ MA, SM; – EL: SA, China
sondaicus J. Polhemus & Keffer – J. Polhemus & Keffer, 1999: 9-12 ‣ NT
tristis (Stål) – Stål, 1854: 241 (*Nepa*) ‣ NG; – EL: AU

Telmatotrephes Stål, 1854
Type species: *Telmatotrephes sculpticollis* Stål, 1854
breddini Montandon – Montandon, 1907a: 329-330 ‣ BO

RANATRINAE Douglas & Scott, 1865

Cercotmetus Amyot & Serville, 1843
Type species: *Cercotmetus asiaticus* Amyot & Serville, 1843
asiaticus Amyot & Serville – Amyot & Serville, 1843: 441
‣ MA, SM, JA, BO; – EL: SA
brevipes Montandon – Montandon, 1909b: 65
‣ MA, SM, JA, BO, LU; – EL: SA, India, China
compositus Montandon – Montandon, 1903: 109-110 ‣ MA, SM; – EL: SA
robustus Montandon – Montandon, 1911b: 92-93 ‣ SL
dissidens Montandon – Montandon, 1911c: 652-654 ‣ NG

Ranatra Fabricius, 1790
Type species: *Nepa linearis* Linnaeus, 1758
akoiachta Nieser – Nieser, 1996: 347-349 ‣ ML
biroi Lundblad – Lundblad, 1933a: 38-40 ‣ NG
gracilis Dallas – Dallas, 1850: 10 ‣ MA; – EL: SA, India
longipes longipes Stål – Stål, 1861: 203 ‣ SM, JA, BO?, SL?
longipes celebensis Lansbury – Lansbury, 1972a: 334-335 ‣ SL
longipes thai Lansbury – Lansbury, 1972a: 334 ‣ MA?; – EL: SA
malayana Lundblad – Lundblad, 1933a: 40-41 ‣ SL
megalops Lansbury – Lansbury, 1972a: 303-306 ‣ NG
natunaensis Lansbury – Lansbury, 1972a: 331-332 ‣ BO
parmata Mayr – Mayr, 1865: 446 ‣ MA, SM, JA, NT; – EL: SA
spinifrons Montandon – Montandon, 1910: 166 ‣ BO, ML
stali Montandon – Montandon, 1903: 390-391 ‣ LU, PH, MI
sulawesii sulawesii Nieser & Chen – Nieser & Chen, 1991: 55-57 ‣ SL; – EL: SA
sulawesii sebui Nieser & Chen – Nieser & Chen, 1996: 173 ‣ MI
varipes Stål – Stål, 1861: 203 ‣ MA, SM, JA; – EL: SA, India

BELOSTOMATIDAE Leach, 1815
BELOSTOMATINAE Leach, 1815

Diplonychus Laporte, 1833
Type species: *Nepa rustica* Fabricius, 1871
molestus (Dufour) – Dufour, 1863: 395 (*Appasus*) ‣ MA, SM; – EL: India
rusticus (Fabricius) – Fabricius, 1871: 333 (*Nepa*)
‣ MA, SM, JA, BO, SL, LU, MI, NG; – EL: SA, HT, India, China, Japan

LETHOCERINAE Lauck & Menke, 1961

Lethocerus Mayr, 1853
Type species: *Belostoma fakir* Gistel, 1848
indicus (Lepeletier & Serville) – Lepeletier & Serville, 1825: 272 (*Belostoma*)
▸ MA, SM, JA, SL, LU; – EL: SA, HT, India, China, Japan
insulanus (Montandon) – Montandon, 1898: 430-431 (*Belostoma*) ▸ NG; – EL: AU, NC

OCHTEROIDEA Kirkaldy, 1815
OCHTERIDAE Kirkaldy, 1815

Ochterus Latreille, 1807
Type species: *Acanthia marginata* Latreille, 1804
alticola Baehr – Baehr, 1990b: 473 ▸ NG
australicus Jaczewski – Jaczewski, 1934: 607-609 ▸ NG, BS; – EL: AU, NC
baltazarae Gapud & San Valentin – Gapud & San Valentin, 1977: 278-283 ▸ LU
barrosoi Gapud – Gapud, 1981: 307-309 ▸ PH
cheesmannae Baehr – Baehr, 1990b: 473-474 ▸ NG
grandiusculus Nieser & Chen – Nieser & Chen, 1992a: 11-12 ▸ SL
gressitti Kormilev – Kormilev, 1971: 438-439 ▸ NG, BS
homorfos Nieser & Chen – Nieser & Chen, 1999: 91-92 ▸ SL
jaczewskii Kormilev – Kormilev, 1971: 439 ▸ NG
kokodae Baehr – Baehr, 1990b: 465-466 ▸ NG
latior Baehr – Baehr, 1990b: 474-475 ▸ NG
louisiadae Baehr – Baehr, 1990b: 470-471 ▸ BS
luzonicus Gapud – Gapud 2003: 100-103 ▸ LU
magnificus Gapud & San Valentin – Gapud & San Valentin, 1977: 283-287 ▸ LU
magnus Gapud & San Valentin – Gapud & San Valentin, 1977: 287-291 ▸ LU
marginatus (Latreille) – Latreille, 1804: 242 (*Acanthia*)
▸ MA, SM, JA, BO, SL, LU, PA, PH, MI; – EL: widespread, from the Mediterranean and most of Africa to China and Japan
nigrinus Baehr – Baehr, 1990b: 464-465 ▸ BS
noualhieri Baehr – Baehr, 1990a: 91-93 ▸ JA, SL
papuasicus Kormilev – Kormilev, 1972: 585-587 ▸ NG
pardalos Nieser & Chen – Nieser & Chen, 1999: 93-94 ▸ MI
paucistriata Baehr – Baehr, 1990b: 468-469 ▸ NG
philippinensis Kormilev – Kormilev, 1971: 436-437 ▸ PH
polhemusi Gapud – Gapud, 1981: 300-301 ▸ LU, PH
surigaoensis Gapud – Gapud, 1995: 42-44 ▸ MI
thienemanni Jaczewski – Jaczewski, 1935: 480-482 ▸ SM, JA, SL, NG
trichotos Nieser & Chen – Nieser & Chen, 1999: 92-93 ▸ SL
xustos Nieser & Chen – Nieser & Chen, 1992a: 12-13 ▸ BO
zetteli Gapud – Gapud 2003: 103-106 ▸ PA

GELASTOCORIDAE Kirkaldy, 1897

Nerthra Say, 1832
Type species: *Naucoris stygicus* Say, 1832

ampliata (Montandon) – Montandon, 1899: 404 (*Mononyx*) ▸ NG
cheesmanae Todd – Todd, 1959: 82-83 ▸ NG
colaticollis Todd – Todd, 1959: 75-78 ▸ NG
conabilis Todd – Todd, 1959: 88-90 ▸ NG
eximia Todd – Todd, 1957a: 155-156 ▸ SM
gurneyi Todd – Todd, 1955a: 419-420 ▸ BS
hamata Todd – Todd, 1957a: 158-159 ▸ NG
hirta Todd – Todd, 1959: 83-84 ▸ NG
improcera Todd – Todd, 1959: 73-74 ▸ NG
infecta Todd – Todd, 1959: 91-92 ▸ NG
laticollis (Guérin-Méneville) – Guérin-Méneville, 1843a: 112 (*Mononyx*) ▸ NG, BS
lobata (Montandon) – Montandon, 1899a: 397-398 (*Mononyx*) ▸ MA, SM, JA
lurida Todd – Todd, 1959: 72-73 ▸ SL
macrostyla Todd – Todd, 1955a: 428-429 ▸ BS
macrothorax (Montrouzier) – Montrouzier, 1855a: 110-111 (*Galgulus*) ▸ BO, SL, PH, NG, BS; – EL: AU, Pacific Islands from Tonga to Comores and Japan
mixta (Montandon) – Montandon, 1899a: 404-405 (*Mononyx*) ▸ NG, BS
mixtella Todd – Todd, 1959: 85-86 ▸ NG
monticola Todd – Todd, 1959: 88 ▸ NG
nieuwenhuisi Todd – Todd, 1957a: 154-155 ▸ BO
occidua Todd – Todd, 1959: 71-72 ▸ SL
omani Todd – Todd, 1955a: 422-423 ▸ BS
petila Todd – Todd, 1959: 90-91 ▸ NG
recta Todd – Todd, 1959: 80-81 ▸ ML
robusta Todd – Todd, 1955a: 429-430 ▸ NG
rugosa (Desjardins) – Desjardins, 1837a: 239 (*Naucoris*) ▸ ?NG
stevensi Todd – Todd, 1959: 87 ▸ NG
tenuistyla Todd – Todd, 1959: 94 ▸ ML
toxopeusi Todd – Todd, 1959: 92-93 ▸ ML

NAUCOROIDEA Leach, 1815
APHELOCHEIRIDAE Fieber, 1851

Aphelocheirus **subgenus** *Aphelocheirus* Westwood, 1833
Type species: *Naucoris aestivalis* Fabricius, 1794

baguio D. Polhemus & J. Polhemus – D. Polhemus & J. Polhemus, 1988: 210-212 ▸ LU
bengkulu D. Polhemus – D. Polhemus, 1994: 74-78 ▸ SM
breviculus Nieser & Chen – Nieser & Chen, 1991: 49-50 ▸ BO

celebensis D. Polhemus & J. Polhemus – D. Polhemus & J. Polhemus, 1988: 188-190 ▸ SL

femoratus D. Polhemus & J. Polhemus – D. Polhemus & J. Polhemus, 1988: 214-216 ▸ MA; – EL: SA

gapudi Zettel – Zettel, 1999b: 110-113 ▸ LU, MI

geros Nieser & Chen – Nieser & Chen, 1996: 163-164 ▸ SL

grik D. Polhemus & J. Polhemus – D. Polhemus & J. Polhemus, 1988: 218-220 ▸ MA

javanicus D. Polhemus & J. Polhemus – D. Polhemus & J. Polhemus, 1988: 197-198 ▸ JA

kedirius Nieser, Chen, Leksawasdi, Thanyakam & Duangsupa – Nieser et al., 2004: 37-39 ▸ JA

kinabalu D. Polhemus & J. Polhemus – D. Polhemus & J. Polhemus, 1988: 192-195 ▸ BO

kodadai Zettel –Zettel, 1999a: 57-59 ▸ BO

malayanus D. Polhemus & J. Polhemus – D. Polhemus & J. Polhemus, 1988: 216-218 ▸ MA

minor D. Polhemus & J. Polhemus – D. Polhemus & J. Polhemus, 1988: 195-197 ▸ BO

palawanensis D. Polhemus & J. Polhemus – D. Polhemus & J. Polhemus, 1988: 199-200 ▸ PA

pallens Horváth – Horváth, 1899: 258 ▸ NG

philippinensis Usinger – Usinger, 1938: 307-309 ▸ LU

robustus Nieser & Chen – Nieser & Chen, 1991: 50-52 ▸ SL

sculpturatus D. Polhemus & J. Polhemus – D. Polhemus & J. Polhemus, 1988: 208-210 ▸ LU

uichancoi Usinger – Usinger, 1938: 305-307 ▸ LU

venus Zettel – Zettel, 1999b: 108-110 ▸ LU

zamboanga zamboanga D. Polhemus & J. Polhemus – D. Polhemus & J. Polhemus, 1988: 201-203 ▸ MI

zamboanga luzonicus D. Polhemus & J. Polhemus – D. Polhemus & J. Polhemus, 1988: 212-214 ▸ LU

zamboanga visayasensis Zettel – Zettel, 1998a: 88-90 ▸ PH

Aphelocheirus **subgenus** *Micraphelocheirus* Hoberlandt & Štys, 1977

Type species: *Aphelocheirus pygmaeus* La Rivers, 1971

kelabitensis Zettel – Zettel, 1993b: 81-84 ▸ BO

NAUCORIDAE Leach, 1815
CHEIROCHELINAE Montandon, 1897
CHEIROCHELINI Montandon, 1897

Coptocatus Montandon, 1909

Type species: *Coptocatus oblongulus* Montandon, 1909

ibanus Zettel – Zettel, 2002a: 173-175 ▸ BO
jaechi Zettel – Zettel, 2002a: 172-173 ▸ BO
kinabalu D. Polhemus – D. Polhemus, 1986: 252-254 ▸ BO
oblongulus Montandon – Montandon, 1909a: 45 ▸ BO
stereos Nieser & Chen – Nieser & Chen, 1991: 52-53 ▸ BO
tebedu D. Polhemus –D. Polhemus, 1968: 254-256 ▸ BO

Gestroiella Montandon, 1897
Type species: *Gestroiella limnocoroides* Montandon, 1897
limnocoroides Montandon – Montandon, 1897d: 371-372 ▸ MA; – EL: SA

SAGOCORINI La Rivers, 1971

Aptinocoris Montandon, 1897
Type species: *Aptinocoris papuus* Montandon, 1897
boikiki D. Polhemus & J. Polhemus – D. Polhemus & J. Polhemus, 2000b: 365-366 ▸ NG
cheesmanae (La Rivers) – La Rivers, 1971a: 44-47 (*Sagocoris*) ▸ NG
fenneri (La Rivers) – La Rivers, 1971a: 49-51 (*Sagocoris*) ▸ NG
minutus D. Polhemus & J. Polhemus – D. Polhemus & J. Polhemus, 2000b: 361-362 ▸ NG
papuus Montandon – Montandon, 1897b: 65-66 ▸ NG
sedlaceki (La Rivers) – La Rivers, 1971a: 47-49 (*Sagocoris*) ▸ NG
sogeri D. Polhemus & J. Polhemus – D. Polhemus & J. Polhemus, 2000b: 366-368 ▸ NG
ziwa D. Polhemus & J. Polhemus – D. Polhemus & J. Polhemus, 2000b: 368-370 ▸ NG

Asthenocoris Usinger, 1938
Type species: *Asthenocoris luzonensis* Usinger, 1938
australis Zettel, Nieser & D. Polhemus – Zettel et al., 1999: 97-99 ▸ MI
luzonensis luzonensis Usinger – Usinger, 1938a: 303-305 ▸ LU, MI
luzonensis leyticus Zettel & Nieser – Zettel et al., 1999: 90-91 ▸ MI
luzonensis paradisianus Zettel & Nieser – Zettel et al., 1999: 88-90 ▸ LU
medius medius Zettel, Nieser & D. Polhemus – Zettel et al., 1999: 94-96 ▸ MI
medius samarensis Zettel & Nieser – Zettel et al., 1999: 96-97 ▸ MI
montanus Zettel, Nieser & D. Polhemus – Zettel et al., 1999: 91-93 ▸ LU

Cavocoris La Rivers, 1971
Type species: *Cavocoris bisulcus* La Rivers, 1971
bisulcus La Rivers – La Rivers, 1971a: 57-59 ▸ NG
ibatiri D. Polhemus & J. Polhemus – D. Polhemus & J. Polhemus, 1989: 78-82 ▸ NG

ismayi D. Polhemus & J. Polhemus – D. Polhemus & J. Polhemus, 1989: 83-85 ▸ NG
minor D. Polhemus & J. Polhemus – D. Polhemus & J. Polhemus, 1989: 82-83 ▸ NG
rotundatus D. Polhemus & J. Polhemus – D. Polhemus & J. Polhemus, 1989: 75-78 ▸ NG

Philippinocoris D. Polhemus & J. Polhemus, 1987
Type species: *Sagocoris usingeri* La Rivers, 1970
sumaldei Zettel, Nieser & D. Polhemus – Zettel et al., 1999: 67-69 ▸ LU
usingeri (La Rivers) – La Rivers, 1970b: 167-169 (*Sagocoris*) ▸ LU

Sagocoris Montandon, 1911
Type species: *Sagocoris biroi* Montandon, 1911
asymmetrica (La Rivers) – La Rivers, 1971a: 51-54 (*Quadricoris*) ▸ NG
biroi Montandon – Montandon, 1911a: 248-250 ▸ NG
flavinotum D. Polhemus & J. Polhemus – D. Polhemus & J. Polhemus, 2000b: 349-351 ▸ NG
gressitti La Rivers – La Rivers, 1971a: 35-37 ▸ NG
intermedius D. Polhemus & J. Polhemus – D. Polhemus & J. Polhemus, 2000b: 346-347 ▸ NG
irianus D. Polhemus & J. Polhemus – D. Polhemus & J. Polhemus, 2000b: 347-349 ▸ NG
lariversae La Rivers – La Rivers, 1971a: 40-42 ▸ NG

Stalocoris La Rivers, 1969
Type species: *Stalocoris breviceps* La Rivers, 1969
bicolanus bicolanus Zettel – Zettel 2003b: 115-117 ▸ LU
bicolanus ruizi Zettel – Zettel 2003b: 117-118 ▸ LU
breviceps La Rivers – La Rivers, 1969: 1-4 ▸ MI
rarosae Zettel – Zettel 2003b: 118-120 ▸ MI
schoedli Zettel, Nieser & D. Polhemus – Zettel et al., 1999: 74-76 ▸ PH
tansiongcoi Zettel, Nieser & D. Polhemus – Zettel et al., 1999: 76-78 ▸ LU, PH
ticaoensis Zettel, Nieser & D. Polhemus – Zettel et al., 1999: 78-79 ▸ PH

Warisia La Rivers, 1971
Type species: *Warisia cavanceps* La Rivers, 1971
cavanceps La Rivers – La Rivers, 1971a: 60-62 ▸ NG

TANYCRICINI La Rivers, 1971

Idiocarus Montandon, 1897
Type species: *Idiocarus elongatus* Montandon, 1897
elongatus Montandon – Montandon, 1897c: 8-9 ▸ NG

intermedius La Rivers – La Rivers, 1971a: 23-25 ▸ NG
isolatus D. Polhemus & J. Polhemus – D. Polhemus & J. Polhemus, 1986a: 43-46 ▸ NG
minor La Rivers – La Rivers, 1971a: 28-31 ▸ NG
papuus D. Polhemus & J. Polhemus – D. Polhemus & J. Polhemus, 1986a: 48-50 ▸ NG
sepikanus D. Polhemus & J. Polhemus – D. Polhemus & J. Polhemus, 1986a: 46-48 ▸ NG

Nesocricos La Rivers, 1971
Type species: *Nesocricos evops* La Rivers, 1971
evops La Rivers – La Rivers, 1971a: 16-19 ▸ NG
kukukuku D. Polhemus & J. Polhemus – D. Polhemus & J. Polhemus, 1985: 198-200 ▸ NG
mion La Rivers – La Rivers, 1974: 14-15 ▸ NG
montanus D. Polhemus & J. Polhemus – D. Polhemus & J. Polhemus, 1985: 200-202 ▸ NG

Tanycricos La Rivers, 1971
Type species: *Tanycricos usingeri* La Rivers, 1971
acumentum La Rivers – La Rivers, 1971a: 5-7 ▸ NG
binarius La Rivers – La Rivers, 1971a: 8-10 ▸ NG
froeschneri D. Polhemus & J. Polhemus – D. Polhemus & J. Polhemus, 1986b: 171-173 ▸ NG
jaetipi D. Polhemus – D. Polhemus, 2000: 171-176 ▸ NG
longiceps La Rivers – La Rivers, 1971a: 10-13 ▸ NG
usingeri La Rivers – La Rivers, 1971a: 13-15 ▸ NG
ziwa D. Polhemus – D. Polhemus, 2000: 176-178 ▸ NG

LACCOCORINAE Stål, 1876

Ctenipocoris Montandon, 1897
Type species: *Ctenipocoris asiaticus* Montandon, 1897
asiaticus Montandon – Montandon, 1897d: 374-376 ▸ MA; – EL: SA

Heleocoris Stål, 1876
Type species: *Naucoris obliquatus* Spinola, 1837
floresensis Nieser & Chen – Nieser & Chen, 1992b: 31-33 ▸ NT
grandis Montandon – Montandon, 1909a: 52-53 ▸ BO
montandoni Lundblad – Lundblad, 1933a: 70-72 ▸ SM, JA
nebulosus Montandon – Montandon, 1909a: 53-55 ▸ SM, JA
obscuratus Montandon – Montandon, 1897b: 56-57 ▸ SM, JA
ovatus Montandon – Montandon, 1897a: 451-452 ▸ MA; – EL: SA

Laccocoris Stål, 1856

Type species: *Naucoris spurcus* Stål, 1856

horvathi Montandon – Montandon, 1897a: 442-443 ▸ BO
lipogonia La Rivers – La Rivers, 1970a: 498-499 ▸ BO
hoogstraali La Rivers – La Rivers, 1970c: 269-273 ▸ MI
maai La Rivers – La Rivers, 1970a: 496-497 ▸ BO
marginatus Montandon – Montandon, 1897b: 54-55 ▸ BO
nervicus Montandon – Montandon, 1897b: 52-53 ▸ MA, SM
staudingeri Montandon – Montandon, 1897a: 440-442 ▸ BO

NAUCORINAE Leach, 1815

Naucoris Geoffroy, 1762

Type species: *Naucoris maculatus* Fabricius, 1798

obscuripennis Stål – Stål, 1854: 239 ▸ LU
pumilus Zettel, Nieser & D. Polhemus – Zettel et al., 1999: 61-63 ▸ LU, PH, MI
rhizomatus J. Polhemus – J. Polhemus, 1984a: 157-158 ▸ MA; – EL: AU
scutellaris Stål – Stål, 1858: 266 ▸ MA, SM, JA, SL; – EL: SA, India
sumatranus Fieber – Fieber, 1851a: 17 ▸ SM

CORIXOIDEA Leach, 1815
CORIXIDAE Leach, 1815
CORIXINAE Leach, 1815

Agraptacorixa Kirkaldy, 1898

Type species: *Agraptacorixa gestroi* Kirkaldy, 1898

eurynome (Kirkaldy) – Kirkaldy, 1897b: 54-55 (*Corixa*) ▸ NG; – EL: AU
hyalinipennis (Fabricius) – Fabricius, 1803: 105 (*Sigara*)
▸ MA, SM, JA, MI, ML, NG; – EL: SA, HT, India
macrops Hungerford – Hungerford, 1953: 38-39 ▸ NG; – EL: AU

Sigara Fabricius, 1775

Type species: *Notonecta striata* Linnaeus, 1758

distorta (Distant) – Distant, 1910a: 343-344 (*Corixa*) ▸ SM; – EL: SA, India, China
insulana Lundblad – Lundblad, 1933a: 79-81 ▸ SM, JA
paivai Lundblad – Lundblad, 1928: 241 ▸ MA, SM, JA; – EL: SA, India, Korea
syn.: *connexa* Lundblad – Lundblad, 1933a: 83-87
papuensis Jansson – Jansson, 1982: 99-100 ▸ NG
tadeuszi Lundblad – Lundblad, 1933a: 81 ▸ NG; – EL: AU, NC
truncatipala (Hale) – Hale, 1922: 314-316 (*Arctocorisa*) ▸ NG; – EL: AU, NC

CYMATIAINAE Walton, 1940

Cnethocymatia Jansson, 1982
Type species: *Cymatia nigra* Hungerford, 1947
nigra (Hungerford) – Hungerford, 1947: 154-155 (*Cymatia*) ▸ NG; – EL: AU

MICRONECTIDAE Jaczewski, 1924

Micronecta Kirkaldy, 1897
Type species: *Notonecta minutissima* Linnaeus, 1758
abra Nieser & Chen – Nieser & Chen, 2003: 132-134 ▸ PA
acuta Lundblad – Lundblad, 1933a: 111-113 ▸ JA
aleksanderi Nieser & Chen – Nieser & Chen, 1999: 85-86 ▸ JA
carbonaria Horváth – Horváth, 1904: 595 ▸ NG
charakta Nieser – Nieser, 2002a: 9-11 ▸ MA; – EL: SA
decorata Lundblad – Lundblad, 1933a: 93-94 ▸ MA, SM, JA
dentifera Nieser – Nieser, 2002a: 11-12 ▸ MA
fugitans Breddin – Breddin, 1905b: 57 ▸ MA, SM, JA, SL; – EL: SA
grisea (Fieber) – Fieber, 1844a: 292 (*Sigara*) ▸ MA, SM, SL; – EL: SA, HT, India
guttatostriata Lundblad – Lundblad, 1933a: 101-102 ▸ MA, JA; – EL: SA
haliploides Horváth – Horváth, 1904: 594-595 ▸ MA, SM, JA; – EL: SA, India
johorensis Fernando – Fernando, 1964a: 18 ▸ MA
kymatista Nieser & Chen – Nieser & Chen, 1999: 82-83 ▸ BO, SL
lemnae Nieser – Nieser, 1999: 281-283 ▸ MA; – EL: SA
leongi Nieser – Nieser, 2002a: 12-15 ▸ MA
ludibunda Breddin – Breddin, 1905b: 57 ▸ MA, JA, NT, SL, NG, BS; – EL: SA, India
maculata Nieser – Nieser, 2002a: 15-16 ▸ MA
malayana Leong – Leong, 1966: 86-88 ▸ MA
melanopardala melanopardala Nieser & Chen – Nieser & Chen, 2003: 134-137 ▸ LU
melanopardala adiaphana Nieser & Chen – Nieser & Chen, 2003: 137-139 ▸ MI
**minajerwi* Tinerella & J. Polhemus – Tinerella & J. Polhemus, 2004: 184-187 ▸ NG
pacheta Nieser & Chen – Nieser & Chen, 2003: 140-141 ▸ MI
pachynychi Nieser & Chen – Nieser & Chen, 1999: 83-85 ▸ SL
polhemusi Nieser – Nieser, 1999: 283-285 ▸ MA; – EL: SA
proxima (Lethierry) – Lethierry, 1877: 101-102 (*Sigara*) ▸ LU
pumilio Lundblad – Lundblad, 1933a: 102-103 ▸ JA
quadristrigata Breddin – Breddin, 1905b: 57
▸ MA, SM, JA, SL, BO, MI; – EL: SA, HT, AU, Iran, India
scutellaris (Stål) – Stål, 1858b: 319 (*Sigara*) ▸ MA; – EL: SA, Africa, Arabia, India
siva (Kirkaldy) – Kirkaldy, 1897a: 240 (*Sigara*) ▸ SM; – EL: SA, India
skutalis Nieser & Chen – Nieser & Chen, 1999: 86-87 ▸ BO, PA
tarsalis Chen – Chen, 1960: 114 ▸ SL; – EL: SA, India
vanduzeei Kirkaldy – Kirkaldy, 1905: 262 ▸ PH
virgata Hale – Hale, 1922: 327 ▸ SL; – EL: AU

Synaptonecta Lundblad, 1933
Type species: *Micronecta issa* Distant, 1910
Synaptonecta issa (Distant) – Distant, 1910a: 350 (*Micronecta*)
‣ MA, JA; – EL: SA, India, Florida (USA)

NOTONECTOIDEA Latreille, 1802
NOTONECTIDAE Latreille, 1802
ANISOPINAE Hutchinson, 1929

Anisops Spinola, 1837
Type species: *Anisops sardeus* Herrich-Schäffer, 1849
barbatus Brooks – Brooks, 1951: 387-389 ‣ MA, SM, JA; – EL: SA, HT, India
biroi Brooks – Brooks, 1951: 454-456 ‣ SL, ML, NG
bouvieri Kirkaldy – Kirkaldy, 1904: 116 ‣ MA, NG; – EL: SA, India
breddini Kirkaldy – Kirkaldy, 1901a: 5-6 ‣ MA, JA, SL, NG; – EL: SA, India
browni Lansbury – Lansbury, 1963: 9-10 ‣ BS
capitatus Lansbury – Lansbury, 1968a: 95-96 ‣ BS
cheesmanae Lansbury – Lansbury, 1963: 6-9 ‣ BS
depressus Lansbury – Lansbury, 1962: 145-147 ‣ NG
elstoni Brooks – Brooks, 1951: 326-327 ‣ BS; – EL: AU
?syn.: *capitatus*; Lansbury – Lansbury, 1978a: 106-107 (may be brachypterous morph of *elstoni*)
exiguus Horváth – Horváth, 1919: 314 ‣ MA, NG; – EL: SA, India
syn.: *exigera* Brooks – Brooks, 1951: 447-449 (misspelling of *exigua*)
fijiensis Brooks – Brooks, 1951: 361-363 ‣ MI; – EL: Fiji
gressitti Lansbury – Lansbury, 1962: 147-149 ‣ NG
kuroiwae Matsumura – Matsumura, 1905a: 109
‣ MA, BO, LU, PH, MI; – EL: SA, HT, India, China, Japan
syn.: *batillifrons* Lundblad – Lundblad, 1933b: 463-464
lansburyi Leong – Leong, 1963: 101-102 ‣ MA
leucothecus Esaki – Esaki, 1926: 76-80 ‣ BS; – EL: Pacific Islands
lihirensis; Lansbury – Lansbury, 1978: 111-113 ‣ BS
lombokensis Brooks – Brooks, 1952: 62-63 ‣ JA, NT
lundbladianus Lansbury – Lansbury, 1962: 149-151 ‣ NG
megalops Lansbury – Lansbury, 1962: 144-145 ‣ NG
minuta Reichart – Reichart, 1985: 235-238 ‣ JA
nasutus Fieber – Fieber, 1851a: 484-485
‣ MA, SL, NG, BS; – EL: AU, India, China, Pacific Islands
nigrolineatus Lundblad – Lundblad, 1933a: 160-163 ‣ MA, JA, PH; – EL: SA, India
niveus (Fabricius) – Fabricius, 1775: 690 (*Notonecta*) ‣ MA, SM; – EL: SA, India
occipitalis Breddin – Breddin, 1905: 152
‣ MA, JA, NT, SL, ML, MI, NG, BS; – EL: AU, NC
paracrinitus Brooks – Brooks, 1951: 329-331 ‣ SM, JA, NT, SL; – EL: AU

philippinensis Brooks – Brooks, 1951: 383-384 ▸ MI, ML, BS, NG; – EL: AU
quadrispinosus Lansbury – Lansbury, 1967: 92-93 ▸ BS
rhomboides Nieser & Chen – Nieser & Chen, 1999: 111-113 ▸ BO, SL, MI
rigoensis Brooks – Brooks, 1951: 456-457 ▸ NG
salibabu Nieser & Chen – Nieser & Chen, 1999: 113-114 ▸ SL
stali Kirkaldy – Kirkaldy, 1904: 113, 132 ▸ JA, NT, SL, MI; – EL: AU
tahitiensis Lundblad – Lundblad, 1934a: 121-123 ▸ MA, NG, BS; – EL: AU, SA, Pacific Islands
thienemanni Lundblad – Lundblad, 1933a: 167-168 ▸ JA; – EL: AU
timorensis Brooks – Brooks, 1951: 343-344 ▸ NT
yanoi Miyamoto – Miyamoto, 1981: 25-26 ▸ LU

NOTONECTINAE Latreille, 1802

Aphelonecta Lansbury, 1965

Type species: *Aphelonecta alexis* Lansbury, 1965
alexis Lansbury – Lansbury, 1965b: 328-330 ▸ BO
gavini Lansbury – Lansbury, 1966: 629-630 ▸ MA; – EL: SA
gigas Zettel – Zettel, 1993a: 662-664 ▸ BO
jaechi Zettel – Zettel, 1995a: 108-109 ▸ SM
nakatae Lansbury – Lansbury, 1965b: 330-331 ▸ BO
philippina Zettel – Zettel, 1995a: 109-110 ▸ PA

Enithares Spinola, 1837

Type species: *Notonecta ciliata* Fabricius, 1798
alexis alexis Lansbury – Lansbury, 1967: 90-92 ▸ BS
alexis lairdi Lansbury – Lansbury, 1968b: 388-389 ▸ BS
amboinensis Lansbury – Lansbury, 1968b: 424-425 ▸ ML
atra Brooks – Brooks, 1948: 48 ▸ NG
bakeri Brooks – Brooks, 1948: 40 ▸ BO, MI, ML, NT, PH, SL
buhleri Brooks – Brooks, 1948: 44-45 ▸ NT
caesaries Nieser & Chen – Nieser & Chen, 1991: 59-62 ▸ SL
charakia Nieser & Chen – Nieser & Chen, 1996: 164-167 ▸ SL
ciliata (Fabricius) – Fabricius, 1798: 524-525 (*Notonecta*) ▸ MA, SM; – EL: SA, India
digitata Lansbury – Lansbury, 1974a: 226-228 ▸ NG
ektakta Nieser & Chen – Nieser & Chen, 1996: 167 ▸ SL
elongata Lansbury – Lansbury, 1974a: 226 ▸ NG
foveatus Lansbury – Lansbury, 1968b: 429-430 ▸ PH
freyi freyi Brooks – Brooks, 1948: 48-49 ▸ PH
freyi quadrispinosus Lansbury – Lansbury, 1967: 94-96 ▸ PA
gantsophora Nieser & Zettel – Nieser & Zettel, 1999: 125-128 ▸ MI
genitalis Lundblad – Lundblad, 1933a: 182-184 ▸ JA
gibbera Brooks – Brooks, 1948: 51-52 ▸ BS

hippokleides Kirkaldy – Kirkaldy, 1898a: 73 ▸ JA
horvathi Kirkaldy – Kirkaldy, 1898a: 72 ▸ SL
intricata Breddin – Breddin, 1905a: 154-155 ▸ SM, JA
lansburyi Nieser & Chen – Nieser & Chen, 1991: 62-63 ▸ SL
loria Brooks – Brooks, 1948: 45-46 ▸ BS, NG; – EL: AU
lombokensis Lansbury – Lansbury, 1968b: 416-418 ▸ NT
maai Lansbury – Lansbury, 1968b: 428-429 ▸ BO
malayensis Brooks – Brooks, 1948: 41 ▸ MA
mandalayensis Distant – Distant, 1910a: 331-332 ▸ MA; – EL: SA
margarethae Nieser & Chen – Nieser & Chen, 1996: 169 ▸ SL
martini Kirkaldy – Kirkaldy, 1898b: 151 ▸ LU, MI
megalops Lansbury – Lansbury, 1968b: 405-406 ▸ NG
metallica Brooks – Brooks, 1948: 420-421 ▸ MA; – EL: SA
nieseri Zettel – Zettel 2003b: 122-123 ▸ MI
nigra Lansbury – Lansbury, 1968b: 395-396 ▸ NG
paramegalops Lansbury – Lansbury, 1968b: 406-408 ▸ ML, NG, SL
phenakismos Nieser & Chen – Nieser & Chen, 1991: 63 ▸ SL
producta Lansbury – Lansbury, 1968b: 368-369 ▸ SL
ripleyana Lansbury – Lansbury, 1968b: 402-403 ▸ ML
skutalis Nieser & Chen – Nieser & Chen, 1991: 63-64 ▸ SL
stansae Nieser & Chen – Nieser & Chen, 1996: 169-170 ▸ SL
stylata Lansbury – Lansbury, 1968b: 425-426 ▸ NG
subparallela Lansbury – Lansbury, 1968b: 408-409 ▸ MI
thienemanni Lundblad – Lundblad, 1933a: 421-423 ▸ SM
timorensis Brooks – Brooks, 1948: 50-51 ▸ NT
uncata Lundblad – Lundblad, 1933a: 179-181 ▸ JA, SM
vicintricata Lansbury – Lansbury, 1968b: 430-432 ▸ BO
vulgaris Lansbury – Lansbury, 1968b: 403-404 ▸ NG

Nychia Stål, 1859
Type species: *Nychia limpida* Stål, 1858
sappho Kirkaldy – Kirkaldy, 1901b: 809-810 ▸ MA, SM, JA, BO, SL, NG; – EL: SA, AU

PLEOIDEA Fieber, 1851
HELOTREPHIDAE Esaki & China, 1927
FISCHEROTREPHINAE Zettel, 1994

Fischerotrephes Zettel, 1994
Type species: *Fischerotrephes depressus* Zettel, 1994
depressus Zettel – Zettel, 1994a: 1071-1072 ▸ BO
jaechi Zettel – Zettel, 1994a: 1073 ▸ MA; EL: SA!
maryatiae Zettel – Zettel, 1997c: 118-119 ▸ BO

TREPHOTOMASINAE Papáček, Štys & Tonner, 1988

Papacekia Zettel, 2005b
Type species: *Papacekia microphtalma* Zettel, 2005
Papacekia microphtalma Zettel – Zettel, 2005b: 84-85 ▸ BO

Platytrephes Zettel, 2005
Type species: *Platytrephes fasciatus* Zettel, 1996a
insulanus (Zettel) – Zettel, 1997a: 120-124 (*Trephotomas*) ▸ BO
kodadai (Zettel) – Zettel, 2000a: 2-3 (*Trephotomas*) ▸ BO

HELOTREPHINAE Esaki & China, 1927
HELOTREPHINI Esaki & China, 1927

Ascetotrephes J. Polhemus & D. Polhemus, 2003
Type species: Ascetotrephes mesilau J. Polhemus & D. Polhemus, 2003
edmundsorum J. Polhemus & D. Polhemus – J. Polhemus & D. Polhemus, 2003: 216-217 ▸ MA
flavomarginatus Zettel – Zettel, 2004f: 1368-1369 ▸ BO
keningau J. Polhemus & D. Polhemus – J. Polhemus & D. Polhemus, 2003: 214-216 ▸ BO
loedli Zettel – Zettel, 2004f: 1369-1370 ▸ BO
mesilau J. Polhemus & D. Polhemus – J. Polhemus & D. Polhemus, 2003: 212-214 ▸ BO
minor Zettel – Zettel, 2004f: 1370 ▸ BO
sumatrensis Zettel – Zettel, 2004f: 1371-1372 ▸ SM
Helotrephes Stål, 1859
Type species: *Helotrephes semiglobosus* Stål, 1859
affinis Zettel & J. Polhemus – Zettel & J. Polhemus, 1998: 132 ▸ MA
australis Zettel & J. Polhemus – Zettel & J. Polhemus, 1998: 119-121 ▸ MA; – EL: SA
ciampori Zettel – Zettel, 2000a: 5-6 ▸ BO
flaviceps Zettel & J. Polhemus – Zettel & J. Polhemus, 1998: 131-132 ▸ MA; – EL: SA
javanicus Zettel & J. Polhemus – Zettel & J. Polhemus, 1998: 128-130 ▸ JA
kodadai Zettel & J. Polhemus – Zettel & J. Polhemus, 1998: 125-127 ▸ BO
mazzoldii Zettel – Zettel, 2000a: 6 ▸ BO
otoeis Nieser & Chen – Nieser & Chen, 1999: 107-108 ▸ BO
recurvatus Zettel – Zettel, 2000a: 4-5 ▸ BO
rompinensis Kovac & Papáček – Kovac & Papáček, 2000: 267-268 (ssp. of flaviceps) ▸ MA; EL: SA!

Hydrotrephes China, 1935
Type species: *Helotrephes bouvieri* (Kirkaldy, 1904)

angulatus (China) – China, 1930: 164-167 (*Helotrephes*) ▸	SM
appendiculatus Zettel – Zettel, 2004f: 1363-1364 ▸	SM
balnearius (Bergroth) – Bergroth, 1918: 125-126 (*Helotrephes*) ▸	LU
bicolanus bicolanus Zettel – Zettel, 2003a: 83-85 ▸	LU
bicolanus seyferti Zettel – Zettel, 2003a: 85-86 ▸	LU
bouvieri (Kirkaldy) – Kirkaldy, 1904a: 129-130 (*Helotrephes*) ▸	SL
busuanganus Zettel – Zettel, 2003a: 58-59 ▸	PA
celebensis J. Polhemus – J. Polhemus, 1997a: 45-47 ▸	SL
corporaali (China) – China, 1930a: 167-169 (*Helotrephes*) ▸	SM
flavus Zettel – Zettel, 2001c: 59-62 ▸	BO
grabenwegeri Zettel – Zettel, 2000a: 14-15 ▸	BO
hybridus Zettel Zettel, 2000a: 7-9 ▸	BO
intermixtus Zettel – Zettel, 2000a: 16-18 ▸	BO
ilocanus Zettel – Zettel, 2004f: 1365-1366 ▸	LU
jani Zettel – Zettel, 2000a: 13-14 ▸	BO
kalimantanensis (Zettel) – Zettel, 2000a: 11-12 (*Helotrephes*) ▸	BO
kamarora J. Polhemus – J. Polhemus, 1997: 47 ▸	SL
luaae Zettel – Zettel, 2001c: 62 ▸	BO
maculatus Papáček & Kovac – Papáček & Kovac, 2001a: 317-318 ▸	MA
mamasanus Zettel – Zettel, 2004f: 1362-1363 ▸	SL
marana J. Polhemus – J. Polhemus, 1997a: 47-49 ▸	SL
martini (Kirkaldy) – Kirkaldy, 1904: 130 (*Helotrephes*) ▸	SM
masbatensis Zettel – Zettel, 2003a: 72-74 ▸	MI
mexon Nieser & Chen – Nieser & Chen, 1999: 101-102 ▸	SL
milanae Zettel – Zettel, 2003a: 78-80 ▸	PH
minutus Zettel – Zettel, 2003a: 60-61 ▸	PA
mixtus Papáček & Kovac – Papáček & Kovac, 2001a: 318-320 ▸	MA
nieseri J. Polhemus – J. Polhemus, 1997: 49-51 ▸	SL
nitidulus Zettel – Zettel, 2001c: 62-63 ▸	BO
ornatus Zettel – Zettel, 2003a: 86-88 ▸	LU
palawanensis Zettel – Zettel, 2003a: 56-58 ▸	PA
pardalos Nieser & Chen – Nieser & Chen, 1999: 103-105 ▸	MI
philippinus Zettel – Zettel, 2003a: 66-68 ▸	LU
polhemi Nieser & Chen – Nieser & Chen, 1999: 102-103 ▸	NT
rarus Zettel – Zettel, 2000a: 15-16 ▸	BO
sallyae Zettel – Zettel, 2003a: 88-90 ▸	MI
samarensis Zettel – Zettel, 2003a: 68-70 ▸	MI
sarawakensis Zettel – Zettel, 2000a: 10-11 ▸	BO
schoedli Zettel – Zettel, 2004f: 1364-1365 ▸	SM
stereoides stereoides Zettel – Zettel, 2003a: 80-82 ▸	LU
stereoides montanus Zettel – Zettel, 2003a: 82 ▸	LU

stereoides mindoroensis Zettel – Zettel, 2003a: 82-83 ▸ LU
stereos Nieser & Chen – Nieser & Chen, 1999: 105-107 ▸ MI
taweli J. Polhemus – J. Polhemus, 1997: 51-52 ▸ SL
variegatus J. Polhemus – J. Polhemus, 1997: 52-53 ▸ SL
viriosus J. Polhemus – J. Polhemus, 1997: 53-54 ▸ SL
visayasensis Zettel – Zettel, 2003a: 70-72 ▸ PH
vulcanus Zettel – Zettel, 2003a: 64-65 ▸ LU
yangae Zettel – Zettel, 2000a: 9-10 ▸ BO
zetteli Nieser & Chen – Nieser & Chen, 1999: 103 ▸ SL

LIMNOTREPHINI J. Polhemus, 1990

Distotrephes J. Polhemus, 1990
Type species: *Distotrephes stysi* J. Polhemus, 1990a

Distotrephes **subgenus** *Distotrephes* J. Polhemus, 1990
zetteli Papáček & Kovac – Papáček & Kovac, 2001b: 1048-1049 ▸ MA; – EL: SA

Distotrephes **subgenus** *Polhemotrephes* Zettel, 1995
Type species: *Distotrephes sarawakensis* Zettel, 1995c
heveli J. Polhemus – J. Polhemus, 1990a: 50-51 ▸ BO
kodadai Zettel – Zettel, 1995c: 166 ▸ BO
maculatus Zettel – Zettel, 2000a: 18 ▸ BO
sarawakensis Zettel – Zettel, 1995c: 165-166 ▸ BO

Idiotrephes Lundblad, 1933
Type species: *Idiotrephes chinai* Lundblad, 1933
asiaticus Papáček & Zettel – Papáček & Zettel, 2000: 205 ▸ MA; – EL: SA
chinai Lundblad – Lundblad, 1933a: 120-125 ▸ MA, SM, BO

Tiphotrephes Esaki & China, 1928
Type species: *Helotrephes indicus* Distant, 1910
indicus (Distant) – Distant, 1910a: 338 (*Helotrephes*) ▸ MA, BO; – EL: SA, India

PLEIDAE Fieber, 1851

Paraplea Esaki & China, 1928
Type species: *Plea pallescens* Distant, 1906
frontalis (Fieber) – Fieber, 1844b: 296 (*Ploa*)
▸ SM, JA, SL, ML; – EL: SA, HT, China, India
horvathi Lundblad – Lundblad, 1933a: 141-142 ▸ NG
liturata (Fieber) – Fieber, 1844b: 297 (*Ploa*)
▸ MA, SM, JA, SL, MI; – EL: SA, AU, NC, India

quinquemaculata Lundblad – Lundblad, 1933a: 135-138 ▸ MA, SM
sobrina (Stål) – Stål, 1859: 268 (*Plea*) ▸ LU

GERROMORPHA Popov, 1971
MESOVELOIDEA Douglas & Scott, 1867
MESOVELIIDAE Douglas & Scott, 1867
MESOVELIINAE Douglas & Scott, 1867

Cryptovelia Andersen & J. Polhemus, 1980
Type species *Cryptovelia terrestris* Andersen & J. Polhemus, 1980
stysi Andersen – Andersen, 2000c: 8-11 ▸ BO

Mesovelia Mulsant & Rey, 1852
Type species *Mesovelia furcata* Mulsant & Rey, 1852
horvathi Lundblad – Lundblad, 1933a: 190-192
▸ MA, SM, JA, BO, NT, SL, PA, MI, ML, NG; – EL: SA, AU, India, Japan
melanesica J. Polhemus & D. Polhemus – J. Polhemus & D. Polhemus, 2000b: 211-215 ▸ NG
subvittata Horváth – Horváth, 1915: 552-553 ▸ ML, NG
stysi J. Polhemus & D. Polhemus – J. Polhemus & D. Polhemus 2000b: 215-217
▸ NG
ujhelyii Lundblad – Lundblad, 1933a: 192-194 ▸ SM, LU
vittigera Horváth – Horváth, 1895: 160
▸ MA, SM, JA, NT, SL!, PA, LU, MI, NG; – EL: SA, HT, AU, Mediterranean, Ethiopian and E. Palearctic regions, India, Oceania

Nereivelia J. Polhemus & D. Polhemus, 1989
Type species *Nereivelia murphyi* J. Polhemus & D. Polhemus, 1989
murphyi J. Polhemus & D. Polhemus – J. Polhemus & D. Polhemus, 1989: 75-77
▸ MA

Phrynovelia Horváth, 1915
Type species *Phrynovelia papua* Horváth, 1915
papua Horváth – Horváth, 1915: 555-556 ▸ NG
**philippinensis* Zettel – Zettel, 2004e: 1353-1358 ▸ LU

HEBROIDEA Amyot & Serville, 1843
HEBRIDAE Amyot & Serville, 1843
HEBRINAE Amyot & Serville, 1843

Hebrus Curtis, 1833
Type species *Lygaeus pusillus* Fallén, 1807
balnearius Bergroth – Bergroth, 1918: 120-121 ▸ LU

bergrothi Horváth – Horváth, 1929: 317 ▸ LU
cruciatus (Distant) – Distant, 1910b: 60 (*Merragata*) ▸ LU, PA!, MI!; – EL: SA
drakei Porter – Porter, 1959: 33-34 ▸ LU
elegans Lundblad – Lundblad, 1933a: 265-269 ▸ MA, SM, JA
haddeni Porter – Porter, 1954: 78-79 ▸ LU, MI!, PH!
harrisi Porter – Porter, 1959: 34-35 ▸ LU
hirsutulus Zettel – Zettel, 2004c: 538-539 ▸ MI
hoberlandti Porter – Porter, 1959: 31-33 ▸ LU
humeralis Horváth – Horváth, 1929: 315 ▸ JA
mangrovensis J. Polhemus & D. Polhemus – J. Polhemus & D. Polhemus, 1989: 79-81 ▸ MA
murphyi Zettel – Zettel, 2004b: 75-77 ▸ PH
nereis J. Polhemus & D. Polhemus – J. Polhemus & D. Polhemus, 1989: 77-79 ▸ MA – EL: SA
nipponicus Horváth – Horváth, 1929: 314 ▸ SM, JA; – EL: Japan
palawanensis Zettel – Zettel, 2004c: 537-538 ▸ PA
papuanus Horváth – Horváth, 1929: 316 ▸ NG
parameralis Zettel – Zettel, 2004c: 536-537 ▸ PA
peculiaris Horváth – Horváth, 1929: 316-317 ▸ NG
timasiformis Zettel – Zettel, 2004c: 5539-541 ▸ LU
ullrichi Zettel – Zettel, 2004c: 549-551 ▸ SM

Merragata Buchanan-White, 1877
Type species *Merragata hebroides* White, 1877
pallescens var *whitei* Lundblad – Lundblad, 1933a: 277-280 ▸ MA, JA; – EL: SA

Timasius Distant, 1909
Type species *Timasius splendens* Distant, 1909
chinai (Lundblad) – Lundblad, 1933a: 283-285 (*Timasiellus*) ▸ JA
jaechi Zettel & Chen – Zettel & Chen, 2000: 1196 ▸ BO
kodadai Zettel & Chen – Zettel & Chen, 2000: 1197 ▸ BO
malayensis Zettel – Zettel, 2004d: 250-251 ▸ MA
schuhi Zettel – Zettel, 2004d: 255 ▸ MA
yangae Zettel – Zettel, 2004d: 257 ▸ MA

HYRCANINAE Andersen, 1981

Hyrcanus Distant, 1911
Type species *Hyrcanus capitatus* Distant, 1911
varicolor Andersen – Andersen, 1981a: 410-411 ▸ SM, JA; – EL: SA

HYDROMETROIDEA Billberg, 1820
HYDROMETRIDAE Billberg, 1820
HETEROCLEPTINAE Villiers, 1948

Heterocleptes Villiers, 1948
Type species *Heterocleptes tuberculatus* Villiers, 1948
spinosus Andersen – Andersen, 1982b: 105-107▸ BO

HYDROMETRINAE Billberg, 1820

Hydrometra Latreille, 1796
Type species: *Cimex stagnorum* Linnaeus, 1801
anamana Hungerford & Evans – Hungerford & Evans, 1934: 68-71
▸ MA!; – EL: SA, HT, China, Japan
**borneensis* Zettel & Yang – Zettel & Yang, 2004: 391-393 ▸ BO
**brevitarsus* Zettel & Yang – Zettel & Yang, 2004: 389-391 ▸ MA, SM, BO
carinata J. Polhemus & D. Polhemus – J. Polhemus & D. Polhemus, 1995a: 17-19
▸ MA, BO
cracens J. Polhemus & D. Polhemus – J. Polhemus & D. Polhemus, 1995a: 19-20
▸ MA, BO
eioana J. Polhemus & Lansbury – J. Polhemus & Lansbury, 1997: 11-12 ▸ NG
gilloglyi J. Polhemus & D. Polhemus – J. Polhemus & D. Polhemus, 1995a: 20-22
▸ MA; – EL: SA
greeni Kirkaldy – Kirkaldy, 1898: 2 ▸ MA, SM; – EL: SA, China, India, Nepal
**heoki heoki* Zettel & Yang – Zettel & Yang, 2004: 393-396▸ BO
**heoki paya* Zettel & Yang – Zettel & Yang, 2004: 396-398 ▸ MA
horvathi Hungerford & Evans – Hungerford & Evans, 1934: 62-65▸ NG, BS
jaczewskii Lundblad – Lundblad, 1933a: 433-435 ▸ SM, JA, BO; – EL: SA
julieni Hungerford & Evans – Hungerford & Evans, 1934: 55-56
▸ MA, BO, LU; – EL: SA
julienoidea J. Polhemus & D. Polhemus – J. Polhemus & D. Polhemus, 1995a: 26-28▸ NT
kelantan J. Polhemus & D. Polhemus – J. Polhemus & D. Polhemus, 1995a: 28-29
▸ MA
kiunga J. Polhemus & Lansbury – J. Polhemus & Lansbury, 1997: 18-20▸ NG
lineata Eschscholtz – Eschscholtz, 1822: 110
▸ MA, SM, JA, BO, SL, LU, PH, PA, MI, ML, NG; – EL: HT, China
lombok J. Polhemus & D. Polhemus – J. Polhemus & D. Polhemus, 1995a: 31-32
▸ JA, NT
longicapitis de la Torre-Bueno – de la Torre-Bueno, 1927: 31-32
▸ MA, SM, JA, BO; – EL: SA
maidli Hungerford & Evans – Hungerford & Evans, 1934: 73-74
▸ MA, JA, NT, SL; – EL: SA, HT

mindoroensis J. Polhemus – J. Polhemus, 1976: 283-284
‣ BO, SL, LU, PH!, PA, MI, NG
orientalis Lundblad – Lundblad, 1933a: 430-433
‣ MA, SM, JA, NT, SL, LU, PA!, MI, NG; – EL: SA, AU
papuana Kirkaldy – Kirkaldy, 1901: 807‣ NG, BS; – EL: AU
strigosa (Skuse) – Skuse, 1893: 43 (*Limnobates*)
‣ BS; – EL: AU, NC, New Zealand, Tahiti

GERROIDEA Leach, 1815
HERMATOBATIDAE Coutière & Martin, 1901

Hermatobates Carpenter, 1892
Type species *Hermatobates haddoni* Carpenter 1892
marchei (Coutière & Martin) – Coutière & Martin, 1901: 215-226 (*Hermatobatodes*)
‣ JA, PA, MI, NG; – EL: AU
singaporensis Cheng – Cheng, 1976: 209-210‣ MA
weddi China – China, 1957: 346-350‣ NG – EL: AU, NC, Japan, Pacific Islands

VELIIDAE Amyot & Serville, 1843
MICROVELIINAE China & Usinger, 1949

Aegilipsicola J. Polhemus & D. Polhemus, 1994
Type species: *Aegilipsicola rapida* J & D. Polhemus, 1994
**auga* D. Polhemus & J. Polhemus – D. Polhemus & J. Polhemus, 2004: 143-145
‣ NG
insularis D. Polhemus & J. Polhemus – D. Polhemus & J. Polhemus, 2000c: 285-287‣ NG
iriana D. Polhemus & J. Polhemus – D. Polhemus & J. Polhemus, 2000a: 104-105
‣ NG
**peninsularis* D. Polhemus & J. Polhemus – D. Polhemus & D. Polhemus, 2004: 141-143‣ NG
rapida J. Polhemus & D. Polhemus – J. Polhemus & D. Polhemus, 1994a: 60-62
‣ NG
robinae D. Polhemus & J. Polhemus – D. Polhemus & J. Polhemus, 2000a: 105-107‣ NG

Baptista Distant, 1903
Type species: *Baptista gestroi* Distant, 1903
collaris (Andersen) – Andersen, 1989c: 379-380 (*Lathriovelia*)‣ MA
femoralis Andersen – Andersen, 1989c: 368-371 ‣ MA; – EL: SA

Brechivelia D. Polhemus & J. Polhemus, 2004

tufi D. Polhemus & J. Polhemus – D. Polhemus & J. Polhemus, 2004: 133-135 ▸ NG

Lathriovelia Andersen, 1989

Type species: *Lathriovelia capitata* Andersen, 1989

capitata Andersen – Andersen, 1989c: 377-379 ▸ MA

rickmersi Kovac & Yang – Kovac & Yang, 2000: 157-161 ▸ MA, SM, BO; – EL: SA

Microvelia Westwood, 1834

Type species: *Microvelia pulchella* Westwood, 1834

albolineolata de la Torre Bueno – de la Torre Bueno, 1927: 33-34 ▸ MA, SM, JA

atroelegans Zettel & Gapud – Zettel & Gapud, 1999: 142-143 ▸ LU

douglasi douglasi Scott – Scott, 1874: 448-449 ▸ MA, SM, JA, BO!, SL, PA; – EL: SA, HT, India, Japan, Pacific Islands

douglasi atrolineata Bergroth – Bergroth, 1918: 122 ▸ LU, MI

genitalis genitalis Lundblad – Lundblad, 1933a: 361-364 ▸ JA; – EL: SA, HT

gestroi Kirkaldy – Kirkaldy, 1901b: 806-807 ▸ NG

insignis (Distant) – Distant, 1903a: 256 (*Rhagovelia*) ▸ MA

jaechi Zettel & Gapud – Zettel & Gapud, 1999: 144-145 ▸ NT

leveillei (Lethierry) – Lethierry, 1877: CI (*Hydroessa*) ▸ MA, SM, JA, BO!, NT, LU, MI; – EL: SA, HT, India

syn.: M *diluta* Distant – Distant, 1909: 500

loriae Kirkaldy – Kirkaldy, 1901b: 807 ▸ NG

magnifica Lundblad – Lundblad, 1933a: 326-328 ▸ JA

plumbea Lundblad – Lundblad, 1933a: 339-341 ▸ MA, JA

somnokrene Zettel & Gapud – Zettel & Gapud, 1999: 138-142 ▸ LU

villosula de la Torre Bueno – de la Torre Bueno, 1927: 34-35 ▸ SM

Neoalardus Distant, 1912

Type species: *Alardus typicus* Distant, 1903

typicus (Distant) – Distant, 1903b: 475 (*Alardus*) ▸ MA, SM; – EL: SA, India, China

Neusterensifer D. Polhemus & J. Polhemus, 2000

Type species: *Neusterinsifer compactus* J & D. Polhemus, 1994

acuminatus D. Polhemus & J. Polhemus – D. Polhemus & J. Polhemus, 2000a: 119-121 ▸ NG

aviavi D. Polhemus & J. Polhemus - D. Polhemus & J. Polhemus, 2004: 185-187 ▸ NG

batantana D. Polhemus & J. Polhemus – D. Polhemus & J. Polhemus, 2000c: 282-283 ▸ NG

bowutu D. Polhemus & J. Polhemus – D. Polhemus & J. Polhemus, 2004: 179-181 ▸ NG

compactus (J. Polhemus & D. Polhemus) – J. Polhemus & D. Polhemus, 1994a: 69 (*Neusterinsifer*) ▸ NG
cyclops (J. Polhemus & D. Polhemus) – J. Polhemus & D. Polhemus, 1994a: 69-70 (*Neusterinsifer*) ▸ NG
**deantrecasteaux* D. Polhemus & J. Polhemus – D. Polhemus & J. Polhemus, 2004: 172-174 ▸ NG, BS
etna D. Polhemus & J. Polhemus – D. Polhemus & J. Polhemus, 2000a: 121-122 ▸ NG
**femoralis* D. Polhemus & J. Polhemus – D. Polhemus & J. Polhemus, 2004: 157-160 ▸ NG, BS
gamensis D. Polhemus & J. Polhemus – D. Polhemus & J. Polhemus, 2000c: 283-285 ▸ NG
gladius (J. Polhemus & D. Polhemus) – J. Polhemus & D. Polhemus, 1994a: 70-71 (*Neusterinsifer*) ▸ NG
**goilala* D. Polhemus & J. Polhemus – D. Polhemus & J. Polhemus, 2004: 181-184 ▸ NG
**hunteri* D. Polhemus & J. Polhemus – D. Polhemus & J. Polhemus, 2004: 160-163 ▸ NG, BS
irianus D. Polhemus & J. Polhemus – D. Polhemus & J. Polhemus, 2000a: 122-123 ▸ NG
**kula* D. Polhemus & J. Polhemus – D. Polhemus & J. Polhemus, 2004: 174-177 ▸ BS
kutubu D. Polhemus & J. Polhemus – D. Polhemus & J. Polhemus, 2000a: 115-117 ▸ NG
**louisiadae* D. Polhemus & J. Polhemus – D. Polhemus & J. Polhemus, 2004: 163-164 ▸ BS
lubu D. Polhemus & J. Polhemus – D. Polhemus & J. Polhemus, 2000a: 115 ▸ NG
**microrivula* D. Polhemus & J. Polhemus – D. Polhemus & J. Polhemus, 2004: 184-185 ▸ NG
**misima* D. Polhemus & J. Polhemus – D. Polhemus & J. Polhemus, 2004: 168-170 ▸ BS
misoolicus D. Polhemus & J. Polhemus – D. Polhemus & J. Polhemus, 2000c: 281-282 ▸ NG
**muyuw* D. Polhemus & J. Polhemus – D. Polhemus & J. Polhemus, 2004: 170-172 ▸ BS
nabire (J. Polhemus & D. Polhemus) – J. Polhemus & D. Polhemus, 1994a: 71-72 (*Neusterinsifer*) ▸ NG
pseudocyclops D. Polhemus & J. Polhemus – D. Polhemus & J. Polhemus, 2000a: 117-119 ▸ NG
**sagrai* D. Polhemus & J. Polhemus – D. Polhemus & J. Polhemus, 2004: 155-157 ▸ NG
sepik (J. Polhemus & D. Polhemus) – J. Polhemus & D. Polhemus, 1994a: 72-74 (*Neusterinsifer*) ▸ NG

**sulcata* D. Polhemus & J. Polhemus – D. Polhemus & J. Polhemus, 2004: 165-168 ▸ BS

**tufi* D. Polhemus & J. Polhemus – D. Polhemus & J. Polhemus, 2004: 177-179 ▸ NG

**yela* D. Polhemus & J. Polhemus – D. Polhemus & J. Polhemus, 2004: 164-165 ▸ BS

Papuavelia D. Polhemus & J. Polhemus, 2000

Type species: *Papuavelia siculifera* D. Polhemus & J. Polhemus, 2000

siculifera D. Polhemus & J. Polhemus – D. Polhemus & J. Polhemus, 2000a: 94-95 ▸ NG

Phoreticovelia D. Polhemus & J. Polhemus, 2000

Type species: *Phoreticovelia rotunda* D. Polhemus & J. Polhemus, 2000

nigra D. Polhemus & J. Polhemus – D. Polhemus & J. Polhemus, 2000a: 98-100 ▸ NG

rotunda D. Polhemus & J. Polhemus – D. Polhemus & J. Polhemus, 2000a: 97-98 ▸ NG; – EL: AU

Polhemovelia Zettel & Sehnal, 2000

Type species: *Polhemovelia septuaginta* Zettel & Sehnal, 2000

johnpolhemi Zettel & Sehnal – Zettel & Sehnal, 2000: 131-132 ▸ BO

septuaginta Zettel & Sehnal – Zettel & Sehnal, 2000: 130-131 ▸ BO

setosa Zettel & Sehnal – Zettel & Sehnal, 2000: 132-133 ▸ BO

Pseudovelia Hoberlandt, 1950

Type species: *Pseudovelia tibialis* Esaki & Miyamoto, 1955

Pseudovelia subgenus *Pseudovelia* Hoberlandt, 1950

Type species: *Pseudovelia tibialis* Esaki & Miyamoto, 1955

aflia Nieser – Nieser, 1995: 72 ▸ SL

afrofila Nieser – Nieser, 1995: 72-74 ▸ SL

argyropardala Nieser – Nieser, 1995: 74-76 ▸ SL, MI

borneensis Andersen – Andersen, 1983: 264 ▸ BO

epimekta Nieser – Nieser, 1995: 76-78 ▸ SL

feuerborni (Lundblad) – Lundblad, 1933: 333-336 (*Microvelia*) ▸ MA, JA – EL: SA

hypodonta (Lundblad) – Lundblad, 1933: 329-331 (*Microvelia*) ▸ SM

kalami Nieser – Nieser, 1995: 78-79 ▸ MI

koutali Nieser – Nieser, 1995: 79 ▸ SL

lata Sehnal – Sehnal, 1999: 152-154 ▸ PA

lundbladi Andersen – Andersen, 1983: 262-264 ▸ MA

mystax Nieser – Nieser, 1995: 80 ▸ SL

pyrokrene Nieser – Nieser, 1995: 80-81 ▸ SL

reiseni J. Polhemus – J. Polhemus, 1976: 272-273 ▸ LU
sangihe Nieser – Nieser, 1995: 81-83 ▸ SL
sexualis (Paiva) – Paiva, 1917: 77-79 (*Microvelia*) ▸ MA, SM; – EL: SA, India, Pakistan
tenuis Sehnal – Sehnal, 1999: 149-152 ▸ PA
tjurupensis (Lundblad) – Lundblad, 1933: 329-331 (*Microvelia*) ▸ SM

**Rheovelia* D. Polhemus & J. Polhemus, 2004
**anomala* D. Polhemus & J. Polhemus – D. Polhemus & J. Polhemus, 2004: 125-127 ▸ BS
**asymmetrica* D. Polhemus & J. Polhemus – D. Polhemus & J. Polhemus, 2004: 128-129 ▸ NG
**basilaki* D. Polhemus & J. Polhemus – D. Polhemus & J. Polhemus, 2004: 130-132 ▸ NG
**fonticola* D. Polhemus & J. Polhemus – D. Polhemus & J. Polehemus, 2004: 129-130 ▸ NG
**insularis* D. Polhemus & J. Polhemus – D. Polhemus & J. Polhemus, 2004: 123-125 ▸ BS
**petrophila* D. Polhemus & J. Polhemus – D. Polhemus & J. Polhemus, 2004: 121-122 ▸ NG
**robinae* D. Polhemus & J. Polhemus – D. Polhemus & J. Polhemus, 2004: 122-123 ▸ BS
**truncata* D. Polhemus & J. Polhemus – D. Polhemus & J. Polhemus, 2004: 127-128 ▸ NG

Tanyvelia J. Polhemus & D. Polhemus, 1994
Type species: *Tanyvelia missim* J. Polhemus & D. Polhemus, 1994
bosavi D. Polhemus & J. Polhemus – D. Polhemus & J. Polhemus, 2000a: 102-104 ▸ NG
**minima* D. Polhemus & J. Polhemus – D. Polhemus & D. Polhemus, 2004: 136-138 ▸ NG
missim J. Polhemus & D. Polhemus – J. Polhemus & D. Polhemus, 1994a: 58-60 ▸ NG
**papuana* D. Polhemus & J. Polhemus – D. Polhemus & J. Polhemus, 2004: 139-141 ▸ NG
**tagulana* D. Polhemus & J. Polhemus – D. Polhemus & J. Polhemus, 2004: 138-139 ▸ BS

Tarsovelia J. Polhemus & D. Polhemus, 1994
Type species: *Tarsovelia alta* J & D. Polhemus, 1994
alta J. Polhemus & D. Polhemus – J. Polhemus & D. Polhemus, 1994a: 64-65 ▸ NG
arfak J. Polhemus & D. Polhemus – J. Polhemus & D. Polhemus, 1994a: 65 ▸ NG
bosavi D. Polhemus & J. Polhemus – D. Polhemus & J. Polhemus, 2000a: 109-111 ▸ NG

dani J. Polhemus & D. Polhemus – J. Polhemus & D. Polhemus, 1994a: 65-67 ▸ NG

kikori D. Polhemus & J. Polhemus – D. Polhemus & J. Polhemus, 2000a: 108-109 ▸ NG

**louisiadensis* D. Polhemus & J. Polhemus – D. Polhemus & J. Polhemus, 2004: 146-149 ▸ NG

rajana D. Polhemus & J. Polhemus – D. Polhemus & J. Polhemus, 2000c: 287-289 ▸ NG

reclusa D. Polhemus & J. Polhemus – D. Polhemus & J. Polhemus, 2000a: 111 ▸ NG

ziwa D. Polhemus & J. Polhemus – D. Polhemus & J. Polhemus, 2000a: 112-113 ▸ NG

Velohebria Štys, 1976
Type species: *Velohebria antennalis* Štys, 1977
antennalis Štys – Štys, 1977: 393-394 ▸ NG

Xiphovelia Lundblad, 1933
Type species: *Xiphovelia ensis* Lundblad, 1933
ensis Lundblad – Lundblad, 1933a: 364-368 ▸ SM
skoteina Nieser – Nieser, 1995: 85-86 ▸ SL

HALOVELIINAE Esaki, 1930

Entomovelia Esaki, 1930
Type species: *Entomovelia doveri* Esaki, 1930
doveri Esaki – Esaki, 1930: 21-22 ▸ MA

Halovelia Bergroth, 1893
Type species: *Halovelia maritima* Bergroth, 1893
abdominalis Andersen – Andersen, 1989b: 184-186 ▸ MA, JA
anderseni Lansbury – Lansbury, 1996a: 18-20 ▸ NG
annemariae Andersen – Andersen, 1989a: 115-116 ▸ NG, BS
bergrothi Esaki – Esaki, 1926: 161-162 ▸ LU, PH, PA, MI, NG, BS; – EL: SA, NC, Pacific Islands
corallia Andersen – Andersen, 1989a: 92-93 ▸ NG; – EL: AU
esakii Andersen – Andersen, 1989a: 107-109 ▸ NT, SL, LU, PH, PA, MI, ML, NG, BS
halophila Andersen – Andersen, 1989a: 92 ▸ BO, NT
hilli China – China, 1957: 352-354 ▸ NG; – EL: AU, NC
lannae Andersen – Andersen, 1989a: 116-118 ▸ MA, JA, BO, PA, MI
malaya Esaki – Esaki, 1930: 18-20 ▸ MA; – EL: SA!
martima Bergroth – Bergroth, 1893: 277-278 ▸ NG; – EL: AU
novoguinensis Andersen – Andersen, 1989a: 112 ▸ NG

solomon Andersen – Andersen, 1989a: 110-112 ‣ BS
sulawesi Andersen – Andersen, 1989b: 183-184 ‣ SL
sumaldei Zettel – Zettel, 1998f: 462-464 ‣ MI, PH!
wallacei Andersen – Andersen, 1989a: 119-120 ‣ SL, NT

Haloveloides Andersen, 1992
Type species: *Halovelia papuensis* Esaki, 1926
anderseni Zettel – Zettel, 2003c: 167-170 ‣ LU
brevicornis Andersen – Andersen, 1992b: 395-396 ‣ SL, PA, ML
browni (Lansbury) – Lansbury, 1989a: 102-104 (*Halovelia*) ‣ NG, BS
christyae Zettel – Zettel, 1998f: 465-468 ‣ LU, PH, PA
cornutus Andersen – Andersen, 1992b: 400-401 ‣ LU
danpolhemi Andersen – Andersen, 1992b: 399-400 ‣ PA
femoralis Andersen – Andersen, 1992b: 401-402 ‣ PH, PA
fluvialis Zettel – Zettel, 2003c: 170-172 ‣ MI
gapudi Zettel – Zettel, 2003c: 172-174 ‣ LU
hirsutus Zettel – Zettel, 2003c: 174-176 ‣ MI
lansburyi Zettel – Zettel, 1998f: 464-465 ‣ MI
papuensis (Esaki) – Esaki, 1926: 160 (*Halovelia*) ‣ NG, BS
sundaensis Andersen – Andersen, 1992b: 398-399 ‣ MA, JA, BO; EL: SA!

Strongylovelia Esaki, 1924
Type species: *Strongylovelia formosa* Esaki, 1924
aberrans Lansbury & Zettel – Lansbury & Zettel, 1997: 65 ‣ BO
albicollis Esaki – Esaki 1926: 156-157 ‣ NG
bukidnonica Lansbury & Zettel – Lansbury & Zettel, 1997: 71-72 (ssp. of *philippinensis*) ‣ MI
cebuana Lansbury & Zettel – Lansbury & Zettel, 1997: 74-76 ‣ PH
connexivum Lansbury & Zettel – Lansbury & Zettel, 1997: 58-60 ‣ BO
esakii Lansbury & Zettel – Lansbury & Zettel, 1997: 60-62 ‣ BO
hirsutula Lansbury & Zettel – Lansbury & Zettel, 1997: 58 ‣ BO
marinduquensis Zettel -Zettel, 2003d: 187-188 ‣ LU
mindoroensis Lansbury & Zettel – Lansbury & Zettel, 1997: 72-74 ‣ LU
palawanensis Lansbury & Zettel – Lansbury & Zettel, 1997: 62-64 ‣ PA
philippinensis philippinensis Lansbury & Zettel – Lansbury & Zettel, 1997: 66-68 ‣ LU
philippinensis boholensis Lansbury & Zettel – Lansbury & Zettel, 1997: 70-71 ‣ MI
philippinensis surigaoensis Zettel – Zettel, 2003d: 188-190 ‣ MI
priori Lansbury – Lansbury, 1993a: 15-18 ‣ BS
samarensis Zettel – Zettel, 2003d: 184-187 ‣ MI
seyferti Zettel – Zettel 2003f: 1135-1137 ‣ SL
sibuyana Lansbury & Zettel – Lansbury & Zettel, 1997: 68-70 ‣ LU

Xenobates Esaki, 1927

Type species: *Microbates seminulum* Esaki, 1927

argentatus Andersen – Andersen, 2000a: 280-281 ▸ MA, BO, MI, PA
caudatus Andersen & Weir – Andersen & Weir, 1999: 330-332 ▸ NG
maculatus Andersen – Andersen, 2000a: 282-283 ▸ BO
mandai Andersen – Andersen, 2000a: 278-280 ▸ MA; EL: SA!
murphyi Andersen – Andersen, 2000a: 277-278 ▸ MA, BO, PH
oculatus (Lansbury) **comb. nov.** – Lansbury, 1989a: 104-105 (Halovelia) ▸ BS
pictus Andersen – Andersen, 2000a: 281-282 ▸ MA, BO
pilosellus Lansbury – Lansbury, 1996a: 21-22 ▸ NG
seminulum (Esaki) – Esaki, 1926: 154-155 (*Microbates*) ▸ NG
singaporensis Andersen – Andersen, 2000a: 275-277 ▸ MA
solomonensis Lansbury – Lansbury, 1989a: 107-109 ▸ NG, BS

PERITTOPINAE China & Usinger, 1949

Perittopus Fieber, 1861

Type species: *Perittopus breddini* Kirkaldy, 1901

asiaticus Zettel – Zettel, 2001a: 110-113 ▸ MA; – EL: SA
borneensis Zettel – Zettel, 2001a: 114-115 ▸ BO
breddini Kirkaldy – Kirkaldy, 1901c: 286 ▸ JA; – EL: SA
schuhi Zettel – Zettel, 2001a: 117-118 ▸ JA
sumatrensis Zettel – Zettel, 2001a: 115-117 ▸ SM
vicarians Breddin – Breddin, 1905a: 129-130 ▸ JA; – EL: SA
webbi Zettel – Zettel, 2001a: 113-114 ▸ MA

RHAGOVELIINAE China & Usinger, 1949

Rhagovelia Mayr, 1865

Type species: *Velia nigricans* Burmeister, 1835

aberrans Andersen – Andersen, 1965: 111-117 ▸ MI
abra Nieser, Zettel & Chen – Nieser et al., 1997: 22-24 ▸ SL
achna Nieser, Zettel & Chen – Nieser et al., 1997: 24-25 ▸ SL
agilis J. Polhemus – J. Polhemus, 1976: 278-279 ▸ LU
akrita J. Polhemus – J. Polhemus, 1976: 280-281 ▸ LU
ambonensis Lundblad – Lundblad, 1936: 30-31 ▸ ML
amnica Lansbury – Lansbury, 1993b: 31-33 ▸ BS
aureospicata Lansbury – Lansbury, 1993: 41-43 ▸ NG
bacanensis J. Polhemus & D. Polhemus – J. Polhemus & D. Polhemus, 1988: 192-193 ▸ ML
bipennicillata Lundblad – Lundblad, 1936: 23-25 ▸ SM
biroi Lundblad – Lundblad, 1936: 13-14 ▸ NG, BS
blogiokommena Nieser & Chen – Nieser & Chen, 1993: 268 ▸ SL

borneensis J. Polhemus & D. Polhemus – J. Polhemus & D. Polhemus, 1988: 183-184 ▸ BO
browni Lansbury – Lansbury, 1993b: 33-36 ▸ BS
caesia Lansbury – Lansbury, 1993b: 50-53 ▸ NG
camiguinana Zettel – Zettel, 1996c: 123-124 ▸ MI
canalaonensis Zettel – Zettel, 1996c: 118-120 ▸ PH
celebensis J. Polhemus & D. Polhemus – J. Polhemus & D. Polhemus, 1988: 199-200 ▸ SL
christenseni J. Polhemus & D. Polhemus – J. Polhemus & D. Polhemus, 1988: 201-202 ▸ BO
chrysomalla Nieser & Chen – Nieser & Chen, 1993: 268-270 ▸ SL
cotabatoensis Hungerford & Matsuda – Hungerford & Matsuda, 1961a: 259-260 ▸ LU, PH!, MI
crinita Lansbury – Lansbury, 1993b: 46-48 ▸ NG
cylindros Nieser, Zettel & Chen – Nieser et al., 1997: 39-40 ▸ SL
daktylophora Nieser & Chen – Nieser & Chen, 1993: 270-271 ▸ SL
doveri Lundblad – Lundblad, 1933a: 295-296 ▸ SM
estrella Zettel – Zettel, 1994c: 44-45 ▸ PA
femorata Dover – Dover, 1928: 66-67 ▸ MA
fischeri Zettel – Zettel, 1999i: 748-751 ▸ LU
fulva Lansbury – Lansbury, 1993b: 36-37 ▸ BS
grayi J. Polhemus & D. Polhemus – J. Polhemus & D. Polhemus, 1988: 207-208 ▸ SL
gyrista Nieser, Zettel & Chen – Nieser et al., 1997: 32-33 ▸ SL
hamdjahi J. Polhemus & D. Polhemus – J. Polhemus & D. Polhemus, 1988: 194-195 ▸ SL
herzogensis Lansbury – Lansbury, 1993b: 43-46 ▸ NG
hirsuta Lansbury – Lansbury, 1993b: 39-41 ▸ NG
hoberlandti Hungerford & Matsuda – Hungerford & Matsuda, 1961a: 260-261 ▸ PA
hoogstraali Hungerford & Matsuda – Hungerford & Matsuda, 1961a: 264-266 ▸ MI
horaia Nieser & Chen – Nieser & Chen, 1993: 271-273 ▸ SL
hungerfordi Lundblad – Lundblad, 1933a: 293-295 ▸ SM
hutchinsoni Lundblad – Lundblad, 1933: 292-293 ▸ JA; – EL: SA
incognita J. Polhemus & D. Polhemus – J. Polhemus & D. Polhemus, 1988: 179-180 ▸ BO
kalami Nieser & Chen – Nieser & Chen, 1993: 273-274 ▸ SL
kastanoparuphe Nieser & Chen – Nieser & Chen, 1993: 274-275 ▸ SL
kawakamii (Matsumura) – Matsumura, 1913: 99 (*Kotovelia*) ▸ BO, LU, PA; – EL: HT
krama Nieser, Zettel & Chen – Nieser et al., 1997: 25-27 ▸ SL
lansburyi Zettel – Zettel, 1995d: 46-47 ▸ PA
leyteensis Zettel – Zettel, 1996c: 124-126 ▸ MI

lorelinduana J. Polhemus & D. Polhemus – J. Polhemus & D. Polhemus, 1988: 208-209 ▸ SL
lugubris Lundblad – Lundblad, 1933a: 296-297 ▸ SM, JA
lundbladi Hungerford & Matsuda – Hungerford & Matsuda, 1961a: 257-259 ▸ MI
luzonica Lundblad – Lundblad, 1937: 6-7 ▸ LU
meikdelyi J. Polhemus & D. Polhemus – J. Polhemus & D. Polhemus, 1988: 202-203 ▸ ML
melanopsis J. Polhemus & D. Polhemus – J. Polhemus & D. Polhemus, 1988: 181-182 ▸ BO
minahasa J. Polhemus & D. Polhemus – J. Polhemus & D. Polhemus, 1988: 205-206 ▸ SL
mindanaoensis Hungerford & Matsuda – Hungerford & Matsuda, 1961a: 263-264 ▸ MI
mindoroensis Zettel – Zettel, 1994b: 226-227 ▸ LU
minuta Lundblad – Lundblad, 1936: 34 ▸ LU
minutissima Hungerford & Matsuda – Hungerford & Matsuda, 1961a: 267-268 ▸ PA
negrosensis Zettel – Zettel, 1995d: 66-67 ▸ PH
nieseri Zettel – Zettel, 1995d: 73-74 ▸ MI
nigra Hungerford – Hungerford, 1933: 2-3 ▸ SM
obi J. Polhemus & D. Polhemus – J. Polhemus & D. Polhemus, 1988: 193-194 ▸ ML
ochra Nieser, Zettel & Chen – Nieser et al., 1997: 27-28 ▸ SL
orientalis Lundblad – Lundblad, 1937: 5-6 ▸ MI
orientaloides Zettel – Zettel, 1995d: 57-60 ▸ MI
palawanensis Zettel – Zettel, 1994c: 40-44 ▸ PA
panayensis Zettel – Zettel, 1995d: 67-68 ▸ PH
papuensis Lundblad – Lundblad, 1936: 28-30 ▸ NG, BS
philippina Lundblad – Lundblad, 1936: 36-37 ▸ LU
phoretica D. Polhemus – D. Polhemus, 1995: 61-63 ▸ MI, ?PH
plychona Nieser, Zettel & Chen – Nieser et al., 1997: 35-37 ▸ SL
potamophila Zettel – Zettel, 1996c: 117-118 ▸ LU
priori Lansbury – Lansbury, 1993b: 29-31 ▸ NG
pruinosa J. Polhemus & D. Polhemus – J. Polhemus & D. Polhemus, 1988: 195-196 ▸ SL
pseudocelebensis Nieser & Chen – Nieser & Chen, 1993: 275 ▸ SL
raddai Zettel – Zettel, 1994b: 228-229 ▸ LU
ranau J. Polhemus & D. Polhemus – J. Polhemus & D. Polhemus, 1988: 180-181 ▸ BO
raymondi Zettel – Zettel, 1995d: 68-69 ▸ PH
ridicula D. Polhemus – D. Polhemus, 1995: 55-60 ▸ ?LU, MI
robina Nieser & Chen – Nieser & Chen, 1993: 275-277 ▸ SL
rudischuhi Zettel – Zettel, 1993c: 39-41 ▸ BO, MA

sabela J. Polhemus & D. Polhemus – J. Polhemus & D. Polhemus, 1988: 191 ▸ ML
sallyae Zettel – Zettel 2003e: 1142-1144 ▸ MI
samardaca J. Polhemus & D. Polhemus – J. Polhemus & D. Polhemus, 1988: 200-201 ▸ BO
samarinda J. Polhemus & D. Polhemus – J. Polhemus & D. Polhemus, 1988: 186-187 ▸ BO, ?SL
sarawakensis J. Polhemus & D. Polhemus – J. Polhemus & D. Polhemus, 1988: 190 ▸ BO
schoedli Zettel – Zettel, 1996c: 120-121 ▸ PH
seyferti Zettel – Zettel, 1995: 61-63 ▸ PH
sibuyana Zettel – Zettel, 1996c: 116-117 ▸ LU
silau J. Polhemus & D. Polhemus – J. Polhemus & D. Polhemus, 1988: 184-185 ▸ BO
simulata J. Polhemus & D. Polhemus – J. Polhemus & D. Polhemus, 1988: 182-183 ▸ BO
singaporensis Yang & D. Polhemus – Yang & D. Polhemus, 1994: 987-993 ▸ BO, MA
skoura Nieser, Zettel & Chen – Nieser et al., 1997: 28-31 ▸ SL
sondaica J. Polhemus & D. Polhemus – J. Polhemus & D. Polhemus, 1988: 188-190 ▸ MA, BO
sterea Nieser, Zettel & Chen – Nieser et al., 1997: 33-35 ▸ SL
sulawesiana J. Polhemus & D. Polhemus – J. Polhemus & D. Polhemus, 1988: 198-199 ▸ SL
sumatrensis Lundblad – Lundblad, 1933a: 287-290 ▸ BO, MA, SM, JA, NT; – EL: SA, HT, India
tablasensis Zettel – Zettel, 1996c: 121-122 ▸ PH
tansiongcoi Zettel – Zettel, 1995d: 69-71 ▸ LU
tawau J. Polhemus & D. Polhemus – J. Polhemus & D. Polhemus, 1988: 187-188 ▸ BO
tebakang J. Polhemus & D. Polhemus – J. Polhemus & D. Polhemus, 1988: 186 ▸ BO
thysanotos Lansbury – Lansbury, 1993b: 48-50 ▸ NG
trichota Nieser & Chen – Nieser & Chen, 1993: 277-278 ▸ SL
tropidata Nieser & Chen – Nieser & Chen, 1993: 278-279 ▸ SL
tsouloufi Nieser, Zettel & Chen – Nieser et al., 1997: 31-32 ▸ SL
unica J. Polhemus & D. Polhemus – J. Polhemus & D. Polhemus, 1988: 195 ▸ SL
usingeri Hungerford & Matsuda – Hungerford & Matsuda, 1961a: 262-263 ▸ MI
wallacei J. Polhemus & D. Polhemus – J. Polhemus & D. Polhemus, 1988: 203-205 ▸ SL
werneri Hungerford & Matsuda – Hungerford & Matsuda, 1961a: 266-267 ▸ MI

Tetraripis Lundblad, 1936
Type species: *Rhagovelia ravana* Kirkaldy, 1901
bituberculatus Zettel & Yang – Zettel & Yang, 2002: 1536-1538 ▸ BO

borneensis Zettel – Zettel, 1995e: 26-27 ▸ BO
doveri Lundblad – Lundblad, 1936: 56-57 ▸ MA
drescheri J. Polhemus & D. Polhemus – J. Polhemus & D. Polhemus, 1999: 124-126 ▸ JA
kodadai Zettel – Zettel, 1995e: 27-29 ▸ BO

VELIINAE Brullé, 1836

Angilia Stål, 1865
Type species: *Velia albidotincta* Stål, 1855

Angilia **subgenus** *Adriennella* Poisson, 1942
Type species: *Angilia schoutedeni* Poisson, 1942
anderseni Zettel & Hecher – Zettel & Hecher, 1998: 338-339 ▸ BO
borneensis Zettel & Hecher – Zettel & Hecher, 1998: 339-340 ▸ BO
mazzoldii Zettel & Hecher – Zettel & Hecher, 1998: 336-338 ▸ BO
orientalis Andersen – Andersen, 1981b: 343-345 ▸ PA; – EL: SA, HT
philippiensis Drake & Hoberlandt – Drake & Hoberlandt, 1953: 223-226 ▸ LU
trispinosa Andersen – Andersen, 1981b: 350-351 ▸ SM

Angilovelia Andersen, 1981
Type species: *Velia y-alba* Paiva, 1918
y-alba (Paiva) – Paiva, 1918: 20 (*Velia*) ▸ MA, ?SL; – EL: SA, HT, India

GERRIDAE Leach, 1815
RHAGADOTARSINAE Lundblad, 1933

Rhagadotarsus Breddin, 1905
Type species: *Rhagadotarsus kraepelini* Breddin, 1905

Rhagadotarsus **subgenus** *Rhagadotarsus* Breddin, 1905
Type species: *Rhagadotarsus kraepelini* Breddin, 1905
anomalus J. Polhemus & Karunaratne – J. Polhemus & Karunaratne, 1993: 103-104 ▸ NG; – EL: AU
borneensis J. Polhemus & Karunaratne – J. Polhemus & Karunaratne, 1993: 106-107 ▸ BO, NG
kraepelini Breddin – Breddin, 1905a: 137-139 ▸ MA, SM, JA, BO, NT, SL, LU, PH, PA, MI, NG, BS; – EL: SA, HT, India, China, Pacific Islands
**palawanensis* Zettel – Zettel, 2004e: 363-364 ▸ PA

TREPOBATINAE Matsuda, 1960

Metrobatini J. Polhemus & D. Polhemus, 1993

Andersenella J. Polhemus & D. Polhemus, 1993
Type species: *Andersenella binotata* J. Polhemus & D. Polhemus, 1993
binotata J. Polhemus & D. Polhemus – J. Polhemus & D. Polhemus, 1993: 248-249 ▸ NG

Ciliometra J. Polhemus & D. Polhemus, 1993
Type species: *Ciliometra kiunga* J. Polhemus & D. Polhemus, 1993
femorata J. Polhemus & D. Polhemus – J. Polhemus & D. Polhemus, 1993: 252 ▸ NG
hirsuta Lansbury – Lansbury, 1996b: 59-60 ▸ NG
kiunga J. Polhemus & D. Polhemus – J. Polhemus & D. Polhemus, 1993: 251-252 ▸ NG
minajerwi J. Polhemus & D. Polhemus – J. Polhemus & D. Polhemus, 2000a: 296-297 ▸ NG
priori Lansbury – Lansbury, 1996b: 55-58 ▸ NG
sepik J. Polhemus & D. Polhemus – J. Polhemus & D. Polhemus, 1993: 252-253 ▸ NG
setosa J. Polhemus & D. Polhemus – J. Polhemus & D. Polhemus, 2000a: 292-295 ▸ NG
waigeo J. Polhemus & D. Polhemus – J. Polhemus & D. Polhemus, 2000a: 292-295 ▸ NG

Iobates J. Polhemus & D. Polhemus, 1993
Type species: *Merobatopsis affinis* Esaki, 1926
affinis (Esaki) – Esaki, 1926: 146-148 (*Metrobatopsis*) ▸ NG
ivimka J. Polhemus & D. Polhemus – J. Polhemus & D. Polhemus, 2002: 287-288 ▸ NG
salawati J. Polhemus & D. Polhemus – J. Polhemus & D. Polhemus, 1993: 255-256 ▸ NG
somare J. Polhemus & D. Polhemus – J. Polhemus & D. Polhemus, 1993: 256-258 ▸ NG

Metrobatoides J. Polhemus & D. Polhemus, 1993
Type species: *Metrobatoides genitalis* J. Polhemus & D. Polhemus, 1993
bifurcatus J. Polhemus & D. Polhemus – J. Polhemus & D. Polhemus, 2002: 284-287 ▸ NG
genitalis J. Polhemus & D. Polhemus – J. Polhemus & D. Polhemus, 1993: 261 ▸ NG

Metrobatopsis Esaki, 1926

Type species: *Metrobatopsis flavonotatus* Esaki, 1926

browni J. Polhemus & D. Polhemus – J. Polhemus & D. Polhemus, 1993: 266-268 ▸ BS

flavonotatus Esaki – Esaki, 1926: 146 ▸ NG

insularis J. Polhemus & D. Polhemus – J. Polhemus & D. Polhemus, 1993: 266 ▸ NG

lannae J. Polhemus & D. Polhemus – J. Polhemus & D. Polhemus, 1993: 269-270 ▸ BS

mussau J. Polhemus & D. Polhemus – J. Polhemus & D. Polhemus, 1993: 269 ▸ NG

solomonensis Hungerford & Matsuda – Hungerford & Matsuda, 1959: 35-36 ▸ BS

Stygiobates J. Polhemus & D. Polhemus, 1993

Type species: *Stygiobates iriana* J. Polhemus & D. Polhemus, 1993

iriana J. Polhemus & D. Polhemus – J. Polhemus & D. Polhemus, 1993: 273-275 ▸ NG

iweka J. Polhemus & D. Polhemus – J. Polhemus & D. Polhemus, 2000a: 297-300 ▸ NG

moratai J. Polhemus & D. Polhemus – J. Polhemus & D. Polhemus, 1993: 275-278 ▸ ML

mubi J. Polhemus & D. Polhemus – J. Polhemus & D. Polhemus, 2000a: 300-301 ▸ NG

rajana J. Polhemus & D. Polhemus – J. Polhemus & D. Polhemus, 2000a: 301-303 ▸ NG

Talaudia J. Polhemus & D. Polhemus, 2002

Type species: *Andersenella nilsi* Chen & Nieser, 2000

nilsi (Chen & Nieser) – Chen & Nieser, 2000: 69-71 (*Andersenella*) ▸ SL

Naboandelini J. Polhemus & D. Polhemus, 1994

Calyptobates J. Polhemus & D. Polhemus, 1994

Type species: *Calyptobates jourama* J. Polhemus & D. Polhemus, 1994

amboina J. Polhemus & D. Polhemus – J. Polhemus & D. Polhemus, 1994b: 337-338 ▸ ML

kamoro J. Polhemus & D. Polhemus – J. Polhemus & D. Polhemus, 2000a: 308-309 ▸ NG

kopi J. Polhemus & D. Polhemus – J. Polhemus & D. Polhemus, 2000a: 306-308 ▸ NG

samarinda J. Polhemus & D. Polhemus – J. Polhemus & D. Polhemus, 1994b: 344-345 ▸ BO *simplex* J. Polhemus & D. Polhemus – J. Polhemus & D. Polhemus, 1994: 345-346 ▸ NG

Naboandelus Distant, 1910
Type species: *Naboandelus signatus* Distant, 1910
borneensis J. Polhemus & D. Polhemus – J. Polhemus & D. Polhemus, 1994b: 350-351 ▸ BO
johorensis J. Polhemus & D. Polhemus – J. Polhemus & D. Polhemus, 1994b: 351 ▸ MA

Trepobatini Matsuda, 1960

Cryptobates Esaki, 1929
Type species: *Gerris raja* Distant, 1910
johorensis J. Polhemus & D. Polhemus – J. Polhemus & D. Polhemus, 1995b: 102-103 ▸ MA
obscurus Miyamoto – Miyamoto, 1967: 253-255 ▸ BO
rufus J. Polhemus & D. Polhemus – J. Polhemus & D. Polhemus, 1995b: 105-107 ▸ MA

Stenobatini J. Polhemus & D. Polhemus, 1993

Pseudohalobates J. Polhemus & D. Polhemus, 1996
Type species: *Pseudohalobates inobonto* J. Polhemus & D. Polhemus, 1996
inobonto J. Polhemus & D. Polhemus – J. Polhemus & D. Polhemus, 1996: 294-299 ▸ SL, MI, ML, NG

Rheumatometroides Hungerford & Matsuda, 1958
Type species: *Rheumatometroides browni* Hungerford & Matsuda, 1958
browni Hungerford & Matsuda – Hungerford & Matsuda, 1958b: 204-206 ▸ NG, BS
drepanephoros Chen & Nieser – Chen & Nieser, 1992: 160 ▸ SL
insularis insularis (J. Polhemus & Cheng) – J. Polhemus & Cheng, 1982: 225-227 (*Stenobates*) ▸ MA
insularis papar J. Polhemus & D. Polhemus – J. Polhemus & D. Polhemus, 1996: 303-304 ▸ BO
kikori J. Polhemus & D. Polhemus – J. Polhemus & D. Polhemus, 1996: 305-307 ▸ NG
sele J. Polhemus & D. Polhemus – J. Polhemus & D. Polhemus, 1996: 307-311 ▸ NG
wabon J. Polhemus & D. Polhemus – J. Polhemus & D. Polhemus, 1996: 311-313 ▸ NG

Stenobates Esaki, 1927
Type species: *Stenometra biroi* Esaki, 1926

australicus J. Polhemus & D. Polhemus – J. Polhemus & D. Polhemus, 1991: 3-5 ▸ NG; – EL: AU
biroi (Esaki) – Esaki, 1926: 119-120 (*Stenometra*) ▸ MA
fakfak J. Polhemus & D. Polhemus – J. Polhemus & D. Polhemus, 1996: 317-319 ▸ NG
kamojo J. Polhemus & D. Polhemus – J. Polhemus & D. Polhemus, 1996: 319-320 ▸ NG
kasim J. Polhemus & D. Polhemus – J. Polhemus & D. Polhemus, 1996: 320-324 ▸ NG
labuha J. Polhemus & D. Polhemus – J. Polhemus & D. Polhemus, 1996: 325-326 ▸ ML
langoban J. Polhemus & D. Polhemus – J. Polhemus & D. Polhemus, 1996: 326-332 ▸ PA
makraitos (Chen & Nieser) – Chen & Nieser, 1992: 159-160 (*Rheumatobates*) ▸ SL
sangihe J. Polhemus & D. Polhemus – J. Polhemus & D. Polhemus, 1996: 333-335 ▸ SL
zamboanga J. Polhemus & D. Polhemus – J. Polhemus & D. Polhemus, 1996: 335-338 ▸ MI

Stenobatopsis J. Polhemus & D. Polhemus, 1996
Type species: *Stenobatopsis stygius* J. Polhemus & D. Polhemus, 1996

**coronensis* Zettel – Zettel, 2004a: 365-367 ▸ PA
stygius J. Polhemus & D. Polhemus – J. Polhemus & D. Polhemus, 1996: 340-341 ▸ ML

Thetibates J. Polhemus & D. Polhemus, 1996
Type species: *Rheumatometroides serena* Lansbury, 1992

matawa (Lansbury) – Lansbury, 1989b: 67-69 (*Rheumatometroides*) ▸ BS
serena (Lansbury) – Lansbury, 1992: 1-6 (*Rheumatometroides*) ▸ NG

GERRINAE Leach, 1815

Aquarius Schellenberg, 1800
Type species: *Cimex najas* De Geer, 1773

adelaidis (Dohrn) – Dohrn, 1860: 408 (*Gerris*) ▸ MA, SM, BO, LU; – EL: SA, China, India
lili D. Polhemus & J. Polhemus – D. Polhemus & J. Polhemus, 1994: 54-57 ▸ NT
philippinensis Zettel & Ruiz – Zettel & Ruiz, 2003: 196-201 ▸ LU

Limnogonus Stål, 1868
Type species: *Hydrometra hyalina* Fabricius, 1803

Limnogonus **subgenus** *Limnogonus* Stål, 1868

Type species: *Hydrometra hyalina* Fabricius, 1803

**anderseni* Zettel – Zettel, 2004e: 371-373 ▸ MI

darthulus (Kirkaldy) – Kirkaldy, 1901b: 805 (*Gerris*) ▸ NG, BS

fossarum fossarum (Fabricius) – Fabricius 1775: 727 (*Cimex*)
▸ MA, SM, BO, LU, PH, PA, MI; – EL: SA, HT, India, Japan

fossarum gilguy Andersen & Weir – Andersen & Weir, 1997: 255-263
▸ JA, BO, SL, NT, NG, BS; – EL: AU, NC, Pacific Islands

syn.: *fossarum skusei* Andersen – Andersen, 1975: 36-39 (*nec* de la Torre- Bueno)

fossarum rennellensis Brown – Brown, 1968: 85-87 ▸ BS

hungerfordi Andersen – Andersen, 1975: 46-47 ▸ MA, SM, BO, SL, LU, PH, PA, NG; – EL: HT, AU, Pacific Islands, Nicobar Isands

luctuosus (Montrousier) – Montrousier, 1865: 242 (*Gerris*)
▸ BS; – EL: AU, NC, Pacific Islands

nitidus (Mayr) – Mayr, 1865: 443 (*Hydrometra*)
▸ MA, SM, JA, BO, NT, SL, LU, PH, PA, MI; – EL: SA, HT, China, India

papuensis Andersen – Andersen, 1975: 48-51 ▸ NG, BS

Limnogonus **subgenus** *Limnogonoides* Poisson, 1965

Type species: *Gerris hypoleuca* Gerstaecker, 1873

pectoralis (Mayr) – Mayr, 1865: 443 (*Hydrometra*) ▸ MA; – EL: SA, India

Limnometra Mayr, 1865

Type species: *Limnometra femorata* Mayr, 1865

annulicornis (Breddin) – Breddin, 1901: 83-85 (*Gerris*) ▸ SL

aploa Nieser & Chen – Nieser & Chen, 1992c: 16-17 ▸ SL

arachnis Nieser & Chen – Nieser & Chen, 1992c: 17 ▸ SL

borneensis Hungerford & Matsuda – Hungerford & Matsuda, 1958a: 407-408 ▸ BO

bruneiensis (Miyamoto) – Miyamoto, 1967: 229-232 (*Tenagogonus*) ▸ BO

ciliata Mayr – Mayr, 1865: 444
▸ MA, SM, JA, BO, SL, LU, PA, MI, ML, NG, BS; – EL: Pacific Islands

ciliodes Andersen & Weir – Andersen & Weir, 1997: 242-246 ▸ NG; – EL: AU

femorata Mayr – Mayr, 1865: 443 ▸ MA, SM, LU; MI, PH; – EL: HT, Japan

genitalis Nieser & Chen – Nieser & Chen, 1992c: 18-19 ▸ SL

grallator D. Polhemus & J. Polhemus – D. Polhemus & J. Polhemus, 1997: 25-30
▸ NG

hysterema Nieser & Chen – Nieser & Chen, 1992c: 23-24 ▸ BS

insularis Hungerford & Matsuda – Hungerford & Matsuda, 1958a: 404-405
▸ MA, SM, JA, BO; – EL: SA

kallisto (Kirkaldy) – Kirkaldy, 1899b: 506 (*Gerris*) ▸ NG, BS

lepta Nieser & Chen – Nieser & Chen, 1992c: 19-20 ▸ SL

lipovskyi Hungerford & Matsuda – Hungerford & Matsuda, 1958a: 399-400 ▸ NG, BS; – EL: AU

matsudai (Miyamoto) – Miyamoto, 1967: 223-226 (*Tenagogonus*) ▸ MA; – EL: SA, China

melanochroa Nieser & Chen – Nieser & Chen, 1992c: 20-21 ▸ SL

monochroma Nieser & Chen – Nieser & Chen, 1992c: 21-22 ▸ BS

nigripennis nigripennis Mayr – Mayr, 1865: 443 ▸ LU, PH, MI

**nigripennis amabilis* Zettel – Zettel, 2004e: 376-377 ▸ MI

**nigripennis bicolana* Zettel – Zettel, 2004e: 375-376 ▸ LU, MI

**nigripennis cebuana* Zettel – Zettel, 2004e: 373-375 ▸ PH

octopunctata Hungerford – Hungerford, 1955: 67-68 ▸ BO, MA, SM, SL

palawanensis Zettel & Chen – Zettel & Chen, 2000: 81-83 ▸ PA

pseudoinsularis Nieser & Chen – Nieser & Chen, 1992c: 21 ▸ BO

pulchra Mayr – Mayr, 1865: 443 ▸ ML

rossi Hungerford & Matsuda – Hungerford & Matsuda, 1958a: 408-409 ▸ LU

spinosa Zettel – Zettel, 2002b: 137-141 ▸ MA, SM, JA, BO

submarginalis (Miyamoto) – Miyamoto, 1967: 226-229 (*Tenagogonus*) ▸ BO

Neogerris Matsumura, 1913

Type species: *Neogerris boninensis* Matsumura, 1913

assimilis Andersen – Andersen, 1975: 90-91 ▸ MA, SM, BO

parvulus (Stål) – Stål, 1859: 265 (*Gerris*) ▸ MA, JA, NT, PH, PA, MI, NG, BS; – EL: SA, HT, Arabia, Iran, India, China, Japan

**philippinensis* Zettel – Zettel, 2004e: 367-370 ▸ LU, MI

Tenagogonus Stål, 1853

Type species: *Tenagogonus albovittatus* Stål, 1853

akanthinos Chen & Nieser – Chen & Nieser, 1992: 150 ▸ BO

bergrothi Hungerford & Matsuda – Hungerford & Matsuda, 1958a: 393-394 ▸ LU, MI

divergens Hungerford & Matsuda – Hungerford & Matsuda, 1958a: 385-386 ▸ SM

kampaspe (Kirkaldy) – Kirkaldy, 1901b: 805 (*Gerris*) ▸ NG, BS

maai Hungerford & Matsuda – Hungerford & Matsuda, 1962: 143-144 ▸ MA, BO

pravipes pravipes Bergroth – Bergroth, 1915: 121-122 ▸ MA, SM, JA, NT, SL

pravipes claggi Hungerford & Matsuda – Hungerford & Matsuda, 1958a: 389 ▸ MI

EOTRECHINAE Matsuda, 1960

Amemboa Esaki, 1925
Type species: *Amemboa fumi* Esaki, 1925

Amemboa **subgenus** *Amemboa* Esaki, 1925
Type species: *Amemboa fumi* Esaki, 1925

brevifasciata Miyamoto – Miyamoto, 1967: 236-238 ▸ MA, SM; – EL: SA
cristata J. Polhemus & Andersen – J. Polhemus & Andersen, 1984: 95-97 ▸ MA; – EL: SA
**heissi* Zettel – Zettel, 2002c: 1531-1534 ▸ MA
incurvata J. Polhemus & Andersen – J. Polhemus & Andersen, 1984: 97-98 ▸ MA, BO
javanica Lundblad – Lundblad, 1933a: 405-408 ▸ MA, JA; – EL: SA
lannae J. Polhemus & Andersen – J. Polhemus & Andersen, 1984: 90 ▸ MA
mathildae Zettel – Zettel, 1995f: 63-64 ▸ BO
philippinensis J. Polhemus & Andersen – J. Polhemus & Andersen, 1984: 99 ▸ MI
prostata J. Polhemus & Andersen – J. Polhemus & Andersen, 1984: 87 ▸ MA

Amemboa **subgenus** *Amemboides* J. Polhemus & Andersen, 1984
Type species: *Amemboa (Amemboides) perlata* J. Polhemus & Andersen, 1984

nodosa J. Polhemus & Andersen – J. Polhemus & Andersen, 1984: 103-104 ▸ MA

Onychotrechus Kirkaldy, 1903
Type species: *Onychotrechus rhexenor* Kirkaldy, 1903

esakii Andersen – Andersen, 1980: 127-128 ▸ MA; – EL: HT, SA
pallidus Andersen – Andersen, 1980: 131 ▸ MA

CYLINDROSTETHINAE Matsuda, 1960

Cylindrostethus Fieber, 1861
Type species: *Cylindrostethus fieberi* Mayr, 1865

malayensis D. Polhemus – D. Polhemus, 1994b: 8-9 ▸ MA, SM
samarinda D. Polhemus – D. Polhemus, 1994b: 6-7 ▸ BO
sumatranus Lundblad – Lundblad, 1933a: 392-394 ▸ SM
persephone Kirkaldy – Kirkaldy, 1899b: 508 ▸ SL
scrutator (Kirkaldy) – Kirkaldy, 1899a: 90 (*Gerris*) ▸ MA, SM, JA, BO, ML; – EL: SA, India
vittipes (Stål) – Stål, 1871: 705 (*Hydrobates*) ▸ LU, MI

PTILOMERINAE Bianchi, 1896

Pleciobates Esaki, 1930
Type species: *Pleciobates tuberculatus* Esaki, 1930
tuberculatus Esaki – Esaki, 1930: 14-15 ‣ MA

Potamometropsis Lundblad, 1933
Type species: *Potamometropsis obnubila* Lundblad, 1933
anomalis Chen & Nieser – Chen & Nieser, 1992: 154-155 ‣ SL
bruneiensis J. Polhemus & Zettel – J. Polhemus & Zettel, 1997: 30-31 ‣ BO
crassifemur J. Polhemus & Zettel – J. Polhemus & Zettel, 1997: 37-38 ‣ BO
fischeri Zettel – Zettel, 1994d: 89-90 ‣ BO
hoogstraali Hungerford – Hungerford, 1957: 125-127 ‣ MI
ikarus Zettel – Zettel, 1994d: 95-97 ‣ LU
kundesan J. Polhemus & Zettel – J. Polhemus & Zettel, 1997: 31-34 ‣ BO
**longipes* Zettel – Zettel, 2004e: 373-378 ‣ LU
luzonica Zettel – Zettel, 1994d: 86-89 ‣ LU
obnubila Lundblad – Lundblad, 1933a: 415-417 ‣ SM
poring J. Polhemus & Zettel – J. Polhemus & Zettel, 1997: 28-30 ‣ BO
sabah J. Polhemus & Zettel – J. Polhemus & Zettel, 1997: 34-36 ‣ BO
sumaldei Zettel – Zettel, 1999g: 156-158 ‣ LU
werneri werneri Hungerford – Hungerford, 1957: 127-130 ‣ MI
werneri aberrans Zettel – Zettel, 1994d: 95 ‣ MI

Ptilomera Amyot & Serville, 1843
Type species: *Gerris laticaudata* Hardwicke, 1825

Ptilomera subgenus *Ptilomera* Amyot & Serville, 1843
Type species: *Gerris laticaudata* Hardwicke, 1825
aello aello Breddin – Breddin, 1906: 9 ‣ NG
aello cheesmanae Hungerford & Matsuda – Hungerford & Matsuda, 1965: 459-462 ‣ NG
arfak D. Polhemus & J, Polhemus – D. Polhemus & J. Polhemus, 2001: 147-149 ‣ NG
biroi D. Polhemus & J. Polhemus – D. Polhemus & J. Polhemus, 2001: 140-143 ‣ NG
bismarckensis D. Polhemus & J. Polhemus – D. Polhemus & J. Polhemus, 2001: 161-163 ‣ BS
breddini Hungerford & Matsuda – Hungerford & Matsuda, 1965: 451-454 ‣ NG
ceramensis Hungerford & Matsuda – Hungerford & Matsuda, 1965: 470-471 ‣ ML
cheesmanae Hungerford & Matsuda – Hungerford & Matsuda, 1965: 459-462 ‣ NG
chiasma Nieser & Chen – Nieser & Chen, 1992b: 36-39 ‣ NT

chinai Hungerford & Matsuda – Hungerford & Matsuda, 1965: 438-440 ▸ BO
dorceus Breddin – Breddin, 1901: 87-88 ▸ SL
dromas Breddin – Breddin, 1901: pl. 1 fig. 14a, b ▸ JA
etna D. Polhemus & J. Polhemus – D. Polhemus & J. Polhemus, 2001: 133-136 ▸ NG
gressitti Hungerford & Matsuda – Hungerford & Matsuda, 1965: 446-448 ▸ BO
harpalos Schmidt – Schmidt, 1926: 66 ▸ SM
hungerfordi Andersen – Andersen, 1967: 262-267 ▸ PA
insularis D. Polhemus & J. Polhemus – D. Polhemus & J. Polhemus, 2001: 156-158 ▸ BS
iriana D. Polhemus & J. Polhemus – D. Polhemus & J. Polhemus, 2001: 124-127 ▸ NG
jimi D. Polhemus & J. Polhemus – D. Polhemus & J. Polhemus, 2001: 109-112 ▸ NG
kirkaldyi Hungerford & Matsuda – Hungerford & Matsuda, 1965: 444-446 ▸ BO
kiunga D. Polhemus & J. Polhemus – D. Polhemus & J. Polhemus, 2001: 117-120 ▸ NG
kutubu D. Polhemus & J. Polhemus – D. Polhemus & J. Polhemus, 2001: 143-147 ▸ NG
laelaps Breddin – Breddin, 1901: 88-89 ▸ SL
lundbladi Hungerford & Matsuda – Hungerford & Matsuda, 1965: 488 ▸ SM
maai Hungerford & Matsuda – Hungerford & Matsuda, 1965: 476-478 ▸ BO
misoolensis D. Polhemus & J. Polhemus – D. Polhemus & J. Polhemus, 2001: 152-156 ▸ NG
morobe D. Polhemus & J. Polhemus – D. Polhemus & J. Polhemus, 2001: 112-115 ▸ NG
novabrittanica D. Polhemus & J. Polhemus – D. Polhemus & J. Polhemus, 2001: 158-161 ▸ BS
nunikanensis Hungerford & Matsuda – Hungerford & Matsuda, 1965: 441-444 ▸ BO
omo D. Polhemus & J. Polhemus – D. Polhemus & J. Polhemus, 2001: 115-117 ▸ NG
oribasus Breddin – Breddin, 1901: 88 ▸ SL
palawanensis D. Polhemus – D. Polhemus, 1998: 261-264 ▸ PA
pamphagus Breddin – Breddin, 1901: 86-87 ▸ SL
papuensis Hungerford & Matsuda – Hungerford & Matsuda, 1965: 454-455 ▸ NG
sarawakensis Hungerford & Matsuda – Hungerford & Matsuda, 1965: 424-426 ▸ BO
sumbaensis Hungerford & Matsuda – Hungerford & Matsuda, 1965: 474-476 ▸ NT
sumatranus Hungerford & Matsuda – Hungerford & Matsuda, 1965: 485-488 ▸ SM
sumizome Esaki – Esaki, 1925: 59-60 ▸ SL
tigrina Uhler – Uhler, 1860: 230 ▸ MA; – EL: SA, HT, China

timika D. Polhemus & J. Polhemus – D. Polhemus & J. Polhemus, 2001: 120-124 ▸ NG
timorensis Hungerford & Matsuda – Hungerford & Matsuda, 1965: 436-438 ▸ NT
waigeo D. Polhemus & J. Polhemus – D. Polhemus & J. Polhemus, 2001: 149-152 ▸ NG
wapoga D. Polhemus & J. Polhemus – D. Polhemus & J. Polhemus, 2001: 127-129 ▸ NG
werneri Hungerford & Matsuda – Hungerford & Matsuda, 1958d: 69-72 ▸ PA
wewak D. Polhemus & J. Polhemus – D. Polhemus & J. Polhemus, 2001: 136-140 ▸ NG
yapenana D. Polhemus & J. Polhemus – D. Polhemus & J. Polhemus, 2001: 129-133 ▸ NG

Rheumatogonus Kirkaldy, 1909
Type species: *Ptilomera luzonica* Kirkaldy, 1909
borneensis Esaki – Esaki, 1927: 267-268 ▸ BO, PA
esakii Chen & Nieser – Chen & Nieser, 2002: 390-393 ▸ BO
intermedius Hungerford – Hungerford, 1933: 3-4 ▸ MA, SM
inusitatus Chen & Nieser – Chen & Nieser, 2002: 395-399 ▸ BO
luzonicus (Kirkaldy) – Kirkaldy, 1909: 389 (*Ptilomera*) ▸ LU, PH!, MI
seyferti Zettel – Zettel, 1994: 79-80 ▸ PH
vantoli Chen & Nieser – Chen & Nieser, 2002: 393-395 ▸ BO, PA
**vietnamensis* Zettel & Chen – Zettel & Chen, 1996: 171-174 ▸ BO; - EL: SA

Rhyacobates Esaki, 1923
Type species: *Rhyacobates takahashii* Esaki, 1923
imadatei Miyamoto – Miyamoto, 1967: 238-241 ▸ BO

Note A cladistic analysis of the genus *Rhyacobates* (Andersen & Chen, 1995) showed that *R. imadatei* does not belong in this genus. So far no new generic assignment has been made.

HALOBATINAE Bianchi, 1896

Halobatini Bianchi, 1896

Asclepios Distant, 1915
Type species: *Asclepios annandalei* Distant, 1915
annandalei Distant – Distant, 1915: 505-506 ▸ MA; – EL: India

Halobates Eschscholtz, 1822
Type species: *Halobates micans* Eschscholtz, 1822

Halobates **subgenus** *Halobates* Eschscholtz, 1822

Type species: *Halobates micans* Eschscholtz, 1822

browni Herring – Herring, 1961: 270-271 ▸ BS
calyptus Herring – Herring, 1961: 285-286 ▸ LU, ?PH, MI, NG
dianae Zettel, 2001b: 1097-1102
esakii Miyamoto – Miyamoto, 1967: 241-243 ▸ BO
flaviventris Eschscholtz – Eschscholtz, 1822: 109
▸ MA, JA, BS; – EL: SA, India, Indo- and West Pacific; East Africa
germanus White – White, 1883: 50-52 ▸ MA, SM, JA, BO, NT, SL, ML, NG, BS; – EL: Red Sea, Indian and Pacific Oceans
hayanus White – White, 1883: 52-54 ▸ BO, MA, JA, PA, ML, NG; – EL: SA, AU, Nicobar Islands, coast of Indo-Pacific Ocean
maculatus Schadow – Schadow, 1922: 2 ▸ LU!, NG, BS
micans Eschscholtz – Eschscholtz, 1822: 107
▸ SM, BO, SL, LU, MI, NG; – EL: tropical oceans
nereis Herring – Herring, 1961: 272-273 ▸ NG; – EL: Palau Islands
peronis Herring – Herring, 1961: 278-279 ▸ LU, BS
princeps White – White, 1883: 44-45
▸ MA, JA, NT, SL, ML, NG; – EL: AU, China Sea
proavus White – White, 1883: 54 ▸ MA, JA, LU!, MI, BS; – EL: Nicobar Islands
regalis Carpenter – Carpenter, 1892: 144-146 ▸ NG
sexualis Distant – Distant, 1903a: 258 ▸ MA
trynae Herring – Herring, 1964: 85-86 ▸ MA; – EL: Bay of Bengal

Metrocorini Matsuda, 1960

Esakia Lundblad, 1933

Type species: *Esakiella ventidioides* Lundblad, 1933

**cenizae* Zettel – Zettel, 2004e: 382-385 ▸ MI
fernandoi Cheng – Cheng, 1966: 16-18 ▸ MA, BO
johorensis Cheng – Cheng, 1966: 18-20 ▸ MA
lundbladi Cheng – Cheng, 1966: 20-22 ▸ MA
**palawanensis* Zettel – Zettel, 2004e: 379-382 ▸ PA
usingeri Hungerford & Matsuda – Hungerford & Matsuda, 1958: 196-197 ▸ LU
ventidioides (Lundblad) – Lundblad, 1933a: 401-405 (*Esakiella*) ▸ SM; – EL: SA

Metrocoris Mayr, 1865

Type species: *Metrocoris brevis* Mayr, 1865

borneensis D. Polhemus – D. Polhemus, 1990: 9-14 ▸ BO
breviculus Chen & Nieser – Chen & Nieser, 1992: 156 ▸ SL
celebensis D. Polhemus – D. Polhemus, 1990: 8-9 ▸ SL
luzonicus D. Polhemus – D. Polhemus, 1990: 15-16 ▸ LU
malayensis Chen & Nieser – Chen & Nieser, 1993: 15-16 ▸ MA

medius Chen & Nieser – Chen & Nieser, 1996: 72-75 ▸ SL
nigrofasciatus Distant – Distant, 1903a: 257 ▸ MA; – EL: SA
philippinensis den Boer – den Boer, 1965: 29-33 ▸ MI
squamifer Lundblad – Lundblad, 1933a: 398-399 ▸ MA, SM
strangulator Breddin – Breddin, 1905a: 134-135 ▸ SM, JA
sunda D. Polhemus – D. Polhemus, 1990: 6-8 ▸ NT
tenuicornis Esaki – Esaki, 1926: 125-126
▸ MA, SM, JA, BO, PA; – EL: SA, India, China, HT
tigrinus D. Polhemus – D. Polhemus, 1990: 14-15 ▸ BO
zetteli D. Polhemus – D. Polhemus, 1998: 264-267 ▸ MI

Ventidius Distant, 1910
Type species: *Ventidius aquarius* Distant, 1910

Ventidius **subgenus** *Ventidius* Hungerford & Matsuda, 1960
Type species: *Ventidius aquarius* Distant, 1910
distanti Paiva, 1918: 25 ▸ MA, SM, JA, BO; – EL: SA, India
= *modulatus* Lundblad **syn. nov.** – Lundblad, 1933a: 339-401
harrisoni Cheng – Cheng, 1965: 155-156 ▸ MA, BO, NT
hungerfordi Cheng – Cheng, 1965: 158-159 ▸ MA; – EL: SA
malayensis Hungerford & Matsuda – Hungerford & Matsuda, 1960: 325-326
▸ MA, BO, NT; – EL: SA
pilosus Chen & Zettel – Chen & Zettel, 1999: 169-170 ▸ NT
polhemorum Chen & Zettel – Chen & Zettel, 1999: 165-167 ▸ BO
usingeri Hungerford & Matsuda – Hungerford & Matsuda, 1960: 326-327
▸ LU, MI
werneri Hungerford & Matsuda – Hungerford & Matsuda, 1960: 330-331 ▸ PA

Ventidius **subgenus** *Ventidioides* Hungerford & Matsuda, 1960
Type species: *Ventidius (Ventidioides) kuiterti* Hungerford & Matsuda, 1960
heissi Chen & Zettel – Chen & Zettel, 1999: 197-199 ▸ BO
kurtokalami Chen & Nieser – Chen & Nieser, 1992: 158-159 ▸ BO
lundbladi Miyamoto – Miyamoto, 1967: 245-247 ▸ MA
nieseri Chen & Zettel – Chen & Zettel, 1999: 196-197 ▸ BO
pulai Cheng – Cheng, 1965: 153-155 ▸ MA
xiphibion Chen & Nieser – Chen & Nieser, 1992: 156-157 ▸ SL
xyele Chen & Nieser – Chen & Nieser, 1992: 157-158 ▸ SL

Ventidius **subgenus** *Ventidiopsis* Miyamoto, 1967
Type species: *Ventidiopsis imadatei* Miyamoto, 1967
imadatei Miyamoto – Miyamoto, 1967: 248-251 ▸ BO
yangae Chen & Zettel – Chen & Zettel, 1999: 203-204 ▸ BO

Glossary

Note: 'in Gerromorpha' and 'in Nepomorpha' indicates that a structure does not occur, or a term is not used, for the other infraorder in this book.

a *abdominal scent gland* – gland lying dorsally in the abdomen in larvae of most Nepomorpha, and in larvae and adults of some taxa of Gerromorpha.

accessory sclerite, see morphology, genitalia (chapter 2).

acetabulum (acetabula) – in Gerromorpha: cup-like structures in which the coxae fit (fig. 4, ac).

aedeagus (aedeagi) – endosoma or penis, the male organ which actually conducts the sperm into the female.

Amphibicorisae – 'amphibious bugs', synonym of Gerromorpha, also spelled Amphibiocorisae.

anal cone or *anal lid* – see proctiger.

antennal socket – antennal tubercle.

antennal tubercle – tubercle and groove on the head where the antenna originates.

antennomere – antennal segment.

anteclypeus – anterior part of the head, just caudally of the labrum.

apomorphy (apomorphic character) – character developed directly from its pre-existing homologue. An apomorphy is consequently characteristic for the species and its descendants which acquire it. When shared by two or more descendants it is termed synapomorphy. Synapomorphies define monophyletic groups or monophyletic taxa.

apterous – wings (in adult) totally absent.

apical sclerite –chitinous part towards the tip.

arolium (arolia) – a pair of structures inserted between the tarsal claws. The dorsal one is usually smaller than the ventral one (fig. 13, va).

Australasia – eastern part of Malesia and Australia, containing the Moluccas, Irian Jaya, New Guinea and Australia.

Australian region – biogeographical region containing Australia, Tasmania, New Guinea, Bismarck Archipelago, and Solomon Islands. It formerly also included New Zealand and Oceania, but these are nowadays usually considered separate regions.

autapomorphy (autapomorphic character) – an apomorphy developed in a single species.

autotomy – self-mutilation of wings in macropterous specimens (fig. 3).

b *basal articulatory apparatus* – basal plate.

basal plate – basal part of the aedeagus.

basal sclerite – the most basally directed chitinous part.

brachypterous – wings in adult reduced to about half their length when fully developed. See morphology, wing (chapter 2).

buccula(e) – ventral lobes of head in Hebridae (figs. 259-263).

c *carina(e)* – keel; narrow chitinous elevation perpendicular to the basal surface, usually ventrally on thorax or abdomen.

cephalic – adjective meaning belonging to the head.

cephalonotum (cephalonota) – fused head and thorax (in Helotrephidae; fig. 235).

chordotonal organ – scolopal organ; sensory organ reacting to changes in pressure. See also 'sensory adaptations' (chapter 2).

cibarium – food pump.

clade – a monophyletic group of species, a natural taxon.

clasper – paramere.

class – taxonomic category above order.

claval suture – the suture between clavus and corium.

clavus (clavi) – proximal part of the hemielytron (fig. 16).

clypeus (clypei) – anterior part of the head between and in front of the eyes.

co-aptatation – structure at the base of the wings to keep the wings in place when closed (see chapter 2 morphology, wings).

conjunctivum (conjunctiva) – see endosoma.

connexivum (connexiva) – combined laterotergites on one side of the abdomen (fig. 2, c).

continental South-east Asia – Myanmar, Thailand, Laos, Cambodia, Vietnam. In addition, peninsular Malaysia and Singapore are usually considered to belong to this area. However, in this book they are not included here when 'continental South-east Asia' is used.

coreaceous – leathery, usually used to describe the hardened basal part of the hemielytron.

corium (coria) – lateral part of the hemielytron (fig. 16).

costal margin – lateral margin of a hemielytron. In most Gerromorpha it is recognizably a vein, the costa, hence the name.In Nepomorpha the vein is not or hardly recognizable. See hemielytron.

coxa(e) – basal segment of a leg (fig. 13).

Cryptocerata – synonym of Nepomorpha.

d *dealate* – macropterous specimen of which the apical part of the hemielytra and wings are torn off (fig. 3).

declivent (declivous) – sloping downward.

dextral – (oriented) to the right. Dextral asymmetry in male Corixoidea is recognized by the strigil being on the right side of the abdomen.

distal – the part the furthest away from the body. See proximal.

dorsal – the back, the upper part of the body.

dorsal sclerite – chitinous part situated at dorsal side. See also Morphology: male genitalia of Gerromorpha.

e *ecdysis*– moult. In Heteroptera there are nearly always five ecdyses, one after each larval instar. Occasionally also used, wrongly, for eclosion.

eclosion – the process of the larva emerging from its egg.

elytraceous – shield-like.

embolar groove – in Corixoidea: the trough-like lateral part of the embolium. See hemielytron.

embolar suture – in Nepomorpha: longitudinal suture on corium separating the embolium from the rest of the corium.

embolium – in Nepomorpha: narrow lateral area of corium, proximally of the nodal furrow. See hemielytron.

endemic – restricted to a particular region.

endosoma – apical part of the aedeagus, in Gerromorpha consisting of a basal part, or conjectivum, and an apical part, or vesica. See Morphology, genitalia.

endosomal sclerites – vesical sclerites.

Entomostraca – name for all Crustacea, not included in the higher Crustacea or subclassis Malacostraca. In freshwater mainly represented by the small planktonic taxa Cladocera and Copepoda, and the Ostracoda.

f *femur* – third segment of a leg. See morphology, legs (chapter 2).

food pump – muscular organ lying internally at the base of the rostrum pumping out saliva and taking in fluid food. Also called cibarium. See morphology, mouthparts (chapter 2).

fovea – a large pit.

frons – anterior part of the head, roughly between the eyes.

g *gena(e)* – lateroventral part of the head in front of the compound eyes.

genal plate – small sclerite anteriorly of the pronotal plate in Helotrephidae (fig. 231).

genital capsule – pygophore, abdominal segment 9 in males, partly (most Gerromorpha) or totally (Nepomorpha) retracted inside the abdomen. The ventral sclerite is strongly developed and forms a capsule around the internal organ, the phallus. Laterally there is usually a pair of hooks, the parameres (figs. 4, py, 20, 433).

genital clasper – paramere.

genital operculum – see subgenital operculum.

genital segments – abdominal segments 8-10, forming the external genital organs (fig. 4, g).

Geocorisae – 'land bugs', the Heteroptera with exclusion of Gerromorpha and Nepomorpha.

genus (genera) – set of related species differing from other similar species by synapomorphies.

gonapophyse – apical part of the female genital segments, attached to the gonocoxae.

gonocoxa(e) – paired basal part of the modified female genital segments. The paired basal part of abdominal segment 8 form the first gonocoxae, those of segment 9 the second. In Nepomorpha they are retracted inside the abdomen. In Gerromorpha they can also be retracted, but usually the first gonocoxae are visible as a pair of apico-ventral valves at the apex of the female abdomen, where they cover the oviposition apparatus. In the text, gonocoxae refers to the first pair unless otherwise stated. See morphology, genitalia (chapter 2).

gonocoxite – gonocoxa.

gula(e) – ventral area of the head between the rostrum and the pronotum.

Gymnocerata – Heteroptera, with exclusion of the Nepomorpha.

h

harpagone – paramere.

hemielytron or hemelytron (hemielytra) – the fore wing of Heteroptera, specifically those with distinctly thickened proximal part and membranous distal part. Strictly speaking, the fore wings of Gerromorpha and some other primitive infraorders are considered not to be true hemielytra. The official term for these fore wings is tegmen (tegmina). Here the term hemielytra is used for both Nepomorpha and Gerromorpha. In this book we consider the basic position of the hemielytra when they are folded over the back, in which case the costal margin becomes laterally oriented. When the wings are expanded, the costal margin is anteriorly oriented.

Hemiptera – either Heteroptera and Homoptera combined or a synonym of Heteroptera.

Heteroptera – order or suborder of insects containing the true bugs.

holotype – the specimen to which the name of a species is fixed.

homonymy – the same name applied to two different species within the same genus.

humeral width – maximum width of pronotum between or close to the wing bases.

Homoptera – order or suborder of insects closely related to the Heteroptera.

Hydrocorisae – literally meaning 'water bugs', used as synonym of Nepomorpha.

hydrofuge – water repellent.

hygropetric – living on wet rocky or stony surface.

i *incrassate* – swollen or thickened.

Indo-Australia – Indian subcontinent and Sri Lanka, continental South-east Asia, Malesia and Australia. Usually South-west China is also included, but not in this book.

infraorder – subdivision of an order, assemblage of families with a few characters in common.

instar – larval stage between two moults

interoculus (interocula) – dorsal part of the head between the eyes.

j *Johnston's organ* – sense organ in the antenna. See Adaptations to aquatic life (chapter 2).

k *keel* – see carina.

l *labium* – lower (or posterior) part of the mouthparts, forming a sheath around the mandibulae and maxillae in Hemiptera, which together form the rostrum (figs. 4-7, r).

labrum – anterior part of head forming a small sclerite, usually lying over the base of the mouth parts (figs. 5, 7, l).

large – see size.

larva – stage between egg and adult.

lateral – side-.

lateral sclerite – sclerites lying to the sides of the animal. The abdominal pleurites (lateral sclerites) are considered to have fused with the sternites, which carry the abdominal tracheal spiracles. See also Morphology (chapter 2).

laterotergite – lateral part of the dorsal sclerites of an abdominal segment. The laterotergites on one side of the abdomen together form a connexivum (fig. 2, c).

lectotype – specimen chosen from the original syntype series to which the name of a species is fixed when the original author did not designate a holotype.

length – in this book used as total body length, unless specified. It is measured – with the animal oriented with its dorsum upward with the main body axis horizontal – from (anterior) tip of head to (posterior) tip of abdomen, except in macropterous Nepomorpha, where it is measured from (anterior) tip of head to (posterior) tip of hemielytra.

lentic – stagnant, water without (perceivable) current, such as a pond or pool.

limnic – fresh (water).

lotic – water with a current: streams and rivers.

m *macro-hair layer* – a less dense layer of longer hairs (macrotrichia) on top of the micro-hair layer in Gerromorpha.

macropterous – both pairs of wings fully developed.

macrotrichia – cuticular outgrowths with a length of 15-100 μ and a thickness 1-5 μ, in each individual of most species usually present in two or three sizes, and much less dense than the microtrichia.

mandibula(e) – first pair of mouth parts, in Heteroptera lying within the labium or rostrum, and used for piercing the food.

maxilla(e) – second pair of the mouth parts, in Heteroptera lying under the labium or rostrum, and used for injecting saliva and sucking in food.

maxillary plate – anterior part of the head lateral of the base of the rostrum. The most important part is the ventral lobe or buccula.

mediotergite – middle part of an abdominal tergite.

meiopterous – with reduced wings.

Melanesia – A number of island groups in the central Pacific including New Guinea, the Torres Straits, Bismarck and Solomon Islands, New Hebrides, New Caledonia and Fiji.

membrane – distal part of the fully developed hemielytron (fig. 16).

meso- – middle-, e.g. mesofemur is the femur of the middle leg.

mesonotum – dorsal sclerite of the mesothorax (fig. 2, ms).

mesopleuron – lateral sclerite(s) of the mesothorax.

mesopostnotum – posterior part of the mesonotum (figs. 376-379).

mesoscutellum, see scutellum.

mesosternum – ventral sclerite of the mesothorax (fig. 4, mu).

mesothorax – second segment of the thorax.

meta- – behind, e.g. metafemur is the femur of the hind leg.

metacetabular groove – suture on the dorsal part of the thorax of Gerridae (figs. 376-379).

metacetabulum (metacetabula) – acetabulum of a hind leg.

metanotum – dorsal sclerite of the metathorax (fig. 2, mt).

metapleuron – lateral sclerite(s) of the metathorax.

metasternal xiphus – triangular sclerite lying mediocaudally on the metasternum (fig. 4, x).

metasternal (scent) gland orifice – opening of the metathoracic (scent) gland.

metasternum – ventral sclerite of the metathorax (fig. 4, mv).

metathoracic (scent) gland – a gland in the metathorax, opening near the posterior margin of the metasternum present in most Heteroptera.

metathorax – third segment of the thorax.

micro-hair layer – a dense layer of tiny and fine cuticular outgrowths (microtrichia) covering part or most of the body of Gerromorpha.

micropyle – opening in the wall of an egg through which the sperm passes.

micropterous – with very short wings in adult individuals.

microtrichia – hairs forming a micro-hair layer. They are 2-10μ long and 0.5-0.6μ wide, and very densely placed.
mixohaline – where freshwater and seawater meet and mix.
monobasis – based on or containing one lower taxon, e.g. a monobasic genus contains only one species.
monophyletic taxon – natural taxon sharing a common ancestor.

n *natural taxon* – monophyletic taxon.
neotype – a specimen to which the name of a species is fixed after the original type(s) has (have) been lost
nodal furrow – in some Nepomorpha: a short transverse suture distally on the corium near the membranal suture, running form the costal margin of the corium inward.
nominate – in subgenus and subspecies: the taxon having the original name, e.g. *Ranatra sulawesii sulawesii*. This was the taxon which was described before other subspecies were recognised.
nymph – larva of an insect with gradual metamorphosis.

o *Oceania* – Pacific Islands east of Malesia and Australia.
ocellus (ocelli) – organ with a small single lens. When present, there is a pair of them posteriorly on the vertex.
omphalium – modified area of the metasternum where the metathoracic gland opens.
operculum – see subgenital operculum.
order – main division within a class combining taxa which have a few general treats in common, e.g. Odonata (damselflies + dragonflies).
Oriental region – the area of tropical Asia west of Weber's line. The inclusion of the Nusa Tenggara east of Bali and Sulawesi is disputable. We follow the concept of Oosterbroek (1998).
orifice – opening, used in connection with glands. See also ostiole.
ostiole – small opening, in Heteroptera especially used for the opening of the metathoracic gland.
ovipositor – sclerites of segments 8 and 9 of the female abdomen, modified for the deposition of eggs.

p *pala(e)* – pala frontalis: the modified anterior tarsus of Corixoidea.
palm – the ventral side of a pala.
Palearctic region – the non tropical northern part of the Old World. The southernmost areas are North Africa, Saudi Arabia, the Himalayas, Myanmar, the area north of Thailand, China north of 30° northern latitude and Japan.

pantropical – occurring throughout the tropical regions of the world.

Papuan subregion – part of the Australian region comprising New Guinea, the Bismarck Archipelago and the Solomon Islands.

parameres – a pair of lateral hooks on the pygophore in males, surrounding the penis.

paraphyletic taxon – taxon which includes a common ancestor and some but not all of its descendants.

paratergite – laterotergite.

paratype – specimen used for the original description of a species other than the holotype.

parasternites – sclerites laterally of the median part of an abdominal sternite.

parempodia – paired structure ventral to the tarsal claws, inserted on the same basal sclerite as the claws, thinner than the claws but sometimes longer. The dorsal one is usually smaller than the ventral one.

pars stridens (stridulitrum) – the, in principle, not moving part of a stridulatory mechanism. See plectrum.

pedunculate – on a stalk.

phallosoma – phallotheca and conjunctivum together.

phallotheca – a hardened capsule connected with the basal plate of the aedeagus to hold the endosoma when not active.

phoresy – female carrying the male on her back for protracted time intervals during which copulation takes place.

phragma(ta) – internal ridge of the chitinous body wall to provide a larger area for muscle attachment.

phylum – taxonomic category above class level, e.g. Arthropoda.

physical gill – breathing system in which gas is exchanged in a gas bubble. Used by subaquatic animals to obtain extra oxygen (see chapter 2: Adaptations to aquatic life).

pilosity – layer of short dense hairs.

plastron – thin film of gas in subaquatic insects, covering an exposed part of the body in which sufficient gas exchange takes place to meet demand of oxygen (see chapter 2: Adaptations to aquatic life).

plectrum – the moving part of a stridulatory mechanism. See pars stridens.

pleurite – lateral sclerite of an abdominal segment.

polyphyletic taxon – taxon of which the most recent common ancestor also belongs to another taxon.

postclypeus – part of the head in front of the eyes.

pretarsus – appendages of the apical tarsal segment: arolia, claws and parempodia.

pro- – anterior, e.g. profemur is the femur of the anterior leg.

proctiger – remnant of abdominal segment 10, together with segment 11 forming a lid posteriorly on the pygophore, containing the anal opening, hence also called anal lid.

prolarva – full grown embryo which after eclosion becomes the first instar larva.
pronotal plate – sclerite ventrally of the cephalonotum in Helotrephidae (fig. 205).
pronotum – dorsal sclerite of the first thoracic segment.
propleuron – lateral sclerite of the first thoracic segment.
prosternum – ventral sclerite of the first thoracic segment.
prothorax – first thoracic segment.
proximal – close to the body (opposite: distal).
pruinose – having a whitish, powdery covering or bloom
pseudo-endocorium (pseudendocorium) – inner part of corium adjacent to the scutellum
pseudomembrane – thickened caudal ridge on a hemielytron connecting the parts of the hemielytra.
pygophore – see genital capsule.

r *reniform* – kidney-shaped.
reservoir breathers – subaquatic species using virtually only the oxygen they take down in an air store (see chapter 2: Adaptations to aquatic life).
respiratory siphon – pair of caudal filaments forming a breathing tube.
rheophilous – living in streams.
rhombic – parallelogram-shaped.
rostral groove – ventral groove on head, usually extending onto thorax, in which the rostrum rests.
rostral prong – in males of *Anisops*: a paired lateral stripe on the rostrum. See also fig. 5.
rostrum – in Heteroptera and Homoptera: sheath around the mouth parts formed by the labium (figs. 5-10).
ruptor or ruptor ovi – chintinous hook-like structure on head of prolarva used to break the egg shell at eclosion.

s *scent gland* – gland producing scent fluid. See under 'Adaptations to aquatic life – secretion grooming' (chapter 2).
sclerite – chitinous component of a body segment.
scolopal organs – sensory organ reacting to changes in pressure. See also 'sensory adaptations' (chapter 2).
scolopophorous – scolopal.
scutellum (mesoscutellum) – triangular medio-posterior sclerite of the mesonotum, visible between the base of the hemielytra in most Heteroptera. It is absent in Gerromorpha, excepting macropterous Mesoveliidae. In Hebridae and Hydrometridae there is a metanotal structure which looks like, and is more or less in the same place as, the scutellum, which in literature is indicated as scutellum or pseudoscutellum (fig. 3, sc).

secretion grooming, see adaptations to aquatic life (Chapter 2).

sessile – attached without a stalk.

siphon, see respiratory siphon.

sinistral – (oriented) to the left. Sinistral asymmetry in male Corixoidea is recognized by the strigil being on the left side of the abdomen.

species – basic unit of taxonomy.

spinules – small spines.

spiracle – opening to the tracheal system.

static organs – in Nepidae: subcircular organs on ventral laterotergites of segments 4-6, looking like large openings of tracheal stigmata, consisting of strong chitinous ring with a central field. In Aphelocheiridae and some Naucoridae static organs are present on thoracic venter. See Nepidae (p. 53).

sternite – ventral sclerite of an abdominal segment (fig. 4, st).

sternopleurite – fused sternite and pleurite of an abdominal segment.

stigma, see tracheal stigma.

strigil – specialized organ of most male Corixidae, situated laterally on the hind margin of abdominal tergite 6, which keeps the subhemielytral air store open during copulation. When the strigil is on the right side it is called dextral abdominal asymmetry, and sinistral abdominal asymmetry when it is on the left side.

stylets – modified mandibulae and maxillae in Heteroptera and Homoptera.

sub- – prefix meaning 'nearly', 'slightly' or 'under', e.g. 'subparallel' means 'nearly parallel', and 'subaquatic' means 'under water'.

subgenital operculum – enlarged abdominal sternite 7 or 8, covering genitalia in ventral view in some Nepomorpha.

subspecies (geographical subspecies) – geographically isolated (allopatric) subgroup(s) of a species, differing in minor but distinguishable aspects from other representatives of the species, with no or incomplete reproductive isolation between/among the subgroups.

Sundaland – continental Malaysia and the islands of Borneo, Sumatra, Java and Palawan, which during the last ice age formed a single land mass (fig. 563).

suture – groove between segments of body and appendages or parts of the hemielytra.

synapomorphy (synapomorphic character) – a homologous character found in two or more taxa, hypothesized to have arisen in the ancestral species of these taxa and not in an earlier ancestor. Hence it is considered true that taxa sharing a synapomorphy form a monophyletic group or natural taxon. See also apomorphy.

synthlipsis – minimum distance between eyes, measured posteriorly in dorsal view.

syntype – specimen of a type series in which no holotype was designated.

systematics – science of the diversity of and relationships among organisms.

t *tarsal formula* – number of tarsal segments of front, middle and hind tarsi respectively, e.g. tarsal formula 1-2-2 means that the fore tarsus has one, and the middle and hind tarsi each have two segments.

tarsus (tarsi) – distal-most part of an insect leg bordering the tibia, in aquatic bugs consisting of one to three segments.

taxon (taxa) – a group of organisms having a recognized taxonomic standing, e.g. species, genus, family.

taxonomy – theory and practice of classifying organisms.

tegmen – see hemielytron.

teneral – just moulted. Used for adults with the reproductive organs not yet productive. Usually the body wall is not yet fully hardened and often not fully coloured.

tergite – dorsal sclerite of an abdominal segment (fig. 2, t).

terminalia – caudal parts of abdomen.

tibia – the fourth leg segment counted from base.

trachea(e) – tube through which gas transport takes place inside insects.

tracheal gill – structure with a thin cuticula, provided with numerous tracheoles for gas exchange between insect and water.

tracheal spiracle – opening of a trachea to the outside for gas exchange between the insect and its environment.

tracheal stigma – tracheal spiracle.

tracheole – narrow tube at the end of trachea where gas exchange between tubes and tissues takes place, or sometimes, in tracheal gills, between tubes and water.

tribe or *tribus* – assemblage of related genera within a subfamily (ending: -ini).

trichobothria – sensory hairs usually placed in a pit. Important in classification are the cephalic trichobothria. See morphology, head of adult (chapter 2).

trochanter – second leg segment from base between coxa and femur.

tumescence – distension, swelling.

tympanal organ – sensory organ for detection of sound.

v *ventral* – belly-; view with the belly towards the observer.

ventral lobe – antero-ventral part of head capsule, laterally of anteclypeus and often covering the base of the rostrum in lateral view. See also buccula.

ventral sclerite – chitinous part situated at ventral side. See also Morphology: vesical sclerites.

vertex – dorsal part of the head capsule (fig. 10, v).

vesica – apical part of the endosoma (apical part of the aedeagus).

vesical sclerites – small chitinous parts lying in vesica.

w *Wallacea* – islands between Sundaland and Australia/New Guinea: Sulawesi, the Moluccas and the Lesser Sunda Islands. Some authors add the Philippines.

Wallace's Line – biogeographical border west of Sulawesi separating Wallacea and the Philippines from Sundaland (fig. 1).

Weber's Line – biogeographical border east of Wallacea separating the Oriental area from Australia and New Guinea (fig. 1).

x *xiphus* – see metasternal xiphus.

References

Ahmad, M. & Khan, N. H., 1977. Bionomics of *Micronecta scutellaris* Stål.– Journal of the Bombay Natural History Society 74: 453-458.
Amyot, C. J. B. & Serville, A., 1843. Histoire naturelles des Insectes. Hémiptères.– Roret, Paris, lxxvi+675 pp.
Andersen, N. M., 1965. A remarkable new species of *Rhagovelia* Mayr from the Philippines (Heteroptera: Veliidae).– Entomologiske Meddelelser 34: 111-117.
Andersen, N. M., 1967. A contribution to the knowledge of Philippine semiaquatic Hemiptera-Heteroptera.– Entomologske Meddelelser 35: 26-282.
Andersen, N. M., 1973. Seasonal polymorphism and developmental changes in organs of flight and reproduction in bivoltine pondskaters (Hem. Gerridae).– Entomologica Scandinavica 4: 1-20.
Andersen, N. M., 1975. The *Limnogonus* and *Neogerris* of the Old World with Character Analysis and a Reclassification of the Gerrinae (Hemiptera: Gerridae).– Entomologica Scandinavica Supplementum 7: 1-96.
Andersen, N. M., 1976. A comparative study of locomotion on the water surface in semiaquatic bugs (Insecta, Hemiptera, Gerromorpha).– Videnskabelige Meddelelser fra Dansk Naturhistorisk Forening 139: 337-396.
Andersen, N. M., 1977a. Fine structure of the body hair layers and morphology of the spiracles of semiaquatic bugs (Insecta, Hemiptera, Gerromorpha) in relation to life on the water surface.– Videnskabelige Meddelelser fra Dansk Naturhistorisk Forening 140: 7-37.
Andersen, N. M., 1977b. A new and primitive genus and species of Hydrometridae (Hemiptera, Gerromorpha) with a cladistic analysis of relationships within the family.– Entomologica Scandinavica 8: 301-316.
Andersen, N. M., 1980. Hygropetric waterstriders of the genus *Onychotrechus* Kirkaldy with description of a related genus (Insecta, Hemiptera, Gerridae).– Steenstrupia 6: 113-146.
Andersen, N. M., 1981a. Semiaquatic bugs: phylogeny and classification of the Hebridae (Heteroptera: Gerromorpha) with revisions of *Timasius*, *Neotimasius* and *Hyrcanus*.– Systematic Entomology 6: 377-412.
Andersen, N. M., 1981b. A new genus of Veliinae and descriptions of new Oriental species of the subfamily (Hemiptera: Veliidae).– Entomologica Scandinavica 12: 339-356.
Andersen, N. M., 1982a. The Semiaquatic Bugs (Hemiptera, Gerromorpha).– Entomonograph 3: 1-455. Klampenborg, Denmark.
Andersen, N. M., 1982b. The first species of *Heterocleptes* Villiers from the Oriental region (Hemiptera: Hydrometridae).– Entomologica Scandinavica 13: 105-108.
Andersen, N. M. 1982c. Semiterrestrial waterstriders of the genera *Eotrechus* Kirkaldy and *Chimarrhometra* Bianchi (Insecta, Hemiptera, Gerridae).– Steenstrupia, 9(1): 1-25.
Andersen, N. M., 1983. The Old World Microveliinae (Hemiptera: Veliidae) I. The status of *Pseudovelia* Hoberlandt and *Perivelia* Poisson, with a review of Oriental species.– Entomologica Scandinavica 14: 253-268.
Andersen, N. M., 1989a. The coral bugs, genus *Halovelia* Bergroth (Hemiptera, Veliidae). I. History, classification, and taxonomy of species except the *H. malaya*-group.– Entomologica Scandinavica 20: 75-120.
Andersen, N. M., 1989b. The coral bugs, genus *Halovelia* Bergroth (Hemiptera, Veliidae). II. Taxonomy of the *H. malaya*-group, cladistics, ecology, biology, and biogeography.– Entomologica Scandinavica 20: 179-273.
Andersen, N. M., 1989c. The Old World Microveliinae (Hemiptera: Veliidae). II. Three new species of *Baptista* Distant and a new genus from the Oriental region.– Entomologica Scandinavica 19: 363-380.

Andersen, N. M., 1990. Phylogeny and taxonomy of waterstriders, genus *Aquarius* Schellenberg (Insecta, Hemiptera, Gerridae), with a new species from Australia.– Steenstrupia 16: 37-81.

Andersen, N. M., 1991a. Cladistic Biogeography of Marine Waterstriders (Insecta, Hemiptera) in the Indo-Pacific.– Australian Systematic Botany 4: 151-163.

Andersen, N. M., 1991b. Marine insects: genital morphology, phylogeny and evolution of sea skaters, genus *Halobates* (Hemiptera: Gerridae).– Zoological Journal of the Linnean Society 103: 21-60.

Andersen, N. M., 1992a. The *Hydrometra longicapitis* Torre-Bueno-group with a new species from Thailand (Heteroptera, Hydrometridae).– Steenstrupia 18: 1-8.

Andersen, N. M., 1992b. A new genus of marine waterstriders (Hemiptera, Veliidae) with five new species from Malesia.– Entomologica Scandinavica 22: 389-404.

Andersen, N. M., 1995a. Cladistic inference, and evolutionary scenarios: locomotory structure, function, and performance in waterstriders.– Cladistics 11: 279-295.

Andersen, N. M., 1995b. Cladistics, historical biogeography, and a checklist of gerrine waterstriders (Hemiptera, Gerridae) of the World.– Steenstrupia 21: 93-123.

Andersen, N.M., 1998. Waterstriders from the Paleogene of Denmark with a review of the fossil record and evolution of semiaquatic bugs (Hemiptera, Gerromorpha).– Biologiske Skrifter 50: 1-157.

Andersen, N. M., 1999a. *Cryptovelia stysi* sp. n. from Borneo, with a reanalysis of the phylogeny of the Mesoveliidae (Heteroptera: Gerromorpha).– Acta Societatis Zoologicae Bohemicae 63: 5-18.

Andersen, N. M., 1999b. The evolution of marine insects: phylogenetic, ecological and geographical aspects of species diversity in marine waterstriders.– Ecography 22: 98-111.

Andersen, N.M., 1999c. *Microvelia polhemi*, n.sp. (Heteroptera: Veliidae) from Dominican amber: the first fossil record of a phytotelmic waterstrider.– Journal of the New York Entomological Society 107: 135-144.

Andersen, N.M., 2000a. The marine Haloveliinae (Hemiptera: Veliidae) of Singapore, Malaysia and Thailand, with six new species of *Xenobates* Esaki.– Raffles Bulletin of Zoology 48: 273-292.

Andersen, N. M., 2000b. Fossil waterstriders in the Eocene Baltic amber (Hemiptera, Gerromorpha).– Insect Systematics and Evolution 31: 257-284.

Andersen, N.M., 2000c [1999]. A new species of *Tetraripis* from Thailand, with a critical assement of the generic classification of the subfamily Rhagovelinae (Hemiptera, Gerromorpha, Veliidae).– Tijdschrift voor Entomologie 142: 185-194.

Andersen, N. M., 2001. Fossil waterstriders in the Oligocene/Miocene Dominican amber (Hemiptera: Gerromorpha).– Insect Systematics and Evolution 31: 411-431.

Andersen, N. M. & Chen, P.-p., 1995. A taxonomic revision of the Ptilomerine genus *Rhyacobates* Esaki (Hemiptera: Gerridae), with five new species from China.– Tijdschrift voor Entomologie 138: 51-67.

Andersen, N. M., Farma, A., Minelli, A. & Piccoli, G., 1994. A fossil *Halobates* from the Mediterranean and the origin of sea skaters (Hemiptera, Gerridae).– Zoological Journal of the Linnean Society 112: 479-489.

Andersen, N. M. & Foster, W. A., 1992. Sea skaters of India, Sri Lanka and the Maldives, with a new species and a revised key to Indian Ocean species of *Halobates* and *Asclepios* (Hemiptera, Gerridae).– Journal of Natural History 26: 533-553.

Andersen, N. M. & Grimaldi, D., 2001. A fossil water measurer from the mid-Cretaceous Burmese amber (Hemiptera: Gerromorpha: Hydrometridae).– Insect Systematics and Evolution 32: 381-392.

Andersen, N. M. & Poinar jr., G. O., 1992. Phylogeny and classification of an extinct water strider genus (Hemiptera, Gerridae) from Dominican amber, with evidence of mate guarding in a fossil insect.– Zeitschrift für Zoologische Systematik und Evolutionsforschung 30: 256-267.

Andersen, N. M. & Poinar jr., G.O., 1998. A marine waterstrider (Hemiptera: Veliidae) from Dominican amber.– Entomologica Scandinavica 29: 1-9.

Andersen, N. M. & Polhemus, J.T., 1976. Waterstriders (Hemiptera: Gerridae, Veliidae etc.).– In: Cheng, L., (ed.). Marine Insects. North Holland Publishing Company, Amsterdam: 187-224.

Andersen, N. M. & Polhemus, J. T., 1980. Four new genera of Mesoveliidae (Hemiptera, Gerromorpha) and the phylogeny and classification of the family.– Entomologica Scandinavica 11: 369-392.

Andersen, N. M., Spence, J. R. & Wilson, M. V. H., 1993. 50 Million years of structural stasis in water striders (Hemiptera: Gerridae).– American Entomologist 39: 174-176.

Andersen, N. M. & Weir, T. A., 1994a. *Austrobates rivularis*, gen. et sp. nov., a freshwater relative of *Halobates* Eschscholtz (Hemiptera: Gerridae), with a new perspective on the evolution of sea skaters.– Invertebrate Taxonomy, 8: 1-15.

Andersen, N.M. & Weir, T. A., 1994b. The Sea Skaters, genus *Halobates* Eschscholtz (Hemiptera: Gerridae), of Australia: Taxonomy, Phylogeny and Zoogeography.– Invertebrate Taxonomy 8: 861-909.

Andersen, N. M. & Weir, T. A., 1997. The Gerrine Waterstriders of Australia (Hemiptera: Gerridae): Taxonomy, Distribution and Ecology.– Invertebrate Taxonomy 11: 203-299.

Andersen, N.M. & Weir, T. A., 1998. Australian waterstriders of the subfamilies Trepobatinae and Rhagadotarsinae.– Invertebrate Taxonomy 12: 509-544.

Andersen, N. M. & Weir, T. A., 1999. The marine Haloveliinae (Hemiptera: Veliidae) of Australia, New Caledonia and southern New Guinea.– Invertebrate Taxonomy 13: 309-350.

Andersen, N. M. & Weir, T. A., 2000. The coral treaders, *Hermatobates* Carpenter (Hemiptera: Hermatobatidae), of Australia and New Caledonia with notes on biology and ecology.– Invertebrate Taxonomy 14: 327-345.

Andersen, N. M. & Weir, T.A., 2001. New genera of Veliidae (Hemiptera: Heteroptera) from Australia, with notes on the generic classification of the subfamily Microveliinae.– Invertebrate Taxonomy 15: 217-258.

Andersen, N. M. & Weir, T.A., 2003a. The genus *Microvelia* Westwood in Australia (Hemiptera: Heteroptera: Veliidae).– Invertebrate Systematics 17: 261-348.

Andersen, N. M. & Weir, T. A., 2003b. A new species of sea skaters, *Halobates* Eschscholtz, from Robinson River, Western Australia (Hemiptera-Heteroptera: Gerridae).– Aquatic Insects 25: 9-18.

Andersen, N. M. & Weir, T. A., 2004. Mesoveliidae, Hebridae, and Hydrometridae of Australia (Hemiptera: Heteroptera: Gerromorpha), with a reanalysis of the phylogeny of semiaquatic bugs.– Invertebrate Systematics 18: 467-522.

Arnett, Jr., R. H., Samuelson, G. A. & Nishida, G. M., 1993. The insect and spider collections of the world. 2nd Ed.– Flora & Fauna Handbooks 11: 310 pp.

Arntz, B., 1972. The hearing capacity of water bugs (Hemiptera, Heteroptera).– Journal of Comparative Physiology 80: 309-311.

Arntz, B., 1975. Das Hörvermogen von *Nepa cinerea* L. Zur Funktionsweise der thorakalen Scolopalorgane.– Journal of Comparative Physiology 96: 53-72.

Aukema, B. & Rieger, C. (Eds.), 1995. Catalogue of the heteroptera of the Palearctic Region 1.– Netherlands Entomological Society, Amsterdam: i-xxvi, 1-222.

Bacon, J. A., 1956. A taxonomic study of the genus *Rhagovelia* of the Western Hemisphere.– The University of Kansas Scientific Bulletin 38: 695-913.

Baehr, M., 1990a. *Ochterus noualhieri* sp. n., a new ochterid from Java [Heteroptera, Ochteridae].– Revue Française d'Entomologie (N. S.) 12: 91-93.

Baehr, M., 1990b. Revision of the genus *Ochterus* Latreille in the Australian Region (Heteroptera: Ochteridae).– Entomologica Scandinavica 20: 449-477.
Baehr, M., 1990c. Revision of the genus *Megochterus* Jaczewski (Insecta: Heteroptera: Ochteridae).– Invertebrate Taxonomy 4: 197-203.
Bakonyi, G., 1978. Contribution to the knowledge of the feeding habits of some water boatman: *Sigara* spp. (Heteroptera: Corixidae).– Folia Entomologica Hungarica (ser. nova) 31: 19-24.
Ban, Y. Shibata, S. & Ishikawa, M., 1988. Life history of the water strider *Metrocoris historio* B. White (Hemiptera: Gerridae) in Aichi Prefecture, Japan.– Internationale Vereinigung für Theoretische und Angewandte Limnologie Verhandlungen 23: 2145-2151.
Bare, C. O., 1928. Haemoglobin cells and other studies of the genus *Buenoa*.– The University of Kansas Science Bulletin 18: 265-349.
Baunacke, W., 1912. Statische Sinnesorganen bei den Nepiden.– Zoologische Jahrbücher Abteilung für Anatomie 34: 179-346.
Benwitz, G., 1956. Der Kopf von *Corixa punctata* Ill. (*geoffroyi* Leach) (Hemiptera-Heteroptera).– Zoologische Jaarbücher Abteilung für Anatomie und Ontogenie der Tiere 75(3): 311-378.
Benzie, J. A.H., 1989. The immature stages of *Plea frontalis* (Fieber, 1844) (Hemiptera, Pleidae), with a description of the adult.– Hydrobiologia 179: 157-171.
Bergroth, E., 1893. On two halophilous Hemiptera.– Entomologist's Monthly magazine (2)4: 277-279.
Bergroth, E., 1915. Some javanese Hemiptera collected by E. Jacobson and Th. H. MacGillavry. –Zoologische Mededeelingen 1: 117-123.
Bergroth, E., 1918. Studies in Philippine Heteroptera, I.– Philippine Journal of Science (D) 13: 120-126.
Billberg, G. J., 1820. Enumeratio Insectorum in Museo Gust. Joh. Billberg: 1-138. Gadelianis, Stockholm.
Blaustein, L. Kotler, B. P. & Ward, D., 1995. Direct and indirect effects of a predatory backswimmer (*Notonecta maculata*) on community structure of desert temporary pools.– Ecological Entomology 20: 311-318.
Bonke, D., 1974. Der Bau und die Antwortcharakteristik des Schirmrezeptors aus dem Statorgansystem von *Nepa cinerea* L. (Hemiptera, Rhynchota).– Verhandlungen des Deutschen Zoologischen Gesellschafts 67: 42-45.
Börner, C., 1934. Űber System und Stammesgeschichte der Schnabelkerfe.– Entomologische Beihefte Berlin-Dahlem 1: 138-144.
Boulard, M. & Coffin, J., 1991. Sur la biologie juvénile d'*Ochterus marginatus* (Latreille, 1804) camouflage et construction (Hemiptera Ochteridae).– EPHE, Travaux du Laboratoire Biologie et Evolution des Insectes 4: 57-68.
Breddin, G., 1898. Studia hemipterologica IV.– Jahresbericht des Naturwissenschaftlichen Vereins in Magdeburg 1896-1897 2: 149-163.
Breddin, G., 1901. Die Hemipteren von Celebes.– Abhandlungen der Naturforschenden Gesellschaft zu Halle 24: 1-213, 1 plate.
Breddin, G., 1903. Neue tropische Wasserläufer der Gattung *Ptilomera* Am.– Societas Entomologica 17: 147-148.
Breddin, G., 1905a. Rhychota Heteroptera aus Java.– Mitteilungen aus dem Naturhistorischen Museum, Hamburg 12: 111-159.
Breddin, G., 1905b. Übersicht der javanischen *Micronecta*-Arten.– Societas Entomologica Zürich 20: 57.
Breddin, G., 1906. Eine neue *Ptilomera* (Rhynchota Heteroptera).– Societas Entomologica 21: 9.
Brooks, G.T., 1948. New species of *Enithares* (Hemiptera, Notonectidae).– Journal of the Kansas Entomological Society 21: 37-54.

Brooks, G. T., 1951. A revision of the genus *Anisops* (Notonectidae, Hemiptera).– The University of Kansas Science Bulletin 34: 301-519.
Brooks, G. T., 1952. Three new species of *Anisops* (Notonectidae, Hemiptera).– Journal of the Kansas Entomological Society 25: 60-64.
Brooks, G. T., 1954. A new species of *Anisops* from Sumba (Het., Notonectidae).– Verhandlungen der Naturforschenden Gesellschaft, Basel 65: 29-30.
Brown, E. S., 1968. The Aquatic Hemiptera (Excluding Notonectidae) of Rennell Island.– Natural History of Rennell Island 5: 83-93.
Bueno, see Torre Bueno, J. R. de la
Carpenter, G.H., 1892. Reports on the zoological collections made in Torres Straits by Professor A. C. Haddon, 888-1889: Rhynchota from Murray Island and Mabuiag.– Proceedings of the Royal Dublin Society 7: 137-146, pls 12, 13.
Cassis, G. & Gross, G. F. 1995. Hemiptera: Heteroptera (Coleorrhyncha to Cimicomorpha). In: Houston, W. W. K. & Maynard, G.V. (eds). Zoological catalogue of Australia 27.3A. Melbourne: CSIRO Australia: i-xv, 1-560.
Chen, L. C., 1960. A study of the genus *Micronecta* of India, Japan, Taiwan and adjacent regions.– Journal of the Kansas Entomological Society 33: 99-118.
Chen, P.-p. & Nieser, N., 1992. Gerridae, mainly from Sulawesi and Pulau Buton (Indonesia).– Tijdschrift voor Entomologie 135: 145-162.
Chen, P.-p., & Nieser, N., 1993. A taxonomic revision of the Oriental waterstrider genus *Metrocoris* Mayr (Hemiptera, Gerridae). Part I, II.– Steenstrupia 19: 1-43, 45-82.
Chen, P.-p. & Nieser, N., 1996. A new species of *Metrocoris* (Heteroptera: Gerridae).– Entomologische Berichten, Amsterdam 56: 72-75.
Chen, P.-p. & Nieser, N., 2000. A new species of *Andersenella* (Heterotera: Gerridae) from Sulawesi (Indonesia).– Entomologische Berichten, Amsterdam, 60: 69-71.
Chen, P.-p. & Nieser, N., 2002. Taxonomic characters of the male endosomal structure in the genus *Rheumatogonus* Kirkaldy (Hemiptera: Gerridae), with descriptions of four new species.– Zoologische Mededelingen, Leiden 76: 371-409.
Chen, P.-p., Nieser, N. & Ho, J.-z., 2004. Review of Chinese Ranatrinae (Hemiptera: Nepidae), with descriptions of four new species of *Ranatra* Fabricius.– Tijdschrift voor Entomologie 147: 81-102.
Chen, P.-p., Nieser, N. & Wattanachaiyingcharoen, W., 2002. A new genus, *Pleciogonus*, and four new species of semiaquatic and aquatic bugs from Thailand (Heteroptera: Gerromorpha, Nepomorpha).– Tijdschrift voor Entomologie 145: 193-212.
Chen, P.-p., Wongsiri, S., Jamyanya, T., Rinderer, T. E., Vonsamanode, S., Matsuka, M., Sylvester, H. A. & Oldroyd, B. P., 1998. Honey-bees and other edible insects used as human food in Thailand.– American Entomologist Spring 1998: 24-29.
Chen, P.-p. & Zettel, H., 1998. Key to the genera and subgenera of Gerridae (Gerromorpha) of Thailand and adjacent countries, with a check-list of species known from Thailand.– Amemboa 2, News and results on Heteroptera of Thailand: 24-41.
Chen, P.-p., & Zettel, H., 1999. A taxonomic revision of the Oriental waterstrider genus *Ventidius* Distant (Hemiptera, Gerromorpha, Gerridae).– Tijdschrift voor Entomologie 141: 137-208.
Cheng, L., 1965. The genus *Ventidius* Distant (Heteroptera: Gerridae) in Malaya, with a description of four new species.– Proceedings of the Entomological Society of London (B) 34: 153-163.
Cheng, L. 1966a. A curious marine insect, *Hermatobates*.– Malayan Nature Journal 19: 283-285.
Cheng, L., 1966b. Three new species of *Esakia* Lundblad (Heteroptera: Gerridae) from Malaya.– Proceedings of the Entomological Society of London (B) 35: 16-22.

Cheng, L., 1967a. Studies on the biology of the Gerridae (Hem., Heteroptera) I: Observations on the feeding of *Limnogonus fossarum* (F.).– Entomologists' Monthly Magazine 102 [1966]: 121-128.
Cheng, L., 1967b. Studies on the biology of the Gerridae (Hem., Heteroptera) II: The life history of *Metrocoris tenuicornis* Esaki.– Entomologists' Monthly Magazine 102 [1966]: 273-282.
Cheng, L., 1976. A new species of *Hermatobates* (Hemiptera: Heteroptera).– The Pan-Pacific Entomologist 52: 209-212.
Cheng, L., 1977. The elusive sea bug *Hermatobates* (Heteroptera).– The Pan-Pacific Entomologist 53: 87-97.
Cheng, L.,1980. Notes on the seabug *Hermatobates hawaiiensis* China (Heteroptera: Hermatobatidae).– Proceedings of the Hawaiian Entomological Society 23: 193-197.
Cheng, L., 1981. *Halobates* (Heteroptera: Gerridae) from Micronesia with notes on a laboratory population of *H. mariannarum*.– Micronesica 17: 97-106.
Cheng, L., 1985. Biology of *Halobates* (Heteroptera: Gerridae).– Annual Review of Entomology 30: 111-135.
Cheng, L., 1989. Biogeography and phylogeny of the sea-skater *Halobates*.– Chinese Journal of Oceanology and Limnology 7: 233-239.
Cheng, L. & Fernando, C. H., 1969. A taxonomic study of the Malayan Gerridae (Hemiptera: Heteroptera) with notes on their biology and distribution.– Oriental Insects 3: 97-160.
Cheng, L. & Fernando, C. H., 1971. The life history and biology of the riffle bug *Rhagovelia obesa* Uhler (Heteroptera: Veliidae) in southern Ontario.– Canadian Journal of Zoology 49: 435-442.
Cheng, L. & Frank, J. H., 1993. Marine insects and their reproduction. – Oceanography and marine biology annual review 31: 479-506.
Cheng, L., Yang, C. M. & Andersen, N. M., 2001. Guide to aquatic Heteroptera of Singapore and Peninsular Malaysia. I. Gerridae and Hermatobatidae.– The Raffles Bulletin of Zoology 49: 129-148.
Cheng, L., Yang, C. M. & Polhemus, J. T., 2001. Guide to aquatic Heteroptera of Singapore and Peninsular Malaysia.– The Raffles Bulletin of Zoology 49: 121-127.
Chesson, J., 1984. Effect of notonectids (Hemiptera: Notonectidae) on mosquitoes (Diptera: Culicidae): predation or selective oviposition?– Environmental Entomology 13: 531-538.
Chesson, J., 1989. The effect of alternative prey on the functional response of *Notonecta hoffmanni*.– Ecology 70: 1227-1235.
China, W. E., 1930. Three new species of the genus *Helotrephes* (Hemiptera-Heteroptera).– Annals and Magazine of Natural History (ser. 10) 5: 164-172.
China, W. E., 1935. New and little known Helotrephidae (Hemiptera, Heteroptera).– Annals and Magazine of Natural History (10) 15: 593-614.
China, W. E., 1940. New South American Helotrephidae (Hemiptera – Heteroptera).– Annals and Magazine of Natural History (10) 17: 527-538.
China, W. E., 1955. The evolution of the water bugs.– Bulletin of the National Institute of Science of India 7: 91-103
China, W. E., 1956. A new species of the genus *Hermatobates* from the Hawaiian Islands (Hemiptera-Heteroptera, Gerridae, Halobatinae).– Annals and Magazine of Natural History (12) 9: 353-357.
China, W.E., 1957. The marine Hemiptera of the Monte Bello Islands, with descriptions of some allied species.– Journal of the Linnean Society of London. Zoology 43: 342-357.
Clark, F., 1992. A study of a population of *Micronecta scutellaris* Stål (Hemiptera: Corixidae) in Lake Naivasha, Kenya.– Hydrobiologia 248: 115-124.
Cloarec, A., 1972. Activité respiratoire de *Ranatra linearis* (Insecte, Hétéroptère aquatique).– Bulletin de la Société Zoologigique de France 97: 729-736.

Cloarec, A., 1989a. Variations of foraging tactics in a water bug, *Diplonychus indicus*.– Journal of Ethology 7: 27-34.
Cloarec, A., 1989b. Variations of predatory tactics of a water bug during development.– Ethology 86: 33-46.
Cloarec, A., 1991a. Handling time and multi-prey capture by a water bug.– Animal Behaviour 42: 607-613.
Cloarec, A., 1991b. Predatory versatility in the water bug *Diplonychus indicus*.– Behavioural Processes 23: 231-242.
Cloarec, A., 1992. The influence of feeding on predatory tactics in a water bug.– Physiological Entomology 17: 25-32.
Cloarec, A. & Joly, D., 1988. Choice of perch by the water stick insect.– Behavioural Processes 17: 131-144.
Cobben, R. A., 1946. Wantsennieuws uit Midden-Limburg. – Natuurhistorisch Maandblad 35: 56-58.
Cobben, R. H., 1965. Egg-life and symbiont transmission in a predatory bug, *Mesovelia furcata* Ms. & Rey (Heteroptera, Mesoveliidae).– Proceedings of the XII International Congress of Entomology, London 1964: 162-163.
Cobben, R.H., 1968. Evolutionary trends in Heteroptera I. Eggs.– Mededelingen Laboratorium voor Entomologie, Wageningen 151: viii, 1-475.
Cobben, R.H., 1978. Evolutionary trends in Heteroptera II. Mouthparts.– Mededelingen Laboratorium voor Entomologie, Wageningen 289: 1-407.
Cobben, R. H. 1981: The hebrid fauna of the Ethiopian Kaffa Province, with considerations on species grouping (Hebridae, Heteroptera).– Tijdschrift voor Entomologie 125: 1-24.
Cobben, R. H., 1987. *Aradus signaticornis* in Nederland met opmerkingen over enkele andere, met *Pinus* geassocieerde Hemiptera (Heteroptera: Aradidae; Homoptera: Cercopidae).– Entomologische Berichten, Amsterdam 47: 33-38.
Coutière, H. & Martin, J., 1901. Sur un nouvel Hémiptère Halophile, *Hermatobatodes marchei*, n.gen., n.sp.– Bulletin du Musée de Histoire Naturelle Paris 5: 214-226.
Dallas, W. S., 1850. Notice of some Hemiptera in the collection of the Hon. East India Company.– Transactions of the Royal Entomological Society of London 1: 4-11.
Damgaard, J., Andersen, N.M., Cheng, L. & Sperling, F.A.H., 2000a. Phylogeny of sea skaters, *Halobates* Eschscholtz (Hemiptera, Gerridae), based on mtDNA sequence and morphology.– Zoological Journal of the Linnean Society 130: 511-526.
Damgaard, J., Andersen, N. M. & Sperling, F. A. H., 2000b. Phylogeny of the waterstrider genus *Aquarius* Schellenberg (Heteroptera: Gerridae) based on nuclear and mitochondrial DNA sequences and morphology.– Insect Systematics and Evolution 31: 71-90.
De Carlo, J. A., 1964. Los 'Ranatridae' de America. – Revista del Museo Argentina de Ciencias Naturales 'Bernardino Rivadavia' Entomologia 1: 133-215.
Decu, V., Gruia, M., Keffer, S. L. & Sarbu, S. M., 1994a. Stygobiotic waterscorpion, *Nepa anophthalma*, n.sp. (Heteroptera: Nepidae), from a sulfurous cave in Romania.– Annals of the Entomological Society of America 87: 755-761.
Deshefy, G. S., 1980. Anti-predator behavior in swarms of *Rhagovelia obesa* (Hemiptera: Veliidae).– The Pan-Pacific Entomologist 56: 111-112.
Desjardins, J.F., 1837. Note sur un insecte nouveau faisant partie de la faune de l'île Maurice. – Annales de la Société Entomologique de France 6: 239-242.
Dickerson, R. E., Merrill, E.D., McGregor, R.C., Schultze, W., Taylor, E.H. & Herre, A. W., 1928. Distribution of Life in the Philippines.– Monographs of the Bureau of Sciences, Manila, 21: 1-322, plates 1-42.
Distant, W.L., 1903a. Report on the Rhynchota. Part I. Heteroptera.– Fasciculi Malayensis (Zoology) 2: 221-271.

Distant, W. L. 1903b. Rhynchota Notes XIX.– Annals and Magazine of Natural History (7) 12: 469-480.
Distant, W. L., 1909. Oriental Rhynchota Heteroptera.– Annals and Magazine of Natural History (8) 3: 491-507.
Distant, W. L., 1910a. The fauna of British India. Rhynchota V (Heteroptera). Appendix: i-xii, 1-362. Taylor & Francis, London.
Distant, W.L., 1910b. Some undescribed Gerrinae.– Annals and Magazine of Natural History 10: 505-513.
Distant, W. L., 1915. A few undescribed Rhynchota.– Annals and Magazine of Natural History (8) 15: 503-507.
Dohrn, A., 1860. Zur Heteropteren Fauna Ceylon's.– Stettiner Entomologische Zeitung 21: 399-409.
Douglas, J.W. & Scott, J., 1865a. The British Hemiptera. Vol. 1 Hemiptera-Heteroptera.– Ray Society, London, 627 pp.
Dover, C., 1928. Notes on a collection of aquatic Rhynchota from the Buitenzorg Museum.– Treubia 10: 65-72.
Drake, C.J. 1917. A survey of the North American species of *Merragata*.– Ohio Journal of Science 17: 101-105.
Drake, C.J. & Hoberlandt, L., 1953. A new species of the genus *Angilia* from the Philippines, with a catalogue of the described species.– The Pan-Pacific Entomologist 30: 223-227.
Dudgeon, D. 1960. Feeding by the aquatic heteropteran, *Diplonychus rusticum* (Belostomatidae): an effect of prey density on meal size.– Hydrobiologia 190: 93-96.
Dufour, L., 1863. Essai monographique sur les Bélostomides.– Annales de la Société Entomologique de France (4) 3: 373-404.
Dupuis, C., 1955. Les génitalia des Hémiptères-Hétéroptères– Mémoires du Muséum National d'Histoire Naturelle Série A. Zoologie 6(4): 183-278.
Duviard, D., 1974. Flight activity of Belostomatidae in central Ivory Coast.– Oecologia 15: 321-328.
Ekblom, T., 1926. Morphological and Biological Studies of the Swedish Families of Hemiptera-Heteroptera. Part 1. The Families Saldidae, Nabidae, Lygaeidae, Hydrometridae, Veliidae and Gerridae.– Zoologiska Bidrag från Uppsala 10: 31-176.
Ekblom, T., 1930. Morphological and Biological Studies of the Swedish Families of Hemiptera-Heteroptera. Part 2. The Families Mesoveliidae, Corizidae and Corixidae.– Zoologiska Bidrag från Uppsala 12: 113-149.
Esaki, T., 1923. An interesting new waterstrider from Formosa.– Philippine Journal of Science 22: 387-391.
Esaki, T., 1924a. On the curious halophilous waterstrider, *Halovelia maritima* Bergroth (Hemiptera, Gerridae).– Bulletin of the Brooklyn Entomological Society 19: 29-34.
Esaki, T., 1924b. On a new genus and species of the Gerridae from Formosa.– Annals of the Entomological Society of America 17: 228-230.
Esaki, T., 1924c. On the genus *Halobates* from Japanese and Formosan coasts (Hemiptera: Gerridae).– Psyche 3: 112-119.
Esaki, T., 1926. The water-striders of the subfamily Halobatinae in the Hungarian National Museum.– Annales Musei Nationalis Hungarici 23: 117-164.
Esaki, T., 1927a. Revision of the *Ptilomera*-group of the Gerridae, with descriptions of three new species.– Eos 3: 251-268.
Esaki, T., 1927b. An interesting new genus and species of Hydrometridae (Hem.) from South America.– The Entomologist 60: 181-184.
Esaki, T., 1928. New or little known Gerridae. I. Ceylonese species.– Annals and Magazine of Natural History (10) 2: 505-513.

Esaki, T., 1929. New or little known Gerridae. II. Indian species.– Annals and Magazine of Natural History (10) 4: 412-419.
Esaki, T., 1930a. New or little-known Gerridae from continental Malaysia.– Journal of the Federated Malay States Museums *16*: 13-24.
Esaki, T., 1930b. Marine waterstriders from the Corean coast. – The Entomologist's Monthly Magazine *66*: 158-161.
Esaki, T., 1937. A new *Asclepios* from Japan.– Mushi 9: 135-136.
Esaki, T. & China, W. E. 1927: A new family of Heteroptera.– Transactions of the Entomological Society London 75: 279-295.
Esaki, T. & China, W. E., 1928. A mongraph of the Helotrephidae, subfamily Helotrephinae (Hem. Heteroptera).– Eos 4: 129-172.
Esaki, T. & Miyamoto, S., 1955. Veliidae of Japan and adjacent territory (Hemiptera-Heteroptera) I. *Microvelia* Westwood and *Pseudovelia* Hoberlandt of Japan.– Sieboldia 1: 169-204.
Eschscholtz, J. F., 1822. Entomographien 1: 1-128. Reimer, Berlin.
Fabricius, J. C., 1775. Systema Entomologiae: i-xxx, 1-823. Kortii, Flensburgi et Lipsiae.
Fabricius, J. C., 1781. Species insectorum exhibentes eorum differentias specificas, synonyma auctorum, loca natalia, metamorphosis adjectis observationibus, descriptionibus 2: 1-517. Bohnii, Hamburgi et Kiloni.
Fabricius, J.C., 1790. Nova insectorum genera.– Skrifter Naturhistorie Selskabet Kjöbenhavn 1: 213-228.
Fabricius, J. C., 1794: Systema Entomologiae 4: i-v, 1-472 pp. Proft, Hafniae.
Fabricius, J. C., 1798. Supplementum Entomologiae Systematicae: 1-572. Profi & Storch, Hafniae.
Fabricius, J. C., 1803. Systema Rhyngotorum: i-vi, 1-315. Reichard, Brunsvigae.
Fernando, C. H., 1964 [1963]. The distribution of aquatic insects in South-east Asia with special references to their dispersal.– Bulletin of the National Museum, Singapore 32: 72-79.
Fernando, C. H., 1964a. A new species of *Micronecta* (Hemiptera: Corixidae) from Malaya.– Proceedings of the Royal Entomological Society of London (B) 33: 17-18.
Fernando, C. H., 1964b. Notes on aquatic insects colonizing an isolated pond in Mawai, Johore.– Bulletin of the National Museum, Singapore 32: 80-89.
Fernando, C. H. & Cheng, L., 1974. A preliminary study on the fauna and distribution of aquatic Hemiptera in Malay and Singapore.– Federation Museum Journal 19: 21-44.
Fernando, C. H. & Leong, C. Y., 1963. Miscellaneous notes on the biology of Malayan Corixidae (Hemiptera: Heteroptera) and a study of the life histories of two species, *Micronecta quadristrigata* Bredd. and Agraptacorixa hyalinipennis (F.).– Annals and Magazine of Natural History (13) 6: 545-558.
Ferrari, E. von, 1888. Die Hemipteren-Gattung *Nepa* Latr. (sens. natur.)– Annalen des Naturhistorischen Museums in Wien 3: 161-194.
Fieber, F. X., 1844a. Monographie der Gattung *Sigara*.– Abhandlungen der Böhmischer Gesellschaft der Wissenschaften 3: 289-293.
Fieber, F.X., 1844b. Monographie der Gattung *Ploa*.–Abhandlungen der Böhmischer Gesellschaft der Wissenschaften 3: 294-297.
Fieber, F. X., 1848. Synopsis aller bisher in Europa entdeckten Arten der Gattung *Corisa*.– Bulletin de la Societé Impériale des Naturalistes de Moscou 21: 505-593.
Fieber, F. X., 1851a. Rhynchotographieen III. Die Gruppe der Notonectae: 45-62. J. G. Calve, Pragae.
Fieber, F. X., 1851b. Genera Hydrocoridum: 1-31. J.G. Calve, Pragae.
Fieber, F. X., 1851c. Species generis *Corisa*: 1-48. J. G. Calve, Pragae.
Flor, G., 1860. Die Rhynchoten Livlands in systematischer Folge beschrieben. 1: 1-826. Schulz, Dorpat.

Foster, W.A., 1989. Zonation, behaviour and morphology of the intertidal coral treader *Hermatobates* (Hemiptera: Hermatobatidae) in the south-west Pacific.– Zoological Journal of the Linnean Society 96: 87-105.

Fox, L.R., 1975a. Some demographic consequences of food shortage for the predator, *Notonecta hoffmanni*. Ecology 56: 868-880.

Fox, L. R., 1975b. Factors influencing cannibalism, a mechanism of population limitation in the predator *Notonecta hoffmanni*. Ecology 56: 933-941.

Frick, K. E., 1949. The biology of *Microvelia capitata* Guérin, 1857, in the Panama Canal Zone and its role as a predator on Anopheline larvae.– Annals of the Entomological Society of America 42: 77-100.

Gagné, W.C. & Howarth, F.G., 1975. The cavernicolous fauna of Hawaiian lava tubes. 6. Mesoveliidae or water treaders (Heteroptera).– Pacific Insects 16: 399-413.

Gapud, V. P., 1981. Contribution to the taxonomy of the genus *Ochterus* Latreille (Hemiptera: Ochteridae).– Kalikasan, Philippine Journal of Biology 10: 300-309.

Gapud, V. P., 1995. A new species of *Ochterus* Latreille (Hemiptera: Ochteridae) from the Philippines.– Asia Life Sciences 4: 41-44.

Gapud, V. P., 2003 [2002]. Two new Philippine *Ochterus* (Insecta: Hemiptera: Ochteridae) and checklist of Philippine species.– Annalen des Naturhistorischen Museum in Wien 104 B: 99-108.

Gapud, V. P. & San Valentin, H. O., 1977. The Ochteridae (Hemiptera) of the Philippines.– Kalikasan, Philippine Journal of Biology 6: 269-300.

Gapud, V. P., Zettel, H. & Yang, C. M. 2003 [2002]. The Hydrometridae (Insecta: Heteroptera) of the Philippine Islands.– Annalen des Naturhistorischen Museum in Wien 104B: 143-162.

Gerstaecker, C. E. A., 1873. Hemiptera.– In: Baron Carl Claus von der Decken's Reisen in Ost Afrika. Wissenschaftliche Ergebnise. Gliederthiere 3: 394-460. Winter, Leipzig, Heidelberg.

Gilbert, J. J. & Burns, C. W., 1999. Some observations on the diet of the backswimmer, *Anisops wakefieldi* (Hemiptera: Notonectidae).– Hydrobiologia 412: 111-118.

Giller, P.S. & McNeil, S., 1981. Predation strategies, resource partitioning and habitat selection in *Notonecta* (Hemiptera/Heteroptera).– Journal of Animal Ecology 50: 789-808.

Gistel, J., 1848. Naturgeschichte des Thierreiches für höhere Schulen: i-xvi, 1-216. Hoffmann, Stuttgart.

Gittelman, S.H., 1975. The ecology of some Costa Rican backswimmers (Hemiptera: Notonectidae).– Annals of the Entomological Society of America 68: 511-518.

Gogala, M., 1996. Waterstriders and their ripple communication.– The Malayan Naturalist 49: 18-20.

Gorai, A. K. & Chaudhuri, D.N., 1962. Food and feeding habits of *Anisops bouvieri* Kirk. (Heteroptera: Notonectidae).– Journal of the Asiatic Society 4: 135-139.

Griffith, M. E., 1945. The environment, life history and structure of the water boatman, *Ramphocorixa acuminata* (Uhler) (Hemiptera, Corixidae).– The University of Kansas Science Bulletin 30: 241-363.

Guérin-Méneville, F. E., 1843a. Note sur la *Naucoris rugosa* de J. Desjardins, formant un nouveau genre d'Hémiptères, et description de plusieurs espèces des genres *Pelogonus* et *Mononyx*.– Revue Zoologique Societé Cuvier 6: 112-114.

Gupta, S. P., Prakash, A., Choudhury, J., Rao, J. & Gupta, A., 1990. New records of bugs infesting paddy fields in Orissa.– Journal of Insect Science 3: 185.

Gupta, Y.C., 1981. A new species of *Ventidius* Distant (Hemiptera: Gerridae) from India.– Oriental Insects 15: 97-102.

Gupta, Y. C., 1985. Bionomics of *Gerris adelaidis* Dohrn (Heteroptera: Gerridae).– Bulletin of Entomology 26: 129-139.

Hale, H.M., 1922. Studies in Australian aquatic Hemiptera No. I. Family Corixidae.– Records of the South Australian Museum 2: 309-330.

Hale, H.M., 1923. Studies in Australian aquatic Hemiptera II. Family Notonectidae.– Records of the South Australian Museum 2: 397-424.

Hale, H.M., 1924. Two new Hemiptera from New South Wales.– Proceedings of the Linnean Society of New South Wales 49: 461-467.

Hall, R., 1998. The plate tectonics of Cenozoic SE Asia and the distribution of land and sea.– In: Hall, R. & Holloway, J. D., (eds). Biogeography and Geological Evolution of SE Asia: 99-131. Backhuys Publishers, Leiden.

Hamilton, M.A., 1931. The morphology of the water scorpion, *Nepa cinerea* Linn. (Rhynchota, Heteroptera).– Proceedings of the Zoological Society of London 1931: 1067-1136, pl. 1-6.

Hecher, C., 1998. Key to the genera of Veliidae (Gerromorpha) of Thailand and adjacent countries, with a check-list of genera and species known from Thailand.– Amemboa 2: 3-7.

Hecher, C. & Zettel, H., 1996. Faunistical and morphological notes on *Limnogonus* subgenus *Limnogonoides* Poisson, 1965 (Heteroptera, Gerridae).– Linzer biologische Beiträge 28: 325-333.

Heming-van Battum, K. E. & Heming, B. S. 1986: Structure, function and evolution of the reproductive system in females of *Hebrus pusillus* and *H. ruficeps* (Hemiptera, Gerromorpha, Hebridae).– Journal of Morphology 190: 121-167.

Heming-van Battum, K. E. & Heming, B. S. 1989: Structure, function and evolution of the reproductive system in males of *Hebrus ruficeps* and *H. pusillus* (Hemiptera, Gerromorpha, Hebridae).– Journal of Morphology 202: 281-323.

Herrich-Schäffer, G.A.W., 1849. Die Wanzenartige Insekten 9(1): 1-44. Lotzbeck, Nürnberg.

Herring, J. L., 1961. The genus *Halobates* (Hemiptera: Gerridae).– Pacific Insects 3: 223-305.

Herring, J. L., 1964. A new species of *Halobates* from the Bay of Bengal.– Proceedings of the Entomological Society of Washington 66: 85-86.

Hinton, H. E., 1961. The structure and function of the egg-shell in the Nepidae (Hemiptera).– Journal of Insect Physiology 7: 224-257.

Hoberlandt, L., 1950. Semiaquatic Heteroptera collected in Lunda, North East Angola (Portuguese West Africa) by Dr. A. de Barros Machado in 1946-1949. – Publicações Culturais da Companhia de Diamantes de Angola 10: 7-49.

Hoberlandt, L. & Štys, P., 1979. *Tampocoris asiaticus* gen. and sp. n. – A new aphelocheirine from Vietnam and further studies on Naucoridae (Heteroptera).– Acta Musei Nationalis Pragae 33(B) (1977): 1-20.

Hoffmann, W.E., 1927. Biological notes on *Laccotrephes* (Hemiptera, Nepidae).– The Lingnan Agricultural Review 4: 77-93.

Hoffmann, W. E., 1933. The life history of a second species of *Laccotrephes* (Hemiptera, Nepidae).– Lingnan Science Journal 12: 245-257.

Holling, C.S., 1966. The functional response of invertebrate predators to prey density.– Memoirs of the Entomological Society of Canada 48: 1-86.

Hope, F. W., 1841. Descriptions of some new insects, collected in Assam by William Griffith.– Transactions of the Linnean Society 18: 435-447.

Horváth, G., 1895. Hémiptères nouveaux de l'Europe et des pays limitrophes.– Revue de l'Entomologie 14: 152-165.

Horváth, G., 1899. Monographia generis *Aphelocheirus*.–Termeszetrajzi Füzeket 22: 256-267.

Horváth, G., 1904. Hydrocorisae tres novae.– Annales Historico-Naturales Musei Nationalis Hungarici 2: 594-595.

Horváth, G., 1915. Monographie des Mésovéliides.– Annales Musei Nationalis Hungarici 13: 535-556.

Horváth, G., 1919. Hemipteren von den Aru und Kei-inseln.– Abhandlungen der Senckenbergischen Naturforschenden Gesellschaft 35: 305-314.
Horváth, G. 1929. Species novae hebridarum (Hem. Het.).– Annales Musei Nationalis Hungarici 26: 313-317.
Hüfner, B., 1939. Die Antennen und antennalen Sinnesorgane der Hydrocores mit besonderer Berücksichtigung der Nepiden.– Zoologica 97: 1-32
Hungerford, H. B., 1920. The biology and ecology of aquatic and semiaquatic Hemiptera.– The Kansas University Science Bulletin 11 (1919): 1-328.
Hungerford, H. B., 1922. Oxyhaemoglobin present in backswimmer, *Buenoa margaritacea* Bueno.– Canadian Entomologist 54: 262-263.
Hungerford, H.B., 1933. Some Aquatic and Semiaquatic Hemiptera from Sumatra.– Miscellanea Zoologica Sumatrana 75: 1-5.
Hungerford, H. B., 1947. A new species of *Cymatia* from Australia (Hemiptera, Corixidae).– Journal of the Kansas Entomological Society 20: 154-157.
Hungerford, H. B., 1948a. The Corixidae of the Western Hemisphere (Hemiptera).– The University of Kansas Science Bulletin 32: 1-827.
Hungerford, H. B., 1948b. The eggs of Corixidae (Hemiptera).– Journal of the Kansas Entomological Society 21: 141-146.
Hungerford, H. B., 1953. A new *Agraptacorixa* from New Guinea (Corixidae-Hemiptera).– Journal of the Kansas Entomological Society 26: 38-39.
Hungerford, H. B., 1955. A new *Limnometra* from Sumatra.– Journal of the Kansas Entomological Society 28: 67-68.
Hungerford, H. B., 1957. Two new Gerrids from the Philippines.– Journal of the Kansas Entomological Society 30: 125-130.
Hungerford, H. B. & Evans, N. E., 1934. The Hydrometridae of the Hungarian National Museum and other studies in the family (Hemiptera).– Annales Musei Nationalis Hungarici 28: 31-112.
Hungerford, H. B. & Matsuda, R., 1957. Descriptions of two species of the genus *Brachymetra* (Gerridae, Hemiptera).– Journal of the Kansas Entomological Society 30: 19-25.
Hungerford, H. B. & Matsuda, R., 1958a. The *Tenagogonus-Limnometra* Complex of the Gerridae.– The University of Kansas Science Bulletin 39: 371-457.
Hungerford, H. B. & Matsuda, R., 1958b. A new genus of Gerridae from the Solomon Islands.– The Pan-Pacific Entomologist 34: 203-206.
Hungerford, H. B. & Matsuda, R., 1958c. The genus *Esakia* Lundblad with two new species (Heteroptera, Gerridae).– Journal of the Kansas Entomological Society 31: 193-195.
Hungerford, H. B. & Matsuda, R., 1958d. A new *Ptilomera* from the Philippines (Hemiptera: Gerridae).– Bulletin of the Brooklyn Entomological Society 53: 69-73.
Hungerford, H. B. & Matsuda, R., 1958e. A new primitive *Ptilomera* from the Himalayas and other notes (Gerridae, Hemiptera).– Bulletin of the Brooklyn Entomological Society 53: 117-123.
Hungerford, H. B. & Matsuda, R., 1959. Concerning the genus *Metrobatopsis* Esaki with description of a new species.– Bulletin of the Brooklyn Entomological Society 54: 29-36.
Hungerford, H.B. & Matsuda, R., 1961a. Some new species of *Rhagovelia* from the Philippines (Veliidae, Heteroptera)– The University of Kansas Science Bulletin 42: 257-279.
Hungerford, H. B. & Matsuda, R., 1961b. A new species of *Tenagogonus* from the Fiji Islands (Gerridae: Hemiptera).– Journal of the Kansas Entomological Society 34: 173-176.
Hungerford, H. B. & Matsuda, R., 1962a. Concerning some species of *Tenagogonus* s.str. (Gerridae: Hemiptera).– Bulletin of the Brooklyn Entomological Society 57: 141-145.
Hungerford, H. B. & Matsuda, R., 1962b. The genus *Cylindrostethus* Fieber from the Eastern Hemisphere.– The University of Kansas Science Bulletin 43: 83-111.

Hungerford, H. B. & Matsuda, R., 1965. The genus *Ptilomera* Amyot & Serville (Gerridae: Hemiptera).– The University of Kansas Science Bulletin 45: 397-515.

Hutchinson, G. E., 1929. A revision of the Notonectidae and Corixidae of South Africa.– Annals of the South African Museum 25: 359-474.

Hutchinson, G. E., 1940. A revision of the Corixidae of India and adjacent regions. Transactions of the Connecticut Academy of Arts and Sciences 33: 339-476, pls. 1-36.

Hutchinson, G. E., 1957. On a new species of *Anisops* (Hemiptera, Notonectidae) from the Moluccas.– Postilla 33: 1-4.

Hynes, H. B. N., 1955. Biological notes on some east African aquatic Heteroptera.– Proceedings of the Royal Entomological Society of London (A) 30: 43-54.

Ichikawa, N., 1988. Male brooding behavior of the giant water bug *Lethocerus deyrollei* Vuillefroy (Hemiptera: Belostomatidae).– Journal of Ethology 6: 121-127.

Ichikawa, N., 1989. Repeated copulations benefit of the female in *Lethocerus deyrollei* Vuillefroy (Heteroptera: Belostomatidae).– Journal of Ethology 7: 113-117.

Ichikawa, N., 1990. Egg mass destroying behavior of the female giant water bug *Lethocerus deyrollei* Vuillefroy (Heteroptera: Belostomatidae).– Journal of Ethology 8: 5-11.

Ichikawa, N., 1991a. Egg mass destroying and guarding behavior of the giant water bug *Lethocerus deyrollei* Vuillefroy (Heteroptera: Belostomatidae).– Journal of Ethology 9: 25-29.

Ichikawa, N., 1991b. Additional benefit of egg mass desctruction by giant water bug *Lethocerus deyrollei* Vuillefroy (Heteroptera: Belostomatidae) females.– Journal of Ethology 9: 34-36.

Ichikawa, N., 1993. Biased operational sex ratio causes the female giant water bug *Lethocerus deyrollei* to destroy egg masses.– Journal of Ethology 11: 151-152.

Ichikawa, N., 1995. Male counterstrategy against infanticide of the female giant water bug *Lethocerus deyrollei* (Hemiptera: Belostomatidae).– Journal of Insect Behaviour 8: 181-188.

Izzard, R. J., 1936. The Hemiptera of Christmas Island.– Annals and Magazine of Natural History (10) 17: 577-600.

Jackson, R. R. & Walls, E. I., 1998. Predatory and scavenging behaviour of *Microvelia macgregory* (Hemiptera: Veliidae), a water-surface bug from New Zealand.– New Zealand Journal of Zoology 25: 23-28.

Jacobs, W. & Renner, M., 1988: Biologie und Őkologie der Insekten.– Gustav Fischer, Stuttgart, New York: 690 pp.

Jaczewski, T., 1934. Notes on the Old World species of Ochteridae (Heteroptera).– Annals and Magazine of Natural History (10) 11: 597-613.

Jaczewski, T., 1935. Die Uferwanzen (Heteroptera: Leptopodidae, Acanthiidae, Ochteridae und Mononychidae) der Deutschen Limnologischen Sunda-Expedition.– Archiv für Hydrobiologie Supplement 13: 474-483.

Jansson, A., 1973. Stridulation and its significance in the genus *Cenocorixa* (Hemiptera: Corixidae).– Behaviour 46: 1-36.

Jansson, A., 1982. Notes on some Corixidae (Heteroptera) from New Guinea and New Caledonia.– Pacific Insects 24: 95-103.

Jansson, A., 1986. The Corixidae (Heteroptera) of Europe and some adjacent regions.– Acta Entomologica Fennica 47: 1-94.

Jansson, A., 1989. Stridulation of Micronectinae (Heteroptera, Corixidae).– Annales Entomologici Fennici 55: 161-175.

Jansson, A., 1995. Family Corixidae Leach, 1815. Water boatmen.– In: Aukema, B. & Rieger, C. (eds). Catalogue of the Heteroptera of the Palaearctic Region. Netherlands Entomological Society, Amsterdam: i-xxvi, 1-222.

Jansson, A. & Meyer-Rochow, V. B., 1990. Sound production in *Synaptonecta issa* (Heteroptera: Corixidae, Micronectinae).– Applied Entomological Zoology 25: 328-331.

Jell, P.A. & Duncan, P.M., *1986*. Invertebrates, mainly insects, from the freshwater, Lower Cretaceous, Koonwarra Fossil Bed (Korumburra group), South Gippsland, Victoria.– Memoires of the Association for Australasian Palaeontology 3: 111-205.
Jordan, K.H.C. 1935: Beitrag zur Lebensweise der Wanzen auf feuchten Böden (Heteropt.).– Stettiner Entomologische Zeitung 96: 1-26.
Jordan, K.H.C. 1952: Wasserläufer.– Die Neue Brehm Bücherei 52: 1-32.
Kanyukova, E.V., 1997. Hebridae of Russia and adjacent countries (Heteroptera).– Zoosystematica Rossica 6(1-2): 223-236.
Keffer, S.L., 1996. Systematics of the New World waterscorpion genus *Curicta* Stål (Heteroptera: Nepidae).– Journal of the New York Entomological Society 104: 117-215.
Keffer, S.L., 2000.– Waterscorpions (Nepidae).– In: Schaefer, C.W. & Panizzi, A.R. (eds). Heteroptera of economic importance. CRC Press, Boca Raton: 583-589.
Kellen, W.R., 1959. Notes on the biology of *Halovelia marianarum* Usinger in Samoa (Veliidae: Heteroptera).– Annals of the Entomological Society of America 52: 53-62.
King, I.M., 1976. Underwater sound production in *Micronecta batilla* Hale (Heteroptera: Corixidae).– Journal of the Australian Entomological Society 15: 35-43.
King, I.M., 1997. Two new species of *Micronecta* Kirkaldy (Heteroptera: Corixidae) from Victoria, Australia.– Australian Entomologist 24: 145 152.
Kirkaldy, G.W., 1897a. Notes on the genus *Sigara* Fabr.– The Entomologist, London 30: 238-240.
Kirkaldy, G.W., 1897b. Aquatic Rhynchota: Descriptions and notes – No. 1.– Annals and Magazine of Natural History (6) 20: 52-60.
Kirkaldy, G.W., 1898a. Deux espèces nouvelles du genre *Enithares* Spin. et notes hémiptérologiques diverses.– Revue d'Entomologie, Caen 17: 72-73.
Kirkaldy, G.W., 1898b. Neue und seltene Notonectiden-Arten.– Wiener Entomologische Zeitung 17: 141-142.
Kirkaldy, G.W., 1898c. Description d'une espèce nouvelle de Notonectidae (Hemiptera) de la collection du Muséum d'Histoire Naturelle de Paris.– Bulletin du Muséum National d'Histoire Naturelle 1898: 151.
Kirkaldy, G.W., 1898d. Note sopra alcuni rincoti acquatici africani del Museo Civico di Storia Naturale di Genova.– Annali dei Museo Civico di Storia Naturale Giacomo Doria 39: 144-145.
Kirkaldy, G.W., 1899a. Sur quelques hémiptères aquatiques nouveaux ou peu connus.– Revue de Entomologie, Caen 18: 85-96.
Kirkaldy, G.W., 1899b. Aquatic Rhynchota in the collection of the Royal Museum of Belgium.– Annales de la Société Entomologique de Belgique 43: 42-47.
Kirkaldy, G.W., 1901a. Miscellania Rhynchotalia.– The Entomologist 34: 5-6, 116-117.
Kirkaldy, G.W., 1901b. On some Rhynchota, principally from New Guinea. (Amphibicorisae and Notonectidae).– Annali dei Museo Civico di Storia Naturale Giacomo Doria 20: 804-810.
Kirkaldy, G.W., 1901c. Notes of the division *Veliiaria* [Rhynchota].– The Entomologist 34: 285-286, 308-310.
Kirkaldy, G.W., 1902. Miscellania Rhynchotalia No. 3.– The Entomologist 35: 136-137.
Kirkaldy, G.W., 1904. Über Notonectiden (Hemiptera).– Wiener Entomologische Zeitung 23: 93-135.
Kirkaldy, G.W., 1905. Five New Species of *Micronecta* Kirkaldy (Aquatic Hemiptera).– Entomological News 16: 260-263.
Kirkaldy, G.W., 1909. Hemiptera old and new, No. 2.– Canadian Entomologist 41: 388-391.
Knowles, J.N., 1974. A Revision of Australian species of *Agraptacorixa* Kirkaldy and *Diaprepocoris* Kirkaldy (Heteroptera: Corixidae).– Australian Journal of Marine and Freshwater Research 25: 173-191.

Kormilev, N. A., 1971. Ochteridae from the Oriental and Australian regions.– Pacific Insects 13: 429-444.
Kormilev, N. A., 1972. A new species of the genus *Ochterus* Latreille, 1807, from New Guinea.– Pacific Insects 14: 585-587.
Kovac, D., 1982. Zur Überwinterung der Wasserwanze *Plea minutissima* Leach (Heteroptera, Pleidae): Diapause mit Hilfe der Plastronatmung.– Nachrichten des Entomologischen Vereins Apollo, Neue Folge 3: 59-76.
Kovac, D., 1987. Untersuchungen zur biologischen bedeutung des Sekretsputzen bei Wasserwanzen und Wasserkäfern.– Dissertation Frankfurt/Main: 1-242.
Kovac, D., 1993. A quantitative analysis of sectretion-grooming behaviour in the water bug *Plea minutissima* Leach (Heteroptera, Pleidae): control by abiotic factors.– Ethology 93: 41-61.
Kovac, D. & Maschwitz, U., 1989. Secretion grooming in the water bug *Plea minutissima*: a chemical defence against microorganisms interfering with the hydrofuge properties of the respiratory region.– Ecological Entomology 14: 403-411.
Kovac, D. & Maschwitz, U., 1990. Sekretputzen bei *Ilyocoris cimicoides* (Heteroptera, Naucoridae).– Nachrichten des Entomologischen Vereins Apollo Frankfurt N. F. 11: 155-164.
Kovac, D. & Maschwitz, U., 1991. The function of the metathoracic scent gland in corixid bugs (Hemiptera, Corixidae): secretion grooming on the water surface.– Journal of Natural History 25: 331-340.
Kovac, D. & Papáček, M., 2000. *Helotrephes steiningeri* sp.n., and notes on two further Helotrephini spp. (Heteroptera: Helotrephidae) from Thailand and West Malaysia.– Linzer biologische Beiträge 32(1): 265-271.
Kovac, D. & Yang, C. M., 1990. A preliminary checklist of the semiaquatic and aquatic Hemiptera (Heteroptera: Gerromorpha and Nepomorpha) of Ulu Kinchin, Pahang, Malaysia.– Malayan Nature Journal 43: 282-288.
Kovac, D. & Yang, C. M., 2000. Revision on the Oriental bamboo-inhabiting bug genus *Lathriovelia* Andersen, 1989 (Heteroptera, Veliidae) with description of *L. rickmersi* new species, and notes on the genus *Baptista* Distant, 1903.– The Raffles Bulletin of Zoology 48: 153-165.
Lahr, J., 2000. Effects of insecticides on invertebrates in temporary ponds in the Sahel.– Thesis Vrije Universiteit te Amsterdam: 1-169.
Lahr, J., Diallo, A. O., Ndour, K. B., Badji, A. & Diouf, P. S., 1999. Phenology of invertebrates living in a sahelian temporary pond.– Hydrobiologia 405: 189-205.
Laird, M., 1956. Studies of mosquitoes and freshwater ecology in the South Pacific.– Bulletin of the Royal Society of New Zealand 6: 1-213.
Lansbury, I., 1962. Notes on the genus *Anisops* in Bishop Museum (Hem.: Notonectidae).– Pacific Insects 4: 141-151.
Lansbury, I., 1963. Notes on the water-bugs of Solomon Islands and New Hebrides.– Pacific Insects 5: 5-10.
Lansbury, I., 1964a. Some observations on the Notonectidae (Hemiptera-Heteroptera) of Viêt-Nam and adjacent countries.– Annales Zoologici Polska Akademia Nauk 22: 203-219.
Lansbury, I., 1964b. The genus *Anisops* in Australia (Hemiptera: Notonectidae) I.– Journal of the Entomological Society of Queensland 3: 52-65.
Lansbury, I., 1965a. Notes on the species of the genus *Anisops* Spin. (Hemiptera-Heteroptera, Notonectidae) of Java.– Annales Zoologici Polska Akademia Nauk 23: 57-68.
Lansbury, I., 1965b. A new tribe and genus of Notonectidae (Heteroptera: Notonectidae) from Borneo.– Pacific Insects 7: 327-332.
Lansbury, I. 1966. Notes on the genus *Aphelonecta* (Hemiptera-Heteroptera: Notonectidae).– Pacific Insects 8(3): 269-230.
Lansbury, I., 1967. Aquatic Heteroptera of the Noona Dan Expedition to the Philippine and Bismarck Islands.– Entomologiske Meddelelser 35: 89-98.

Lansbury, I., 1968a. Notonectidae (Hemiptera-Heteroptera) of Rennell Island.– The Natural History of Rennel Island, British Solomon Islands 5: 95-98.
Lansbury, I., 1968b. The *Enithares* (Hemiptera-Heteroptera: Notonectidae) of the Oriental Region.– Pacific Insects 10: 353-442.
Lansbury, I., 1969. The genus *Anisops* in Australia.– Journal of Natural History 3: 433-458.
Lansbury, I., 1970. Revision of the Australian *Sigara* (Hemiptera-Heteroptera, Corixidae).– Journal of Natural History 4: 39-54.
Lansbury, I., 1972a. A review of the Oriental species of *Ranatra* Fabricius (Hemiptera-Heteroptera: Nepidae).– Transactions of the Royal Entomological Society London 124: 287-341.
Lansbury, I., 1972b. A revision of the genus *Telmatotrephes* Stål (Hemiptera-Heteroptera: Nepidae).– Zoologica Scripta 1: 271-286.
Lansbury, I., 1973a. A review of the genus *Cercotmetus* Amyot & Serville, 1843 (Hemiptera-Heteroptera: Nepidae).– Tijdschrift voor Entomologie 116: 83-106.
Lansbury, I., 1973b. *Montonepa* gen. n. from India with notes on the genus *Borborophyes* Stål (Hemiptera-Heteroptera, Nepidae).– Zoologica Scripta 2: 111-118.
Lansbury, I., 1974a. Notes on the genus *Enithares* Spinola (Hem., Notonectidae).– Entomologist's Monthly Magazine 109 (1973): 226-231.
Lansbury, I. 1974b. Notes on *Ranatra (Amphischizops) compressicollis* Montandon with a review of its systematic position within the American *Ranatra* (Hemiptera-Heteroptera, Nepidae).– Zoologica Scripta 3: 23-30.
Lansbury, I., 1974c. A new genus of Nepidae from Australia with a revised classification of the family (Hemiptera: Heteroptera).– Journal of the Australian Entomological Society 13: 219-227.
Lansbury, I., 1978a. Comments on the species concept in some Australian *Anisops* Spinola (Hemiptera: Notonectidae).– Memoirs of the Natural Museum of Victoria 39: 101-115.
Lansbury, I., 1978b. A review of *Goondnomdanepa* Lansbury (Heteroptera: Nepidae).– Australian Journal of Marine and Freshwater Research 29: 117-126.
Lansbury, I., 1983. Notes on the Australasian species of *Cymatia* Flor s.l. (Insecta, Heteroptera, Corixidae).– Transactions of the Royal Society of South Australia 107: 51-57.
Lansbury, I., 1984. Some Nepomorpha (Corixidae, Notonectidae and Nepidae) (Hemiptera-Heteroptera) of North-West Australia.– Transactions of the Royal Society of South Australia 108: 39-40.
Lansbury, I., 1985. Notes on the identity of *Nychia* Stål (Hemiptera-Heteroptera: Notonectidae) in Australia.– The Beagle, Occasional Papers of the Northern Territory Museum of Arts and Science 2: 1-9.
Lansbury, I., 1988. Gelastocoridae (Hem.-Het.) in the Hope Entomological Collections, University Museum Oxford.– Entomologist's Monthly Magazine 124: 243-250.
Lansbury, I., 1989a. Notes on the Haloveliinae of Australia and the Solomon Islands (Insecta, Hemiptera, Heteroptera, Veliidae).– Reichenbachia 26: 93-109.
Lansbury, I., 1989b. Notes on marine *Rheumatometroides* from the Solomon Islands (Hem., Gerridae).– Entomologist's Monthly Magazine 125: 63-71.
Lansbury, I., 1992. Notes on the marine-freshwater gerrid genus *Rheumatometroides* (Hemiptera, Gerridae) of Papua New Guinea.– Tijdschrift voor Entomologie 135: 1-10.
Lansbury, I., 1993a. *Strongylovelia* (Veliidae) and *Metrobatopsis* (Gerridae) and associated pleuston Hemiptera of West New Britain.– Tijdschrift voor Entomologie 136: 15-22.
Lansbury, I., 1993b. *Rhagovelia* of Papua New Guinea, Solomon Islands and Australia (Hemiptera-Veliidae).– Tijdschrift voor Entomologie 136: 23-54.
Lansbury, I., 1995a. Notes on the Corixidae and Notonectidae (Hemiptera: Heteroptera) of southern Western Australia.– Records of the Western Australian Museum 17: 181-189.

Lansbury, I., 1995b. Notes on the genus *Anisops* Spinola (Hemiptera-Heteroptera, Notonectidae) of the Northern Territory and Western Australia.– The Beagle, Records of the Museums and Art Galleries of the Northern Territory 12: 65-74.

Lansbury, I., 1996a. Notes of the marine veliid genera *Haloveloides*, *Halovelia* and *Xenobates* (Hemiptera-Heteroptera, Veliidae) of Papua New Guinea.– Tijdschrift voor Entomologie 139: 17-28.

Lansbury, I., 1996b. Two new species of *Ciliometra* Polhemus (Hem., Gerridae) from Papua New Guinea.– Entomologist's Monthly Magazine 132: 55-60.

Lansbury, I. & Zettel, H., 1997. New species and subspecies of the genus *Strongylovelia* Esaki (Insecta: Heteroptera: Veliidae) from Borneo and the Philippines.– Annalen des Naturhistorischen Museums in Wien 99B: 51-77.

Laporte, F.L. de, 1832-1833. Essai d'une classification systematique de l'ordre des Hémiptères.– Magasin de Zoologie 2 (suppl.): 1-88.

La Rivers, I., 1951. A revision of the genus *Ambrysus* in the United States (Hemiptera: Naucoridae).– University of California Publications in Entomology 8: 277-332.

La Rivers, I., 1953. New Gelastocorid and Naucorid records and miscellaneous notes, with a description of the new species *Ambrysus amargosus* (Hemiptera: Naucoridae).– The Wasmann Journal of Biology 11: 83-96.

La Rivers, I., 1969. A new genus and species of Naucorid from the Philippines.– Occasional Papers of the Biological Society of Nevada 18: 1-4.

La Rivers, I., 1970a. Two new species of *Laccocoris* from British North Borneo.– Proceedings of the Entomological Society of Washington 72: 496-499.

La Rivers, I., 1970b. A new Philippine *Sagocoris*.– The Pan-Pacific Entomologist 46: 167-169.

La Rivers, I., 1970c. A new species of *Laccocoris* from the Philippines (Hemiptera: Naucoridae).– The Wasmann Journal of Biology 28: 269-273.

La Rivers, I., 1971a. The Naucoridae of New Guinea (Hemiptera).– Memoirs of the Biological Society of Nevada 2: 1-63, 101-120.

La Rivers, I., 1971b. Catalogue of Taxa Described in the Family Naucoridae (Hemiptera).– Memoirs of the Biological Society of Nevada 2: 65-99.

La Rivers, I., 1971c. Descriptions and notes concerning some oriental *Aphelocheirus* (Hemiptera: Aphelocheiridae).– Bulletin of the South California Academy of Sciences 70: 69-72.

La Rivers, I., 1974. Catalogue of taxa described in the family Naucoridae (Hemiptera) Supplement No. 1.– Occasional Papers of the Biological Society of Nevada 38: 1-17.

La Rivers, I., 1976. Catalogue of taxa described in the family Naucoridae (Hemiptera).– Occasional papaers of the biological society of Nevada 41, Suppl. no. 2: 1-17.

Larsén, O., 1924. Zur Kenntnis von *Aphelocheirus aestivalis* F.– Arkiv för Zoologi 16 No 16: 1-21.

Larsén, O., 1927. Über die Entwicklung und Biologie von *Aphelocheirus aestivalis* Fabr.– Entomologisk Tidskrift 48: 181-206.

Larsén, O., 1932. Beiträge zur Ökologie und Biologie von *Aphelocheirus aestivalis* Fabr.– Internationale Revue der Gesamten Hydrobiologie und Hydrografie 26: 1-19.

Larsén, O., 1938. Untersuchungen über den Geschlechtsapparat der aquatilen Wanzen.– Opuscula Entomologica 1: 1-388.

Larsén, O., 1949. Die Ortsbewegungen von *Ranatra linearis* L.– Lunds Universitets Årsskrift N.F. Avdelning 2, 15(6): 1-82

Larsén, O., 1957.– Truncale scolopalorgane in den pterothorakalen und bei den beiden ersten abdominalen Segmenten der aquatilen Heteropteren. – Acta Universitatis Lundensis (n.s.) 53(1): 1-67.

Latreille, P.A., 1796. Précis des caractères génériques des insectes, disposés dans un ordre naturel: i-xiii, 1-208. Brive, Bordeaux.

Latreille, P. A., 1802. Histoire naturelle, générale et particulière, des crustacés et des insectes. Vol. 3: i-xii, 1-467. Dufart, Paris.

Latreille, P. A., 1804. Histoire naturelle, générale et particulière, des crustacés et des insectes. Vol. 12: 1-424. Dufart, Paris.

Latreille, P. A., 1807. Genera crustaceorum et insectorum secundum ordinem naturalem in familias dispisita iconibus exemplisque plurimis explicata 3: 1-259. Koenig, Parisiis et Argentorati.

Lauck, D. R. & Menke, A. S., 1961. The higher classification of the Belostomatidae (Heteroptera).– Annals of the Entomological Society of America 54: 644-657.

Leach, W. E., 1815. Entomology. In: Brewster, D. (ed.). The Edinburgh Encyclopaedia 9: 57-172.

Leach, W. E., 1817. On the classification of the natural tribe of insects Notonectides with descriptions of British species.– Transactions of the Linnean Society of London 12: 10-18.

Lebrun, D., 1960. Recherches sur la biologie et l'éthologie de quelques Hétéroptères aquatiques.– Annales de la Société Entomologique de France 129: 179-199.

Leong, C. Y., 1962. The life-history of *Anisops breddini* Kirkaldy (Hemiptera, Notonectidae).– Annals and Magazine of Natural History (13) 5: 377-383.

Leong, C.Y., 1963. *Anisops lansburyi* sp. n. (Hemiptera: Notonectidae) from Malaya. – Proceedings of the Royal Entomological Society of London (B) 32: 101-102.

Leong, C. Y., 1966. A taxonomic study of the Malayan Corixidae (Hemiptera-Heteroptera) with the description of *Micronecta malayana* sp. nov.– Bulletin of the National Museum Singapore 12: 83-90.

Lepeletier, A.L. M. & Serville, J. G. A., 1825. Encyclopédie Méthodique 10: 344 pp.

Lethierry, L., 1877. Séance du 13 juin 1877.– Bulletin de la Société Entomologique de France 1877: C-CII [100-102].

Linnaeus, C., 1758. Systema Naturae. 10th Ed.: i-v, 1-823. Salvii, Holmiae.

Lundblad, O., 1928. Studien über schwedische Corixiden VII.– Entomologisk Tidsckrift 49 9-17, 3 pls.

Lundblad, O., 1933a. Zur Kenntnis der aquatilen und semiaquatilen Hemipteren von Sumatra, Java und Bali.– Archiv für Hydrobiologie Supplement Band 12: 1-195, 263-488, pls. 1-21.

Lundblad, O., 1933b. Some new or little known Rhynchota from China.– Annals and Magazine of Natural History (10) 12: 449-464.

Lundblad, O., 1934a. Aquatic and semiaquatic Heteroptera of Tahiti.– Bishop Museum Bulletin 113: 121-126.

Lundblad, O., 1934b. Schwedisch-chinesische wissenschaftliche Expedition nach den nordwestlichen Provinsen Chinas. 28-1 Wasserhemipteren.– Arkiv för Zoologi 27A-14: 1-31, 2 pls.

Lundblad, O., 1936. Die altweltlichen Arten der Veliidengattungen *Rhagovelia* und *Tetraripis*.– Arkiv för Zoologi 28A-21: 1-63, pls. 1-12.

Lundblad, O., 1937. Einige neue oder wenig bekannte ostasiatische *Rhagovelia*-Arten.– Entomologisk Tidskrift 58: 1-9.

Macan, T. T., 1954a. The Corixidae (Hemipt.) of some Danish lakes.– Hydrobiologia 6: 44-69.

Macan, T. T., 1954b. A contribution to the study of the ecology of Corixidae (Hemipt.).– Journal of Animal Ecology 23: 115-141.

Mahner, M., 1993. Systema Cryptoceratorum Phylogeneticum (Insecta, Heteroptera).– Zoologica 143: i-ix, 1-302.

Malipatil, M. B. & Monteith, G. B., 1983. One new genus and four new species of terrestrial Mesoveliidae (Hemiptera: Gerromorpha) from Australia and New Caledonia.– Australian Journal of Zoology 31: 943-955.

Markl, H. & Wiese, K., 1969. The sensitivity of *Notonecta glauca* L. to ripples of water surface.– Zeitschrift für Vergleichende Physiologie 62: 413-420.
Marks, E. P., 1958. A comparative study of the foregut of several aquatic Hemiptera.– Journal of the Kansas Entomological Society 31: 138-153.
Matsuda, R., 1960. Morphology, evolution and a classification of the Gerridae (Hemiptera-Heteroptera).– The University of Kansas Science Bulletin 41: 25-632.
Matsumura, S., 1905. Die Wasser-Hemipteren Japans.– Journal of the Sapporo Agricultural College 2: 53-66.
Matsumura, S., 1913. Thousand Insects of Japan. Additamenta 1: 1-184.
Mayr, E. & Ashlock, P.D., 1969. Principles of systematic zoology. 2nd Edition: i-xx, 1-475. McGraw Hill, New York.
Mayr, G. L., 1853. Zwei neue Wanzen aus Kordofan.– Verhandlungen des Zoologisch-Botanischen Vereins in Wien 2: 14-18, pls. 1-2.
Mayr, G.L., 1865. Diagnosen neuer Hemipteren 2.– Verhandlungen der Zoologisch-Botanischen Gesellschaft in Wien 15: 429-446.
Menke, A. S., 1960. A review of the genus *Lethocerus* (Hemiptera: Belostomatidae) in the Eastern Hemisphere with the description of a new species from Australia.– Australian Journal of Zoology: 285-288.
Menke, A. S. & Stange, L. A., 1964. A new genus of Nepidae from Australia with notes on the higher classification of the family.– Proceedings of the Royal Society of Queensland 75: 67-72.
Merritt, R. W. & Cummins, K. W. [eds.], 1984. An introduction to the aquatic insects of North America (2 edition).– Kendall / Hunt Publishing Company, Dubuque, Iowa, pp.722.
Messner, B., Lunk, A., Groth, I., Subklew, H.-J. & Taschenberger, D., 1981. Neue Befunde zum Atmungssystem der Grundwanze *Aphelocheirus aestivalis* Fab. (Heteroptera, Hydrocorisae). I. Imagines.– Zoologische Jahrbücher für Anatomie 105: 474-496.
Michel, K., 1977. Feinstrukturelle Besonderheiten im mesothorakalen Tympanalorgan der Ruderwanze *Callicorixa praeusta* (Heteroptera: Corixidae).– Entomologica Germanica 3: 316-323.
Midak, Z., 1965. 'Plukwiaki róznoskrzydle (Hemiptera-Heteroptera) wód okoloc Gorzowa Wielkopolskiego.' [Heteroptera of the waters in the surroundings of Gorzów Wielkopolski.] – Badania Pizjograficzne Nad Polska Zachodna 16: 7-25.
Miller, N. C. E., 1971. The biology of the Heteroptera. (2 edition): xi + 206 pp. E. W. Classey Ltd., Hampton.
Miller, P. L., 1964. The possible role of haemoglobin in *Anisops* and *Buenoa* (Hemiptera: Notonectidae).– Proceedings of the Royal Entomologicl Society of London (A) 39: 166-175.
Miyahala, K. & Oho, N., 1957. On the resistance of *Microvelia douglasi* Scott to insecticides.– Proceedings of the Association for Plant protection Kyushu 3: 62-64.
Miyamoto, S., 1937. Observations on a marine water-strider, *Asclepios coreanus miyamotoi* Esaki.– Mushi 9: 137-142.
Miyamoto, S., 1952. Biology of *Helotrephes formosanus* Esaki et Miyamoto, with descriptions of larval stages.– Sieboldia 1: 1-10.
Miyamoto, S., 1953. Biology of *Microvelia diluta* Distant, with descriptions of its brachypterous form and larval stages.– Sieboldia 1: 113-133, pls. 15-18.
Miyamoto, S., 1961. Insecta Japonica Series 1 Part 3 Hemiptera: Gerridae: 1-39.– Hokuryukan Publishing Co., Tokyo.
Miyamoto, S., 1964. Veliidae of the Ryukyus (Hemiptera, Heteroptera).– Kontyû 32: 137-150.
Miyamoto, S., 1967. Gerridae of Thailand and North Borneo taken by the Joint Thai-Japanese Biological Expedition 1961-1962.– Nature and Life in South-east Asia 5: 217-257.

Miyamoto, S., 1981. *Anisops yanoi* sp.n.– In: Yano, K., Miyamoto, S. & Gabriel B. P. Faunal and ecological studies on the insects of paddy fields in Asia. 4. Aquatic and semiaquatic Heteroptera from the Philippines. Esakia 16: 25-26.
Möller, H., 1921. Über *Lethocerus uhleri* Mont.– Zoologische Jahrbücher Abteilung für Anatomie und Ontogenie 42: 43-90.
Montandon, A. L., 1897a. Hemiptera Cryptocerata. Fam. Naucoridae. – Sous-fam. Laccocorinae.– Verhandlungen der Kaiserlich-Königlichen Zoologisch-Botanischen Gesellschaft in Wien 47: 435-454.
Montandon, A. L., 1897b. Hémiptères-Hétéroptères exotiques. Notes et descriptions.– Annales de la Société Entomoligique de Belgique An. : 50-66.
Montandon, A. L., 1897c. Hemiptera Cryptocerata. Fam. Naucoridae. – Sous-fam. Cryptocricinae.– Verhandlungen der Kaiserlich-Königlichen Zoologisch-Botanischen Gesellschaft in Wien 47: 6-23.
Montandon, A. L., 1897d. Viaggio di Leonardo Fea in Birmania e regioni vicine LXXV Hemiptera Cryptocerata.– Annali del Museo Civico di Storia naturale di Genova 37: 365-377.
Montandon, A. L., 1898. Hemiptera Cryptocerata. Notes et descriptions d'espèces nouvelles II. Fam. Belostomidae.– Bulletin de la Société des Sciences de Bucarest 7: 430-432.
Montandon, A. L., 1899. Hemiptera Cryptocerata. Sousfamille Mononychinae I, II.– Bulletin de la Société des Sciences de Bucarest 8: 392-407, 774-780.
Montandon, A. L., 1900. Notes sur quelques Hemiptères Hétéroptères et descriptions d'espèces des collections du Musée Civique de Gênes.– Annali del Museo Civico di Storia Naturale di Genova 40: 531-541.
Montandon, A. L., 1903. Hémiptères aquatiques.– Bulletin de la Société des Sciences de Bucarest 12: 97-121.
Montandon, A. L., 1905. Espèces nouvelles ou peu connues du genre *Ranatra*.– Bulletin de la Société des Sciences de Bucarest 14: 389-398.
Montandon, A. L., 1907. Hémiptères Hétéroptères, espèces nouvelles ou peu connues.– Bulletin de la Société des Sciences de Bucarest 15: 308-331.
Montandon, A. L., 1909a. Naucoridae, descriptions d'espèces nouvelles.– Bulletin de la société des Sciences de Bucarest 18: 43-61.
Montandon, A.L., 1909b. Nepidae et Belostomidae. Notes diverses et descriptions d'espèces nouvelles.– Annales Historico-Naturales Musei Nationalis Hungarici 7: 59-70.
Montandon, A.L., 1909c. Belostomatidae et Nepidae. Notes diverses et descriptions d'espèces nouvelles.– Bulletin de la Société des Sciences de Bucarest 18: 137-147.
Montandon, A. L., 1910. Nepidae. Notes et descriptions d'espèces nouvelles.– Annales Historico-Naturales Musei Nationalis Hungarici 8: 163-171.
Montandon, A.L., 1911a. Deux genres nouveaux d'Hydrocorises.– Annales Historico-Naturales Musei Nationalis Hungarici 9: 244-250.
Montandon, A. L., 1911b. Nouvelles espèces d'hydrocorises appartenant aux collections du Muséum Entomologique de Berlin.– Bulletin de la Société des Sciences de Bucarest 20: 83-93.
Montandon, A. L., 1911c. Nepidae (Hemipt.). Nouvelles contributions.– Bulletin de la Société des Sciences de Bucarest 20: 648-656.
Montandon, A. L., 1913. Nepidae et Belostomatidae. Descriptions de deux espèces nouvelles.– Bulletin de la Société des Sciences de Bucarest 22: 122-125.
Montrousier, P., 1865. Essai sur la faune entomologique de Kanala (Nouvelle Calédonie) et description de quelques espèces nouvelles ou peu connues.– Annales de la Société Linnéenne de Lyon 11: 46-257.
Mulsant, E. & Rey, C., 1852. Description de quelques Hémiptères Hétéroptères nouveaux ou peu connues.– Annales de la Société Linnéenne de Lyon 1852: 76-141, 311.

Muraji, M. & Tachikawa S., 2000. Phylogenetic analysis of waterstriders (Hemiptera: Gerridae) based on partial sequences of mitochondrial and nuclear ribosomal RNA-genes.– Entomological Science 3(4): 615-626.

Murphey, R. K., 1971. Sensory aspects of the control of orientation to prey by the waterstrider, *Gerris remigis*.– Zeitschrift für Vergleichende Physiologie 72: 168-185.

Nakasuji, F. & Dyck, V. A., 1984. Evaluation of the role of *Microvelia douglasi atrolineata* (Bergroth) as predator of the brown planthopper *Nilaparvata lugens* Stål.– Researches on Population Ecology 26: 134-149.

Nel, A. & Paicheler, J.-C., 1992. Les Heteroptera aquatiques fossils, état actuel des connaissances (Heteroptera: Nepomorpha et Gerromorpha).– Entomologica Gallica 3: 159-182.

Nel, A. & Paicheler, J.-C., 1993a. Les Heteroptera aquatiques fossils, état actuel des connaissances (Heteroptera: Nepomorpha et Gerromorpha) (suite).– Entomologica Gallica 4: 15-21.

Nel, A. Paicheler, J.-C., 1993b. Les Heteroptera aquatiques fossils, état actuel des connaissances (Heteroptera: Nepomorpha et Gerromorpha) (suite et fin).– Entomologica Gallica 4: 79-89.

Nel, A. & Popov, Y. A., 2000. The oldest known fossil Hydrometridae from the lower Cretaceous from Brazil (Heteroptera: Gerromorpha).– Journal of Natural History 34: 2315-2322.

Nieser, N., 1977. A revision of the genus *Tenagobia* Bergroth (Heteroptera: Corixidae).– Studies on Neotropical Fauna and Environment 12: 1-56.

Nieser, N., 1982. De Nederlandse water- en oppervlaktewantsen (Heteroptera: Nepomorpha en Gerromorpha).– Wetenschappelijke Mededelingen van de Koninklijke Nederlandse Natuurhistorische Vereniging 155: 1-103.

Nieser, N., 1994. A new species and a new status in *Neogerris* Matsumura (Heteroptera: Gerridae) with a key to American species.– Storkia 3: 27-37.

Nieser, N., 1995. Nine new species of *Pseudovelia* and a new *Xiphovelia* (Heteroptera: Veliidae) from Sulawesi (Indonesia) and Mindanao (Philippines).– Tijdschrift voor Entomologie 138: 69-87.

Nieser, N., 1996. Two new species of *Ranatra* Fabricius from South-east Asia (Insecta: Heteroptera: Nepidae).– Annalen des Naturhistorischen Museums in Wien 98B: 347-351.

Nieser, N., 1997. A new species of *Ranatra* from Thailand (Insecta: Heteroptera: Nepidae).– Annalen des Naturhistorischen Museums in Wien 99B: 79-82.

Nieser, N., 1998. Introduction to the Notonectidae (Nepomorpha) of Thailand.– Amemboa 2: 10-13.

Nieser, N., 2000 [1999]. Three new species of *Micronecta* from Thailand with a key to SE Asian species (Heteroptera: Corixidae).– Journal of the New York Entomological Society 107: 277-288.

Nieser, N., 2002a. Four new species of *Micronecta* (Micronectidae) from continental Malaysia.– The Raffles Bulletin of Zoology 50: 9-16.

Nieser, N. 2002b. Guide to aquatic Heteroptera of Singapore and Peninsular Malaysia. IV. Corixoidae.– The Raffles Bulletin of Zoology 50: 263-274.

Nieser, N., 2004. Guide to aquatic Heteroptera of Singapore and Peninsular Malaysia. III. Pleidae and Notonectidae.– The Raffles Bulletin of Zoology 52: 79-96.

Nieser, N. & Chen, P.-p., 1991. Notes on Malesian aquatic bugs (Heteroptera) I. Naucoridae, Nepidae and Notonectidae, mainly from Sulawesi and Pulau Buton (Indonesia).– Tijdschrift voor Entomologie 134: 47-67.

Nieser, N. & Chen, P.-p., 1992a. Notes on Gelastocoridae and Ochteridae (Heteroptera) with the description of five new species.– Storkia 1: 2-13.

Nieser, N. & Chen, P.-p., 1992b. Notes on Indonesian water bugs (Nepomorpha & Gerromorpha).– Storkia 1: 30-40.

Nieser, N. & Chen, P.-p., 1992c. Revision of *Limnometra* Mayr (Gerridae) in the Malay Archipelago.– Tijdschrift voor Entomologie 135: 11-26.
Nieser, N. & Chen, P.-p., 1993. The *Rhagovelia* (Heteroptera: Veliidae) of Sulawesi (Indonesia).– Tijdschrift voor Entomologie 136: 259-281.
Nieser, N. & Chen, P.-p., 1996. Six new taxa of Nepomorpha from Sulawesi and Mindanao.– Tijdschrift voor Entomologie 139: 161-174.
Nieser, N. & Chen, P.-p., 1999. Sixteen new species of Nepomorpha (Heteroptera) mainly from Sulawesi.– Tijdschrift voor Entomologie 142: 77-123.
Nieser, N., & Chen, P.-p., 2002. Six new species of *Neotrephes* China, 1936 (Heteroptera: Helotrephidae) from Brazil, with a a key to Neotropical Helotrephidae.– Lundiana 3: 31-40.
Nieser, N., & Chen, P.-p., 2003 [2002]. Four new taxa of *Micronecta* (Heteroptera: Micronectidae) from the Philippines.– Annalen des Naturhistorischen Museums in Wien 104B: 131-142.
Nieser, N., Chen, P.-p., Leksawasdi, P., Thanyakam, A. & Duangsupa, C., 2004. Five new species of Nepomorpha (Heteroptera) from Southeast Asia.– Tijdschrift voor Entomologie 147: 29-40.
Nieser, N. & Lopez-Ruf, M., 2001. A review of *Limnocoris* Stål (Heteroptera: Naucoridae) in southern South America east of the Andes.– Tijdschrift voor Entomologie 144: 261-328.
Nieser, N., Pelli, A. & Lane de Melo, A., 1999. Two new Ambrysinae (Heteroptera: Naucoridae) from Minas Gerais, Brazil.– Acta Societatis Zoologicae Bohemicae 63: 157-163.
Nieser, N. & Zettel, H., 1999. The *Enithares* Spinola, 1873 (Insecta: Heteroptera: Notonectidae) of the Philippines, with descriptions of two new taxa.– Annalen des naturhistorisches Museums in Wien 101 B: 123-133.
Nieser, N. & Zettel, H., 2001. First record of *Paranisops* Hale, 1924 (Insecta: Heteroptera: Notonectidae) from South-east Asia, with description of *P. leucopardalos* sp.n.– Annalen des Naturhistorischen Museums in Wien 103 B:243-247.
Nieser, N., Zettel, H. & Chen, P.-p., 1997. Ten new species of *Rhagovelia* (Heteroptera: Veliidae) from Sulawesi (Indonesia).– Tijdschrift voor Entomologie 140: 17-41.
Nummelin, M., 1988a. The territorial behaviour of four Ugandan waterstrider species (Heteroptera, Gerridae): A comparative study.– Annales Entomologici Fennici 54: 121-134.
Nummelin, M., 1988b. Waterstriders (Het.: Gerridae) as predators of hatching mosquitoes.– Biocovas Proceedings 1: 121-125.
Oosterbroek, P., 1998. The families of Diptera of the Malay Archipelago.– Fauna Malesiana Handbook 1: i-xii, 1-227. Brill, Leiden, Boston, Köln.
Paiva, C. A., 1918. Aquatic Rhynchota from the Inlé Lake.– Records of the Indian Museum 14: 19-32, pl. 8.
Papáček, M., 1987. The ventrolateral thoracic region and thoracico-abdominal junction of *Plea minutissima* (Heteroptera, Pleidae).– Věstník Československé Společnosti Zoologické 51: 199-213.
Papáček, M., 1993. K problematice morfologie a bionomie vodnich plostic nadceledi Pleoidea a Notonectoidea (Heteroptera: Nepomorpha).– Thesis, University of South Bohemia, Česke Budejoviče: 1-232.
Papáček, M., 1994. *Idiotrephes maior* sp.n., a new species of water bug from Vietnam with morphological notes on *I. chinai* (Heteroptera: Helotrephidae).– European Journal of Entomology 91: 419-428.
Papáček, M., 1995. *Idiotrephes meszarosi* sp.n., a new helotrephid (Heteroptera: Helotrephidae) from Vietnam.– Aquatic Insects 17(2): 105-111.
Papáček, M., 2000. Small aquatic bugs (Nepomorpha) with slight or underestimate economic importance.– In: Schaefer, C. W. & Panizzi, A. R. (eds). Heteroptera of economic importance. CRC Press, Boca Raton: 591-600.

Papáček, M., 2001. Small aquatic and ripicolous bugs (Heteroptera: Nepomorpha) as predators and prey: The question of economic importance.– European Journal of Entomology 98: 1–12.

Papáček, M. & Kovac, D., 2001a. Three new species of the genera *Helotrephes* and *Hydrotrephes* (Heteroptera: Nepomorpha: Helotrephidae) from Thailand.– Linzer biologische Beiträge 33: 1047-1055.

Papáček, M. & Kovac, D., 2001b. Three new species of Limnotrephini (Heteroptera, Nepomorpha, Helotrephidae) from Peninsular Malaysia.– Linzer biologische Beiträge 33: 1047-1055.

Papáček, M. & Štys, P., 1985. Fossil Clypostemmatinae compared with modern Notonectidae (Heteroptera).– Acta Entomologica Bohemoslovaca 82: 28-48.

Papáček, M., Štys, P. & Tonner, M., 1988. A new subfamily of Helotrephidae (Heteroptera, Nepomorpha) from South-east Asia.– Acta Entomologica Bohemoslovaca 85: 120-152.

Papáček, M., Štys, P. & Tonner, M., 1989. A new genus and species of Helotrephidae from Afghanistan and Iran (Heteroptera, Nepomorpha).– Věstník Československé Společnosti Zoologické 53: 107-122.

Papáček, M., Štys, P. & Tonner, M., 1990. Ventrolateral thoracic region and thoracico-abdominal junction of some Helotrephidae (Heteroptera, Nepomorpha).– Acta Societatis Zoologicae Bohemoslovaca 54: 107-129.

Papáček, M. & Zettel, H., 2000. A revision of the Oriental genus *Idiotrephes* Lundblad, 1933 (Heteroptera, Nepomorpha: Helotrephidae).– European Journal of Entomology 97: 201-211.

Papáček, M. & Zettel, H., 2001. Notes on the Helotrephidae (Insecta: Heteroptera) of South India, with description of *Nanotrephes* gen.n. and five new species.– Annalen des Naturhistorichen Museums in Wien 103B: 283-300.

Papáček, M. & Zettel, H., 2003. On the species taxonomy of the limnotrephine genera *Limnotrephes*, *Mixotrephes*, and *Tiphotrephes* (Hemiptera: Helotrephidae).– Tijdschrift voor Entomologie 146: 219-234.

Papáček, M. & Zettel, H., in press. *Tiphotrephes immaculatus* sp.n. (Heteroptera: Helotrephidae: Limnotrephini), a new species from Viet Nam.- Linzer biologische Beiträge 37.

Parsons, M. C., 1959. Skeleton and musculature of the thorax of *Gelastocoris oculatus* (Fabricius) (Hemiptera - Heteroptera).– Bulletin of the Museum of Comparative Zoology at Harvard College, Cambridge 122: 299-357.

Parsons, M. C., 1964. The origin and development of the Hemipteran cranium.- Canadian Journal of Zoology 42: 409-432.

Parsons, M. C., 1967. Respiratory significance of the thoracic and abdominal morphology of three Corixidae, *Diaprepocoris*, *Micronecta* and *Hesperocorixa* (Hemiptera: Heteroptera: Corixidae).– Psyche 83: 132-179.

Parsons, M. C., 1969a. Skeletomusculature of the pterothorax and first abdominal segment in micropterous *Aphelocheirus aestivalis* F. (Het: Naucoridae).– Transactions of the Royal Entomological Society 121: 1-39.

Parsons, M. C., 1969b. The food pump of *Aphelocheirus aestivalis* as compared with that of typical Naucoridae (Heteroptera).– Journal of Morphology 129: 17-30.

Parsons, M.C., 1970. Respiratory significance of the thoracic and abdominal morphology of the three aquatic bugs *Ambrysus*, *Notonecta* and *Hesperocorixa* (Insecta, Heteroptera).– Zeitschrift für Morphologie der Tiere 66: 242-298.

Parsons, M. C., 1972a. Respiratory significance of the thoracic and abdominal morphology of *Belostoma* and *Ranatra* (Insecta, Het.).– Zeitschrift für Morphologie der Tiere 73: 163-194.

Parsons, M. C., 1972b. Morphology of the three anterior pairs of spiracles of *Belostoma* and *Ranatra* (Aquatic Het., Belostomatidae, Nepidae).– Canadian Journal of Zoology 50: 865-876.

Parsons, M. C., 1973. Morphology of the eight abdominal spiracles of *Belostoma* and *Ranatra* (Aquatic Heteroptera: Belostomatidae, Nepidae).– Journal of Natural History 7: 255-265.
Parsons, M. C., 1976. Respiratory significance of the thoracic and abdominal morphology of three Corixidae, *Diaprepocoris*, *Micronecta*, and *Hesperocorixa* (Hemiptera: Heteroptera: Hydrocorisae).– Psyche 83: 132-179.
Peckarsky, B. L., 1984. Predator-prey interactions among aquatic insects. In: Resh, V. H. & Rosenberg, D. M., (eds). The ecology of aquatic insects. Praeger, New York: 196-254.
Pemberton, R.W., 1988. The use of the Thai giant waterbug, *Lethocerus indicus* (Hemiptera: Belostomatidae), as human food in California.– Pan-Pacific Entomologist 64: 81-82.
Poisson, R., 1926. L'*Anisops producta* Fieb. (Hemiptère Notonectidae). Observations sur son anatomie et sa biologie.– Archives de Zoologie exptl. et génerale, Paris 65: 181-208.
Poisson, R. A., 1938. Les Hémiptères aquatiques Sandaliorrhyncha de la faune française. II. Micronectinae.– Annales de la Société Entomologique de France 107: 81-120.
Poisson, R., 1944. Contributions á la connaissance des espèces africaines du genre *Hebrus* Curtis, 1833 (Hem. Gymnocerata) (Missions de Ch. Alluaud et R. Jeannel en Afrique orientale).– Revue Francaise Entomologique 10: 89-112.
Poisson, R., 1952. Contribution á l'étude des Hydrocorises de Madagascar (missions J. Millot et R. Paulian 1949).– Mémoires de l'Institut Scientifique de Madagascar (E) 1: 23-70.
Poisson, R., 1956a. Quelques Hydrocorises nouveaux de l'Afrique du Sud (Mission Suédoise Brinck et Rudebeck).– Bulletin de la Société Scientifique de Bretagne 31: 49-59.
Poisson, R., 1956b. Contribution á l'étude des Hydrocorises de Madagascar (4e Mémoire).– Mémoires de l'Institut Scientifique de Madagascar (E) 7: 243-265.
Poisson, R. A., 1965. Catalogue des Hétéroptères Hydrocorises africano-malgaches de la famille des *Nepidae* (Latreille) 1802.– Bulletin de l'Institut Français d'Afrique Noire 27: 229-269.
Polhemus, D. A., 1986. A review of the genus *Coptocatus* Montandon (Hemiptera: Naucoridae).– The Pan-Pacific Entomologist 62: 248-256.
Polhemus, D. A., 1987. *Heleocoris brasiliensis* De Carlo is a *Ctenipocoris* (Hemiptera: Naucoridae).– The Pan-Pacific Entomologist 63: 370.
Polhemus, D.A., 1990. A revision of the genus *Metrocoris* Mayr (Heteroptera: Gerridae) in the Malay Archipelago and the Philippines.– Entomologica Scandinavica 21: 1-28.
Polhemus, D.A., 1994a. A new species of *Aphelocheirus* from Sumatra, and addenda to the world checklist (Heteroptera: Naucoridae).– Journal of the New York Entomological Society 102: 74-78.
Polhemus, D. A., 1994b. Taxonomy, Phylogeny, and Zoogeography of the Genus *Cylindrostethus* Fieber in the Palaeotropical Region (Heteroptera: Gerridae).– Bishop Museum Occasional Papers 38: 1-34.
Polhemus, D. A., 1995. Two new species of *Rhagovelia* from the Philippines, with a discussion of zoogeographical relationships between the Philippines and New Guinea (Heteroptera: Veliidae).– Journal of the New York Entomological Society 103: 55-68.
Polhemus, D. A., 1997. Systematics of the genus *Rhagovelia* Mayr (Heteroptera: Veliidae) in the Western Hemisphere (exclusive of the *angustipes* complex).– Thomas Say Publications in Entomology, Monograph: i-ii, 1-386.
Polhemus, D. A., 1998. Two new species of waterstriders (Heteroptera: Gerridae) from the Philippines.– Proceedings of the Entomological Society of Washington 100: 261-268.
Polhemus, D. A., 2000. Naucoridae (Heteroptera) of New Guinea. 5. A review of *Tanycricos* La Rivers in Irian Jaya, with description of two new species.– Journal of the New York Entomological Society 107(2-3): 171-180.

Polhemus, D.A., 2001. A review of the genus *Ptilomera* (Heteroptera: Gerridae) in Indochina, with descriptions of two new species.– Journal of the New York Entomological Society 109: 214-234.

Polhemus, D. A. & Polhemus, J.T., 1985. Naucoridae (Hemiptera) of New Guinea. I. A review of the genus *Nesocricos*, with descriptions of two new species.– International Journal of Entomology 27: 197-203.

Polhemus, D. A. & Polhemus, J.T., 1986a. Naucoridae (Hemiptera) of New Guinea. II. A review of the genus *Idiocarus* Montandon, with descriptions of three new species.– Journal of the New York Entomological Society 94: 39-50.

Polhemus, D.A. & Polhemus, J. T., 1986b. Naucoridae (Hemiptera) of New Guinea. III. A review of the genus *Tanycricos* La Rivers, with the description of a new species.– Journal of the New York Entomological Society 94: 163-173.

Polhemus, D. A. & Polhemus, J. T., 1987. A new genus of Naucoridae (Hemiptera) from the Philippines with comments on zoogeography.– The Pan-Pacific Entomologist 63: 265-269.

Polhemus, D. A. & Polhemus, J. T., 1988. The Aphelocheirinae of Tropical Asia (Heteroptera: Naucoridae).– The Raffles Bulletin of Zoology 36: 167-300.

Polhemus, D. A. & Polhemus, J. T., 1989. Naucoridae (Hemiptera) of New Guinea. IV. A revision of the genus *Cavocoris* with descriptions of four new species.– Journal of the New York Entomological Society 97: 73-86.

Polhemus, D. A. & Polhemus, J. T., 1994. A new species of *Aquarius* (Heteroptera: Gerridae) from Timor, with notes on Timorese zoogeography.– Proceedings of the Entomological Society of Washington 96: 54-62.

Polhemus, D.A. & Polhemus, J.T., 1997. A review of the genus *Limnometra* Mayr in New Guinea, with the description of a very large new species (Heteroptera: Gerridae).– Journal of the New York Entomological Society 105: 24-39.

Polhemus, D.A. & Polhemus, J.T., 2000a. Additional new genera and species of Microveliinae (Heteroptera: Veliidae) from New Guinea and adjacent regions.– Tijdschrift voor Entomologie 143: 91-123.

Polhemus, D.A. & Polhemus, J. T., 2000b. Naucoridae (Heteroptera) of New Guinea. 6. A revision of the genera *Sagocoris* and *Aptinocoris*, with descriptions of new species.– Journal of the New York Entomological Society (1999)107: 331-371.

Polhemus, D.A. & Polhemus, J. T., 2000c. New species of Microveliinae (Heteroptera: Veliidae) from the Raja Ampat Islands.– Tijdschrift voor Entomologie 143: 279-289.

Polhemus, D.A. & Polhemus, J. T., 2001. A revision of the genus *Ptilomera* (Heteroptera: Gerridae) on New Guinea and nearby islands.– Journal of the New York Entomological Society 109: 81-166.

Polhemus, D.A. & Polhemus, J. T., 2003. A review of the Veliinae of Vietnam (Heteroptera: Veliidae) with description of a new *Velia* species.– Journal of the New York Entomological Society 111: 29-40.

***Polhemus**, D.A. & Polhemus, J. T., 2004. Two new genera and thirty new species of Microveliinae (Heteroptera: Veliidae) from the East Papua Composite Terrane, far eastern New Guinea.– Tijdschrift voor Entomologie 147: 113-189.

Polhemus, J.T., 1970. A new genus of Veliidae from Mexico.– Proceedings of the Entomological Society of Washington 72: 443-448.

Polhemus, J. T., 1974. The *austrina* group of the genus *Microvelia*.– The Great Basin Naturalist 34: 207-217.

Polhemus, J.T., 1982. Marine Hemiptera of the Northern Territory, including the first fresh-water species of *Halobates* Eschscholtz (Gerridae, Veliidae, Hermatobatidae and Corixidae).– Journal of the Australian Entomological Society 21: 5-11.

Polhemus, J. T., 1984a. A review of the Naucorinae of Australia (Heteroptera: Naucoridae).– Journal of the Australian Entomological Society 23: 157-160.
Polhemus, J. T., 1984b. Aquatic and Semiaquatic Hemiptera.– In: Merritt, R. W. & Cummins, K. W. (eds). An introduction to the aquatic insects of North America. Kendall/Hunt Publishing Company, Dubuque, Iowa: 231-260.
Polhemus, J. T., 1990a. A new tribe, a new genus and three new species of Helotrephidae (Heteroptera) from South-east Asia, and a world checklist.– Acta Entomologica Bohemoslovaca 87: 45-63.
Polhemus, J. T., 1990b. Miscellaneous studies on the genus *Rhagovelia* Mayr (Heteroptera: Veliidae) in south-east Asia and in the Seychelles islands, with keys and descriptions of new species.– The Raffles Bulletin of Zoology 38: 65-75.
Polhemus, J. T., 1990c. Surface wave communication in waterstriders; field observations of unreported taxa (Heteroptera: Gerridae, Veliidae).– Journal of the New York Entomological Society 98: 383-384.
Polhemus, J. T., 1991. A new and primitive genus of Cryphocricinae (Heteroptera: Naucoridae). Pan-Pacific Entomologist 67: 119-123.
Polhemus, J.T., 1994a. The identity and synonymy of the Belostomatidae (Heteroptera) of Johann Christian Fabricius 1775-1807.– Proceedings of the Entomological Society of Washington 96: 687-695.
Polhemus, J. T., 1994b. Stridulatory mechanisms in aquatic and semiaquatic Heteroptera.– Journal of the New York Entomological Society 102: 270-274.
Polhemus, J. T., 1995. Nomenclatural and synonymical notes on the genera *Diplonychus* Laporte and *Appasus* Amyot and Serville (Heteroptera: Belostomatidae).– Proceedings of the Entomological Society of Washington 97: 649-653.
Polhemus, J. T., 1997. Seven new species of *Hydrotrephes* China (Helotrephidae: Heteroptera) from Sulawesi.– Tijdschrift voor Entomologie 140: 43-54.
Polhemus, J. T. & Andersen, N. M., 1984. A revision of *Amemboa* Esaki with notes on the phylogeny and ecological evolution of eotrechine waterstriders (Insecta, Hemiptera, Gerridae).– Steenstrupia 10: 65-111.
Polhemus, J. T. & Cheng, L., 1982. Notes on marine water-striders with descriptions of new species.– Pacific Insects 24: 219-227.
Polhemus, J. T. & Copeland, R. S., 1996. A new genus of Microveliinae from treeholes in Kenya (Heteroptera: Veliidae).– Tijdschrift voor Entomologie 139: 73-77.
Polhemus, J. T. & Hogue, C. L., 1972. Two new *Microvelia* from crabholes in Costa Rica (Hemiptera: Veliidae).– Contributions in Science 224: 1-6.
Polhemus, J. T. & Karunaratne, P. B., 1993. A review of the genus *Rhagadotarsus*, with descriptions of three new species (Heteroptera: Gerridae).– The Raffles Bulletin of Zoology 41: 95-112.
Polhemus, J. T. & Keffer, S. L., 1999. Notes on the genus *Laccotrephes* Stål (Heteroptera: Nepidae) in the Malay Archipelago, with the description of two new species.– Journal of the New York Entomological Society 107(1): 1-13.
Polhemus, J. T. & Kerzhner, I. M., 1995. *Nepa rustica* Fabricius, 1781 and *Zaitha stollii* Amyot & Servile, 1843 (currently *Diplonychus rusticus* and *Belostoma stollii*: Insecta, Heteroptera): proposed conservation of the specific names. Case 2941.– Bulletin of Zoological Nomenclature 52: 40-43.
Polhemus, J. T. & Lansbury, I., 1997. Revision of the genus *Hydrometra* Latreille in Australia, Melanesia and the Southwest Pacific (Heteroptera: Hydrometridae).– Bishop Museum Occasional Papers 47: 1-67.

Polhemus, J. T. & Lindskog, P., 1994. The stridulatory mechanism of *Nerthra* Say, a new species, and synonymy (Heteroptera: Gelastocoridae).– Journal of the New York Entomological Society 102: 242-248.
Polhemus, J. T. & Polhemus, D.A., 1984. Notes on neotropical Veliidae (Hemiptera) VI. Revision of the genus *Euvelia* Drake.– The Pan-Pacific Entomologist 60: 55-62.
Polhemus, J.T. & Polhemus, D. A., 1987. Terrestrial Hydrometridae (Heteroptera) from Madagascar, and the remarkable thoracic polymorphism of a closely related species from south-east Asia.– Journal of the New York Entomological Society 95: 509-517.
Polhemus, J.T. & Polhemus, D.A., 1988. Zoogeography, ecology and systematics of the genus *Rhagovelia* Mayr (Heteroptera: Veliidae) in Borneo, Celebes, and the Moluccas.– Insecta Mundi 2: 161-230.
Polhemus, J. T. & Polhemus, D. A., 1989. A new mesoveliid genus and two new species of *Hebrus* (Heteroptera: Mesoveliidae, Hebridae) from intertidal habitats in South-east Asian mangrove swamps.– The Raffles Bulletin of Zoology 37: 73-82.
Polhemus, J. T. & Polhemus, D. A., 1990. Zoogeography of the aquatic Heteroptera of Celebes: regional relationships versus insular endemism.– In: Knight, W. J. & Holloway, J. D. (eds). Insects and the rain forests of South east Asia (Wallacea). The Royal Entomological Society of London: 73-86.
Polhemus, J. T. & Polhemus, D. A., 1991. Three new species of marine waterstriders from the Australasian region, with notes on other species (Gerridae: Halobatinae, Trepobatinae).– The Raffles Bulletin of Zoology 39: 1-13.
Polhemus, J. T. & Polhemus, D. A., 1993. The Trepobatinae (Heteroptera: Gerridae) of New Guinea and surrounding regions, with a review of the world fauna. Part 1. Tribe Metrobatini.– Entomologica Scandinavica 24: 241-284.
Polhemus, J.T. & Polhemus, D. A., 1994a. Four new genera of Microveliinae (Heteroptera) from New Guinea.– Tijdschrift voor Entomologie 137: 57-74.
Polhemus, J. T. & Polhemus, D. A., 1994b. The Trepobatinae (Heteroptera: Gerridae) of New Guinea and surrounding regions, with a review of the world fauna. Part 2. Tribe Naboandelini.– Entomologica Scandinavica 25: 333-359.
Polhemus, J. T. & Polhemus, D.A., 1995a. Revision of the genus *Hydrometra* Latreille in Indochina and the Western Malay Archipelago (Heteroptera: Hydrometridae).– Bishop Museum Occasional Papers 43: 9-72.
Polhemus, J. T. & Polhemus, D. A., 1995b. The Trepobatinae (Heteroptera: Gerridae) of New Guinea and surrounding regions, with a review of the world fauna. Part 3. Tribe Trepobatini.– Entomologica Scandinavica 26: 97-118.
Polhemus, J. T. & Polhemus, D. A., 1996. The Trepobatinae (Heteroptera: Gerridae) of New Guinea and surrounding regions, with a review of the World fauna. Part 4. The marine tribe Stenobatini.– Entomologica Scandinavica 27: 279-346.
Polhemus, J. T. & Polhemus, D. A., 1999. Notes on Asian Veliinae (Heteroptera: Veliidae) with descriptions of three new species.– Journal of the New York Entomological Society 106: 117-131.
Polhemus, J. T. & Polhemus, D. A., 2000a. The Trepobatinae (Gerridae) of New Guinea and surrounding regions, with a review of the world fauna. Part 5. Taxonomic and distributional addenda.– Insect Systematics and Evolution 31: 291-316.
Polhemus, J. T. & Polhemus, D.A., 2000b. The genus *Mesovelia* Mulsant & Rey in New Guinea (Heteroptera: Mesoveliidae).– Journal of the New York Entomological Society 108: 205-230.
Polhemus, J. T. & Polhemus, D. A., 2002. The Trepobatinae (Gerridae) of New Guinea and surrounding regions, with a review of the world fauna. Part 6. Phylogeny, biogeography, world checklist, bibliography and final taxonomic addenda.– Insect Systematics and Evolution 33: 253-290.

Polhemus, J. T. & Polhemus, D. A., 2003. A new genus of Helotrephinae from insular Malaysia and Borneo (Heteroptera: Helotrephidae).– Tijdschrift voor Entomologie 146: 209-218.
Polhemus, J. T. & Reisen, W.K., 1976. Aquatic Hemiptera of the Philippines.– Kalikasan, Philippine Journal of Biology 5: 259-294.
Polhemus, J.T. & Rutter, R. P., 1997. *Synaptonecta issa* (Heteroptera, Corixidae), first new world record of an Asian water bug in Florida.– Entomological News 108: 300-304.
Polhemus, J. T. & Zettel, H., 1997. Five new *Potamometropsis* species (Insecta: Heteroptera: Gerridae) from Borneo.– Annalen des Naturhistorischen Museums in Wien 99B: 21-40.
Popham, E. J., 1960. On the respiration of aquatic Hemiptera Heteroptera with special reference to the Corixidae.– Proceedings of the Zoological Society of London 135: 209-242.
Popham, E. J., 1961. The function of Hagemann's organ in *Corixa punctata* (Illig.) (Hemiptera: Corixidae).– Proceedings of the Royal Entomological Society of London (A) 36: 119-125.
Popham, E. J., Bryant, M.T. & Savage, A. A., 1984a. The function of the abdominal strigil in male corixid bugs.– Journal of Natural History 18: 441-444.
Popham, E.J., Bryant, M.T. & Savage, A. A., 1984b. The role of front legs of British corixid bugs in feeding and mating.– Journal of Natural History 18: 445-464.
Popov, Y. A., 1964. Novoe podsemeystvo vodnykh klopov (Heteroptera) iz mezozoya Zabaykal'ya.– Paleontologicheskij Zhurnal 2: 63-71.
Popov, Y.A., 1971a. Origin and main evolutionary trends of Nepomorpha bugs.– Proceedings of the 13th International Congress of Entomology, Moscow. 1: 282-283.
Popov, Y. A., 1971b. Istoritsheskoe razvitie poluzhestkokrylych infraotrjada Nepomorpha (Heteroptera).– Trudy Paleontologichekogo Instituta 129: 1-230.
Popov, Y. A., 1986. Nasjekomye v rannjemjejelovykh ekosistjemakh zapadnoj Mongolii.– Sovmjestnaja sovjetsko-mongol'skaja paleontologicheskaja ekspjeditsija 28: 50-83.
Popov, Y. A. & Wootton, R. J., 1977. The Upper Liassic Heteroptera of Mecklenburg and Saxony.– Systematic Entomology 2: 333-351.
Porter, T. W., 1954. A new species of *Hebrus* Curtis from the Philippine Islands (Hemiptera, Hebridae).– Journal of the Kansas Entomological Society 27: 78-79.
Porter, T.W., 1959. Three new species of *Hebrus* Curtis from the Philippine Islands (Hemiptera: Hebridae).– Journal of the Kansas Entomological Society 32: 31-35.
Prager, J., 1973. Die Hörschwelle des mesothoracalen Tympanalorgan von *Corixa punctata* Ill. (Heteroptera, Corixidae).– Journal of Comparative Physiology 86: 55-58.
Prager, J., 1976. Das mesothorakale Tympanalorgan von *Corixa punctata* Ill. (Heteroptera, Corixidae).– Journal of Comparative Physiology 110: 33-50.
Prager, J. & Streng, R., 1981. Schallrezeption und ihre Anpassung in die aquatile Lebensweise bei Corixiden.– Mitteilungen der Deutschen Gesselschaft für Allgemeine und Angewandte Entomologie 3: 239-242.
Rabe, W., 1953. Beiträge zum Orientierungsproblem der Wasserwanzen.– Zeitschrift für Vergleichende Physiologie Berlin 35: 300-325.
Radinovsky, S., 1964. Cannibal of the pond.– Natural History 73: 16-25.
Rawat, B. L., 1939. Notes on the anatomy of *Naucoris cimicoides* L. (Hemiptera-Heteroptera).– Zoologische Jahrbücher Abteilung für Anatomie 65: 535-600.
Reichart, C. V., 1985. A new species of *Anisops* from Java (Hemiptera: Notonectidae).– International Journal of Entomology (Honolulu) 27: 235-238.
Rensing, L., 1962. Beiträge zur vergleichenden Morphologie, Physiologie und Ethologie der Wasserlaufer (Gerroidea).– Zoologische Beiträge (N. F.) 7: 447-485.
Rieger, C., 1973. Der basale Vorderflügelrostralrand der Hydrocorisae und seine Bedeutung für die Stellung der Corixidae.– Zoologische Anzeiger 191: 343-347.
Rieger, C., 1976. Skelett und Muskulatur des Kopfes und prothorax von *Ochterus marginatus* Latreille.– Zoomorphologie 83: 109-191.

Rieger, C., 1977. Neue Ochteriden aus der Alten Welt.– Deutsche Entomologische Zeitschrift (N. F.) 24: 213-217.
Ruiter, L. de, Wolvekamp, H. P., Tooren, A. J. van & Vlasblom, A., 1952. Experiments on the efficiency of the 'physical gill' (*Hydrous piceus* L., *Naucoris cimicoides* L., *Notonecta glauca* L.).– Acta Physiologica et Pharmacologica Neerlandica 2: 180-213.
Sailer, R.I., 1948. The genus *Trichocorixa* (Corixidae, Hemiptera).– In: Hungerford, H. B. (The Corixidae of the Western Hemisphere (Hemiptera). The University of Kansas Science Bulletin 32.): 289-407.
Schadow, W., 1922. Die Gattung *Halobates* Eschsch. und die Beziehung der Meeres- und Süsswasserhydrometriden zueinander.– Auszug Hamburg Universität Dissertationen 1922: 1-5.
Schmidt, E. T., 1915. Das Genus *Cylindrostethus* Fieber.– Stettiner Entomologische Zeitung 76: 361-364.
Schmidt, E. T., 1926. Neue *Ptilomera*-Arten aus dem Stettiner Museum.– Entomologische Mitteilungen 15: 63-66.
Schuh, R. T. 1975. The structure, distribution and taxonomic importance of trichobothria in the Miridae.– American Museum Novitates 2585: 1-26.
Schuh, R. T. & Slater, J. A., 1995. True Bugs of the World: i-xiv, 336 pp. Cornell University Press, New York.
Scott, J., 1874. On a collection of Hemiptera-Heteroptera from Japan. Description of various new genera and species.– Annals and Magazine of Natural History (4) 14: 426-452.
Scudder, G. G.E., 1959. The female genitalia of the Heteroptera: morphology and bearing on classification.– Transactions of the Royal Entomological Society of London 111: 405-467.
Sehnal, C., 1999. Two new species of *Pseudovelia* Hoberlandt, 1950 (Insecta: Heteroptera: Veliidae) from Palawan and Busuanga.– Annalen des Naturhistorischen Museums in Wien 101B: 147-154.
Selvanayagam, M. & Rao, T. K. R., 1986. Feeding behaviour in Gerridae (Hemiptera: Heteroptera).– Proceedings of the II Oriental Symposium , 21-24 February 1984, Association for Advancement of Entomology, Trivandrum, India: 155-164.
Sites, R. W., 2000. Creeping water bugs (Naucoridae).– In: Schaefer, C. W. & Panizzi, A. R. (eds). Heteroptera of economic importance. CRC Press, Boca Raton: 571-576.
Sites, R. W. & Nichols, B. J., 1999. Egg architecture of Naucoridae (Heteroptera): internal and external structure of the chorion and micropyle.– Proceedings of the Entomological Society of Washington 101: 1-25.
Sites, R. W., Nichols, B. J. & Permkam, S., 1997. The Naucoridae (Heteroptera) of southern Thailand.– The Pan-Pacific Entomologist 73: 127-134.
Sites, R. W. & Polhemus, J. T., 1994. Nepidae (Hemiptera) of the United States and Canada.– Annals of the Entomological Society of America 87: 27-42.
Sites, R. W. & Polhemus, J. T., 2000. A new species of *Telmatotrephes* (Heteroptera: Nepidae) from Thailand, with distributional notes on congeners.– Aquatic Insects 23: 333-340.
Smith, R. L., 1997. Evolution of parental care in the giant water bugs (Heteroptera: Belostomatidae). In: Choe, J. C. & Crespi, B. J. (eds). The evolution of social behaviour in insects and arachnids. Cambridge University Press, Cambridge, New York & Oakleigh: 116-149.
Soós, Á., 1963. Heteroptera VIII.– Fauna Hungariae 68 (XVII-8): 1-48.
Spence, J.R. & Andersen, N. M., 1994. Biology of Waterstriders: Interactions between Systematics and Ecology.– Annual Review of Entomology 39: 101-128.
Spence, J. R. & Andersen, N. M., 2000. Semiaquatic bugs (Gerromorpha).– In: Schaefer, C. W. & Panizzi, A. R. (eds). Heteroptera of economic importance. CRC Press, Boca Raton: 601-606.
Spinola, M., 1837a. Essai sur les genres d'insectes appartenants à l'ordre des Hémiptères, Lin. ou Rhyngotes, Fab. et à la section des Hétéroptères, Dufour: 383 pp. Graviers, Genes.

Sprague, I.B., 1956. The biology and morphology of *Hydrometra martini* Kirkaldy.– The University of Kansas Science Bulletin 38: 579-693.
Staddon, B.W., 1972. On the suggestion that the secretion from the metathoracic scent glands of a surface dwelling insect, *Gerris najas* (De Geer) (Heteroptera, Gerridae) has a waterproofing function.– Journal of Experimental Biology 57: 765-769.
Staddon, B.W., 1979. The scent glands of Heteroptera.– Advances in Insect Physiology 14: 351-419.
Stål, C., 1854. Nya Hemiptera.– Öfversigt af Kongliga Vetenskaps-Akademiens Förhandlingar 11: 231-255.
Stål, C., 1856. Hemiptera samlade af Victorin i Caplandet.– Öfversigt af Kongliga Vetenskaps-Akademiens Förhandlingar 13: 193-199.
Stål, C., 1858. Orthoptera och Hemiptera från södra-Afrika.– Öfversigt af Svenska Vetenskaps-Akademiens Förhandlingar 15: 307-320.
Stål, C., 1859. Hemiptera. Konigliga Svenska Fregaten Eugenies Resa omkring Jorden. Zoologi 1: 219-298.
Stål, C., 1861. Nova methodus familias quasdam Hemiptera disponendi.– Öfversigt af Svenska Vetenskaps-Akademiens Förhandlingar 18: 195-212.
Stål, C., 1866. Hemiptera Africana. 3 (1865): 1-200. Norstedtiana, Holmiae.
Stål, C., 1871. Hemiptera insularum Philippinarum.– Öfversigt af Svenska Vetenskaps-Akademiens Förhandlingar 27 (1870): 607-776.
Stål, C., 1876. Enumeratio Hemipterorum.– Kungliga Svenska Vetenskaps-Akademiens Handlingar 14: 1-162.
Štys, P., 1976. *Velohebria antennalis* gen.n., sp.n. – a primitive terrestrial Microveliine from New Guinea, and a revised classification of the family Veliidae (Heteroptera).– Acta Entomologica Bohemoslovaca 73: 388-403.
Štys, P. & Davidová-Vilimová, J., 1989. Unusual numbers of instars in Heteroptera: a review.– Acta Entomologica Bohemoslovaca 86: 1-32.
Štys, P. & Jansson, A., 1988. Checklist of recent family group and genus-group names of Nepomorpha (Heteroptera) of the world.– Acta Entomologica Fennica 50: 1-44.
Štys, P. & Kerzhner, I., 1975. The rank and nomenclature of higher taxa in recent Heteroptera.– Acta Entomologica Bohemoslovaca 72: 65-79.
Szabó-Patay, J., 1924. Sur la morphologie et la fonction de l'appareil respiratoire des *Aphelocheirus*.– Annales Musei Nationalis Hungarici 21: 33-55.
Takahashi, R., 1941. Observations on the biology of insects in Siam and Indo-China.– Botanica et Zoologica Tokyo 9: 795-802.
Takahashi, R.M., Steward, R.J., Schefer, C.H. & Sjogren, R.D., 1979. An assessment of *Plea striola* (Hemiptera: Pleidae) as a mosquito control agent in California.– Mosquito news 39: 514-519.
Thirumalai, G., 1992. A new record of *Pleciobates tuberculatus* Esaki from India with a key to the known species of *Pleciobates* Esaki (Gerridae: Heteroptera: Insecta).– Hexapoda 4: 170-174.
Thorpe, W.H., 1966. The habitat of *Aphelocheirus aestivalis* (F.) (Hem.-Het., Aphelocheiridae).– Entomologist's Monthly Magazine 102(1965): 251-253.
Thorpe, W.H. & Crisp, D.J., 1947a. Studies in plastron respiration I. The biology of *Aphelocheirus* [(Hemiptera), Aphelocheiridae (Naucoridae)].– Journal of Experimental Biology 24: 227-269.
Thorpe, W.H. & Crisp, D.J., 1947b. Studies in plastron respiration II. The respiratory efficiency of the plastron of *Aphelocheirus*.– Journal of Experimental Biology 24: 270-303.
Thorpe, W.H. & Crisp, D.J., 1947c. Studies in plastron respiration III. The orientation responses of *Aphelocheirus*.– Journal of Experimental Biology 24: 310-328.

Tindale, N. B., 1961. The chlorocresol method for field collecting.– Journal of the Lepidopterists' Society 15: 195-197.
***Tinerella**, P.P. & Polhemus, J. T., 2004. A new species of *Micronecta* Kirkaldy (Heteroptera: Corixidae: Micronectinae) from western New Guinea.– Journal of the New York Entomological Society 112: 183-189.
Todd, E. L., 1955. A taxonomic revision of the family Gelastocoridae.– The University of Kansas Science Bulletin 37: 277-475.
Todd, E. L., 1957. Five new species of Gelastocoridae with comments on other species (Hemiptera).– Proceedings of the Entomological Society of Washington 59: 145-162.
Todd, E. L., 1959. Results of the Archibold Expeditions. The Gelastocoridae of Melanesia (Hemiptera).– Nova Guinea (N.S.) 10: 61-94.
Torre Bueno, J.R. de la, 1905. The genus *Notonecta* in America North of Mexico.– Journal of the New York Entomological Society 13: 143-167.
Torre Bueno, J. R. de la, 1907. On *Rhagovelia obesa* Uhler.– Canadian Entomologist 39: 61-64.
Torre Bueno, J.R. de la, 1927. Fauna Sumatrana – aquatic Heteroptera.– Bulletin of the Brooklyn Entomological Society 12: 29-35.
Tran, A. D. & Yang, C. M., 2004. The aquatic bugs (Heteroptera) of the Anambas and Natuna Islands, Indonesia.– The Raffles Bulletin of Zoology Supplement 11: 55-59.
Travers, S. E., 1993. Group foraging facilitates food finding in a semi aquatic Hemipteran, *Microvelia austrina* Bueno (Hemiptera: Veliidae).– Pan-Pacific Entomologist 69: 117-121.
Usinger, R. L., 1937. Notes on the biology of *Hydrotrephes balnearius* (Helotrephidae, Hemiptera-Heteroptera).– Entomologist' Monthly Magazine 23: 179-180.
Usinger, R.L., 1938. The Naucoridae of the Philippine Islands (Hemiptera).– Philippine Journal of Science 64: 299-311, pl. 1.
Usinger, R. L., 1941. Key to the subfamilies of Naucoridae with a generic synopsis of the new subfamily Ambrysinae (Hemiptera).– Annals of the Entomological Society of America 34: 5-15.
Usinger, R. L., 1956. Hemiptera. In: Usinger, R. L. (ed.), Aquatic Insects of California. University of California Press, Berkeley & Los Angeles: 182-228.
Venkatesan, P., 1983. Male brooding behaviour of *Diplonychus indicus* Venk. & Rao (Hemiptera: Belostomatidae).– Journal of the Kansas Entomological Society 56: 80-87.
Venkatesan, P., 2000. Giant water bugs (Belostomatidae).– In: Schaefer, C. W. & Panizzi, A. R. (eds). Heteroptera of economic importance. CRC Press, Boca Raton: 577-582.
Venkatesan, P. & D'Silva, T., 1990. Influence of prey size on choice by the water bug, *Diplonychus indicus* Venk. & Rao (Hemiptera: Belostomatidae)– Journal of Entomological Research 14: 130-138.
Venkatesan, P. & Rao, T. K. R., 1980. Description of a new species and a key to Indian species of Belostomatidae.– Journal of the Bombay Natural History Society 77: 299-303.
Vepsäläinen, K., 1974. The life cycles and wing lengths of Finnish *Gerris* Fabr. species (Heteroptera, Gerridae).– Acta Zoologica Fennica 141: 1-73.
Villiers, A., 1948. Hémiptères Réduviidés de l'Afrique noire.– Faune de l'Empire Français 9: 1-488.
Vuillefroy, F. de, 1864. Hémiptères nouveaux.– Annales de la Societé Entomologique de France 33: 141-142.
Wachmann, E., 1989. Wanzen beobachten – Kennenlernen.– Neumann-Neudamm, Melsungen: 274 pp.
Walton, G.A., 1940. Classification of the Corixidae. – In: Hutchinson G. E. (ed). A revision of the Corixidae of India and adjacent regions. Transactions of the Connecticut Academy of Arts and Sciences 33: 343-345.
Weber, H., 1930. Biologie der Hemipteren. Julius Springer, Berlin: i-vii, 1-543 pp.

Weber, M., 1902. Der Indo-Australische Archipel und die Geschichte seiner Tierwelt. Fischer, Jena: I, 46 pp.

Wefelscheid, H., 1912. Über die Biologie und Anatomie von *Plea minutissima* Leach.– Zoologische Jahrbücher Abteilung für Systematik 32: 389-467, pls. 14, 15.

Wesenberg-Lund, C., 1943. Biologie der Süsswasserinsekten. Springer Verlag, Berlin, Wien: i-x, 682 pp.

Westwood, J. O., 1833. On the connecting links between the Geocorisae and Hydrocorisae of Latreille, or the land and water bug tribes.– Magazine of Natural History 6: 228-229.

Westwood, J. O., 1834. Mémoire sur les genres *Xylocoris*, *Hylophila*, *Microphysa*, *Leptopus*, *Velia*, *Microvelia* et *Hebrus*; avec quelques observations sur les Amphibicorisae de M. Dufour et sur l'état imparfait, mais identique de certaines espèces.– Annales de la Société Entomologique de France 3: 637-653, plate 6.

Westwood, J. O., 1871. On a species of *Corixa*, destructive to the ova of fishes in India.– Proceedings of the Entomological Society of London 1871: iii-iv.

White, F. B., 1833. Report on the pelagic Hemiptera procured during the voyage of H.M.S. Challenger in the years 1873-1876.– Voyage of H.M.S. Challenger, Scientific Reports, Zoology 7: 1-80.

White, F. B., 1879. List of the Hemiptera collected in he Amazons by Prof. J. W. Trail, M.S., M.D. in the years 1873-1875, with Descriptions of the New Species.– Transactions of the Entomological Society of London 4: 482-489.

Wichard, W., Arens, W. & Eisenbeis, G., 1995.– Atlas zur Biologie der Wasserinsekten. Gustav Fischer, Stuttgart, Jena, New York. 2.5 Ordnung: Hemiptera – Schnabelkerfe: i-xii, 1-338 pp.

Wilcox, R. S., 1972. Communication by surface wave: mating behaviour of a water-strider (Gerridae).– Journal of Comparative Physiology 80: 255-266.

Wilcox, R. S., 1975. Sound-producing mechanisms of *Buenoa macrotibialis* Hungerford (Hemiptera: Notonectidae).– International Journal of Insect Morphology and Embryology 4: 169-182.

Wilcox, R. S., 1980. Ripple communication.– Oceanus 23: 61-68.

Wilcox, R. S. & Di Stefano, J., 1991. Vibratory signals enhance mate-guarding in a water-strider (Hemiptera: Gerridae).– Journal of Insect Behaviour 4: 43-50.

Williamson, K. B., 1949. Naturalistic measures of Anopheline Control. In: M. F. Boyd (ed.), Malariology II. Saunders, Philadelphia: 1360-1384.

Wilson, C.A., 1958. Aquatic and semiaquatic Hemiptera of Mississippi.– Tulane Studies in Zoology 6: 115-170.

Wise, K. A. J., 1965. An annotated list of the aquatic and semiaquatic insects of New Zealand.– Pacific Insects 7: 1891-216.

Wotzel, F., 1933. Zur Kenntnis der Tympanalorgane der Hydrocores.– Zeitschrift für Wissenschaftliche Zoologie 143: 241-262.

Wróblewski, A., 1939. Pluskwiaki róznoskrzydle wodne (Hemiptera – Heteroptera aquatica) okolic Poznania.– Gragments Faunistica Musei Zoological Polonici 4: 107-141.

Wróblewski, A., 1952. Pluskwiaki róznoskrzydle (Hemiptera – Heteroptera) wód wyspy Wolina (Pomorze Zachodnia).– Prace Kimosji Biologicznej Poznan 13: 341-361.

Wróblewski, A., 1962a. Notes on Micronectinae from Viêt-Nam (*Heteroptera, Corixidae*).– Bulletin de l'Académie Polonaise des Sciences Cl.II (Sciences Biologiques) 10: 175-180.

Wróblewski, A., 1962b. Notes on Micronectinae from Melanesia (Heteroptera, Corixidae).– Bulletin de l'Académie Polonaise des Sciences Cl.II (Sciences Biologiques) 10: 319-324.

Wróblewski, A., 1967. Further notes on Micronectinae from Viêt-Nam (Heteroptera, Corixidae).– Polskie Pismo Entomologiczne 37: 229-251.

Wróblewski, A., 1968. Notes on Oriental Micronectinae (Heteroptera, Corixidae).– Polske Pismo Entomologiczne 38: 753-779.

Wróblewski, A., 1972a. Further notes on Micronectinae from Ceylon (Heteroptera, Corixidae).– Polske Pismo Entomologiczne 42: 1-52, pls. 1-3.
Wróblewski, A., 1972b. Supplementary notes on Australian Micronectinae (Heteroptera, Corixidae).– Polske Pismo Entomologiczne 42: 517-526.
Yang, C. M. & Polhemus, D. A., 1994. Notes on *Rhagovelia* Mayr (Hemiptera: Veliidae) from Singapore, with description of a new species.– The Raffles Bulletin of Zoology 42: 987-993.
Yang, C. M., Wong, T. C. M., Lua, H. K. & Koh, L M., 1999. A checklist of aquatic and semiaquatic bugs (Insecta: Hemiptera: Heteroptera) from Pulau Tioman, Peninsular Malaysia.– The Raffles Bulletin of Zoology, Supplement No. 6: 277-288.
Yano, K., Miyamoto, S. & Gabriel, B. P., 1981. Faunal and biological studies on the insects of paddy fields in Asia. IV. Aquatic and semiaquatic Heteroptera from the Philippines.– Esakia 16: 5-32.
Young, E.C., 1965. Flight muscle polymorphism in British Corixidae, ecological observations.– Journal of Animal Ecology 34: 353-390.
Zettel, H., 1993a. *Apheloneta gigas* n.sp. aus Sarawak (Borneo) mit Bemerkungen zur Stellung der Gattung in einem phylogenetischen System (Heteroptera: Notonectidae).– Linzer Biologische Beiträge 25: 661-667.
Zettel, H., 1993b. Zur Kenntnis der *Aphelocheirus*-Arten in Borneo (Sarawak) und Sulawesi (Heteroptera: Naucoridae).– Zeitschrift der Arbeitsgemeinschaft Österreichischer Entomologen 45: 81-86.
Zettel, H., 1993c. Eine neue *Rhagovelia* Mayr aus Westmalaysia (Heteroptera: Veliidae).– Entomological Problems 24: 39-42.
Zettel, H., 1994a. *Fischerotrephes* gen. n., eine orientalische Gattung der Familie Helotrephidae, welche eine neue Unterfamilie repräsentiert (Heteroptera: Helotrephidae).– Linzer Biologische Beiträge 26: 1069-1079.
Zettel, H., 1994b. Zwei neue *Rhagovelia*-Arten aus Mindoro, Philippinen (Heteroptera, Veliidae).– Entomofauna 15: 225-234.
Zettel, H., 1994c. Revision der philippinischen Arten der Gattung *Rhagovelia* Mayr, 1. Teil (Heteroptera: Veliidae).– Entomological Problems 25: 33-48.
Zettel, H., 1994d. Revision der Gattung *Potamometropsis* Lundblad.– Annalen des Naturhistorischen Museums in Wien 96B: 75-98.
Zettel, H., 1995a. Neubeschreibungen von zwei Arten der Gattung *Aphelonecta* Lansbury aus Indonesien und den Philippinen sowie ein Bestimmungsschlüssel zu den Arten (Heteroptera: Notonectidae).– Linzer Biologische Beiträge 27: 107-113.
Zettel, H., 1995b. Zwei neue Arten der Gattung *Helotrephes* Stål aus China (Heteroptera: Helotrephidae).– Tijdschrift voor Entomologie 138: 291-295.
Zettel, H., 1995c. Neue Arten der Gattung *Distotrephes* Polhemus, 1990, aus China und Borneo sowie faunistische Notizen zu anderen Limnotrephini.– Annalen des Naturhistorischen Museums in Wien 97B: 159-168.
Zettel, H., 1995d. Revision der philippinischen Arten der Gattung *Rhagovelia* (Heteroptera: Veliidae) 2. Teil.– Entomological Problems 26: 43-78.
Zettel, H., 1995e. Zwei neue Arten der Gattung *Tetraripis* Lundblad aus Borneo (Heteroptera: Veliidae).– Zeitschrift der Arbeitsgemeinschaft Österreichischer Entomologen 47: 25-30.
Zettel, H., 1995f. Zwei neue *Amemboa* Esaki, 1925, aus Borneo und Vietnam (Heteroptera: Gerridae).– Annales Historico Naturales Musei Nationalis Hungarici 87: 63-66.
Zettel, H., 1996a: Beitrag zur Kenntnis der Gattung *Trephotomas* mit Beschreibung von zwei neuen Arten aus China und Laos (Heteroptera: Helotrephidae).– Linzer Biologische Beiträge 28(2): 1061-1071.
Zettel, H., 1996b. *Chenevelia stridulans* gen.n. et sp.n. from North Thailand.– Annalen des Naturhistorischen Museums in Wien 98B: 353-359.

Zettel, H., 1996c. Revision der philippinischen Arten der Gattung *Rhagovelia* 3. Teil (Heteroptera: Veliidae).– Entomological Problems 27: 111-140.
Zettel, H., 1997a: One new genus and two new species of Helotrephidae (Insecta: Heteroptera) from India, with notes on the phylogeny of the family.– Annalen des Naturhistorischen Museums in Wien 99B: 83-95.
Zettel, H., 1997b: *Limnotrephes minutissimus* sp.n. (Heteroptera: Helotrephidae) aus Indien.– Linzer Biologische Beiträge 29(2): 683-687.
Zettel, H., 1997c. Notes on Helotrephidae (Insecta: Heteroptera) from Borneo, with descriptions of two new species of the genera *Fischerotrephes* and *Trephotomas*.– Entomological Problems 28: 117-126.
Zettel, H., 1998a. Neue Taxa der Gattung *Aphelocheirus* Westwood, 1833 (Insecta: Heteroptera: Aphelocheiridae) aus der orientalische Region sowie Bemerkungen zu einigen beschriebenen Arten und zu den Raubbeinen der Naucoroidea.– Annalen des Naturhistorischen Museums in Wien 100 B: 77-97.
Zettel, H., 1998b. Four new species of *Hydrotrephes* (Heteroptera: Helotrephidae) from Thailand and Laos.– Entomological Problems 29: 129-137.
Zettel, H., 1998c. Introduction to the Helotrephidae (Nepomorpha) in Thailand and adjacent countries.– Amemboa 2: 15-18.
Zettel, H., 1998d. Revision der Gattung *Hyrcanus* Distant, 1910 (Heteroptera: Hebridae) mit Beschreibung neuer Arten aus Indien, Thailand, Laos und China.– Stapfia 55: 585-606.
Zettel, H., 1998e. A remarkable new *Timasius* Distant 1909 (Heteroptera: Hebridae) from Taiwan.– Linzer Biologische Beiträge 30: 601-604.
Zettel, H., 1998f. Notes on marine Haloveliinae from the Philippine Islands, with descriptions of three new species (Heteroptera, Veliidae).– Entomofauna 19: 461-472.
Zettel, H., 1998g. On the Oriental genus *Neoalardus* Distant 1912 (Heteroptera: Veliidae).– Linzer Biologische Beiträge 30: 595-599.
Zettel, H., 1999a. *Aphelocheirus kodadai* sp.n. (Heteroptera: Aphelocheiridae) aus Borneo.– Entomological Problems 30: 57-59.
Zettel, H., 1999b. Two new Philippine *Aphelocheirus* Westwood, 1833 (Heteroptera: Aphelocheiridae) and a key to the Philippine species.– Annalen des Naturhistorischen Museums in Wien 101B: 107-121.
Zettel, H., 1999c. Addition to the key to the genera of Helotrephidae (Nepomorpha) by Zettel (1999, Amemboa 2).– Amemboa 3: 13.
Zettel, H., 1999d. Notes on the genus *Distotrephes* (Heteroptera: Helotrephidae): new species from Thailand, Laos, and China, a key to species, and first description of the macropterous morph.– Acta Societatis Zoologicae Bohemicae 63: 251-266.
Zettel, H., 1999e. *Nieserius* gen.n., a new genus of the subfamily Hyrcaninae (Heteroptera: Hebridae) from Thailand, Laos, and Nepal, with the first known subaquatic species of Gerromorpha.– Aquatic Insects 21: 39-52.
Zettel, H., 1999f. The genus *Merragata* Buchanan-White, 1877 (Gerromorpha: Hebridae) in Thailand, with notes on the Oriental taxa.– Amemboa 3: 6-8.
Zettel, H., 1999g. Notes on Philippine *Potamometropsis* Lundblad, 1933 (Insecta: Heteroptera: Gerridae). – Annalen des Naturhistorischen Museums in Wien 101B: 155-161.
Zettel, H., 1999h. An illustrated key to the genera of Hebridae (Gerromorpha) in Thailand, with notes on the Oriental taxa.– Amemboa 3: 22-26.
Zettel, H., 1999i. *Rhagovelia fischeri* sp.n. (Heteroptera: Veliidae), erster Nachweis der *Rhagovelia borneensis* Gruppe für die Philippinen.– Linzer biologische Beiträge 31: 747-753.
Zettel, H., 2000a. The Helotrephidae (Heteroptera) of Borneo.– Entomological Problems 31: 1-22.

Zettel, H., 2000b. Neue und wenig bekannte Hebridae (Insecta: Heteroptera) aus Süd-Indien mit Neubeschreibungen von sechs Arten aus den Gattungen *Hyrcanus* Distant, 1910, *Timasius* Distant, 1909 und *Hebrus* Curtis, 1833.– Annalen des Naturhistorischen Museums in Wien 102B: 97-110.
Zettel, H., 2000c. Erster Nachtrag zur Bearbeitung der Helotrephini Thailands (Heteroptera: Helotrephidae).– Linzer biologische Beiträge 32: 145-148.
Zettel, H., 2001a. Five new species of *Perritopus* Fieber, 1861 (Hemiptera: Veliidae) from South-east Asia.– The Raffles Bulletin of Zoology 49: 1-11.
Zettel, H., 2001b. *Halobates dianae* sp.n. (Heteroptera: Gerridae), a new sea skater from the Philippines.– Linzer biologische Beiträge 33: 1097-1102.
Zettel, H., 2001c. First notes on the Helotrephidae (Heteroptera) of Kalimantan Barat, Indonesia: descriptions of three new species of *Hydrotrephes* China, 1935, and first records of *Tiphotrephes* Esaki & China, 1928, from Borneo.– Entomological Problems 32: 59-64.
Zettel, H., 2001d. *Helotrephes papaceki* sp.n. und Erstnachweise vier weiterer *Helotrephes* Arten aus Yunnan, China (Heteroptera: Helotrephidae).– Linzer biologische Beiträge 33: 97-102.
Zettel, H., 2001e. *Perittopus ceylanicus* sp.n., first record of the subfamily Perittopinae (Insecta: Heteroptera: Veliidae) from Sri Lanka.– Annalen des Naturhistorischen Museums in Wien 103B: 269-271.
Zettel, H., 2001f: Die Indische Zwergschwimmwanze *Nanonaucoris* gen.n., eine neue Gattung der Naucorini (Heteroptera: Naucoridae) aus Südindien.– Linzer biologische Beiträge 33: 1085-1095.
Zettel, H., 2002a. Beschreibung von zwei neuen Arten der Gattung *Coptocatus* Montandon (Heteroptera: Naucoridae: Cheirochelinae) aus Sarawak (Borneo), mit einem Bestimmungsschlüssel zu den Artren der Gattung.– Entomological Problems 32: 171-178.
Zettel, H., 2002b. *Limnometra spinosa*, new species (Heteroptera: Gerridae), a rare waterstrider from Indonesia and Malaysia.– The Raffles Bulletin of Zoology 50: 137-141.
Zettel, H., 2002c. Description of *Amemboa heissi* sp. n. (Heteroptera: Gerridae) from South Thailand.– Linzer Biologische Beiträge 34: 1531-1534.
Zettel, H., 2003a [2002]. The Helotrephidae (Insecta: Heteroptera) of the Philippine Islands.– Annalen des Naturhistorischen Museums in Wien 104B: 45-97.
Zettel, H., 2003b [2002]. Additional notes on the Aphelocheiridae, Naucoridae, and Notonectidae (Insecta: Heteroptera: Nepomorpha) of the Philippine Islands.– Annalen des Naturhistorischen Museums in Wien 104B: 109-130.
Zettel, H., 2003c [2002]. Notes on the marine Haloveliinae (Insecta: Heteroptera: Veliidae) of the Philippines, with descriptions of four new species, identification key, and classification of all species of *Haloveloides* Andersen, 1992.– Annalen des Naturhistorischen Museums in Wien 104B: 163-181.
Zettel, H., 2003d [2002]. New species, subspecies, and records of *Strongylovelia* Esaki, 1924 (Insecta: Heteroptera: Veliidae) from the Philippines.– Annalen des Naturhistorischen Museums in Wien 104B: 183-193.
Zettel, H., 2003e. Eine neue *Rhagovelia* (Heteroptera: Veliidae) mit 'phoretischen Männchen' von der Insel Samar, Philippinen.– Linzer biologische Beiträge 35: 1141-1146.
Zettel, H., 2003f. *Strongylovelia seyferti* sp.n. (Heteroptera: Veliidae: Haloveliinae) from Waleakodi, Kepulauan Togian, first record of the genus for the Sulawesi area.– Linzer biologische Beiträge 35: 1135-1140.
Zettel, H., 2004a [2003]. Neue Wasserläufer (Insecta: Heteroptera: Gerridae) von den Philippinen.– Annalen des Naturhistorischen Museums in Wien 105B: 361-387.
Zettel, H., 2004b. *Hebrus murphyi*, new species (Heteroptera: Hebridae) from an intertidal mangrove habitat in Burias Island, Philippines.– The Raffles Bulletin of Zoology 52: 75-77.

Zettel, H., 2004c. Neue Arten der Gattung *Hebrus* Curtis 1833 (Heteroptera: Hebridae) aus Südostasien.– Linzer biologische Beiträge 36: 533-558.

Zettel, H., 2004d. Revision of the *Timasius chinai* species group (Insecta: Heteroptera: Hebridae), with descriptions of ten new species.– Insect Systematics & Evolution 35: 241-260.

Zettel, H., 2004e. *Phrynovelia philippinensis* sp.n. (Heteroptera: Mesoveliidae) von der Insel Polillo, Philippinen.– Linzer biologische Beiträge 36: 1353-1358.

Zettel, H., 2004f. Weitere neue Helotrephini (Heteroptera: Helotrephidae) aus China, Indonesien und Malaysien sowie von den Philippinen.– Linzer biologische Beiträge 36: 1359-1382.

Zettel, H., 2004g [2003]. *Baptista hoedli* sp.n. (Insecta: Heteroptera: Veliidae) from Laos, and definition of the *B. collaris* species group.– Annalen des Naturhistorischen Museums in Wien 105B: 441-445.

Zettel, H., 2004h [2003]. Helotrephidae (Insecta: Heteroptera) aus den chinesischen Provinzen Guangdong, Yünnan und Guizhou, mit Beschreibungen neuer Arten der Gattungen *Helotrephes* und *Distotrephes*.– Annalen des Naturhistorischen Museums in Wien 105B: 397-410.

Zettel, H., 2005a [2004]. Notes on the Helotrephini (Insecta: Heteroptera: Helotrephidae) from Thailand and Vietnam, with descriptions of three new species.– Annalen des Naturhistorischen Museums in Wien 106B: 67-80.

Zettel, H., 2005b [2004]. Neue Gattungen, Arten und Nachweise aus der Unterfamilie Trephotomasinae (Insecta: Heteroptera: Helotrephidae). – Annalen des Naturhistorischen Museums in Wien 106B: 81-92.

Zettel, H. & Chen, P.-p., 1996. Beitrag zur Taxonomie und Faunistik der Gerridae Vietnams mit Neubeschreibungen der Gattung *Andersenius* gen. nov. aus der Unterfamilie Ptilomerinae und weitere Arten (Insecta: Heteroptera: Gerridae).– Entomologische Abhandlungen 57: 149-182.

Zettel, H. & Chen, P.-p., 2000a. *Limnometra palawanensis* spec. nov. (Heteroptera: Gerridae), and a synopsis of the Philippine species of *Limnometra*.– Entomologische Berichten, Amsterdam 60: 73-83.

Zettel, H. & Chen, P.-p., 2000b. The genus *Timasius* Distant, 1909 (Heteroptera: Hebridae) in Borneo, with description of two new species.– Linzer Biologische Beiträge 32: 1195-1201.

Zettel, H. & Gapud, V. P., 1999. A new species group of Oriental *Microvelia* s.l. (Insecta: Heteroptera: Veliidae), with descriptions of three new species.– Annalen des Naturhistorischen Museums in Wien 101B: 135-146.

Zettel, H. & Hecher, C., 1998. Notes on the Veliidae (Heteroptera) of Borneo: 1. Three new species of *Angilia* Stål, 1865, with a key to the Oriental species.– The Raffles Bulletin of Zoology 46: 335-344.

Zettel, H., Nieser, N. & Polhemus, D. A., 1999. The Naucoridae (Insecta: Heteroptera) of the Philippine Islands.– Annalen des Naturhistorischen Museums in Wien 101B: 43-105.

Zettel, H. & Papáček, M., 2004 [2003]: A review of the genus *Esakiella* China, 1932 (Insecta: Heteroptera: Helotrephidae) in Madagascar, with descriptions of five new species.- Annalen des Naturhistorischen Museums in Wien 105B: 327-348.

Zettel, H. & Polhemus, J. T., 1998. A revision of the genus *Helotrephes* Stål, 1860 (Insecta: Heteroptera: Helotrephidae) with descriptions of twelve new taxa from the Oriental Realm.– Annalen des Naturhistorischen Museums in Wien 100B: 99-136.

Zettel, H. & Ruiz, R.B., 2003 [2002]. *Aquarius philippinensis* sp.n., a large endemic waterstrider (Insecta: Heteroptera: Gerridae) from ancient crater lakes in South Luzon, Philippines.– Annalen des Naturhistorischen Museums in Wien 104B: 195-202.

Zettel, H. & Sehnal, P. 1995: Description of male genitalia of the Australian species *Hebrus nourlangiei* Lansbury, 1990 (Heteroptera: Hebridae).– Linzer Biologische Beiträge 27: 1085-1087.

Zettel, H. & Sehnal, C., 2000 [1999]. Notes on the Veliidae (Heteroptera) of Borneo: 2. *Polhemovelia* n. gen. with description of three new species.– Journal of the New York Entomological Society 107: 124-134.

Zettel, H. & Thirumalai, G., 2001a. *Stridulobates anderseni*, a new genus and species of ptilomerine Gerridae (Hemiptera): Heteroptera) with 'stridulatory devices' from South India.– Insect Systematics and Evolution 31: 432-438.

Zettel, H. & Thirumalai, G., 2001b. Re-establishment of the ptilomerine genus *Jucundus* Distant, 1910 (Insecta: Heteroptera: Gerridae), with redescription of the type species *Jucundus custodiendus* Distant, 1910 from South India and notes on *J. vittatus* (Esaki, 1928) comb.n. from Sri Lanka.– Annalen des Naturhistorischen Museums in Wien 103 B: 273-282.

Zettel, H. & Yang, C.M., 2002. To the knowledge on Southeast Asian species of the genera *Tetraripis* and *Angilia* (Heteroptera: Veliidae).– Linzer biologische Beiträge 34: 1535-1544.

***Zettel**, H. & Yang, C. M. 2004. New Oriental species and subspecies of *Hydrometra* Latreille (Heteroptera: Hydrometridae).- The Raffles Bulletin of Zoology 52: 389-398.

Zimmermann, G., 1984. Heteroptera aus dem Nepal-Himalaya. *Geovelia* n. gen., eine Gattung terrestrischer Microveliinae.– Senckenbergiana Biologica 65: 65-74.

Maps

FIGURE 562

Map of areas with abbreviations as used in the checklist. Abbreviations for geographical areas: **BO**: Borneo (Kalimantan, Sabah, Sarawak and Brunei); **BS**: Bismarck and Solomon Archipelagos; **JA**: Java (including Madura and Bali); **LU**: Philippines, Luzon; **MA**: southern part of Thailand, peninsular Malaysia and Singapore; **MI**: Philippines, Mindanao (including Bohol, Leyte, Samar & Sulu Islands); **ML**: Moluccas; **NG**: New Guinea (including adjacent islands, notably the Raja Empat Islands); **NT**: Nusa Tenggara (smaller Sunda Islands); **PA**: Philippines, Palawan; **PH**: remaining Philippine Islands; **SL**: Sulawesi; **SM**: Sumatra. Extralimital areas: **AU**: Australia; **HT**: China: Hainan, Hong Kong, Macao and Taiwan; **NC**: New Caledonia; **SA**: South-east Asia (Myanmar, northern part of Thailand, Laos, Vietnam and Cambodia).

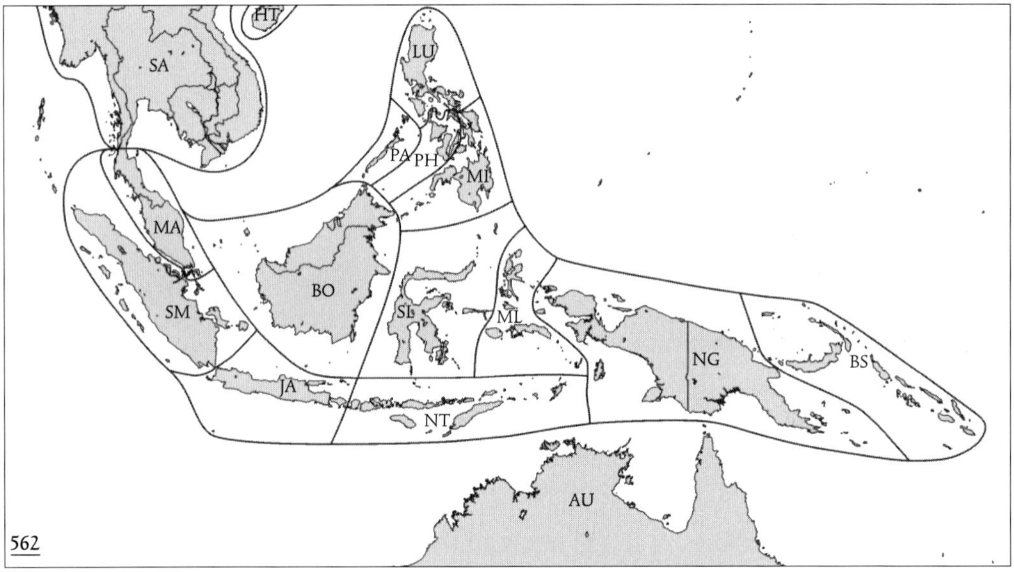

FIGURE 563

Map illustrating Sundaland. The dry area on the continental shelfs during glacial periods is hatched. Expansion of dry areas around isolated islands as a rule is too small to be shown. Especially the eastern Philippines were apparently a single dry area.

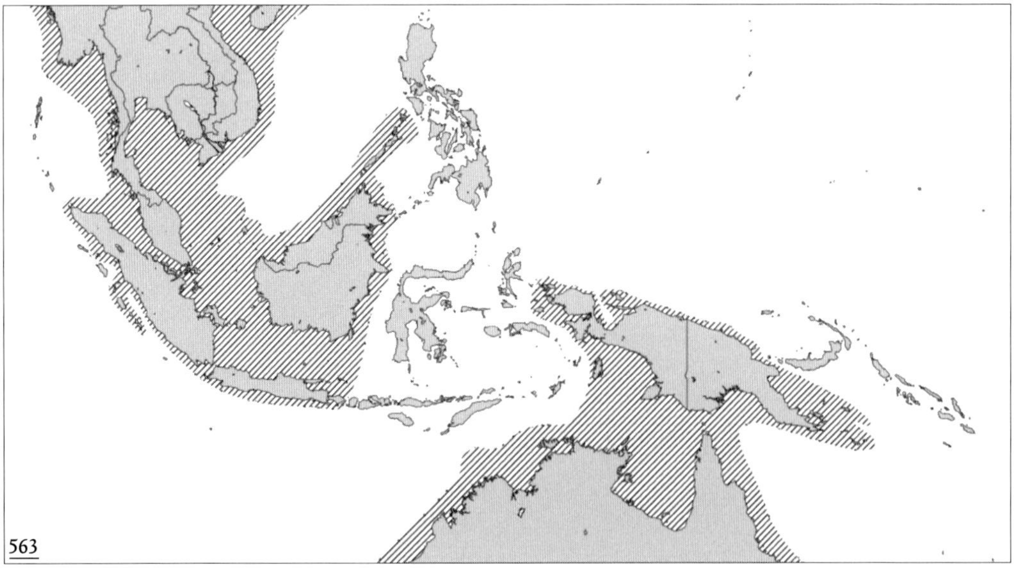

563

FIGURE 564

Range of various genera of aquatic and semiaquatic Heteroptera in Malesia, illustrating the paths of expansion of genera originating in continental Asia (partly based on J. Polhemus & D. Polhemus, 1990): 1 = *Gestroiella*, 2 = *Hyrcanus*, 3 = *Amemboa*, 4 = *Cylindrostethus*, 5 = *Metrocoris*, 6 = *Pseudovelia*.

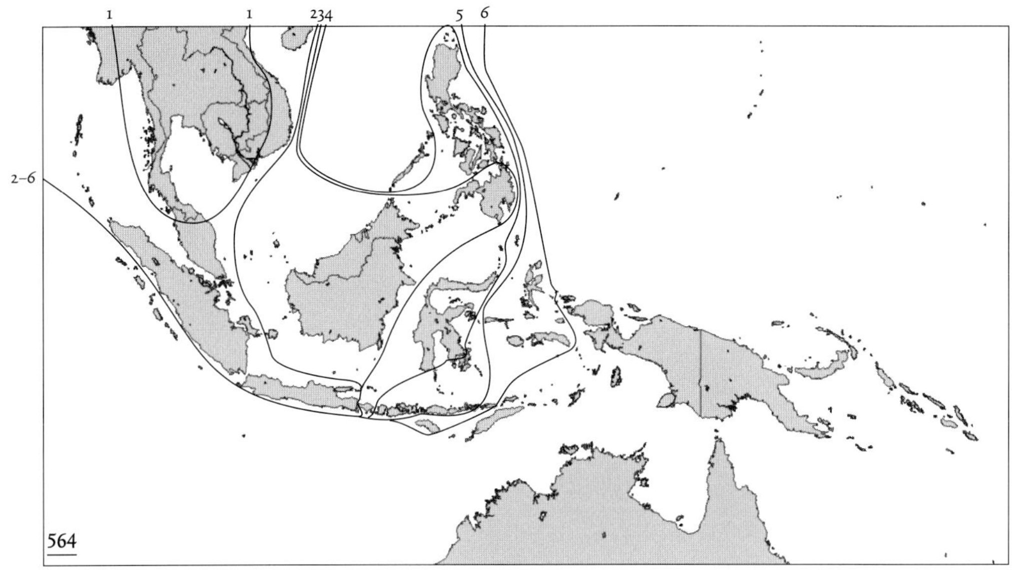

FIGURE 565

Distribution of Nepomorpha. Solid line indicates eastern margin of distribution of *Diplonychus*, *Lethocerus*, *Laccotrephes*, *Cercotmetus*, *Ranatra* and *Aphelocheirus* subgenus *Aphelocheirus*; horizontal hatching = eastern part of the distribution of *Heleocoris*; vertical hatching = *Stalocoris*; oblique hatching + diamond = *Aptinocoris*; triangle = *Borborophyes*; circle = *Telmatotrephes*.

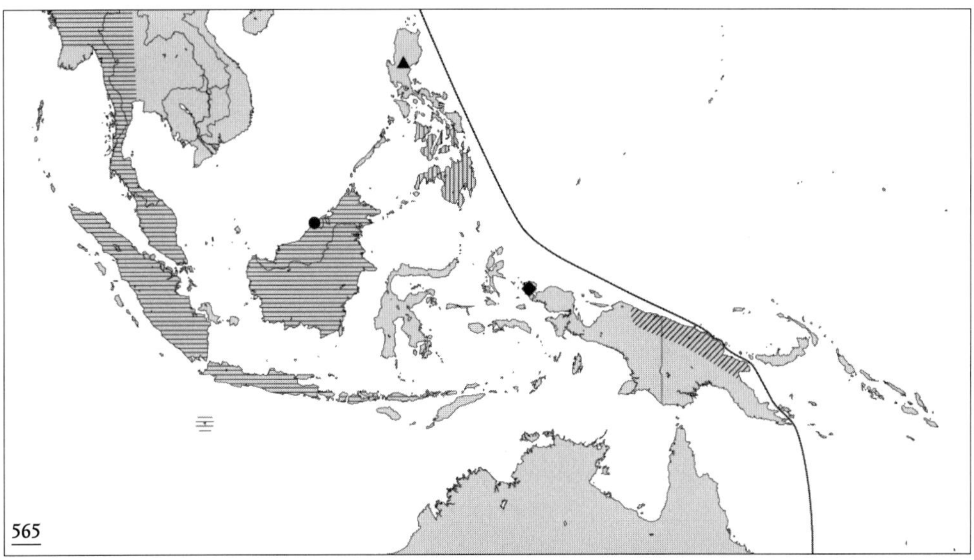

FIGURE 566
Distribution of Aphelocheiridae and Naucoridae. Solid line = eastern margin of *Naucoris*; horizontal hatching = eastern part of distribution of *Gestroiella*; vertical hatching = south-eastern part of distribution of *Aphelocheirus* subgenus *Micraphelocheirus*; oblique hatching to the left = *Asthenocoris*; oblique hatching to the right = *Coptocatus*; horizontal broken hatching = *Sagocoris*; triangle = *Aphelocheirus* (*Micraphelocheirus*) *kelabitensis*.

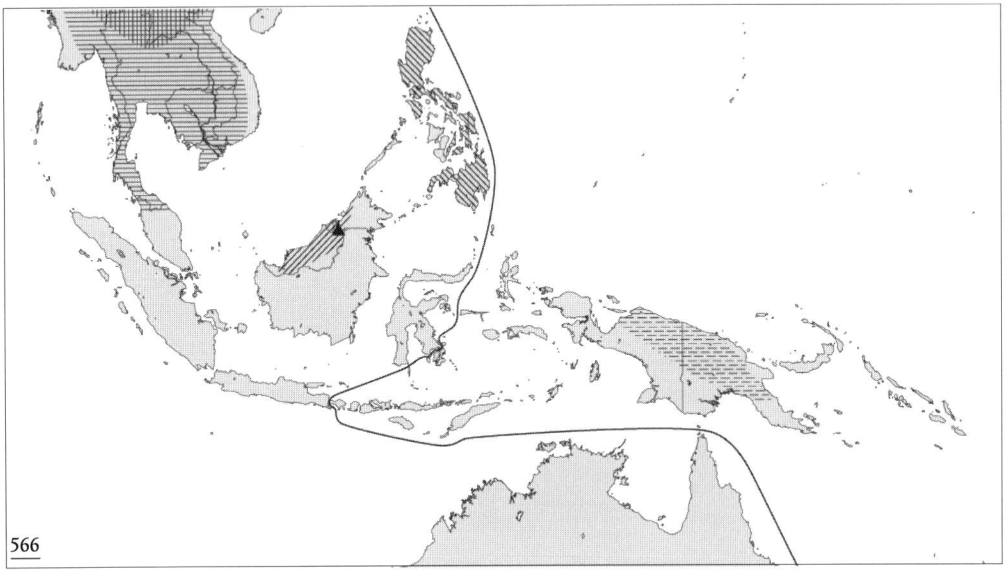

FIGURE 567
Distribution of Naucoridae. Horizontal hatching = *Laccocoris*; vertical hatching = *Philippinocoris*; black = *Cavocoris*.

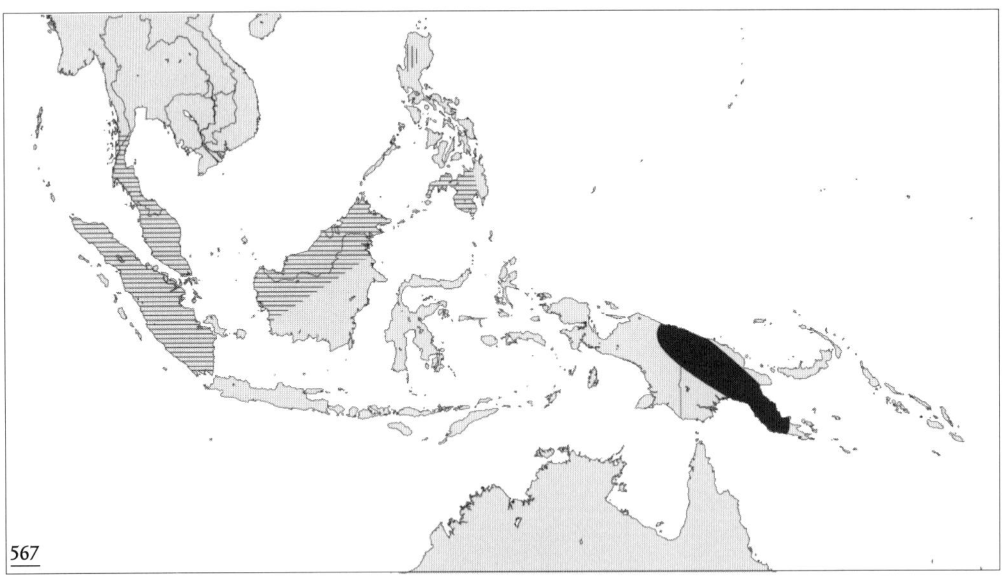

FIGURE 568
Distribution of Naucoridae. Horizontal hatching = *Ctenipocoris*; vertical hatching = *Warisia*.

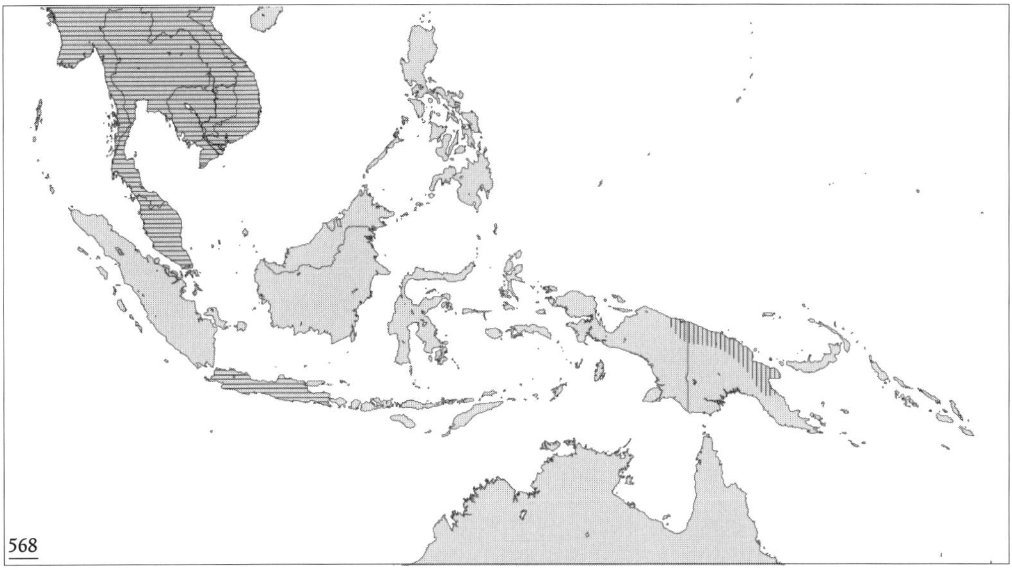
568

FIGURE 569
Distribution of Naucoridae, *Idiocarus*.

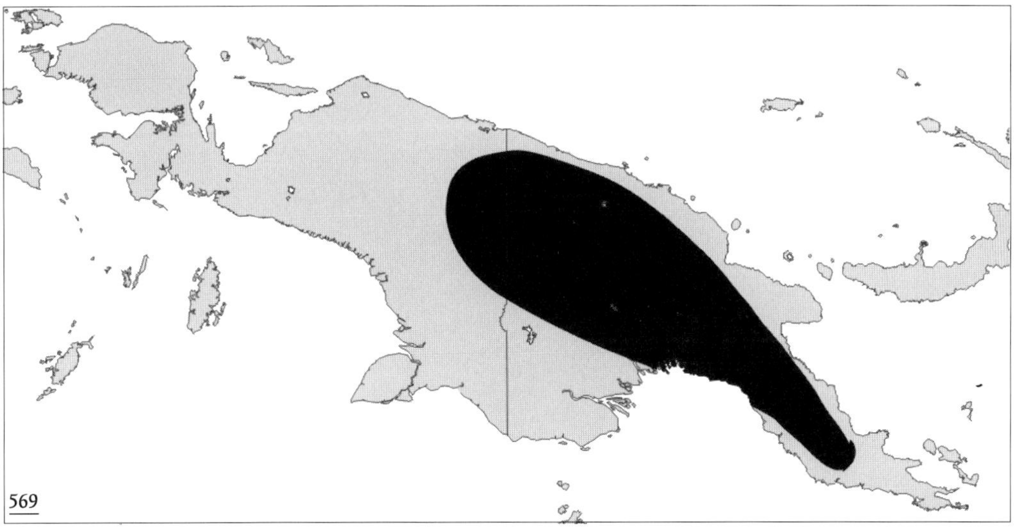
569

FIGURE 570
Distribution of Naucoridae, *Nesocricos*.

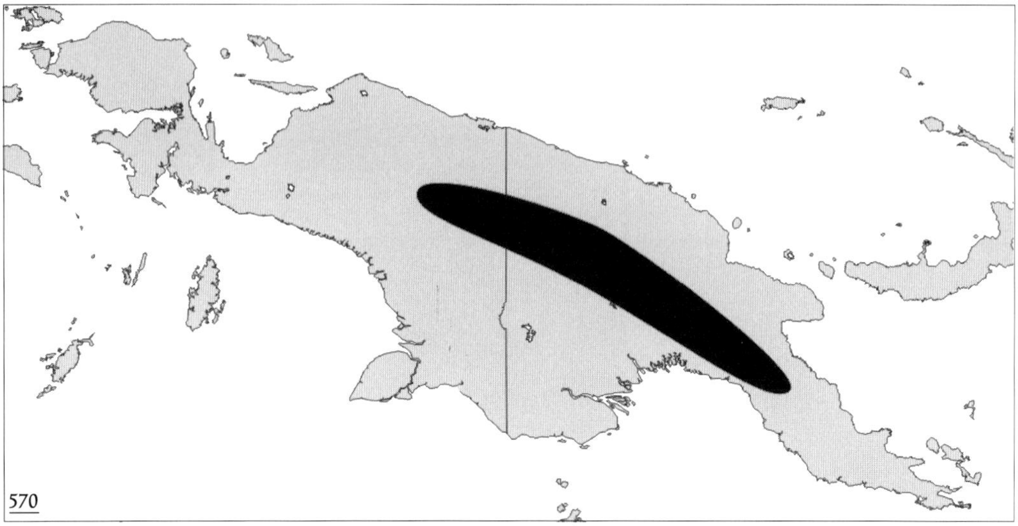

FIGURES 571
Distribution of Naucoridae, *Tanycricos*.

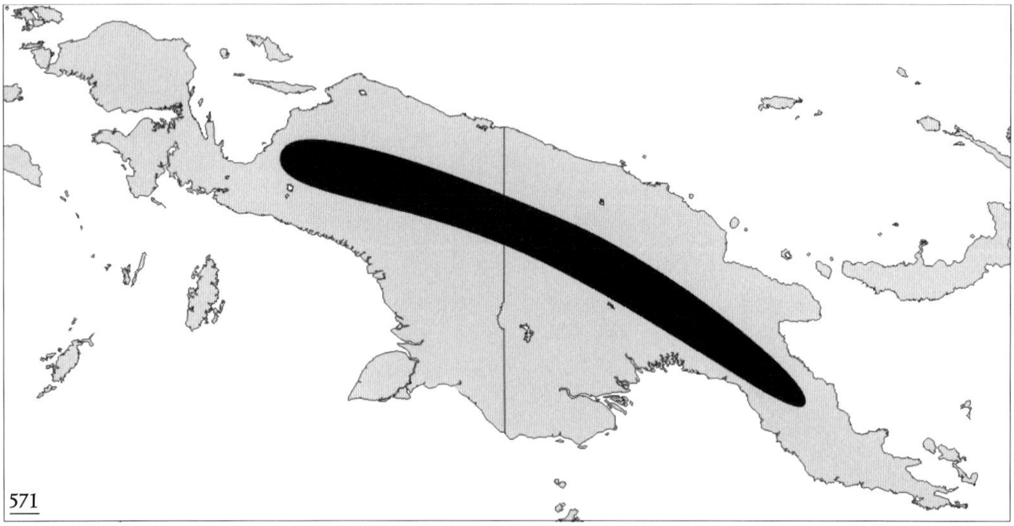

FIGURE 572
Distribution of Corixidae. Horizontal hatching + circle = *Agraptacorixa*; vertical hatching = *Cnethocymatia*; line = eastern margin of *Sigara* subgenus *Tropocorixa*.

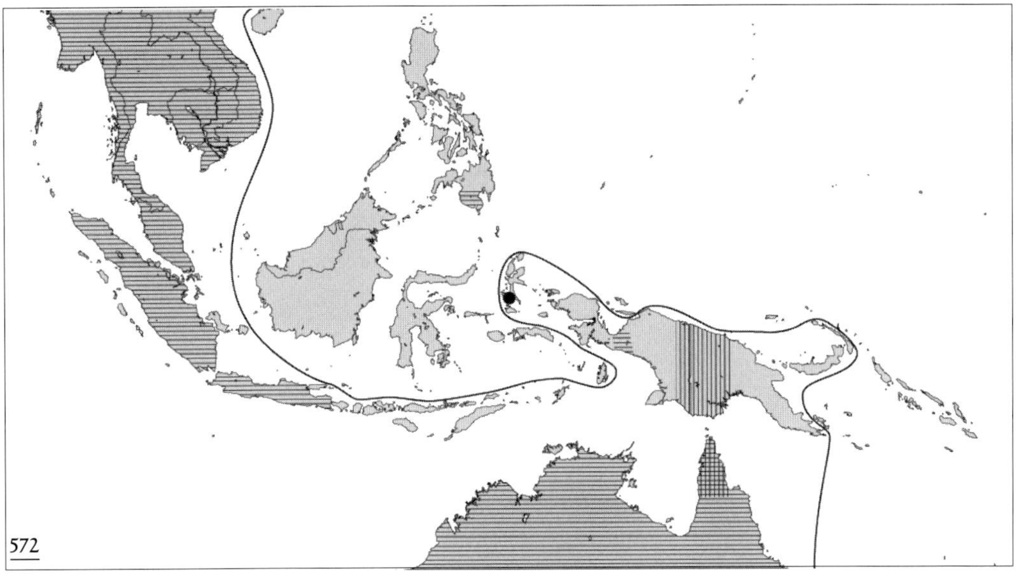

FIGURE 573
Distribution of Micronectidae and Notonectidae. Horizontal hatching = *Aphelonecta*. Lines represent eastern margin of distribution of 1 *Synaptonecta*; 2 *Nychia*.

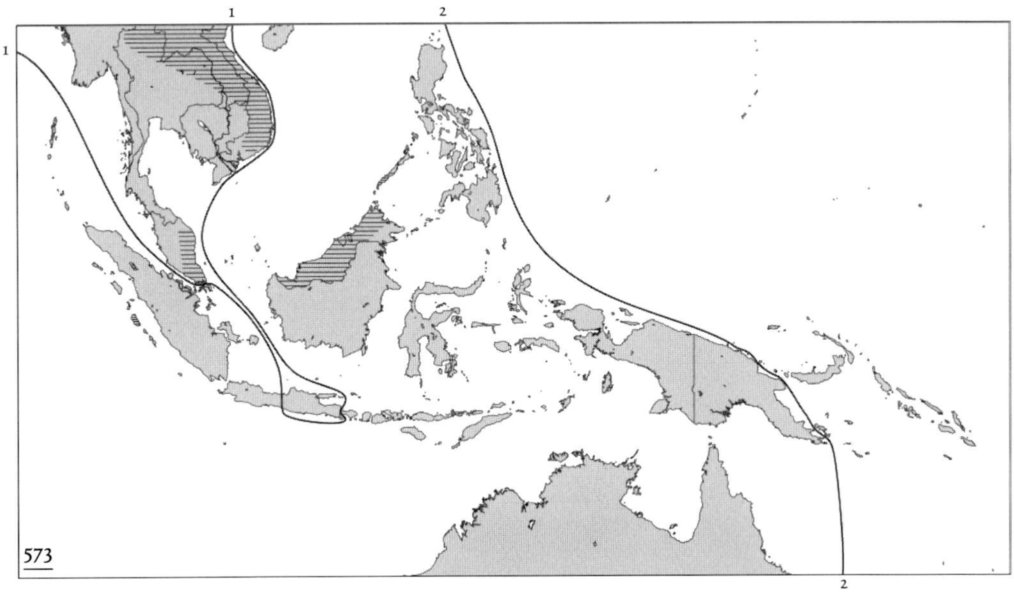

FIGURE 574
Distribution of Helotrephidae and Pleidae. Horizontal hatching = *Trephotomasinae*; vertical hatching = *Fischerotrephes*; line = eastern margin of *Paraplea*.

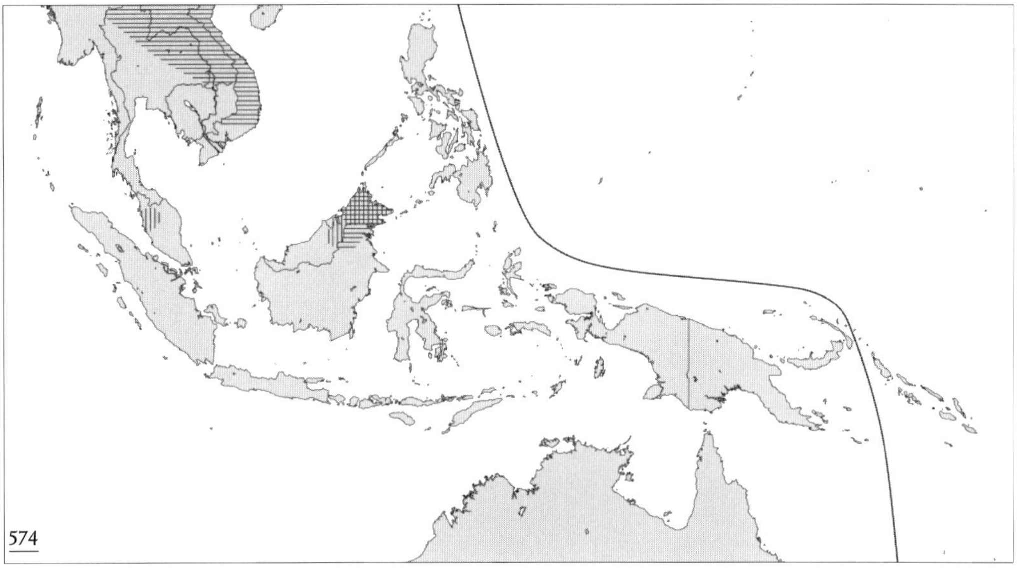

FIGURE 575
Distribution of Helotrephidae. Horizontal hatching = *Distotrephes* s.str.; vertical hatching = *Distotrephes* subgenus *Polhemotrephes*; line = extension of *Helotrephes* into Malesia.

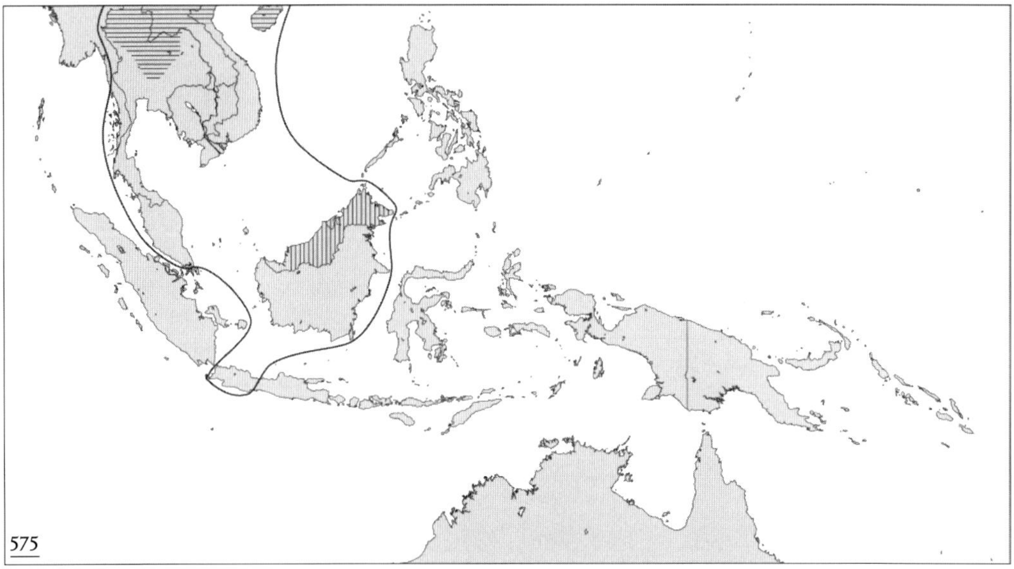

FIGURE 576
Distribution of Helotrephidae. Horizontal hatching = *Tiphotrephes*; vertical hatching = *Idiotrephes*; line = extension of *Hydrotrephes* into Malesia.

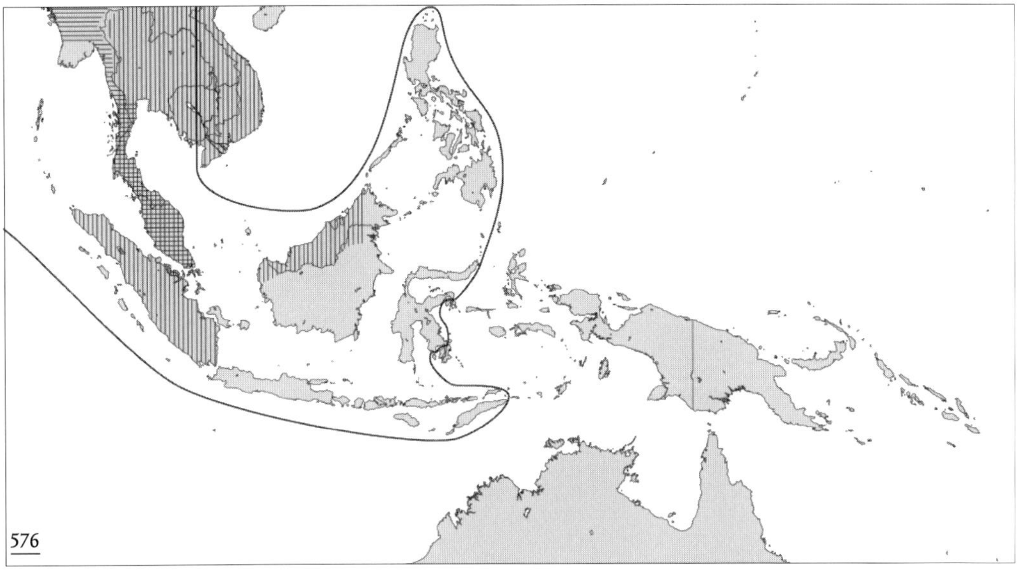

FIGURE 577
Distribution of Mesoveliidae and Hebridae. Line 1 = extension of *Timasius* into Malesia (based on J. Polhemus & D. Polhemus, 1990); line 2 = eastern margin of *Hebrus*; triangle = *Cryptovelia*; circles = *Nereivelia*; square = *Phrynovelia*; diamonds = *Merragata*.

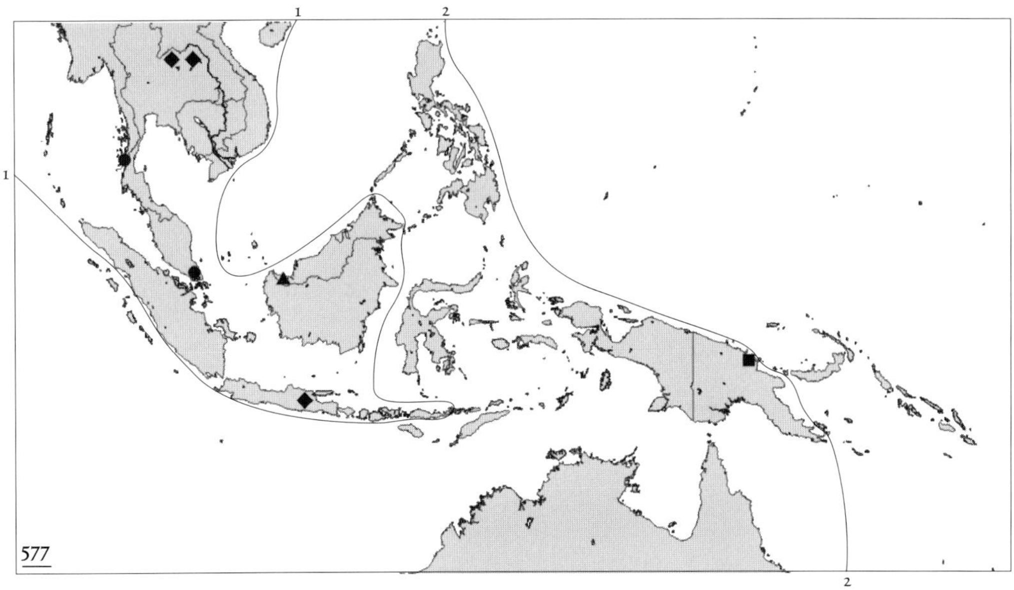

FIGURE 578
Distribution of Gerromorpha. Horizontal hatching and circles = *Hermatobates*; vertical hatching = *Aegilipsicola*; line = extension of *Hyrcanus* into Malesia; triangle = *Heterocleptes*.

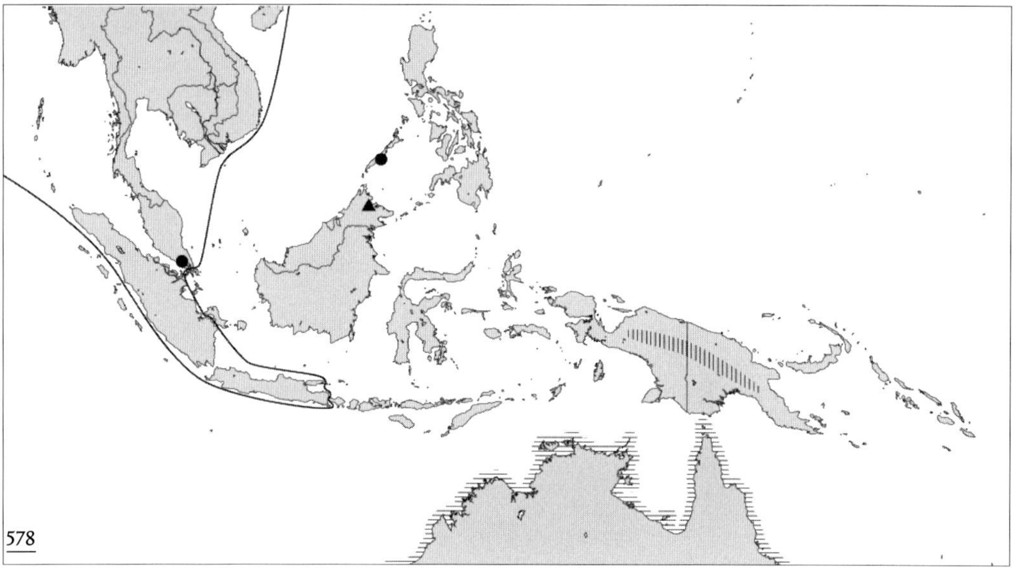

578

FIGURE 579
Distribution of Microveliinae. Horizontal hatching = *Baptista*; vertical hatching = *Lathriovelia*; oblique hatching = *Neusterensifer*.

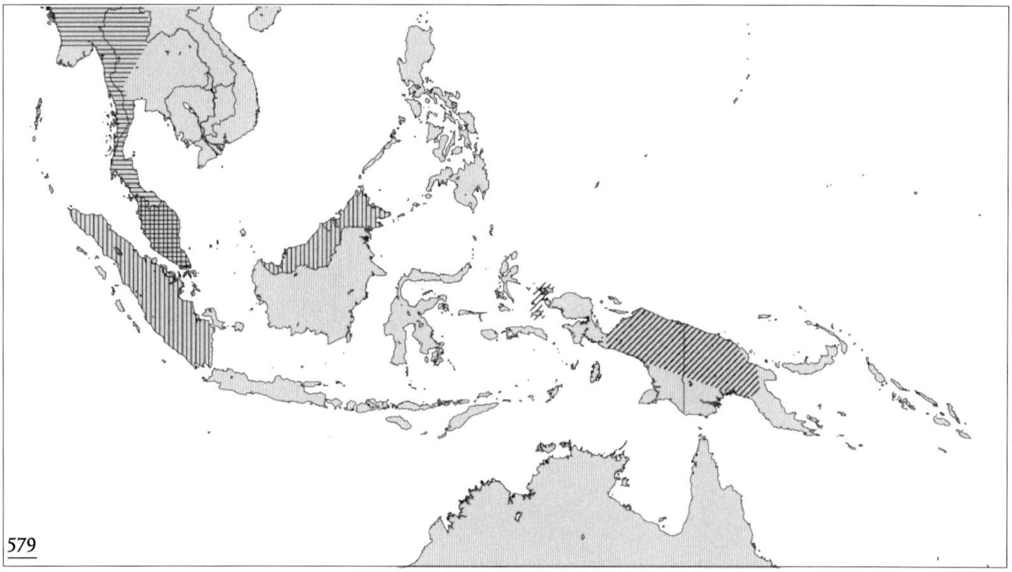

579

FIGURE 580

Distribution of Microveliinae. Horizontal hatching = *Polhemovelia*; vertical hatching = *Phoreticovelia*; oblique hatching = *Xiphovelia*. Lines represent extension into Malesia: 1 = *Pseudovelia*; 2 = *Neoalardus* (after J. Polhemus & D. Polhemus, 1990); circle = *Velohebria*.

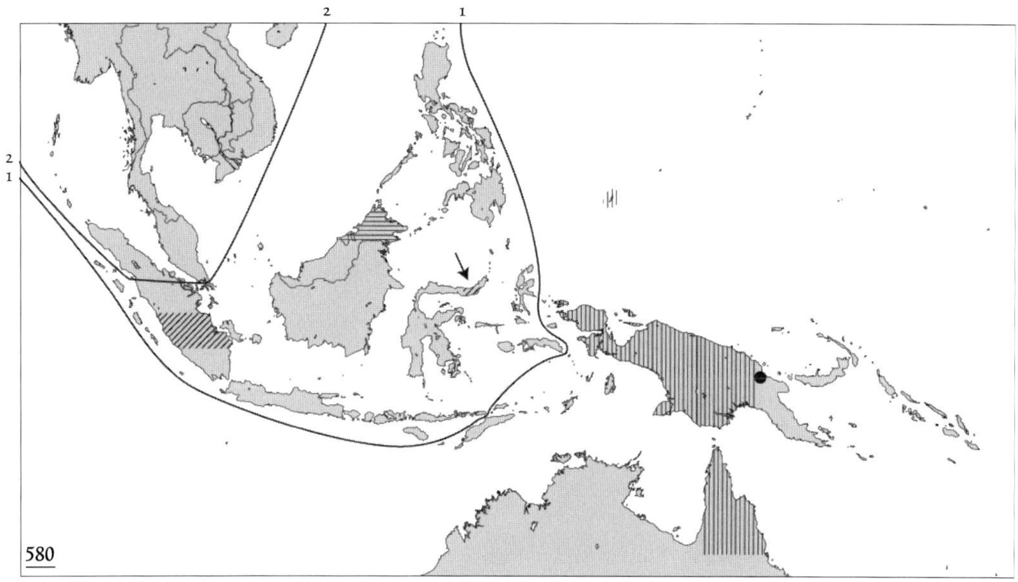

FIGURE 581

Distribution of Microveliinae. Horizontal hatching = *Tarsovelia*; triangles = *Papuavelia*; circles = *Tanyvelia*.

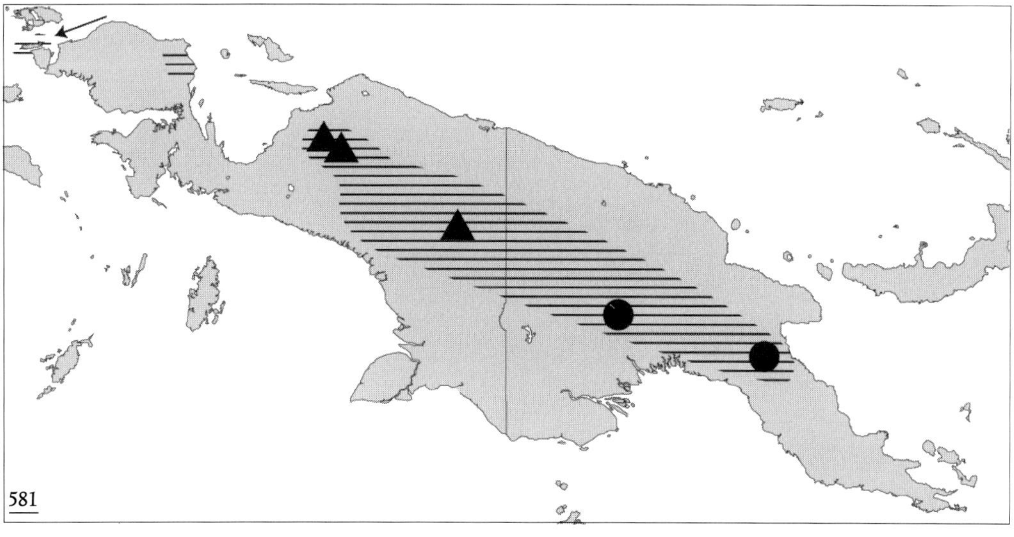

FIGURE 582
Distribution of Haloveliinae. Horizontal hatching = *Entomovelia*; line = *Halovelia*.

FIGURE 583
Distribution of *Haloveloides* (after Andersen, 1992b).

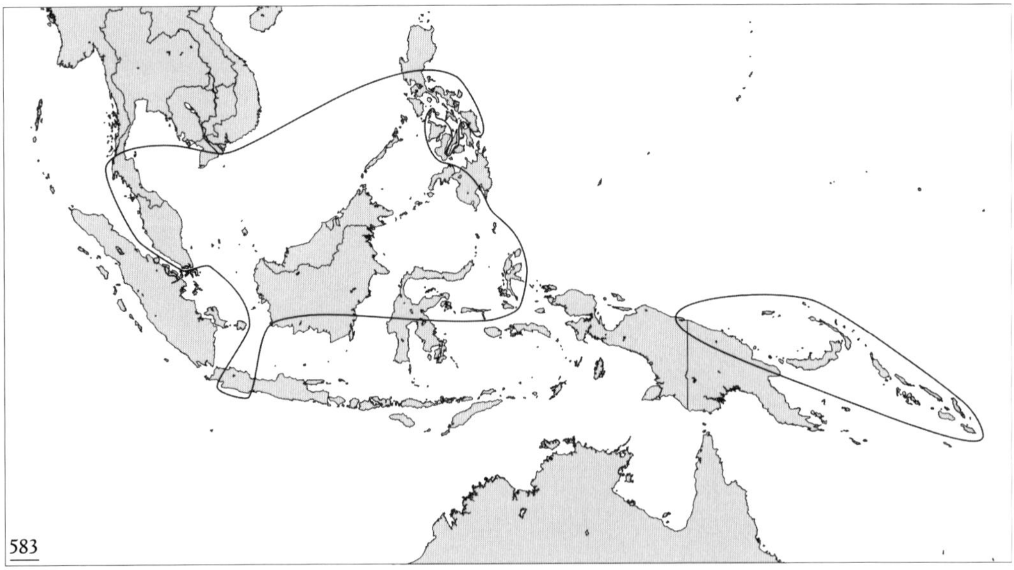

FIGURE 584
Distribution of Haloveliinae and Rhagoveliinae. Horizontal hatching = *Tetraripis*; line = distribution of *Strongylovelia* (after Andersen, 1982a).

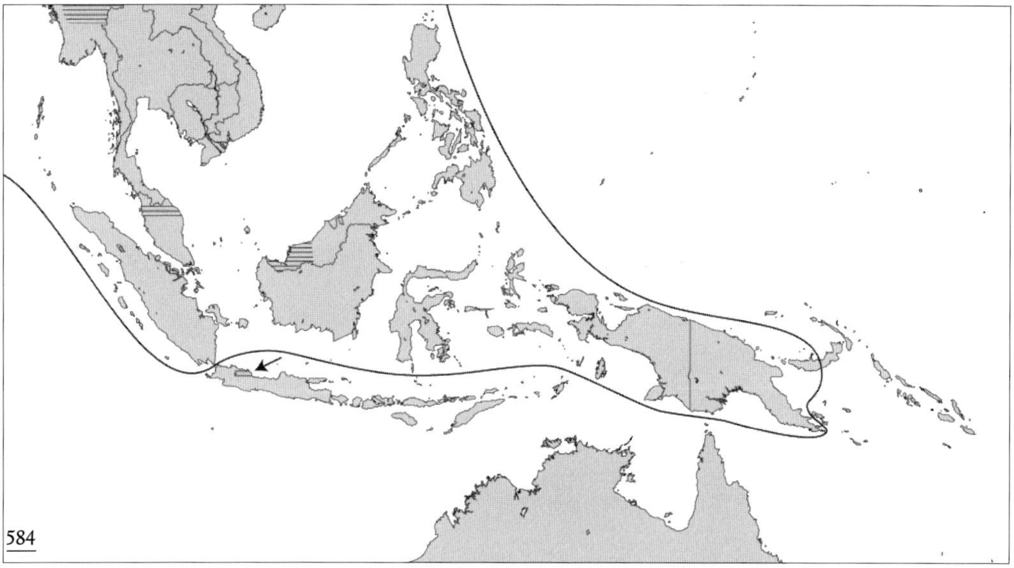

584

FIGURE 585
Distribution of Haloveliinae and Perittopinae. Horizontal hatching = *Perittopus*; line = distribution of *Xenobates* (based on Andersen, 1989b).

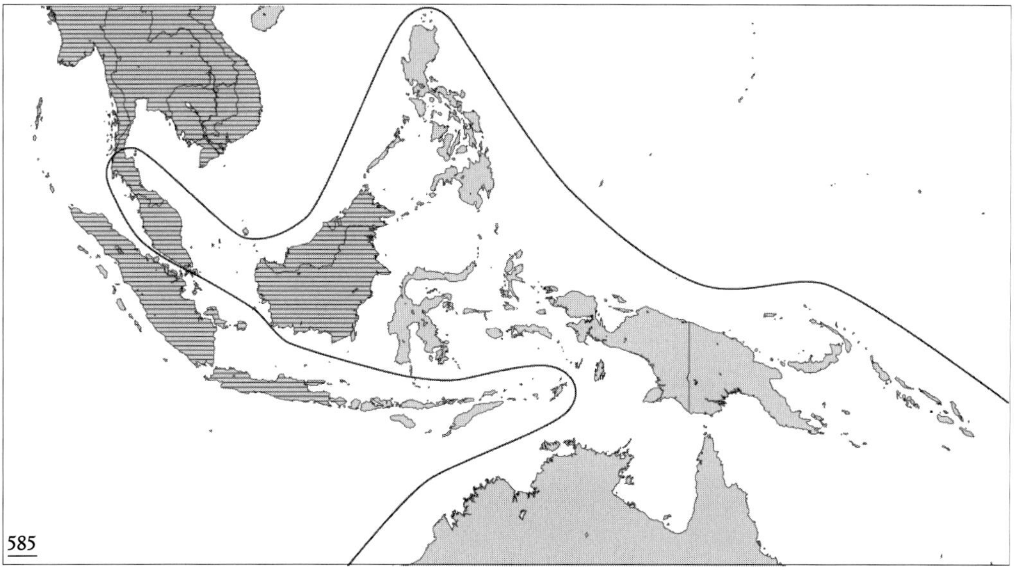

585

FIGURE 586
Distribution of Veliinae. Line = *Angilia*; triangles = *Angilovelia*.

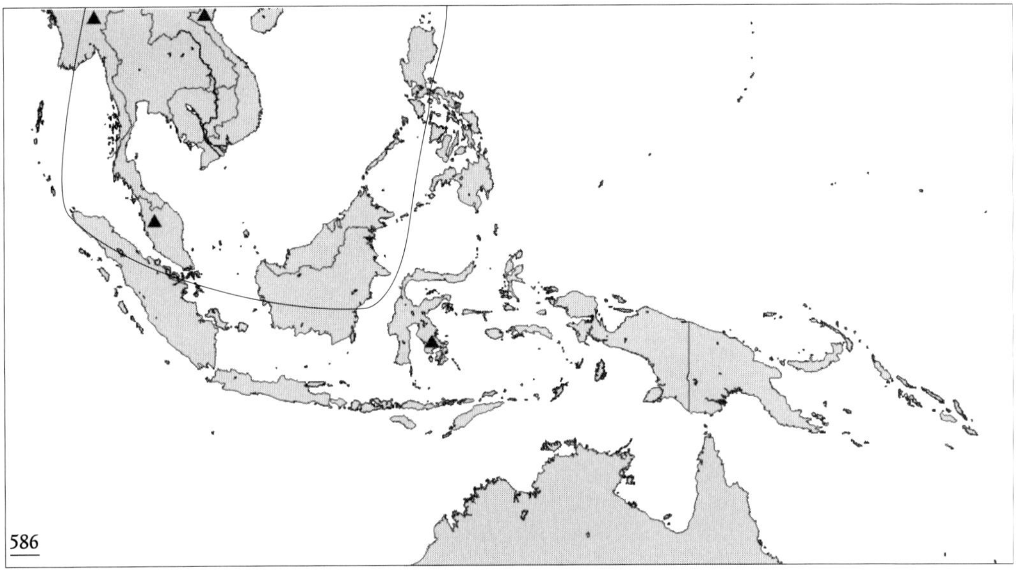

FIGURE 587
Distribution of Rhagadotarsinae and Trepobatinae. Line = eastern margin of distribution of *Rhagadotarsus*; triangles = *Andersenella*; circles = *Metrobatoides*; star = *Talaudia*.

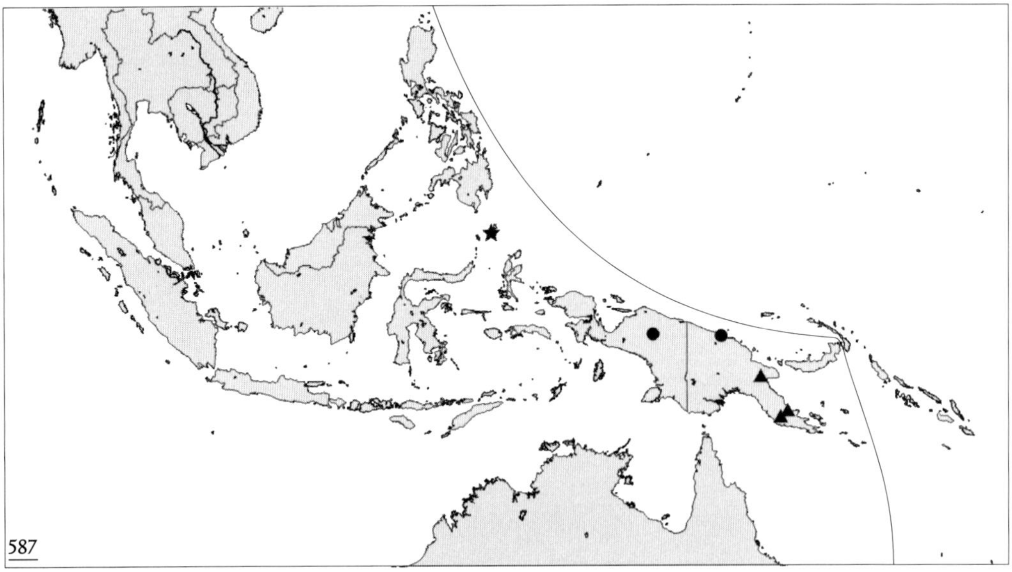

FIGURE 588
Distribution of Trepobatinae, *Ciliometra* (adapted from J. Polhemus & D. Polhemus, 1993).

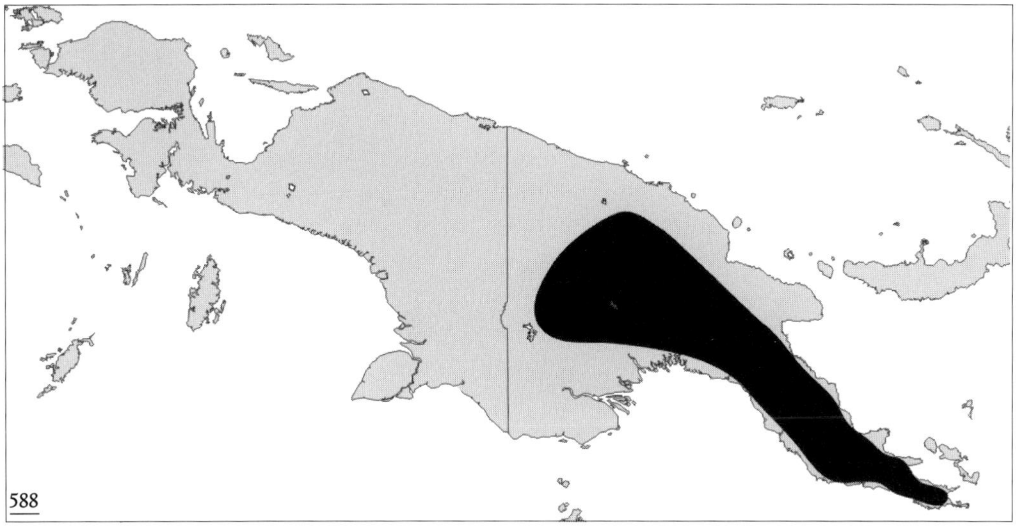

588

FIGURE 589
Distribution of Trepobatinae, *Iobates* (adapted from J. Polhemus & D. Polhemus, 1993).

589

FIGURE 590
Distribution of Trepobatinae. Vertical hatching and circles = *Metrobatopsis*; horizontal hatching = *Stygiobates* (adapted from J. Polhemus & D. Polhemus, 1993).

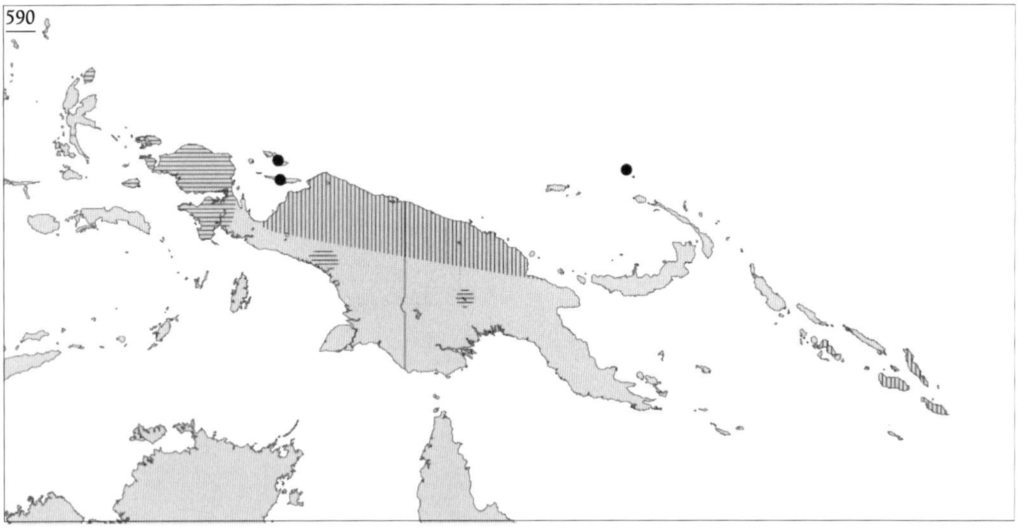

FIGURE 591
Distribution of Trepobatinae. Line = *Naboandelus*; triangles = *Calyptobates* (adapted from J. Polhemus & D. Polhemus, 1994b).

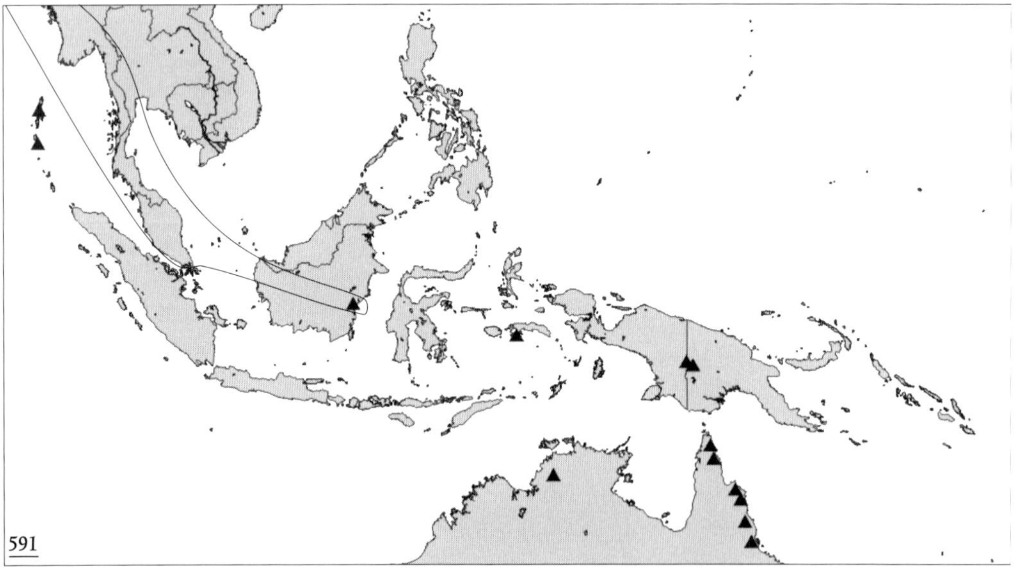

FIGURE 592

Distribution of Trepobatinae. Line = *Cryptobates*; horizontal hatching = *Pseudohalobates* (adapted from J. Polhemus & D. Polhemus, 1995b, 1996).

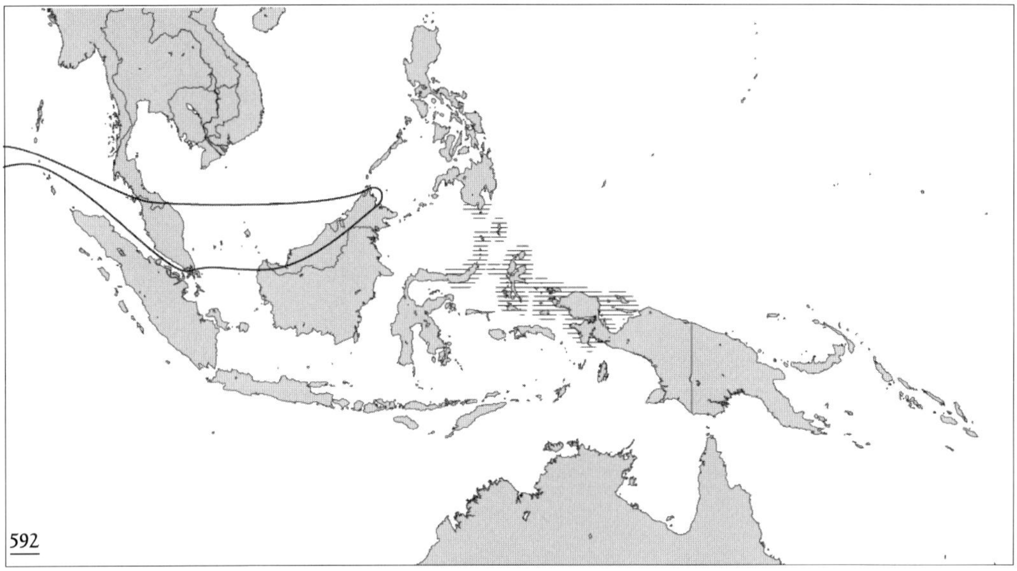

FIGURE 593

Distribution of Trepobatinae. Line = *Rheumatometroides*; triangles = *Thetibates* (adapted from J. Polhemus & D. Polhemus, 1995b, 1996).

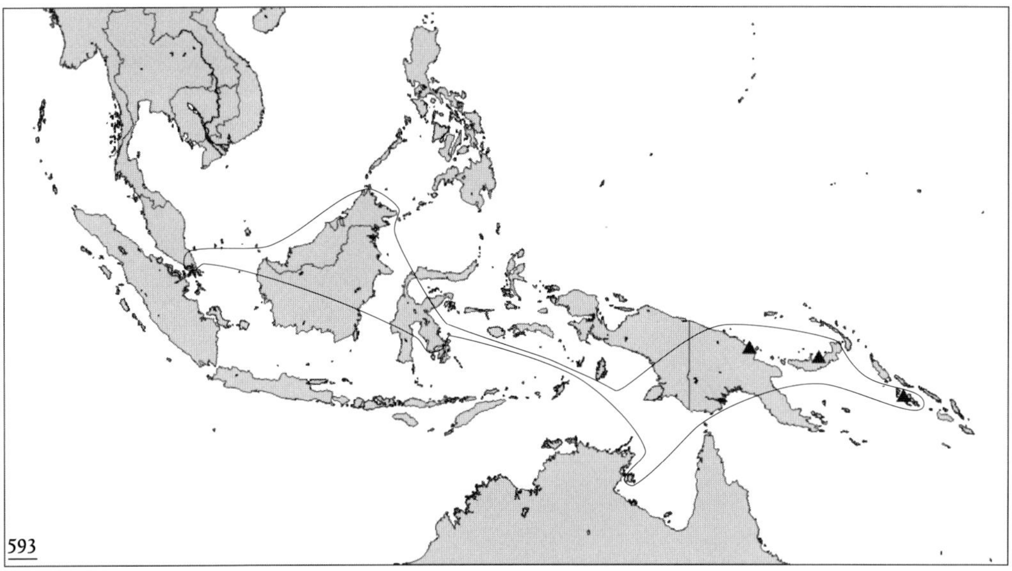

FIGURE 594
Distribution of Trepobatinae. Line plus circle = *Stenobates*; triangles = *Stenobatopsis* (adapted from J. Polhemus & D. Polhemus, 1996).

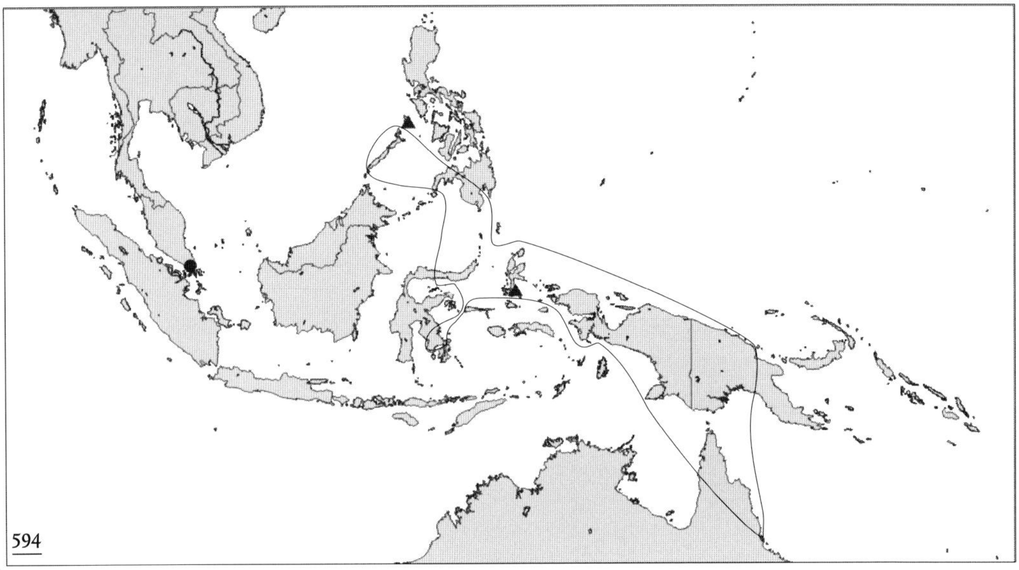

FIGURE 595
Distribution of Gerrinae. Lines and circle = *Aquarius*; triangle = *Limnogonus* subgenus *Limnogonoides*.

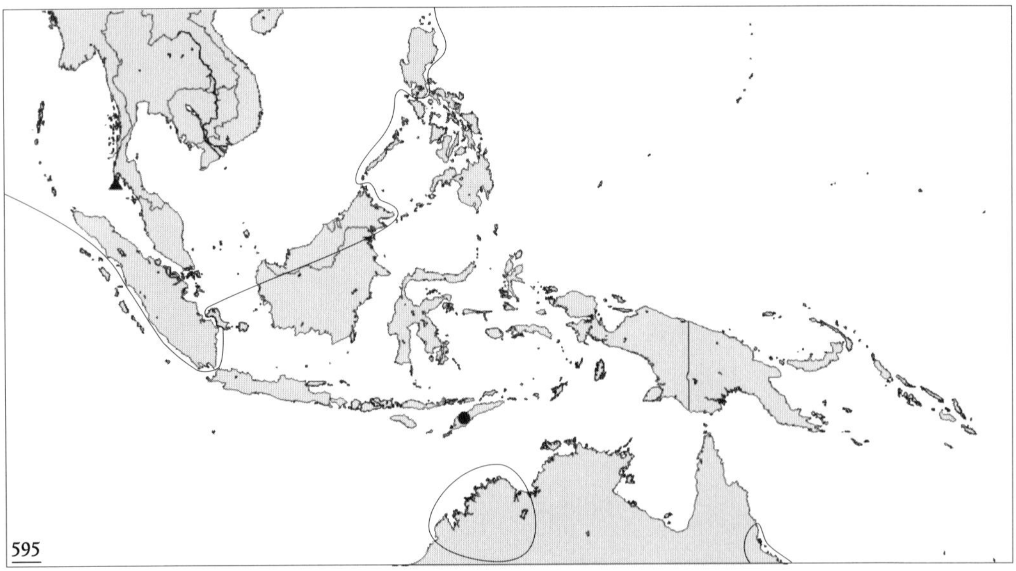

FIGURE 596
Distribution of Eotrechinae. Horizontal hatching = *Onychotrechus*; line = *Amemboa* subgenus *Amemboa*; triangles = *Amemboa* subgenus *Amemboides*.

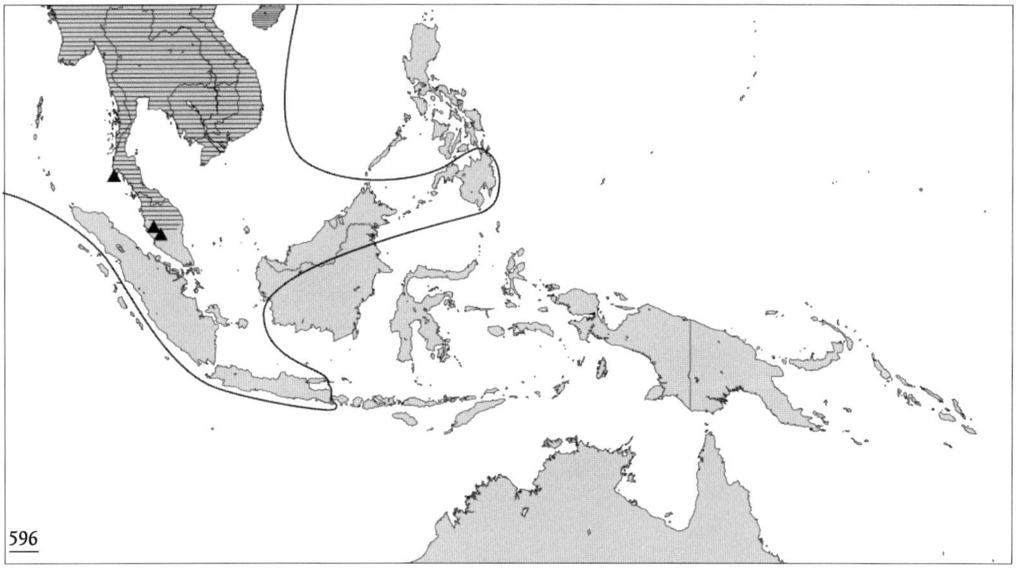

596

FIGURE 597
Distribution of Cylindrostethinae and Ptilomerinae. Horizontal hatching = *Pleciobates*; vertical hatching = '*Rhyacobates*' *imadatai*; line = *Cylindrostethus*.

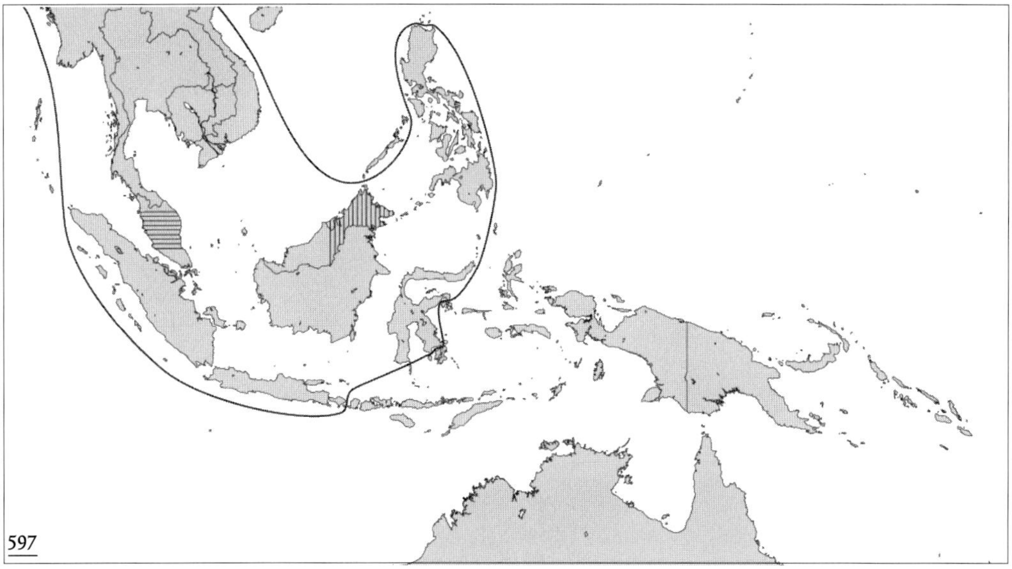

597

FIGURE 598
Distribution of Ptilomerinae. Horizontal hatching = *Potamometropsis*; line = *Ptilomera*.

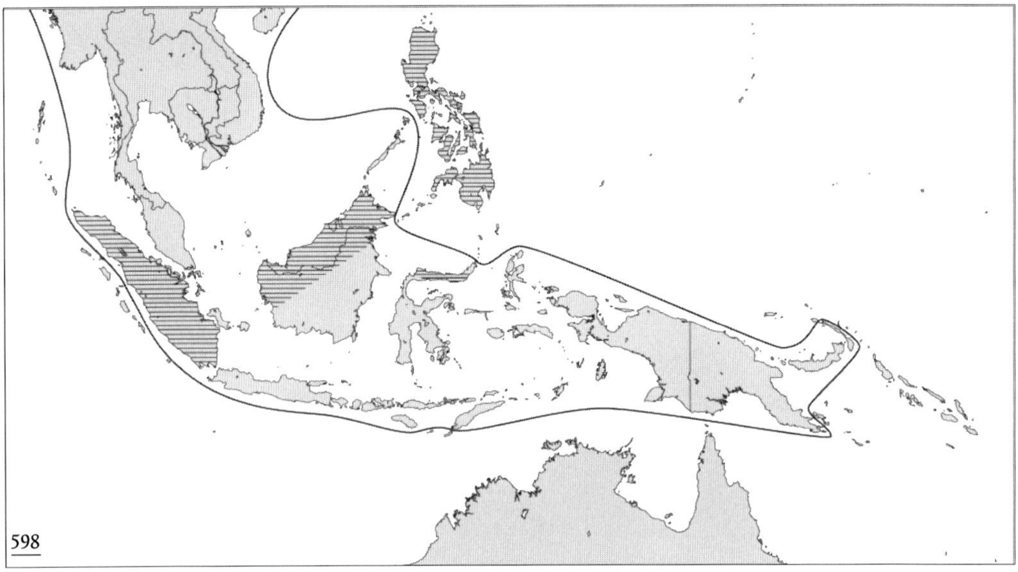

FIGURE 599
Distribution of Ptilomerinae and Halobatinae. Line = *Rheumatogonus*; triangles = *Asclepios*.

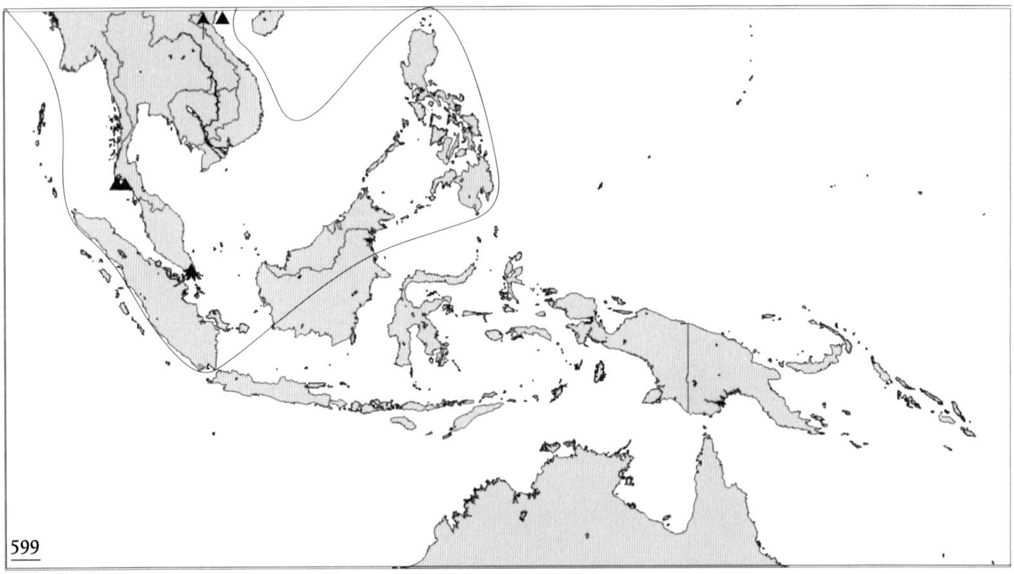

FIGURE 600
Distribution of Halobatinae. 1 = *Esakia*; 2 = *Metrocoris*.

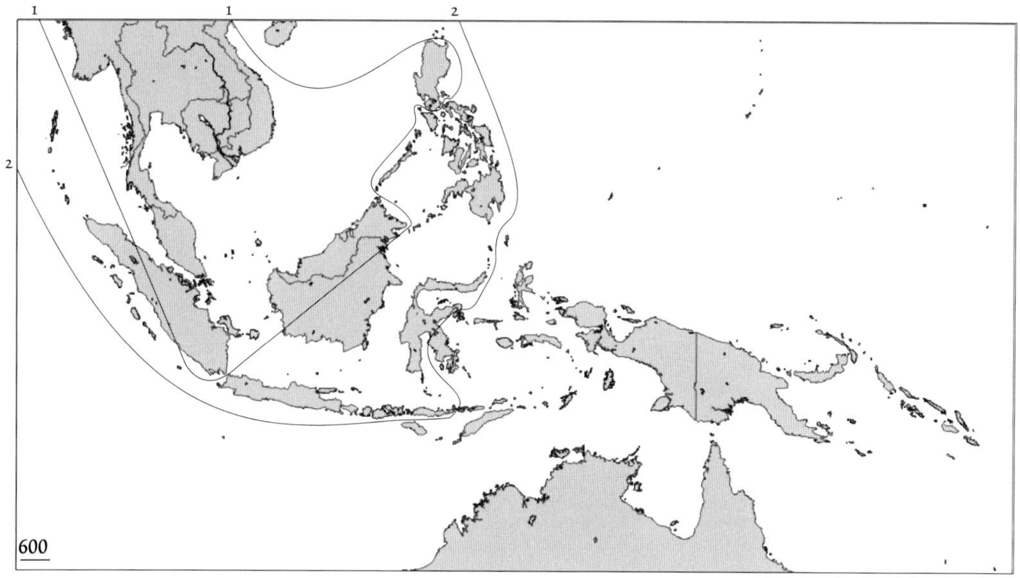

FIGURE 601
Distribution of the subgenera of *Ventidius*. Horizontal hatching = *Ventidius* subgenus *Ventidiopsis*; line 1 = *Ventidius* subgenus *Ventidius*; line 2 = *Ventidius* subgenus *Ventidioides*.

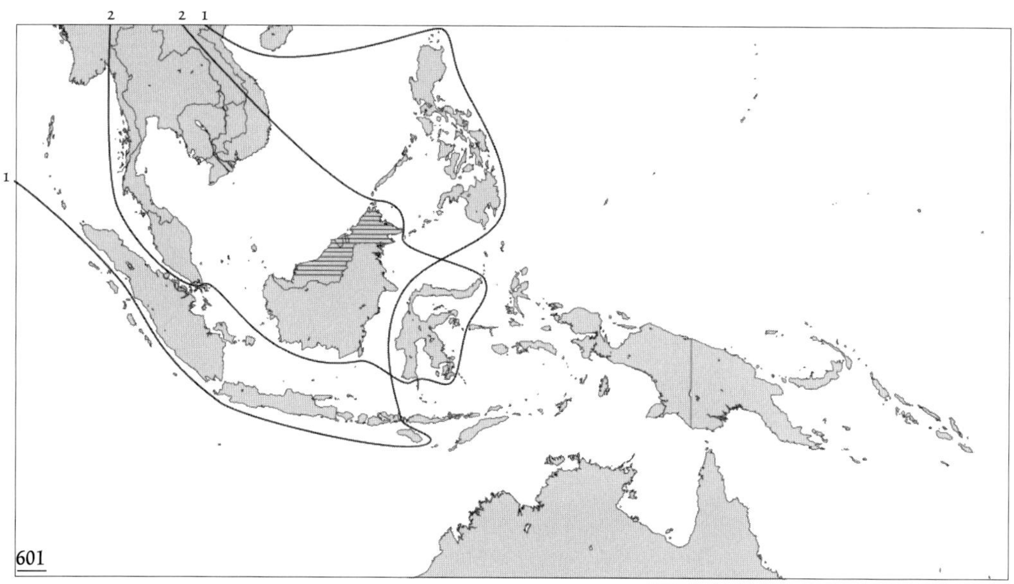

Appendix

FAMILY IN MALESIA *subfamily in Malesia*	GENERA WORLDWIDE INCLUDING FOSSILS	GENERA WORLDWIDE EXCLUDING FOSSILS	DESCRIBED SPECIES WORLD-WIDE INCLUDING FOSSILS	DESCRIBED SPECIES WORLD-WIDE EXCLUDING FOSSILS	GENERA IN MALESIA	DESCRIBED SPECIES IN MALESIA	RANGE
(rounded) total worldwide	356	289	4150	4030	113	957	
Nepidae	15	14	255	255	5	31	
Nepinae	11	10	90	90	3	14	cosmopolitan except polar regions
Ranatrinae	4	4	165	165	2	17	cosmopolitan except polar regions
Belostomatidae	20	9	155	140	2	4	tropics and subtropics
Belostomatinae	7	7	110	110	1	2	tropics and subtropics
Lethocerinae	1	1	25	25	1	2	tropics and subtropics
Ochteridae	3	3	60	60	1	28	tropics to temperate regions
Gelastocoridae	3	3	99	99	1	28	tropics and subtropics
Nerthrinae	1	1	87	87	1	28	tropics and subtropics
Aphelocheiridae	2	1	65	65	1	23	Old World except polar regions and New Zealand
Naucoridae	48	34	370	350	16	76	cosmopolitan except polar regions and New Zealand
Cheirochelinae	13	13	63	63	12	57	Indo-Australia
Laccocorinae	7	7	65	65	3	14	tropics
Naucorinae	16	7	70	60	1	5	cosmopolitan except polar regions and New Zealand
Corixidae	50	32	490	450	3	10	cosmopolitan
Corixinae	28	27	425	420	2	9	cosmopolitan
Cymatiainae	2	2	6	6	1	1	northern temperate area and Papuan subregion
Micronectidae	3	3	140	140	2	32	cosmopolitan except polar regions and Nearctis
Notonectidae	14	10	360	350	4	86	cosmopolitan except polar regions
Anisopinae	4	4	185	185	1	35	tropics and subtropics
Notonectinae	6	6	170	165	3	51	cosmopolitan except polar regions
Helotrephidae	19	19	175	175	8	68	tropics, just extending into subtropics
Fischerotrephinae	1	1	6	6	1	3	India, Borneo
Trephotomasinae	3	3	7	7	3	3	Asian tropics
Helotrephinae	10	10	120	120	6	72	Old World tropics, just extending into subtropics

FAMILY IN MALESIA *subfamily in Malesia*	GENERA WORLDWIDE INCLUDING FOSSILS	GENERA WORLDWIDE EXCLUDING FOSSILS	DESCRIBED SPECIES WORLD-WIDE INCLUDING FOSSILS	DESCRIBED SPECIES WORLD-WIDE EXCLUDING FOSSILS	GENERA IN MALESIA	DESCRIBED SPECIES IN MALESIA	RANGE
Pleidae	3	3	40	40	1	5	cosmopolitan except polar regions and New Zealand
Mesoveliidae	12	11	44	43	4	9	cosmopolitan except polar regions
Hebridae	9	8	163	163	4	19	cosmopolitan except polar regions and New Zealand
Hebrinae	7	6	180	180	3	27	cosmopolitan except polar regions and New Zealand
Hyrcaninae	2	2	13	13	1	1	tropical Asia
Hydrometridae	14	7	134	125	2	20	cosmopolitan except polar regions
Heterocleptinae	3	2	6	5	1	1	tropics
Hydrometrinae	5	5	120	120	1	19	cosmopolitan
Hermatobatidae	1	1	9	9	1	3	tropics
Veliidae	59	56	850	850	24	257	cosmopolitan except polar regions
Microveliinae	37	37	350	350	14	75	cosmopolitan except polar regions
Haloveliinae	5	5	81	80	5	56	Old World tropics and subtropics
Perittopinae	1	1	11	11	1	7	tropical Asia
Rhagoveliinae	3	3	300	300	2	112	tropics and subtropics
Veliinae	11	9	130	130	2	7	cosmopolitan except polar regions, New Zealand and Australia
Gerridae	80	74	725	700	34	258	cosmopolitan
Rhagadotarsinae	2	2	38	38	1	3	Old World tropics and New World except polar regions
Trepobatinae	26	26	131	130	15	57	Old World tropics and New World except polar regions
Gerrinae	18	14	179	169	5	44	cosmopolitan except polar regions
Eotrechinae	5	5	60	60	2	11	Oriental region
Cylindrostethinae	3	3	32	32	1	6	tropics
Ptilomerinae	12	12	120	120	5	86	tropical Asia and Madagascar
Halobatinae	10	10	152	152	5	53	Old World tropics and subtropics, Caribbean, in East Asia reaching into temperate region

Index

This index lists all scientific names, and a few English ones, referred to in the text. Of the taxon names mentioned in the checklist (p. 409), only those above the species level are included. The principal treatment of a taxon starts on the pages underlined; illustrations of a particular taxon are in **bold**.

O

P